The Exploration of Northern Canada

The Exploration of Northern Canada

500 to 1920
A CHRONOLOGY

Alan Cooke
and Clive Holland

The Arctic History Press

The Arctic History Press
18 Waterman Avenue,
Toronto, Ontario.
M4B 3G3

This Edition has been limited to 1,100 copies
and will not be reprinted in this form.

The research for this work was financially supported by the National Museum of Man., one of the National Museums of Canada

CANADIAN CATALOGUING IN PUBLICATION DATA

Cooke, Alan, 1933-
The exploration of northern Canada, 500 to 1920

Bibliography: p.
Includes indexes.
ISBN 0-7710-2265-4

1. Canada, Northern – Discovery and exploration – Chronology. 2. Canada, Northern – Exploring expeditions – Chronology. I. Holland, Clive. II. Title.

FC3956.C66 971.01 C78-001084-1
F1060.7.C66

Contents

Introduction

We began work on this chronological list in 1968 at the Scott Polar Research Institute, University of Cambridge, with financial support from the History Division of the National Museum of Man, a part of the National Museums of Canada, Ottawa. By 1970, we had carried the work far enough to begin publishing parts of it in *Polar Record*, the journal of the Scott Polar Research Institute. This form of preliminary publication continued in *Polar Record* in 11 consecutive chapters over a period of three years.

We had hoped that preliminary publication would bring additions and corrections from our readers. In fact, although we have managed to include much new and amended material in this revised text, we received few responses from readers other than comments upon the usefulness of such a list. We had suspected, of course, that the list would be useful, for it is modelled on a similar work of demonstrated usefulness, *Chronological list of Antarctic expeditions* by Dr. Brian Roberts, our colleague at the Institute. His list was published in *Polar Record*, Vol. 9, Nos. 59 and 60, 1958, pp. 97-134 and 191-239. We are confident that, once this book has been published, other northern nations will feel the need for similar listings of their own explorations and historical events.

We chose 1920 as the terminal date for this list because the introduction of aircraft to northern Canada at about that date brought very many more visitors into the region. The list, to be continued after that date, would have to be conducted on different principles.

The word "expedition" is used in a broad sense; it includes geographical discoveries, the collection of scientific information, and even many journeys of no particular importance that, nevertheless, resulted in some form of publication. "Events" include political actions, dates of trading posts and of some settlements, epidemics and other disasters, etc. The region implied by "northern Canada" changes with time; its southern boundary retreats westward and northward with progressive occupation of the land by pioneer European settlement.

Because of the importance of the North-west Passage in the history of northern Canada, we have listed all known North-west Passage expeditions, including some that went no farther north than Newfoundland or Vancouver Island, and some that are apocryphal. For brevity and clarity we have used modern place-names and some other anachronisms. The dates of the establishment of many trading posts caused doubt and trouble. In general, we have listed a post under the first date by which a more or less permanent building is known to have been constructed. When an important post has been abandoned for any length of time, its re-establishment is noticed.

Another cause of confusion lies in the difference between the Julian or Old Style (OS) calendar and the Gregorian or New Style (NS) calendar, OS being 10 days behind NS dates. France had used the NS calendar from 1582, but England continued to use the OS calendar until 1752. Dates in entries related to English activities can be assumed to be OS, with NS used for French activities. When a date from an English source appears in an entry for a French activity, and vice versa, it is marked according to its provenance.

Each entry is laid out in the same way: across the top, from left to right, are details of the expedition's date, its nature and national, commercial or other association, its leader, captain and, frequently, other senior members, and its ship – or, in a few cases, its sledge. Hudson's Bay Company ships of the same name are distinguished by Roman numerals in square brackets, as with *Dering* [I], *Dering* [II], etc. At the beginning of the text of each entry, there is a statement of the expedition's points of departure and return and the dates of its duration, if known. We then give, as succinctly as possible, some account of what the expedition attempted and what it achieved. The references at the end of each entry refer to the bibliography at the end of the book. An asterisk (*) indicates a comprehensive narrative (or, at least, the most comprehensive published narrative available), and a dagger (†) indicates a scientific paper. The two symbols together indicate a comprehensive narrative with scientific information included or appended.

In addition to the main body of the list, the reader will find not only maps, bibliography, and index, but also a roster of members of the expeditions listed, most of whom are not mentioned in the main entries.

We owe a particular debt of gratitude to Mr. Fred J. Thorpe, then as now Chief of the History Division, National Museum of Man, for his patient support and encouragement during the nine years that this work has slowly grown. We must also mention with special gratitude the archivists of the Hudson's Bay Company with whom we have worked: Miss Alice

Johnson and her successor, Mrs. Joan Craig, and their assistant, Miss G. Kemp, and the present archivist, Mrs. Shirlee A. Smith. Through the kind permission of the Governor and Committee of the Hudson's Bay Company, we have been able to include in this work many hitherto unpublished facts taken from their archives.

It is above all a pleasure to acknowledge, on the one hand, the extraordinary wealth of the Scott Polar Research Institute's library and, on the other, the interest and warm encouragement that we received from its staff: Dr. G. de Q. Robin, Director; Dr. Terence Armstrong, Deputy Director; Mr. Harry King, Librarian; Dr. Brian Roberts, Research Associate; and Mr. L.M. Forbes, formerly editor of *Polar Record.* They generously helped in many aspects of our labour. Polar specialists from all over the world visit the Institute and, while the work was in progress, many of them, especially the Canadian visitors, offered useful suggestions and stimulating conversation.

Inevitably there will be – there must be – mistakes and omissions in this work. We shall be grateful to any reader kind enough to draw our attention to them.

Alan Cooke, formerly Editor, *Polar Record*, Scott Polar Research Institute, Cambridge, and now Consultant, Centre for Northern Studies and Research, McGill University, Montreal

Clive Holland, Curator and Assistant Librarian, Scott Polar Research Institute, Cambridge.

Chronology

Irish voyages | St Brendan | ? **c500**

At this time the Irish began to visit islands in the North Atlantic Ocean, and, according to tradition, even reached North America, a discovery often attributed to St Brendan. So far, the claim is supported only by the unreliable evidence of semi-legendary accounts of his voyages.

OLESON, 1963, pp. 100-01; DCB, Vol. 1, p. 675.

Discovery of Baffin Island | Eirik Thorvaldsson (Eric the Red) | ? **982**

The theory recently advanced that Eric spent this summer exploring the Cumberland Peninsula, Baffin Island, is supported by a brief reference in one of the sagas: "That summer Eric explored the western wilderness, remaining there a long time and giving many local names there." Most scholars, however, believe the reference is to west Greenland.

MOWAT, 1966, pp. 58-66.

The Vinland voyages **c1000**

Voyages to Vinland are recounted in two medieval Icelandic sagas, *Groenlendinga saga* and *Eirik's saga.* Scholars now agree that they are authentic records of the Norse discovery of America about 1000, a belief reinforced by the publication in 1965 of a map of Vinland made about 1440, and by the discovery that same year of Norse artifacts and ruins in northernmost Newfoundland. Doubt has since been cast on the authenticity of the Vinland map. However, the sagas conflict in several important details. According to *Groenlendinga saga,* Bjarni (see below) discovered Vinland and Leif explored it; according to *Eirik's saga,* Leif discovered and explored it. Such differences make the voyages difficult to date precisely. The table below shows approximate dates suggested by some of the principal scholars: Gathorne-Hardy (1921) and Oleson (1963) combined elements of both sagas; Storm (1889) treated each saga independently.

Voyage	Oleson	Gathorne-Hardy	Storm *Groenlendinga saga*	Storm *Eirik's saga*
Bjarni	986	986	985	—
Leif	1001-02	1002-03	1003-04	1000
Thorvald	1004-06	1004-07	1005-06	—
Thorstein	1007	1008	1007	1001
Karlsefni	1011-13	1020-23	1009-11	1003-06
Freydis	1014-15	c1024	1012-13	—

STORM, 1889; GATHORNE-HARDY, 1921; OLESON, 1963; *SKELTON AND OTHERS, 1965; *MAGNUSSON AND PALSSON, 1966; INGSTAD, 1969; JONES, GWYN, *1964; 1968; WALLIS AND OTHERS, 1974.

Discovery of America | Bjarni Herjulfsson | ? **c1000**

On present evidence, Bjarni appears to have been the discoverer of the mainland of America. According to *Groenlendinga saga*, he was driven off course westward in a storm when sailing from Iceland to Greenland and sighted a wooded, hilly land. Sailing north, he sighted a second land that was low and wooded, a third that was mountainous and icy, and then he reached Greenland. The coasts were probably those of Newfoundland, Labrador, and Baffin Island. *Eirik's saga* does not mention this voyage.

MAGNUSSON AND PALSSON, 1966, pp. 51-54.

c1000 Vinland voyage | Leif Eiriksson | ?

Eirik's saga relates that Leif discovered Vinland accidentally, when sailing from Norway to Greenland, and that he landed only for a short time to take samples of the things he found there.

According to *Groenlendinga saga,* Leif sought the coasts that Bjarni had seen. He came first to the icy coast, probably Baffin Island or northern Labrador, which he named Helluland. He made a second landing in flat, wooded country, perhaps a part of Labrador, which he named Markland. Finally he reached Vinland, a warm land, where grapes grew wild. Here he built a house and wintered. The next spring, he returned to Greenland with a cargo of grapes, vines, and trees.

MAGNUSSON AND PALSSON, 1966, pp. 54-56, 84-88.

c1000 Vinland voyage | Thorvald Eiriksson | ?

Thorvald led a colonizing expedition from Greenland to Leif's wintering site in Vinland, wintered there, and, in the following spring and summer, explored the coast by ship. At one point he went ashore and was killed in a fight with the natives (probably Indians). The rest of his company returned to Vinland, wintered a second time, then returned to Greenland with grapes and vines. This voyage is not mentioned in *Eirik's saga.*

MAGNUSSON AND PALSSON, 1966, pp. 59-61.

c1000 Vinland voyage | Thorstein Eiriksson | ?

Thorstein also attempted a colonizing expedition from Greenland to Vinland, but he ran into bad weather and returned home. This voyage is related in both sagas.

MAGNUSSON AND PALSSON, 1966, pp. 61-62, 87.

c1000 Vinland voyage | Thorfinn Thordarsson (Karlsefni) | ?

According to *Groenlendinga saga,* Karlsefni sailed from Greenland intending to colonize Vinland. In the second year, natives came to trade furs. When they were caught stealing weapons, a fight broke out that ended in the Indians' defeat. In that same year, Karlsefni's wife, Gudrid, bore a son, Snorri, the first European child born in America. The third year they returned to Greenland with grapes, vines and furs.

According to *Eirik's saga,* Karlsefni was the discoverer of Helluland, Markland, and a place named Straumfjord, where they wintered. Here Thorhall the Hunter argued about where to find Vinland, then left in his own ship for Ireland, where he was killed. Meantime Karlsefni's party spent a second winter in Vinland where, in addition to the events related in *Groenlendinga saga,* they met a Uniped. They wintered once more in Straumfjord, then sailed back to Greenland.

MAGNUSSON AND PALSSON, 1966, pp. 64-67, 93-103.

Vinland voyage — Freydis, Helgi, Finnbogi — ? — c1000

Freydis was daughter to Eric the Red and half-sister to his sons, Thorvald and Thorstein, mentioned above. Helgi and Finnbogi were Norwegian brothers. They sailed from Greenland to Vinland on a colonizing expedition, but an argument arose between the brothers and Freydis because she had taken more men than had been agreed. The expedition split into two camps. In the winter, Freydis had the brothers and their company slain, and returned with her own party to Greenland.

MAGNUSSON AND PALSSON, 1966, pp. 67-70.

Voyage to America — Gudleif Gudlaugsson — ? — 1025

Gudleif was supposedly blown off course when sailing from Dublin to Iceland and eventually landed on the coast of America. He was captured by the natives but later freed by Bjorn Asbrandsson and returned to Iceland. This improbable story is backed by little evidence, but it finds support among some present-day scholars. Some suggest that Bjorn, like Bjarni (see c1000) had been blown off course and had been shipwrecked on the coast of America some 20-30 years earlier.

MOWAT, 1966, pp. 296-97.

Vinland voyage — Jon — ? — 1059

There is a doubtful record of Jon, who sailed on a missionary voyage to Vinland, where he was later reported to have been murdered.

MOWAT, 1966, pp. 297-98.

Vinland voyage — Bishop Eirik Gnupsson — ? — 1121-?

It has long been known that in this year Gnupsson left Greenland to search for Vinland. Until recently, nothing more has been known of him. However, a legend on the recently discovered Vinland map suggests that he remained a long time in Vinland, then returned to Greenland.

SKELTON, 1965, pp. 140-41; 223-26; PAINTER, 1965, pp. 255-62; GATHORNE-HARDY, 1921, p. 283.

1347	Voyage to Markland	?	?

A ship from Greenland that had been to Markland [Labrador?], probably for timber, was driven off course and eventually reached Iceland. This record indicates that the Norse continued to visit North America for many years after the voyages of discovery, c1000.

SKELTON, 1965, pp. 226-27.

c1360	Voyage to America	Paul Knutsson	?

The King of Norway commissioned an expedition to Greenland to restore to Christianity the inhabitants who had fallen into paganism, but there is no good evidence to show that the expedition ever set out. However, the discovery in Minnesota of a runestone known as the Kensington Stone (of doubtful authenticity) that records the visit of Norsemen in 1362 has led some scholars to conclude that Knutsson's expedition somehow found its way through Hudson Strait, down to James Bay or perhaps the Nelson River, and thence southward to Minnesota.

c1380	Voyage to America	Nicolo Zeno Antonio Zeno	?

In 1558, one of the Zenos, a Venetian family, published a claim that his ancestors Nicolo and Antonio had sailed across the Atlantic about 1380, visited an island in mid-ocean called Frisland, and from there sailed to parts of North America named Estotiland and Drogeo. Since the 19th century, it has been generally accepted that no such voyage occurred, but at the time the narrative and the map accompanying it were believed to be genuine, and they misled seamen and scholars for many years. Mercator included many of their fictional features on his 1569 map (*qv*), and Frobisher apparently thought in 1576 that the east coast of Greenland was Frisland.

MAJOR, 1873; LUCAS, 1898; OLESON, 1963, pp. 108-10; DCB Vol. 1, pp. 680-81.

c1476	Voyage to America	John Scolvus	?

The only real evidence of Scolvus's visit to America is a Latin inscription between Labrador and Greenland on Gemma Frisius's globe of about 1536, which reads: *"Quij populi ad quos Ioés Scoluus danus peruenit circa annum* 1476" (Quij, the people to whom John Scolvus, a Dane, penetrated about the year 1476). No details of his voyage are known. He could have accompanied Pining and Pothorst on an expedition to Greenland sent out by Christian I of Denmark, or he might have sailed to Labrador in one of the ships that still made regular trips from Greenland to America for timber.

LARSEN, 1925; COLLINSON, 1867, p. 4; DCB, Vol. 1, p. 679.

Voyages from Bristol

c1480-95

During these years Bristol merchants sent a number of ships across the North Atlantic to search for the Island of Brasil. Evidence suggests some of them might have reached the east coast of Canada or islands lying off it.

WILLIAMSON, 1962, pp. 19-32.

British North-west Passage expedition | John Cabot | *Matthew*

1497

From Bristol; 2 (?) May - 6 (?) August; sponsored by King Henry VII. Little is known of Cabot's course or the location of the landfall he made on 24 June: Maine and Nova Scotia have been suggested as possible sites, and this is supported by the fact that he made a long coasting voyage eastwards before departing for home from, possibly, Cape Race or Cape Breton. He reported large shoals of fish in the area of his discovery. This observation brought many European fishing vessels to the cod fishery off Newfoundland and Labrador in the years following.

WILLIAMSON, *1962; 1937; DODGE, 1961, pp. 15-18; JACKSON, 1963; OLESON, 1963, pp. 128-40; DCB, Vol. 1, pp. 146-52; HOFFMAN, 1961, pp. 6-12.

British North-west Passage expedition | John Cabot | 5 ships

1498

From Bristol; early May - (?); sponsored by King Henry VII. Cabot probably intended to return to his landfall of 1497, then to coast southward in search of Japan. It is not known whether or not he reached the American coast and he may have died during the voyage. There is some evidence to suggest that he may have explored southwards from his discoveries of 1497, but he never returned to England.

WILLIAMSON, *1962; 1937; DODGE, 1961, pp. 15-18; OLESON, 1963, pp. 128-40; DCB, Vol. 1, pp. 146-52; HOFFMAN, 1961, pp. 6-12.

Portuguese North-west Passage expedition | Gaspar Corte-Real | 2 ships

1500

From Lisbon; June - (?); sponsored by King Emmanuel to search for a North-west Passage and possibly to establish Portuguese sovereignty over the land discovered by Cabot in 1497. Corte-Real sighted a coast which he took for Asia: it was probably Greenland, but may have been Newfoundland.

LARSEN, 1921-22; BIGGAR, 1903; OLESON, 1963, p. 143; DCB Vol. 1, pp. 234-36; HOFFMAN, 1961, pp. 26-29.

Bristol voyages

1500-05

Documentary evidence shows that voyages of discovery and trade from Bristol to North America continued for some years after Cabot's 1497 discovery.

WILLIAMSON, 1962, pp. 116-44.

1501

Portuguese North-west Passage expedition | Gaspar Corte-Real | 3 ships

From Lisbon; 15 May - 11 October; sponsored by King Emmanuel to search for a North-west Passage and claim for Portugal the land discovered by Cabot in 1497. He made a landfall, probably on the Labrador coast, and from there sailed south, possibly visiting Hamilton Inlet. The three ships separated in or near the Strait of Belle Isle and Corte-Real and his ship were never seen again. The other two ships returned with about 50 captive Eskimos.

LARSEN, 1921-22; BIGGAR, 1903; OLESON, 1963, p. 143; DCB, Vol. 1, pp. 234-36; HOFFMAN, 1961, pp. 26-29.

1502

Portuguese expedition in search of Corte-Real | Miguel Corte-Real | 3 ships

From Lisbon; May - (?); sponsored by King Emmanuel to search for Miguel's brother, Gaspar, lost during his 1501 expedition. The fleet reached Newfoundland in June. The ships then separated, intending to rendezvous on 20 August, but Corte-Real's ship, like that of his brother, disappeared and was never heard of again.

BIGGAR, 1903; OLESON, 1963, p. 144; DCB, Vol. 1, p. 236; HOFFMAN, 1961, pp. 26-29.

1504

French fishing voyages

Fishing vessels from Brittany began to visit the cod fishery on the Grand Banks.

COLLINSON, 1867, pp. vii-viii.

1506

French fishing voyage | Jean Denys | 1 ship

From Honfleur. First known Norman fishing voyage to Newfoundland.

EMMANUEL, 1959, p. 55; DCB, Vol. 1, p. 256.

1506

Portuguese fishing voyages

Portuguese vessels began fishing off Newfoundland on or before this date.

GOSLING, 1910, p. 35.

1508-09

British North-west Passage expedition | Sebastian Cabot | ?

Whether or not this expedition ever took place is disputed. By his own account, Cabot entered a strait, which opened between lats 61° and 64°N, in search of a North-west Passage, and continued westward for about 10° of longitude. He then turned south into a large sea, which he believed to be the Pacific Ocean. The position and size of this strait correspond approximately to Hudson Strait, and some consider that Cabot discovered Hudson Strait and Hudson Bay on this voyage.

*WILLIAMSON, 1962, pp. 145-72; DODGE, 1961, pp. 18-19; OLESON, 1963, pp. 132-35; DCB, Vol. 1, pp. 152-58; HOFFMAN, 1961, pp. 16-25.

1510

French fishing voyage | ? | *Jacquette*

This ship arrived in Rouen to sell fish caught off Newfoundland.

GOSLING, 1910, p.36.

1513

Spanish overland expedition | Vasco Nuñez de Balboa

Balboa crossed the Isthmus of Panama and discovered the Pacific Ocean. He thereupon claimed for Spain all the territory washed by the waters of this ocean "for all time past, present or to come, without contradiction . . . north and south, . . . from the Pole Arctic to the Pole Antarctic." In the latter part of the 18th century, Spain based her rights to the west coast of Canada on this declaration.

MARQUIS, 1914.

1517

Fishing voyages

By this date, at least 50 Spanish, French, and Portuguese ships were employed cod fishing off the Grand Banks.

COLLINSON, 1867, p. viii; GOSLING, 1910, p. 36.

1520

Portuguese expedition | João Alvarez Fagundes | ?

Visited the vicinity of Newfoundland, Nova Scotia, Cape Breton, St Pierre and Miquelon. On return to Portugal, Fagundes was given rights to these islands, and he may later have tried to found a colony.

HOFFMAN, 1961, pp. 33-35, 100-01; DCB, Vol. 1, pp. 303-04.

1524

French North-west Passage expedition | Giovanni da Verrazzano | *Dauphine*

From Madeira; 17 January - early July; sponsored by King François I. He probably landed near Cape Fear, North Carolina, and sailed northward until he reached about lat 50°N off Newfoundland.

HOFFMAN, 1961, pp. 105-12; DCB, Vol. 1, pp. 657-60.

1524-25

Spanish North-west Passage expedition | Estevão Gomes, Pedro de Luna | *La Anunciada*

From La Coruña; 24 September 1524 - 21 August 1525; sponsored by King Charles V. Coasted northward from Cuba and Florida to Cape Race, Newfoundland.

VIGNERAS, 1970; DCB, Vol. 1, pp. 342-43.

1527

British voyage	?	*Dominus Vobiscum* 1 other ship

From London; 9 May - early October; sponsored by King Henry VIII. All that is known of this voyage is that one ship was lost in the Strait of Belle Isle and the other returned to England. The only known account of this voyage is Hakluyt's. Very probably, his information was wrong; this may be no more than a garbled account of Rut's voyage (see below, 1527-28).

HAKLUYT, 1965, Vol. 2, p. 517; DODGE, 1961, p. 30.

1527-28

British North-west Passage expedition	John Rut ?	*Mary of Gilford* *Samson*

From Plymouth; 10 June 1527 - (?) 1528; sponsored by King Henry VIII. The ships were separated in a storm in the Atlantic on 1 July, and *Samson* was probably lost. Rut may have reached as far as lat 64°N before turning south to explore the coast of Labrador. He landed in one inlet, possibly St Lewis Inlet. He then visited St John's harbour, Newfoundland, and continued southward to the West Indies before returning to England.

PURCHAS, 1906, Vol. 14, pp. 304-05; DODGE, 1961, pp. 30-31; DCB, Vol. 1, pp. 585-86; HOFFMAN, 1961, pp. 119-21.

1534

French North-west Passage expedition	Jacques Cartier	2 ships

From St Malo; 20 April – 1 September; sponsored by King François I. They reached Cape Bonavista, Newfoundland, on 10 May, then sailed toward the Labrador coast, passed through the Strait of Belle Isle, sailed southward along the unexplored west coast of Newfoundland, discovered Magdalen Islands and Prince Edward Island and entered Baie de Gaspé, where they remained for some time. They then returned to St Malo, having explored most of the shores of the Gulf of St Lawrence without having noticed the mouth of the St Lawrence River.

BIGGAR, 1924; 1930; DODGE, 1961, pp. 25-26; DCB, Vol. 1, pp. 165-71; HOFFMAN, 1961, pp. 131-37.

1535-36

French North-west Passage expedition	Jacques Cartier Macé Jalobert Guillaume le Breton Bastille	*Grande Hermine* *Petite Hermine* *Émérillon*

From St Malo; 19 May 1535 - 16 July 1536; sponsored by King François I to continue exploration of the Gulf of St Lawrence. They passed southward through the Strait of Belle Isle and, following the directions of Indians, ascended the St Lawrence River to the first rapids, the present site of Montreal. They lost 25 men during a harsh winter at Stadacona (Quebec) and left for France on 20 May 1536.

BIGGAR, *1924; 1930; DODGE, 1961, pp. 26-27; DCB, Vol. 1, pp. 165-71; HOFFMAN, 1961, pp. 137-42.

British expedition	Richard Hore ?	*Trinity* *William*	1536

From Gravesend; end of April - early October. The object of the voyage is obscure; it may have been a fishing voyage, a voyage of exploration, or a sightseeing trip. They reached Cape Breton, then continued to Penguin Island off the south coast of Newfoundland. Here the provisions gave out and they apparently resorted to cannibalism before they managed to capture an unidentified French ship that was well supplied with food, which enabled them to return to England.

HAKLUYT, 1965, Vol. 2, pp. 517-19; TAYLOR, 1931; DODGE, 1961, pp. 33-34; DCB, Vol. 1, pp. 371-72.

French expedition	Jacques Cartier ? ?	*Grande Hermine* *Émérillon* 3 other ships	1541-42

From St Malo; 23 May 1541 - (?) 1542; sponsored by King François I to seek precious metals and to consider sites for colonization along the St Lawrence River. They built a fort at Cap Rouge on the north shore of the river and wintered there. The following year they loaded their ships with worthless pyrites and quartz, thinking they contained gold and diamonds. Returning, they met Roberval (see 1542-43), Cartier's commanding officer, who ordered them to return to the St Lawrence. However, Cartier slipped away in the night and continued to France.

*BIGGAR, 1924; 1930; HOFFMAN, 1961, pp. 142-47; DCB, Vol. 1, pp. 168-69.

French expedition	Jean François de la Rocque de Roberval ? ?	*Valentine* *Anne* *Lèchefraye*	1542-43

From La Rochelle; 16 April 1542 - (?) 1543; sponsored by King François I to establish a colony on St Lawrence River. Roberval was to have accompanied Cartier's 1541-42 expedition, but his departure was delayed. He met Cartier on 8 June 1542 at St John's, Newfoundland, and tried unsuccessfully to force him to return to the St Lawrence. He sailed up the St Lawrence to Cap Rouge, where Cartier had built a fort, then attempted to explore the Saguenay River in search of precious metals. The colony soon ran into difficulty and he sent to France for help. The entire company returned to France in 1543 in ships sent out by François I.

BIGGAR, 1930; HOFFMAN, 1961, pp. 146-47, 183-86; DCB, Vol. 1, pp. 423-24.

French North-west Passage expedition	Jean Alfonse	?	c1543

Alfonse was pilot to Roberval's 1542-43 expedition. He probably made his search for a North-west Passage while returning to France from the St Lawrence. He sailed north through the Strait of Belle Isle, probably entered Davis Strait, and was back in La Rochelle in May 1543.

HOFFMAN, 1961, pp. 168-71, 184, 213; DCB, Vol. 1, pp. 309-10.

1555	Portuguese North-west Passage expedition	Martin Chake	?

An English sailor, Thomas Cowles, stated in 1579 that, six years before, he had been in Lisbon, where a Portuguese mariner named Martin Chake (also spelled Chacke or Chaque) told him that he had discovered a North-west Passage in 1555. His story was that he had sailed from India for Portugal in a small ship and in the company of four other vessels. His ship was separated from the others in a storm and, after passing many islands, sailed through a gulf from the Pacific to the Atlantic and emerged in lat 59°N near Newfoundland. Another version of the story states that he sailed from the Atlantic to the Pacific: in either case, the story is plainly a hoax, but it contains the first mention of the Strait of Anian dividing America from Asia.

PURCHAS, 1906, Vol. 14, pp. 414-15; FORSTER, 1786, pp. 461-62.

1558 Publication of the Zeno map and narrative (see 1380).

1569 Mercator map

What appears to be Ungava Bay is represented on a map by Gerhardus Mercator, published in 1569: Hudson Strait and Ungava Bay may, therefore, have already been discovered by Europeans. Portuguese fishermen, from whom Mercator obtained much of his information, may have made these discoveries, but the first known visit to Hudson Strait was Frobisher's voyage of 1578. Mercator's map also includes several fictional features taken from the Zeno map (see 1380 and 1558).

DCB, Vol. 1, pp. 680-81.

1576	British North-west Passage expedition	Martin Frobisher Christopher Hall Gryffyn Owen	 *Gabriel* *Michael*

From Deptford; 7 June - 2 October; sponsored by the Muscovy Company. Owen returned home in *Michael* after a storm in the Atlantic in early July, but Frobisher and Hall sailed on in *Gabriel* and discovered Frobisher Bay on 11 August. They remained there 15 days, during which time Frobisher took possession of Hall Island in the name of Queen Elizabeth I, fought with Eskimos who captured five of his men, and took specimens of plants. A piece of stone he carried back to England was analysed and reported to contain gold, which prompted further expeditions to Frobisher Bay.

HAKLUYT, 1965, Vol. 2, pp. 615-22; *COLLINSON, 1867; *STEFANSSON, 1938; DCB, Vol. 1, pp. 316-19.

1577

British North-west Passage and mining expedition	Martin Frobisher	*Ayde*
	Edward Fenton	*Gabriel*
	Gilbert Yorke	*Michael*

From Blackwall; 25 May - 23 September; sponsored by the Cathay Company and Queen Elizabeth I to mine gold in Frobisher Bay and to search for a North-west Passage. They entered Frobisher Bay on 19 July and remained for five weeks. They met more Eskimos, but saw no sign of the men captured in the previous year. They returned to England with about 200 tons of useless ore.

HAKLUYT, 1965, Vol. 2, pp. 622-30; *COLLINSON, 1867; *STEFANSSON, 1938; HULTON, 1961; DCB, Vol. 1, pp. 316-19.

1578

British mining expedition	Martin Frobisher	*Ayde*
	Gilbert Yorke	*Thomas Allen*
	Edward Fenton	*Judith*
	Captain Kynersley	*Michael*
	George Beste	*Anne Francis*
	Henry Carewe	*Hopewell*
	Richard Philpott	*Beare*
	William Tanfield	*Thomas*
	Captain Courtney	*Emanuell*
	Thomas Moyles	*Frances*
	Captain Upcot	*Moone*
	Captain Newton	*Emanuell (Buss)*
	Hugh Randal	*Solomon*
	Captain Kendall	*Dennis*
	Edward Harvey	*Gabriel*

From Harwich; 30 May - 1 October; sponsored by the Cathay Company and Queen Elizabeth I to leave 100 men in Frobisher Bay to mine gold during the winter. Frobisher's plans were upset during the outward voyage by the sinking of *Dennis*, which carried part of the prefabricated house in which the winterers were to have lived. During the outward voyage, the fleet entered Hudson Strait, which Frobisher named Mistaken straytes. He was eager to explore it in search of a North-west Passage, but his orders were this time for mining only. A small house was built on Kodlunarn (Warwick) Island in Frobisher Bay, and the ships returned to England carrying over 1,000 tons of useless ore. During the return, the officers of *Emanuell (Buss)* sighted an island – never seen again – which they named Buss. Seamen searched for this island during the next two centuries. It may have been an ice island.

HAKLUYT, 1965, Vol. 2, pp. 630-35; *COLLINSON, 1867; *STEFANSSON, 1938; BLACKADAR, 1967; DCB, Vol. 1, pp. 316-19; KENYON, 1975.

1578

Fishing voyages

By this date, about 150 French vessels and 200 ships from other nations were fishing off Newfoundland and Labrador, and in addition there were some 30 Biscayan whalers.

LOW, 1896, p. 9L.

1583

British colonizing expedition	Sir Humphrey Gilbert	
	William Winter	*Delight*
	Walter Raleigh	*Raleigh*
	Edward Hayes	*Golden Hind*
	Maurice Browne	*Swallow*
	William Andrewes	*Squirrel*

From Plymouth; 11 June - 22 September; to search for a site for a colony on the east coast of America. Raleigh turned back shortly after sailing. On 23 July, the ships were separated in fog off Newfoundland, but regathered in St John's harbour. On 5 August, Gilbert formally took possession of Newfoundland and all lands for 200 leagues to the north and south. *Swallow* returned to England with the sick, while the three other ships sailed to Sable Island. Here *Delight* ran aground and was lost. The remaining two ships thereupon set sail for England but *Squirrel*, with Gilbert on board, sank on 9 September. *Golden Hind* returned to Falmouth alone.

HAKLUYT, 1965, Vol. 2, pp. 679-97; DCB, Vol. 1, pp. 331-36.

1585

British North-west Passage expedition	John Davis	*Sunneshine*
	William Bruton	*Mooneshine*

From Dartmouth; 7 June - 30 September; financed by London and Devon merchants. They followed the west coast of Greenland northward to Godthaab Fjord, crossed Davis Strait to Exeter Sound, and entered Cumberland Sound.

*MARKHAM, A.H., 1880, pp. 1-14; MARKHAM, C.R., 1889, pp. 34-43; DODGE, 1961, pp. 91-94; DCB, Vol. 1, pp. 251-52.

1586

British North-west Passage expedition	John Davis	*Mermayde*
	?	*Mooneshine*
	Richard Pope	*Sunneshine*
	?	*Northstarre*

From Dartmouth; 7 May - 14 October; private enterprise. *Sunneshine* and *Northstarre* went to east Greenland, while Davis followed the west coast of Greenland to Disko Island. *Mermayde* returned home from here; *Mooneshine*, with Davis on board, crossed Davis Strait to Baffin Island at lat 67°N and sailed south to lat 54°30'N off Labrador. They failed to notice Hudson Strait. After two of his men had been killed by Labrador Eskimos, Davis returned home.

*MARKHAM, A.H., 1880, pp. 15-38; MARKHAM, C.R., 1889, pp. 43-52; DODGE, 1961, pp. 94-98; DCB, Vol. 1, pp. 251-52.

1587

British North-west Passage expedition	John Davis	*Elizabeth*
	?	*Sunneshine*
	John Churchyard	*Ellen*

From Dartmouth; 19 May - 15 September; sponsored by London merchants. Davis returned to the west coast of Greenland and reached lat 72°12'N before crossing Davis Strait to Baffin Island.

He again explored Cumberland Sound, noticed the entrances to Frobisher Bay and to Hudson Strait, and named Cape Chidley on the southern side of Hudson Strait. Sailing southward along the coast of Labrador, he entered Davis Inlet and reached the vicinity of Hamilton Inlet before turning home.

*MARKHAM, A.H., 1880, pp. 39-59; MARKHAM, C.R., 1889, pp. 53-71; DODGE, 1961, pp. 98-101; DCB, Vol. 1, pp. 251-52.

Spanish North-west Passage expedition — Lorenzo Ferrer Maldonado — 1588

The report of this voyage was a hoax. Maldonado probably intended to profit from Spain's growing concern with English attempts to find a North-west Passage. He described entering Davis Strait at lat 60°N and following it north-westward 280 leagues to lat 75°N. He then continued westward for 790 leagues through the Strait of Anian, which divided Asia from America, to emerge into the Pacific Ocean. After exploring the west coast of America southward to about lat 55°N, he returned by the same route to Spain.

WAGNER, 1931; DODGE, 1961, pp. 183-84; DCB, Vol. 1, p. 676.

Spanish North-west Passage expedition — Juan de Fuca — ? — 1592

In 1596, Michael Lok, one of the chief supporters of Frobisher's expeditions, 1576-78, was told by de Fuca, an old Greek pilot in Venice, that in 1592 he had been instructed by the Viceroy of Mexico to sail north along the California coast in search of the Strait of Anian. He said he had sailed into a broad inlet, which he had found between lat 47° and 48°N, and that he had reached the Arctic Ocean before turning back. His story is unlikely, but the strait that separates Vancouver Island from the mainland bears his name.

PURCHAS, 1906, Vol. 14, pp. 415-21; WAGNER, 1931; WILLIAMS, 1962, pp. 273-76; DODGE, 1961, p. 182; DCB, Vol. 1, p. 677.

French colonizing expedition	Pierre de Chauvin de Tonnetuit	*Don de Dieu*	1600-02
	?	*Espérance*	
	?	*Bon Espoir*	
	?	*Saint Jean*	

From Honfleur; a private venture to trade furs and establish a colony on the St Lawrence. Chauvin established the first trading post in Canada at Tadoussac in 1600. He returned to France with all his ships in the same year, leaving behind 16 men. In 1601, *Espérance* returned to Tadoussac and carried home the five surviving colonists. In 1602, Chauvin again sailed to Tadoussac, in *Don de Dieu* and *Espérance*, with trade his sole object. He remained in Canada only a few months and left no colonists. He died before he could make a further voyage. Although his venture was a failure, it encouraged others to search for more suitable sites for trade and colonization in Canada. One of them, Samuel de Champlain, established Quebec (see 1608).

BIGGAR, 1901, pp. 42-43; DCB, Vol. 1, pp. 209-10.

1602

British North-west Passage expedition | George Weymouth, John Drewe | *Discovery*, *Godspeed*

From Redcliffe, on the Thames; 2 May - 5 August; sponsored by the East India Company. On 28 June they made a landfall at lat 64°N on the Baffin Island coast. From there they allegedly reached lat 69°N before the crew forced them back. Coasting southward, they entered Frobisher Bay, Hudson Strait, and Ungava Bay.

*PURCHAS, 1906, Vol. 14, pp. 306-18; DODGE, 1961, pp. 103-07; *STEVENS, 1886; DCB, Vol. 1, pp. 667-68; FOXE, 1635, pp. 47-50.

1606

British North-west Passage expedition | John Knight | *Hopewell*

From Gravesend; 18 April - 24 September; sponsored by the East India Company and the Muscovy Company. They battled through heavy ice off Greenland and reached the Labrador coast on 19 June, where the ship was damaged on rocks in about lat 57°N. On 26 June, Knight and three other men went ashore and were never seen again. They were probably killed by Eskimos, who two days later tried to attack the ship. The survivors sailed to the Bay of Fogo, Newfoundland, where repairs were made with the help of fishermen, and they returned home.

*MARKHAM, C.R., 1877, pp. 279-94; *PURCHAS, 1906, Vol. 14, pp. 353-65; DODGE, 1961, pp. 111-12; DCB, Vol. 1, p. 408.

1608

Establishment of Quebec | Samuel de Champlain | *Don de Dieu*

Since 1603, Champlain had made several voyages to the trading post at Tadoussac (see 1600-02), and had explored the lower St Lawrence River region and the Gulf of St Lawrence. On 13 April 1608, he left France with *Don de Dieu*, reached Tadoussac on 3 June, then went upriver in a bark and established a fort at Quebec on 3 July. He continued to visit Quebec and to encourage its growth as a trading post and a settlement until his death there in 1635.

DCB, Vol. 1, pp. 186-99.

1610-11

British North-west Passage expedition | Henry Hudson | *Discovery*

From Gravesend; 17 April 1610 - (?) September 1611; sponsored by the Northwest Company. Hudson explored the southern shores of Hudson Strait, and followed the east coast of Hudson Bay south to James Bay where, having been frozen in during November, he wintered. Provisions ran short and, on 22 June 1611, the crew mutinied and set Hudson and eight others adrift in a boat. During the return, four of the mutineers were killed by Eskimos in Hudson Strait, and one more died of starvation before they reached Britain. No trace was ever found of the men set adrift. This is the first certain discovery of Hudson Bay by Europeans, although others may have been there earlier.

*ASHER, 1860; *POWYS, 1927; NEATBY, 1958, pp. 14-29; DODGE, 1961, pp. 114-26; *FOXE, 1635, pp. 71-117; DCB, Vol. 1, pp. 374-79.

1612-13

British North-west Passage expedition	Thomas Button John Ingram	*Resolution* *Discovery*

From London; 14 April 1612 - 27 September 1613; sponsored by the Northwest Company and the Prince of Wales. Button sought a North-west Passage through Hudson Strait. In his instructions, there was no mention of a search for Hudson (see 1610-11), who may have been alive in James Bay. Button named the island at the entrance to Hudson Strait after his ship *Resolution*, steered southwestward across Hudson Bay, discovered Coats Island, and wintered at Port Nelson, where he lost many men during a hard winter. In June 1613, he explored the west coast of Hudson Bay, discovered Churchill River, and reached lat 65°N in Roes Welcome Sound. During the return to England, he discovered Mansel Island.

*FOXE, 1635, pp. 117-36; NEATBY, 1958, pp. 32-34; DODGE, 1961, pp. 130-34; DCB, Vol. 1, pp. 144-45.

1614

British North-west Passage expedition	William Gibbons	*Discovery*

From England; March - (?); sponsored by the Northwest Company. Unusually heavy ice in the North Atlantic forced the ship away from the entrance to Hudson Strait and down the Labrador coast, where they lay icebound for ten weeks in a bay, named Gibbons Hole (possibly Saglek Bay), at about lat 58°N. When they broke out, the season was so far gone they returned to England with nothing accomplished.

*FOXE, 1635, p. 137; DODGE, 1961, p. 134, DCB, Vol. 1, p. 329.

1615

British North-west Passage expedition	Robert Bylot William Baffin	*Discovery*

From England; 15 March - 8 September; sponsored by the Northwest Company. They landed at Resolution Island, charted the south coast of Baffin Island, discovered Mill Island, and spent a month in the vicinity of Salisbury and Nottingham islands, Foxe Channel, and Southampton Island, observing tides. They entered Frozen Strait and concluded correctly that there was no possibility of a North-west Passage through Hudson Bay. They stopped at

Digges Islands during the return to Plymouth. Baffin was the first to reckon longitude at sea. Observations of the variation of the compass and of tides in Hudson Bay were found by Parry in 1821-23 to be extremely accurate.

*PURCHAS, 1906, Vol. 14, pp. 379-86; *FOXE, 1635, pp. 137-49; *MARKHAM, C.R., 1881, pp. 103-37; DODGE, 1961, pp. 134-37; NEATBY, 1958, pp. 34-35; DCB, Vol. 1, pp. 74-75.

1616

British North-west Passage expedition	Robert Bylot William Baffin	*Discovery*

From Gravesend; 26 March - 30 August; sponsored by the North-west Company. They explored the west coast of Greenland north to Smith Sound and discovered Cary Islands. Turning south, they discovered the entrances to Jones Sound, where they landed, and Lancaster Sound. They then explored the Baffin Island coast southward to about lat 65°40'N, in the region already explored by Davis, 1585-87, and concluded there was no passage to be found by way of Davis Strait. Baffin had carefully charted all the coasts of Baffin Bay but their discoveries came to be regarded as fantastic and were later removed from maps. They were not confirmed until 1818, when John Ross rediscovered Baffin Bay.

*PURCHAS, 1906, Vol. 14, pp. 401-11; *FOXE, 1635, pp. 153-59; *MARKHAM, C.R., 1881, pp. 138-49; DODGE, 1961, pp. 140-42; NEATBY, 1958, pp. 35-38; DCB, Vol. 1, pp. 74-75.

1619-20

Danish North-west Passage expedition	Jens Munk Jens Hendrichsen	*Enhiörningen* *Lamprenen*

From Denmark; 16 May 1619 - 21 September 1620; sponsored by King Christian IV. Sent to search for a North-west Passage by way of Hudson Strait, they entered Frobisher Bay by mistake and reached Digges Islands only on 20 August. They crossed to the west coast of Hudson Bay at lat 63°20'N, then coasted south to the mouth of Churchill River, where they wintered. Scurvy broke out and, of the original 65 men, only Munk and two others survived to the next summer. These three set out on 16 July 1620 and managed to sail *Lamprenen* to Norway.

*GOSCH, 1897; *HANSEN, 1970; DODGE, 1961, pp. 144-53; DCB, Vol 1, pp. 514-15; YOUNG, D., 1973.

1625

British North-west Passage expedition	William Hawkeridge ?	*Lion's Whelp* 1 other ship

From England; June - (?); sponsored by the Northwest Company. They passed through Hudson Strait in July, keeping to its northern shore. Nothing is known of their course in Hudson Bay, except that they kept among the islands in the north of the bay, near the entrance to Foxe Channel, and accomplished nothing.

*FOXE, 1635, pp. 166-71; *CHRISTY, 1894, Vol. 1, pp. 248-59; *CHRISTY, 1927; ROYAL GEOGRAPHICAL SOCIETY, 1927; RUNDALL, 1849, pp. 150-52; GOSCH, 1897, pp. lxxxvi-xciv; DODGE, 1961, pp. 153-55; DCB, Vol. 1, p. 362.

1631

British North-west Passage expedition | Luke Foxe | *Charles*

From Deptford; 5 May - 31 October; sponsored by London merchants. He entered Hudson Strait on 22 May, then followed the coasts of Coats and Southampton islands to Cape Fullerton in Roes Welcome Sound, worked down the western shore of Hudson Bay to Port Nelson, and then explored the south-western shore of the bay to Cape Henrietta Maria, thus linking the discoveries of Hudson in 1610 with those of Button in 1612-13. He met James (see 1631-32) during this stage of his exploration. Returning northward, he discovered the Sleeper Islands and explored the Baffin Island shore of Foxe Basin to lat 60°47'N. Here scurvy broke out, so he returned to England.

*FOXE, 1635; *CHRISTY, 1894; DODGE, 1961, pp. 156-64; MERBS, 1969; 1971; DCB, Vol. 1, pp. 311-12.

1631-32

British North-west Passage expedition | Thomas James | *Henrietta Maria*

From Bristol; 3 May 1631 - 22 October 1632; sponsored by Bristol merchants. James was not a trained seaman, and he encountered many obstacles in passing through Hudson Strait and crossing Hudson Bay to Churchill, where he arrived on 11 August 1631. He then explored the south-western shore of the bay from Port Nelson to Cape Henrietta Maria, near which he fell in with Foxe (see 1631). He continued down the western shore of James Bay, discovering several islands. The ship ran aground in shoal water near Charlton Island, and, to prevent her from breaking up, James decided to sink her, and wintered on the island. Scurvy became serious in February 1632. In May they found vetches and sorrel to eat and cure it, and they managed to refloat the ship. Later, James was nearly killed when, in an attempt to attract the attention of the natives, he lit a fire, which spread beyond control. On 1 July he sailed north, visited Nottingham Island, and returned to England.

*CHRISTY, 1894; *BODILLY, 1928; TOMLINSON, 1951; DODGE, 1961, pp. 165-77; DCB, Vol. 1, pp. 384-85; MacINNES, 1967.

1640

Spanish North-west Passage expedition | Bartholomew de Fonte | *St Spiritus*
| Diego Pennelossa | *St Lucia*
| Pedro de Barnarda | *Rosario*
| Philip de Ronquillo | *King Philip*

In 1708, a London magazine published a detailed account of the discovery of a North-west Passage by a fictitious explorer, Bartholomew de Fonte. The editor of the magazine, James Petiver, is now believed to have invented the story, but it was widely credited during the 18th century, and several expeditions were sent to search for de Fonte's passage. The expedition allegedly left Callao, Peru, on 3 April 1640, with orders to search for a North-west Passage.

They sailed up the west coast of America to lat 53°N, where they discovered an opening they called Rio de los Reyes. From here de Barnarda explored a channel leading in a north-easterly direction towards northern Baffin Bay, but which had no outlet. De Fonte, however, explored a channel leading eastward, through Lake Belle and Lake de Fonte. He did not sail all the way through the passage, but he met a Boston ship that had come from the east. He then returned home.

WILLIAMS, 1962, pp. 277-82; WAGNER, 1931; DODGE, 1961, pp. 182-83; DCB, Vol. 1, p. 677.

1647

French overland expedition — Jean de Quen

From Tadoussac; 11 July - (?). De Quen, in charge of the Jesuit mission to the Montagnais Indians at Tadoussac, learned that some neophytes on their way to the mission had fallen ill at Lake St John. With two Montagnais guides, he covered the 200 km to the lake in five days. He returned, having discovered not only Lake St John, but also the route leading into the interior of the Saguenay region, which the Indians had until then kept secret.

DCB, Vol. 1, pp. 558-59.

1651

French overland expedition — Jacques Buteux

From Trois Rivières; 27 March - 18 June. Buteux explored St Maurice River northward to its source, and there met Indians who had had no contact with white men.

DCB, Vol. 1, pp. 142-43.

1652

French overland expedition — Jacques Buteux

From Trois Rivières; 4 April - 8 June. Wishing to extend the explorations he had begun in 1651, Buteux set out northward again in the first of several unsuccessful attempts by the French to reach Hudson Bay overland. After five weeks of forced marches with some 60 Attikamegue Indians, the party was attacked by Iroquois, and Buteux was killed on 10 May. Thomas Tsondoutannen, a Huron member of the party, returned with this news to Trois Rivières on 8 June.

DCB, Vol. 1, pp. 142-43.

1657

French expedition — Jean Bourdon — ?

From Quebec; 2 May - 11 August; sponsored by the Conseil Souverain de Québec to take possession of Hudson Bay for the French. He was apparently forced to turn back near the mouth of Kaipokok River on the Labrador coast in lat 55°N, where his Huron guides had been killed by Eskimos. The French later tried to claim

Hudson Bay by right of this voyage, asserting that Bourdon had in fact reached the bay and formally taken possession of it.

DCB, Vol. 1, pp. 111-13.

1658

La Traite de Tadoussac or King's Posts

The King of France granted to Sieur Demaure a lease, La Traite de Tadoussac, which gave him exclusive trading, hunting and fishing rights from Isle aux Coudres along the north shore of the St Lawrence River to near Sept-Îles, and north to the heads of the rivers draining that coast. This territory was known as the Domaine du Roi; trading posts established within it as Postes du Roi. In 1733, the northern boundary of the Domaine was redefined to include Lake Mistassini and the upper waters of Eastmain, Fort George, Kaniapiskau, Hamilton and Rupert rivers. By this time, the inland posts in the Domaine included Chicoutimi (established c 1650), Lake St John (c 1650), Ashuapmouchuan (1690), Mistassini (c 1673), Nemiscau (1685), Nikabau (before 1700) and Nichicun (before 1725).

The lease on the Postes du Roi passed to the Compagnie des Postes du Roi about 1700. About 1788, it was granted by the British Government to the North West Company; it passed to the Hudson's Bay Company in 1821, and the monopoly finally expired by limitation in 1859.

VOORHIS, 1930, pp. 12-13; GREAT BRITAIN, PRIVY COUNCIL, 1927, pp. 3125-3131.

1659-60

French overland expedition — Pierre Esprit Radisson, Médard Chouart des Groseilliers

In August 1659, Radisson and Groseilliers left Quebec and travelled up the St Lawrence and Ottawa rivers to the Great Lakes on a fur-trading expedition. They wintered among the Indians south of Lake Superior. In 1660, they crossed Lake Superior to its north shore and then, according to Radisson, travelled northward to Hudson Bay. However, it is now generally agreed that they did not have sufficient time to make a trip to Hudson Bay before returning to Quebec that same summer, and that Radisson invented this part of his narrative in order to win support for his plans to extend the fur trade to Hudson Bay.

NUTE, 1943; DCB, Vol. 1, pp. 224-25, Vol. 2, p. 536.

1661

Seigneurie de Mingan

On 25 February, the Compagnie des Cent-Associés granted to François Byssot de la Rivière a tract of land along the north shore of the St Lawrence from the eastern limit of the Domaine du Roi (see 1658) eastward to Goynish River. He also had fishing rights over much of the Quebec-Labrador peninsula. Byssot set up a trading post at Mingan, but otherwise he seems to have made little use of the grant.

GREAT BRITAIN, PRIVY COUNCIL, 1927, pp. 3144-64; VOORHIS, 1930, pp. 114-15; DCB, Vol. 1, pp. 145-46.

1661

French overland expedition | Claude Dablon, Gabriel Druillettes

The second unsuccessful attempt by the French to reach Hudson Bay overland; their object was geographical exploration and the extension of missionary activities to the north-west of the St Lawrence. They ascended Saguenay River to Lake St John and continued overland to Nikabau. Here their Montagnais guides took fear of the Iroquois and refused to continue, so the expedition had to return to Quebec. Later the French, in their dispute with the English over Hudson Bay, claimed that they had reached the mouth of Rupert River and had erected a small trading post.

DCB, Vol. 1, pp. 244, 281-82.

1663

British fur-trading voyage | Pierre Esprit Radisson, Médard Chouart des Groseilliers, (captain ?) | ?

From Massachusetts; sponsored by Boston merchants; to open a sea route to Hudson Bay for the purpose of fur trading. The ship left so late that the captain decided to turn back when faced with the hazards of Hudson Strait.

RICH, 1958, p. 29.

1663

French overland expedition | Guillaume Couture, Pierre Duquet de la Chesnaye, Jean Langlois

From Quebec; mid-May - (?); to find an overland route to Hudson Bay. In this third unsuccessful attempt by the French to reach Hudson Bay overland from the St Lawrence, Couture and his companions travelled by way of Saguenay River, Lake Mistassini, and Rupert River to Lake Nemiscau, where their Indian guides refused to continue. They traded with the Indians there and handed out presents, then returned to Quebec.

MARTIJN AND ROGERS, 1969, p. 73; DCB, Vol. 2, pp. 157-59.

1668-69

British fur-trading voyage | Zachariah Gillam, William Stannard | *Nonsuch*, *Eaglet*

From London; 5 June 1668 - (?) October 1669; sponsored by Prince Rupert and London merchants.

After their abortive attempt to reach Hudson Bay from Boston (see 1663), Radisson and Groseilliers turned to London for support in their attempt to open a sea route to Hudson Bay for the fur trade. Radisson was aboard *Eaglet,* which turned back to Plymouth in August 1668. *Nonsuch* reached Hudson Bay with Groseilliers on board. They built Fort Charles (renamed Rupert House in 1670) at the mouth of Rupert River and, while wintering there, carried on a successful fur trade with the Indians. They returned to England in 1669. The success of this voyage was largely responsible for the establishment of the Hudson's Bay Company.

NUTE, 1943; RICH, 1958, pp. 36-43; NUTE, 1945.

British fur-trading voyage | William Stannard, Pierre Esprit Radisson | *Wivenhoe*

1669

Forced to turn back on their previous attempt to sail to Hudson Bay (see 1668-69), Stannard and Radisson set out from England again in May 1669, hoping to join their companions who were wintering in the bay on *Nonsuch.* They again failed in their attempt, although they probably reached Hudson Strait before turning back.

RICH, 1958, p. 65.

Incorporation of the Hudson's Bay Company

1670

Following the success of the voyage of *Nonsuch* in 1668-69, King Charles II granted a charter on 2 May to the 18 adventurers who had subscribed to support that voyage. They incorporated under the name of "Governor and Company of Adventurers of England trading into Hudson's Bay." Prince Rupert was nominated governor, and the company was granted the "sole Trade and Commerce of all those Seas Streightes Bayes Rivers Lakes Creekes and Soundes in whatsoever Latitude they shall bee that lye within the entrance of the Streightes commonly called Hudsons Streightes together with all the Landes and Territoryes upon the Countryes Coastes and confynes of the Seas Bayes Lakes Rivers Creekes and Soundes aforesaid that are not actually possessed by the Subjectes of any other Christian Prince or State." This territory, called Rupert's Land, was to be considered as a colony and the company held mineral and fishing rights as well as the exclusive trade.

RICH, 1958, pp. 52-60.

Hudson's Bay Company fur-trading voyage; establishment of Rupert House | Zachariah Gillam, Médard Chouart des Groseilliers, Charles Bayly | *Prince Rupert*
| Robert Newland, Pierre Esprit Radisson | *Wivenhoe*

1670-71

From London; 31 May 1670 - 26 October 1671; the first voyage under the charter of the new Hudson's Bay Company, to establish permanent posts in James Bay and at Nelson River. *Wivenhoe* was nearly wrecked on Mansel Island but continued to Nelson River to build a post. No Indians came and, after Newland's death, his men abandoned the site and went to Rupert River.

Prince Rupert reached Rupert River on 8 September. Fort Charles (see 1668) was still standing, but Governor Bayly immediately began to build another dwelling, eventually known as Rupert House.

1671: Radisson and Bayly on an exploratory excursion traded with the Indians near the mouth of Moose River. Bayly's recommendation of that region led to the establishment of Moose Factory (see 1673). Bayly could not persuade enough men to stay at Rupert House a second winter, so they all sailed home on 1 July.

On return, they explored parts of the east coast of Hudson Bay and found relics of James's wintering on Charlton Island (see 1631-32).

RICH, 1958, p. 66; 1948, pp. 348-54.

1671-72

French overland expedition — Charles Albanel SJ, Paul Denis de St Simon, Sébastien Pennara

From Quebec; 6 August 1671 - (?) 1672. Jean-Baptiste Talon, Intendant of New France, sent this expedition to take possession of the country around Hudson Bay for the King of France and to persuade the Indians of James Bay to trade with the French, rather than with the Hudson's Bay Company. Albanel met his guides, St Simon and Pennara, at Tadoussac. At Lake St John, Indians informed him of two English ships in James Bay (see 1670-71). He therefore sent to Quebec for passports, which arrived too late for him to continue the journey that year.

1672: They set out on 1 June with 16 Indians, passed through lakes Mistassini and Nemiscau, and down Rupert River to Rupert House. There they found two empty houses and sighted one of the English ships, but they failed to meet the Englishmen and set out homeward on 6 July.

*ROUSSEAU, 1950; 1950a; RICH, 1958, pp. 74-75; DCB, Vol. 1, pp. 47-50.

1672-75

Hudson's Bay Company fur-trading voyage	Zachariah Gillam	*Prince Rupert*
	Samuel Cole	*Employ*
	Robert Morris	*Messenger*
	Charles Bayly	

From London; June 1672 - (?) 1675; the second attempt to establish permanent posts in James Bay and at Nelson River (see 1670-71). Bayly was again sent as governor.

1673: After a successful trade at Rupert River, *Prince Rupert* and *Messenger* returned to England. Bayly, from Rupert House, built a post at Moose River (see 1673, below), and made a second unsuccessful attempt to establish a post at Nelson River. The men sent there found relics of Button's expedition (see 1612-13), which were later taken to England.

1674: Bayly made trading expeditions to Moose, Albany and Severn rivers (see 1674, below). Thomas Gorst explored Nottaway River to its first falls. In September, *Prince Rupert* and *Shaftesbury* arrived with fresh provisions (see 1674-75).

1675: All the ships returned home, leaving Bayly with three men at Rupert House.

RICH, 1958, pp. 70-79.

1673

Establishment of Moose Factory

Following Radisson and Bayly's visit to Moose River in 1671 (see 1670-71), the Hudson's Bay Company decided to build a post there for occasional occupation. This was done in 1673, and the post became known as Fort Moose. Bayly visited Fort Moose again in summer 1674 and carried on a successful trade (see below, 1674). He returned to pass winter 1674-75 there and may have made it his headquarters, in preference to Rupert House, subsequently. When John Nixon replaced Bayly as governor in 1679, he was encouraged to keep his main strength at Moose to prevent French encroachments on the west coast of James Bay. During the English-French conflict in Hudson Bay, Fort Moose exchanged hands a number of times, until the English demolished it (see 1686 and 1692-93).

RICH, 1948, pp. 357-61; *MOCCASIN TELEGRAPH*, 1970, pp. 105-06.

1674

Hudson's Bay Company fur-trading voyage | Charles Bayly | [*Employ?*]

In late spring, Bayly sailed from Rupert House to Moose River, where the company had built an outpost (see 1673). He made a successful trade with Indians who had come to meet him from Albany River. He then proceeded to Albany River, where he stayed 18-21 July, but found little fur. He promised the Indians he would return the next year, but it is not known if he did so. He then searched for Thomas James's Severn River (see 1631-32), which he apparently found, but did not trade there. He intended to continue to Nelson River in a third attempt to establish a post there, but ice conditions forced him to return to Rupert House. On return he was driven ashore on Charlton Island, but got off again.

RICH, 1948, pp. 345, 354, 358, 361.

c1674

French fur-trading expedition | ?

According to Thomas Gorst (see 1672-75), the French had a trading post not more than eight days' journey up Rupert River (probably at Lake Mistassini) to intercept Indians going down to trade at Rupert House. Governor Bayly considered moving his headquarters to Moose Factory to avoid such interception (see 1673).

RICH, 1948, pp. xxxviii, 350-51.

1674-75

French overland expedition | Charles Albanel SJ

From Tadoussac; 13 January 1674 - (?) 1675; second French attempt to persuade the Indians of James Bay to trade with them, rather than with the English, and also to urge Chouart des Groseilliers, a Frenchman employed by the Hudson's Bay Company, to change his allegiance to the French. An accident delayed Albanel some months at Lake St John, and he reached Rupert House on 30 August 1674. There the English took him prisoner and sent him to London the next spring. He was soon freed, continued to France, then returned to Quebec.

ROUSSEAU, 1950; DCB, Vol. 1, pp. 47-50.

1674-75	Hudson's Bay Company supply voyage	Zachariah Gillam William Lydall Thomas Shepard	*Prince Rupert* *Shaftesbury*

From London; they sailed to Rupert House to relieve Bayly's wintering party (see 1672-75) and brought out a new governor, William Lydall. After spending the winter at Rupert House, Lydall decided to return to England, and left Bayly, with three men, to continue as governor.

RICH, 1958, p. 78.

1676	Hudson's Bay Company supply voyage	Joseph Thompson	*Shaftesbury*

From London. Routine voyage to James Bay and back.

RICH, 1958, p. 80.

1677	Hudson's Bay Company supply voyage	Joseph Thompson	*Shaftesbury*

From London. Routine voyage to James Bay and back.

RICH, 1958, p. 82; 1945, p. 342.

1678	Hudson's Bay Company supply voyage	Richard Power Joseph Thompson	*Prince Rupert* *Shaftesbury*

From London. *Prince Rupert* was ordered to winter in James Bay. *Shaftesbury*, returning with the furs in 1678, was wrecked off the Scilly Isles. *Prince Rupert* remained in James Bay until 1681.

RICH, 1945, pp. 340-42; 1958, p. 82; DCB, Vol. 1, pp. 553, 642.

1679	Hudson's Bay Company supply voyage	Nehemiah Walker John Nixon James Tatnam	*John and Alexander* *Colleton*

From London; May-December. *Colleton* turned back early on the outward voyage, whilst Captain Tatnam happened to be on board *John and Alexander*, which continued with him to Rupert House. John Nixon replaced Governor Bayly, who had been in James Bay continuously since 1672.

RICH, 1958, p. 90; DCB, Vol. 1, p. 666.

1679	French overland expedition	Louis Jolliet Zacharie Jolliet Etienne Lessard Antoine Silvy SJ	

From Quebec; 13 April - 25 October; sponsored by Comte de Frontenac, Governor of New France, probably to spy on English trading activities in James Bay. In addition, Silvy intended to found a mission at Lake Nemiscau. Jolliet's party travelled from Tadoussac to Chicoutimi, where Silvy joined them. Leaving Silvy at Lake Nemiscau, the rest continued to Rupert House, where they met Governor Bayly and enjoyed his hospitality. A few days later, they returned up Rupert River, through lakes Mistassini, Albanel and St John, to Quebec.

Jolliet built a trading post at Lake Mistassini, leaving his brother Zacharie in charge. Whether or not this post was the first on the lake is not clear. Certainly there had been intermittent trade there before this date (see 1674), and, very likely, at Nemiscau as well. According to Radisson, the French had built a post by 1681 less than one day's journey above Rupert House (probably at Nemiscau, which is about three days' journey upriver).

BURGESSE, 1947; DELANGLEZ, 1944; DCB, Vol. 1. pp. 392-98.

c1679

Establishment of Fort Albany

Governor Bayly, during a voyage of trading and discovery around James Bay, had first visited Albany River (see 1674). It is not certain when a post was built but, before the end of 1679, the company had a "house of some strength" there. In 1686, the French captured the post and renamed it Fort Sainte-Anne. In 1696, after the post had changed hands a number of times, the English recaptured it. Albany was the only post the English held from that date until the Treaty of Utrecht (see 1713).

RICH, 1948, pp. 5, 345-48; KENYON AND TURNBULL, 1971, pp. 117-26; KENYON, 1961; 1965; 1973.

1680-81

Hudson's Bay Company supply voyage; establishment of Charlton Depot	Richard Greenway Thomas Draper Walsall Cobbie	*Prudent Mary* *Albemarle* *Colleton*

From London; 2 June 1680 - (?) 1681; the ships were ordered to Charlton Island, where the company had decided to build a warehouse and a depot for the other posts on James Bay. *Albemarle* and *Colleton* apparently wintered there; *Prudent Mary,* returning to London in 1680 with the fur returns, was wrecked in Hudson Bay. Governor Nixon did not favour Charlton Island as a depot, but it continued so until 1686, when the company ordered it abandoned. Before these orders could be carried out, the post was captured by the French (see 1686).

1681: *Albemarle* sailed to London with *Prince Rupert* (which had been in James Bay since 1678); *Colleton* remained permanently in James Bay.

RICH, 1948, pp. 354-57; KENYON, 1974.

1681-82

Hudson's Bay Company supply voyage	Nehemiah Walker	*Diligence*

From London; June 1681 - September 1682. After wintering at Charlton Island with Governor Nixon, Walker returned to London in 1682 with the fur returns.

RICH, 1948, p. 30n, 355; DCB, Vol. 1, p. 666.

1682-83

Hudson's Bay Company supply voyage	Zachariah Gillam	*Prince Rupert*
	Ezbon Sandford	*Albemarle*
	William Bond	*Craven*
	Samuel Edwards	*Friendship*
	Nicholas Reymer	*Lucy*

From London. *Craven, Friendship* and *Lucy* were ordered to make the regular voyage to Charlton Depot, and *Prince Rupert* and *Albemarle* to sail to Nelson River with John Bridgar in a fourth attempt to establish a trading post there (see 1670-71, 1672-75, and 1674). *Friendship* and *Lucy* returned to London in 1682, without furs. *Craven* remained permanently in James Bay.

Captain Sandford died on the outward voyage, so *Albemarle* sailed to James Bay and *Prince Rupert* went alone to Nelson River, where she sank on 21 October. Zachariah Gillam and nine crew were lost. The survivors were taken prisoner by Radisson and Groseilliers' interloping expedition and, in 1683, carried in Benjamin Gillam's ship to Quebec and there set free (see 1682-83, below).

Albemarle sailed from James Bay to Nelson River in 1683, was also captured by the French, and kept there until 1695, when they sailed her to Europe (see 1694-95).

RICH, 1958, p. 133.

1682-83

New England fur-trading voyage	Benjamin Gillam	*Bachelor's Delight*

From Boston; 21 June 1682 - 20 October 1683; private enterprise. Knowing the Hudson's Bay Company's intention to establish a trading post at Nelson River, Gillam (Zachariah Gillam's son) decided to establish a rival post there. He reached Nelson River on 18 or 19 August. Soon after, the French interlopers arrived (see 1682-83, below), captured his house and ship and took him and his crew prisoner. In 1683, the French took him, his ship and crew, and the survivors of Zachariah Gillam's expedition (see 1682-83, above) to Quebec and set them free.

RICH, 1958, p. 135; DCB, Vol. 2, p. 247.

1682-83

French fur-trading voyage; establishment of Fort Bourbon	Pierre Esprit Radisson	*St Pierre*
	Médard Chouart des Groseilliers	*Ste Anne*

From Quebec; (?)1682 - 20 October (OS) 1683; private enterprise. Radisson and Groseilliers had deserted the Hudson's Bay Company and, like Benjamin Gillam (see 1682-83, above), knew the Hudson's Bay Company intended to establish a post at Nelson River and decided to rival them. They reached Nelson River about 20 August (OS) and captured Gillam, his crew, and the Hudson's Bay Company's servants. They then removed to the mouth of Hayes River and built Fort Bourbon.

1683: *St Pierre and Ste Anne* were damaged in the spring; the former was abandoned, and some of the Hudson's Bay Company prisoners were allowed to take the latter to James Bay. Radisson and his men returned to Quebec with *Bachelor's Delight.* Jean-Baptiste Chouart, Radisson's nephew, and a small party occupied Fort Bourbon until they were captured by the English (see 1684).

NUTE, 1943, pp. 188-97; RICH, 1948, p. 103; 1958, p. 135.

Hudson's Bay Company supply voyage; establishment of Fort Hayes	Nehemiah Walker John Abraham (out) William Bond (home) Henry Sergeant	*Diligence* *George*	1683

From London; they captured the interloping ship *Charles* (see 1683, below) and took her to Charlton Depot. Sergeant replaced John Nixon, governor since 1679. Mrs Sergeant and her companion or servant Mrs Maurice were the first English women known to have gone to Hudson Bay. Abraham then took *George* and *Albemarle* (see 1682-83 for *Albemarle's* arrival) to Nelson River, where he found the French in possession (see 1682-83). He built Fort Hayes near Fort Bourbon and competed for trade with the Indians. *Diligence* and *George* returned to London in October.

RICH, 1948, p. 389; 1958, pp. 146-51.

British fur-trading voyage	Richard Lucas	*Charles*	1683

From Dartmouth; private enterprise. Lucas hoped to trade surreptitiously in Hudson's Bay Company territory, but was captured on the outward voyage (see 1683, above). He and his crew were taken prisoner to James Bay, then returned to England on company vessels. *Charles* was wrecked later by Nehemiah Walker on Charlton Island.

RICH, 1946, pp. 263-85, 330; 1958, pp. 149-50; DCB, Vol. 1, pp. 476-77.

Hudson's Bay Company supply voyage; establishment of York Factory	Leonard Edgcombe William Bond John Outlaw Pierre Esprit Radisson George Geyer	*John and Thomas* *Happy Return* *Lucy*	1684

From London. *John and Thomas* made the regular voyage to Charlton Depot and back. *Happy Return* and *Lucy* sailed to Fort Hayes (see 1683). Radisson, who had now rejoined the Hudson's Bay Company, persuaded the Frenchmen (led by his nephew), who had occupied Fort Bourbon on Hayes River since 1682, to surrender their furs to the company, abandon their fort, and to return to Europe on the ship *Lucy.* Geyer began constructing a new post on Hayes River. Later known as York Factory, this post changed hands a number of times until taken by the French in 1697. It was restored to the Hudson's Bay Company with others in 1713.

RICH, 1938, pp. 426-27; *1948, pp. 363-69; 1958, pp. 148, 154, 201-02; DCB, Vol. 1, p. 301.

1684-85	French fur-trading voyage	Claude de Bermen de la Martinière Pierre l'Allemand P. Moisan Antoine Silvy SJ	2 ships

From Isle-aux-Coudres; 12 July 1684 - autumn 1685; sponsored by the Compagnie du Nord to supply Fort Bourbon (see 1682-83). They reached the post on 22 September to find it had been taken by the English (see 1684). They established a post near by and spent the winter quarrelling with their neighbours. In spring 1685, they traded some furs from the Indians and sailed for Quebec in July.

TYRRELL, J.B., 1931.

1685	Hudson's Bay Company supply voyage	Richard Lucas William Bond John Outlaw Edward Hume	*Owner's Goodwill* *Happy Return* *Success* *Perpetuana Merchant*

From London; the French captured *Perpetuana Merchant* at Digges Islands on 27 July during the outward voyage. *Happy Return* made the routine voyage to York Factory and back; *Owner's Goodwill* and *Success* to Charlton Island and back. *Success* sank during her return voyage on 11 December.

RICH, 1958, pp. 202-03; DCB, Vol. 1, pp. 107, 476-77, 529.

1685 Establishment of Fort Severn

The Hudson's Bay Company had ordered Governor Nixon to build a post at Severn River in 1680. After occasional visits to the location the post, first known as Churchill Fort, was built by 1685. The English burned it in 1690 to prevent its capture by the French. In 1691, the French built a new post near the site, which they held most of the time until 1713, when the Treaty of Utrecht restored it to the English. Fort Severn was abandoned in 1713, but was reestablished in 1759 *(qv)*.

RICH, 1948, pp. 361-63.

1686	Hudson's Bay Company supply voyage	Robert Porten William Bond	*Abraham and Robert* *Happy Return*

From London; *Happy Return*, carrying instructions to abandon Charlton Depot, sank in Hudson Strait on 10 July on her way there. *Abraham and Robert*, having saved the crew of *Happy Return*, continued to York Factory, then returned to England in October.

RICH, 1958, p. 219; DCB, Vol. 1, p. 107.

1686	Hudson's Bay Company exploring expedition	Michael Grimington John Abraham	*Hayes*

From York Factory; 25 June - 25 July. Rediscovered Churchill River (see 1612-13). They recommended the site for a trading post, and an unsuccessful attempt to establish one was made in 1688. Although they saw signs of natives they met none. Pieces of iron they found were thought to have been left by Jens Munk (see 1619-20).

*RICH, 1948, p. 340; 1958, p. 243; KENNEY, 1932, p. 18; DAVIES, 1965, p. 370.

1686

French overland expedition — Pierre de Troyes, Pierre Le Moyne d'Iberville

From Montreal; 20 March - early October; sponsored by the Compagnie du Nord to capture Hudson's Bay Company posts in James Bay in retaliation for the loss of Fort Bourbon (see 1684). The detachment of 100 men travelled in 35 canoes by the Ottawa River and lakes Timiskaming and Abitibi and reached James Bay on 20 June. They took Moose Fort on 21 June, Rupert House on 3 July, and Albany on 17 July, all by surprise, and changed their names to St-Louis (Moose), St-Jacques (Rupert) and Ste-Anne (Albany). De Troyes left for Quebec on 19 August, leaving Iberville as governor with 40 men. Fifty-two Englishmen were sent in the ship *Colleton* to York Factory, the rest were kept prisoner in James Bay over the winter.

*KENYON AND TURNBULL, 1971; RICH, 1958, pp. 212-27; MATHEWS, 1966.

1686

Treaty of Neutrality

This agreement between Louis XIV and James II in November provided for the continuance of peace between the French and English dominions in America, even in the event of war in Europe, and for a commission to settle claims. Clause IV provided that each king should continue to hold what he held at that date. The English forts captured by de Troyes in summer 1686, an accomplishment unknown to the negotiators of the treaty, remained, therefore, in French hands.

RICH, 1948, pp. 229-30, n.2.

1687

English prisoners sent overland

The prisoners not sent to York Factory in *Colleton* (see 1686, French overland expedition) were obliged to shift for themselves in the woods around Fort St-Jacques (formerly Rupert House). In June they were sent up to Nemiscau to fish, then, in late July, carried in French canoes overland to Quebec, kept on short rations, and made to carry at the portages. They were the first Englishmen to cross overland from Hudson Bay to St Lawrence River.

RICH, 1958, pp. 242-43.

1687-88

Hudson's Bay Company supply voyage		
	William Bond	*Dering* [I]
	Richard Smithsend	*Huband*
	Leonard Edgcombe	*John and Thomas*

From London; because the posts in James Bay were held by the French, the ships sailed direct to York Factory. *Huband* wintered there, the others returned to London in autumn 1687. In spring 1688, *Huband* sailed to Charlton Island and was captured by the French.

RICH, 1958, p. 228; HUDSON'S BAY COMPANY (2).

1688	British fur-trading voyage	John Abraham John Outlaw	*Mary*

From England; private enterprise by two former Hudson's Bay Company employees intent on rivalry. In Hudson Strait, when their ship was near sinking, they were taken prisoner aboard *Churchill* (see 1688-89) and carried to James Bay, where they were all captured by the French.

RICH, 1958, p. 237.

1688-89	Hudson's Bay Company supply voyage	James Young Leonard Edgcombe William Bond John Simpson	*Dering* [I] *John and Thomas* *Churchill* *Yonge*

From London; 6 June 1688 - (?) 1689. *John and Thomas* and *Dering* sailed to York Factory; the former returned in 1688, the latter in 1689, narrowly escaping capture by Iberville (see 1688-89, below). *Churchill* and *Yonge* sailed to Albany to trade but, during the winter, Iberville captured Bond and others and, in spring 1689, he captured the ships. He then sent Le Moyne de Sainte-Hélène (his brother) overland to Quebec with Bond and several other prisoners. The other English survivors were given *Huband* (see 1687-88) for their return to England. Iberville sailed *Churchill* to Quebec, nearly capturing *Dering* on the way.

RICH, 1958, p. 250; HUDSON'S BAY COMPANY (2).

1688-89	French fur-trading voyage	Pierre Le Moyne d'Iberville ?	*Soleil d'Afrique* 2 other ships

From Quebec; June 1688 - (?) 1689; sponsored by the Compagnie du Nord (also commonly known as the Compagnie de la Baie d'Hudson). Iberville reached Charlton Island, which the French were using as a depot, on 9 September, so late that the furs had been taken back from there to Fort St-Louis (Moose Factory). He retrieved the furs, which *Soleil d'Afrique* carried to La Rochelle that autumn. He then went to Albany and there met *Churchill* and *Yonge*, which he captured by a series of ruses (see 1688-89, above). He returned to Quebec in the captured *Churchill* in 1689, having sent some English prisoners to Canada overland and others out of Hudson Bay by ship.

RICH, 1958, pp. 244-48.

French overland expedition — Jacques Le Moyne de Sainte-Hélène — 1689

From Quebec; 5 July - November; sponsored by the Compagnie du Nord. Sainte-Hélène crossed overland to Fort Ste-Anne (Albany) with 38 men to supply his brother, Iberville, and to bring back his fur returns in ships captured from the Hudson's Bay Company. His plans were changed on arrival at Fort Ste-Anne, and, instead of returning by ship, he escorted a party of English prisoners overland back to Quebec (see 1688-89).

DCB, Vol. 1, pp. 465-66; RICH, 1958, p. 247.

First founding of Churchill	James Young	*Hopewell*	1689
	Amos Dumuresq	*Albemarle*	
	Thomas Savage		
	Ekie Grimard		
	Henry Kelsey		

The Hudson's Bay Company had ordered the establishment of a post on Churchill River in 1687, following Abraham's description of the place (see 1686), which had mentioned thousands of white whales in the river. The ships left York Factory separately, in May and April, and, while the house was being built on Churchill River, Kelsey set out northward to find Indians to trade there (see 1689, below). However, a fire accidentally destroyed the unfinished house on 1 or 2 August, and they all returned to York. No further attempt to establish the post was made until 1717, although the nine tons of whale blubber sent home got a high price.

KENNEY, 1932.

Hudson's Bay Company expedition	Henry Kelsey		1689
	James Young	*Hopewell*	

From Churchill; 17 June - 28 July. On 27 June, because of ice conditions only 100 km north of Churchill, Kelsey and an Indian boy went ashore and walked at least 200 km further north in search of Indians to trade at Churchill (see 1689, above). They met none and the boy, afraid of meeting Eskimos, refused to continue. Kelsey saw a musk-ox on this journey, the first white man recorded to do so.

KENNEY, 1932, pp. 21-22.

French fur-trading voyage	Pierre Le Moyne d'Iberville	*Ste Anne*	1690
	?	*St François-Xavier*	
	?	*Armes de la Compagnie*	

From Quebec; sponsored by the Compagnie du Nord, Iberville, now commissioned as governor of the posts in Hudson Bay, attempted to capture York Factory, but was driven off. He then attacked Fort Severn (see 1685), but the men burned the post to prevent its capture and travelled overland to York. It is not clear whether or not Iberville wintered in Hudson Bay. York Factory was now the Hudson's Bay Company's single post.

RICH, 1958, p. 289.

1690-91

Hudson's Bay Company supply voyage	James Young	*Dering* [II]
	Leonard Edgcombe	*Royal Hudson's Bay*
	Michael Grimington	*Prosperous*

From London; June - October. Routine voyage to York Factory. The fire-ship *Prosperous* (formerly *Dering* [I]) remained in Hudson Bay until 1691, the other two ships returned to London.

RICH, 1958, pp. 264, 276-77; HUDSON'S BAY COMPANY (2).

1690-92

Hudson's Bay Company overland expedition	Henry Kelsey	

From York Factory; 12 June 1690 - (?) 1692. In an effort to offset losses to the French, Kelsey was sent inland with a group of Assiniboine Indians to bring interior Indians to trade at York. By 10 July he had gone about 950 km south-west of York Factory and set up a base camp, which he named Dering's Point, somewhere north of Lake Winnipeg. From here he travelled further inland, approximately 930 km, to the Gros Ventres Indians, and evidently passed the winter of 1691-92 among them. He returned to York with "a good Fleet of Indians" in 1692. He is the first European on record to see the Canadian prairies, grizzly bears, and bison.

MORTON, 1939, pp. 111-13; RICH, 1958, pp. 297-99.

1691

Hudson's Bay Company supply voyage	Leonard Edgcombe	*Royal Hudson's Bay*
	James Young	*Loyalty*

From London; 29 May - (?) October. Routine voyage to York Factory. They returned to London the same year in company with *Prosperous,* which had gone to York in 1690.

RICH, 1958, pp. 279-82; HUDSON'S BAY COMPANY (2).

1692-93

Hudson's Bay Company supply voyage; reoccupation of Albany	James Knight	
	Michael Grimington	*Royal Hudson's Bay*
	Henry Baley	*Prosperous*
	Charles Cotesworth	*Pery*
	Leonard Edgcombe	*Dering* [III]

From London. All the ships made the regular voyage to York Factory. *Dering* returned to London with the fur returns in 1692. The other ships wintered at Gilpin Island, off Eastmain River, and, in 1693, Knight used them to capture Fort Ste-Anne (Fort Albany). The French had already abandoned the post, except for a small garrison. Not having been supplied since 1690, they were starving and mutinous and retreated overland to Canada. Knight then entered Moose and Rupert rivers and burned Fort St-Louis and Fort St-Jacques, without attempting to reoccupy them. The French now held no posts on Hudson Bay, but continued their rivalry from Nemiscau (see 1679) and other inland posts. Two of Knight's ships then returned to London, but *Pery* remained in Hudson Bay until 1697.

RICH, 1958, p. 286; 1948, p. 336, n. 1.

Hudson's Bay Company supply voyage — Leonard Edgcombe, James Young — *Dering* [III], *Supply* — 1693

From London. The ships made the regular voyage to York Factory and Edgcombe, assured that Albany was in English hands, carried goods there. Both ships returned to London in October with *Prosperous* and *Royal Hudson's Bay* (see 1692-93).

RICH, 1958, p. 314; HUDSON'S BAY COMPANY (2).

French fur-trading voyage — ? — *Ste Anne* — 1693

From Quebec; July-November; sponsored by the Compagnie du Nord. Despite annual attempts, the French posts on Hudson Bay had not been supplied since 1690. James Knight had just recaptured Fort Ste-Anne (Albany) when the French ship arrived. *Ste Anne* escaped an English vessel sent in pursuit and, presumably, evacuated Fort St-Louis and Fort St-Jacques because, with Albany back in English hands, they were untenable.

RICH, 1958, p. 304.

Hudson's Bay Company supply voyage — Henry Baley, Michael Grimington — *Dering* [III], *Royal Hudson's Bay* — 1694

From London. The two ships made the regular voyage to York Factory, which post they left in August, before Iberville arrived (see 1694-95). They stopped at Albany, then returned to England.

RICH, 1958, p. 329; HUDSON'S BAY COMPANY (2).

French fur-trading and exploring voyage — Louis Jolliet — ? — 1694

From Quebec; 28 April - mid-October. He stopped at Mingan to trade and to reconstruct buildings burned by the English. On 9 June he turned northward along the Labrador coast. About 9 July he passed Cape Charles in lat 52°10'N and entered unknown waters. On 15 August, in lat 56°8'N, near present-day Zoar, he turned back to Quebec with the first detailed account and map of the Labrador coast between Cape Charles and Zoar.

EMMANUEL, 1959, pp. 122-24; DCB, Vol. 1, pp. 397-98.

1694-95

French fur-trading voyage; occupation of York Factory	Pierre Le Moyne d'Iberville	*Poli*
	?	*Salamandre*

From Quebec; c 10 August 1694 - October 1695; private enterprise. With ships from the French Royal Navy, Iberville sailed to Hayes River, which he reached on 24 September, and, in October, laid siege to York Factory, which soon surrendered. Iberville wintered fhere, making most of the Englishmen, including Henry Kelsey, shift for themselves in the woods.

1695: After trading with the Indians in the spring, Iberville waited through the summer hoping to catch the English supply ships. None had been sent, and he left on 7 September for La Rochelle, leaving Gabriel Testard de La Forest and 70 Canadians to occupy the post, again named Fort Bourbon (see 1682-83). The English were transported to prisons in France in *Albemarle,* a ship that had been in Hudson Bay since 1682.

RICH, 1958, p. 329.

1696

Hudson's Bay Company supply voyage; reoccupation of York Factory	Michael Grimington	*Royal Hudson's Bay*
	Nicholas Smithsend	*Knight*
	Henry Baley	*Dering* [III]
	William Allen	HMS *Bonaventure*
	Captain Watkins (or Grange?)	HMS *Seaford*

From London. The Hudson's Bay Company had sent no supply ship in 1695 and learned of the conquest of York Factory from the English prisoners, after their release from gaol in France (see 1694-95). Now, with the assistance of two Admiralty men-of-war, which served as convoy into Hudson Bay, the English laid siege to Fort Bourbon, which surrendered on 5 September on terms that were later altered and disputed. The French were taken prisoners back to England.

RICH, 1958, p. 336; HUDSON'S BAY COMPANY (2).

1696

French fur-trading voyage	Joseph Le Moyne de Serigny	2 ships

From France. Serigny, Iberville's brother, reached Fort Bourbon a few hours after the Hudson's Bay Company ships with the two men-of-war (see 1696, above). As his ships were small and the English were in greater force and in possession of Hayes River, he returned to France, losing one ship on the way.

DOUGLAS AND WALLACE, 1926, p. 27.

1697

French fur-trading voyage; reoccupation of York Factory	Pierre Le Moyne d'Iberville	*Pélican*
	Joseph Le Moyne de Serigny	*Palmier*
	?	*Esquimo*
	Captain Dugué	*Profond*
	Captain Chatrie	*Weesp*

From La Rochelle; sponsored by the Compagnie du Nord. *Esquimo* sank on the outward voyage. The others continued to Hayes River, having separated in Hudson Strait. *Pélican* arrived off York Factory first, but was soon followed by the English ships (see 1697, below). Against great odds, Iberville managed to sink HMS *Hampshire,* forced *Royal Hudson's Bay* to surrender, and put *Dering* to flight. The other French ships soon arrived and they besieged the fort. The English capitulated for the second time. Iberville left his brother, Serigny, in charge and Nicholas Jérémie (who was to spend 20 years in Hudson Bay) second in command. Then, in November, he sailed for France with the English fur returns.

RICH, 1958, p. 350.

1697

Hudson's Bay Company supply voyage	Nicholas Smithsend	*Royal Hudson's Bay*
	Michael Grimington	*Dering* [III]
	John Fletcher	HMS *Hampshire*
	Captain Lloyd	HMS *Owner's Love*

From London. *Owner's Love* was lost on the outward voyage. The others sailed to York Factory and there found Iberville in *Pélican.* They joined battle, which the English lost (see 1697, above). On 13 September, York Factory surrendered to the French, although not without a struggle. In this year, also, *Pery,* Captain Thomas Mann, returned to England from Albany after five years in Hudson Bay, with James Knight aboard.

RICH, 1958, p. 349; HUDSON'S BAY COMPANY (2).

1697

Treaty of Ryswick

Signed on 20 September, the treaty made peace between Great Britain and France and, among other things, attempted to settle the dispute between the Hudson's Bay Company and the French in Hudson Bay. Clause VII restored to each party the possessions each had held before the war. Clause VIII provided for the appointment of commissioners to settle the boundaries between Rupert's Land and Canada. The commissioners were limited by the agreement that "Possession of those Places which were taken by the French during the Peace that preceded this present War, and were retaken by the English during this War, shall be left to the French by virtue of the foregoing Article." In fact the terms of the treaty were never enforced in Hudson Bay and the dispute dragged on until the Treaty of Utrecht, 1713.

RICH, 1958, pp. 343-54.

1698-1700

Hudson's Bay Company supply voyage	Michael Grimington	*Dering* [III]
	Henry Baley	*Pery*
	James Knight	

From London. Both ships sailed to Albany, the only post still held by the Hudson's Bay Company after the Treaty of Ryswick (see

1697). They took out James Knight as governor. *Dering* returned to London the same year, *Pery* returned in 1700.

RICH, 1958, pp. 370-72; DCB, Vol. 2, pp. 42, 266.

1699

Hudson's Bay Company supply voyage | Michael Grimington | *Dering* [III]

From London. Routine voyage to Albany and back.

DAVIES, 1965, p. 335.

1701

Hudson's Bay Company supply voyage | Michael Grimington | *Pery*

From London. Routine voyage to Albany and back.

RICH, 1958, p. 376.

1702

French settlement of Labrador | Augustin Le Gardeur de Courtemanche

On 17 October 1702, Courtemanche was granted an exclusive 10-year trading concession to the region between Kégashka River on the north shore of the St Lawrence to Hamilton Inlet, then known as Kessessakiou or Baye des Esquimaux, on the coast of Labrador. He seems to have visited the region about 1704, and he wrote a glowing report of its riches. In 1704 also he moved his first post from Old Fort Bay to Baye de Phélypeaux (Brador Bay) and built a new post, Fort Pontchartrain. In 1712, Courtemanche's concession was renewed, although confined to the neighbourhood of Baye de Phélypeaux, and he was appointed commander for the King to adjudicate disputes related to the fisheries of Labrador. In 1716, he proposed to build a fort at Kessessakiou, but he died the next year. The concession was continued by his widow and his step-son, François Martel de Brouague. Courtemanche, just before his death, captured two Eskimo girls in a skirmish. They lived for two years with Mme de Courtemanche and Brouague, and they began to learn each other's language. Brouague's account of their conversations must therefore be the first detailed description of Eskimo culture in North America.

DCB, Vol 2, pp. 7-10, 383-84; GOSLING, 1910, pp. 131-53; GREAT BRITAIN, PRIVY COUNCIL, 1927, pp. 3175-3179. EMMANUEL, 1959, p. 126.

1702-03

Hudson's Bay Company supply voyage | Joseph Davis | *Hudson's Bay* [II]

From London; 21 July 1702 - November 1703. The ship was late leaving London and, on the assumption that she would be unable to return home before the onset of winter, Grimington was ordered to winter at Gilpin Island, off the mouth of Eastmain River, before visiting Albany. She returned to London in 1703 with a good cargo of furs from Albany.

RICH, 1958, p. 377.

1705-06

Hudson's Bay Company supply voyage — Michael Grimington — *Hudson's Bay* [II]

From London. He made the routine voyage to Albany, but there he fell ill, and his crew failed to get clear of Albany River before the ice set in. They therefore wintered there and returned to London in 1706.

RICH, 1958, p. 380.

1706-07

Hudson's Bay Company supply voyage — Joseph Davis — *Pery*

From London. Davis sailed with orders to go first to Albany, then to the east coast of James Bay, where he was to begin trading under Henry Kelsey. However, he arrived at Albany only on 1 October, too late to continue the voyage. He wintered at Albany and returned direct to London in 1707.

RICH, 1958, p. 383.

1708

Hudson's Bay Company supply voyage — Michael Grimington — *Hudson's Bay* [II]; Joseph Davis — *Eastmain*

From London. Both ships sailed to Albany. *Hudson's Bay* returned home the same year. *Eastmain* remained permanently in Hudson Bay.

DAVIES, 1965, p. 336.

1708

Publication of the de Fonte narrative

In 1708, the London magazine *Monthly miscellany, or memoirs for the curious* published a fictional account of the discovery of a North-west Passage in 1640 by the explorer Bartholomew de Fonte (see 1640). This narrative, written in the form of a letter from de Fonte, is now believed to have been composed by the magazine's editor, James Petiver. The account was widely credited during the 18th century, and did much to revive interest in the North-west Passage, which had been almost forgotten since the voyages of Foxe and James (see 1631 and 1631-32). Several expeditions were sent in search of de Fonte's North-west Passage, and it took George Vancouver's meticulous examination of the west coast of Canada and southern Alaska in 1791-95 to remove all remaining belief in the existence of such a passage.

WILLIAMS, 1962, pp. 277-82.

1710-11

Hudson's Bay Company supply voyage — Michael Grimington, Joseph Davis — *Hudson's Bay* [II]

From London. Grimington died at the beginning of the voyage, and Joseph Davis took over command. He made the regular voyage to Albany and returned to London in 1711.

RICH, 1958, p. 387; DAVIES, 1965, p. 336; DCB, Vol. 2, p. 267.

1711-12 Hudson's Bay Company supply voyage — Richard Ward — *Pery*

From London. The voyage took an exceptionally long time and the ship was wrecked in September when entering Albany River. Ward and his crew returned to London in 1712 on *Knight*.

RICH, 1958, p. 388; DAVIES, 1965, p. 336.

1712 Hudson's Bay Company supply voyage — Joseph Davis — *Hudson's Bay* [II]

From London. Routine voyage to Albany and back.

RICH, 1958, p. 388; DAVIES, 1965, p. 336.

1713 Treaty of Utrecht

Signed in March, the treaty between France and Great Britain reversed the terms of the Treaty of Ryswick (see 1697) with respect to the French-English conflict in Hudson Bay and, among other things, provided for the "restoration" of Hudson Bay to Great Britain and compensation to the Hudson's Bay Company. "Restoration" implied official recognition of the Hudson's Bay Company's title to Rupert's Land, a point that, until then, had always been in dispute.

RICH, 1954; 1958, pp. 416-25.

1713 French fur-trading voyage — ? — *Providence*

From France; June - (?) November; sent to evacuate Fort Bourbon (York Factory), now restored to the Hudson's Bay Company by the Treaty of Utrecht. Although the French had presumably made many voyages to Hudson Bay after the Treaty of Ryswick (see 1697), this last voyage is the only one of which any details are known. *Providence* left York Factory 20 September.

HUDSON'S BAY COMPANY (2).

1713-14 Hudson's Bay Company supply voyage — Joseph Davis, Peter Clements — *Hudson's Bay* [II], *Prosperous*

From London; 13 June 1713 - (?) 1714; to reoccupy the forts restored to the Hudson's Bay Company by the Treaty of Utrecht. The ships arrived in Hudson Bay very late in the season, too late to reach York Factory. *Prosperous* reached Albany 8 September 1713 and remained in Hudson Bay until 1721. *Hudson's Bay* also visited Albany before returning to London in 1714.

RICH, 1958, pp. 428-33.

1714 Hudson's Bay Company supply voyage — James Belcher, Richard Harle, James Knight — *Port Nelson*, *Union*

From London; 6 July - November. *Port Nelson* made the routine voyage to Albany and was back at London on 2 November, *Union* carried James Knight to reoccupy York Factory, which the French Governor, Nicholas Jérémie, had allowed to become very dilapidated. *Union* arrived on 7 September, left on the 20th, and reached London on 23 November.

HUDSON'S BAY COMPANY (2); RICH, 1958, pp. 431-32; DAVIES, 1965, pp. 34-41.

1715-16

Hudson's Bay Company supply voyage	Joseph Davis James Belcher	*Hudson's Bay* [III] *Port Nelson*

From London. Davis reached the vicinity of York Factory on 24 August and, until 11 September, sailed up and down the coast, unable to find the entrance of Hayes River. He finally gave up and returned to London without supplying the post. Belcher made the routine voyage to Albany. Soon after arrival, on 30 August, he lost all his ship's boats in a storm and was obliged to winter there. He returned to England in 1716.

HUDSON'S BAY COMPANY (2), note 28; RICH, 1958, pp. 438-39; DAVIES, 1965, p. 40.

1715-16

Hudson's Bay Company overland expedition	William Stewart

From York Factory; 27 June 1715 - 7 May 1716. The Chipewyan Indians had not yet traded with the Hudson's Bay Company because of warfare with the Crees. To make peace between them and to allow the Chipewyans to descend unmolested to Hudson Bay, Stewart set out with a Chipewyan known as the Slave Woman as a guide and interpreter and about 150 Crees. He reached Churchill on 28 August, then travelled north-westward across the Barren Grounds until "he Imagined he was in the Latitude of abt. 63." He continued westward and came to "a very plentifull Country for Beasts." Here he made a peace between the tribes and persuaded ten young Chipewyans to accompany him back to York Factory. He claimed to have travelled 1,600 km inland. Modern scholars agree that he went 1,100 km inland, to the area south of Great Slave Lake and east of Slave River.

MORTON, 1939, pp. 131-33; RICH, 1958, p. 435; DAVIES, 1965, pp. 410-17.

1716

Hudson's Bay Company supply voyage	George Berley Richard Ward	*Albany* *Hudson's Bay* [III]

From London. Berley made the routine voyage to Albany and back. Ward sailed to York Factory, remained there 4 - 21 September, and arrived in London on 25 October.

HUDSON'S BAY COMPANY (2); DAVIES, 1965, p. 337.

1717

Hudson's Bay Company supply voyage	George Berley Peter Clements James Belcher Richard Ward	*Albany* *Diligence* *Port Nelson* *Hudson's Bay* [III]

From London. *Albany* and *Diligence* sailed to Albany. *Albany* returned to London in 1717, but *Diligence* remained permanently in Hudson Bay. *Port Nelson* made the routine voyage to and from York Factory, where she remained 14 August - 8 September. *Hudson's Bay* sailed to Churchill, where James Knight was establishing a post (see 1717, below), then returned to London.

HUDSON'S BAY COMPANY (2); DAVIES, 1965, p. 337.

1717

Second founding of Churchill	James Knight David Vaughan Michael Grimington Jr John Carruthers William Stewart Richard Norton	 *Good Success* *Prosperous*

The Hudson's Bay Company had earlier failed to establish a post at Churchill River (see 1689). In 1717 Knight sent an advance party (Carruthers, Stewart, the boy Norton, and six workmen) from York Factory to build a post at Churchill River and bring Indians to trade. They left York in a small boat on 10 June, but were delayed and had made little progress when Knight arrived on 14 July with the ships. Knight rejected the site occupied in 1689, but chose the spot where he believed Munk had wintered (see 1619-20). The wooden fort he built, later named Prince of Wales Fort (see 1719), was eventually replaced by a great stone fort of the same name (see 1731). Knight remained at Churchill until 7 or 8 September 1718, when he returned to York Factory, then England, on *Albany. Success* and *Prosperous* remained at Churchill and were later used for northward discovery and trade.

KENNEY, 1932; DAVIES, 1965.

1717-18

Hudson's Bay Company overland expedition	Richard Norton	

From Churchill; 18 July 1717 - (?) 1718; sent by Governor James Knight to search for Chipewyan Indians, to winter with them "to divert 'em from going too warr, and to desire 'em to go to trade at York Fort." Norton set out with a Chipewyan Indian and the Slave Woman, travelled north-west to about 60°N, then "took a great sweep to the Southward of the West," met the Chipewyans and took some of them to trade at York. Exactly where Norton went is not certain: he was only a young apprentice and untrained in survey. He underwent great hardship, but established a peace among the Indians west of Churchill that lasted for many years and enabled the Indians from the region of Great Slave Lake and Lake Athabasca to trade in safety with the posts on Hudson Bay.

RICH, 1958, p. 436; 1967, p. 98; DAVIES, 1965, pp. li-liv.

1718

Hudson's Bay Company supply voyage	James Belcher Richard Ward George Berley	*Mary* [I] *Hudson's Bay* [III] *Albany*

From London. *Mary* made the routine voyage to Albany and back. *Hudson's Bay* reached York Factory on 25 August and left for home on 16 September, but, having run aground in a storm, had to return for repairs. She left again on 26 September and reached London on 13 November. *Albany* carried supplies for the new post at Churchill River (see 1717), remained there 21 August - 7 or 8 September, then returned to London with James Knight.

HUDSON'S BAY COMPANY (2); DAVIES, 1965, p. 337; KENNEY, 1932, pp. 77-78.

1719

Hudson's Bay Company supply voyage	James Belcher Richard Ward	*Mary* [I] *Hudson's Bay* [III]

From London; 5 June - 18 October. *Mary* made the routine voyage to Albany, remained there 17 August - 4 September, and was back at London on 18 October. Ward carried supplies to both Churchill and York Factory. He reached Churchill on 17 August, and on behalf of the Hudson's Bay Company, proclaimed British sovereignty and named the fort after the Prince of Wales. He left on 21 August, but was wrecked *en route* to York Factory on 24 August. Ward and his crew finally reached York safely and returned to London in 1720.

HUDSON'S BAY COMPANY (1) and (2); RICH, 1958, p. 445; WILLIAMS, 1962, p. 19; DAVIES, 1965, p. 338.

1719

Hudson's Bay Company trading and mining expedition	Henry Kelsey John Hancock	*Prosperous* *Good Success*

From Churchill; 2 July - 9 August; to search for copper. It was also the first of several attempts by Kelsey to develop the Eskimo trade north of Churchill. He seems to have coasted at least as far north as 62°40'N, beyond Marble Island, before returning to York Factory.

KENNEY, 1932, p. 81; WILLIAMS, 1962, pp. 20-21.

1719-21

Hudson's Bay Company North-west Passage and mining expedition	James Knight George Berley David Vaughan	 *Albany* *Discovery*

From London; 5 June 1719 - (?) 1721; to search the west coast of Hudson Bay north of 64°N for gold, copper, and the Strait of Anian. They sailed with the Hudson's Bay Company supply ships *Mary* [I] and *Hudson's Bay* [III] but, after parting company with them, were never seen again. Nothing was known of their fate until 1767, when *Success*, Captain Joseph Stevens, found the remains of a house and of Knight's ships on Marble Island. In 1769, Samuel Hearne learned from an old Eskimo that the expedition had reached the island safely, but had died of sickness and famine. In the second winter, 1720-21, only 20 of more than 30 men were living. The Eskimos fed them when they could but, by 1721, only five men survived, and they died within the year.

WILLIAMS, 1962, pp. 1-30; KENNEY, 1932, pp. 75-89; SMITH, 1971; 1972; SMITH AND BARR, 1971; ROSS AND BARR, 1972; ZACHARCHUK, 1973.

1720 | Hudson's Bay Company supply voyage | James Belcher
Ingram Gofton | *Mary* [I]
Hannah

From London; (?) - 26 October. *Mary* made the routine voyage to Albany, stopping there 31 August - 14 September. *Hannah* made the routine voyage to York Factory, stopping there 3 - 18 September, and returned to London with the captain and crew of *Hudson's Bay,* wrecked near York Factory in 1719.

HUDSON'S BAY COMPANY (1); DAVIES, 1965, p. 338; RICH, 1958, p. 449.

1720 | Hudson's Bay Company trading expedition | John Hancock
James Napper | *Prosperous*
Good Success

From Churchill; 19 July - 9 August; to trade for copper, oil, and whalebone with the Eskimos of the west coast of Hudson Bay as far north as 61°N. He carried orders for Knight's expedition (see 1719-21) to return to Churchill. *Good Success* was wrecked soon after sailing. Hancock did not go far enough to meet Knight's expedition, but he learned from Eskimos south of Marble Island that Knight had "winter'd where wee had been last summer [see 1719, Kelsey and Hancock] and had traded with those Indians and Spoiled our Trade." Hancock returned to York Factory on 9 August.

WILLIAMS, 1962, p. 21; KENNEY, 1932, p. 82; DODGE, 1961, p. 190.

1721 | Hudson's Bay Company supply voyage | James Belcher
Ingram Gofton | *Mary* [I]
Hannah

From London; (?) - 13 November (*Hannah*). *Mary* made the routine voyage to Albany and back. *Hannah* sailed to York Factory, remained there 24 August - 12 September, and departed in company with *Whalebone* (see 1721-22), intending to accompany her to Churchill. Bad weather forced *Hannah* to return to England, but *Whalebone* continued to Churchill.

HUDSON'S BAY COMPANY (2); DAVIES, 1965, p. 338; KENNEY, 1932, p. 83.

1721 | Hudson's Bay Company trading and mining expedition | Henry Kelsey
Richard Norton | *Prosperous*

From Churchill; 13 July - 16 August; to trade with Eskimos north of Churchill, and he took a Northern Indian "to show me ye copper." He saw Eskimos three times, on 21 and 23 July and on 1 August, but apparently did not trade with them. Unfavourable winds forced him to give up a search for Knight's lost expedition (see 1719-21) and to turn back on 9 August. He had, however, found some articles belonging to Knight's ships.

KENNEY, 1932, p. 82; DAVIES, 1965, pp. lvi-lvii; WILLIAMS, 1962, pp. 21-22.

1721-22 | Hudson's Bay Company exploring expedition | John Scroggs
Richard Norton | *Whalebone*

From London; instructed to winter at Churchill in order to start early in spring 1722 on an exploration of the west coast of Hudson Bay as far as 66°30'N and to search for copper mines. He reached York Factory on 1 September 1721 and left for Churchill on the 12th. Having wintered at Churchill, he started north on 22 June 1722, trading with Eskimos for whalebone along the way. At Chesterfield Inlet, Norton went ashore and reported a clear passage to the west, but Scroggs refused to investigate it. He probably also visited Daly Bay. On return, he sent a boat ashore at Marble Island, where traces of Knight's ships (see 1719-21) were found. After calling at Churchill, he arrived at York Factory on 16 August. This was the Hudson's Bay Company's last exploration northward from Churchill until 1737.

WILLIAMS, 1962, pp. 22-26; KENNEY, 1932, pp. 82-84; RICH, 1949, pp. xliii-xlv.

Hudson's Bay Company supply voyage	George Spurrell	*Hudson's Bay* [IV]	1722
	James Belcher	*Mary* [I]	
	Ingram Gofton	*Hannah*	

From London. *Hudson's Bay* made the routine voyage to Albany and back. *Mary* sailed to York Factory, arrived on 10 August, left on 1 September, and reached London on 31 October. *Hannah* visited Churchill first, then York Factory, where she remained 16 August - 1 September, and then returned to London.

HUDSON'S BAY COMPANY (1) and (2); DAVIES, 1965, p. 338.

Hudson's Bay Company supply voyage	Ingram Gofton	*Hannah*	1723
	George Spurrell	*Hudson's Bay* [IV]	
	James Belcher	*Mary* [I]	
	George Kennedy	*Beaver*	

From London. *Hannah* made the routine voyage to Albany and back; *Hudson's Bay* to York Factory and back. *Beaver* sailed to Albany, where she remained. *Mary* visited Churchill 17 - 23 August and York Factory 31 August - 3 September, and arrived back at London on 10 October.

HUDSON'S BAY COMPANY (1) and (2); DAVIES, 1965, p. 338.

Establishment of Eastmain post	Joseph Myatt	1723-24

The history of the Hudson's Bay Company's activities on Eastmain River before 1723 is not clear. Beginning in 1684, the London Committee had several times considered, and sometimes ordered, the establishment of a trading post there, but it appears that no permanent, regularly staffed building was ever constructed. In the early 18th century, there were frequent slooping voyages from Albany to Eastmain River, and Eastmain was regularly used as a wintering place for ships attached to Albany because it provided a better harbour than Albany River. In 1719 the London Committee again authorized the building of a house there, but this was reported in 1722 as "none of the best fortifications." Finally, in the winter of 1723-24, Joseph Myatt built a permanent post at the mouth of Eastmain River. This post has been in almost constant operation ever since. It was rebuilt in 1739-40 and probably again in 1764, and from 1732 to 1821 was the headquarters of Hudson's Bay Company operations on the east coast of Hudson Bay.

RICH, 1948, pp. 121-22; DAVIES, 1963, p. xviii; DAVIES, 1965, p. xlii; VOORHIS, 1930, p. 64.

1724

Hudson's Bay Company supply voyage	James Belcher	*Mary* [I]
	George Spurrell	*Hudson's Bay* [IV]

From London. *Mary* made the routine voyage to Albany and arrived on 14 August. She left for London on the 26th, but was wrecked the next day off Weston Island. *Hudson's Bay* made the routine voyage to York Factory, remained there 8 - 18 August, and was back at London on 12 October.

HUDSON'S BAY COMPANY (1) and (2); DAVIES, 1965, pp. 338-39.

1725

Hudson's Bay Company supply voyage	George Spurrell	*Hudson's Bay* [IV]
	Christopher Middleton	*Hannah*

From London. *Hudson's Bay* made the routine voyage to Albany, remained there 13-25 August, and arrived back at London on 2 October. *Hannah* made the routine voyage to Churchill and remained there 10-16 August. She stopped at York Factory 20-26 August and returned to London with *Whalebone,* which had been lying at York since returning from her expedition of 1721-22. *Hannah* arrived back at London on 6 October.

HUDSON'S BAY COMPANY (1) and (2); DAVIES, 1965, p. 339.

1726

Hudson's Bay Company supply voyage	George Spurrell	*Hudson's Bay* [IV]
	Christopher Middleton	*Hannah*

From London. *Hudson's Bay* made the routine voyage to Albany, stayed there 8-13 August, and arrived back at London on 5 October. *Hannah* called at Churchill, then stopped at York Factory 20 - 29 August. She arrived back at London on 19 October.

HUDSON'S BAY COMPANY (1) and (2); DAVIES, 1965, p. 339.

1727

Hudson's Bay Company supply voyage	George Spurrell	*Hudson's Bay* [IV]
	Christopher Middleton	*Hannah*
	William Coats	*Mary* [II]

From London. *Hudson's Bay* made the routine voyage to Albany and back. *Hannah* stopped at Churchill 3-7 August, then at York Factory 12-20 August. She arrived back at London on 5 October. *Mary,* carrying Thomas Macklish, the new Governor of Hudson Bay, was wrecked on 26 June off Cape Farewell (Kap Farvel) on her voyage to York Factory. Macklish, with everyone on board, was saved and taken by *Hannah* to Churchill.

HUDSON'S BAY COMPANY (1) and (2); DAVIES, 1965, p. 339.

Hudson's Bay Company supply voyage — Christopher Middleton, *Hudson's Bay* [IV]; George Spurrell, *Mary* [III]; William Coats, *Hannah* 1728

Hudson's Bay Company supply voyage	Christopher Middleton	*Hudson's Bay* [IV]
	George Spurrell	*Mary* [III]
	William Coats	*Hannah*

From London. *Hudson's Bay* sailed to Albany, was there 1-8 August, and back at London on 17 September. *Mary* called at York Factory 1-10 August and was back at London on 18 September. *Hannah* was at Churchill 30 July - 5 August and reached London on 10 September.

HUDSON'S BAY COMPANY (1) and (2); DAVIES, 1965, p. 339.

Hudson's Bay Company supply voyage — William Coats, *Hannah*; Christopher Middleton, *Hudson's Bay* [IV]; George Spurrell, *Mary* [III] 1729

Hudson's Bay Company supply voyage	William Coats	*Hannah*
	Christopher Middleton	*Hudson's Bay* [IV]
	George Spurrell	*Mary* [III]

From London. *Hannah* sailed to Albany, remained there 27 July - 5 August, and reached London on 17 September. *Hudson's Bay* called at York Factory 24 July - 1 August. *Mary* sailed to Churchill and back.

HUDSON'S BAY COMPANY (1) and (2); DAVIES, 1965, p. 339.

New England whaling voyage — Henry Atkins, *Whale* 1729

New England whaling voyage	Henry Atkins	*Whale*

From Boston. Atkins sailed to Davis Strait. On return, he landed at several places on the Labrador coast south of Davis Inlet. At 53°40'N he met natives who had apparently never before seen Europeans. He traded "some trifles" for whalebone.

BERNARD, 1792; GOSLING, 1910, pp. 321-23.

Hudson's Bay Company supply voyage — Christopher Middleton, *Hudson's Bay* [IV]; William Coats, *Hannah*; George Spurrell, *Mary* [III]; Richard Griffies, *Moose River* [I] 1730

Hudson's Bay Company supply voyage	Christopher Middleton	*Hudson's Bay* [IV]
	William Coats	*Hannah*
	George Spurrell	*Mary* [III]
	Richard Griffies	*Moose River* [I]

From London. *Hudson's Bay* and *Moose River* sailed to Albany. *Hudson's Bay* remained there 22 August - 1 September, then returned to London. *Moose River* stayed permanently in Hudson Bay. *Hannah* sailed to York Factory, was there 10-18 August, and back at London on 16 October. *Mary* sailed to Churchill and back.

HUDSON'S BAY COMPANY (1) and (2); DAVIES, 1965, p. 339.

Hudson's Bay Company supply voyage — Christopher Middleton, *Hudson's Bay* [IV]; George Spurrell, *Mary* [III]; William Coats, *Hannah* 1731

Hudson's Bay Company supply voyage	Christopher Middleton	*Hudson's Bay* [IV]
	George Spurrell	*Mary* [III]
	William Coats	*Hannah*

From London. *Hudson's Bay* made the routine voyage to Albany and back. *Mary* called at Churchill 29-31 July, then at York Factory, where she stayed 2-8 August. *Hannah* visited Churchill 29 July - 14 August and reached London on 23 September.

HUDSON'S BAY COMPANY (1); DAVIES, 1965, p. 340.

1731

Construction of
Prince of Wales Fort begun

James Knight had built a wooden fort at Churchill (see 1717), which was officially named Prince of Wales in 1719 (*qv*). In 1731, the Hudson's Bay Company sent Richard Norton to erect at Churchill a stone fortress capable of repulsing a naval attack. His ships, *Mary* and *Hannah,* carried masons, labourers, and construction equipment. Work began on the foundations of the fort at Eskimo Point, near the mouth of Churchill River, in August. Through the winter, the men gathered stone and timber, and building began to rise in 1732. By 1736, the south and east bastions were completed; by 1746, the ramparts were finished and a large house had been built inside them. Construction and repair progressed slowly, and in 1771, when 42 cannons had been mounted, the work was apparently barely completed. The fort was about 100 m long on its north and south sides; walls were about 12 m thick at base and 2 m thick on top. Despite this elaborate preparation, Comte de la Pérouse captured and destroyed the fort in 1782 (*qv*). Its remains are now a Historic Memorial Site.

KENNEY, 1932, pp. 93-97; DAVIES, 1965.

1732

Hudson's Bay Company supply voyage	Christopher Middleton	*Hudson's Bay* [IV]
	George Spurrell	*Mary* [III]
	William Coats	*Hannah*
	George Green	*Moose River* [II]

From London. *Hudson's Bay* visited first Albany, then Moose Factory. *Moose River* [II] sailed to Moose Factory and remained there permanently, replacing *Moose River* [I] (see 1731), which was renamed *Churchill* and sent to Prince of Wales Fort. *Mary* sailed to York Factory, was there 3-19 August, and returned to London on 28 September. *Hannah* visited Churchill 11-25 August, and was back at London on 8 October.

HUDSON'S BAY COMPANY (1) and (2); DAVIES, 1965, p. 340.

1733

Hudson's Bay Company supply voyage	Christopher Middleton	*Hudson's Bay* [IV]
	George Spurrell	*Mary* [III]
	William Coats	*Hannah*

From London. *Hudson's Bay* visited Albany and Moose Factory and returned to London. *Mary* visited Churchill 4-6 August, York Factory 11-17 August, and was back at London on 30 September. *Hannah* sailed to Churchill and back.

HUDSON'S BAY COMPANY (1) and (2); DAVIES, 1965, p. 340.

1733

Boundaries of the Domaine
du Roi redefined

Pierre Carlier, who had taken up La Traite de Tadoussac (see 1658), petitioned the Intendant of New France, Gilles Hocquart, to redefine the limits of the Domaine du Roi. In 1731, Hocquart had ordered Sieur Aubert de la Chenaye to make a survey of the interior and, in 1732, he ordered Sieur Joseph Normandin to continue it. On the basis of their reports and map, Hocquart issued an Ordinance, dated 23 May 1733, which indicated that the French had a good knowledge of the rivers draining into Lake St John and St Lawrence River and that the domaine supported many posts. The boundaries ran north from Les Eboulements and Cape Cormorant on St Lawrence River. The western boundary included lakes St John and Mistassini and extended to Hudson Bay. The eastern ran north to the height of land and included Moisie River and lakes Pletipi and Ashuanipi.

HOCQUART, 1733; VOORHIS, 1930, pp. 12-13.

1734

Hudson's Bay Company supply voyage	Christopher Middleton George Spurrell William Coats	*Seahorse* [I] *Mary* [III] *Hudson's Bay* [IV]

From London. *Hudson's Bay* made the routine voyage to Churchill and back; *Seahorse* to Moose Factory and back. *Mary* sailed to Churchill, remained there 28 July - 1 August, then continued to York Factory and remained there 4-13 August. She arrived back at London on 14 September.

HUDSON'S BAY COMPANY (1) AND (2); DAVIES, 1965, p. 340.

1735

Hudson's Bay Company supply voyage	Christopher Middleton William Coats George Spurrell	*Seahorse* [I] *Hudson's Bay* [IV] *Mary* [III]

From London. *Seahorse* visited Moose Factory 3-30 August, and was back at London on 13 October. *Hudson's Bay* visited York Factory 1-7 August and reached London on 17 September. *Mary* was at Churchill 30 July - 19 August, and reached London on 20 September.

HUDSON'S BAY COMPANY (1) and (2); DAVIES, 1965, p. 340.

1736

Hudson's Bay Company supply voyage	Christopher Middleton William Coats George Spurrell	*Seahorse* [I] *Hudson's Bay* [IV] *Mary* [III]

From London. *Hudson's Bay* was lost 3 July in Hudson Strait on the voyage to York Factory. *Mary* visited Churchill 12-19 August, York 24-28 August and arrived back at London on 8 October. *Seahorse* visited Moose Factory 15-28 August, Albany 31 August - 4 September. She reached London on 11 October.

HUDSON'S BAY COMPANY (1) and (2); DAVIES, 1965, pp. 340-41.

1736-37

Hudson's Bay Company trading, mining, and North-west Passage expedition	James Napper Robert Crow	*Churchill* *Musquash*

From London. At the instigation of Arthur Dobbs, who soon became an energetic campaigner for the renewal of the search for a

North-west Passage, the Hudson's Bay Company sent two ships on an exploring voyage northward from Churchill to survey Roes Welcome Sound, to open trade with the Eskimos, to dig for minerals, and to search for a North-west Passage. The ships reached Churchill too late in 1736 to begin the voyage that year and were, in any case, needed for transport between Churchill and York Factory owing to the loss of *Hudson's Bay* (see 1736, above). The ships left Churchill on 4 July 1737 but went only as far as Whale Cove, in lat 62°10'N, where they established friendly relations with the Eskimos and promised to return the following year. Napper died at Whale Cove. *Churchill* arrived back at Churchill on 18 August and *Musquash* on 22 August. They had discovered nothing and had made a negligible trade. Probably the company would not have sent another voyage northward had Napper not promised the Eskimos to return the next year.

RICH, 1949, pp. xlviii-xlix; WILLIAMS, 1962, p. 44; DAVIES, 1965, pp. 241-42; DODGE, 1961, pp. 192-93; RICH, 1958, p. 560.

1737

Hudson's Bay Company supply voyage	Christopher Middleton	*Hudson's Bay* [V]
	George Spurrell	*Seahorse* [I]
	William Coats	*Mary* [III]

From London. *Seahorse* visited Moose Factory 7-24 August and was back at London on 29 September. *Mary* visited Churchill 27 July - 23 August, and reached London on 23 September. *Hudson's Bay* called at Churchill 20-27 July and at York Factory 29 July - 7 August. She returned to London on 20 September.

HUDSON'S BAY COMPANY (1) and (2); DAVIES, 1965, p. 341.

1738

Hudson's Bay Company supply voyage	Christopher Middleton	*Hudson's Bay* [V]
	William Coats	*Mary* [IV]
	George Spurrell	*Seahorse* [I]

From London. *Hudson's Bay* made the routine voyage to Moose Factory. *Mary* called first at Churchill, then stopped at York Factory 24 August - 1 September. She reached London on 29 October. *Seahorse* visited Churchill 27 July - 17 August, and returned to London on 20 September.

HUDSON'S BAY COMPANY (1) and (2); DAVIES, 1965, p. 341.

1738

Hudson's Bay Company trading voyage	Francis Smith	*Churchill*

From Churchill; 5-15 August; James Napper had promised the Eskimos of Whale Cove (see 1736-37) that a ship would return with trade goods. This voyage sailed to honour that promise but no Eskimos were sighted, and the ship returned to Churchill. Even after this disappointment, the Hudson's Bay Company ordered annual trading voyages northward from Churchill.

DAVIES, 1965, pp. 253, 258.

1739

Hudson's Bay Company supply voyage — William Coats, *Mary* [IV]; George Spurrell, *Seahorse* [I]; Christopher Middleton, *Hudson's Bay* [V]

From London. *Mary* called at Moose Factory 5-19 August and returned to London on 3 October. *Seahorse* was at Churchill 12-21 August, and at York Factory 25 August - 1 September. She reached London on 29 October. *Hudson's Bay* was at Churchill 12-27 August. She arrived back at London on 12 October.

HUDSON'S BAY COMPANY (1) and (2); DAVIES, 1965, p. 341.

1739

Hudson's Bay Company trading voyage — (Francis Smith?), *Churchill*

From Churchill; 4-9 August. Traded with Eskimos at Whale Cove.

HUDSON'S BAY COMPANY, ARCHIVES, B42/a/19/2d-3.

1740

Hudson's Bay Company supply voyage — Christopher Middleton, *Hudson's Bay* [V]; William Coats, *Mary* [IV]; George Spurrell, *Seahorse* [I]

From London. *Hudson's Bay* called at Moose Factory 14-20 August, at Albany 22-28 August, and arrived back at London on 6 November. *Mary* visited York Factory 20-30 July and reached London on 20 September. *Seahorse* visited Churchill 16 July - 11 August and reached London on 28 September.

HUDSON'S BAY COMPANY (1) and (2); DAVIES, 1965, p. 341.

1740

Hudson's Bay Company trading voyage — Francis Smith, *Churchill*

From Churchill; 5-24 July. He traded with Eskimos along the coast north of Churchill. Despite instructions that these northern trading voyages should be made annually (see 1738) Richard Norton, chief at Churchill, cancelled the voyage for 1741.

RICH, 1949, pp. 336-37.

1741

Hudson's Bay Company supply voyage — William Coats, *Mary* [IV]; George Spurrell, *Seahorse* [I]

From London. *Mary* sailed to Moose Factory, where she remained 26 August - 1 September, then to Albany, where she remained 3-8 September. *Seahorse* visited Churchill 22-31 July, and York Factory 2-13 August. She reached London on 3 October.

HUDSON'S BAY COMPANY (1).

1741-42

British North-west Passage expedition — Christopher Middleton, HMS *Furnace*; William Moor, HMS *Discovery*

From London; 8 June 1741 - 13 October 1742; sent by the Admiralty, at the suggestion of Arthur Dobbs, to search for a North-west Passage in Roes Welcome Sound. They made slow progress across the Atlantic Ocean and decided to winter at Churchill. They spent an uncomfortable winter, during which ten men died of scurvy. On 1 July 1742 they left Churchill, but their progress was repeatedly hampered by ice. On 12 July, they passed Whalebone Point (65°N), the most northerly point sighted by Scroggs (see 1721-22), and entered unknown waters. The same day they reached Cape Dobbs and discovered Wager Bay. Further progress north was delayed by ice driving into Wager Bay, and Middleton took the opportunity to send four boats to explore it. The boatmen observed many whales in the bay, that the flood tides came from the east, and that at the most westerly point they reached, the water was nearly fresh. Middleton rightly concluded, therefore, that there was no hope of a North-west Passage in Wager Bay. On 4 August, they continued north again and discovered Repulse Bay and the entrance to Frozen Strait, but Middleton decided, again rightly, that here, too, there was no hope of a passage. On the return, they explored the west coast of Hudson Bay from Cape Dobbs south to Marble Island, then returned to England.

MIDDLETON, 1743; DOBBS, 1744; WILLIAMS, 1962, pp. 58-78; DODGE, 1961, pp. 194-210.

1742

Hudson's Bay Company supply voyage	William Coats George Spurrell	*Mary* [IV] *Seahorse* [I]

From London; 7 May - 1 October (*Mary*). *Mary* visited Moose Factory 27 July - 4 August and Albany 6 - 13 August, then returned to London. *Seahorse* called at Churchill, then at York Factory, where she stopped 3 - 15 August, then returned to London.

HUDSON'S BAY COMPANY (1).

1742

Hudson's Bay Company trading voyage	Francis Smith	*Churchill*

From Churchill. He sailed northward from Churchill with Christopher Middleton's expedition (see 1741-42), accompanied Middleton for part of his journey (perhaps as far as Whale Cove in lat 62°10'N), and "Made But a poor Trade" with the Eskimos along the coast.

RICH, 1949, pp. 336-37; 1958, p. 574.

1743

Hudson's Bay Company supply voyage	William Coats George Spurrell	*Mary* [IV] *Seahorse* [I]

From London; 7 May - 2 October (*Mary*). *Mary* called at Moose Factory 3 - 14 August and at Albany 15 - 21 August, then returned to London. *Seahorse* called at Churchill 2 - 8 August and at York Factory 10 - 19 August, and was back at London on 30 September.

HUDSON'S BAY COMPANY (1) and (2).

Hudson's Bay Company trading voyage | Francis Smith | *Churchill* **1743**

From Churchill. Traded with Eskimos north of Churchill.

RICH, 1949, pp. lxxii, 337; 1958, pp. 574-75.

Establishment of Henley House **1743**

The post, about 160 km up Albany River, was first established in 1743 to combat "the encroachment of the French Pedlars." It was twice ambushed by Indians. In 1755, the master, William Lamb, and four men were killed; in 1759, the master, George Clark, was killed, but the men escaped to Albany. It was reestablished in 1766.

RICH, 1951, p. xv; 1959, p. 92; WILLIAMS, 1969, pp. 253-55.

French exploration of Hamilton Inlet | Louis Fornel **1743**

From Quebec; 16 May - 25 August. Fornel, who had been trading for some years in Chateau Bay, Strait of Belle Isle, applied in 1742 for a trading concession in the region of Hamilton Inlet – Lake Melville, then known as Esquimaux Bay or Kessessakiou (see 1702). The French do not appear to have been active on the coast north of the strait owing to the number and fierceness of the Eskimos, but the outline of the coast was well known to fishermen. Fornel reached Esquimaux Bay on 10 July and, after a very brief reconnaissance, turned home. He left behind Jean Pilot and his son, who, with some Montagnais Indian families, travelled overland during the winter to Mecatina. Fornel died in 1745, but in 1749 his widow and a company she had formed were granted a 12-year trading concession to Esquimaux Bay.

GREAT BRITAIN. PRIVY COUNCIL, 1927, pp. 3131-3143, 3275-3355; ROY, 1942, pp. 201-261.

Hudson's Bay Company supply voyage **1744**

	William Coats	*Mary* [IV]
	George Spurrell	*Prince Rupert* [II]
	Jonathan Fowler	*Hudson's Bay* [V]

From London; 9 May - 20 November. *Mary* called at Churchill 8 - 18 August then, although not instructed to do so, sailed to York Factory to accompany *Prince Rupert* back to London. *Prince Rupert* was at Churchill 8 - 10 August and at York 13 - 28 August. *Hudson's Bay* visited Moose Factory 12 - 21 August, and Albany 23 - 30 August.

HUDSON'S BAY COMPANY (1) and (2).

1744	Hudson's Bay Company exploring expedition	Thomas Mitchell John Longland	*Eastmain* *Phoenix*

From Moose Factory; 26 June - 14 August; the company's first attempt to explore the Eastmain or east coast of Hudson Bay – little known before that time because ice conditions had forced the annual supply ships well out into the bay. Indians had reported in 1739 that three great lakes in lat 60°N occupied much of the interior of Labrador. This news gave rise to hopes that a waterway might lead into the interior and allow the company to open trade with the Indians of the Quebec-Labrador peninsula. Leaving Moose Factory, *Phoenix* stopped at Strutton Island in James Bay, where she was joined by *Eastmain.* The ships visited the mouths of Fort George, Great Whale, and Little Whale rivers, and made a hasty exploration of Richmond Gulf before returning home. They were back at Albany on 14 August. At Little Whale River they had picked up minerals that they thought might be diamonds, but otherwise they were discouraged by the coast. Their rough sketch maps of the coast they had explored between Cape Jones and Richmond Gulf were of little use. There was no further attempt to explore this coast until 1749.

RICH, 1958, pp. 617-18; DAVIES, 1963, pp. xix-xx.

1744	Hudson's Bay Company trading voyage	Francis Smith	*Churchill*

From Churchill; he sailed northward to trade with Eskimos of the west coast of Hudson Bay. On 8 July he discovered and named Knapp Bay, just north of Eskimo Point in lat 61°07'N. This bay afterwards became a regular stopping place for ships on trading voyages from Churchill. He then continued northward to Whale Cove, but in general, the trade was poor; he remarked that he had seen three times as many Eskimos on his 1743 voyage. He arrived back at Churchill on 25 July.

RICH, 1949, p. 337; HUDSON'S BAY COMPANY, ARCHIVES, B42/a/26-38.

1744-45	Hudson's Bay Company voyage	John Fowler	*Seahorse* [I]

From London. Because England was at war with France and Spain, *Seahorse* was equipped with 50 men and 18 months' provisions to defend Churchill and York Factory against possible attacks. About 10 men were left for the defence of York Factory, and *Seahorse* with the others wintered at Churchill. In 1745, she called at York Factory, then left for London, where she arrived on 28 September.

HUDSON'S BAY COMPANY (1) and (2).

1745	Hudson's Bay Company supply voyage	William Coats George Spurrell Jonathan Fowler	*Mary* [IV] *Prince Rupert* [II] *Hudson's Bay* [V]

From London. *Mary* called at Moose Factory 21 - 29 July and at Albany 1 - 4 August. *Hudson's Bay* called at Albany 22 July - 4 August. *Prince Rupert* visited Churchill 17 - 27 July, then York Factory 29 July - 15 August. All three ships were back at London on 28 September.

HUDSON'S BAY COMPANY (1) and (2).

1746

Hudson's Bay Company supply voyage	John Fowler	*Seahorse* [I]
	William Coats	*Mary* [IV]
	George Spurrell	*Prince Rupert* [II]
	Jonathan Fowler	*Hudson's Bay* [V]

From London. *Mary* called at Albany 17 - 27 August and arrived back at London 8 October. *Prince Rupert* visited York Factory 3 - 16 August, and reached London 16 September. *Hudson's Bay* called at Moose Factory 17 - 27 August and reached London 8 October. *Seahorse* called at Churchill 27 July - 6 August, at York Factory 7 - 17 August, and was back at London 19 September.

HUDSON'S BAY COMPANY (1) and (2).

1746-47

British North-west Passage expedition	William Moor	*Dobbs*
	Francis Smith	*California*

From London; 20 May 1746 - 14 October 1747; sponsored by the North West Committee. Arthur Dobbs, dissatisfied by the results of Middleton's expedition (see 1741-42) arranged this private venture to reexamine Wager Bay, where he was sure a passage would be found. The expedition reached Marble Island on 26 August and, having explored the island, sailed to York Factory. They were received coldly by the Hudson's Bay Company employees there, but were given clothing and allowed to remain for the winter. On 24 June 1747, they resumed their voyage, discovered Chesterfield Inlet, and reached Douglas Harbour in Wager Bay on 29 July. The two captains, in longboats, went to the head of Wager Bay, and rightly concluded that there was no North-west Passage there, but, to be certain, one of the boats re-surveyed the north shore of Wager Bay. Having decided against exploring Repulse Bay and Chesterfield Inlet, they turned homeward and reached Yarmouth 14 October.

SWAINE, 1748; ELLIS, 1748; 1946a; 1946b, WILLIAMS, 1962, pp. 79 108; DODGE, 1961, pp. 200-10; EAVENSON, 1949; 1950.

1747

Hudson's Bay Company supply voyage	William Coats	*Mary* [IV]
	George Spurrell	*Prince Rupert* [II]
	Jonathan Fowler	*Hudson's Bay* [V]
	John Fowler	*Seahorse* [I]

From London. *Mary* called at Moose 12 - 22 August and reached London 16 October. *Seahorse* visited Albany 13 - 22 August and reached London 17 October. *Prince Rupert* called at York Factory 8 - 22 August. *Hudson's Bay* visited Churchill 6 - 15 August, then York Factory 18 - 22 August. She returned to London with *Prince Rupert*.

HUDSON'S BAY COMPANY (1) and (2).

1748

Hudson's Bay Company supply voyage | William Coats | *Mary* [IV]
| George Spurrell | *Prince Rupert* [II]
| Jonathan Fowler | *Hudson's Bay* [V]
| John Fowler | *Seahorse* [I]

From London. *Mary* called at Churchill 17 August - 6 September and reached London 7 October. *Prince Rupert* called at York Factory 18 - 30 August and returned to London. *Hudson's Bay* visited Moose Factory 20 - 29 August and reached London 7 October. *Seahorse* called at Albany 16 - 27 August, at Moose Factory 28 - 29 August, and returned to London 6 October.

HUDSON'S BAY COMPANY (1) and (2).

1749

Hudson's Bay Company supply voyage | George Spurrell | *Prince Rupert* [II]
| Jonathan Fowler | *Hudson's Bay* [V]
| John Fowler | *Seahorse* [I]

From London. *Prince Rupert* called at York Factory 28 July - 14 August. *Hudson's Bay* called at Churchill 26 July - 5 August, then joined *Prince Rupert* at York Factory. Both ships reached London 15 September. *Seahorse* called at Moose Factory 27 July - 6 August and at Albany 8 - 17 August. She reached London on 20 September.

HUDSON'S BAY COMPANY (1) and (2).

1749

Hudson's Bay Company exploring expedition | William Coats | *Mary* [IV]
| Thomas Mitchell | *Success*

From London. Coats was ordered to continue Mitchell's exploration of the east coast of Hudson Bay (see 1744) by charting the coast southward from Cape Digges. He was also to locate a suitable site for a trading post, to observe currents and tides, and to collect minerals and specimens of flora and fauna. Between Cape Digges and Richmond Gulf, he found only one reasonable harbour, probably Mosquito Bay. He frequently went ashore in boats, but found the coast bleak, barren, and devoid of wildlife. The region of Richmond Gulf seemed much more favourable to prospects of trade, it had a safe harbour, and he recommended it as the best site for a post. He returned to London with detailed maps of Richmond Gulf and of the coast between Cape Digges and Little Whale River.

WILLIAMS, 1963; DAVIES, 1963, pp. xx-xxii; HUDSON'S BAY COMPANY (2).

1749

Parliamentary inquiry into the Hudson's Bay Company

In March, partly in response to the criticisms of Arthur Dobbs and others, the House of Commons set up a committee to "inquire into the State and Condition of the Countries adjoining to Hudson's Bay, and of the Trade carried on there." The report was printed on 26 April and presented on 1 May. Although the company's charter and monopolistic rights had been severely attacked, the

House decided that there was no case for annulling the charter nor for throwing the trade open to others.

RICH, 1958, pp. 653-54.

Hudson's Bay Company supply voyage	William Coats	*King George* [I]	1750
	George Spurrell	*Prince Rupert* [II]	
	Jonathan Fowler	*Hudson's Bay* [V]	
	John Fowler	*Seahorse* [I]	

From London. *King George* supplied Richmond Fort (see 1750, below), and stayed there 14 August - 3 September. She reached London 1 November. *Prince Rupert* made the routine voyage to York Factory, was there 18 August - 3 September, and reached London on 24 October. *Hudson's Bay* visited Churchill 17 - 29 August, and reached London on 12 October. *Seahorse* was at Albany 17 - 26 August, at Moose 27 August - 4 September, and reached London on 24 October.

HUDSON'S BAY COMPANY (1).

Establishment of Richmond Fort	John Potts		1750
	Thomas Mitchell	*Eastmain*	
	?	*Success*	

The Hudson's Bay Company decided to establish a trading post at Richmond Gulf, the site recommended by Coats and Mitchell (see 1749), and sent three miners to work mineral deposits that Coats had located between Richmond Gulf and Little Whale River. The sloops sailed from Albany on 20 June, and the post was built by 25 July. Potts arrived later in the year to take command of the post, and William Coats arrived on 14 August with supplies from London (see above, 1750). Potts soon realized that the post would not meet expectations: the miners were sent home after the mines had produced nothing of value, and in the first six years, the post received only about £100 worth of furs. At first a few Eskimos visited the post, but they stopped coming after 1754, when two Eskimos and a young company servant were killed in a scuffle. A dozen Naskapi Indians visited in 1755, but only five returned in 1756. In 1756, the company moved the post to Little Whale River, where they had been whaling in summer. Trade continued to be poor, and the post was abandoned in 1759. There is no record of any other post at Richmond Gulf until 1923.

DAVIES, 1963, pp. xxi-xxv, xxxv-xxxvi, 275, 277; MORTON, 1939, p. 229; RICH, 1958, pp. 619-20; VOORHIS, 1930, p. 148.

Hudson's Bay Company exploring expedition	James Walker	*Churchill*	1750

From Churchill; (?) - 28 July; sent to find a site for a trading post near Whale Cove, then to sail as far north as possible, examining all the inlets and observing the tides. He reached Whale Cove on 20 July, remained there five days to mend nets and scrape the sloop, then returned to Churchill, having accomplished nothing.

From this date, trading voyages northwards from Churchill to Knapp Bay, Nevill Bay and Whale Cove, which had hitherto been only sporadic, were made annually.

WILLIAMS, 1962, p. 123; ROSS, W.G., 1973, p. 5.

1750

Establishment of Fort Paskoyac

During the 1730's and 1740's, French fur traders, led by Pierre Gaultier de Varennes, Sieur de la Vérendrye, had explored northwestward from Lake Superior, and had established trading posts on Lake Winnipeg. Towards the end of the 1740's, one of La Vérendrye's sons discovered the Saskatchewan River and, in 1750, the first fur trading post on the Saskatchewan, Fort Paskoyac, was built near the present site of The Pas. La Vérendrye's son Pierre probably ordered the building of the post, and it was occupied by Chevalier de Niverville towards the end of the year. In spring 1751, de Niverville sent a party of men to explore upstream in canoes, and they are said to have ascended the Saskatchewan as far as the Rocky Mountains, where they built Fort La Jonquière. The French are known to have built other forts on the Saskatchewan between Fort Paskoyac and the Forks at about this time, but neither the precise date of their establishment nor their exact location is known. They were named forts Nipawi, St-Louis, des Prairies, and La Corne. Fort Paskoyac was visited and described by Anthony Henday in 1754-55 (*qv*). Its ruins were seen by Alexander Henry in 1808.

BURPEE, 1935, Vol. 1, pp. 266-87; MORTON, 1939, pp. 231, 236-37; VOORHIS, 1930, p. 134.

1751

Hudson's Bay Company supply voyage	William Coats	*King George* [I]
	George Spurrell	*Prince Rupert* [II]
	Joseph Spurrell	*Hudson's Bay* [V]
	Jonathan Fowler	*Seahorse* [I]

From London. *King George* supplied Richmond Fort (see 1750), remained there 26 July - 16 August, and returned to London on 21 September. This appears to have been Coats' 25th and last voyage to Hudson Bay. *Prince Rupert* visited York Factory 25 July - 10 August and reached London 14 September. *Hudson's Bay* was at Churchill 26 July - 23 August, and reached London 30 September. *Seahorse* was at Albany 26 July - 3 August, at Moose Factory 4 - 10 August, and reached London 14 September.

HUDSON'S BAY COMPANY (1).

1751

Hudson's Bay Company exploring and trading expedition	James Walker	*Churchill*

From Churchill; Walker's second attempt (see 1750) to develop northern trade and to explore the coast north of Whale Cove. He visited the Eskimos of Whale Cove, then sailed north for two days, reaching lat 64°N, but never went close to the shore, and brought back no new information about the coastline.

WILLIAMS, 1962, p. 124.

Hudson's Bay Company supply voyage	William Norton George Spurrell Jonathan Fowler Joseph Spurrell	*Hudson's Bay* [V] *Prince Rupert* [II] *Seahorse* [I] *King George* [I]	1752

From London. *Hudson's Bay* visited Richmond Fort 26 July - 15 August and reached London on 29 September*. *Prince Rupert* visited York Factory 24 July - 9 August and reached London on 21 September. *Seahorse* was at Churchill 22 July - 11 August and reached London on 21 September. *King George* visited Albany 2 - 10 August, then Moose Factory 13 - 21 August, and reached London on 23 October.

HUDSON'S BAY COMPANY (1).

Hudson's Bay Company exploring and trading expedition	James Walker	*Churchill*	1752

From Churchill; Walker's third trading voyage northward to Whale Cove. In particular he was to search for an inlet in about lat 64°N, called by the Indians Kish-Stack-Ewen, but he did not attempt to carry out this part of his instructions. During this voyage, Walker made one of the company's infrequent visits to Marble Island. He fired guns at the island to attract natives, but without success.

WILLIAMS, 1962, pp. 124-25; 1969, p. 333; ROSS AND BARR, 1972, p. 31.

British missionary and trading voyage to Labrador	Captain Madgeson Elijah Goff John Christian Erhardt	*Hope*	1752

From London; 18 May - late November; sponsored by the London merchants Nisbet, Grace, and Bell to trade with Eskimos and to establish a mission on the Labrador coast. Erhardt, with four other Moravian Brethren, undertook the missionary activity. They reached southern Labrador on 11 July and coasted northward, visiting Sandwich Bay and Hamilton Inlet. They first met Eskimos on 29 July; on 31 July, they found an anchorage in about lat 55°10'N, which they thought a suitable site for a mission, and named it Nisbet's Harbour. In August, they built a house and traded with Eskimos. On 5 September, the ship departed, taking Erhardt, but leaving behind the four missionaries. Ten days later, the ships returned, commanded by the mate, Goff, who reported that Madgeson, Erhardt, and five others had gone ashore at Davis Inlet to trade with Eskimos and had not been seen again. They

*In this year, England adopted the New Style calendar. Dates are given in Old Style for 1752 and in New Style for 1753 onwards.

were presumably killed by Eskimos. Goff and the missionaries tried to return to Davis Inlet, but bad weather forced them back to Nisbet's Harbour. The missionaries then decided to return to London with the ship, intending to return in 1753. They left on 20 September, called at St John's, and reached London at the end of November.

GOSLING, 1910, pp. 252-53; WHITELEY, 1964, p. 30; PEACOCK, 1960; HILLER, 1966.

1753

Hudson's Bay Company supply voyage	William Norton George Spurrell Joseph Spurrell Jonathan Fowler	*Hudson's Bay* [V] *Prince Rupert* [II] *King George* [I] *Seahorse* [I]

From London, *Hudson's Bay* visited Richmond Fort 20 August - 2 September, and reached London on 5 October. *Prince Rupert* visited York Factory 27 August - 8 September, Churchill 12 September, and reached London 10 October. *King George* was at Churchill 23 August - 8 September, and back at London 12 October. *Seahorse* called at Albany 27 August - 2 September, and at Moose Factory 4 - 11 September. She reached London 10 October.

HUDSON'S BAY COMPANY (1).

1753

Hudson's Bay Company trading voyage	James Walker	*Churchill*

From Churchill; routine trading voyage northward to Whale Cove. Three hundred Eskimos tried to attack his ship off the coast south of Marble Island, but he managed to escape.

WILLIAMS, 1962, p. 27; 1969, pp. 219-20.

1753

New England voyage to Labrador	Henry Atkins Captain Prebble	?

From Boston. Prebble went ashore at Fortune Bay, an unidentified part of the Labrador coast. He traded with Indians and observed that they had never before "left their own country to trade with the French."

BERNARD, 1792; GOSLING, 1910, p. 326.

1753

American North-west Passage expedition	Charles Swaine	*Argo*

From Philadelphia; 4 March - 10 November; sponsored by Benjamin Franklin and others, under the name of the Northwest Company. The ship reached the mouth of Hudson Strait on 25 June, but was stopped by drifting ice. On 14 July, she met the Hudson's Bay Company's supply vessels (see 1753, above), tried to accompany them through the strait, but lost them on 19 July, in lat 62°N. After several more unsuccessful attempts, they turned back, and decided to explore the Labrador coast thoroughly between 56° and 55°N. From this close inspection and having sailed to the heads of six inlets, they concluded that no waterway connected the Labrador coast with Hudson Bay south of Hudson Strait. The Hudson's Bay Company had once suspected that such a strait might exist (*eg,* see 1744, Mitchell). They also discovered a coastal fishing bank between 57° and 54°N.

SOLIS-COHEN, 1943; 1952; DODGE, 1961, pp. 216-18; EAVENSON, 1949, pp. 46-51.

British trading voyage to Labrador	Elijah Goff	*Hope*	1753

During his North-west Passage expedition, Charles Swaine (see 1753, above) fell in with the brig *Hope,* Captain Elijah Goff, who was presumably making a trading voyage similar to the one made in *Hope* in 1752 (*qv*). Goff met Swaine in a harbour on the Labrador coast on 5 August and told of his meeting with Captain Taylor (see below, 1753). On 9 August, Swaine's ship left, and, on 13 August, Goff found the remains of the men lost during the 1752 voyage. He then sailed for New York.

SOLIS-COHEN, 1943; 1952; EAVENSON, 1949, pp. 48-50; DODGE, 1961, pp. 218-19.

New England North-west Passage expedition	John Taylor	?	1753

From Rhode Island. Nothing more is known of this expedition than what Goff told Swaine (see 1753, above). Apparently, Taylor had left Rhode Island in a sloop of about 35 tons to search for a North-west Passage or, failing that, to explore the Labrador coast. Goff and Taylor met on 9 July, agreed to collaborate, and remained together until early August.

SOLIS-COHEN, 1943, p. 27; 1952, p. 160; DODGE, 1961, pp. 218-19.

Hudson's Bay Company supply voyage	William Norton George Spurrell Jonathan Fowler Joseph Spurrell	*Hudson's Bay* [V] *Prince Rupert* [II] *Seahorse* [I] *King George* [I]	1754

From London. *Hudson's Bay* visited Richmond Fort 1-11 September and reached London 17 October. *Prince Rupert* called at York Factory 1-11 September. *Seahorse* called at Churchill 27 August - 7 September. *King George* called at Moose Factory 2-15 September, then at Albany 16-21 September.

HUDSON'S BAY COMPANY (1) and (2).

Hudson's Bay Company trading voyage	James Walker	*Churchill*	1754

From Churchill. Routine trading voyage north to Whale Cove. Having failed to accomplish the exploratory work he was ordered to do since 1750, Walker was replaced by John Bean in 1755.

WILLIAMS, 1962, p. 125.

1754	American North-west Passage expedition	Charles Swaine	*Argo*

From Philadelphia; c 2 May - 20 October; sponsored by Benjamin Franklin and others. Following the failure of Swaine's first attempt (see 1753), he set out once more, but returned after three men, including the mineralogist, John Patten, had been killed by Eskimos while looking for a mine on the Labrador coast.

SOLIS-COHEN, 1943; 1952; EAVENSON, 1949, pp. 59-61; DODGE, 1961, pp. 219-20.

1754-55	Hudson's Bay Company overland expedition	Anthony Henday	

From York Factory; 26 June 1754 - 20 June 1755; sent by James Isham to make peace with the Blackfeet Indians, who had not yet traded with the Hudson's Bay Company, to teach them to catch beaver and paddle canoes, and to encourage them to return with him to trade at York Factory. He left York in the company of trusted Indians and followed Hayes River and its tributaries to a portage that leads to Cross Lake on Nelson River; he left Cross Lake by Minago River and portaged to Moose Lake. He then followed Summerberry River to the French Fort Paskoyac (see 1750) on Saskatchewan River. He arrived here on 22 July. He next followed the South Saskatchewan River to near the site of present-day Saskatoon, crossed the prairies north-westward to the elbow of the North Saskatchewan, and followed the valley of Battle River to Manito Lake, where he arrived on 15 September. From here he travelled south-westward, crossed Red River on 24 October, then turned north-westward to the country east of Clearwater River. Here, within sight of Rocky Mountains, he remained trapping furs and hunting buffalo until 23 December, when he began the homeward journey. He returned by the North Saskatchewan River, again visited Fort Paskoyac, then continued to York with a flotilla of about 70 Indian canoes carrying furs.

BURPEE, 1907; MORTON, 1939, pp. 244-50; RICH, 1958, pp. 632-41.

1755	Hudson's Bay Company supply voyage	William Norton	*Hudson's Bay* [V]
		George Spurrell	*Prince Rupert* [II]
		Jonathan Fowler	*Seahorse* [I]
		Joseph Spurrell	*King George* [I]

From London. *Hudson's Bay* called at Richmond Fort 30 August - 9 September, and reached London 31 October. *Prince Rupert* called at York Factory 23 August - 4 September, and reached London 29 October. *Seahorse* called at Moose Factory 28 August - 7 September, at Albany 9-17 September, and was back at London 31 October. *King George* visited Churchill 2-17 September, and reached London 6 September.

HUDSON'S BAY COMPANY (1).

Hudson's Bay Company exploring and trading expedition	John Bean Moses Norton	*Churchill*	1755

From Churchill. Bean replaced James Walker as master of *Churchill* for the annual voyages to trade at Whale Cove and to explore the coast farther north—particularly to search in about lat 64°N for Chesterfield Inlet, discovered by Moor and Smith (see 1746-47), but not seen since. It was hoped that this inlet might prove to be an entrance to a North-west Passage. Bean carefully explored the coast northward from 64°N. Not surprisingly, he failed to find Chesterfield Inlet, which is south of that latitude.

WILLIAMS, 1962, p. 126.

Hudson's Bay Company supply voyage	William Norton George Spurrell Jonathan Fowler Joseph Spurrell	*Hudson's Bay* [V] *Prince Rupert* [II] *Prince Rupert* [III] *Seahorse* [I]	1756

From London. *Hudson's Bay* called at Richmond 5-21 August and reached London 14 October. *Prince Rupert* [II] called at York Factory 3-15 August, at Churchill 19-25 August, and reached London 14 October. This appears to have been the 35th and last of Spurrell's voyages to Hudson Bay, which began in 1722. *Prince Rupert* [III] called at Churchill 2-25 August and reached London 16 October. *Seahorse* called at Albany 1-14 September, at Moose Factory 16-23 September, and reached London 10 November.

HUDSON'S BAY COMPANY (1).

Hudson's Bay Company trading and exploring expedition	John Bean	*Churchill*	1756

From Churchill. Bean made a second attempt to locate Chesterfield Inlet north of 64°N. Again failing, he was convinced that the inlet did not exist.

WILLIAMS, 1962, p. 126.

Hudson's Bay Company overland expedition	Joseph Smith Joseph Waggoner		1756-57

From York Factory; 23 August 1756 - (?) 1757; sent to find a route from Cross Lake on Nelson River to lakes Winnipeg and Winnipegosis and Cedar Lake. They hoped by this means to guide Indians past the French traders, who had become very active in that area (see 1750). Leaving York Factory with a band of Indians, they followed Henday's route (see 1754-55) to Cross Lake, then ascended East River to Little Playgreen Lake and, on 22 September, reached Lake Winnipeg. They crossed Lake Winnipeg to Saskatchewan River and, on 31 October, reached the French Fort

Bourbon on Cedar Lake, but met no Frenchmen. Continuing overland and on lake ice south-westward towards Porcupine Hills, they began to meet the Frenchmen wintering among the Indians. On 30 November, they crossed Assiniboine River, began to hunt buffalo to make pemmican for the return journey, and were joined by a Frenchman who elected to winter with them. In March 1757, they retraced their steps to Swan River, built canoes and, on 5 May, set out back towards York Factory. They canoed down Swan River, through Swan and Winnipegosis lakes, and crossed Mossy Portage back to Cedar Lake, where they had a cordial meeting with the Frenchmen at Fort Bourbon. They returned to York Factory by way of Lake Winnipeg, Little Playgreen Lake, and Echimamish and Hayes rivers, thus becoming the first Englishmen to use what later became the main route between Hudson Bay and the interior.

MORTON, 1939, pp. 250-51.

1757

Hudson's Bay Company supply voyage	William Norton	*Seahorse* [I]
	Jonathan Fowler Sr	*Prince Rupert* [II]
	Jonathan Fowler Jr	*Hudson's Bay* [V]
	Joseph Spurrell	*Prince Rupert* [III]

From London. *Seahorse* visited Albany 11 - 21 August, Moose Factory 24 August - 1 September, and reached London on 28 October. *Prince Rupert* [II] called at York Factory 10 - 22 August and reached London on 26 October. *Hudson's Bay* called at Richmond Fort 10 - 25 August and reached London on 27 October. *Prince Rupert* [III] called at Churchill 11 - 22 August and reached London on 28 October.

HUDSON'S BAY COMPANY (1).

1757

Hudson's Bay Company trading voyage	John Bean	*Churchill*

From Churchill. Having convinced himself that Chesterfield Inlet, for which he had twice searched (see 1755 and 1756), did not exist, Bean did not attempt any exploration during this voyage, but simply conducted the routine trade with the Eskimos northward to Whale Cove. Bean landed on Marble Island on 28 July.

WILLIAMS, 1962, p. 126; ROSS AND BARR, 1972, p. 31.

1757-58

Hudson's Bay Company overland expedition	Joseph Smith
	Joseph Waggoner

From York Factory. A week after returning from their successful exploratory journey of 1756-57, Smith and Waggoner set out again into the interior to bring Indians to York Factory over the route they had discovered. Again they hunted buffalo during the winter on the prairies south of Assiniboine River, and in 1758 returned to York Factory with a fleet of 57 canoes of Assiniboine Indians.

MORTON, 1939, p. 252.

Hudson's Bay Company overland expedition	Isaac Batt George Potts	1757-58

From York Factory. Continuing the company's new policy of sending men inland to persuade Indians in regions formerly dominated by French traders to take their furs to York Factory, Batt and Potts followed Smith and Waggoner's new route to Lake Winnipeg (see 1756-57), wintered on Saskatchewan River west of Cumberland Lake, and returned to York Factory in 1758 with 64 canoes of Sturgeon Indians.

MORTON, 1939, p. 252.

Hudson's Bay Company supply voyage	William Norton Jonathan Fowler Sr Jonathan Fowler Jr Joseph Spurrell	*Seahorse* [I] *Prince Rupert* [II] *Hudson's Bay* [V] *Prince Rupert* [III]	1758

From London. *Seahorse* called at Moose Factory 16 September - 2 October. *Prince Rupert* [II] visited York Factory 7 - 17 September. *Hudson's Bay* called at Richmond Fort 2 - 8 September. *Prince Rupert* [III] called at Churchill 10 - 19 September. All ships reached London on 27 November.

HUDSON'S BAY COMPANY (1).

Hudson's Bay Company trading voyage	James Wood	*Churchill*	1758

From Churchill. Wood replaced John Bean as master of *Churchill*, and made a routine trading voyage northward to Whale Cove. There was no trading voyage in *Churchill* in 1759.

WILLIAMS, 1962, p. 126.

New England trading voyage to Labrador	Henry Atkins	?	1758

From Boston. This was Atkins' third recorded voyage to the Labrador coast (see also 1729 and 1753). He left in early June, landed on the Labrador coast in lats 57°N, 56°40'N, and 53°50'N. At this last stop, he found a "delightful Inlet," and made observations on game, fish, and trees. He left it on 29 September, and continued searching the coast and trading with natives during the return south; he visited Fortune Bay, Labrador, and St Peter's Bay, Newfoundland, then returned to Boston.

BERNARD, 1792; GOSLING, 1910, p. 324.

Hudson's Bay Company supply voyage	William Norton Jonathan Fowler Sr Jonathan Fowler Jr Joseph Spurrell	*Seahorse* [I] *Prince Rupert* [II] *Hudson's Bay* [V] *Prince Rupert* [III]	1759

From London. *Seahorse* visited Albany 26 August - 2 September, Moose Factory 4 - 11 September, and reached London 23 October. *Prince Rupert* [II] called at York 20 August - 4 September, and reached London 22 October. *Hudson's Bay* called at Richmond Fort 17 August - 1 September, and reached London 23 October. *Prince Rupert* [III] called at Churchill 21 - 30 August and reached London 24 October.

HUDSON'S BAY COMPANY (1).

1759

Re-establishment of Fort Severn

The original Fort Severn had been established by the Hudson's Bay Company in 1685 (*qv*), held by the French under the name of Fort Ste-Thérèse, from 1691 until 1713, when, after its restoration to the Hudson's Bay Company under the Treaty of Utrecht, it was abandoned. It was re-established by the company in 1759 under the name of James Fort. The name was changed to Severn House in 1761, and it has been in continuous operation ever since.

MACFIE, 1970; VOORHIS, 1930, p. 162; RICH, 1952, p. xxiv.

1759

Closure of Richmond Fort

Richmond Fort, which had been established in 1750 (*qv*), was closed in 1759.

FOR BIBLIOGRAPHY, SEE UNDER 1750.

1759-60

Hudson's Bay Company overland expedition — Isaac Batt

From York Factory. Batt repeated his journey of 1757-58, and visited the Indians of Saskatchewan River west of Cumberland Lake to take them to trade at York Factory.

MORTON, 1939, p. 252.

1759-60

Hudson's Bay Company overland expedition — Anthony Henday, Joseph Smith

From York Factory. Henday returned to the country at the foot of the Rocky Mountains that he had visited in 1754-55, and brought back 61 canoes of Blackfeet Indians to trade at York Factory. After this journey the Bloods, a tribe of the Blackfeet, began to trade regularly at York Factory.

MORTON, 1939, p. 252.

1760

Hudson's Bay Company supply voyage	William Norton	*Seahorse* [I]
	Jonathan Fowler Jr	*Prince Rupert* [II]
	Joseph Spurrell	*Prince Rupert* [III]

From London. *Seahorse* called at Moose Factory 26 August - 16 September and reached London 23 October. *Prince Rupert* [II] called at Churchill 23 August - 2 September. *Prince Rupert* [III] called at York Factory 26 August - 11 September. Both ships reached London 24 October.

HUDSON'S BAY COMPANY (1).

Hudson's Bay Company trading voyage | James Wood | *Churchill*

1760

From Churchill. Routine trading voyage north to Whale Cove.

WILLIAMS, 1962, p. 126.

Hudson's Bay Company overland expedition | Joseph Waggoner

1760-61

From York Factory. Travelled inland, probably to Lake Winnipeg region, and returned to York Factory with 65 canoes of Indians.

MORTON, 1939, p. 252.

Hudson's Bay Company supply voyage

William Norton — *Prince Rupert* [III]
Jonathan Fowler Jr — *Seahorse* [I]
Joseph Spurrell — *King George* [II]

1761

From London. *Prince Rupert* called at Albany 21 - 29 August, at Moose Factory 29 August - 5 September, and reached London 19 November. *Seahorse* called at Churchill 15 - 26 August, and reached London 13 November. *King George* called at York Factory 16 August - 1 September, and reached London on 21 November.

HUDSON'S BAY COMPANY (1).

Hudson's Bay Company exploring expedition | William Christopher | *Churchill*

1761

From Churchill. Christopher was instructed, as previous masters of *Churchill* had been, to trade with Eskimos at Whale Cove, then to search for Chesterfield Inlet in about lat 64°N. The inlet had been discovered in 1747 by Moor and Smith (see 1746-47), but further searches through the 1750's had failed to relocate it. In this year, Christopher found the inlet and sailed up it some 150 km, until forced by contrary winds to return. This voyage renewed hope that a North-west Passage might be found by that route, and that the inlet might give easy access to minerals and furs in the interior.

WILLIAMS, 1962, p. 128.

Hudson's Bay Company overland expedition | Henry Pressick

1761-62

From York Factory. Pressick repeated Henday's journey of 1759-60, and visited the Blackfeet Indians at the foot of the Rocky Mountains to encourage them to trade at York Factory.

MORTON, 1939, p. 252.

Hudson's Bay Company supply voyage

William Norton — *Prince Rupert* [III]
Jonathan Fowler Jr — *Seahorse* [I]
Joseph Spurrell — *King George* [II]

1762

From London. *Prince Rupert* called at Moose Factory 2 - 12 September and reached London 10 November. *Seahorse* called at Churchill

22 August - 1 September and reached London 2 November. *King George* called at York Factory 27 August - 6 September, and reached London on 10 November.

HUDSON'S BAY COMPANY (1).

1762

Hudson's Bay Company exploring expedition	William Christopher	*Churchill*
	?	*Strivewell*
	Moses Norton	

From Churchill. Christopher returned to complete his exploration of Chesterfield Inlet, which he had rediscovered in 1761 (*qv*). Christopher sailed from Churchill on 13 July, reached Chesterfield Inlet two weeks later, and explored it as far as the head of Baker Lake, where he discovered that there was no hope of a North-west Passage by this route.

WILLIAMS, 1962, p. 129.

1763

Hudson's Bay Company supply voyage	William Norton	*Prince Rupert* [III]
	Jonathan Fowler Jr	*Seahorse* [I]
	Joseph Spurrell	*King George* [II]

From London. *Prince Rupert* called at Albany 18 - 27 August, at Moose Factory 29 August - 5 September, and reached London 14 October. *Seahorse* called at Churchill 28 August - 8 September, and reached London 30 October. *King George* called at York Factory 29 August - 9 September, and reached London 5 November.

HUDSON'S BAY COMPANY (1).

1763

Hudson's Bay Company trading voyage	William Christopher	*Churchill*

From Churchill. Routine trading voyage northward to Whale Cove and Marble Island. When he touched at Marble Island, Christopher noticed whales sporting in the harbour. His report of this encouraged the Committee in London to issue instructions for whaling gear to be taken on these annual voyages. These instructions were first implemented in 1765 (*qv*).

WILLIAMS, 1962, p. 131; ROSS, W.G., 1973, p. 5.

1763

Treaty of Paris

The Seven Years' War (1756-63), although mainly fought in Europe, extended to the French and English colonies. In 1759, the English captured Canada and, in 1763, at the Treaty of Paris, France ceded Canada to Great Britain. Even before the Cession, English and Scottish merchants had begun to follow the river-and-lake routes that the French had explored from Montreal into the west and north-west, where, in the heart of the fur country, during the next half century they competed aggressively with the Hudson's Bay Company.

MORTON, A.S., 1939, pp. 257-58; RICH, 1967, pp. 130-34.

Hudson's Bay Company overland fur-trading expeditions	Joseph Smith Isaac Batt		1763-64

From York Factory. Smith travelled by Hayes and Fox rivers to Split Lake on Nelson River. He continued by way of Grass River to Athapapuskow Lake and River, then by way of Goose and Namew Lakes to Cumberland Lake on Saskatchewan River. He was the first European known to have followed this route from Hudson Bay into the interior. He then ascended Saskatchewan River to a point above Nipawi. He died in spring 1764, while returning with 71 canoes of Indians to trade at York Factory. Batt travelled with Smith to Fox River, then probably continued by way of Leaf River, Cross Lake, and Minago River to the region of The Pas. He returned in 1764 with 39 canoes of Indians to trade.

MORTON, 1939, p. 274.

Hudson's Bay Company supply voyage	Jonathan Fowler Jr Joseph Richards John Horner	*King George* [II] *Seahorse* [II] *Prince Rupert* [III]	1764

From London. King George called at York Factory 13-25 August, and reached London 6 October. *Seahorse* visited Churchill 13-26 August, and reached London 5 October. *Prince Rupert* visited Albany 11-20 August, Moose Factory 21-28 August, and reached London 8 October.

HUDSON'S BAY COMPANY (1) and (2).

Moravian missionary expedition to Labrador	Jens Haven	?	1764

From England. After the Moravians' failure to establish a mission in Labrador in 1752 (*qv*), Haven became interested in the project. In 1764, he set out to consider possible sites. He encountered some difficulty in crossing from Newfoundland to Labrador, for the English had only just begun to trade along that coast, and few ships went there. At Quirpon he met Captain James Cook, who arranged for an Irish fishing vessel to take him to Labrador. On 24 August, Haven landed at Chateau Bay, but met no Eskimos. He returned to Quirpon, where he did meet some Eskimos and preached to them. He returned to England in the frigate *Lark* with a recommendation from Sir Hugh Palliser to the Board of Trade, which enabled him to make a second voyage in 1765 (*qv*).

PEACOCK, 1960; WHITELEY, 1964; PUBLIC RECORD OFFICE, CO 194/16, pp. 59-62 [Haven's manuscript journal].

Hudson's Bay Company trading and exploring expedition	Magnus Johnston	*Churchill*	1764

From Churchill. Johnston succeeded William Christopher as master of *Churchill* on her annual voyage to trade with Eskimos north of Churchill. He also took over the task of exploring the coast in search of a North-west Passage. On this voyage, he explored the coast opposite Marble Island, but discovered nothing new.

WILLIAMS, 1962, p. 131.

1765	Hudson's Bay Company supply voyage	Jonathan Fowler Jr Joseph Richards John Horner	*King George* [II] *Prince Rupert* [III] *Seahorse* [II]

From London. *King George* visited York Factory 6-16 September, and reached London 20 October. *Prince Rupert* visited Churchill 4-10 September, and reached London 17 October. *Seahorse* visited Moose Factory 4-20 September, and reached London 3 November.

HUDSON'S BAY COMPANY (1) and (2).

1765	Moravian missionary expedition to Labrador	Jens Haven Christian Drachardt John Hill Andrew Schloezer

From Spithead; May - (?). They sailed to Croque, Newfoundland, in *Lark,* in company with the British fishing fleet, then to Chateau Bay, Labrador, in HMS *Niger,* arriving on 17 July. Haven and Schloezer then coasted northward in the *Hope* schooner, searching for Eskimos and for a suitable site for a mission. However, they met no Eskimos, the crew refused to sail north of Davis Inlet, and they were back at Chateau Bay by early September. Meanwhile, Drachardt and Hill had had a very successful meeting with Eskimos at Chateau Bay, during which they recorded information on the Eskimos' mode of living, dwelling places, numbers, trading activities, and preferences in trade goods.

PEACOCK, 1960; WHITELEY, 1964; PUBLIC RECORD OFFICE, CO 194/16, pp. 223-45.

1765	Hudson's Bay Company whaling and trading voyage	Magnus Johnston	*Churchill*

From Churchill. Johnston was sent on the routine trading voyage northward from Churchill but also, for the first time, he carried whaling gear and was given instructions to attempt whaling in the vicinity of Marble Island. He remained at Marble Island 4-10 August, but killed no whales.

ROSS, W.G., 1973, pp. 5-6, 11.

1765-66	Hudson's Bay Company overland fur-trading expeditions	Louis Primeau Isaac Batt

From York Factory. Primeau wintered inland with the Assiniboine Indians, and returned to York Factory with 20 canoes of

Indians. Batt also wintered inland, probably near The Pas, and returned with 33 canoes of Indians.

MORTON, 1939, p. 275.

Hudson's Bay Company supply voyage	Jonathan Fowler Jr Joseph Richards John Horner	*King George* [II] *Prince Rupert* [III] *Seahorse* [II]	1766

From London. *King George* visited York Factory 11-28 August, and returned to London 17 October. *Prince Rupert* visited Churchill 7-27 August, and reached London 10 October. *Seahorse* visited Albany 17-27 August, Moose Factory 29 August - 6 September, and reached London 20 October.

HUDSON'S BAY COMPANY (1) and (2).

British naval expedition to Labrador	Sir Thomas Adams Constantine Phipps Joseph Banks	HMS *Niger*	1766

From England; 7 April - 17 November; to investigate the fisheries of Newfoundland and Labrador. They visited St John's during late May and early June, and Chateau Bay during August and September. Banks and Phipps made several excursions ashore, during which Banks collected plants and about 70 species of birds, the first comprehensive biological collection from that region. They returned to Croque, Newfoundland, on 7 October, and thence continued to England.

*LYSAGHT, 1971; CAMERON, 1952, p. 4.

Hudson's Bay Company whaling and trading voyage	Magnus Johnston John Garbut	*Churchill* *Success*	1766

From Churchill. Johnston was sent on the routine trading voyage northward to Whale Cove; Garbut sailed direct to Marble Island to attempt whaling. As on the previous year's voyage (*qv*), no whales were caught, though Garbut did confirm that whales were abundant around Marble Island.

ROSS, W.G., 1973, pp. 6, 11.

Hudson's Bay Company overland fur-trading expeditions	Louis Primeau Edward Lutit James Dearing James Allen Isaac Batt William Pink		1766-67

Primeau was sent to "the famous . . . Beaver River," which had not yet been seen by white men. (He probably ascended Churchill River.) Lutit went far up the North Saskatchewan River with Assiniboine Indians. Dearing travelled inland with Attikasish,

Henday's guide in 1745-55 (*qv*). They probably wintered near Red Deer River, Alberta, within sight of the Rocky Mountains. Allen travelled far up North Saskatchewan River with a well known trading Indian, Cabosh, where they wintered. Batt's wintering-ground among the Indians is not recorded: it may have been on Saskatchewan River near Pasquia or on Sturgeon River, with which he was later recorded to be familiar. Pink appears to have revisited the region of Nipawi (see Smith 1763-64) and probably hunted buffalo west of South Saskatchewan River. On 25 June 1767, Pink, Batt, Allen and Lutit arrived at York Factory with 156 canoes of Indians to trade.

MORTON, 1939, pp. 275-76.

1767

Hudson's Bay Company supply voyage	Jonathan Fowler Jr Joseph Richards John Horner	*King George* [II] *Prince Rupert* [III] *Seahorse* [II]

From London. *King George* visited York Factory 29 August - 8 September, and reached London 22 October. *Prince Rupert* visited Churchill 28 August - 11 September, and reached London 22 October. *Seahorse* visited Moose Factory 4 - 16 September, and reached London 1 November.

HUDSON'S BAY COMPANY (1) and (2).

1767

Hudson's Bay Company whaling and trading voyage	Joseph Stevens Magnus Johnston Samuel Hearne	*Success* *Churchill*

From Churchill. *Success* was sent whaling at Marble Island, but they caught only one small whale. Johnston made the routine voyage north to trade with Eskimos. When *Success* was anchored at Marble Island, members of her crew stumbled on relics of James Knight's North-west Passage expedition (see 1719-21), the first firm evidence regarding the fate of that expedition. A few days later, Johnston arrived at the island in *Churchill* and joined the search for more relics. Hearne, mate of *Churchill*, saw the hulks of Knight's two ships underwater. In 1769 he returned to search the island again, and to learn from an old Eskimo the full story of Knight's fate (*qv*).

WILLIAMS, 1962, pp. 26-27; GLOVER, 1958, pp. lxi-lxiii; KENNEY, 1932, pp. 85-86; ROSS, W.G., 1973, p. 6.

1767-68

Hudson's Bay Company overland fur-trading expeditions	Louis Primeau Isaac Batt James Dearing James Allen Edward Lutit William Pink

From York Factory. Primeau made the routine journey to winter among trading Indians on Saskatchewan River. On his return, he reported the presence there of a pedlar (private trader) from

Montreal, François le Blanc (also called Saswe), and urged the company to establish an inland post to prevent the Montreal pedlars from capturing the trade. Batt wintered on the upper reaches of Assiniboine River. Dearing, Allen and Lutit wintered at different places on Saskatchewan River. Pink travelled by the Hayes River – Cross Lake – Minago River route to the Saskatchewan. He left the North Saskatchewan at Shell River and travelled overland towards the north-west, to reach the upper waters of Beaver River, near its junction with Muriel River, on 19 August. He spent part of the winter on Beaver and Amisk rivers, then made his way southward, back to North Saskatchewan River, where Blackfeet Indians visited him. He set out homeward on 8 May 1768 and fell in with Allen and Dearing on the 10th. They reached York Factory on 3 July.

MORTON, 1939, pp. 276-79.

1767-68

Hudson's Bay Company overland expedition — William Tomison

From Fort Severn; to investigate reports that pedlars from Montreal had built trading posts in the hinterland of Fort Severn. He ascended Severn River and descended Winnipeg River to its mouth, where he met Indians who were waiting for pedlars, who, Indians told him, had several forts in the region. On 2 October 1767, he met François le Blanc (*cf* 1767-68, Primeau, above). Tomison wintered near Lake Winnipeg, but failed to persuade the Indians there to change their allegiance to the Hudson's Bay Company. Tomison was the first Hudson's Bay Company trader to view the site of present-day Winnipeg.

MORTON, 1939, pp. 276-77.

1768

Hudson's Bay Company supply voyage	Jonathan Fowler Jr	*King George* [II]
	Joseph Richards	*Prince Rupert* [III]
	John Horner	*Seahorse* [II]

From London. *King George* visited York Factory 12 - 28 August, and reached London 9 October. *Prince Rupert* visited Churchill 9 August - 1 September, and reached London 10 October. *Seahorse* visited Albany 20 - 31 August, Moose Factory 2 - 6 September, and reached London 31 October.

HUDSON'S BAY COMPANY (1) and (2).

1768

Hudson's Bay Company whaling and trading voyage	Magnus Johnston	*Churchill*
	Joseph Stevens	*Success*
	?	*Speedwell*

From Churchill. *Churchill* made the routine coastal trading voyage northward to Whale Cove; *Success* and *Speedwell* were sent whaling off Marble Island. They remained there 2 - 18 August but, as in previous years, no whales were caught.

ROSS, W.G., 1973, p. 9.

1768-69

Hudson's Bay Company overland fur-trading expeditions

William Pink
Isaac Batt
James Allen
Edward Lutit
James Dearing

From York Factory. Pink followed the Grass River route discovered by Joseph Smith (see 1763-64) to Saskatchewan River, then followed his own course of 1767-68 (*qv*) to Frog Creek, Alberta, which he reached on 30 September, and to Mooswa Creek. He wintered with Indians in the region of present-day Vermilion. There he met Batt. When building canoes at Mooswa Creek in 1769, he met Allen with his Indians; Lutit and Dearing built their canoes nearby. The five men and their Indians returned together to York Factory. At Nipawi, a post on Saskatchewan River maintained by François le Blanc (see 1767-68, both Primeau and Tomison), they met James Finlay, the first British trader from Montreal to reach Saskatchewan River. Finlay led a party of 12 Frenchmen. Pink had learned earlier that Finlay had an outpost at Pasquia, so they avoided that post by returning to York Factory by way of the Grass River route.

MORTON, 1939, pp. 279-80.

1768-69

British scientific expedition to Hudson Bay

William Wales
Joseph Dymond

From London; sponsored by the Royal Society, which sent scientists to many parts of the world to observe the transit of Venus on 3 June 1769. They sailed to Churchill in 1768 in *Prince Rupert* (*qv*). Wales kept a journal and, with Dymond, a meteorological record (published in 1771). Wales used his leisure at Churchill to compute tables of the equations to equal altitudes for facilitating the determination of time, an aid in the determination of longitude. They returned to London in autumn 1769 in *Prince Rupert* (*qv*). It was probably at Wales's suggestion that the Royal Society requested the Hudson's Bay Company to supply specimens of the natural history of Hudson Bay, which resulted in a great advance in scientific knowledge of the region.

*WALES, †1770; *1771; †DYMOND AND WALES, 1771; WILLIAMS, 1969, pp. xx-xxi.

1769

Hudson's Bay Company supply voyage

Hudson's Bay Company supply voyage	Jonathan Fowler Jr	*King George* [II]
	Joseph Richards	*Prince Rupert* [IV]
	John Horner	*Seahorse* [II]

From London. *King George* visited York Factory 14 - 27 August and arrived back at London 29 September. *Prince Rupert* visited Churchill 14 August - 8 September, and reached London 31 October. *Seahorse* visited Albany 18 - 30 August, Moose Factory 1 - 8 September, and reached London 30 October.

HUDSON'S BAY COMPANY (1) and (2).

Hudson's Bay Company whaling voyage — Joseph Stevens, Samuel Hearne — *Charlotte*

1769

From Churchill. Routine whaling voyage to Marble Island. Hearne, the ship's mate, made a second visit to the harbour on Marble Island where relics of James Knight's North-west Passage expedition (see 1719-21) had been found in 1767 (*qv*). While he was there, an old Eskimo told him the full story of the fate of Knight's ships. Attempts at whaling on this voyage were, as in previous years, unsuccessful.

WILLIAMS, 1962, pp. 26-27; GLOVER, 1958, pp. lxi-iii; KENNEY, 1932, pp. 86-87; ROSS. W.G., 1973, p. 9.

Hudson's Bay Company first expedition to Coppermine River — Samuel Hearne

1769

From Churchill; 6 November - 11 December; to search for copper deposits and to determine whether a low-latitude North-west Passage could be found north of Churchill. He travelled overland and had gone only about 320 km north-west of Churchill, when his Cree guide, "Captain" Chawchinahaw, deserted him and he had to return to Churchill. He made a second attempt in 1770 (*qv*).

*HEARNE, 1795; *GLOVER, 1958; SPECK, 1963.

Hudson's Bay Company overland fur-trading expeditions — William Pink, James Dearing, James Allen, Thomas Haddle, Louis Primeau

1769-70

From York Factory. They made the routine journey inland towards Saskatchewan River to assemble Indians to trade at York Factory. Pink travelled by way of Minago River to the Saskatchewan and continued to Eagle Hills, probably intending to observe the activities of the Montreal pedlars based at Pasquia. He wintered among Assiniboine, Blood, and Blackfeet Indians on the plains beyond Manitou Lake, where he met Dearing, Allen and Haddle. The pedlars seem not to have returned to Saskatchewan River that season, so the company men had no difficulty in persuading the Indians to go to York Factory.

MORTON, 1939, pp. 280-81.

Hudson's Bay Company overland fur-trading expedition — William Tomison

1769-70

From Fort Severn. He reached Lake Winnipeg by Berens (Pigeon) River, crossed the lake, continued by Dauphin River to Lake Manitoba, then passed east of Lake Dauphin to Assiniboine River. Near present-day Birtle, Manitoba, he turned north-eastward back to Lake Dauphin, where he built canoes for his return. On

Mossy River, near Lake Winnipeg, he heard a false report that Fort Severn had been abandoned. He therefore turned toward York Factory, travelling by lakes Manitoba and Winnipeg, Little Playgreen Lake, East River, and Cross Lake, and reached the post with only four canoes of Indians.

MORTON, 1939, pp. 281-82.

1770

Hudson's Bay Company supply voyage	Jonathan Fowler Jr Joseph Richards William Christopher	*King George* [II] *Prince Rupert* [IV] *Seahorse* [II]

From London. *King George* visited Churchill 24 August - 7 September, and reached London 15 October. *Prince Rupert* visited York Factory 25 August - 8 September, and reached London 16 October. *Seahorse* visited Moose Factory 24 August - 11 September, and reached London 17 October.

HUDSON'S BAY COMPANY (1) and (2).

1770

Moravian missionary expedition to Labrador	Francis Mugford Jens Haven Christian Drachardt Stephen Jensen	*Jersey Packet*

From Exmouth; 17 May - (?) November. Haven (see 1764 and 1765) had obtained permission from the Board of Trade to establish a mission on the Labrador coast. After visiting southern Labrador and Byron Bay, they continued north. On 1 August, they met and traded with a large number of Eskimos, to whom Drachardt and Haven preached and from whom they purchased 30 km^2 of land on which to build a mission (see 1771). They then sailed south to Chateau Bay, participated in the fisheries for a time, then returned to England.

WHITELEY, 1964, p. 43; GOSLING, 1910, p. 266.

1770

Hudson's Bay Company whaling voyage	Joseph Stevens	*Charlotte*

From Churchill. Sent on the routine whaling voyage to Marble Island, Stevens encountered unusually heavy pack ice and turned back to Churchill after having reached only 59°34'N.

ROSS, W.G., 1973, p. 9.

1770

Hudson's Bay Company second expedition to Coppermine River	Samuel Hearne

From Churchill; 23 February - 25 November. On return from his first unsuccessful expedition (see 1769), Hearne began to plan a second attempt. He left Churchill with five Indians, including a Chipewyan guide, Conne-e-quese, crossed Seal River (probably at the western end of Shethanei Lake), then travelled north. About 1

June, he passed Baralzon Lake, crossed the Barren Grounds to Yathkyed Lake, crossed Kazan River, and continued northward. By the time he reached Dubawnt River, he had been joined by about 600 Indians. He possibly discovered Dubawnt Lake, and passed close to Aberdeen Lake. Somewhere north-west of Dubawnt Lake, he accidentally broke his quadrant, and he felt obliged to turn back. His return journey cannot be accurately traced. He probably crossed Dubawnt River where it flows into Dubawnt Lake (about 63°N), then turned south-east and travelled overland to Kazan River, which he reached just above Angikuni Lake. He continued south-eastwards to Thlewiaza River, again crossed Seal River, and reached Churchill. Although he failed in the main object of his journey, Hearne had penetrated farther into sub-Arctic Canada than any European before him—600 km north-west from Churchill and 500 km inland from Hudson Bay. He also recorded many observations of natural history.

*HEARNE, 1795; *GLOVER, 1958; SPECK, 1963.

1770-71

Hudson's Bay Company overland fur-trading expedition — Louis Primeau

From Churchill. Primeau, with three Hudson's Bay Company servants whose names are unrecorded, made the routine journey inland to meet Indians on Saskatchewan River. They returned in 1771 with 131 canoes of Indians.

MORTON, 1939, p. 282.

1770-72

Hudson's Bay Company third expedition to Coppermine River — Samuel Hearne

From Churchill; 7 December 1770 - 30 June 1772. After two failures (see 1769 and 1770), Hearne set out again, this time with a Chipewyan guide, Matonabbee. They reached Nueltin Lake on 31 December, then continued westward by Kasba Lake to Snowbird Lake, where they remained for some time in a Chipewyan encampment. From here they travelled north-westward in company with a large party of Chipewyans, reached Clowey Lake on 3 May 1771, Clinton-Colden Lake on 27 May, then passed by Lake Aylmer, Lac de Gras, Lac du Sauvage, and Contwoyto Lake. They finally reached Coppermine River near Sandstone Rapids, and descended it from there to the Arctic Ocean. On 17 July, at Bloody Fall, near the mouth of the river, Hearne's Indian companions massacred a group of Eskimos fishing there. The following day Hearne reached the sea, the first white man to stand on the north coast of America. Some writers have doubted that Hearne actually reached the sea, suggesting that the Indians, knowing his disapproval of the intended massacre, had left him some distance behind, and that Hearne had pieced out details of the massacre and the coastal scenery from what the Indians had told him. His dead-reckoning position of the river's mouth was about 320 km to the north in error, but John Franklin confirmed his description of it 50 years later (see 1819-22). He returned to Churchill by way of Point Lake, Great Slave Lake (which he saw on 24 December), Wholdaia Lake, Ennadai Lake, and Kazan River. He brought back a sample of copper, but the company made no attempt to exploit the copper deposits. His journey demonstrated that there was no hope of a low-latitude North-west Passage.

*HEARNE, 1795; *GLOVER, 1958; SPECK, 1963; MOWAT, 1958; RICHARDSON, 1836; GLOVER, 1951; WILSON, J.T., 1949; BURPEE, 1946; WRAY, 1934.

1770-86

Cartwright's residence in Labrador

George Cartwright hunted, trapped, and traded from his base at Cape Charles for 16 years. He was a keen observer of natural history, and he corresponded on scientific subjects with Sir Joseph Banks (see 1766). In 1773, he took some Eskimos with him on a visit to England, and they created a sensation. His journal (published in 1792) was the standard reference to Labrador for many years.

GOSLING, 1910, pp. 222-250; TOWNSEND, 1911; LYSAGHT, 1971, pp. 71-89.

1771

Hudson's Bay Company supply voyage	Jonathan Fowler Jr	*King George* [II]
	Joseph Richards	*Prince Rupert* [IV]
	William Christopher	*Seahorse* [II]

From London. *King George* visited Churchill 24 August - 7 September, and reached London 17 October. *Prince Rupert* visited Moose Factory 25 August - 14 September, and reached London 29 October. *Seahorse* visited York Factory 25 August - 7 September, and reached London 17 October.

HUDSON'S BAY COMPANY (1) and (2).

1771

Moravian missionary expedition to Labrador; establishment of Nain	Francis Mugford	*Amity*
	Jens Haven	
	Christian Drachardt	
	Christopher Braasen	
	John Schneider	
	Stephen Jensen	

From London; 8 May - (?); to establish a mission on the site selected by Haven in 1770 (*qv*). They reached Nain on 9 August, and the Moravians began at once to build a mission house, which was completed on 22 September. Two days later, the ship sailed for England, leaving Haven, Braasen, Schneider, and their wives, Drachardt, Jensen, and six other Moravian Brethren at Nain. The mission has been in operation ever since. *Amity,* Captain Mugford, made annual supply voyages from London to the mission station until 1777, when she was replaced by *Good Intent. Amity* was again used between 1780 and 1786. Between 1787 and 1802, the first of several ships named *Harmony* made the annual voyage. In 1782, Captain James Fraser replaced Captain Mugford.

WHITELEY, 1964; HILLER, 1971.

Hudson's Bay Company whaling voyage	Thomas Robinson	*Charlotte*	1771

From Churchill. Routine whaling voyage to Marble Island. He remained at the island 10 July - 11 August, and killed three whales.

ROSS, W.G., 1973, pp. 10-11.

Hudson's Bay Company overland fur-trading expedition	Louis Primeau Isaac Batt		1771-72

From York Factory. They made the routine journey inland to meet trading Indians on Saskatchewan River, and to conduct them back to York Factory.

MORTON, 1939, p. 282.

Hudson's Bay Company supply voyage	Jonathan Fowler Jr Joseph Richards William Christopher	*King George* [II] *Prince Rupert* [IV] *Seahorse* [II]	1772

From London. *King George* visited York Factory 21 August - 5 September, and reached London 3 October. *Prince Rupert* visited Churchill 19 August - 6 September, and reached London 8 October. *Seahorse* visited Albany 20 August - 3 September, Moose Factory 5-11 September, and reached London 17 October.

HUDSON'S BAY COMPANY (1) and (2).

Hudson's Bay Company whaling voyage	Thomas Robinson	*Charlotte*	1772

From Churchill; routine whaling voyage to Marble Island. He remained at the island 13 July - 13 August, but killed only one small whale. After this voyage, the Hudson's Bay Company abandoned their attempts at whaling in Hudson Bay. In the eight years since their first whaling voyage to Marble Island, in 1765, only five whales had been caught.

ROSS, W. G., 1973, pp. 10-11.

Hudson's Bay Company overland fur-trading expedition	Matthew Cocking Louis Primeau John Cole		1772-73

From York Factory. Routine journey inland to meet trading Indians in the Saskatchewan River region. In 1773, Primeau deserted to the Montreal traders (pedlars), and did not return to York Factory (see 1773-74). Cole did return, but the details of his journey are not recorded. Cocking left York Factory on 27 June 1772, probably travelled by way of Hayes River, Deer Lake, Cross Lake, Nelson and Minago rivers to reach Saskatchewan River at The Pas. He continued past Cumberland Lake, crossed the prairie north of Duck Lake to North Saskatchewan River, continued to Eagle Hills, then turned southward. On 1 December, he met the

Gros-Ventres Indians and tried unsuccessfully to make them return with him to trade their furs at York Factory. In spring 1773, he assembled the Indians who were to return with him, and they set out towards York Factory. However, when returning along Saskatchewan River they stopped at Nipawi, a trading post belonging to the rival Montreal traders François le Blanc and James Finlay (see also 1768-69), where many of Cocking's Indians traded their furs. They met more Montreal traders on their route, with the result that, on arrival at York Factory, there were only 16 canoes of Indians left with Cocking, carrying largely inferior furs.

BURPEE, 1908.

1773

Hudson's Bay Company supply voyage	Jonathan Fowler Jr	*King George* [II]
	Joseph Richards	*Prince Rupert* [IV]
	William Christopher	*Seahorse* [II]

From London. *King George* visited York Factory 23 August - 5 September, and reached London 10 October. *Prince Rupert* visited Churchill 20 August - 7 September, and reached London 13 October. *Seahorse* called at Moose Factory 26 August - 2 September, Albany 3-10 September, and reached London 13 October.

HUDSON'S BAY COMPANY (1) and (2).

1773-74

Hudson's Bay Company overland expedition — Joseph Hansom

From Churchill. Pedlars from Montreal were gradually robbing York Factory of its trade from the Saskatchewan River region (*eg*, see 1772-73, Cocking), so Moses Norton, Governor of Churchill, sent Hansom inland to investigate the effect of the pedlars on the trade at Churchill. Hansom travelled by way of Southern Indian Lake, Kississing Lake, Athapapuskow Lake, and Cumberland Lake to Saskatchewan River, then returned to winter at the house built by a pedlar, Joseph Frobisher, on Cumberland Lake (see below, 1773-74). Here he learned of the great extent of the pedlars' invasion. In spring 1774, Hansom assembled Indians to go with him to Churchill, but the Indians traded many of their furs with Frobisher, who had gone to plant himself on their route (see 1774-75).

Only one man, John Cole, made the routine journey inland from York Factory in 1773, but he deserted to the pedlars and did not return.

MORTON, 1939, pp. 289-90.

1773-74

Montreal pedlars' fur-trading expedition	Joseph Frobisher
	Louis Primeau

Frobisher arrived on Saskatchewan River from Montreal for the first time in 1773 and Primeau, who had recently deserted from the Hudson's Bay Company (see 1772-73), joined him. Together, they built a house on the south side of Cumberland Lake and wintered there. Joseph Hansom of the Hudson's Bay Company visited them during the winter. In spring 1774, they left their house and travelled north to intercept Indians on Churchill River (see 1774-75).

TYRRELL, J.B., 1934, pp. 107, 120, 122.

Hudson's Bay Company supply voyage	Jonathan Fowler Jr Joseph Richards William Christopher	*King George* [II] *Prince Rupert* [IV] *Seahorse* [II]	1774

From London. *King George* visited York Factory 18 August - 2 September, and reached London 6 October. *Prince Rupert* visited Moose Factory 5-21 September, and reached London 28 October. *Seahorse* visited Churchill 18-28 August, and reached London 2 October.

HUDSON'S BAY COMPANY (1) and (2).

Establishment of Cumberland House	Samuel Hearne		1774

Because of severe competition from the pedlars inland, the Hudson's Bay Company decided in 1772 to establish its first post well inland, an action that put an end to the company's "sleep by the Frozen Sea" and marked the beginning of an active and finally successful competition with its Montreal-based rivals. In 1774, after a year's delay, Hearne led a party to Cumberland Lake, a strategic location on the routes to both Churchill and Saskatchewan rivers, not far from the old French Fort Paskoyac (see 1750). The post has operated ever since. In 1780, the North West Company opened a rival post nearby, which was closed after the coalition of the two companies in 1821.

RICH, 1951, pp. xii, xiv, xxiv, xxv; 1938, p. 415.

Moravian missionary expedition to Labrador	Alexander Wilson Jens Haven John Hill Christopher Braasen Christian Lister Gottfried Lehman	*George*	1774

From St John's, Newfoundland; to find a site for a second mission station, to serve Eskimos who had been travelling long distances to Nain (see 1771). The ship sailed first to Nain. In early September, with four Moravian missionaries from Nain as passengers, Wilson sailed north along the Labrador coast as far as 59°N, where he turned back because the Eskimo pilot, Auluk, did not know the coast farther north. On 14 September, about 57°N, *George* ran aground and sank with the loss of Braasen and Lehman. The survivors repaired the ship's boat and, on 18 September, continued to Nain. They then returned to the wreck to recover the two bodies and to salvage what they could.

NEATBY, 1966.

1774

Spanish expedition to the Northwest Coast | Juan Josef Perez Hernandez, Esteban José Martinez | *Santiago*

From San Blas, Mexico; 25 January - (?). The Spanish, who claimed the entire Pacific coast of North America (see 1513, Balboa), were troubled by reports of Russian activities on the Northwest Coast and by England's powerful hold on eastern North America, so they sent an 88-man expedition to renew their claim to the coast north of California, which they had not yet visited. On 18 July, about 55°N, they reached the Queen Charlotte Islands. Because of storms and sickness, they disobeyed orders to sail to 60°N, and turned back. About 54°N, they met Haida Indians, of whom the Franciscan, Father Crespi, a member of the expedition, published the earliest account. On 18 August, they found a good anchorage, which they named San Lorenzo, near the entrance of the body of water that Cook, in 1778, named Nootka Sound (see 1776-80, Cook). During the "Nootka Affair" (see 1789, Martinez), Spain based her claim to the sole trading rights along the coast from California to Alaska and her claim to occupy Nootka on this voyage.

MARQUIS, 1914; ESPINOSA Y TELLO, 1802; BANCROFT, 1890, pp. 195-97.

1774-75

Montreal pedlars' fur-trading expedition; establishment of Frog Portage Fort | Joseph Frobisher, Louis Primeau

Frobisher and Primeau, who had spent the winter of 1773-74 on Cumberland Lake (*qv*), travelled north from there in spring 1774. They probably took the route by Amisk (Beaver), Mirond, Pelican and Wood lakes to Frog Portage, which divides Churchill River from the Nelson River system. There they built a temporary trading post, known as Frog Portage Fort or Fort la Traite, at which they wintered. They successfully intercepted Indians travelling from the west to trade at the Hudson's Bay Company post at Churchill (*eg*, see 1773-74, Hansom). This was Frobisher's first visit to Churchill River, which he named English River. In 1775, Frobisher returned to Saskatchewan River, whilst Primeau remained on Churchill River (see 1775-76). Frog Portage Fort was occupied by Joseph and Thomas Frobisher and Alexander Henry in spring 1776 and, perhaps, by Thomas Frobisher in spring 1777. After that, it was probably abandoned.

TYRRELL, 1934, pp. 106-07, 120, 158, 339.

1775

Hudson's Bay Company supply voyage | Jonathan Fowler Jr, Joseph Richards, William Christopher | *King George* [II], *Prince Rupert* [IV], *Seahorse* [II]

From London. *King George* visited York Factory 14-29 August, and reached London 8 October. *Prince Rupert* visited Churchill 13-26 August, and reached London 8 October. *Seahorse* visited Albany

16-22 August, Moose Factory 23 August - 6 September, and reached London 14 October.

HUDSON'S BAY COMPANY (1) and (2).

Establishment of Okak mission	Jens Haven Stephen Jensen	?	1775

Haven and Jensen laid out the boundary stones of Okak, the second Moravian missionary station on the Labrador coast, about 240 km north of Nain. The mission house was built in the following year. The mission was closed in 1919.

GOSLING, 1910, p. 272; WHITELEY, 1964; HILLER, 1971.

Spanish North-west Passage expedition	Bruno de Heceta Juan Josef Perez Hernandez Juan Francisco de la Bodega y Quadra Francisco Antonio Maurelle	*Santiago* *Felicidad* or *Sonora*	1775

From San Blas, Mexico; 16 March - 20 November; to search for the Strait of Juan de Fuca, which a Spanish sailor had claimed to have discovered in about 47° or 48°N on the west coast of America in 1592 (*qv*), and which was still considered a possible entrance to the North-west Passage. The expedition landed at Point Grenville, 47°20'N, where they took possession in the name of the King of Spain. Seven seamen from *Felicidad* were massacred by Indians at this spot. At about 50°N, Heceta separated from his convoy on 30 July and turned back to Monterey, where he arrived on 29 August. He discovered Columbia River on the way. Meanwhile, Quadra, Perez and Maurelle continued northward in *Felicidad,* sighted Mount Edgecumbe and continued north to nearly 58°N in the vicinity of Cross Sound. On 24 August, they landed on Prince of Wales Island and claimed it and surrounding lands for the King. On 28 August, distressed by storms and scurvy, they turned back. Perez died during their return.

ESPINOSA Y TELLO, 1802; BANCROFT, 1890, pp. 197-202; *MAURELLE, 1781.

Montreal pedlars' fur-trading expedition; establishment of Fort Beaver Lake	Joseph Frobisher Thomas Frobisher Alexander Henry Charles Patterson J.-B. Cadotte Peter Pond		1775-76

Henry and Cadotte left Sault Ste Marie on 10 June 1775. They met Pond on the way and travelled together north along the east shore of Lake Winnipeg. On 7 September Joseph Frobisher, who had recently returned from his wintering on Churchill River (see 1774-75), his brother Thomas, and Patterson overtook them. The party now included 30 canoes and 130 men. They reached the mouth of Saskatchewan River on 1 October, Cedar Lake on 3 October, and Cumberland House on 26 October, where they broke up. Henry, Cadotte, and the Frobishers travelled north to Amisk (Beaver) Lake, where they built a fort and wintered. Pond wintered south of Lake Winnipegosis on Dauphin Lake; the rest wintered along the Saskatchewan to the west, effectively intercepting trade before it reached the Hudson's Bay Company posts at Cumberland House and on Hudson Bay. It was such loose associations of free traders as these that led to the formation of the North West Company in 1779 (*qv*).

TYRRELL, J. B., 1934, p. 120; INNIS, 1930, pp. 68-73.

1775-76

Montreal pedlars' fur-trading expedition — Louis Primeau

In 1774-75, Primeau had wintered with Joseph Frobisher at Frog Portage on Churchill River (*qv*). In 1775, when Frobisher went south (see 1775-76, above), Primeau ascended Churchill River, probably to Primeau Lake, where he built a house and wintered. In 1776, he left this post, descended Churchill River, joined Thomas Frobisher and Alexander Henry, who had gone up from Amisk Lake to meet him, and returned with them to Fort Beaver Lake (see 1776-77).

TYRRELL, J. B., 1934, pp. 41-42, 121-22, 354.

1776

Hudson's Bay Company supply voyage	Jonathan Fowler Jr	*King George* [II]
	Joseph Richards	*Prince Rupert* [IV]
	William Christopher	*Seahorse* [II]

From London. *King George* visited York Factory 24 August - 3 September, and reached London 14 October. *Prince Rupert* visited Churchill 24 August - 4 September, and reached London 13 October. *Seahorse* visited Moose Factory 29 August - 17 September, and reached London 23 October.

HUDSON'S BAY COMPANY (1) and (2).

1776

British naval expedition to Davis Strait and Labrador — Richard Pickersgill — HMS *Lyon*

From Deptford; 25 May - late October; sent by the Admiralty to make a preliminary exploration of the coasts of Baffin Bay, in preparation for a more detailed examination in 1777 (*qv*), and so that guides might be sent to Captain Cook to assist him through the bay, should his attempt on the North-west Passage be successful (see 1776-80). Pickersgill was also to protect the British whaling fleet in Davis Strait from American privateers. He reached Cape Farewell (Kap Farvel) on 7 July, and, three days later, a Whitby whaler told him that the whaling fleet had left Disco Island for home more than a month previously. He followed the Greenland coast to 68°26'N, where he turned back because the season was late and many of his crew were ill. On return, he stopped at Porcupine Harbour on the Labrador coast 4-27 September, then returned to England. One of *Lyon's* officers, Michael Lane, later accused Pickersgill of drunkenness during the expedition, which caused the latter to be dismissed from the navy.

BARROW, 1818; WILLIAMS, 1962, pp. 184-90.

1776-77

Montreal pedlars' fur-trading expeditions; establishment of Ile-à-la-Crosse post	Thomas Frobisher Alexander Henry Louis Primeau

In March 1776, Frobisher and Henry left Fort Beaver Lake on Amisk Lake (see 1775-76) with eight men "hauling rum for the two canoes residing in the Thopiscow country." This no doubt means they were sent north to Churchill River, where Louis Primeau was wintering with two canoes (see 1775-76). They probably went first to Fort Frog Portage (see 1774-75) to put the buildings in order so they could again be occupied. They then met Primeau and returned with him to Fort Beaver Lake. Henry returned from there to Montreal with Joseph Frobisher, whilst Thomas Frobisher and Primeau returned to Churchill River, which they ascended to Lac Ile-à-la-Crosse. There they built a post, the first of several on that lake, on a peninsula on its western side. Frobisher then returned south to winter on Saskatchewan River; Primeau wintered at the post on Lac Ile-à-la-Crosse. He returned to Montreal in autumn 1777. He probably returned to winter again at Ile-à-la-Crosse in 1778-79.

TYRRELL, J. B., 1934, pp. 41-57; BURPEE, 1935, Vol. 2, pp. 319, 323; INNIS, 1930, pp. 78-80; HENRY, 1969.

1776-80

British naval North-west Passage expedition	James Cook Charles Clerke	HMS *Resolution* HMS *Discovery*

From Plymouth; 12 July 1776 - October 1780; sent by the Admiralty to search for a North-west Passage on the west coast of America north of 65°N. They were ordered to join the American coast at 45°N, but to make no detailed exploration until they reached 65°N. They rounded Cape of Good Hope in 1776. In 1777 they visited New Zealand, Tahiti, the Society Islands and the Hawaiian Islands. They sighted the west coast of North America in 44°33'N on 7 March 1778, then coasted northward, landing at Cape Flattery and again at Nootka, where they spent most of April making repairs. After leaving Nootka, they did not sight land until 55°20'N, and therefore failed to examine the coast in 53°N, where, it was still commonly believed, the entrance to de Fonte's apocryphal North-west Passage (see 1640) might be found. Cook was later criticized for this omission. On 10 May, they sighted Mount St Elias in 60°N, and were dismayed to find that the land then trended west and south. They examined Prince William Sound and Cook Inlet, rounded Alaska Peninsula, then headed east, only to find themselves in Bristol Bay and forced to retreat west again. On 9 August, they reached Cape Prince of

Wales, crossed Bering Strait to East Cape, then headed north-east with high hopes of finding a North-west Passage to Baffin Bay. However, in mid-August they encountered the impenetrable barrier of the Arctic pack ice and had to turn back at Icy Cape. They visited Cape North on the western side of Bering Strait, then returned to the American coast, hoping to find a North-west Passage south of Cape Prince of Wales. They investigated Norton Sound, then sailed south to make repairs at Unalaska, where they met Russian traders, then sailed to Hawaii, where they arrived in January 1779. After nearly three weeks in Hawaii, they left, only to return a few days later because one of *Resolution's* masts had broken. During this visit, relations with the natives deteriorated, and Cook was killed in a scuffle on the beach at Kealakekua Bay on 14 February. Despite Cook's death, the two ships made a second attempt to find a North-west Passage. They reached Kamchatka in late April, passed through Bering Strait, but were again turned back by the pack ice. Captain Clerke died on the way back to Kamchatka. They then sailed to Canton, where they sold furs they had collected on the North American coast for such high prices that the crew, who wanted to return to the American coast for more furs, nearly mutinied. However, they continued the voyage and reached England in October 1780.

*BEAGLEHOLE, 1967; WILLIAMS, 1962, pp. 184-211.

1777

Hudson's Bay Company supply voyage	Jonathan Fowler Jr	*King George* [II]
	Joseph Richards	*Prince Rupert* [IV]
	William Christopher	*Seahorse* [II]

From London. *King George* visited Moose Factory 13-26 September, and reached London 20 November. *Prince Rupert* visited York Factory 18 August - 1 September, and reached London 18 October. *Seahorse* visited Churchill 17-25 August, York Factory 28 August - 1 September, and returned to London with *Prince Rupert.*

HUDSON'S BAY COMPANY (1) and (2).

1777

Establishment of Wapiscogamy House — Thomas Atkinson

Atkinson built Wapiscogamy House on the north bank of Missinaibi River, a tributary of Moose River, near the junction of Opasatika River, for the Hudson's Bay Company. It was renamed Brunswick House about 1783.

RICH, 1959, p. 94.

1777

Establishment of Gloucester House

The Hudson's Bay Company established Gloucester House about 160 km upstream from Henley House (see 1766).

RICH, 1959, p. 93; 1952, p. xxx.

British naval expedition to Davis Strait	Walter Young	HMS *Lyon*	1777

From the Nore; 23 March - 26 August; sent by the Admiralty to complete the survey of Baffin Bay that Pickersgill had begun in 1776 (*qv*), to investigate whether or not the sea discovered by Samuel Hearne (see 1770-72) could be reached from Baffin Bay, and to see if the natives of Baffin Bay used copper. His instructions also referred to Captain Cook (see 1776-80), whose expedition in search of a North-west Passage was expected to emerge into Baffin Bay. Young reached Disco Island on 5 June and attained 72°42'N three days later. Then, for no known reason, he turned around and sailed back to England.

BARROW, 1818; WILLIAMS, 1962, pp. 184-90.

Hudson's Bay Company supply voyage	Jonathan Fowler Jr	*King George* [II]	1778
	Joseph Richards	*Prince Rupert* [IV]	
	William Christopher	*Seahorse* [II]	
	Joseph Tunstall	*Charlotte*	

From London. *King George* visited York Factory 24 August - 6 September, and reached London 25 December. On 9 August, Fowler noticed that there was more ice toward the western end of Hudson Strait than he had ever seen before, and he encountered heavy ice in the strait on his return. The ice and contrary winds considerably delayed his return to London. *Prince Rupert* visited Moose Factory 28 August - 17 September, and reached London 23 December. *Seahorse* visited Churchill 23-31 August then, under instructions to accompany *King George* back to London, visited York Factory 3-6 September. She reached London 22 December. *Charlotte* visited Moose Factory 29 August - 18 September, and reached London 21 December.

HUDSON'S BAY COMPANY (1) and (2).

Establishment of Hudson House	Robert Longmoor William Tomison	1778-79

Upper Hudson House, a Hudson's Bay Company outpost built by Longmoor and Tomison, was about 470 km from Cumberland House (see 1774) on the west bank of North Saskatchewan River. In 1779, Tomison built Lower Hudson House about 22 km farther downstream.

WILLIAMS, 1969, p. 207n; RICH, 1951, pp. 309-11; 1952, p. 69; 1959, p. 79.

Montreal syndicate's fur-trading expedition; establishment of Old Pond Fort	Peter Pond	?	1778-80

Encouraged by the success of Frobisher and Primeau at Ile-à-la-Crosse (see 1776-77), a group of Montreal-based traders along Saskatchewan River joined in 1778 to provide Pond with four canoes and goods to carry into the Athabasca country. He thus had the advantage of an early start from Saskatchewan River, rather than from Grand Portage, where the pedlars were usually obliged to collect their trade goods before setting out for the north. He probably took the well-known route from Cumberland Lake to Ile-à-la-Crosse: by Amisk, Mirond and Wood lakes to Frog Portage, and then up Churchill River. No trader is known to have travelled north beyond Ile-à-la-Crosse before Pond, so he can probably be credited with the discovery of Churchill Lake, Peter Pond Lake, La Loche River and Lac La Loche. He certainly discovered Portage La Loche (Methy Portage), the watershed between Hudson Bay and Arctic Ocean drainage. This pass was the main route of access into the Northwest Territories for many years. Beyond the portage he discovered Clearwater and Athabasca rivers, which he descended to within 45 km of Lake Athabasca, where he built a house and wintered. This house, Old Pond Fort or the Old Establishment, was visited frequently by Pond and other traders until it was replaced by Fort Chipewyan (see 1788). Pond was the first to carry trade goods to the Chipewyan and Cree Indians of this fur-rich country. They were delighted to be spared the arduous journey to Churchill and were willing to pay high prices for trade goods. He brought out more than 80,000 fine beaver skins, but had traded still more, which he left behind *en cache.* He returned to Cumberland Lake in July 1779, then went to Grand Portage. He appears to have returned directly to bring out the furs he had left in Athabasca, spent a second winter there, come south again late in 1780, and spent winter 1780-81 at Mackinac.

*INNIS, 1928; 1930, pp. 80-82; CAMPBELL, M. W., 1957, p. 9; RICH, 1959, pp. 116-17; MORTON, 1939, p. 328; HUDSON'S BAY COMPANY, ARCHIVES, B39/e/3/20-21d.

1779

Hudson's Bay Company supply voyage	Jonathan Fowler Jr	*King George* [II]
	Joseph Richards	*Prince Rupert* [IV]
	William Christopher	*Seahorse* [II]

From London. *King George* sailed towards York Factory but, as in 1778 (*qv*), Fowler encountered unusually thick ice in Hudson Bay, which prevented him from reaching York Factory. Instead, he delivered his trade goods to Churchill, where he remained 12-19 September, and reached London 17 November. *Seahorse* visited Churchill 11-19 September, and returned with *King George. Prince Rupert* sailed to Moose Factory and back to London.

HUDSON'S BAY COMPANY (1) and (2).

1779

Formation of the
North West Company

Various fur traders from and merchants in Montreal and London had, from about 1760, joined in temporary partnerships, combinations, or syndicates, to trade with Indians north-west and south-west of the Great Lakes. Frobisher's success in 1776-77 (*qv*) and Pond's in 1778-79 (see 1778-80) had been supported by such combinations. In 1779, seven small partnerships and two individual merchants joined in a copartnership styled the North West Company. It was organized into 16 shares, which were held by nine partnerships, and it lasted only one year. The traders included Benjamin and Joseph Frobisher, William Holmes, Robert Grant, John Ross, Etienne Waden, Venant St-Germain, and Peter Pond; the merchants included Isaac Todd, Patrick Small, James and John McGill, George McBeath, Simon McTavish, Forrest Oakes, and Lawrence Ermatinger. In 1780, essentially the same combination of men signed a new, revised agreement, intended to run for three years, but it was abandoned after two. A third agreement was worked out in 1783-84 (*qv*).

INNIS, 1927, p. 312; 1930, p. 88; CAMPBELL, M. W., 1957, pp. 18-19; RICH, 1966, pp. 74-78; 1959, p. 117; DAVIDSON, 1918, pp. 1-31.

1779

Spanish expedition to the Northwest Coast	Ignacio Arteaga Juan Francisco de la Bodega y Quadra Alfrez Antonio Maurelle Fernando Quiros	*Princesa* *Favorita*

From San Blas, Mexico; 11 February - 21 November; to explore the west coast of America northward to 70°N. They reached Port Bucareli in Prince of Wales Archipelago on 4 May. They spent several weeks there, replenishing their provisions. They continued northward to within sight of Mount St Elias, then turned back to San Blas.

*TERO, 1973; ESPINOSA Y TELLO, 1802.

1779-82

North West Company overland fur-trading expedition; establishment of post at Lac la Ronge	Etienne Waden

The company sent Waden to take over the trade in the Athabasca region that Pond had opened up (see 1778-80). He built a house on Lac la Ronge, possibly on Waden Bay, and spent two winters there, although he may have returned south for a short time in summer 1780. When the agreement of the North West Company expired in 1780, the fur traders along Saskatchewan River made a new agreement, from which Waden was excluded. He stayed at Lac la Ronge a third winter, 1781-82, to represent another, smaller group of traders, during which he was murdered (see 1781-82, Pond).

INNIS, 1930, pp. 90, 93-97; RICH, 1966, pp. 76-77; 1959; pp. 117-18; MORTON, 1939, p. 334.

1780

Hudson's Bay Company supply voyage	Jonathan Fowler Jr Joseph Richards William Christopher	*King George* [II] *Prince Rupert* [IV] *Seahorse* [II]

From London. *King George* visited Churchill 1-8 September and York Factory 15-17 September. *Seahorse* visited York Factory 30

August - 17 September, and, with *King George*, reached London 10 November. *Prince Rupert* visited Moose Factory 5-29 September, and returned to London.

HUDSON'S BAY COMPANY (1) and (2).

1781

Hudson's Bay Company supply voyage	Jonathan Fowler Jr	*King George* [III]
	Joseph Richards	*Prince Rupert* [IV]
	William Christopher	*Seahorse* [II]

From London. *Prince Rupert* visited Churchill 20-28 August, and York Factory 1-8 September. *Seahorse* visited York Factory 20 August - 8 September, and, with *Prince Rupert*, reached London 31 October. *King George*, making her first voyage, visited Moose Factory 28 August - 22 September, and reached London on 1 November.

HUDSON'S BAY COMPANY (1) and (2).

1781-82

Smallpox epidemic

In 1781, a smallpox epidemic, which fur traders believed had begun along Mississippi River, began to spread north and west towards Saskatchewan River. It reached Hudson House in October and Cumberland House in December 1781. By 1782, it had spread to the Indians of the Athabasca region, and then to the Chipewyans of the Barren Grounds. Few Europeans caught the disease, but the Indians died in great numbers: one estimate stated that three-fifths of the prairie Indians had perished, and Samuel Hearne considered that nine-tenths of the Chipewyans had died during the epidemic. These deaths considerably damaged the fur trade of both the Hudson's Bay Company and the Montreal companies.

RICH, 1959, p. 82; 1952, pp. lvii-lxii.

1781-82

Montreal syndicate's overland fur-trading expedition	Peter Pond

From Grand Portage; sponsored by the larger of the groups of fur traders that had formed after the lapse of the first North West Company agreement (see 1779). Pond wintered at the post on Lac la Ronge, which Waden had established for the smaller group of traders in 1779 (see 1779-82), and where he still traded. During the winter, the two men quarrelled and, in March, Waden was killed. After a trial in Montreal, probably in 1784, Pond was acquitted of Waden's murder, but the suspicion of it lingered. This tragedy further disrupted the informal alliances of the Montreal traders and encouraged the major interests to establish a new North West Company on a more formal basis (see 1783-84). Pond probably returned to Grand Portage in summer 1782.

INNIS, 1930, pp. 93-98; RICH, 1959, p. 118; 1966, pp. 76-77.

Hudson's Bay Company supply voyage | Jonathan Fowler Jr, William Christopher, Joseph Richards | *King George* [III], *Prince Rupert* [IV], *Seahorse* [III] — 1782

Hudson's Bay Company supply voyage	Jonathan Fowler Jr	*King George* [III]	1782
	William Christopher	*Prince Rupert* [IV]	
	Joseph Richards	*Seahorse* [III]	

From London. *King George* reached York Factory on 15 August. While she was there, the French squadron commanded by la Pérouse (see below, 1782) appeared. On 24 August, the day on which York Factory was surrendered, *King George* slipped away and reached London on 27 November. *Prince Rupert* reached Churchill to find the fort already destroyed by the French. She returned to London at once, and arrived on 26 November. *Seahorse*, on her first voyage, visited Moose Factory 24 August - 25 September, and reached London on 26 November. On 18 June, during the outward voyage, Captain Richards died in the Orkney Islands.

HUDSON'S BAY COMPANY (1) and (2).

French attack on Hudson's Bay Company posts	Jean-François Galaup, Comte de la Pérouse	*Sceptre*	1782
	Marquis de Lajaille	*Engageante*	
	P.-A. de Langle	*Astrée*	

From St Domingo; 31 May - 12 October. The ships arrived off Churchill on 9 August and, early on the 10th, Samuel Hearne surrendered Prince of Wales Fort to them without resistance. The French left Churchill on 13 August, having partly demolished it, and on the 17th, arrived off York Factory, where the governor, Humphrey Marten, surrendered without resistance on 22 August, and the French burned it down. On 2 September, the French ships, with Hearne, Marten, William Falconer (master of Severn House), and about 30 other Hudson's Bay Company servants on board, left York Factory. They had in tow the sloop *Severn*, which had been used for voyages between York Factory and Severn House. On 10 September, at the eastern end of Hudson Strait, the Hudson's Bay Company's servants were transferred to *Severn* and set free; they reached Stromness on 15 October. The French ships sailed to Cadiz, and arrived on 12 October.

GLOVER, 1951a; RICH, 1959, pp. 83-88; HUDSON'S BAY COMPANY (2), note 74.

Establishment of Hopedale mission	1782

Following the establishment of missions at Nain and at Okak (see 1771 and 1775), the Moravians received a third grant of land from the Privy Council to build Hopedale, a mission about 240 km south of Nain.

HILLER, 1971, p. 843; GRENFELL AND OTHERS, 1909, p. 34; GOSLING, 1910, p. 272.

Montreal pedlars' overland fur-trading expedition	?	1782-83

It appears that two rival groups of canoemen, each supplied by Montreal merchants, set out for the Athabasca district in 1782.

One of the groups, probably led by Peter Pond, failed to reach Athabasca River and wintered at Lac Ile-à-la-Crosse. The other, however, crossed Portage La Loche and wintered on Athabasca River. There, however, they found the Indian population greatly diminished by the smallpox epidemic (see 1781-82). They returned south in 1783 with only seven packs of beaver.

INNIS, 1930, pp. 98-99; MORTON, 1939, pp. 334-35.

1783

Hudson's Bay Company supply voyage	William Christopher	*King George* [III]
	Joshua Tunstall	*Prince Rupert* [IV]
	John Richards	*Seahorse* [III]
	William Falconer	*Severn*

From London. *King George*, delayed on the outward voyage by adverse winds and unusually heavy ice, visited York Factory 14-25 September, and reached London 3 November. *Prince Rupert* visited Churchill 15-27 September, and reached London 9 November. *Seahorse* visited Moose Factory 17 September - 5 October, and reached London 24 November. *Severn*, captured by the French and sent to England in 1782 (*qv*), returned to York Factory with *King George*, then, on 25 September, continued to her station at Severn House.

HUDSON'S BAY COMPANY (1) and (2).

1783-84

Formation of the North West Company

The fur traders of Montreal, by now very active on Saskatchewan, Churchill, and Athabasca rivers, had made various temporary alliances (see 1779), and now combined more formally in a new North West Company. The company was neither a chartered corporation nor a limited liability company; there was no common capital, but each partner contributed to a common stock of trade goods. The stock was divided into 16 shares, which included six shares divided evenly between the two houses of Simon McTavish and Benjamin and Joseph Frobisher. Peter Pond, offered only one share in the company, was dissatisfied, and refused to sign the agreement until 1785. A number of smaller traders were left out of the agreement. Some of them joined Gregory, McLeod and Company (see 1784) as rivals of the North West Company.

RICH, 1959, pp. 118-19; DAVIDSON, 1918, pp. 10-11.

1783-84

Montreal pedlars' overland fur-trading expedition	Peter Pond

Pond returned to Old Pond Fort on Athabasca River for the third time, probably direct from his wintering place on Lac Ile-à-la-Crosse (see 1782-83). It was probably on the explorations of this journey that he based his famous map of 1785. He seems to have partly explored Lake Athabasca, but he probably did not go below it. The Indians told him of the approximate locations of Great Slave Lake and Great Bear Lake and that Mackenzie River emptied into the Arctic Ocean. He met some of the Indians who had accompanied Hearne to Coppermine River (see 1770-72), and some Mackenzie River Indians, who had traded with the Hudson's Bay Company before Churchill was destroyed (see 1782), but were now willing to deal with the Montreal traders. In 1784, Pond returned to Grand Portage and Montreal.

INNIS, 1930, pp. 99-102.

Hudson's Bay Company supply voyage	William Christopher	*King George* [III]	1784
	Joshua Tunstall	*Prince Rupert* [IV]	
	John Richards	*Seahorse* [III]	

From London. *King George* visited Moose Factory 4-20 September, and reached London 27 October. *Prince Rupert* visited Churchill 2-12 September, and reached London 30 October. *Seahorse* visited York Factory 1-18 September, and reached London 2 November.

HUDSON'S BAY COMPANY (1) and (2).

Gregory, McLeod and Company 1784

Peter Pangman, one of the fur traders excluded from the new North West Company (see 1783-84), persuaded the Montreal firm of Gregory, McLeod and Company, which had formerly been trading at Detroit and Michilimackinac, to enter the trade of Saskatchewan, Churchill, and Athabasca rivers. Members of this company who later traded in the north-west were Alexander and Roderick Mackenzie, Laurent Leroux, and John Ross. In 1787, this company amalgamated with the North West Company.

INNIS, 1930, pp. 106-09; MORTON, 1939, pp. 336-43.

Hudson's Bay Company supply voyage	William Christopher	*King George* [III]	1785
	Joshua Tunstall	*Prince Rupert* [IV]	
	John Richards	*Seahorse* [III]	

From London. *King George* visited Moose Factory 8-23 September, and reached London 3 November. *Prince Rupert* visited Churchill 21 August - 3 September, and reached London 14 October. *Seahorse* visited York Factory 25 August - 16 September, and reached London 19 October.

HUDSON'S BAY COMPANY (1) and (2).

Fur-trading voyage to the Northwest Coast	James Hanna	*Harmon* or *Sea Otter*	1785

The publication in 1784 of the narrative of Cook's last voyage (see 1776-80) encouraged many traders to sail to the Northwest Coast of North America in search of the valuable furs he described. Hanna was the first of them. He sailed from Macao on 15 April, reached Nootka on 8 August, and gathered 560 otter skins from the Indians, which he profitably sold at Canton. Many other traders followed him to the Northwest Coast and, as well as trading for furs, they made geographical discoveries. These voyages will not, however, be included in this list.

HOWAY, 1930.

1785

Establishment of Frederick House — Philip Turnor

From Moose Factory. In May-June 1782, Turnor surveyed Abitibi River and Lake for the Hudson's Bay Company. His superiors were impressed by the possibilities of the region and ordered him to establish a post there. In 1784, he built a temporary post at the junction of Abitibi and Frederick House rivers, but, in 1785, he built another post about 80 km farther south on Lake Waratowaha. This was the company's first regular post on Abitibi waters.

TYRRELL, J.B., 1934, pp. 75-85.

1785-87

Gregory, McLeod and Company overland fur-trading expeditions — Peter Pangman, Alexander Mackenzie, Roderick Mackenzie, John Ross, Laurent Leroux

From Grand Portage. In 1785, Gregory, McLeod and Company (see 1784) sent out its first fur-trading brigades to the north-west. Pangman went to South Saskatchewan River, where he built Fort des Isles. Alexander Mackenzie went to Churchill River and wintered on Lac Ile-à-la-Crosse, close to Patrick Small, who represented the North West Company. Roderick Mackenzie spent winter 1785-86 in charge of the company's depot at Grand Portage, then joined his cousin Alexander at Ile-à-la-Crosse. In 1786-87, he wintered at an outpost he built at Lac des Serpents, close to, and on very friendly terms with, William McGillivary, whom the North West Company had sent to winter on the same lake. Ross and Leroux travelled to Athabasca River in 1785, and built a post close to Old Pond Fort (see 1778-80), where Peter Pond himself was in residence (see 1785-88). Nothing is known of Ross and Leroux's activities during winter 1785-86. In 1786, Leroux left to build a post on Great Slave Lake, close to a post built at the same time by Cuthbert Grant for the North West Company (see 1786). During winter 1786-87, relations between Ross and Pond worsened and, in spring 1787, Ross was killed in a scuffle. In June 1787, Roderick Mackenzie, then at Ile-à-la-Crosse, learned of the death and returned to report it to his partners who were assembled at Grand Portage. Because of this and other reports of violence between the servants of the two companies, they decided, in July 1787, to amalgamate.

INNIS, 1930, pp. 106-09; MORTON, 1939, pp. 336-42; MACKENZIE, 1801, pp. xviii-xx.

North West Company overland fur-trading expedition	Peter Pond Cuthbert Grant	1785-88

From Grand Portage. Pond returned to the post he had established on Athabasca River (see 1778-80), where he wintered alongside John Ross and Laurent Leroux of Gregory, McLeod and Company (see 1785-87). Nothing is known of his activities in winter 1785-86, but his map of 1790 shows that he made excursions in the summers of 1786 and 1787. In 1786, he sent Grant to establish a post on Great Slave Lake (*qv*), and he probably spent the summer organizing the district so that the canoes could return to Grand Portage without difficulty and trade and exploration could be developed. In 1787 one of his servants shot John Ross in a scuffle (see 1785-87) and two of his men, François Nadeau and Eustache le Compte, were sent to Quebec to be tried for murder. They were acquitted. Pond appears to have visited Great Slave Lake in summer 1787, for he remarked on ice conditions in the lake in July of that year. He may have crossed the lake to Yellowknife Bay, but this is not certain. He did, however, learn that Mackenzie River flows from the western end of the lake, information that proved useful to Alexander Mackenzie in 1789 (*qv*). In 1787, Alexander Mackenzie came to succeed Pond. They spent the winter together, and Pond probably gave Mackenzie the information that he needed to discover Mackenzie River. Pond left the Athabasca district for the last time in 1788.

INNIS, 1930, pp. 106-09; RICH, 1967, p. 174; MORTON, 1939, pp. 337-42.

French North-west Passage expedition	Jean-François Galaup, Comte de la Pérouse P.-A. de Langle	*Boussole* *Astrolabe*	1785-88

From Brest; August 1785 - (?) 1788; sent by the French Government to carry out extensive exploration in the Pacific Ocean and to search the American coast between Monterey in California and Mount St Elias for a North-west Passage to Hudson Bay. The ships passed Cape Horn on 1 February 1786, visited Conception, Easter, and the Hawaiian islands, and sighted Mount St Elias on 23 June. On 2 July the ships anchored in Lituya Bay, Alaska, in about 58°30'N. Here, six officers and 15 men were lost in a storm. La Pérouse then sailed southward, charting the main features of the coast. He entered Queen Charlotte Sound, sailed by the west coast of Vancouver Island, and reached Manila on 14 September 1786.

1787: the ships reached Macao from Manila on 2 January, then sailed by way of Formosa, Korea, and the Sea of Japan to reach Petropavlovsk on 30 August. Here they received instructions to sail for New Zealand, and they sent Barthélemy de Lesseps overland to Europe with documents. They left Petropavlovsk on 30 September and visited the Samoan Islands where 12 men, including de Langle, were killed by natives.

1788: they reached Botany Bay on 26 January, left there at the end of February, and were never seen again.

DUNMORE, 1965; *LA PÉROUSE, 1797.

1786

Hudson's Bay Company supply voyage — William Christopher, *King George* [III]; Joshua Tunstall, *Prince Rupert* [IV]; John Richards, *Seahorse* [III]

From London. *King George* carried supplies for both Churchill and York Factory. She visited Churchill 13-16 August, York Factory 18-30 August, and reached London 4 October. *Prince Rupert* and *Seahorse* visited Moose Factory 16 August - 9 September, and reached London 17 October.

HUDSON'S BAY COMPANY (1) and (2).

1786

Hudson's Bay Company exploring expedition — Alexander Brand — *Moose*

From Moose Factory; 22 July - (?). Since closing Richmond Fort in 1759 (*qv*), the Hudson's Bay Company had paid no attention to the possibilities of trade in the interior of the Quebec-Labrador peninsula, but, in the early 1780's, an Indian report of a large river north of Richmond renewed hope of easy access to the Indians in the interior. Brand, sent to search for this river, visited Great and Little Whale rivers, meeting Indians at the latter place and Eskimos farther north. In 60°40'N, he reached Mosquito Bay (probably discovered earlier by William Coats, see 1749), but failed to find the large river in question and returned to Moose Factory. This was the company's last serious effort to explore the interior of the Quebec-Labrador peninsula by a strait leading inland.

DAVIES, 1963, pp. xxv-xxvi, 275-76.

1786

Establishment of Great Slave Lake post — Laurent Leroux; Cuthbert Grant

Leroux, representing Gregory, McLeod and Company, and Grant, representing the North West Company, were both sent north in 1786 from Athabasca River (see 1785-87 and 1785-88) to establish trading posts on Great Slave Lake. Five of Grant's men were drowned at Portage des Noyés on Slave River. Both built posts east of Slave River delta. The site chosen is close to present-day Fort Resolution, a much later name. The original name was possibly Slave Fort. Leroux wintered there in 1786-87. In 1787, after the amalgamation of the two companies, Alexander Mackenzie ordered Leroux to abandon his post but, apparently, it was Grant who ceased to trade on Great Slave Lake, whilst Leroux returned from Old Pond Fort on Athabasca River to winter there a second time in 1787-88. During one of these winterings, he sent the English Chief, a Chipewyan who accompanied Alexander Mackenzie on his exploration of 1789, to encourage the Indians north of Great Slave Lake to trade with him. He sent James Sutherland on a similar mission. It is not known where Sutherland went, but he apparently persuaded many Indians to return with him to the fort. Leroux is reported to have established Fort Providence on the north side of the lake between 1786 and 1788. He did visit the north side of the lake in 1788-89 (*qv*), but it was not until 1789 that a permanent post was built there. Leroux's post east of Slave River was moved, a few years later, to Moose Deer Island in Slave River delta, opposite the present site of Fort Resolution. Moose Deer Island post remained there until after the coalition of the North West Company with the Hudson's Bay Company in 1821. It was then abandoned in favour of Fort Resolution, built by the Hudson's Bay Company in 1819 (*qv*), about eight km west of Slave River.

INNIS, 1930, p. 108; BURPEE, 1935, Vol. 2, p. 422; MASSON, 1960, Vol. 1, p. 94 [Wentzel's letters].

1787

Hudson's Bay Company supply voyage	William Christopher Joshua Tunstall	*King George* [III] *Seahorse* [III]

From London. *King George* visited Moose Factory 27 August - 10 September, and reached London on 16 October. *Seahorse* visited Churchill 14-25 August, York Factory 27 August - 9 September, and reached London 16 October.

HUDSON'S BAY COMPANY (1) and (2).

1788

Hudson's Bay Company supply voyage	William Christopher Joshua Tunstall John Richards	*King George* [III] *Seahorse* [III] *Beaver*

From London. *King George* visited Churchill 8-14 August, York Factory 18-31 August, and reached London 4 October. *Seahorse* visited Moose Factory 16 August - 10 September, and reached London 7 November. *Beaver* visited Eastmain 18-29 August, and reached London 18 November.

HUDSON'S BAY COMPANY (1) and (2).

1788

Establishment of Fort Chipewyan	Roderick Mackenzie

On the amalgamation of the North West Company with Gregory, McLeod and Company in 1787, Alexander Mackenzie left Ile-à-la-Crosse (see 1785-87) to replace Peter Pond (see 1785-88) at Old Pond Fort and to take charge of the Athabasca district. In 1788, Roderick Mackenzie joined him at Old Pond Fort, then went to build a post on Lake Athabasca to replace Old Pond Fort. He chose Old Fort Bay, about 13 km east of the Athabasca River delta. He named the post Fort "Chipewean". In 1804, the post was removed to its present site on the north shore of the lake. Rival companies built several posts nearby: about 1800, the XY Company had a post about one km north of it and two posts on Lake Claire, but they were soon abandoned. In 1802, the Hudson's Bay Company built Nottingham House near the present site of Fort Chipewyan (see 1802-03), and, in 1815, built Fort Wedderburn on Coal Island opposite it (see 1815-16). Fort Wedderburn was abandoned in favour of Fort Chipewyan after the amalgamation of the North West Company and the Hudson's Bay Company in 1821.

MASSON, 1960, Vol. 1, pp. 26-28 [Mackenzie's reminiscences]; RICH, 1938, pp. 414-15.

1788

Establishment of the Old Establishment on Peace River — Charles Boyer

Under Alexander Mackenzie's instructions, Boyer built for the North West Company the first trading post on Peace River. It was on the north bank, about 10 km above the mouth of Boyer River, close to the present site of Fort Vermilion. At first, the post was evidently of some importance: Alexander Mackenzie, then in charge of the Athabasca district, wrote in 1789 that "the greatest dependence of this place at present is on the Peace River." Boyer left the post in 1789, replaced by a Mr Vandreil (Vandrieul or Vaudreuil), who surveyed Peace River from Lake Athabasca to this point. The history of this post after 1789 is not known, but it was probably soon abandoned in favour of other posts on Peace River. The buildings were still standing when Alexander Mackenzie visited them in 1792 (see 1792-93), but members of his party accidentally burned down the main building.

MORTON, 1939, p. 413; BURPEE, 1935, Vol. 2, p. 451; MASSON, 1960, Vol. 1, p. 30 [Mackenzie's reminiscences]; MACKENZIE, 1801, p. 124; MACGREGOR, 1952, pp. 65-66.

1788

Spanish expedition to the Northwest Coast	Esteban José Martinez	*Princesa*
	Gonzalo Lopez de Haro	*San Carlos*

From San Blas; 8 March - 5 December; to investigate reports that the Russian American Fur Company intended to establish trading posts at Nootka and Prince William Sound, over which the Spanish claimed sovereignty. They reached Prince William Sound on 17 May, traded with Indians there, and found one Russian establishment. On 23 June, they reached Cook Inlet and, a week later, discovered a Russian colony, where they learned that Russian activities in the area were already widespread. In August, they visited a Russian post on Unalaska, then returned to San Blas with the news that the Russians intended to establish a post at Nootka and that they claimed the Northwest Coast down to California. This news encouraged the Viceroy of New Spain to send a strong fleet to establish a post at Nootka in 1789 (*qv*). *San Carlos* arrived back at San Blas on 22 October, and *Princesa* on 5 December.

ESPINOSA Y TELLO, 1802; BANCROFT, 1890, pp. 270-73; MARQUIS, 1914.

1789

Hudson's Bay Company supply voyage	Joshua Tunstall	*King George* [III]
	John Richards	*Seahorse* [III]
	J. Williamson	*Beaver*

From London. *King George* visited Churchill 16-22 August, York Factory 25 August - 10 September, and reached London 10 October. *Seahorse* visited Moose Factory 15 August - 14 September, and reached London 21 October. *Beaver* arrived at Moose Factory 15 September, delivered cargo to Eastmain 17-27 August, to Albany 30 August - 8 September, returned to Moose Factory until 14 September, and reached London 30 October.

HUDSON'S BAY COMPANY (1) and (2).

Spanish expedition to the Northwest Coast; the Nootka affair	Esteban José Martinez ?	*Princesa* *San Carlos*	1789

From San Blas; 17 February - 6 December; to confirm Spanish sovereignty at Nootka by building a trading post there. The ships reached Nootka on 5 May, and found the trading vessels *Iphigenia* and *Columbia* at anchor. Martinez plundered *Iphigenia,* captured another trading vessel, *North-west America,* when it arrived on 8 June, and took possession of the country on 24 June. He captured two more ships, *Argonaut* and *Princess Royal,* when they arrived in July, and sent them and their crews to San Blas. *Princesa* then returned to San Blas; *San Carlos* remained at Nootka, where a post was built. Martinez's handling of this situation sparked the Nootka affair, which threatened to cause war between England and Spain. However, it terminated without bloodshed, when the Nootka Convention was signed on 28 October 1790. Spain relinquished her claim to the north Pacific coast on the understanding that no permanent settlement at Nootka was to be made by either side. After 1795, when Spain ceded possession of the coast to England, 30 years passed before the coast was again settled.

ESPINOSA Y TELLO, 1802; MARQUIS, 1914.

Establishment of Fort Providence	Laurent Leroux	1789

Leroux had traded on the south side of Great Slave Lake for the North West Company since 1786 (*qv*). Early in 1789, he crossed to the north shore of the lake, where he met and traded with Chipewyan and Slave Indians. Exactly where he went is not recorded, but it was probably to Yellowknife Bay. The Indians promised to meet him again that summer and he returned to fulfill his promise. He accompanied Alexander Mackenzie (see 1789, below) from Fort Chipewyan to Yellowknife Bay on the North Arm of Great Slave Lake, where they parted on 25 June. During the summer, he went north to Lac la Martre, where he traded with Slave and Beaver Indians. When Mackenzie rejoined him on 24 August 1789, he had built a house on Yellowknife Bay, later known as Fort Providence, and there he spent the winter. The post was in operation until 1823, when the Hudson's Bay Company closed it. It has been suggested that Leroux had opened Fort Providence for one or two seasons before 1789, perhaps in 1786-88, and had re-established it in 1789, but there is little evidence for this view. Leroux apparently built another post, on Lac la Martre, which was occupied intermittently until other posts on Mackenzie River were opened.

STAGER, 1962, pp. 38-39; MASSON, 1960, Vol. 1, p. 95 [Wentzel's letters]; MACKENZIE, 1801, pp. 18, 112-15; PERRY AND CLARK, 1971.

North West Company exploration of Mackenzie River	Alexander Mackenzie The English Chief	1789

From Fort Chipewyan; 3 June - 12 September; to explore the river that, according to Indian report, flowed northward from the west end of Great Slave Lake, which Mackenzie hoped would carry him to the Pacific Ocean. Mackenzie left Fort Chipewyan in a birch-bark canoe with four Canadian boatmen, a German, and the Chipewyan guide called the English Chief. Laurent Leroux accompanied him to Great Slave Lake to trade with Indians there (see 1789, above). They reached the lake on 9 June, then waited five days at Leroux's post east of Slave River for ice on the lake to clear. They got to the north shore on 22 June, coasted westward to the North Arm, and passed Yellowknife Bay on 23 June. Near here, Mackenzie hired a Copper Indian as guide and, on 25 June, left Leroux behind. They found the river on 29 June. In his progress downstream, Mackenzie met Slave, Dogrib, and Hare Indians, who told terrifying tales of the river's great length and of the lack of provisions at its northern end, and he barely managed to prevent his guides from deserting. He passed Great Bear River on 5 July and reached the head of the delta on 12 July. On 14 July, he camped on Whale Island. The next morning, the rising tide having swamped his baggage, he realized that he had reached the Arctic Ocean, not the Pacific Ocean. On 16 July, he began his return up what he now called the River of Disappointment. On his way back, he questioned Indians about lands to the west and learned vaguely of Yukon River. He rejoined Leroux at Great Slave Lake on 22 August. Here he paid off his guides and returned to Fort Chipewyan on 12 September, a round trip of nearly 4,800 km in 102 days.

*MACKENZIE, 1801; *LAMB, 1970; WOOLLACOTT, 1927; McDONALD, T.H., 1966; DANIELLS, 1969.

1790

Hudson's Bay Company supply voyage; North-west Passage expedition	John Richards John Turner Henry Hanwell Charles Duncan	*King George* [III] *Queen Charlotte* *Seahorse* [III]

From London. *King George,* carrying cargo for Fort Albany, visited Moose Factory 29 August - 21 September, and reached London 8 December. *Queen Charlotte,* carrying cargo for Eastmain, visited Moose Factory 14-23 September, and reached London 6 December. *Seahorse* visited York Factory 24 August - 11 September, Churchill 16-24 September, and reached London 8 December. She carried Captain Charles Duncan RN to Churchill and, after his first attempt to embark on a search for a North-west Passage had failed, back to London (see also 1791-92).

HUDSON'S BAY COMPANY (1) and (2); GOLDSON, 1793, pp. 52-53; WILLIAMS, 1962, pp. 245-46.

1790

Establishment of McLeod's Fort on Peace River	Alexander McLeod

Alexander Mackenzie sent McLeod to build for the North West Company a trading post farther up Peace River than the previous post (see 1788, Boyer). He built a post at the mouth of Whitemud

River, which was abandoned after two years in favour of Fork Fort, a post still farther upstream (see 1792).

MACGREGOR, J. G., 1952, pp. 68, 126, 385; WALLACE, J.N., 1929, p. 29.

1790

Spanish North-west Passage expedition	Manuel Quimper	*Princess Royal*

From San Blas, Mexico. Quimper sailed to Nootka to return *Princess Royal* to the trader from whom it had been taken by Martinez in 1789 (*qv*). Before returning the ship, he used her to explore the west coast of Vancouver Island and to search for the apocryphal North-west Passages of de Fuca (see 1592) and de Fonte (see 1640). He charted the entrance to Juan de Fuca Strait.

ORMSBY, M.A., 1958, p. 21.

1790-92

Hudson's Bay Company overland surveying expedition	Philip Turnor Peter Fidler Malcolm Ross

From Cumberland House; 13 September 1790 - 17 July 1792; to survey the North West Company's route to Lake Athabasca and Great Slave Lake, which had previously been used only by that company. They travelled north from Cumberland House by way of Amisk Lake, Dog Portage, Corneille Lake, Pine Portage, Miron, Pelican and Wood lakes to Churchill River, which they ascended to Lac Ile-à-la-Crosse, arriving on 6 October. They wintered there at the house of Patrick Small of the North West Company (see 1785-87).

1791: Leaving Ile-à-la-Crosse on 30 May, they crossed Methy Portage (Portage la Loche) to Clearwater and Athabasca rivers. On 28 June, they reached Fort Chipewyan (see 1788), and left on 1 July for Great Slave Lake, arriving on 19 July. They spent a week surveying part of the southern shore of the lake, then returned to Fort Chipewyan and surveyed eastward to the end of Lake Athabasca. On 2 September, Indians came to the fort to invite Fidler to winter among them. He agreed, and they set out on the 4th. He spent some time with Indians west of Great Slave Lake and returned to Fort Chipewyan on 4 April 1792. The whole party left on 9 May for York Factory, arriving on 17 July.

*TYRRELL, J.B., 1934, FIDLER, 1968.

1791

Hudson's Bay Company supply voyage	John Richards John Turner Henry Hanwell	*King George* [III] *Queen Charlotte* *Seahorse* [III]

From London. *King George,* carrying cargo for Albany, visited Moose Factory 7-28 September, and reached London on 17 November. *Queen Charlotte,* carrying cargo for Eastmain, visited Moose Factory 31 August - 7 September. *Seahorse* visited Churchill 1-7 September, York Factory 8 September - 4 October, and reached London 15 November.

HUDSON'S BAY COMPANY (1) and (2).

1791 | Hudson's Bay Company trading voyage | William Robinson | ?

From Eastmain; 29 June - 14 July; the first expedition to trade for oil with the Indians of Fort George River. Previous coasting voyages northward from Eastmain had been to Great and Little Whale rivers. The voyage was unsuccessful: he returned with only three gallons of seal oil.

DAVIES, 1963, pp. 263, 278.

1791 | Spanish North-west Passage expedition | Alejandro Malaspina, José de Bustamente | *Descubierta*, *Atrevida*

From Acapulco; 1 May - 9 October; to investigate Maldonado's claim to have discovered a North-west Passage (see 1588). They coasted northward to 57°59'N, which they reached on 25 June. After some further exploration of the Alaskan coast, they returned to Nootka on 13 August, and were back at San Blas on 9 October.

*VIANA, 1849; ESPINOSA Y TELLO, 1802; DODGE, 1961, pp. 183-84.

1791 | Spanish Northwest Coast survey | Francisco Eliza, José Maria Narvaez | ?

From Nootka. They were the first to sail into Juan de Fuca Strait, which had long been considered a possible entrance to a North-west Passage, and explored Haro Strait, the Strait of Georgia, and Texada Island.

ORMSBY, 1958, p. 21.

1791-92 | British North-west Passage expedition | Charles Duncan, George Taylor | *Beaver*

From London; sent by the Admiralty to search the coast of Hudson Bay north of Churchill and to pay particular attention to the possibilities of Corbett and Chesterfield inlets. Duncan had first attempted this expedition in 1790 (*qv*), when he had sailed to Churchill only to find that the sloop *Churchill,* which had been intended for the expedition, was unfit for service. On 2 May 1791, he left London on a second attempt. Ice held him in Hudson Strait for seven weeks, he was further delayed at Marble Island, and he completed only the first part of his instructions, the exploration of Corbett Inlet, before he was forced to return south to winter at Churchill. He set out again on 15 July 1792, and reached Chesterfield Inlet early in August. He sailed part way up the inlet, then continued in a longboat to Beaver Lake at its head. Here he entered the small river that Christopher and Norton had examined in 1762 (*qv*). The mate, Taylor, walked 30 km up the river, but found it neither became navigable nor did it trend in the direction of Dubawnt Lake, as had been hoped. Duncan therefore gave up the search, returned to Churchill at the end of August, then sailed for England.

GOLDSON, 1793; pp. 52-54; WILLIAMS, 1962, pp. 241-48; RICH, 1959, pp. 130-32; *HUDSON'S BAY COMPANY, ARCHIVES, C. 1/204, 205 [Journal of George Taylor].

1791-95

British naval North-west Passage expedition	George Vancouver William Robert Broughton	*Discovery* *Chatham*

From England; 1 April 1791 - 12 September 1795. Fur-trading voyages to the Northwest Coast of America, stimulated by James Cook's expedition of 1776-80, had confirmed speculation, based largely on the apocryphal voyages of de Fuca and de Fonte, that there were deep inlets in parts of the coast not explored by Cook that might be the entrance to a low-latitude North-west Passage. Also, Peter Pond's map of 1785 had shown Lake Athabasca and Great Slave Lake as having possible outlets to the Pacific Ocean, which would facilitate the trade in furs to China. Vancouver was sent to investigate these inlets. He sailed by way of the Cape of Good Hope, and sighted the west coast of America at 39°20'N on 18 April 1792. Later in April, off the Strait of Juan de Fuca, Vancouver decided to prevent further speculation about the possibility of a North-west Passage by this route by surveying the strait in detail and by examining each of its inlets in boats. He explored Puget Sound and Hood Canal, then continued up the Strait of Georgia. Here, to his surprise, he met the two Spanish ships commanded by Galiano and Valdés (see 1792), who were performing precisely the same task, and who informed him that Eliza and Narvaez had preceded him in the discovery of the Strait of Georgia (see 1791). On 9 August, Vancouver passed out of the strait through Queen Charlotte Sound, visited Nootka, where there was a Spanish establishment, then sailed to the Hawaiian Islands, where he wintered.

1793: He returned to the northernmost point of his explorations of the previous year, then continued northward on his meticulous examination of the coast. He discovered the outlet of Bella Coola River at the beginning of June, only seven weeks before Alexander Mackenzie reached the same spot (see 1792-93), and he discovered the mouth of Skeena River. By the time he had reached 56°N, he had disproved all theories based on the apocryphal discoveries of de Fonte and de Fuca. He wintered again at the Hawaiian Islands.

1794: He explored both branches of Cook Inlet, Alaska, in April, then explored southward, examining Prince William Sound and Cross Sound, until, on 19 August, he reached 56°N, the northernmost point of his explorations of 1793. His survey thus completed, he returned home by way of Cape Horn, and reached the River Shannon on 12 September 1795.

*VANCOUVER, 1798; WILLIAMS, 1962, pp. 248-68.

1792

Hudson's Bay Company supply voyage	John Richards John Turner Henry Hanwell Thomas Williamson	*King George* [III] *Queen Charlotte* *Seahorse* [III] *Nimble*

From London. *King George* visited Moose Factory 17 August - 16 September, and reached London 21 October. *Queen Charlotte* visited Moose Factory 12-18 September, and reached London 20 October. *Seahorse* visited York Factory 20 August - 17 September, and reached London 22 October. *Nimble* visited Churchill 16

August - 3 September, and reached London 21 October.

HUDSON'S BAY COMPANY (1) and (2).

1792

Spanish Northwest Coast Survey | Dionisio Galiano, Cayetano Valdés | *Sutil*, *Mexicana*

From Acapulco; 8 March - 23 September; to complete the detailed examination of Juan de Fuca Strait and the Strait of Georgia that Eliza had begun in 1791 (*qv*). They sailed first to Nootka, where they remained from mid-May to early June. They then began their exploration of Juan de Fuca Strait and sailed into the Strait of Georgia where, at the end of June, they met Vancouver's expedition (see 1791-95). Vancouver was examining the strait in exactly the same manner as themselves, and the two expeditions had duplicated many of each other's surveys. They worked together for a short time, then separated. Galiano continued his exploration northward, passed through Queen Charlotte Strait into the Pacific Ocean, and returned to Nootka, arriving on 30 August, having circumnavigated Vancouver Island. They left Nootka after a short stay, and sailed back to Monterey, arriving on 23 September.

*ESPINOSA Y TELLO, 1802; ORMSBY, 1958, pp. 21-24.

1792

Hudson's Bay Company trading voyage | Robert Robinson | ?

From Eastmain Factory; 3 July - 19 August; he was sent by Bartholomew Nelson, chief at Eastmain Factory, on a second voyage to trade for oil along the Eastmain coast. His instructions ordered him to Great and Little Whale rivers, but it is not known if he visited them.

DAVIES, 1963, p. 279.

1792

Overland expedition to the interior of the Quebec-Labrador peninsula | André Michaux

From Quebec; 31 July - (?). Michaux, a French naturalist, had arrived in North America in 1785 to collect plants for the Jardin du Roi in Paris. In 1792, he decided to travel overland from Quebec to Hudson Bay. He left Quebec on 31 July and went by way of Tadoussac and Saguenay River to Chicoutimi, where he loaded two canoes for his journey. He started up Chicoutimi River on 13 August, crossed Lake St John, and entered Mistassini River on 21 August. His route from Mistassini River to Lake Mistassini has long been a subject of speculation. He may have ascended Mistassini River to Swan Lake, portaged north-west by way of the head of Nestowkanow River to Lac Albanel, crossed this lake, and entered Lake Mistassini on 4 September. He may then have followed the north-western shore of this lake to Rupert River. After only one day on Rupert River he decided to retrace his course because of the lateness of the season, and was back at Lake St John on 12 September. According to another theory, he left Mistassini River at Swan Lake, portaged north-westward through a series of small lakes and rivers to the southern extremity of Lake Mistassini, then followed its south-western shore to Rupert River, where he turned back. Still other theories hold that he travelled nearly to the head of Mistassini River before crossing to Lake Mistassini. Michaux was the first naturalist to penetrate into the interior of the Quebec-Labrador peninsula.

LEPAGE AND DUTILLY, 1945; ROUSSEAU, 1948a.

Establishment of Fork Fort on Peace River	Alexander McLeod John Finlay Alexander Mackenzie	1792

In autumn 1792, McLeod sent two men from Fort McLeod (see 1790) to build a house near the confluence of Peace and Smoky rivers. Mackenzie, on his expedition toward the Pacific Ocean, arrived later in the autumn, completed the house, and traded there during the winter. When he left in spring 1793, McLeod took charge and may have occupied the post until 1799, when he left Peace River. It is possible also that he may have moved the post from the south to the northern side of Peace River, where a post, sometimes known as Fort McLeod, existed in the 1790's. The post seems to have been occupied intermittently until 1805, when it was abandoned in favour of Fort Dunvegan (see 1805). The XY Company (see 1799) built a post nearby in 1803, but it was probably occupied only one year.

MACGREGOR, 1952, pp. 126-33; WALLACE, J.N., 1929, pp. 31-33, 47; VOORHIS, 1930, p. 69.

Spanish North-west Passage expedition	Jacinto Caamaño	*Aranzazu*	1792

From San Blas; 20 March - 22 October; sent to examine the Northwest Coast of America from Bucareli to Nootka in search of de Fonte's Northwest Passage (see 1640). They reached Nootka on 14 May, began their search on 23 May, and reached Bucareli on 12 June. From here, they sent out boats to investigate nearby channels, then sailed north to search more channels. They returned to Nootka on 7 September, and were back at Monterey on 22 October.

ESPINOSA Y TELLO, 1802.

North West Company overland expedition to the Pacific Ocean	Alexander Mackenzie	1792-93

From Fort Chipewyan; 10 October 1792 - September 1793; to discover an overland route from Fort Chipewyan to the Pacific. He ascended Peace River to the mouth of Smoky River, where men sent by A. McLeod had built a house in preparation for his arrival (see 1792, Fork Fort) and where he wintered. He left it on 9 May 1793 and by 17 May was in sight of the Rocky Mountains. On 31 May, he reached Finlay Forks, where the north and south branches of Peace River meet. He chose to follow the south branch, Parsnip River, to its source at 54°50'N, 121°50'W; he crossed the height of land to Bad River (James Creek), followed the latter to Herrick Creek, and so reached Fraser River. He descended it to the site of present-day Alexandria, where, on the advice of Carrier Indians, he decided to strike westward overland. He therefore returned north to the mouth of West Road River, which he reached on 3 July, paddled some distance up it, then set off across country with Carrier guides to Dean River. He crossed Dean River, reached Bella Coola River on 17 July, and continued down it in the canoes of Coast Indians to its mouth on North Bentinck Arm, arriving on 20 July. He reckoned the end of his journey in Dean Channel was at 52°20'48"N. After one night there, he returned by the same route to Fork Fort, arriving on 24 August. He then continued downstream to Fort Chipewyan, where he wintered.

*MACKENZIE, 1801; DANIELLS, 1969, pp.104-60; BURPEE, 1935, Vol. 2, pp. 449-91; *LAMB, 1970; WOOLLACOTT, 1927.

1793	Hudson's Bay Company supply voyage	John Richards	*King George* [III]
		Henry Hanwell	*Prince of Wales* [I]
		John Turner	*Queen Charlotte*
		Thomas Williamson	*Nimble*
		George Taylor	*Beaver*

From London. *King George* visited Moose Factory 23 August - 21 September, and reached London 7 November. *Prince of Wales* visited York Factory 12 August - 7 September, and reached London 8 November. *Queen Charlotte* visited Churchill 8-20 August, York Factory 23 August - 7 September, and reached London 4 November. *Nimble* visited Eastmain 24 August - 8 September, Moose Factory 10-21 September, and reached London 4 November. *Beaver* remained in Hudson Bay and voyaged for many years between York and Severn House.

HUDSON'S BAY COMPANY (1) and (2)

1793	Establishment of Little Whale River post	George Jackman

Bartholomew Nelson, chief at Eastmain Factory, was instructed to establish a post at Little Whale River. A party under the surgeon, Jackman, left Eastmain on 11 September and reached Little Whale River on the 27th. They went a few miles upstream and found on the north shore the remains of an old building, presumably Richmond Fort, which had been moved there from Richmond Gulf in 1756, then abandoned in 1759 (see 1750). A short time after building his post, Jackman and his party were murdered by a group of about 20 Eskimos, although the massacre was at first thought to have been the work of Indians. The first news of the tragedy reached Eastmain on 22 August 1794, when a party of Indians from near Little Whale River visited. In 1795, an expedition was sent from Eastmain to examine the site of the massacre, and ships visited the river for the purpose of whaling and trading in 1802, 1804 (*qv*), and in each year between 1807-12. When Great Whale River post was opened in 1813 (*qv*), a ship was sent each year from that post to Little Whale River for whaling, but no permanent post was built there until 1851.

DAVIES, 1963, p. 280; HUDSON'S BAY COMPANY, ARCHIVES, B77/a/14/53.

Establishment of Neoskweskau post	John Clarke	1793

In June 1793, Clarke left Eastmain to build a trading post for the Hudson's Bay Company on Lake Mistassini in an attempt to prevent the encroachment of Canadian traders. His party paddled up Eastmain River; they found the journey difficult and, after 43 days' travelling, decided not to continue to Lake Mistassini but to build their post on Neoskweskau Lake, close to Eastmain River, where they wintered. In 1794, Clarke took the returns of trade back to Eastmain by way of Rupert River which, being more navigable than Eastmain River, became the main supply route for the post. In 1796, the post was destroyed by an explosion, but it was rebuilt the same year. It remained in operation until 1822, when it was closed. Neoskweskau was also known as St John's House and Carlton House.

DAVIES, 1963, pp. 301-13.

Hudson's Bay Company survey of Seal River	Peter Fidler	1793

Fidler surveyed Seal River as a possible route inland from Churchill to Lake Athabasca. He reported it to be a fine route for flat-bottomed boats.

RICH, 1959, p. 174-75.

Hudson's Bay Company supply voyage	John Richards Henry Hanwell John Turner	*King George* [III] *Prince of Wales* [I] *Queen Charlotte*	1794

From London. *King George* visited Moose Factory 18 August - 13 September, and reached London 10 November. *Prince of Wales* visited Churchill 16-28 August, York Factory 5-26 September, and reached London 12 November. *Queen Charlotte* visited Eastmain 19-31 August, Moose Factory 1-13 September, and reached London on 9 November.

HUDSON'S BAY COMPANY (1) and (2).

1794 Establishment of Nemiscau — James Foggett

French traders had operated posts on Lake Nemiscau intermittently since the 17th century (see 1658 and 1679, Jolliet), but little is known of the history of posts on that lake until 1794, when the Hudson's Bay Company built there. Foggett, a carpenter, left Eastmain on 27 May to establish a post at Lac au Goéland. He was delayed on his journey inland and managed to reach only Lake Nemiscau. The house he built there was, at first, regarded as a temporary establishment, but it continued as a trading post until 1799, when a post at Lac au Goéland was finally opened (see 1799, Cheashquacheston). The latter post closed in 1802, and Nemiscau reopened. Nemiscau suffered opposition from a nearby North West Company post in 1804-06 (see 1803-06). After that, the trade was so poor that it closed again in 1809, and the Indians were instructed to go to Rupert House. It is not known how long it remained closed, nor whether it operated at all in the 19th century.

DAVIES, 1793, pp. 298-301.

1795 Hudson's Bay Company supply voyage

Hudson's Bay Company supply voyage	John Richards	*King George* [III]
	John Turner	*Queen Charlotte*
	Henry Hanwell	*Prince of Wales* [I]

From London. *King George* visited York Factory 27 August - 21 September, and reached London 25 November. *Queen Charlotte* visited Churchill 28 August - 9 September, York Factory 11-21 September, and reached London 24 November. *Prince of Wales* visited Moose Factory 29 August - 26 September, and reached London 26 November.

HUDSON'S BAY COMPANY (1) and (2).

1795 Establishment of Fairford House — Malcolm Ross

In this year, Ross built Fairford House for the Hudson's Bay Company on Churchill River, below the mouth of Reindeer River. It was apparently abandoned the next year in favour of Bedford House on Reindeer Lake (see 1796, Thompson).

GLOVER, 1962, p. 108; VOORHIS, 1930, p. 66.

1796 Hudson's Bay Company supply voyage

Hudson's Bay Company supply voyage	John Richards	*King George* [III]
	John Turner	*Queen Charlotte*
	Henry Hanwell	*Prince of Wales* [I]

From London. *King George* visited York Factory 20 August - 13 September, and reached London 12 November. *Queen Charlotte* visited Churchill 16-30 August, York Factory 2-13 September, and reached London 12 November. *Prince of Wales* visited Moose Factory 20 August - 21 September, and reached London 13 November.

HUDSON'S BAY COMPANY (1) and (2).

Hudson's Bay Company exploring expedition; establishment of Bedford House	David Thompson		1796

From Fairford House on Churchill River (see 1795); 10 June - 21 July; sent, at his own request, to explore the unknown country between the confluence of Churchill and Reindeer rivers and the eastern end of Lake Athabasca. Accompanied by two Chipewyan Indians, he followed Reindeer River northward to Reindeer Lake, canoed along its west side, then continued by a small river and by portages to Wollaston Lake. He crossed the northern end of Wollaston Lake, reached the head of Fond du Lac River on 23 June, and descended it, discovering Hatchet Lake and Black Lake on his way, and reached the eastern end of Lake Athabasca on 2 July, a point reached by Philip Turnor in 1791 (see 1790-92). He returned by the same route to Fairford House, arriving on 21 July. Soon after, Thompson and Malcolm Ross returned to Reindeer Lake, built a trading post, Bedford House, on its western shore and wintered there.

GLOVER, 1962, pp. 108-20.

Establishment of Livingston's Fort	Duncan Livingston		1796

Livingston built for the North West Company the first trading post on Mackenzie River, about 130 km downstream from Great Slave Lake. He maintained the post until 1799, when he was murdered during an expedition down Mackenzie River (*qv*). In 1800, his replacement, John Thompson, abandoned the post and divided its trade between Rocky Mountain Fort and Great Bear Lake post (*qv*).

MASSON, 1960, Vol. 1, p. 95.

Hudson's Bay Company supply voyage	John Richards John Turner Henry Hanwell	*King George* [III] *Queen Charlotte* *Prince of Wales* [I]	1797

From London. *King George* visited York Factory 24 August - 15 September, and reached London 4 December. *Queen Charlotte* visited Churchill 19 August - 3 September, York Factory 6-15 September, and reached London 4 December. *Prince of Wales* visited Moose Factory 28 August - 23 September, and reached London 4 December.

HUDSON'S BAY COMPANY (1) and (2).

North West Company exploration of Finlay River	John Finlay		1797

Finlay, a North West Company agent in the Peace River district, explored part of Finlay River, perhaps as far as Ingenika River. No details of his journey are known. It was not until 1824 that Samuel Black explored the river to its head *(qv)*.

RICH, 1955, pp. xii-xiv.

1798

Hudson's Bay Company supply voyage	John Richards	*King George* [III]
	John Turner	*Queen Charlotte*
	Henry Hanwell	*Prince of Wales* [I]

From London; 6 June - November. *King George* visited York Factory 6-26 September, and reached London 23 November. *Queen Charlotte* visited Churchill 31 August - 15 September, York Factory 17-26 September, and reached London 22 November. *Prince of Wales* visited Moose Factory 6-30 September, and reached London 22 November.

HUDSON'S BAY COMPANY (1) and (2).

c1799

Establishment of
Waswanipi post

The first trading post on Waswanipi Lake was built by the North West Company at least as early as 1799, for in that year Andrew Moar of the Hudson's Bay Company built Cheashquacheston post only 30 km away, frequently met traders from Waswanipi, and once visited that post (see below, 1799). The post continued in operation throughout the early part of the 19th century. In 1819, the Hudson's Bay Company built a rival post nearby (*qv*), but that post was apparently closed after the amalgamation of 1821, and the Hudson's Bay Company continued to operate the North West Company's post.

DAVIES, 1963, pp. 318-26.

1799

Hudson's Bay Company supply voyage	John Richards	*King George* [III]
	John Turner	*Queen Charlotte*
	Henry Hanwell	*Prince of Wales* [I]

From London; 29 May - November. *King George* visited York Factory 30 August - 21 September, and reached London 15 November. *Queen Charlotte* visited Churchill 31 August - 11 September, York Factory 13-21 September, and reached London 14 November. *Prince of Wales* visited Moose Factory 31 August - 1 October, and reached London 16 November.

HUDSON'S BAY COMPANY (1) and (2).

1799

Formation of the
XY Company

The formation of the North West Company in 1783 (*qv*) had amalgamated some of the Montreal-based fur-trading companies in rivalry with the Hudson's Bay Company, but many small concerns had been left out of that and subsequent agreements. In 1798, two of these small concerns – Forsyth, Richardson and Company and Leith, Jamieson and Company – formed a co-partnership. In 1799, the partnership became known as the XY Company, and it was soon joined by Parker, Gerrard and Ogilvy and Company, John Mure and Company, Phyn, Inglis and Company, and by Alexander Mackenzie. During four years, the new company (also known as the New North West Company) traded in fierce competition with the North West Company, and maintained posts in the regions of Lake Athabasca, Great Slave Lake, and Mackenzie River. It amalgamated with the North West Company in 1804.

RICH, 1959, pp. 212-13.

1799

North West Company exploration of Mackenzie River — Duncan Livingston, James Sutherland

Livingston, who operated a trading post on Mackenzie River (see 1796) and four men made the first long journey down the river since its discovery in 1789 (*qv*). It has been generally stated that the entire party was killed by Eskimos about eight days' march downstream from the post, but the recent discovery of a contemporary account of the event indicates that Livingston was murdered by his Indian guides.

WENTZEL, 1823, pp. 78-79; YERBURY, 1977.

1799

Establishment of Cheashquacheston post — Andrew Moar

The Hudson's Bay Company had long wished to establish a trading post at Lac au Goéland, where the North West Company and other Canadian traders had for many years operated posts which adversely affected trade at Eastmain. They had sent men to establish a post there in 1793 and 1794, but both ventures had failed, and Nemiscau had been established instead (see 1794). In 1799, however, Andrew Moar, master at Nemiscau, was instructed to close Nemiscau and to move to Lac au Goéland. He arrived there on 21 July. North West Company traders from Waswanipi (see c1799) were settled nearby, and Moar's share of the trade was poor. The post was known as Windsor House in the first year, but then became known as Cheashquacheston. In 1802, the post was closed, and Nemiscau reopened. The Hudson's Bay Company did not trade again at Lac au Goéland until 1820, when they built a house there as an outpost to Waswanipi (see 1819). That house was closed in 1821.

DAVIES, 1963, pp. 271-75.

c1800

Establishment of Fond du Lac

According to Peter Fidler, the North West Company had a post of this name at the east end of Lake Athabasca in 1800, and it operated at least until 1821, after which it seems to have closed. In 1851, the Hudson's Bay Company ordered a new post to be established there, and it has been in operation ever since.

RICH, 1938, p. 416; RICH, 1956, pp. 120, 225-26.

c1800

Establishment of Great Bear Lake post

The North West Company built this post at the head of Great Bear River on the site later occupied by Fort Franklin (see 1825-27). John Franklin thought it was built in 1799, and Livingston may have established it in that year during his expedition down Mackenzie River (*qv*). More probably, it was established in 1800 by John Thompson, Livingston's successor, who closed Livingston's Fort on Mackenzie River (see 1796) and diverted its trade to Great Bear Lake post and to Rocky Mountain House (see 1800, below). The XY Company (see 1799) appears to have built a rival post on Great Bear Lake between 1800 and 1804, for, in 1804, the master of the North West Company post was killed in a dispute over Indian allegiance. The North West Company post continued in operation until 1815, when, in common with other trading posts in the Mackenzie district, it was closed. The site remained unoccupied until Franklin's expedition arrived in 1825.

STAGER, 1962, pp. 39-40; MASSON, 1960, Vol. 2, pp. 96-127.

1800

Hudson's Bay Company supply voyage	John Richards	*King George* [III]
	John Turner	*Queen Charlotte*
	Henry Hanwell	*Prince of Wales* [I]

From London; 28 May - 5 December. *King George* visited Moose Factory 8 September - 8 October. *Queen Charlotte* visited Churchill 4-16 September, and York Factory 18-23 September. *Prince of Wales* visited York Factory 8-23 September. All ships reached London on 5 December.

HUDSON'S BAY COMPANY (1) and (2).

1800

Establishment of Rocky Mountain House on Mackenzie River — John Thompson

In this year Thompson, North West Company agent in charge of Mackenzie River district, abandoned Livingston's Fort (see 1796) and built instead Great Bear Lake post (see c1800, above) and Rocky Mountain House. The latter was on the right bank of Mackenzie River at Camsell Bend, opposite the mouth of North Nahanni River. It was abandoned before 1805, when Alexander Mackenzie spoke of it as "Old Rocky Mountain Fort," probably in 1803, when Fort of the Forks was built (*qv*).

STAGER, 1962, pp. 39-40.

1801

Hudson's Bay Company supply voyage	John Turner Henry Hanwell	*Prince of Wales* [I] *King George* [III]

From London; 20 May - December. *Prince of Wales* visited Moose Factory 23 August - 14 September, and reached London 16 November. *King George* visited Churchill 16 August - 4 September, York Factory 10-21 September, and reached London 4 December.

HUDSON'S BAY COMPANY (1) and (2).

1802

Hudson's Bay Company supply voyage	John Turner Henry Hanwell Thomas Ramsey	*King George* [III] *Prince of Wales* [I] *Ceres*

From London; 26 May - October. *King George* visited York Factory 26 August - 18 September, and reached London 23 October. *Prince of Wales* visited Moose Factory 19 August - 10 September. From Moose, she carried goods to Albany 21-23 August. She reached London 23 October. *Ceres* visited Churchill 25 August - 7 September, York Factory 9-18 September, and reached London 24 October.

HUDSON'S BAY COMPANY (1) and (2).

1802-03

Hudson's Bay Company expedition to Lake Athabasca; establishment of Nottingham House	Peter Fidler Thomas Swain

From Cumberland House; the company's first attempt to break the North West Company's monopoly of trade in the Athabasca district, and only the second time they had sent men to that region (see 1790-92). They left in summer 1802, reached Lake Athabasca in late September, and built Nottingham House on the present site of Fort Chipewyan and not far from where Fort Chipewyan then stood (see 1788). Fidler wintered there and sent Swain up Peace River to establish Mansfield House close to present-day Fort Vermilion. Swain abandoned Mansfield House in January 1803 because the Indians refused to trade with him. On 19 May 1803, Fidler left for York Factory with the few furs they had collected and returned to Nottingham House later that year (see 1803-04).

TYRRELL, J. B., 1913.

c1803

Establishment of Mistassini post

Little is known of the early history of trading posts on Lake Mistassini. French traders had been there at least as early as 1679 (*qv*), it was included in the Domaine du Roi in 1733 (*qv*), and independent Canadian traders may have been active in the region throughout the 18th century. In the latter part of that century, the Hudson's Bay Company sent several expeditions from James Bay to Lake Mistassini to explore the route there, to investigate the possibility of establishing a post there, and to observe the activities of the Canadian traders. James Robertson made such journeys in 1778 and 1779; Thomas Buchan in 1780; John Clarke in 1790 and 1793. On the last of these journeys, Clarke established Neoskweskau post about 3 days' journey from Lake Mistassini (see 1793) and, in consequence, the Hudson's Bay Company's interest in establishing a post on the lake itself died down for several years. The North West Company was thus the first of the two major trading companies to build a permanent post on Lake Mistassini. This was probably done in 1803, when Angus Shaw, who managed the Postes du Roi on behalf of the North West Company, was reported to be on his way to build a post there (see also 1803-06). James McKenzie described the North West Company's post on Lake Mistassini in 1808; it was situated at the south-eastern end of the lake, about four days' journey from Neoskweskau post. The Hudson's Bay Company built a post nearby in 1812 as an outpost to Neoskweskau, but this was closed in 1816 after fierce competition from the North West Company had rendered it of little value. In 1817, however, the North West Company also abandoned their post. The Hudson's Bay Company therefore returned in 1818 and built a post about 80 km from their former one. In 1821, this post was moved again, and rebuilt close to the site of the old North West Company post. This post has been maintained by the Hudson's Bay Company ever since.

DAVIES, 1963, pp. 288-98.

1803

Hudson's Bay Company supply voyage	John Turner	*King George* [III]
	Henry Hanwell	*Prince of Wales* [I]
	Thomas Ramsey	*Ceres*

From London; 4 June - November. *King George* visited York Factory 15 August - 3 September, and reached London 26 November. *Prince of Wales* visited Albany 22-24 August, Moose Factory 26 August - 11 September, and reached London 27 November. *Ceres* visited Churchill 11-26 August, York Factory 29 August - 3 September, and reached London 29 November.

HUDSON'S BAY COMPANY (1) and (2).

1803

Establishment of Fort of the Forks (Fort Simpson)	James Porter

In 1803-04, Porter of the North West Company wintered at a post at the confluence of the Liard and Mackenzie rivers. As there is no record of a post having been there previously, he may be presumed to have established it. In 1804, W. F. Wentzel took charge of the post, which operated until 1815 (*qv*), when it was abandoned. It was reopened, possibly in 1821, when Fort Alexander (see 1817) was apparently moved upstream from Willowlake River to the Forks. In 1822, Wentzel built a new post on an island at the Forks, downstream from the old buildings and close to the present site. By October 1823, it was known as Fort Simpson, after George Simpson, and it has been in operation ever since.

FLEMING, 1940, p. 44; *MOCCASIN TELEGRAPH*, 1970, pp. 122-23.

1803

Establishment of Great River Post (Fort George) on James Bay	Jacob Atkinson George Atkinson	

When William Bolland, chief at Eastmain Factory, learned that the North West Company intended to send a party to winter on Fort George River, he decided to send the Atkinson brothers to build a rival post there. They left with one Indian on 23 September and built a post, which was at first known as Big or Great River post. They succeeded in taking nearly all the trade of the region. George Atkinson managed the post in 1804-05. Thomas Alder joined him in 1805-06, but, because their supply of trade goods was poor, they made an agreement with the North West Company agent, Duncan McDougall, by which they took two-thirds and McDougall one-third of the trade. In 1806, the North West Company burned their post and left the area. The Hudson's Bay Company's post was evacuated in favour of Great Whale River post in 1813 (*qv*), reopened in 1816, closed again in 1824, and reopened in 1837 after Eastmain Factory closed. The name Fort George came into use at this time, and it has been in operation ever since.

DAVIES, 1963, pp. 263-71.

1803-04

Hudson's Bay Company expedition to Lake Athabasca	Peter Fidler Thomas Swain	

On 12 September, Fidler arrived back at Nottingham House, which he had established on Lake Athabasca in 1802 (see 1802-03). Four days later, he sent Swain to establish Chiswick House on Slave River, about 40 km above Great Slave Lake. On 21 May 1804, they left Nottingham House with a poor trade for Cumberland House.

TYRRELL, J. B., 1913.

1803-06

North West Company invasion of James Bay	John George McTavish Captain Richards Angus Shaw Duncan McDougall Alexander Fraser Robert Folster	*Eddystone*

In these years, the North West Company defied the Hudson's Bay Company's charter (see 1670) by establishing posts in and around James Bay. In late August or early September, McTavish arrived at Charlton Island in *Eddystone* and built a storehouse, named Fort

St Andrews. He was soon joined by Shaw, the North West Company agent in charge of the Postes du Roi (see 1658), who had come overland from Quebec. Henry Hanwell of the Hudson's Bay Company (see 1803, above) attempted to dislodge the invaders, but failed. In September, Shaw left Charlton Island and returned up Eastmain River, presumably to Lake Mistassini, where the North West Company was apparently building a post (see c1803). McDougall built a post at Fort George River, but soon found his trade blocked by a rival post. In November, Fraser descended Rupert River to join McTavish at Charlton Island.

1804: In spring, Folster opened a post at Nemiscau, near the Hudson's Bay Company's post there (see 1794), then left to manage another post near Rupert House. In July, Jack Hester led a party of Indians from Fort St Andrews to reinforce McDougall at Fort George River.

1805: McDougall's trade improved, and he forced George Atkinson, his rival, to concede to him one-third of the trade at Fort George River.

1806: The North West Company abandoned their posts on James Bay, but their post at Lake Mistassini continued in operation until 1817.

DAVIES, 1963, pp. xxix, 263-65, 281-82, 293, 299-300, 305-06, 333; RICH, 1959, p. 257; WALLACE, 1947.

1804

Hudson's Bay Company supply voyage	John Turner Henry Hanwell Thomas Ramsey	*King George* [III] *Prince of Wales* [I] *Ceres*

From London; 1 June - November. *King George* visited York Factory 6-27 August, and reached London 28 November. *Prince of Wales* visited Moose Factory 5 August - 21 September, delivered cargo to Albany 9-17 September, and reached London 28 November. *Ceres* visited Churchill 14 August - 1 September, York Factory on 4 September, and reached London 27 November.

HUDSON'S BAY COMPANY (1) and (2).

1804

Amalgamation of North West Company and XY Company

The XY Company, formed in 1799 (*qv*), amalgamated with the North West Company in November 1804, after a period of violent competition, 1800-04.

RICH, 1959, p. 230.

1804

Establishment of Fort Good Hope

The North West Company established Fort Good Hope on the left bank of Mackenzie River, opposite the mouth of Hare Indian River. It continued in operation until 1815 (*qv*) when, with the other posts on Mackenzie River, it was closed. It reopened from 1818 until 1823, when the Hudson's Bay Company, with which the North West Company had amalgamated in 1821 (*qv*), moved it downstream about 160 km and built a post at the mouth of Trading River at the request of Loucheux Indians, who wished to shorten their journey upstream. It was hoped that the Eskimos would also visit the new post, but none came; the Loucheux trade failed to meet expectations, so, in 1826, the company moved back upstream to build a post near the original site, possibly on Manitou Island. In 1836, spring floods washed away the entire establishment, and the post was rebuilt on the right bank, just south of Hare Indian River, where it still stands.

STAGER, 1962, pp. 40-42.

1804

Hudson's Bay Company voyage — Thomas Alder

From Rupert House; 24 June - 22 August. Alder visited the company's new post at Fort George River (see 1803), then continued northward to Great Whale River, arriving on 4 July. On 9 July, he there met agents of the North West Company, who had come up from their post on Fort George River (see 1803-06). On 18 July, he left to examine the remains of the Hudson's Bay Company's old posts at Little Whale River (see 1793) and Richmond Gulf (see 1750). He found evidence of frequent Eskimo visits to Richmond Gulf, returned south on 8 August, and reached Eastmain Factory on the 22nd. He recommended that men should be sent north each year to kill whales instead of continuing to trade with the Indians for whale oil. He also brought back samples of lead ore from Little Whale River.

DAVIES, 1963, p. 281.

1804-05

Hudson's Bay Company expedition to Lake Athabasca — Peter Fidler, Thomas Swain

This was Fidler and Swain's third season at Nottingham House (see 1802-03 and 1803-04). They left Cumberland House on 3 August and arrived on 11 September. On the 16th, Swain left to build the Hudson's Bay Company's first post on Great Slave Lake, at Moose Island, about 16 km west of Slave River delta. Fidler left for Cumberland House on 24 May 1805, leaving Swain behind, and returned to rejoin him later the same year.

TYRRELL, J.B., 1913.

1805

Hudson's Bay Company supply voyage — John Turner, *King George* [III]; Henry Hanwell, *Prince of Wales* [I]

From London; 6 June - 25 November. *King George* visited Churchill 16-25 August, and York Factory 28 August - 7 September. *Prince of Wales* visited Moose Factory 17 August - 13 September, and carried cargo to Albany 28 August - 8 September. Both ships reached London on 25 November.

HUDSON'S BAY COMPANY (1) and (2).

1805

Establishment of Fort Dunvegan — Archibald Norman McLeod

McLeod began to build Fort Dunvegan for the North West Company in the summer of 1805 on Peace River, near the mouth of Muddy Creek. During winter 1805-06, it was visited by Fraser and McDougall who were engaged in exploration farther up Peace River (see 1805-06). The post continued in existence for many years as one of the main establishments on the river. It was closed in 1823 by the Hudson's Bay Company (who had control of it after the amalgamation of 1821 [*qv*]) to punish Beaver Indians for the massacre of men at Fort St John (see 1806), but it was reopened in 1828. It remained in operation until 1918, when it was finally closed.

MACGREGOR, 1952, pp. 151-68, 200-30.

1805

Establishment of Fort Liard

The early history of trading posts on Liard River is obscure. It has been generally supposed that between about 1805 and 1813 the North West Company operated two posts concurrently there—Fort Liard and Fort Nelson. But an unpublished account by W. F. Wentzel gives the impression that these two posts were, until 1821 at least, one and the same place. During winter 1804-05, Wentzel, who had charge of Fort of the Forks (see 1803), enquired among the Indians about the country up Liard River. Encouraged by their response, he sent a man about 130 km upstream in 1805 to build a temporary store. This store is referred to on Wentzel's map as "Old Fort 1805." It was on the right bank, just above the entrance of Nahanni River. Wentzel visited it later in 1805 and, soon after, built a permanent post on Liard River, to which George Keith was sent as manager in 1806. He seems then to have replaced the storehouse of 1805 with a post much farther upstream, opposite the mouth of Fort Nelson River. This post, known as Fort Nelson, was destroyed during winter 1812-13 by Indians and its inhabitants massacred. The North West Company's trade in the Mackenzie district was then declining, so, for the next seven years there was no post at all on Liard River. In 1820, the company built a new post on the left bank about halfway between the storehouse of 1805 and Fort Nelson of 1806-13, not far from the present site of Fort Liard. Wentzel continued to call this post Fort Nelson; George Simpson called it "Wintzells Fort"; but it soon became generally known as Fort Liard.

WENTZEL, 1821; RICH, 1938, pp. 393-94; STAGER, 1962.

Hudson's Bay Company expedition to Lake Athabasca — Peter Fidler — **1805-06**

Fidler returned to Nottingham House, where he had left Thomas Swain in charge, for the fourth consecutive season. He found that servants of the North West Company from Fort Chipewyan had raided the post and had set up a guard to keep the Indians away. In 1806, therefore, he and Swain abandoned the post, leaving the Athabasca district once again entirely in the hands of the North West Company.

TYRRELL, J.B., 1913.

Establishment of Rocky Mountain Portage House and Fort McLeod — Simon Fraser, John Stuart, James McDougall — **1805-06**

The North West Company decided to extend its trade across the Rocky Mountains to the region Mackenzie explored in 1792-93 (*qv*). In 1805, Fraser ascended Peace River with 20 men. He built a house at Rocky Mountain Portage, left Stuart in charge there, then continued up Peace and Parsnip rivers to Pack River, which Mackenzie had not noticed. He followed it up to McLeod Lake, where he built Fort McLeod. Leaving a man named La Malice in charge here, he returned to Peace River, where he, Stuart, and McDougall wintered at Rocky Mountain Portage House and at Fort Dunvegan. During winter 1805-06, La Malice and the two men with him at Fort McLeod quarrelled, and they returned to Rocky Mountain Portage House. By April 1806, McDougall had made a reconnaissance westward from McLeod Lake to Stuart Lake, a route Fraser and Stuart surveyed in more detail in 1806-07 (*qv*).

LAMB, 1960, pp. 15-16; RICH, 1959, p. 233.

Hudson's Bay Company supply voyage — John Turner, Henry Hanwell — *King George* [III], *Prince of Wales* [I] — **1806**

From London; 4 June - November. *King George* visited Churchill 9-17 August, York Factory 18-30 August, and reached London 22 November. *Prince of Wales* visited Moose Factory 10 August - 13 September, carried cargo to Albany 31 August - 3 September, and reached London 21 November.

HUDSON'S BAY COMPANY (1) and (2).

Establishment of Fort St John on Peace River — **1806**

The North West Company probably built this post near the confluence of the Beatton and Peace rivers, some 25 km below its present site. A new post was built nearby after 1810. In 1823, Indians massacred the master, Guy Hughes, and servants. Fort Dunvegan nearby (see 1805) was closed for a few years to punish the Indians. The site of Fort St John was unoccupied for at least 20 years until, during the 1840's or 1850's, the Hudson's Bay Company built a new post of the same name not far from the original site. The post was rebuilt in 1874, again in 1923, and is still in operation.

MACGREGOR, 1952, pp. 169-99, 385; WALLACE, J.N., 1929, p. 76.

1806-07

North West Company overland expedition — Simon Fraser, John Stuart

Continuing the North West Company's policy of developing trade west of the Rocky Mountains (see 1805-06), Fraser and Stuart left Rocky Mountain Portage House on 20 May and canoed up Peace, Parsnip, and Pack rivers to Fort McLeod. They spent a fortnight there building canoes and resumed the journey on 23 June. They returned down Pack River, then followed Mackenzie's route to the confluence of Fraser and Nechako rivers (see 1792-93), then followed Nechako and Stuart rivers to Stuart Lake (which McDougall had visited in 1805-06 [*qv*]). They arrived there on 26 July and began to build Fort St James. Leaving Fraser there, Stuart set out to explore Fraser Lake, of which he gave such a favourable report that Fraser himself went to see it and built Fort Fraser there. Stuart returned to Fort St James on 18 December. In autumn 1807, he built Fort George, a supply post at the mouth of Nechako River.

LAMB, 1960.

1807

Hudson's Bay Company supply voyage	John Turner	*King George* [III]
	Henry Hanwell	*Prince of Wales* [I]
	Thomas Ramsey	*Eddystone*
	John Davison	*Mainwaring*

From London. *King George* visited Moose Factory 4-24 September, and reached London 15 November. *Prince of Wales* visited Churchill 6-15 September, York Factory 19 September - 4 October, and reached London 15 November. *Eddystone* visited Eastmain Factory 7-18 September, Moose Factory 21-24 September, and reached London 15 November. *Mainwaring,* sent for permanent service in Hudson Bay, was towed by the *King George* and *Prince of Wales* for a considerable part of the outward voyage. She arrived at York Factory on 8 September, and left for Moose Factory on 14 September, commanded by George Taylor. She served in the Bottom of the Bay until 1820, when she was broken up.

HUDSON'S BAY COMPANY (1) and (2).

1807

Hudson's Bay Company surveying expedition — Peter Fidler

From Cumberland House; 2 June - 4 August; to find a practical route to the eastern end of Lake Athabasca, so the company might renew its campaign to win the trade of the Athabasca district (see

1805-06, Fidler). From Cumberland House, he canoed along the well known route by Namew Lake, Sturgeon-Weir River, Wood Lake, and Portage du Traite to Churchill River, then continued up Reindeer River to Reindeer Lake. Here, because the lake was still icebound, he made a detour westward by Oliver Lake and Vermilion River and rejoined Reindeer Lake near Bedford House, the post David Thompson had built in 1796 (*qv*). He then followed Thompson's route toward Lake Athabasca. He reached (and named) Wollaston Lake on 19 June, descended Fond du Lac River, and on 29 June reached the easternmost point that he and Turnor had visited when surveying Lake Athabasca eastward from Fort Chipewyan in 1791 (*qv*). He then retraced his route to Churchill River and descended it to Churchill, arriving 4 August. Fidler was not convinced of the value of the route, of which he had made a careful survey, and the company made no immediate attempt to exploit it.

TYRRELL, J.B., 1913.

1808

Hudson's Bay Company supply voyage	John Turner	*King George* [III]
	Henry Hanwell	*Prince of Wales* [I]
	Thomas Ramsey	*Eddystone*

From London; 20 May - 24 November. *King George* visited York Factory 21 August - 4 September. On the outward voyage, the ship was delayed by ice in Hudson Bay; on 10 August, Turner recorded that he had never seen the ice so close and heavy during 41 years' experience in the bay. *Prince of Wales* visited Moose Factory 14 August - 21 September. She carried cargo to Eastmain 26 August - 2 September, and to Albany 12-15 September. *Eddystone* visited Churchill 12-24 August, York Factory 26 August - 4 September, and returned to London in company with *King George.* All ships reached London on 24 November.

HUDSON'S BAY COMPANY (1) and (2).

1808

North West Company exploration of Fraser River

Simon Fraser
Jules Maurice Quesnel
John Stuart

From Fort George, at the mouth of Nechako River (see 1806-07); 28 May - 6 August; to explore Fraser River, which Mackenzie had discovered in 1792 (see 1792-93), and which Fraser believed to be Columbia River — if true, the fact would have proved most useful to the fur trade. He descended the river in badly built canoes in six weeks, reaching its mouth on 2 July, then returned to Fort George.

LAMB, 1960.

1809

Hudson's Bay Company supply voyage	John Turner	*King George* [III]
	Henry Hanwell	*Prince of Wales* [I]

From London; 31 May - 22 November. *King George* visited Churchill 27 August - 2 September, and York Factory 8-22 September. *Prince of Wales* visited Moose Factory 31 August - 30 September, and carried cargo to Albany 9-12 September. Both ships reached London on 22 November. *Eddystone,* chartered by the British Government for a voyage to Davis Strait, did not visit Hudson Bay in this year.

HUDSON'S BAY COMPANY (1) and (2).

1809

North West Company exploration of Mackenzie River	(John?) Clarke	

The third attempt by a European to descend Mackenzie River to the sea (see 1789 and 1799). Clarke, presumably an employee at one of the North West Company posts on Mackenzie River, apparently reached the head of the delta before being obliged to return by the "menacing attitude" of Eskimos, who appeared on both banks of the river.

WENTZEL, 1823.

c1810

Establishment of Fort Norman

The North West Company built Fort Norman on the right bank of Mackenzie River, nearly opposite Redstone River, in or before 1810. It has been suggested that it may have been built as early as 1804 because Fort of the Forks (established in 1803 [*qv*]) and Fort Good Hope (established in 1804 [*qv*]) were about 725 km apart, and it would have been logical to have a post between them. In common with other posts on Mackenzie River, it was closed in 1815 and reopened about 1821. In 1823, when Fort Good Hope was moved about 160 km downstream, Fort Norman was also moved to a site about 50 km south of Great Bear River to accommodate Hare Indians, who had until then traded at Fort Good Hope. In 1851, it was moved to its present site on the right bank near the mouth of Great Bear River and has operated ever since.

STAGER, 1962, p. 40; ROBINSON AND ROBINSON, 1946; *MOCCASIN TELEGRAPH,* 1970, p. 122.

1810

Hudson's Bay Company supply voyage	John Turner Henry Hanwell Thomas Ramsey	*King George* [III] *Prince of Wales* [I] *Eddystone*

From London; 31 May - 5 December. *King George* visited Churchill 8-15 August, and York Factory 19 August - 4 September. *Prince of Wales* visited Moose Factory 10 August - 24 September, and carried cargo to Albany 9 - 14 September. *Eddystone* visited Churchill 8-16 August, Moose Factory 28 August - 24 September. All ships reached London on 5 December.

HUDSON'S BAY COMPANY (1) and (2).

British naval expedition to interior Newfoundland | David Buchan | *Adonis* | 1811

From Bay of Exploits; 13-30 January; to establish whether there were Indians who resided permanently in Newfoundland, or whether those who had been seen on the coast were merely occasional visitors from the mainland. This is the first expedition known to have penetrated far into the interior of Newfoundland. Leaving his ship, Buchan travelled inland up Exploits River with 24 of his crew, two white guides, and 12 sledges. On 24 January, about 200 km upstream, he discovered an encampment of about 75 Indians, with whom he stayed for several days and established friendly relations. He then returned to his ship. He wrote a short account of these Indians and of the general geography of the interior.

BARROW, 1818, Appendix 1.

Moravian missionary voyage to Ungava Bay | Benjamin Kohlmeister, George Kmoch | ? | 1811

From Okak; 24 June - 4 October; to visit the Eskimos of Ungava Bay and to explore the region. Accompanied by four Eskimo families, they sailed north along the Labrador coast, around Cape Chidley, into Ungava Bay. They named and mapped many features as far west as Koksoak River, which they ascended about 40 km to the present site of Fort Chimo. The favourable report they published of the area was noticed by the Hudson's Bay Company, who were thereby encouraged to extend their trade to Ungava Bay (see 1830).

*KOHLMEISTER AND KMOCH, 1814; COOKE, 1964, pp. 142-43; 1969, pp. 12-13; LEECHMAN, 1947.

Grant of Assiniboia | 1811

On 12 June, the Hudson's Bay Company granted an area of some 185,600 km^2 of its territories to Thomas Douglas, fifth Earl of Selkirk, for the formation of a colony at Red River. The grant, which extended from Lake Winnipeg, Winnipeg and Rainy rivers, southwestward to the height of land, was opposed by many Hudson's Bay Company traders and, of course, by the North West Company, for it interrupted not only their line of communication between Athabasca and Montreal but also their supply of pemmican, without which they could not make such long and rapid canoe trips. The arrival of settlers precipitated a bloody conflict sometimes called the Pemmican War or the Battle for Athabasca. Government concern over the conflict, together with the heavy financial losses suffered by both companies, eventually made a coalition a necessity (see 1821).

MORTON, A.S., 1939, pp. 534-37; STEFANSSON, 1956, pp. 204-24.

1811-12

Hudson's Bay Company supply voyage	Henry Hanwell	*Prince of Wales* [I]
	Thomas Ramsey	*Eddystone*
	Thomas Gull	*Edward and Ann*
	C. Campbell	HMS *Rosamond*

From London. *Prince of Wales* visited Moose Factory 25 September - 10 October. She was obliged to turn back on her homeward voyage on 25 October, and wintered in Strutton Sound. In 1812, *Rosamond* arrived to convoy her home. They set out on 2 August and reached London on 4 December 1812. *Eddystone* was bound for Churchill and York Factory but she arrived in Hudson Bay so late in the season that Ramsey decided not to call at Churchill. He visited York Factory 26 September - 6 October, and arrived back at Yarmouth 28 November. *Edward and Ann* carried men and goods for York Factory and was to call also at Moose Factory to take on timber. However, she too arrived so late in Hudson Bay that, after calling at York Factory, she returned direct to England with *Eddystone*.

HUDSON'S BAY COMPANY (1) and (2).

1812

Hudson's Bay Company supply voyage	John Turner	*King George* [III]
	Thomas Ramsey	*Eddystone*
	John Davison	*Robert Taylor*

From London. *King George* visited Churchill 23-28 August, York Factory 31 August - 14 September, and reached London 1 December. *Eddystone* visited Moose Factory 26 August - 28 September, and reached London 1 December. *Robert Taylor* visited York Factory 26 August - 4 September, sailed to Moose Factory, then returned to London. She carried colonists, led by Owen Keveny, for the Earl of Selkirk's Red River Settlement (see 1811). All the colonists had left York Factory by 9 September and had arrived at the settlement by 27 October.

HUDSON'S BAY COMPANY (1) and (2); *McKEEVOR, 1819.

c1813

Establishment of the Rock Depot

In 1813, the Hudson's Bay Company sent out instructions for the building of a depot on Hayes River, 190 km above York Factory, partly as a result of Lord Selkirk's wish for a "Winter Road" from York Factory to the Red River settlement. The depot, which was not intended as a trading post, was used as an unloading point, where cargo was transferred from the heavy boats which had come up from York Factory to the canoes which were used on the less easily navigable rivers of the interior.

RICH, 1938, pp. 423-24.

1813

Hudson's Bay Company supply voyage	Thomas Ramsey	*Eddystone*
	John Turner	*Prince of Wales* [I]
	James Stirling	HMS *Brazen*

From London. *Eddystone* visited Moose Factory 4 - 27 September, and reached London 27 November. *Prince of Wales* visited Churchill 19 August - 20 September, York Factory 26 September - 6 October, where she landed colonists bound for the Red River settlement, and reached London 26 November. *Brazen* escorted the two ships to and from Hudson Bay.

HUDSON'S BAY COMPANY (1) and (2).

1813

Establishment of Great Whale River post — George Atkinson

Atkinson, in charge of Fort George (see 1803), had conducted several trading expeditions to Great and Little Whale rivers (see 1793). He now received instructions to evacuate Fort George and to remove to Great Whale River. The new post did not entirely replace Fort George, which was reopened in 1816. Whaling was carried on at Great Whale River post for a time, but with little success, and a vessel was sent each year to whale at Little Whale River. In 1824, Fort George, then the headquarters of the district, was closed, and Great Whale River was presumably closed as well.

DAVIES, 1963, pp. 284-88.

1814

Hudson's Bay Company supply voyage	Henry Hanwell	*Prince of Wales* [I]
	John Turner	*Eddystone*
	Edward Stopford	HMS *Rosamond*
	Edward Chappell	

From London; 15 May - 25 November. *Prince of Wales,* carrying passengers bound for the Red River settlement, and her escort *Rosamond,* visited York Factory 2 - 28 September. On arrival, they learned that the Churchill post had burned in November 1813. They returned to London on 25 November. *Eddystone* visited Moose Factory 3 - 20 September, then returned to London.

HUDSON'S BAY COMPANY (1) and (2);*CHAPPELL, 1817; PUBLIC RECORD OFFICE, ADM. 51/2741/10 [log of HMS *Rosamond*].

1815

Abandonment of Mackenzie River district

Prior to 1815, the North West Company's trade in the Mackenzie district declined, for reasons that W. F. Wentzel ascribed to mismanagement. In 1815, no longer able to meet the expenses of the district, the company evacuated all its posts: Fort of the Forks (see 1803), Fort Good Hope (see 1804), Great Bear Lake post (see c1800), and Fort Norman (see c1810). Fort Nelson (see 1805, Fort Liard) had already been destroyed by Indians in 1813. The Hudson's Bay Company had no posts in Mackenzie River district, so there were no European traders in the region at all for a short time, a fact that caused hardship and anger among the Indians. In 1816, however, Wentzel canoed as far as Fort Good Hope with a small supply of trade goods, and promised the Indians that he would try to have the posts re-established. In 1817, Charles Grant made a similar trading voyage downstream and obtained good returns. He possibly established Fort Alexander during this journey (see 1817). By 1818, the company had returned to the region and, by 1820, three posts operated: Fort Alexander, Fort Good Hope, and Fort Nelson. Fort of the Forks and Fort Norman—but not Great Bear Lake post—were reoccupied in or about 1821.

WENTZEL, 1821; STAGER, 1962, p. 41; MASSON, 1889, Vol. 1, p. 123.

1815-16

Hudson's Bay Company supply voyage

Henry Hanwell — *Prince of Wales* [I]
John Turner — *Eddystone*
John Davison — *Hadlow*
Captain Andrews — HMS *Dee*

From London; 25 May 1815 - (?) 1816. *Prince of Wales* visited York Factory 27 August - 23 September, and reached London 22 November 1815. The convoy ship *Dee* returned to London with her. *Eddystone* visited Moose Factory 23 August - 24 September. She was obliged to turn back during her return voyage, and wintered at Strutton Sound. She set off again for London on 12 August 1816. *Hadlow* visited York Factory 26 August - 7 September and Moose Factory 14 - 24 September. She, too, turned back during her return voyage and wintered at Strutton Sound. She left for London on 26 July 1816.

HUDSON'S BAY COMPANY (1) and (2).

1815-16

Hudson's Bay Company expedition to Lake Athabasca; establishment of Fort Wedderburn

John Clarke

The Hudson's Bay Company's first attempt to break the North West Company's monopoly of the fur trade in the Athabasca district had failed (see 1805-06, Fidler). Colin Robertson, a former North West Company trader, had suggested to the Hudson's Bay Company, in 1810 and after, that they could successfully compete with their rivals in the Athabasca region using French Canadian canoemen and a shorter supply route from Hudson Bay. The company now decided to act on his suggestion. While Robertson was occupied at Red River, they sent Clarke with a body of men from Cumberland House to Lake Athabasca. They built Fort Wedderburn on Coal Island, at the western end of the lake and near Fort Chipewyan, and other posts at Great and Lesser Slave lakes, Green Lake, and Peace River. Clarke mistakenly dispersed his men during the winter, leaving them short of provisions, and at Peace River 16 died of starvation. Aulay McAulay at Great Slave Lake was forced to surrender his post after only a few weeks' operation for want of provisions. Fort Wedderburn and Lesser Slave Lake post were little affected by starvation and, when he returned south in 1816, Clarke was convinced the campaign could succeed and he returned to continue it (see 1816-17).

RICH, 1959, pp. 333-34.

Hudson's Bay Company exploration of Great Whale River	George Atkinson John Stewart		1816

From Great Whale River post (see 1813); 11 June - 12 July; the first step in the company's long-discussed plan to explore the interior of the Quebec-Labrador peninsula. They canoed some 560 km up Great Whale River to a point some distance beyond Lake Bienville. They turned back on 5 July and gave a depressing report of an uninhabited country that was destitute of fish, beaver, and other game.

DAVIES, 1963, pp. xxxix, 338.

Establishment of Nichicun post	John Isbister Thomas Isbister		1816

When reports reached Neoskweskau (see 1793) in 1813-14 that Indians in the region of Nichicun Lake were willing to trade with the company, James Clouston, chief at Neoskweskau, determined on establishing an outpost there. The two Isbisters left Neoskweskau with six Indians on 15 August 1816, built a house at the southeast corner of Nichicun Lake, and reported in the following year that the new post promised a fair trade. Clouston surveyed the route between Neoskweskau and Nichicun in August 1817. The route, by Eastmain River, was extremely difficult, and the company tried, without success, to find an easier and cheaper route. John Isbister remained in charge of the post until 1822, when it was closed. It was reopened in 1834 (*qv*).

DAVIES, 1963, pp. 313-16.

Hudson's Bay Company supply voyage	Henry Hanwell Benjamin Bell	*Prince of Wales* [I] *Emerald*	1816-17

From London; 12 May 1816 - 8 November 1817. *Prince of Wales* visited York Factory 22 September - 3 October 1815. During her return voyage, she was obliged by ice conditions to turn back, and wintered at Charlton Island. She left there on 3 August 1817, visited Moose Factory, then sailed for London, where she arrived 8 November 1817. *Emerald* visited Moose Factory 20 September - 4 October 1816. She wintered with *Prince of Wales* at Charlton Island, left there 3 August 1817, visited Moose Factory, and arrived at London 19 October.

HUDSON'S BAY COMPANY (1) and (2).

Hudson's Bay Company expedition to Lake Athabasca	John Clarke		1816-17

Clarke returned to Fort Wedderburn with 80 men to continue the campaign to win control of the Athabasca fur trade (see 1815-16). On arrival he found that A. N. McLeod of the North West Company had arrested the small party of men he had left there in summer 1816. A scuffle ensued, and Clarke himself was arrested, but later released. In January 1817, the North West Company servants captured Fort Wedderburn and arrested Clarke again. They also captured the rival post under François Decoigne at Peace River and the post at Green Lake. Clarke was taken to Great Slave Lake, then Fort Vermilion, arriving in September.

RICH, 1959, pp. 341-42.

1817	British whaling voyage	Captain Muirhead ?	*Larkins* *Elizabeth*

British whalers had been making voyages to Davis Strait for many years before 1817 but, in general, they had kept close to the Greenland coast, and few or none had penetrated into the high latitudes of Baffin Bay or crossed to the Canadian shore. These two ships, *Larkins* from Leith and *Elizabeth* from Aberdeen, followed the Greenland coast to 77°N (75°15'N, according to Ross), crossed to the Canadian shore, then they apparently returned along the Greenland shore. Only after John Ross's expedition of 1818 (*qv*) did the whalers adopt the routine of sailing north into Baffin Bay along the Greenland coast, then returning south along the Canadian coast.

LUBBOCK, B., 1937, p. 204; ROSS, JOHN, 1819, p. 48.

c1817	Establishment of Fort Alexander (Willow Lake River post) on Mackenzie River	Charles Grant?	

This post was apparently built for the North West Company in 1817, possibly by Grant, who made a trading journey down Mackenzie River in 1817 and was the only European known to have been on the river that year (see 1815). On Wentzel's 1821 and 1822 maps, it is on the left bank of Willowlake River, just about its mouth, but recent archaeological investigations have shown that its actual situation was on the right bank of the river, about 3 km upstream from the confluence. Wentzel called it "Willow Lake river" post and listed it as one of the three posts on the Mackenzie River operating in 1821. But it was apparently abandoned in that year in favour of Fort of the Forks (see 1803), then just reopened.

GORDON, B.C., 1973; RICH, 1938, pp. 394-95; STAGER, 1962, p. 40; WENTZEL, 1821; 1822 (map).

1817-18	Hudson's Bay Company supply voyage	John Edman John Davison	*Britannia* *Eddystone*

From London; 9 May 1817 - (?) 1818. *Eddystone* visited Moose Factory 26 August - 14 September, and reached London 24 November 1817. *Britannia* visited York Factory, left there 1 October 1817, but soon met extremely bad weather and heavy ice conditions. Edman tried to return to Churchill, then to Charlton Island, but was finally driven to the shore near Severn River, where the ship was frozen in. In April 1818, while the men were unloading cargo, the ship was destroyed by fire. Edman was taken by boat to York Factory and returned to London later in 1818.

HUDSON'S BAY COMPANY (1) and (2).

Hudson's Bay Company supply voyage	John Davison Benjamin Bell Captain Rennie	*Prince of Wales* [I] *Eddystone* *Levant*	1818

From London; 10 May - November. *Prince of Wales* visited York Factory 14 August - 8 September, and reached London 14 October. *Eddystone* visited Moose Factory 17 August - 11 September, and reached London 11 November. *Levant,* which was towed by *Prince of Wales* for part of the outward voyage, visited York Factory 15 August - 8 September, then returned to London. The ships were delayed by ice in Hudson Strait during the outward voyage.

HUDSON'S BAY COMPANY (1) and (2).

British naval Northwest Passage expedition	John Ross William Edward Parry	HMS *Isabella* HMS *Alexander*	1818

From London; 18 April - 16 November; sent by the Admiralty to search for a passage by way of Baffin Bay. Ross's ships passed Kap Farvel on 26 May, and followed the Greenland coast of Davis Strait, accompanied, in the early stages of the passage north, by a fleet of about 40 whalers. They passed through and named Melville Bay, then, in the region of Thule, located a tribe of Eskimos who had never before seen white men. Their next task was to examine Smith, Jones, and Lancaster sounds, discovered by Bylot and Baffin (see 1616). Ross made only a brief examination of Smith Sound, named the capes on each side of its entrance after his two ships, and decided that it offered no opportunity of a passage. He then turned south-west towards Jones Sound, discovered Coburg Island at its entrance, but, deciding that the sound was only a small inlet enclosed by mountains, made no attempt to explore it. He entered Lancaster Sound on 30 August, but soon after decided that it, too, was enclosed by mountains. Parry did not see the mountains, and his belief that Lancaster Sound offered the best opportunity of a passage to the west resulted in his being sent to examine it again, without Ross (see 1819-20). Returning out of Lancaster Sound, Ross made two landings on Bylot and Baffin islands, followed the coast of Baffin Island southward, but made no attempt to examine Cumberland Sound (discovered by Davis in 1585 [*qv*]) or Frobisher Bay (see 1576). They then returned to England with meteorological and magnetic observations and zoological and geological collections. The voyage confirmed the discoveries of Bylot and Baffin, which had been removed from many 18th century maps as unfounded, and also encouraged whalers to follow his route to the North Water of Baffin Bay and to fish along the Baffin Island coast.

*†ROSS, JOHN, 1819; *FISHER, A., 1819; SABINE, †1819a, †1819b, 1819c; DODGE, 1973, pp. 43-93.

1818

Hudson's Bay Company exploration of Little Whale River — George Atkinson

From Fort George; 19 June - 4 September; sent to explore Little Whale River. Accompanied by a band of Indians, he followed the coast from Fort George northward to Richmond Gulf, then travelled inland by way of Clearwater River and Lake to the region of Lac d'Iberville and Upper Seal Lake. Here he met Indians who told him there were no other traders in the region and that the lakes emptied into Ungava Bay. He continued north-westward for a short distance, but had to turn back on 13 August, when his Indian guide refused to go on. He then travelled south overland to the source of Little Whale River, and surveyed it to the sea. He reached Fort George on 4 September.

DAVIES, 1963, pp. xl-xli, 339-40.

1818-19

Hudson's Bay Company expedition to Lake Athabasca — Colin Robertson, John Clarke

Following Clarke's two abortive attempts to break the North West Company's monopoly of the Athabasca district (see 1815-16 and 1816-17), Robertson made a third attempt. He left Norway House with 26 officers and 160 men in 27 canoes. Their main object was Lake Athabasca, but men were also sent to Ile-à-la-Crosse and Lesser Slave Lake. Clarke took charge of Peace River. The party reached Fort Wedderburn on 17 September. They were again mistreated by the North West Company men, who took Robertson prisoner. Even so, he managed to carry on a fairly successful trade, until he was taken south in June 1819. He escaped at Cumberland House. Meanwhile Clarke had had two posts built in the Peace River area: Joshua Halcro built Colvile House on Loon River, and Jaques Chastellaine built Fort St Mary on Smoky River. These successes gave the company a firm foothold in the Athabasca district, and dealt a severe blow to the North West Company.

RICH, 1959, pp. 347-52.

1819

Establishment of Waswanipi post

News of the success of the North West Company's post on Waswanipi Lake (see c1799) occasionally reached the Hudson's Bay Company, who became eager to establish their own post there. In September 1817, they sent John Pitt Greely inland from Rupert House to survey the route to Waswanipi. Greely wintered at Lake Evans, where he and his party died of starvation. In 1818, another party was sent inland to build a post on Lake Evans. This post was known as Big Lake House, and James Clouston wintered there before continuing south to survey the route to Waswanipi Lake in 1819. After a brief visit to the North West Company post there, he returned to Rupert House. Soon after, a party led by Patrick Flynn left Rupert House to build a post at Waswanipi. This post soon became district headquarters for Big Lake House and Cheashquacheston (see 1799). After the amalgamation in 1821, the Hudson's Bay Company post on Waswanipi Lake was apparently abandoned in favour of the North West Company post. Big Lake House was abandoned in 1822; Waswanipi has continued in operation until the present day.

DAVIES, 1963, pp. 318-26.

Establishment of Fort Resolution | Robert McVicar | 1819

The North West Company had operated a post on the south shore of Great Slave Lake since 1786 (*qv*), and Aulay McAulay had opened the Hudson's Bay Company's first post there in 1815-16 (*qv*), but it had lasted only a few weeks. He again tried unsuccessfully in 1818-19. In 1819, however, Robert McVicar succeeded in building Fort Resolution, and was reported to have 19 men there. The fort was situated about 8 km west of Slave River delta, very close to the North West Company's post on Moose Deer Island, which, however, was abandoned in favour of Fort Resolution after the coalition in 1821. McVicar was in charge of Fort Resolution almost continuously until 1827. The fort has remained in about the same place to the present.

RICH, 1938, p. 416.

Establishment of Harrisons House | George Peter Andries | 1819

Andries built Harrisons House for the Hudson's Bay Company near the eastern end of Lake Athabasca. The North West Company had had a post near there previously, but had abandoned it after some of their men had been murdered by Indians. They re-established their post there in 1819, but abandoned it in 1821. It is not known how long the Hudson's Bay Company post at Harrisons House operated after 1821.

RICH, 1938, pp. 416-17.

Establishment of Berens House | Joseph Greill | 1819

Greill built Berens House for the Hudson's Bay Company on the right bank of Athabasca River, near present-day Fort MacKay, and close to the North West Company's Pierre au Calumet post. Aulay McAulay took charge of Berens House in autumn 1819, but was forced to abandon it during the winter because of starvation. It was re-established in 1820. It was still in existence in 1825, but it is not known how much longer it remained.

RICH, 1938, pp. 413-14.

1819-20	Hudson's Bay Company supply voyage	John Davison Benjamin Bell Thomas Thompson	*Prince of Wales* [I] *Eddystone* *Wear*

From London; 21 May 1819 - November 1820. The ships met very heavy seas on the outward voyage. *Prince of Wales,* carrying John Franklin's Arctic expedition (see 1819-22) and German settlers for Red River, struck the rocks in thick fog off Resolution Island, and was pumped constantly until she reached York Factory, where she was repaired. She remained at York Factory 30 August - 27 September, and reached London 30 October 1819. *Eddystone* visited Moose Factory 29 August - 19 September, and reached London 7 November 1819. *Wear* visited York Factory, then sailed to Moose Factory, carrying North West Company prisoners, whom she was to carry to Quebec. She remained at Moose 20-30 September, left for England, but was unable to penetrate the ice in Hudson Strait and wintered at York Factory. In 1820, she visited Quebec, and reached London in November.

HUDSON'S BAY COMPANY (1) and (2).

1819-20	Hudson's Bay Company overland exploring expedition	James Clouston

From Rupert House; 29 September 1819 - 22 August 1820; sent to explore the interior of the Quebec-Labrador peninsula, then virtually unknown. He ascended Rupert River to the company's post, Neoskweskau, where he wintered until 6 March 1820. He then left for the company's house (see 1816) on Nichicun Lake, arriving on 19 March. He set out again on 17 April to explore northward. By early June he had reached Lake Kaniapiskau, known to him by Indian report. Attempting to continue north, he found Kaniapiskau River scarcely navigable and decided to make a long detour eastward, following a route that cannot now be identified, and rejoined Kaniapiskau River about 60 km above its junction with Larch River and about 160 km from Ungava Bay. Here he turned back. He ascended Kaniapiskau River to Châteauguay River, then followed the latter westward to Lac Châteauguay and continued westward by an unknown route to Clearwater Lake. He then crossed to Hudson Bay at Richmond Gulf and coasted south to Rupert House, completing the longest journey until then made through the peninsula.

DAVIES, 1963, pp. 29-68.

1819-20	British naval Northwest Passage expedition	William Edward Parry Matthew Liddon	HMS *Hecla* HMS *Griper*

From London; 12 May 1819 - mid-November 1820. Parry and John Ross had disagreed about the prospect of finding a Northwest Passage through Lancaster Sound after their voyage of 1818 (*qv*), so Parry was sent by the Admiralty to explore the sound again. The ships followed Ross's route northward through Davis Strait and Baffin Bay, keeping close to the Greenland coast, then crossed to the western shore near Lancaster Sound. Entering the sound, Parry kept close to the Devon Island coast until Maxwell Bay, where he turned south to explore Prince Regent Inlet. He got almost to Cape Kater, where ice conditions forced him back. He again followed the northern shore of Lancaster Sound westward, and discovered the south coasts of Cornwallis, Bathurst, Byam Martin and Melville islands. On 6 September, he crossed 110°W, thus winning a reward of £5000 offered by a King's order in council. Ice conditions prevented his continuing, so he turned back and entered winter quarters at Winter Harbour on the south coast of Melville Island on 26 September. On 1 June 1820, he set out northward across Melville Island with a party of 12 men pulling a cart and, during this journey, explored the south shore of Hecla and Griper Bay, Sabine Bay, and the eastern end of Liddon Gulf. He returned to Winter Harbour on 15 June. He left the harbour with the ships on 1 August, and tried to continue westward, but was again stopped by ice at the south-western point of Melville Island, where he reached his farthest west, 113°48'W. From here, he sighted land to the south-west, which he named Banks Island. Returning eastward towards Baffin Bay, he followed the southern shore of Barrow Strait and Lancaster Sound, charted the north coast of Somerset Island, and discovered the entrances to Admiralty and Navy Board inlets. He then followed the west coast of Baffin Bay southward and returned to England. He carried back magnetic and meteorological observations and zoological and geological collections.

*†PARRY, W.E., 1821; *ANONYMOUS, 1821; *FISHER, A., 1821; SABINE, 1821; †SABINE AND OTHERS, 1824; PARRY, A., 1963, pp. 44-70.

1819-22

British Arctic Land Expedition

John Franklin
John Richardson
George Back
Robert Hood

From London 23 May 1819 - autumn 1822; sent by the Admiralty to explore the north coast of America eastward from the mouth of Coppermine River to Hudson Bay. They sailed in *Prince of Wales* (see 1819-20) to York Factory, where they were advised by officers of both the Hudson's Bay and North West companies to winter at Cumberland House, then to travel to Coppermine River by way of Athabasca River and Lake, Slave River, and Great Slave Lake. They reached Cumberland House on 22 October. Franklin and Back set out on 18 January 1820 on snowshoes for Fort Chipewyan, leaving Richardson and Hood to follow with the supplies and equipment later in spring. On 18 July 1820, after Richardson and Hood had caught them up, they continued down Slave River and across Great Slave Lake to Fort Providence, where W. F. Wentzel of the North West Company was in charge. Wentzel undertook to accompany the expedition to the Arctic coast and he recruited Indians to guide them to Coppermine River and to hunt.

They followed Yellowknife River north into unknown country and eventually reached Winter Lake, where they built Fort Enterprise and wintered. Two Eskimos sent from Hudson Bay to serve as guides and interpreters now joined them, and Back made a snowshoe journey back to Fort Chipewyan to get more supplies. Having made a preliminary survey of the route from Fort Enterprise to Point Lake on Coppermine River, they left the fort on 14 June 1821, descended the difficult river, and reached the sea on 18 July, thus verifying Hearne's discovery of the Arctic coast (see 1770-72). Wentzel and the Indians now left them, and they continued eastward along the coast in two canoes, charting and naming features in Coronation Gulf, Bathurst Inlet, and Melville Sound. At Turnagain Point, on the north shore of Kent Peninsula, they turned back because of short provisions, unrest among the French Canadian boatmen, and the approach of winter. The canoes were too badly damaged to retrace the coast, so they returned to Fort Enterprise overland. They ascended Hood River, then went overland past Kathawachaga Lake to Contwoyto Lake (another of Hearne's discoveries), past Point Lake to the fort. During this journey, Hood was murdered by an Iroquois member of the party, whom Richardson then executed. About 10 of the boatmen died of starvation. They found the fort bare of provisions and had to seek help from Indians. On 11 December 1821, they reached Fort Providence, but soon moved on to Moose Deer Island post on Great Slave Lake. In spring 1822, they returned to York Factory, then to London in the Hudson's Bay Company's supply ship. In addition to making a careful survey of their route, they made extensive scientific observations and collections of geological, botanical, and zoological interest, recorded meteorological and magnetic data, and studied the aurora borealis.

*†FRANKLIN, J., 1823; MACLEOD AND GLOVER, 1971; RICH, 1938, passim; MASSON, 1960, Vol. 1, pp. 130-50; NEATBY, 1958, pp. 56-81; LEA, 1917-18; †HOOKER, W.J., 1829-40; HOUSTON, *1974; 1974a.

1820

Hudson's Bay Company supply voyage	John Davison Benjamin Bell Robert Knox	*Prince of Wales* [I] *Eddystone* *Luna*

From London; 12 May - (?). *Prince of Wales* visited Moose Factory 6 August - 11 September, and reached London 26 October. *Eddystone* visited York Factory 15 August - 9 September, and reached London 26 October. She carried out to York Factory John West (see 1820-23) and some Red River colonists. *Luna* visited York Factory 15 August - 3 September, called at Quebec, then returned to London.

HUDSON'S BAY COMPANY (1) and (2).

1820

British naval voyage to the Labrador coast	Hercules Robinson	HMS *Favorite*

From St John's; 18 June - September; *Favorite,* a fisheries protection vessel on the Newfoundland station, sailed to examine the state of the fisheries on the Labrador coast. Robinson reached Niger Sound on 22 June, then coasted north as far as Black and Grady islands, reaching his farthest north at about 53°46'N. He then turned back and followed the coast southward as far as L'Anse-au-Loup, crossed to Conception Bay, Newfoundland, then returned to St John's. He made notes on weather, physiography and fisheries along the Labrador coast, and reported that 530 vessels were active in the Labrador fisheries that year, of which 49 were operating north of Cape Charles and as far north as Sandwich Bay.

ROBINSON, H., 1834.

1820

British whaling voyages	Captain Johnson	*Cumbrian*
	Captain Cooper	*Earl Fauconberg*
	Captain McBride	*Friendship*
	Captain Hurst	*Ariel*
	?	*Truelove*
	?	*Duckinfield Hall*

This was the first small fleet of whaling vessels to follow the example set by *Larkins* and *Elizabeth* in 1817 (*qv*) by sailing northward beyond the usual whaling grounds off the south-west coast of Greenland to the North Water of Baffin Bay and then to cross to the Canadian coast. *Cumbrian* of Hull sailed about 130 km into Lancaster Sound, observed that the sound continued west at least 30 km farther, but saw no trace of Parry's expedition (see 1819-20), which was wintering farther up the sound at that time. Johnson returned to Hull in September 1820, a month before Parry's return, so he was the first man to report that Lancaster Sound was indeed an open strait and not, as John Ross had stated in 1818 (*qv*), a closed inlet. Cooper, in *Earl Fauconberg* of Grimsby, was also near the entrance to Lancaster Sound, but apparently did not enter it. When returning southward along the east coast of Baffin Island, he noticed recently inhabited Eskimo huts on the shore in about 73°20'N, probably on the east coast of Bylot Island. The other four ships, all from Hull, also fished off the Canadian coast in northern Baffin Bay. The whole fleet probably followed Ross's example of 1818 and returned south along the Baffin Island coast, rather than the Greenland coast. If so, they were the first whaling ships to follow this anti-clockwise course around Baffin Bay, a course that became routine for whalers in the years following.

LUBBOCK, B., 1937, p. 217; *LIVERPOOL MERCURY*, 1820.

1820-21

Arrival of George Simpson

In August 1818, William Williams arrived at York Factory as the new Governor-in-Chief of Rupert's Land. Angered by news of Robertson's treatment at Fort Chipewyan (see 1818-19), he resolved on direct action and, with a party of armed men, intercepted a canoe brigade of the North West Company at Grand Rapid on 21 June 1819 and arrested some of that company's principal traders. The ensuing legal disputes made the Governor and Committee of the Hudson's Bay Company fear that Williams might be sent to England for trial. Therefore, on 26 February 1820, they appointed George Simpson, an accountant in London whose shrewdness and integrity they valued, to serve as Governor-in-Chief *locum tenens* should Williams be removed. Probably only Williams, in Canada, knew of Simpson's special mission and, as he was not removed, he sent Simpson to replace Robertson, who had been arrested, and to take charge of Fort Wedderburn. During winter 1820-21 and despite the gravest difficulties, Simpson managed to improve the trade of the district and, in the shortest possible time, seems to have grasped the whole business of the fur trade, knowledge that was very soon to prove of great advantage to himself and to his employers.

RICH, 1959, pp. 371-75.

1820-23

Church Missionary Society expedition to Hudson Bay	John West	

West, a clergyman in the Church of England, was the first Protestant missionary to visit the shores of Hudson Bay (see 1671-72 and 1674-75). With support from the Church Missionary Society and the Hudson's Bay Company, he sailed to York Factory in *Eddystone* (see 1820) intending to establish a mission in the Red River colony. He arrived on 15 August and stayed some time to preach at York Factory before continuing to Red River, and he returned to the post for the same reason during the next two summers. In 1822, he met Franklin's party on its way to England (see 1819-22). Franklin strongly urged him to extend his missionary work north to the Eskimos at Churchill and, when West returned to York Factory in summer 1823, he did journey on foot to Churchill to open a temporary mission. On 11 September, he left York Factory in *Prince of Wales* (see 1823) for England. He seems to have intended to return to Hudson Bay, but never did so.

BOON, 1962, pp. 1-21; GOULD, 1917, pp. 83-97.

1821

Hudson's Bay Company supply voyage	John Davison	*Prince of Wales* [I]
	Benjamin Bell	*Eddystone*
	James Fulbister	*Lord Wellington*

From London; 22 May - 8 November. *Prince of Wales* visited York Factory 22 August - 14 September. *Eddystone* visited Moose Factory 19 August - 16 September. Both ships reached London 8 November. *Lord Wellington* was chartered by the executors of Lord Selkirk to carry Swiss settlers from Rotterdam to York Factory, *en route* to Red River. She also carried some Hudson's Bay Company cargo to York Factory. She remained there 18 August - 1 September, and returned to London.

HUDSON'S BAY COMPANY (1) and (2).

Coalition of the North West Company and the Hudson's Bay Company

1821

On 26 March 1821, representatives of both companies agreed in London to unite "into one Concern, to be from the 1st day of June next carried on solely by and in the name of the Hudson's Bay Company." The management was to be shared in favour of the older company, although the chief officers in the fur trade itself were mainly chosen from among the wintering partners of the North West Company. The Hudson's Bay Company's charter was once again vindicated and the company's trade was greatly extended, for the Government authorized a 21-year licence "for the sole and exclusive privilege of trading with the Indians" over all of British North America outside of the Canadian provinces and of Rupert's Land itself.

MORTON, A.S., 1939, pp. 623-25; RICH, 1938, pp. l-li; 1959, pp. 385-400; 1967, p. 240.

British naval North-west Passage expedition	William Edward Parry George Francis Lyon	HMS *Fury* HMS *Hecla*

1821-23

From London; 29 April 1821 - mid-October 1823; sent by the Admiralty to search for a passage along the west coast of Foxe Basin, northward from Repulse Bay. Parry sailed north of Southampton Island instead of south, the route preferred by previous navigators, and thus became the first to sail through Frozen Strait and confirmed Middleton's discovery of Repulse Bay (see 1741-42). In late August, having found no passage through Repulse Bay, he explored the coast northward and named, among other features, Haviland Bay, Bushnan and Vansittart islands, Gore Bay, and Lyon and Hoppner inlets. He examined the last two inlets by boats, sailing as far as Ross Bay.

On 1 February 1822, some Eskimos visited the ships' winter quarters on Winter Island and spent much of the winter in the expedition's company, enabling Parry and his men to make detailed observations on their customs and language. On 2 July, they left winter quarters and explored north along the east coast of Melville Peninsula equipped with an Eskimo map of the region. They met other Eskimos at Iglooik and, farther north, found the entrance to Fury and Hecla Strait. Ice prevented their passing through it, but overland journeys confirmed that it led westward to a sea. Having discovered a route that promised to provide a much shorter North-west Passage than Lancaster Sound (see 1819-20), Parry returned to winter quarters at Igloolik in October and stayed there until 12 August 1823, again with the company of Eskimos. A second attempt to sail through Fury and Hecla Strait was unsuccessful and, fearing scurvy, they abandoned the effort and returned home.

†*PARRY, 1824; *LYON 1824; ROSS, W.G., 1960; DODGE, 1961, pp. 263-67; PARRY, A., 1963, pp. 71-82.

1822

Hudson's Bay Company supply voyage | John Davison, Benjamin Bell | *Prince of Wales* [I], *Eddystone*

From London; 27 May - 15 November. *Prince of Wales* visited York Factory 10 August - 7 September, and, carrying members of Franklin's expedition (see 1819-22), returned to London on 21 October. *Eddystone* visited Moose Factory 13 August - 12 September, and reached London 15 November.

HUDSON'S BAY COMPANY (1) and (2).

1822

First crossing of Newfoundland | William Epps Cormack

From Trinity Bay; early September - early November. Cormack walked across the island from Random Island, Trinity Bay, east to St George's Harbour, St George's Bay, with one Micmac Indian.

*CORMACK, 1824; 1928.

1823

Hudson's Bay Company supply voyage | John Davison, Benjamin Bell | *Prince of Wales* [I], *Eddystone*

From London; 16 May - 5 November. *Prince of Wales,* carrying a party of Scottish settlers bound for Red River, visited York Factory 15 August - 11 September, and reached London 25 October. *Eddystone* visited Moose Factory 18 August - 16 September, and reached London 5 November.

HUDSON'S BAY COMPANY (1) and (2).

1823

Hudson's Bay Company expedition to South Nahanni River | Alexander R. McLeod

From Fort Simpson; 9 February - 6 March; to locate and to establish trading relations with the Nahanni Indians, known to the company only by Slave Indian report. His report of the expedition is lost. He probably ascended Liard River to the mouth of South Nahanni River. He suffered considerable hardship, having no experienced hunters with him, and he failed to meet the Nahannis.

PATTERSON, 1961, p. 40.

1823

Hudson's Bay Company expedition to South Nahanni River | John M. McLeod

From Fort Simpson; 5 June - 10 July; a second attempt to make contact with the Nahanni Indians, following A. R. McLeod's failure earlier in the year (*qv*). He canoed up Liard River with a party of Indians, and entered South Nahanni River on 10 June. Soon after, finding navigation too difficult, he continued his journey westward on foot. His route is difficult to trace, for he recorded his progress only by the number of mountain ranges he had crossed. He turned back after the ninth range, when he was probably near the banks of Caribou River. He retraced his steps to the fourth range of mountains, where he finally met some Nahanni Indians on 3 July. He promised to come to them the following year with trade goods (see 1824, McLeod), then returned to Fort Simpson.

*PATTERSON, 1961; MORTON, 1939, pp. 705-06.

Hudson's Bay Company expedition to Liard River **1823**

The company intended to send P. W. Dease to explore Finlay River (see 1797) to its source, but the expedition was abandoned until 1824 (*qv*, Black). News of this change did not reach Fort Liard before a party of seven Indians had set off to meet Dease. They left in June, reached the "Main West Branch" of Liard River by an indirect route, followed it for eight days, then turned back to reach Fort Liard on 19 August. They failed, also, to find any Nahanni Indians, whom they were to persuade to return with them to trade.

RICH, 1955, p. 171, note.

Massacre of Chipewyan Indians **1823**

The Dogrib Indians, who had long suffered from arrogant treatment by the Chipewyan or Copper Indians, plotted to destroy the whole tribe before the traders could stop them. They ambushed seven families after freeze-up between Great Bear and Great Slave lakes, killing all but four young men, who escaped to report the event. The traders seemed generally to have been sympathetic to the Dogribs, but they quickly ended the feud. Hostility between the tribes affected the fur trade during the next several years.

HUDSON'S BAY COMPANY, ARCHIVES, B200/a/3/35-36; B200/a/4/1-3d.

Hudson's Bay Company supply voyage	John Davison Benjamin Bell	*Prince of Wales* [I] *Camden*	**1824**

From London; 28 May - November. *Prince of Wales* visited York Factory 20 August - 12 September, and reached London 1 November. She carried stores, boats, and men for Franklin's second expedition (see 1825-27). *Camden* visited Moose Factory 29 August - 22 September, and reached London 9 November.

HUDSON'S BAY COMPANY (1) and (2).

British naval exploring expedition	George Francis Lyon	HMS *Griper*	**1824**

From London; 10 June - 10 November; to sail to Repulse Bay, cross Melville Peninsula, and explore the north coast of America westward to Turnagain Point on Kent Peninsula, the most easterly point reached by Franklin in 1821 (see 1819-22). They landed on Coats Island (then thought to be part of Southampton Island)

and reported on the Eskimos whom they met there. In Roes Welcome Sound, they were twice nearly shipwrecked and were forced to turn home before reaching Repulse Bay. They reached Portsmouth on 10 November.

*†LYON, 1825; *PATON, 1825; DODGE, 1961, pp. 268-69; DUNBABIN, 1963.

1824

Hudson's Bay Company expedition to South Nahanni River — John M. McLeod

From Fort Simpson; 8 June - 1 July; to fulfil his promise of 1823 (*qv*) to trade with the Nahanni Indians. He canoed up Liard River, reached the mouth of South Nahanni River on 12 June, and, on 15 June, set off westward overland. On 18 June, he rejoined Nahanni River at Deadmen's Valley and camped there five days. He then searched for Indians, whom he found near Meilleur Creek. They had little to trade, but McLeod persuaded their chief to go with him back to Fort Simpson. There the chief described his country and promised to return with furs in spring 1826.

PATTERSON, 1961, pp. 44-47.

1824

Hudson's Bay Company expedition to Liard and Beaver rivers — Murdoch McPherson

From Fort Liard. McPherson left on 7 July, ascended Liard River to the mouth of Beaver River, then followed the latter until 18 July, when he turned back to reach Fort Liard on 22 July. This is the first known ascent of Liard River by a European above the mouth of Fort Nelson River.

RICH, 1955, p. 171, note.

1824

Hudson's Bay Company exploration of Finlay River — Samuel Black

From Rocky Mountain Portage House (Peace River); 13 May - 16 September; to explore Finlay River and extend trade, for the company feared that Russian traders might enter the area. Black began to ascend Finlay River on 22 May; on 1 June, he came to the forks of Finlay and Fox rivers, and followed the Finlay branch. He reached Thutade Lake on 23 June, and explored it to its western end. On 8 July, he cached his canoes at the eastern end of the lake, and set out northward on foot. He crossed Toodoggone, Chukachida, Pitman, and Tucho rivers, and reached Turnagain River on 16 August. He followed this eastward for a short distance, and reached his farthest point at 58°70'N, 128°W, on 17 August. He then retraced his steps to Rocky Mountain Portage House. He reported the river to be hazardous and Indians scarce, and the company did not expand into the area.

*RICH, 1955; WALLACE, J.N., 1928.

British naval North-west Passage expedition | William Edward Parry, Henry Parkyns Hoppner | HMS *Hecla*, HMS *Fury*

1824-25

From London; 8 May 1824 - October 1825; to reach the north coast of America by Prince Regent Inlet, discovered by Parry in 1819-20 (*qv*), then to follow the coast westward. The expedition was delayed by bad weather and ice, and reached Lancaster Sound on 10 September. They wintered at Port Bowen in Prince Regent Inlet 1 October 1824 - 20 July 1825. In early August, *Fury* was driven ashore by ice at Fury Point, Somerset Island, and abandoned. Both crews returned to England in *Hecla*. They made important meteorological, astronomical, geological, zoological, botanical and magnetical observations.

*†PARRY, 1826; MOGG, 1964; PARRY, A., 1963, pp. 83-93.

Hudson's Bay Company supply voyage | John Davison, Benjamin Bell | *Prince of Wales* [I], *Camden*

1825

From London; 30 May - 1 November. *Prince of Wales* visited York Factory 15 August - 2 September, and reached London 25 October. *Camden* visited Moose Factory 16 August - 18 September, and reached London 1 November.

HUDSON'S BAY COMPANY (1) and (2).

Anglo-Russian Treaty

1825

In 1821, Russia had attempted to exclude foreigners from the Bering Sea and Pacific coasts of Russian America by ukase. Both the United States and Great Britain protested, and Russia drew up a treaty with the former in 1824 and the latter in 1825. According to the convention agreed with the British, Russia held only a narrow strip of the Pacific north-west coast north of 54°40', but held all of the main part of Alaska and its islands west of 141°W. In the negotiations, the Russian Government was guided by representations of the Russian-American Company and the British Government by the Hudson's Bay Company. So matters stood at the Purchase of Alaska by the United States (see 1867). The boundary was not surveyed until 1903 (*qv*).

NICHOLSON, 1954, pp. 27-28, TOMPKINS, 1945; BAILEY, 1937; BILLMAN, 1971.

British Arctic Land Expedition | John Franklin, John Richardson, George Back, Edward Nicholas Kendall, Thomas Drummond

1825-27

From Liverpool; 16 February 1825 - 26 September 1827; to explore the north coast of America from the Mackenzie River delta: westward to Icy Cape, Alaska, where Captain Frederick William Beechey in HMS *Blossom* was expected to meet them (see

1825-28), and eastward to Coppermine River. Stores and men had already been sent to Cumberland House (see 1824, Davison); the officers sailed to New York on the packet-ship *Columbia* in 1825. From there, they travelled by canoe, boat, and land to Cumberland House, where they arrived 15 June. There, Drummond left the main party and set out for the Rocky Mountains, where he remained, making botanical and zoological collections, until 1827. The rest of the party continued over Methy Portage and down the Athabasca-Slave-Mackenzie river system to Great Bear Lake, where they wintered at Fort Franklin.

While the others built their house, Franklin and Kendall made a preliminary trip down Mackenzie River to observe sea-ice conditions at its mouth, and Richardson surveyed the north shore of Great Bear Lake eastward to the mouth of Dease River to facilitate his return from Coppermine River in 1826. They wintered at Fort Franklin until 22 June 1826, when they set out down Great Bear and Mackenzie rivers in four eight-metre boats, *Lion, Reliance, Dolphin*, and *Union*. On 4 July, the parties divided at Point Separation at the head of the Mackenzie delta. Franklin and Back went westward in *Lion* and *Reliance,* Richardson and Kendall eastward in *Dolphin* and *Union.* After meeting hostile Eskimos near the mouth of the delta on 7 July, Franklin and Back made slow progress westward, hampered by poor weather and ice conditions. They discovered and named many features along the coast, including Herschel Island, Camden Bay, and Prudhoe Bay. After six weeks, they were still only halfway to Icy Cape, so, with winter approaching, Franklin decided to abandon the planned rendezvous with *Blossom* and to turn back at Return Reef, 148°52'W. (He later learned that he was within 250 km of a boat sent from *Blossom* to meet him.) They arrived back at Fort Franklin on 21 September. In the meantime, Richardson and Kendall in the boats *Dolphin* and *Union* coasted eastwards from the Mackenzie delta, discovered Liverpool, Franklin and Darnley bays, part of Victoria Island, which they named Wollaston Land, Dolphin and Union Strait, and many other features, and reached the mouth of Coppermine River on 8 August, where they corrected Hearne's latitude (see 1770-72). They then travelled overland and by Coppermine and Dease rivers back to Great Bear Lake, and reached Fort Franklin on 1 September. During winter 1826-27, the whole party travelled back to New York, and reached Liverpool on 26 September 1827. During the expedition, they had made important observations on topology, geology, meteorology, solar radiation, the aurora, and magnetic variations.

†*FRANKLIN, 1828; †RICHARDSON, 1829-37; †BIRD, 1967; NEATBY, 1958, pp. 92-98; LEA, 1917-18; †HOOKER, W.J., 1829-40.

1825-28

British naval expedition to Bering Strait | Frederick William Beechey | HMS *Blossom*

From Spithead; 19 May 1825 - September 1828; to await the arrival of the expeditions of Parry (see 1824-25) and Franklin (see

1825-27) in Bering Strait. Beechey sailed around Cape Horn and, after some exploration in the Pacific, reached Kotzebue Sound, Alaska, on 22 July 1826. He continued north to Icy Cape, then sent his mate, Thomas Elson, north-eastward along the coast as far as Point Barrow in the hope of meeting, or finding some trace of one of the expeditions. Elson came within 250 km of the farthest point reached by Franklin in the same year. Beechey spent the winter of 1826-27 in the Pacific, then, in 1827, returned to Bering Strait. He again saw no sign of either expedition, so he set off homeward in October 1827, and reached England in September 1828.

*†BEECHEY, 1831; †HOOKER AND ARNOTT, 1841; HUISH, 1836; *GOUGH, 1973.

Hudson's Bay Company supply voyage	John Davison Benjamin Bell	*Prince of Wales* [I] *Camden*	1826

From London; 30 May - October. *Prince of Wales* visited York Factory 7 - 20 September, and reached London 26 October. *Camden* visited Moose Factory 30 August - 10 September. She then sailed to York to take on cargo from *Prince of Wales*, which had been damaged during the outward voyage, but the latter proved to be still seaworthy, and no help was needed. *Camden* reached London 16 October.

HUDSON'S BAY COMPANY (1) and (2).

Retirement of William Williams — 1826

William Williams, appointed Resident Governor of the Hudson's Bay Company's territories in 1818, Joint Governor with George Simpson in 1821, and Governor of the Southern Department in 1822, was forcibly retired in 1826, and replaced by George Simpson, who now assumed full authority over both Northern and Southern departments. Simpson was not, however, officially named Governor-in-Chief of the company's territories until 1839.

RICH, 1959, p. 438.

Hudson's Bay Company supply voyage	John Davison Benjamin Bell	*Prince of Wales* [I] *Prince Rupert* [V]	1827

From London; 28 May - 15 October. The ships encountered very heavy ice in Hudson Strait and Bay during the outward voyage. *Prince of Wales* visited York Factory 23 August - 15 September. *Prince Rupert* visited Albany 11 August - 9 September. Both ships reached London 15 October.

HUDSON'S BAY COMPANY (1) and (2).

Hudson's Bay Company supply voyage	Henry Hanwell Jr Benjamin Bell	*Prince of Wales* [I] *Prince Rupert* [V]	1828

From London; 3 June - 5 November. *Prince of Wales* visited Moose Factory 29 August - 17 September. *Prince Rupert* visited York Factory 17 August - 21 September. Both ships reached London on 5 November.

HUDSON'S BAY COMPANY (1) and (2).

1828

Hudson's Bay Company expedition to Ungava Bay	William Hendry George Atkinson	

From Moose Factory; 12 March - 6 September; to locate the mouth of Koksoak River and to observe tides before the sloop *Beaver* arrived later in the year. Hendry snowshoed from Moose to Eastmain and Moar Bay, where he met Atkinson on 9 April. On 1 June, they started north to Richmond Gulf by canoe. Crossing the gulf, they canoed up Deer River to Clearwater Lake, then continued along Lower Seal Lake and Larch River to Koksoak River. On 15 July, within sight of the sea, Hendry determined their position and made tidal observations. They then returned by the same route, and reached Richmond Gulf on 27 July. While waiting there in vain for the sloop to arrive, Atkinson made a short journey to Great Whale River, where he met more than 100 Indians. They then returned to their respective posts. Hendry arrived back at Moose Factory on 6 September, and learned that *Beaver* had not been sent north because she was not seaworthy.

DAVIES, 1963, pp. 69-99.

1829

Hudson's Bay Company supply voyage	Benjamin Bell Henry Hanwell Jr Robert Royal	*Prince Rupert* [V] *Prince of Wales* [I] *Montcalm*

From London; 3 June - November. *Prince Rupert* visited York Factory 11 August - 15 September, and reached London 22 October. *Prince of Wales* visited Moose Factory 31 August - 16 September, and reached London 7 November. *Montcalm* visited York Factory 9-25 August to take home any goods that *Prince Rupert* could not stow. She then sailed to St Lawrence River, and thence to England.

HUDSON'S BAY COMPANY (1) and (2).

1829

Establishment of Fort Halkett	John Hutchison	

From Fort Simpson. Hutchinson with four men left on 27 June and began building a post on the upper Liard River on 22 July. The original site of the fort is not known, but, in about 1832, it was moved to the mouth of Smith River. The new site was chosen partly at the request of local Indians, who were well pleased with the traders' arrival and left furs with them on account. The post was pillaged by an armed band of about 60 Nahanni Indians in summer 1836. Hutchison thought that Russian traders had encouraged them to the act. It was pillaged again during winter

1842-43, when the trader was off hunting. The post was abandoned in 1865 on the establishment of Fort Nelson.

HUDSON'S BAY COMPANY, ARCHIVES, A12/1/343; B200/b/5/fo 5 and 12; B200/c/1/fo 4-5; B200/b/19/fo 7d.

1829-33

British North-west Passage expedition | John Ross, James Clark Ross | *Victory*

From London; 23 May 1829 - October 1833; a private venture sponsored by Sir Felix Booth, to discover a North-west Passage by way of Lancaster Sound and Prince Regent Inlet. Despite considerable difficulty with their steam engine, they had a fairly easy passage through Baffin Bay, entered Lancaster Sound on 6 August and Prince Regent Inlet on the 11th. In the inlet, they visited the wreck of HMS *Fury* (see 1824-25 Parry), sailed past Bellot Strait without noticing it, and put into winter quarters at Felix Harbour in southern Boothia Peninsula.

1830: John and James Clark Ross made several journeys across Boothia Isthmus, and J. C. Ross discovered King William Island, which he then believed to be part of the mainland. (For further details of these journeys, see 1830, below.) The ship was icebound all summer, and they spent a second winter in Felix Harbour.

1831: The Rosses made more overland journeys, during one of which J. C. Ross located the North Magnetic Pole on 1 June. (For further details, see 1831, below.) During summer, they moved the ship only 25 km before she was again beset by ice, and they were forced to spend a third winter in the Arctic.

1832: Ross decided to abandon *Victory* during the winter and, in spring, they pulled their boats to Fury Point to make use of the provisions left behind by Parry in 1825, and to use *Fury's* boats on their retreat to Baffin Bay. They set out on 1 August, but heavy ice kept them in Prince Regent Inlet, and they were forced to turn back and spend their fourth winter in the Arctic at Fury Point.

1833: They set out again in boats on 15 August, made rapid progress out of Prince Regent Inlet, and were picked up by the whaler *Isabella* on 26 August in Lancaster Sound. This ship landed them safely at Hull in October.

ROSS, JOHN, *1835; †1835a; 1835b; *HUISH, 1835; BRAITHWAITE, 1849; FRASER, 1957; DODGE, 1973, pp. 112-49.

1830

British overland exploring expeditions | James Clark Ross

From Felix Harbour, Boothia Peninsula, winter quarters of *Victory* (see 1829-33). Between 5 April and mid-April, Ross, with Thomas Blanky and two Eskimos, sledged along the north shore of Lord Mayor Bay west to Netsiksiuvik Inlet, crossed Boothia Isthmus by way of Middle Lake to Spence Bay, sighted land to the north-west which he named Cape Isabella, then returned by the same route to Felix Harbour. This crossing of the isthmus was the first by Europeans. In late April, Ross made another short journey to see if he could cross the isthmus by Sagvak Inlet, the more southerly of the two inlets at the head of Lord Mayor Bay (Netsiksiuvik being the more northern). He followed it to the eastern end of Middle Lake, a point he had visited during his first journey, thereby proving, contrary to the Eskimos' assurances, that the route was also practicable. Between 27 April and 4 May, Ross sledged northward across Thom Bay and along the east coast of Boothia Peninsula to an Eskimo camp at Agvitutsiak, just north of Elizabeth Harbour, hoping to find another passage across Boothia Peninsula reported by the Eskimos (probably Bellot Strait). Between 17 May and mid-June, Ross explored west of Spence Bay, which he had visited in early April. He followed his previous route across Boothia Peninsula to Spence Bay, continued up the north shore of the bay to Josephine Bay, then crossed James Ross Strait by way of the north coasts of Matty and Tennent islands to King William Island. He traced its north shore, which he believed to be part of the mainland, to Victory Point, then returned to the north of Tennent Island, went down Wellington Strait to Cape Norton, King William Island, and crossed St Roch Basin back to Spence Bay. After a short visit to Willersted Inlet, he returned to Felix Harbour.

ROSS, JOHN, 1835; *FRASER, 1957.

1830

British overland exploring expeditions — John Ross

From Felix Harbour, Boothia Peninsula, winter quarters of *Victory* (see 1829-33); May-June. In support of J. C. Ross's expedition to King William Island (see 1830, above), John Ross crossed Boothia Peninsula by way of Netsiksiuvik Inlet, Middle Lake, and Spence Bay (which he named during this journey). He then visited Netsilik Lake and returned to Felix Harbour by way of Middle Lake, Sagvak Inlet, and Lord Mayor Bay. Between 23 June and 3 July, the two Rosses made a fishing trip north to Thom Bay, where Eskimos showed them the mouth of Lord Lindsay River. They returned to the ship with fish purchased from the Eskimos.

ROSS, JOHN, 1835; *FRASER, 1957.

1830

Establishment of Fort Chimo — Nicol Finlayson, Erland Erlandson

The party left Moose Factory on 10 June and followed Hendry's route (see 1828) to reach the mouth of Koksoak River on Ungava Bay on 2 August. They then spent a week searching for a site for a trading post, finding those selected by Kohlmeister and Kmoch (see 1811) and Hendry unsuitable. They finally chose a site about 50 km from the mouth of the Koksoak and wintered there. *Montcalm* arrived with trading goods for the post in September (see 1830-31). Fort Chimo was closed in 1843 (see 1842-43) and reopened in 1866 (*qv*).

DAVIES, 1963, pp. 100-76.

Establishment of Hebron — John C. Beck, Jonathan Mentzel, F. Kruth, August Freytag — **1830**

On 13 April, Beck and Mentzel, with six Eskimos, set out from Okak by dog sled with stores and the framework of a house to build the Moravians' fourth mission station on the Labrador coast. *Harmony* and *Oliver* left London with supplies for the new post and the other stations on 14 June but, delayed by ice, they did not reach Hebron until 28 July. Mentzel returned with the ships to Okak. Work on the buildings continued until January 1831, when the missionaries visited Okak for 12 weeks. They returned in April to occupy the station permanently.

PERIODICAL ACCOUNTS RELATING TO THE MISSIONS OF THE UNITED CHURCH..., Vol 11, 1829, pp. 363-76; Vol 12, 1831, pp. 67-69, 115-18.

Hudson's Bay Company supply voyage — Benjamin Bell, Henry Hanwell Jr, Robert Royal — *Prince Rupert* [V], *Prince of Wales* [I], *Montcalm* — **1830-31**

From London; 2 June 1830 - (?) 1831. *Prince Rupert* visited York Factory 15 August - 17 September, and reached London 29 October 1830. *Prince of Wales* visited Moose Factory 30 August - 22 September. On return, she was beset in ice off Mansfield Island and obliged to winter at Charlton Island, where she remained 26 October 1830 - 2 August 1831. She then sailed to London. *Montcalm* visited York Factory 15 - 27 August, then sailed to Ungava Bay to land goods for the newly established Fort Chimo (see 1830). She arrived there 13 September, left a week later, sailed to St Lawrence River, and then to England.

HUDSON'S BAY COMPANY (1) and (2).

Hudson's Bay Company supply voyage — Henry Hanwell Jr, Benjamin Bell, Robert Royal — *Prince of Wales* [I], *Prince Rupert* [V], *Montcalm* — **1831**

From London; 2 June - November. *Prince of Wales* visited Moose Factory 2 - 7 August, and reached London 22 September. *Prince Rupert* visited Moose Factory 27 August - 27 September, and reached London 23 November. *Montcalm* visited York Factory 25 August - 19 September, and reached London 25 October.

HUDSON'S BAY COMPANY (1) and (2).

British overland exploring expeditions; discovery of the North Magnetic Pole — James Clark Ross, John Ross — **1831**

From Felix Harbour, Boothia Peninsula, winter quarters of *Victory* (see 1829-33); April (?). Ross made a second attempt to find a passage across Boothia Peninsula from Agvitutsiak, near Elizabeth

Harbour (see 1830). Having examined the small inlets in that area, he rightly concluded there was no route across the peninsula south of Brentford Bay, a feature much farther north that he had seen from *Victory* in 1829. However, not until 1852 did Kennedy and Bellot thoroughly examine Brentford Bay and discover Bellot Strait (see 1851-52). Between May - 13 June, the Rosses travelled north to Thom Bay, then crossed Boothia Peninsula by way of Krusenstern, Jeckyll, and Hansteen lakes to Josephine Bay. From here, John Ross returned across Boothia Isthmus to the ship. James Ross continued up the west coast of Boothia Peninsula to Cape Adelaide, near which he located the North Magnetic Pole on 1 June. He then returned to the ship by the way he had come.

ROSS, JOHN, 1835; *FRASER, 1957.

1831

Hudson's Bay Company exploration of George River, Ungava Bay — William Taylor

From Fort Chimo (see 1830); 13 - 24 July; sent out by Nicol Finlayson to bring Eskimos from George River in to trade. Travelling by boat, Taylor sailed along the coast of Ungava Bay, and examined George River for about 30 km upstream. He met a band of Eskimos, who, however, had already taken their furs to the Moravian mission at Okak (see 1775).

DAVIES, 1963, pp. 148, 151.

1832

Hudson's Bay Company supply voyage	Benjamin Bell	*Prince Rupert* [V]
	Henry Hanwell Jr	*Prince of Wales* [I]

From London; 5 June - November. *Prince Rupert* visited York Factory 22 August - 21 September, and reached London 25 October. *Prince of Wales* visited Moose Factory 25 August - 20 September, and reached London 8 November. The ships met unusually heavy ice in Hudson Strait.

HUDSON'S BAY COMPANY (1) and (2).

1832-33

Establishment of South River House — Erland Erlandson

Erlandson left Fort Chimo (see 1830) on 23 June 1832 to establish a trading post on Swampy Bay River. He travelled up Koksoak and Kaniapiskau rivers to the mouth of Swampy Bay River but, at this point, his Indian guide refused to continue. Erlandson persuaded him to go a little further, but the Indian absolutely refused to continue on 28 June, so Erlandson decided to build his post at that spot, a short distance above the river's mouth. Erlandson wintered at the post, but the trade was so poor that he abandoned it on 25 June 1833, and returned to Fort Chimo.

DAVIES, 1963, pp. 199-224.

British whaling voyages	Richard W. Humphreys	*Isabella*	1833
	?	*William Lee*	

These ships entered Prince Regent Inlet in search of traces of John Ross's missing North-west Passage expedition (see 1829-33). They saw no sign of Ross then, but on 26 August, at the entrance of Lancaster Sound, *Isabella* picked up Ross and his crew, who were retreating towards Baffin Bay in a boat.

ROSS, JOHN, 1835, p. 720.

British whaling voyage; exploration of Exeter Sound	George Simpson William Penny	*Traveller*	1833

From Peterhead; routine whaling voyage to northern Baffin Bay. When returning south along the coast of Baffin Island with a full ship of 20 whales, Captain Simpson consulted Eskimos at Exeter Bay, 66°35'N, about the possibility of new whaling grounds in the vicinity. The Eskimos' information encouraged him to send his mate, Penny, to search for an unexplored bay to the south. Penny located Exeter Sound and sailed about 50 km into it before contrary winds forced him to return, having discovered only a few Eskimo winter huts by the shore. This was the first of a number of enquiries made among the Eskimos of southern Baffin Island about new whaling grounds and unexplored inlets there, which culminated in 1840 with Penny's re-discovery of Cumberland Sound (see 1839 and 1840).

**NAUTICAL MAGAZINE*, 1840; HOLLAND, 1970.

Hudson's Bay Company supply voyage	Benjamin Bell (out) John C. Grave (home)	*Prince Rupert* [V]	1833-34
	Henry Hanwell Jr (out) Henry Baker (home)	*Prince of Wales* [I]	

From London; 6 June 1833 - October 1834. *Prince Rupert* visited York Factory 6 - 27 September. *Prince of Wales* visited Moose Factory 7 - 24 September. Both ships met extremely heavy ice during the outward voyage. During the return voyage, they again met very heavy ice in Hudson Strait, and the two captains agreed to winter in Hudson Bay. *Prince Rupert* wintered at Churchill 18 October 1833 - 20 August 1834, during which time Captain Bell fell ill, several of the crew had scurvy, and one died. The ship visited York Factory 22-31 August 1834, then returned to London under the command of Captain Grave, and arrived 20 October. *Prince of Wales* wintered at Charlton Island until 24 July 1834, visited Moose Factory 25 July - 12 August, and reached London under the command of Captain Baker on 16 September, Henry Hanwell having died in Hudson Bay.

HUDSON'S BAY COMPANY (1) and (2).

1833-35

British expedition in search of Ross | George Back, Richard King

From Liverpool; 17 February 1833 - 8 September 1835; private venture, sponsored by public subscription, to search for John Ross's missing North-west Passage expedition (see 1829-33) by descending Thlew-ee-choh or Great Fish (now Back) River, which, according to Indian report, had its source east of Great Slave Lake and flowed north-eastward into the Arctic Ocean, and by searching the sea coast from its mouth. The expedition sailed to New York on the packet-ship *Hibernia,* then followed well-known routes north to Fort Resolution on Great Slave Lake, where they arrived on 8 August. From there, Indian guides showed Back the way to the river, which he located on 29 August. He then returned to the eastern end of Great Slave Lake, where Fort Reliance was being built as the expedition's winter quarters. On 25 April 1834, news reached them of Ross's safe return to England, but Back decided to explore the Back River anyway. They left Fort Reliance on 7 June and travelled by way of Artillery, Clinton-Colden, and Aylmer lakes to Back River, which they reached on 28 June. They spent one month making the difficult descent of the river, naming Pelly, Garry, Franklin lakes, and other features on the way, then spent three weeks exploring the sea coast more extensively, and they returned up Back River to Fort Reliance, where they arrived on 27 September. They spent a second winter there, then, on 21 March 1835, Back set off back to New York, and arrived at Liverpool on the packet-ship *North America* on 8 September. King followed later with the men and equipment, and arrived back at London on a Hudson's Bay Company supply ship in October (see 1835). King, who accompanied the expedition as naturalist, made extensive observations and collections of botanical and zoological interest, and the expedition also made observations of geology, meteorology, the aurora, and magnetic variations. Back River has been descended very few times since this expedition, and Back's maps of the river remained the only ones in use until as late as 1948.

*†BACK, 1836; *KING, 1836; WILSON, 1938; NEATBY, 1958, pp. 106-15; 1963.

1834

Hudson's Bay Company supply voyage | Daniel Friend, Robert Royal | *Prince George*, *Ganymede*

From London; 4 June - November. *Prince George* visited York Factory 22 August - 12 September, and reached London 21 October. *Ganymede* visited Moose Factory 25 August - 8 September, Fort Chimo 27 September - 6 October, and reached London 9 November.

HUDSON'S BAY COMPANY (1) and (2).

1834

Hudson's Bay Company exploring expedition | Erland Erlandson

From Fort Chimo; 6 April - 17 July; to find a communication route between Fort Chimo and Mingan, and to explore the interior of the Quebec-Labrador peninsula with a view to selecting sites for trading posts. He canoed up Whale River with four Indians, left it at about 57°03'N, 66°45'W to travel eastwards overland, but rejoined the river at Ninawawe Lake. He continued upstream through Lac Jeannin to Rivière Savalette, left this river near Elsie Lake, and continued southward overland to Attikamagen Lake, where he arrived on 13 May. On 21 May he reached Petitsikapau Lake, where his guides told him that they intended to turn eastwards and not, therefore, to travel towards Mingan. They continued probably through Dyke, Birch, and Sandgirt lakes, lakes Michikamau and Mackenzie, Seal Lake, then down Naskaupi River to North West River Post on Lake Melville, where he arrived on 22 June. Erlandson was the first European to make an overland journey from Hudson Strait to the Atlantic coast. After two days at North West River Post (which, he was disappointed to learn, was not a Hudson's Bay Company post and had no communication with Mingan nor anybody to guide him there), he retraced his steps to Petitsikapau Lake, then descended Swampy Bay River back to Fort Chimo.

DAVIES, 1963, pp. 247-59.

Hudson's Bay Company exploration of upper Liard River	John McLeod	1834

From Fort Halkett on Liard River (see 1829). McLeod ascended Liard River, then followed its tributary, Dease River, to its source, Dease Lake. He then struck out westward over the height of land to reach Tuya River, a tributary of Stikine River, at Terror Bridge, a frail bridge built by Indians. He crossed it and learned that the location was a meeting place used by Indians from the coast who came inland to trade. At this point, he turned back to Fort Halkett.

MORTON, 1939, p. 706.

Re-opening of Nichicun post	James Kellock	1834

Nichicun post was first opened in 1816, but was closed again in 1822. In 1834, Kellock arrived with a small party from Rupert House and re-occupied the old house. An outpost to Nichicun was also built that year on Kaniapiskau Lake which, it was hoped, would facilitate overland contact with Fort Chimo. Kaniapiskau post remained in operation until about 1870; Nichicun, now known as Nitchequon, still exists.

DAVIES, 1963, p. 316; HUDSON'S BAY COMPANY, ARCHIVES, B147/a/1/3.

Hudson's Bay Company supply voyage	Robert Royal	*Prince of Wales* [I]	1835
	John C. Grave	*Prince Rupert* [V]	
	William Williams	*Nonpareil*	
	J. Duncan	*Esquimaux*	

From London; 7 June - October. *Prince of Wales* visited Moose Factory 18 August - 7 September, and reached London 15 October.

Prince Rupert visited York Factory 23 August - 24 September, and reached London 29 October, carrying among her passengers Richard King, second-in-command of George Back's expedition (see 1833-35). *Nonpareil* left London on 22 June, visited York Factory 14-24 September, then sailed to New Brunswick. *Esquimaux* left Bristol 14 June, stopped at Akpatok Island in Ungava Bay, visited Fort Chimo 6 - 21 September, and reached London 19 October.

HUDSON'S BAY COMPANY (1) and (2).

1835-36

Whaling ships
beset in Davis Strait

In autumn 1835, 11 whalers from British ports were beset in Davis Strait. *Alfred, Harmony, Duncombe, Norfolk* and *Grenville* managed to extricate themselves during the autumn and return to port, but *Middleton* and *Dordon* were crushed and abandoned in October, and *William Torr* was lost with all hands in December. *Abram, Jane,* and *Viewforth* were carried south during the winter until released by the ice. The ships were not provisioned for such emergencies, and there was much suffering and many deaths. Admiralty sent HMS *Cove,* Captain James Clark Ross, to assist the distressed ships, but those that survived made their own way home.

COOKE AND ROSS, 1969; JONES, A.G.E., 1950.

1836

Establishment of Fort Smith (North West River Post) — Simon McGillivray

The first post in the region of Hamilton Inlet, known to the fur traders as Esquimaux Bay, was built in 1743 by Louis Fornel. The region seems to have been continuously occupied since then by French, then French and English, traders. In 1834 (*qv*), Erland Erlandson had visited two posts there, owned by Nathaniel Jones, a private trader from Quebec who had bought the posts from William Lampson in 1832. In 1835, Jones sold the posts to another merchant from Quebec, David Ramsay Stewart. In 1836, the Hudson's Bay Company sent McGillivray with men to Hamilton Inlet to compete with Stewart. The post they built was first named after William Smith, secretary of the company, but after 1840 called North West River. In 1837, Governor Simpson decided Fort Chimo (see 1830) should be supplied from Esquimaux Bay and ordered John McLean to find a practicable canoe route between the two places. McLean's journeys (see 1838 and 1839) demonstrated the impossibility of such a route, but he did find a route from North West River to Fort Nascopie (see 1841). Fort Chimo was therefore abandoned (see 1842-43) and North West River became the focus of forts Nascopie, Manuan, Michikamau, Winokapau, and Sandy Banks in the interior, and Aillik, Rigolet, and Kaipokok on the coast. The history of most of these posts is patchy. Most of them were established about 1840 and occupied intermittently. North West River and Rigolet are still in existence today.

COOKE, 1969, p. 96 et seq.; DAVIES, 1963, p. 231.

Hudson's Bay Company supply voyage	Henry Baker John C. Grave Charles Humphreys William Taylor Butterwick	*Prince of Wales* [I] *Prince Rupert* [V] *Eagle* *Esquimaux*	1836-37

From London; 2 June 1836 - 13 October 1837. *Prince of Wales* visited Moose Factory 2 - 18 September, and reached London 6 November. *Prince Rupert* was beset in the ice in Hudson Bay 23 August - 19 September, then failed in several attempts to put into York Factory. In early October, Captain Grave decided to return to England with his cargo unshipped, having landed only his passengers and a few articles at York Factory. He reached London 12 November. *Eagle* was stranded on the rocks off Button Islands during her outward voyage, 27-28 July. She reached York Factory on 7 September, too late to begin the homeward journey, and wintered in Hayes River. She set out for Fort Chimo 28 August 1837, and reached London 13 October. *Esquimaux* reached York Factory 24 September, much damaged by ice. She took on cargo for Fort Chimo, but, in early October, it was decided that it was too late to go there, and she was sent to winter at Churchill. She set out on 9 October, turned back five days later, and was driven ashore on 16 October. The ship was abandoned on 20 October and the crew walked back to York Factory, where they were told to winter at Ten Shilling Creek.

HUDSON'S BAY COMPANY (1) and (2).

British naval exploring expedition	George Back	HMS *Terror*	1836-37

From London; 14 June 1836 - 3 September 1837; to sail to Wager Bay, from there to carry boats overland to the north coast of America and to chart that coast between the western end of Fury and Hecla Strait and Turnagain Point, thus completing the survey of the north coast (see 1819-22, 1821-23, 1825-27, and 1829-33). Back sailed *Terror* into Foxe Channel, intending to pass round the northern side of Southampton Island as Parry had done, but was beset by ice when approaching Frozen Strait. *Terror* spent the winter drifting south-east along the coast of Southampton Island, until she was finally released in July 1837, near the west end of Hudson Strait. She was now badly damaged and leaking, so Back decided to abandon the expedition and return home. He managed to land her at Lough Swilly, Ireland, on 3 September. Although the expedition had failed in its object, they did chart some parts of the north-eastern coast of Southampton Island between Seahorse Point and Smyth Harbour.

BACK, 1837; *1838; NEATBY, 1958, pp. 116-21; 1963.

Whaling ships beset in Davis Strait	1836-37

In early October 1836, the whalers *Swan, Dee, Grenville Bay, Norfolk, Thomas,* and *Advice* were beset in Davis Strait, just as 11 ships had been in 1835-36 (*qv*). *Thomas* was crushed in the southward drift, but the others, some with many deaths, escaped. On *Dee,* 47 men out of 60 died. These events had some influence on the course of whaling: some owners began to supply warmer clothing and anti-scorbutics. James Clark Ross proposed the establishment of shore stations on the west side of Baffin Bay and Davis Strait and a more active programme of exploration and mapping. Penny's discovery of Cumberland Sound (see 1840) may be seen as a result of this trend of thought.

LUBBOCK, 1937, pp. 328-41; COOKE AND ROSS, 1969, p. 590.

1836-37

Establishment of Dease Lake post	Robert Campbell

Following McLeod's discovery of Dease Lake in 1834 (*qv*), a party of Hudson's Bay Company men went up Liard River in spring 1836 to establish a trading post there. However, fearing strange Indians, they turned back near Fort Halkett, at the junction of Smith and Liard rivers. Therefore, Campbell himself left Fort Simpson in autumn 1836 to establish the post. He wintered at Fort Halkett, and arrived at Dease Lake in July 1837. Leaving some men to build the post, Campbell explored westward, crossed Tuya River at Tuya Bridge, where McLeod had turned back in 1834, and found a large encampment of Indians who traded with the Russians at the mouth of Stikine River. He returned with this news to Fort Simpson in mid-August. Dease Lake post was apparently burned down by Indians soon after its establishment, but rebuilt. Its history later in the 19th century is unclear, and it was apparently closed on several occasions, but it may have been re-opened about 1910.

*WILSON, 1970, pp. 15-22; PARNELL, 1942; MORTON, 1939, p. 706.

1837

Hudson's Bay Company supply voyage	Henry Baker	*Prince of Wales* [I]
	John C. Grave	*Prince Rupert* [V]
	Daniel Friend	*Prince George*
	Captain Cooper	*Aurora*

From London; 1 June - (?) November. *Prince of Wales* visited Moose Factory 27 August - 13 September, and reached London late in November. *Prince Rupert* visited York Factory 28 August - 16 September, and reached London 29 October. *Prince George* visited York Factory 4-21 September; the date of her arrival back at London is not known. The details of the voyage of *Aurora* are not known.

HUDSON'S BAY COMPANY (1) and (2).

1837-39

Hudson's Bay Company exploring expedition	Peter Warren Dease Thomas Simpson

From Fort Chipewyan; 1 June 1837 - 14 October 1839; to explore those parts of the north coast of America that had not already been explored by Franklin (see 1819-22 and 1825-27): between

Return Reef and Point Barrow, Alaska, and between Turnagain Point and Fury and Hecla Strait. They canoed to the mouth of Mackenzie River, which they reached on 9 July 1837, then followed the coast westward, and passed Return Reef on 23 July. Simpson completed the last four days of the journey to Point Barrow on foot. They then returned to Great Bear Lake, where they established their winter quarters, Fort Confidence. On 6 June 1838, they set out along Dease, Kendall, and Coppermine rivers to the sea. They were hampered by ice during their journey eastward along the coast, and were beset for 22 days at Turnagain Point. From there, Simpson continued on foot to Cape Alexander on Kent Peninsula before turning back, during which he sighted land to the north, which he named Victoria Land. They spent a second winter at Fort Confidence, then, in June 1839, returned down Coppermine River to the sea. Before they began their coasting journey, Simpson explored Richardson River and met an encampment of Eskimos at its mouth. They then explored the coast eastwards to Castor and Pollux Bay. Returning, they surveyed the south coasts of King William and Victoria islands, then ascended Coppermine River back to Fort Confidence, which they reached on 24 September. They then continued to Fort Simpson, arriving on 14 October.

SIMPSON, T., 1839; †*1843; DEASE AND SIMPSON, 1840; SIMPSON, A., 1845; NEATBY, 1958, pp. 122-29.

Hudson's Bay Company supply voyage	Robert Royal John C. Grave	*Prince of Wales* [I] *Prince Rupert* [V]	1838

From London; 29 May - 25 October. *Prince of Wales* visited Moose Factory 5 - 19 September, and reached London 25 October. *Prince Rupert,* after meeting heavy ice in Hudson Strait, visited York Factory 28 August - 17 September, and reached London 21 October.

HUDSON'S BAY COMPANY (1) and (2).

Hudson's Bay Company exploring expedition	John McLean	1838

From Fort Chimo; 2 January - 20 April; a second attempt to establish a practicable route of communication between Fort Chimo and the Atlantic coast (see 1834, Erlandson), made in consequence of Governor Simpson's decision to supply Fort Chimo overland from Fort Smith. McLean followed Koksoak River to Highfall River, possibly followed the latter to its head, then took an overland route to Erlandson Lake, which he discovered and named. From there, he struck out south-eastward, crossed Wheeler River to Whale River, then again travelled overland until he reached Lake Michikamau on 3 February. From there, he followed Naskaupi River down to Lake Melville, where he arrived at Fort Smith (see 1836) on 16 February. He remained there for some weeks before returning to Fort Chimo, probably by a similar route. He, like Erlandson, reported that the route between Fort Chimo and Lake Melville was impracticable.

WALLACE, W.S., 1932, pp. 202-20; COOKE, 1969, pp. 102-07.

1838 Establishment of Fort Nascopie — Erland Erlandson

The first major step in establishing a chain of posts ordered by Governor Simpson to be built in the interior of the Quebec-Labrador peninsula. Erlandson left Fort Chimo on 18 June with three small canoes and probably followed the Koksoak-Swampy Bay route to the west arm of Lake Petitsikapau, which he reached 10 July. The post he built there was closed 1839-40. In 1841, it was supplied from North West River and, after the abandonment of Fort Chimo in 1843, it was the chief focus of trade in the interior. The post was closed about 1870, probably as a result of the re-establishment of Fort Chimo in 1866 (see 1830).

WALLACE, W.S., 1932, p. 221; MATTOX, 1964.

1838 Establishment of Fort Siveright (George River Post), Ungava

Following favourable Indian reports about George River, about 130 km east of Fort Chimo, John McLean sent a party of men from Fort Chimo to examine the river's suitability as a site for a trading post. They also reported favourably. In September, McLean sent a second party to build a post there. This post was the first built by the Hudson's Bay Company to trade exclusively with Eskimos, and it remained in operation until 1842 when, like Fort Chimo, it was abandoned (see 1842-43).

WALLACE, W.S., 1932, pp. 222-23; COOKE, 1969, p. 109; VOORHIS, 1930, p. 73.

1839 Hudson's Bay Company supply voyage — Robert Royal, *Prince of Wales* [I]; David J. Herd, *Prince Rupert* [V]

From London; 28 May - 16 October. *Prince of Wales* visited Moose Factory 22 August - 7 September. *Prince Rupert* visited York Factory 14 August - 11 September. Both ships reached London 16 October.

HUDSON'S BAY COMPANY (1) and (2).

1839 Hudson's Bay Company exploring expedition; discovery of Churchill Falls; establishment of Fort Trial — John McLean

From Fort Chimo; 21 June - 18 September; McLean's second attempt to find a practicable route between Fort Chimo and Fort Smith on Lake Melville (see 1836 and 1838). From Ungava Bay, he ascended George River past Fort Siveright (see 1838) to Indian House Lake, where he left most of his party to establish Fort Trial. He continued up Rivière de Pas to Fort Nascopie (see 1838) on Lake Petitsikapau, then, with Erland Erlandson, master in charge of Fort Nascopie, he searched for a large river that, according to the Indians at Fort Smith, rose close to Lake Petitsikapau, and flowed into Lake Melville. They had little trouble in finding Churchill River (formerly the Grand or Hamilton River), an outlet of Lobstick Lake, and began to descend it, only to be halted on reaching Churchill (formerly Grand) Falls. They therefore retraced their steps back to Fort Siveright, where Erlandson remained whilst McLean continued to Fort Chimo. After this journey, Governor Simpson abandoned his resolve to supply Fort Chimo overland from Fort Smith, and decided instead to send a ship from Quebec to Fort Chimo in alternate years.

COOKE, 1969, pp. 110-16; WALLACE, W.S., 1932, pp. 225-33.

Hudson's Bay Company exploring expedition	John Bell	1839

From Fort Good Hope; to make a preliminary investigation of Peel River (first discovered and explored for a short distance by Franklin in 1826, see 1825-27), and to tell the Indians of the company's intention to build a post there in 1840 (see 1840-41). Bell spent two months on Peel River and explored it for an unknown distance, then returned to the Mackenzie delta to explore Rat River. This he explored as far as McDougall Pass, where he met and bartered with Indians. He then returned to Fort Good Hope.

ISBISTER, 1845; STEFANSSON, 1960, pp. 172-73.

British whaling voyage	William Penny	*Neptune*	1839

From Aberdeen; routine whaling voyage to Baffin Bay. Penny crossed to the western side of Baffin Bay at the end of July and anchored at Durban Island (at about 67°N), where he enquired among the Eskimos about a large whaling ground reputed to lie in the unexplored waters farther south, in an inlet known to the Eskimos as Tenudiakbeek. He left Durban Island in a boat with four Eskimos to search for this inlet on 4 September, entered Exeter Sound (which he had visited previously in 1833, *qv*) and explored it thoroughly, but failed to locate Tenudiakbeek. Back at Durban Island, he met an Eskimo, Eenoolooapik, who drew for him a map of Tenudiakbeek. Penny returned to Aberdeen with Eenoolooapik on board in the hope of returning with him the following year to continue the search (see 1840).

*McDONALD, 1841; HOLLAND, 1970.

Hudson's Bay Company supply voyage	Robert Royal David J. Herd	*Prince of Wales* [I] *Prince Rupert* [V]	1840

From London; 3 June - 25 October. *Prince of Wales* visited Moose Factory 12-30 August, and reached London 9 October. *Prince Rupert* visited York Factory 9 August - 6 September, and reached London 25 October.

HUDSON'S BAY COMPANY (1) and (2).

1840

Hudson's Bay Company overland expedition — John McLean

From Fort Chimo; 18 June - 1 September; McLean's third overland journey from Fort Chimo towards Fort Smith on Lake Melville (see 1838 and 1839), this time to meet a ship carrying supplies for Fort Chimo. He probably travelled by way of George River, Fort Nascopie, and Naskaupi River. The ship arrived late at Fort Smith, and he had to abandon a plan to send a small outfit overland to Fort Nascopie. He returned to Fort Chimo on board the ship, *Marten*.

WALLACE, W.S., 1932, pp. 238-40; COOKE, 1969, pp. 118-19.

1840

Hudson's Bay Company exploring expedition — Robert Campbell

From Fort Halkett; to explore Liard River to its source, then to cross the Rocky Mountains and search for any river flowing west into the Pacific. Ascending Liard River, he discovered one of its northern branches, Frances River, and followed this to Frances Lake. He left his canoes there, continued overland up Finlayson River to Finlayson Lake, crossed the watershed between the Liard and Yukon river drainage systems, and discovered Pelly River at Pelly Banks. He then retraced his steps to Fort Halkett, where he arrived about 10 September. He returned to explore Pelly River in 1842-44 (*qv*).

WILSON, C.P., 1970, pp. 41-45; CAMPBELL, 1967; BURPEE, 1945.

1840

Methodist missionary expedition — George Barnley

From Montreal, Barnley travelled overland to Moose Factory where, at the request of the Hudson's Bay Company, he established a mission for the Wesleyan Methodist Missionary Society. This was Moose Factory's first Christian mission, was only the second Protestant mission to be established on the shores of Hudson Bay (see 1820-23, West), and was the most northerly mission in Rupert's Land at the time of its establishment. Barnley remained there for eight years, then returned to England. In the same year, 1840, the Wesleyan Missionary Society sent three other missionaries to Rupert's Land: James Evans went to Norway House, Robert T. Rundle went to Edmonton, and William Mason went to Rainy Lake.

BOON, 1962, pp. 74-75.

1840

Hudson's Bay Company exploring expedition — W.H.A. Davies

From Fort Smith (North West River); 11-26 June; to explore Churchill (Hamilton) River in the hope of finding a practicable supply route to Fort Nascopie (see 1838). He appears to have ascended the river as far as Churchill (Grand) Falls, then turned back. He reported that goods could be carried by canoe to that point and this news, together with John McLean's knowledge of the river above the falls (see 1839), and Indian reports that a portage route around the falls existed, encouraged McLean to explore the whole length of the river in 1841 (*qv*).

HUDSON'S BAY COMPANY, ARCHIVES, B153/a/3/23d; B153/b/1/45; COOKE, 1969, p. 125.

1840

British whaling voyage	William Penny	*Bon Accord*

From Aberdeen; 1 April - 11 October; secondary object, if time after whaling permitted, was to search for the large inlet in southeastern Baffin Island, known to the Eskimos as Tenudiakbeek, after which he had enquired in 1839 (*qv*). He had a guide, the Eskimo Eenoolooapik, whom he had taken to Aberdeen in 1839. The ships were unable this year to cross to the Baffin Island coast by the usual route, the open North Water of Baffin Bay, so Penny retreated to 65°N, where he was able to cross. About the end of July, he located Tenudiakbeek. He entered it with several other whalers, including *Lady Jane,* Captain Harrison, of Newcastle, and *Lord Gambier,* Captain Wareham, and *Truelove,* Captain Parker, of Hull, and named the inlet "Hogarth's Sound." They spent the next month exploring the sound thoroughly, caught some whales, then, leaving Eenoolooapik behind, returned home. They had in fact rediscovered Cumberland Sound, first entered by John Davis in 1585 (*qv*), but not since revisited. In the years that followed, Cumberland Sound became one of the major whaling grounds in the Baffin Bay area.

*McDONALD, 1841; HOLLAND, 1970.

1840-41

Hudson's Bay Company exploring expedition; establishment of Fort McPherson	John Bell Alexander Kennedy Isbister

From Fort Good Hope; 3 June 1840 - (?) 1841; to establish a trading post on Peel River, at a site Bell had selected in 1839 (*qv*), a short distance up Peel River. It was first known as Peel's River post or Fort Bell, but later called Fort McPherson. During summer and autumn 1840, Isbister spent six months surveying Peel River and the surrounding area. He also explored Rat River to the Rat-Bell portage, McDougall Pass, as Bell had done in 1839. In 1840, he travelled overland from the post to Arctic Red River, and descended the latter to the Mackenzie. During his surveys, Isbister made observations on plants, animals, and geology, which he later published.

ISBISTER, 1845; STEFANSSON, 1960, pp. 173-78.

1841

Hudson's Bay Company supply voyage	Robert Royal David J. Herd Henry Boulton	*Prince of Wales* [I] *Prince Rupert* [VI] *Prince Albert*

From London; 2 June - 19 October. *Prince of Wales* visited Moose Factory 30 August - 12 September, and reached London 19 October. *Prince Rupert* visited York Factory 18 August - 13 September, and reached London 17 October. *Prince Albert* visited York Factory 18 August - 13 September, and reached London 19 October.

HUDSON'S BAY COMPANY (1) and (2).

1841

Hudson's Bay Company exploring expedition — John McLean

From Fort Chimo; 3 June - 6 September; to explore Churchill (Hamilton) River as a possible route for supplying Fort Nascopie. McLean had already explored the river above Churchill (Grand) Falls (see 1839), W.H.A. Davies had explored it below the falls (see 1840), and only the portage route around the falls, on which Indians reported favourably, remained untested. His route to Churchill River is unrecorded, though he probably travelled by way of George River and Fort Nascopie. He descended Churchill River with ease, found Big Hill Portage, and on arrival at Fort Smith, sent a supply of goods to Fort Nascopie and himself returned to Fort Chimo by the same route. As a result of this journey, McLean recommended to the Governor and Committee of the company that Fort Chimo should be closed, and that the interior posts, which had drawn their supplies from Fort Chimo, should be supplied from Fort Smith by way of Churchill River. Previous overland journeys had amply demonstrated the extreme difficulty of communication between Fort Chimo and the interior, and the impossibility of establishing a practicable supply route to Fort Chimo from Fort Smith. In 1842, the company acted on his suggestion and ordered the closure of Fort Chimo (see 1842-43).

WALLACE, W.S., 1932, pp. 245-47; COOKE, 1969, pp. 125-30.

1841-47

Robert Ballantyne's apprenticeship

Ballantyne served a six-year apprenticeship to the Hudson's Bay Company, at posts along the St Lawrence River, Red River, Norway House, and York Factory. His first book, *Hudson Bay, or, life in the wilds of North America* (1848) was a great success. He then wrote *Snowflakes and sunbeams, or, the young fur traders* (1856) and *Ungava, a tale of Eskimo land* (1857), the latter book based on a reminiscence he had commissioned Nicol Finlayson to write about the founding and first years of Fort Chimo (see 1830). These books had an immense influence on the popular view of northern Canada and the Hudson's Bay Company.

PARNELL, 1941; DNB SUPPLEMENT, Vol. 1, 1901, pp. 122-23.

1842

Hudson's Bay Company supply voyage	David J. Herd	*Prince Rupert* [VI]
	Robert Royal	*Prince Albert*

From London; 1 June - 29 October. *Prince Rupert* visited York Factory 16 August - 16 September, and reached London 23 October.

Prince Albert visited Moose Factory 29 August - 10 September, and reached London 29 October.

HUDSON'S BAY COMPANY (1) and (2).

1842

Hudson's Bay Company exploring expedition — John Bell

From Fort McPherson (see 1840-41); 23 June - 24 July; to open communication with an unknown Indian tribe west of the post, from whom he received beaver by way of middlemen. After a hard five days' march across the Richardson Mountains, he discovered Bell River, which he descended for five days. He may have reached Porcupine River, into which it flows. He returned to the fort, believing that the overland route was far too difficult to make communication with Bell River practicable. In fact both he, in 1839 (*qv*), and Isbister, in 1840 (*qv*), had been on the verge of discovering a much easier route, across the Rat-Bell portage, but that route was not finally discovered until 1872 (*qv*).

STEFANSSON, 1960, pp. 179-81; HUDSON'S BAY COMPANY, ARCHIVES, B157 /a/2/7-7d.

1842-43

Abandonment of Fort Chimo

Following John McLean's suggestion in 1841 (*qv*), the Governor and Committee of the Hudson's Bay Company ordered Fort Chimo to be closed in 1842. McLean received the instructions at Fort Smith in 1842. Fort Siveright (see 1838) and Fort Trial (see 1839), outposts of Fort Chimo, were closed in 1842, but Fort Chimo was not closed until 1843. So ended the company's Ungava venture, an attempt to open up the whole of the Quebec-Labrador peninsula to the fur trade which had begun in 1830 with the opening of Fort Chimo. The venture was beginning to show a profit when it was abandoned. It had proved extremely difficult to supply the interior from Fort Chimo, and fewer inland posts had been established than Governor George Simpson, the instigator of the project, had wished. Fort Nascopie on Lake Petitsikapau, and its outposts remained open after 1843, for they could now be supplied with relative ease from the Atlantic coast by way of Churchill River, a route opened up by John McLean in 1841. Fort Chimo was re-opened in 1866.

WALLACE, W.S., 1932; DAVIES, 1963, pp. lxxv-lxxvi; COOKE, 1969.

1842-44

Hudson's Bay Company exploring expedition, establishment of Frances Lake post — Robert Campbell

From Fort Simpson; to establish a trading post on Frances Lake, which Campbell had discovered in 1840 (*qv*), and to explore Pelly River. The party reached the lake on 16 August, built a post, and wintered there. In June 1843, Campbell and six men followed his route of 1840 to Pelly Banks, and continued down Pelly River. Below the mouth of Lewes River, they met a party of Knife Indians, who urged them to turn back on account of a tribe of hostile Indians farther downstream. Campbell was obliged by his Indians to accept this advice and they returned to winter at Frances Lake. In 1844, the party returned to Fort Simpson.

WILSON, 1970, pp. 53-79; BURPEE, 1945; PARNELL, 1942.

1843

Hudson's Bay Company supply voyage	Robert Royal David J. Herd J. C. Mallett (?)	*Prince Albert* *Prince Rupert* [VI] *Marten*

From London; 1 June - 24 November. *Prince Albert* visited Moose Factory 24 September - 30 October, and reached London 9 November. *Prince Rupert* visited York Factory 9 August - 21 September, and reached London 9 November. *Marten* left London on 1 July, and sailed to Quebec. She then visited North West River Post 20-21 July, Fort Chimo 21 August - 1 September, Rigolet 21 September - 1 November, and reached London 24 November.

HUDSON'S BAY COMPANY (1) and (2).

1843

Hudson's Bay Company exploration of Keele (Gravel) River	Adam McBeath

From Fort Norman; 3 - (?) July; to explore the country west of Mackenzie River and seek a site suitable for a trading post. McBeath left with five men in a large canoe and, finding the river very swift and full of islands, they left their damaged canoe and walked. On 13 July they turned up a tributary and climbed several mountains to see the country better. On 17 July they turned back and reached the canoe on the 20th. Then they descended in four hours a distance they had taken six days to ascend. McBeath reported no animals but goats, but he thought there might be rich country behind the mountains.

HUDSON'S BAY COMPANY, ARCHIVES, B200/b/19/30-35.

1843

Hudson's Bay Company overland expedition	James Pruden John Lee Lewes

In 1842, John Bell had searched unsuccessfully for a practicable route westward from Fort McPherson on Peel River (*qv*). In June 1843, Pruden led a second attempt to find a route across the Richardson Mountains, having been told by an Indian that he knew a better route than that followed by Bell. However, the Indian abandoned the party, and they were obliged to return.

HUDSON'S BAY COMPANY, ARCHIVES, B157/a/3/6d-7d.

Magnetic survey of northern Canada	John Henry Lefroy		1843-44

From Toronto; 13 April 1843 - 5 December 1844; sent by the British Government at the instigation of the Royal Society to make a magnetic survey of the far northern territories of the Hudson's Bay Company. After visits to Fort William, Fort Garry, Norway House, and York Factory, Lefroy continued to Fort Chipewyan, where he wintered. He left there on 5 March 1844 and descended Slave and Mackenzie rivers to Fort Good Hope, where he arrived on 29 May. He then returned to Fort Chipewyan, ascended Peace River to Dunvegan, travelled on horseback to Edmonton, then continued by canoe back to Toronto. During this journey, Lefroy made magnetic observations at many hundred locations.

*STANLEY, 1955; WALLACE, W.S., 1938; †LEFROY AND RICHARDSON, 1855.

Hudson's Bay Company supply voyage	Robert Royal David J. Herd T. Dunn	*Prince Albert* *Prince Rupert* [VI] *Marten*	1844

From London; May-November. *Prince Albert* left London 29 May, visited Moose Factory 28 August - 10 September, and reached London 14 October. *Prince Rupert* left London 29 May, visited York Factory 11 August - 15 September, and reached London 18 October. *Marten* left London 15 May, visited North West River Post 17-23 July, Mingan 25-26 August, Quebec 30 August - 6 September, Rigolet 25 September - 12 October, and reached London 10 November.

HUDSON'S BAY COMPANY (1) and (2).

Hudson's Bay Company supply voyage	David J. Herd Robert Royal	*Prince Rupert* [VI] *Prince Albert*	1845

From London; 5 June - 4 November. *Prince Rupert* visited York Factory 10 August - 11 September, and reached London 13 October. *Prince Albert* visited Moose Factory 6-24 September, and reached London 4 November.

HUDSON'S BAY COMPANY (1) and (2).

Hudson's Bay Company exploring expedition	John Bell		1845

From Fort McPherson; 28 May - July; a third attempt to extend the company's trade west from Fort McPherson (see also 1842, Bell, and 1843, Pruden). He crossed the Richardson Mountains from Peel River to Bell River by the five-day portage he had used in 1842, then, after building canoes, descended Bell River to its confluence with Yukon River on 16 June. Unaware of the importance of his discovery, he turned homeward six days later. He was back at the portage on 9 July having, in the meantime, met Indians who assured him that there were no Russian traders in the area. The company decided to establish a post in the region of Bell's discoveries in 1846, but lack of boats prevented that until 1847 (*qv*).

STEFANSSON, 1960, pp. 184-86; HUDSON'S BAY COMPANY, ARCHIVES, B200/b/21/2d.

1845

British whaling voyages	Robert Martin	*Enterprise*
	Captain Dannett	*Prince of Wales*

From Peterhead (*Enterprise*) and Hull (*Prince of Wales*). The crews of these ships were the last white men to see the members of Franklin's expedition (see 1845-48) alive. They first met Franklin's ships on 14 July, off the coast of Greenland, 75°12'N, 61°6'W. They met again on 26 July in Melville Bay, 74°48'N, 66°13'W. On this occasion, some of Franklin's officers dined aboard Martin's ship. A few days later, Franklin's ships and the whalers lost sight of each other.

JONES, A.G.E., 1969; GREAT BRITAIN, PARLIAMENT, 1848, p. 51.

1845-48

British naval North-west Passage expedition	Sir John Franklin	HMS *Erebus*
	Francis Rawdon Moira Crozier	HMS *Terror*

From London, 19 May 1845 - (?)1848; to sail through Lancaster Sound and Barrow Strait, then to search for a passage in the unexplored region between Cape Walker on the south side of Barrow Strait and the north coast of America. The ships were last seen in late July 1845 in northern Baffin Bay, heading for Lancaster Sound, by the whalers *Enterprise* and *Prince of Wales* (see above, 1845), but, after that, the members of the expedition were never again seen by white men. The search for the missing expedition began in 1847 and continued until 1880, during which period the main facts regarding the fate of the expedition and the course pursued by it were established, and a few relics found. After entering Lancaster Sound, the expedition sailed into Wellington Channel as far as 77°N, then returned around the west side of Cornwallis Island to Barrow Strait, and wintered at Beechey Island, southwest Devon Island, where the graves of three members of the expedition were later found. In 1846 they entered Peel Sound, passed through Franklin Strait and were beset to the north of King William Island on 12 September 1846. They remained beset until 22 April 1848 (during which period Franklin died, on 11 June 1847), when the survivors (105 men, from the original total of 129), led by Crozier, abandoned the ships and set out towards Back River. They landed at Victory Point, King William Island, where they left records detailing the above sequence of events; there is no written record of the events that followed, they can be reconstructed only from the evidence of skeletons and relics found along the route, and from the accounts of Eskimos who saw the party. They followed the west coast of King William Island round to Cape Herschel where they met Eskimos, who later reported that they had met about 40 members of the expedition (others had already died or had, apparently, returned towards the ship for further provisions) and had sold them some seal meat. The Eskimos' accounts confirm that the party was starving and suffering badly from scurvy. After parting from the Eskimos, the men continued along the coast, some of them dying along the route, then crossed Simpson Strait to the mainland where, in Starvation Cove on Adelaide Peninsula, the last surviving members of the expedition, about 30 or 35 men, died.

The first news of the fate of the expedition was brought back by Dr John Rae (see 1853-54), and other important relics were found by McClintock (see 1857-59), Hall (see 1864-69), Schwatka (see 1878-80) and some 20th century expeditions. No journals of the expedition have ever been found, although expeditions in search of them or other records of the expedition, have continued to visit the area until the present day.

This was the last of the series of North-west Passage expeditions that the British Admiralty had been sending out since 1818 (*qv*); by discovering a channel of communication between known points in Barrow Strait and on the north coast, Franklin's men had, effectively, discovered the passage, although it was not navigated from end to end until Amundsen sailed through it in 1903-06 (*qv*).

CYRIAX, *1939; 1951; 1958; 1959; 1969; CYRIAX AND JONES, 1954; JONES AND CYRIAX, 1953; WRIGHT, *1959; 1949; NEATBY, 1970; GIBSON, 1937; CONINGHAM, n.d.; TRAILL, 1896.

1846

Hudson's Bay Company supply voyage — David J. Herd, *Prince Rupert* [VI]; Robert Royal, *Prince Albert*

From London; 3 June - 21 October. *Prince Rupert* visited York Factory 6 August - 19 September, and reached London on 31 October. *Prince Albert* visited Moose Factory 9-31 August and reached London on 5 October.

HUDSON'S BAY COMPANY (1) and (2).

1846-47

Hudson's Bay Company exploring expedition — John Rae

From York Factory; 13 June 1846 - autumn 1847; to complete the exploration of the north coast of America by surveying the region between Fury and Hecla Strait (Parry's farthest west in 1821-23, *qv*) and Castor and Pollux Bay (Dease and Simpson's farthest east in 1837-39, *qv*). He travelled from York Factory to Repulse Bay in two small boats, crossed Rae Isthmus, and attempted to sail through Committee Bay. He made little progress through the ice, so he returned across Rae Isthmus to Fort Hope, which had been built as his winter quarters.

1847: He set out across Rae Isthmus by sledge on 5 April, and surveyed the coast north-westward to Lord Mayor Bay (first discovered by Ross in 1829-33, *qv*). He returned to Fort Hope on 5 May. On 13 May, he crossed the coast to within about 30 km of Fury and Hecla Strait, where lack of provisions forced him to turn back. He reached Fort Hope on 9 June. On 12 August, he left Fort Hope and sailed back to York Factory. The only remaining unexplored part of the north coast of America was now the west coast of Boothia Peninsula, and he returned to explore it in 1853 (see 1853-54).

RAE, 1850; BARRATT, 1976.

1846-48

British military expedition to Red River	Captain Lowe Captain Mallinson Captain Webster ?	HMS *Crocodile* *Blenheim* *Lady Fitzherbert* *General Palmer*

In 1846, the British Government sent a force of 346 officers and men of the Sixth Regiment of Foot, the Royal Artillery, and the Royal Sappers and Miners to the Red River settlement in connexion with the controversy, then at its height, with the United States over the question of the Oregon boundary. *Crocodile* and the hired transport *Blenheim* left Cork with the troops on 26 June, three days before news of a peaceful settlement of the dispute reached London. *Blenheim* was at York Factory on 9 August and *Crocodile* on 13 August; they both left on 26 August. In 1848, the government chartered *Lady Fitzherbert* to bring these troops home. The Hudson's Bay Company had, meantime, asked to have the troops replaced, so in 1848 the War Office sent out a contingent of Chelsea out-pensioners, 56 men and 99 women and children, in *General Palmer*. They were landed on 16 August. *General Palmer* left on 30 August with some of the returning troops; *Lady Fitzherbert* followed on 12 September with the rest, and reached London 10 October.

RICH, 1960, pp. 540-46; MORTON, A.S., 1939.

1847

Hudson's Bay Company supply voyage	David J. Herd Robert Royal Forbes Michie	*Prince Rupert* [VI] *Prince Albert* *Westminster*

From London; 2 June - 29 October. *Prince Rupert* visited York Factory 25 August - 24 September, and reached London on 29 October. *Westminster*, carrying men and stores for Richardson's expedition (see 1847-49), visited York Factory 30 August - 28 September, then left for Nova Scotia. *Prince Albert* visited Moose Factory 24 August - 9 September, and reached London on 19 October.

HUDSON'S BAY COMPANY (1) and (2).

1847

Establishment of Fort Yukon	Alexander Hunter Murray

Following John Bell's explorations in 1842 (*qv*), the Hudson's Bay Company decided to establish a trading post at the confluence of the Porcupine and Yukon rivers. Murray left Fort McPherson on 11 June and reached La Pierre's House on 14 June. He then travelled by boat down Bell and Porcupine rivers, and reached Yukon River on 25 June. He began to build Fort Yukon on 1 July, and Indians came to trade soon after. In 1864, a new post was built about one kilometre downriver. In 1869, following the Alaska Purchase in 1867, the company moved its trading activity from Fort Yukon to Rampart House on Porcupine River. It, too, was later found to be in United States territory and moved another 20 km upriver into present-day Yukon Territory.

*BURPEE, 1910; STEFANSSON, 1960, pp. 190-220; WATSON, 1929; WILSON, C.P., 1947; ANDREWS, 1947; SHIPLEY, 1967; DCB, VOL. 10, pp. 540-41.

1847

British whaling voyage and Franklin search — William Penny — *St Andrew*

From Aberdeen. At the end of July and in early August, Penny attempted to sail into Lancaster Sound with the dual object of whaling and searching for Sir John Franklin's missing expedition (see 1845-48). Contrary winds and a strong swell forced him to give up the attempt after he had reached only 75°W. This is the first ship known to have attempted to make contact with the missing expedition.

*GREAT BRITAIN, PARLIAMENT, 1848, pp. 52-53; HOLLAND, 1970, pp. 34-35.

1847

Hudson's Bay Company expedition to Belcher Islands — Thomas Wiegand

In March of this year, Wiegand made a trip over the sea ice from Fort George to the Belcher Islands, with the purpose of ascertaining the islands' location and opening up trade with the Eskimos who lived there. He made a second journey to the islands in 1848. These two trips are the only known visits by a white man to the Belcher Islands during the 19th century; indeed, it seems probable that no white man had ever landed there before, though their existence was certainly known to Captain William Coats of the Hudson's Bay Company in the 18th century, and they may have been sighted as early as 1610, by Henry Hudson. Although, later in the 19th century, Hudson's Bay Company officers at Fort George and Great Whale River still knew of the Belcher Islands and received annual visits from their Eskimo population, who travelled over the sea ice in February and March each year to trade, no further attempt was made to visit or explore them. It was not until the arrival of Robert Flaherty in 1910 that the Belcher Islands were properly explored (see 1910-11; 1911-12; 1913-16).

FLAHERTY, 1918, pp. 447-48.

1847-49

British overland Franklin search expedition — Sir John Richardson, John Rae, John Bell

They were sent by the Admiralty to search the coast between Mackenzie and Coppermine rivers and the southern shores of Wollaston Peninsula, Victoria Island. The four boats and 20 men were sent to York Factory in 1847 in the Hudson's Bay Company's supply vessels *Prince Rupert* and *Westminster* (*qv*), and were met by Bell. Richardson and Rae travelled by steamer to New York in 1848, and caught up the rest of the party at Methy Portage. They reached the mouth of Mackenzie River on 3 August 1848, searched the coast in boats to the Coppermine River, then returned to winter at Fort Confidence, which Bell had been building at the north-eastern corner of Great Bear Lake in the meantime.

1849: Richardson returned to England by steamer from Boston, whilst Rae set out on 7 June to continue the search. He reached the coast, but ice prevented him from crossing to Wollaston Peninsula, and he returned to Fort Simpson. He resumed the search under the direction of the Hudson's Bay Company in 1850 (see 1850-51). During this expedition, Richardson made botanical, zoological, meteorological, magnetic, geological, physiographical, and ethnological observations.

*†RICHARDSON, 1851; GREAT BRITAIN, PARLIAMENT, 1848, pp. 19-21; 1849a; 1849b, pp. 8-10; 1850a, pp. 1-8; †HOOKER, 1857; †LEFROY AND RICHARDSON, 1855.

1848

Hudson's Bay Company supply voyage	David J. Herd Robert Royal	*Prince Rupert* [VI] *Prince Albert*

From London; 31 May - 14 October. *Prince Rupert* visited York Factory 18 August - 11 September, and reached London on 14 October. *Prince Albert* visited Moose Factory 10-31 August, and reached London on 6 October.

HUDSON'S BAY COMPANY (1) and (2).

1848

Establishment of Fort Selkirk	Robert Campbell

Campbell, who had spent the winters of 1846-47 and 1847-48 at his post on Frances Lake (see 1842-44), set out in spring 1848 to build a post for the Hudson's Bay Company at the junction of Pelly and Lewes rivers. They arrived there in June, built their post, and met Chilkat Indians in August. The post remained in existence under Campbell's command until 1852 when it was pillaged and burned down by Indians. It was not rebuilt until 1938.

WILSON, C.P., 1970, pp. 94-110.

1848

British whaling voyage	Thomas Lee	*Prince of Wales*

From Hull. This ship sailed some 240 km into Jones Sound then, having failed to sight any whales, returned to Baffin Bay. It is the first ship known to have sailed any great distance into Jones Sound, although other whalers may have been there before. Later in the voyage, on 1 October, off Cape Hooper, Baffin Island, Lee picked up a cask containing a letter that had been thrown overboard from HMS *Investigator* (see 1848-49). This was the latest news brought home from that ship until her re-emergence from Lancaster Sound in 1849.

*OSBORN, 1852, pp. 242-43; GREAT BRITAIN, PARLIAMENT, 1852, p. 128; LUBBOCK, 1937, p. 348.

1848

British naval Franklin search expedition | Henry Kellett | HMS *Herald*

Herald left England in 1845 on a surveying voyage of the Pacific Ocean, but on three occasions (1848, 1849, and 1850) was called to Bering Strait to assist *Plover* (see 1848-52) in the search for Franklin. On this first occasion, she left Panama on 9 May 1848 and reached Kotzebue Sound on 14 September. Two weeks later, when *Plover* had failed to appear, she returned south.

Botanical observations were made throughout the six-and-a-half-year voyage.

*SEEMANN, 1853, Vol. 2, pp. 1-70; GREAT BRITAIN, PARLIAMENT, 1848, pp. 16-18; 1849b, pp. 11-19.

1848-49

British naval Franklin search expedition | Sir James Clark Ross
Edward Joseph Bird | HMS *Enterprise*
HMS *Investigator*

From London; 12 May 1848 - 3 November 1849. Following the route that Franklin had been ordered to take, they passed through Lancaster Sound and reached Port Leopold, north-east Somerset Island, on 11 September, wintered there, and remained beset until 28 August 1849. During spring 1849, they made four sledge journeys along the coasts of Somerset Island and across Barrow Strait and Prince Regent Inlet (see 1849). When the ships were freed, Barrow Strait to the west was impassable. They were carried by ice eastwards into Baffin Bay and abandoned the search, having found no trace of Franklin's expedition.

*GILPIN, 1850; GREAT BRITAIN, PARLIAMENT, 1848a; *1849b, pp. 1-7; *1850a, pp. 58-64; BROWNE, 1850; BURFORD, 1850; MARKHAM, C.R., 1909, pp. 49-67; DODGE, 1973, pp. 230-34; CYRIAX, 1942.

1848-52

British naval Franklin search expedition | Thomas Edward Laws Moore | HMS *Plover*

From Plymouth; 30 January 1848 - 30 June 1852 (when Moore was replaced by Rochfort Maguire, see 1852-54). They arrived at Bering Strait in October 1848, too late to reach Kotzebue Sound and rendezvous with *Herald* (see 1848) as planned. They wintered on the Asiatic side of the strait, where they had friendly dealings with the natives.

1849: They sailed to Kotzebue Sound where they met *Herald* (see 1849) and *Nancy Dawson* (see 1849), and proceeded with them to Wainwright Inlet. An expedition in boats to Mackenzie River set out from here (see 1849-51). *Plover* wintered in Kotzebue Sound.

1850: They sailed to Icy Cape, and Moore continued in boats to Point Barrow. They then returned to winter at Port Clarence.

1851: They sailed to St Michael, returned to Port Clarence, and there met *Daedalus* (see 1851) with a replacement crew. Moore and some of his officers again wintered on board *Plover* at Port Clarence, and were finally replaced in 1852.

GREAT BRITAIN, PARLIAMENT, 1848, pp. 7-11; 1849b, pp. 18-21; *1850a, pp. 34-41; *1851, pp. 28-40; *1852, pp. 89-91, 201-15; *1852b, pp. 42-43; *HOOPER, 1853; SMITH, D.M., 1877, pp. 429-34; *ILLUSTRATED LONDON NEWS*, 1848.

1849

Hudson's Bay Company supply voyage — David J. Herd, *Prince Rupert* [VI]; Robert Royal, *Prince Albert*; William Howard, *Graham*

From London; 31 May - 25 October. *Prince Rupert* visited York Factory 14 August - 11 September, and reached London on 25 October. *Prince Albert* visited Moose Factory 16 August - 9 September, and reached London on 24 October. *Graham,* a chartered vessel, sank in ice on 3 August near Mansel Island. Captain Howard and seven of the crew made their way in a long boat to the Labrador coast and returned to England in the Moravian missionaries' supply ship. The rest of the crew were last seen in Hudson Strait making for shore in a jolly boat.

HUDSON'S BAY COMPANY (1) and (2).

1849

Consecration of the first Bishop of Rupert's Land

David Anderson was nominated first Bishop of Rupert's Land by the Archbishop of Canterbury in early spring 1849, and was consecrated on 25 May. Soon after, he sailed to York Factory on the supply ship *Prince Rupert* (see above, 1849). He left York Factory on 29 August and travelled to the Red River settlement, where he remained.

BOON, 1962, pp. 64-67.

1849

British whaling voyage — John Gravill, *William Ward*

From Hull. Gravill took his ship into Jones Sound and landed on the south coast of Ellesmere Island. This is the first record of a landing on that island, although other whalers may have landed there before. On his return home, Gravill reported to the Admiralty that he had seen no other ships in the sound; this was in reply to their request for information on the whereabouts of HMS *North Star* (see below, 1849-50).

*GREAT BRITAIN, PARLIAMENT, 1850a, pp. 69-70; LOTZ, 1962, p. 151.

British Franklin search expedition | Sir James Clark Ross, Francis Leopold McClintock | 2 sledges | 1849

From *Enterprise* and *Investigator* at Port Leopold, Somerset Island (see 1848-49); 15 May - 22 June. They searched along the north coast of Somerset Island and down the west coast to 72° 38'N.

GILPIN, 1850, pp. 164, 166.

British Franklin search expedition | John James Barnard | sledge | 1849

From *Enterprise* and *Investigator* at Port Leopold, Somerset Island (see 1848-49); 31 May - 5 June. He crossed Barrow Strait to Cape Hurd, Devon Island.

GILPIN, 1850, p. 165.

British Franklin search expedition | William Henry James Browne | sledge | 1849

From *Enterprise* and *Investigator* at Port Leopold, Somerset Island (see 1848-49); (?) May - (?) June. He crossed Prince Regent Inlet and returned to the ships after an absence of eight days.

GILPIN, 1850, p. 165.

British Franklin search expedition | Frederick Robinson | sledge | 1849

From *Enterprise* and *Investigator* at Port Leopold, Somerset Island (see 1848-49); (?) June - 17 June. He searched the east coast of Somerset Island southward, and turned back at Cresswell Bay, a short distance beyond Fury Beach.

GILPIN, 1850, p. 165.

British naval Franklin search expedition | Henry Kellett | HMS *Herald* | 1849

They left Honolulu on 19 May on a second attempt to join *Plover* (see 1848-52) in a search for Franklin by way of Bering Strait. They met *Plover* in Kotzebue Sound and sailed with her to Wainwright Inlet. From there, *Herald* sailed west, sighted Ostrov Vrangelya (Wrangel Island), discovered Ostrov Geral'da (Herald Island), then returned to Mexico for the winter. *Herald* returned to Bering Strait for the last time in 1850 (*qv*).

*SEEMANN, 1853, Vol. 2, pp. 91-129; *GREAT BRITAIN, PARLIAMENT, 1850a, pp. 9-44.

British whaling voyage and Franklin search | William Penny, Robert Anstruther Goodsir | *Advice* | 1849
| Captain Parker | *Truelove* |

Goodsir, brother of Harry Goodsir in Franklin's *Erebus*, joined *Advice* hoping for an opportunity to enter Lancaster Sound in search of Franklin's ships. The two vessels lay whaling off Pond Inlet, Baffin Bay, until an Eskimo report that Franklin and his men had been seen alive was brought to Parker. They then entered Lancaster Sound, but ice prevented their passing beyond Brodeur Peninsula. They turned back, leaving depots on Wollaston Island and at Cape Hay, Bylot Island.

*GOODSIR, 1850; GREAT BRITAIN, PARLIAMENT, 1850a, pp. 70-74.

1849

British Franklin search expedition | Robert Shedden | *Nancy Dawson*

Shedden sailed to Bering Strait to assist *Herald* (see above, 1849) in the search for Franklin during a circumnavigation of the globe in this schooner from the Royal Thames Yacht Club. He first met *Herald* at Petropavlovsk in June, rejoined her and *Plover* (see 1848-52) in Kotzebue Sound on 17 July, and offered his help. He sailed with them to Wainwright Inlet, and accompanied Pullen and Hooper on their boat voyage from *Plover* (see 1849-51) until about 80 km east of Point Barrow, where he left a cache of stores at Refuge Inlet. Soon after, on 12 August, his crew began to mutiny, and Shedden was obliged to turn south. He met *Herald* on 24 August, handed over three of the mutineers, and took on board an officer of *Herald*, who was ordered by Captain Kellett to sail *Nancy Dawson* to Mazatlan to await *Herald*. The ship reached Mazatlan on 13 November, where Shedden died shortly after.

*JONES, A.G.E, 1958; *GREAT BRITAIN, PARLIAMENT, 1850a, pp. 9-11, 19, 22, 27-33.

1849-50

British naval supply voyage | James Saunders | HMS *North Star*

From London; 16 May 1849 - 28 September 1850; to resupply Sir James Clark Ross's searching expedition (see 1848-49), then to examine Smith and Jones sounds in search of traces of Franklin's expedition. The ship was detained by bad ice conditions in Melville Bay and forced to winter in Wolstenholme Sound, north-west Greenland. Four of the crew died of scurvy during the winter. When they were released from the ice on 1 August 1850, they progressed westward through Lancaster Sound to Prince Regent Inlet, but were unable to land provisions at Port Bowen or Port Leopold as intended. They therefore turned homeward, leaving a depot in Navy Board Inlet.

GREAT BRITAIN, PARLIAMENT, 1849; *1850a, pp. 64-68; *1851, pp. 56-68; *CYRIAX, 1964.

1849-51

British naval Franklin search expedition | William John Samuel Pullen, William Hulme Hooper

From HMS *Plover* in Wainwright Inlet, Alaska (see 1848-52); 25 July 1849 - early October 1851; to search the coast between Wainwright Inlet and Mackenzie River from two boats. They were accompanied by *Nancy Dawson* (see 1849) as far as Point Barrow. They then continued westward along the coast and ascended Mackenzie and Great Bear rivers as far as Fort Franklin on Great Bear Lake, where Hooper wintered. Pullen continued up Mackenzie River to winter at Fort Simpson.

1850: Hooper joined Pullen at Fort Simpson in the spring, and they set out down the Mackenzie on 11 July intending to search the coasts of Banks Island and Wollaston Peninsula. However, they reached only Cape Bathurst before turning back. They discovered Pullen and Hooper islands, then re-ascended Mackenzie River to Fort Simpson, where they arrived on 5 October. They wintered there, then travelled overland to York Factory and sailed to England on the Hudson's Bay Company supply ship *Prince of Wales* (see 1851) on 9 September. Hooper later recorded his observations on the flora, fauna, and ethnology of Arctic North America and the Chukotski Peninsula.

*HOOPER, 1853; *GREAT BRITAIN, PARLIAMENT, 1850a, pp. 25-33; 1851, pp. 51-56; *PULLEN, H.F., 1954; CALDERWOOD, 1935; PULLEN, W.J.S., 1947; SMITH, D.M., 1877, pp. 460-67; NEATBY, 1970, pp. 110-14.

Hudson's Bay Company supply voyage	David J. Herd	*Prince of Wales* [II]	1850
	Robert Royal	*Prince Albert*	
	William B. Mannock	*Prince Rupert* [VI]	
	Captain May	*Flora*	
	John Midwood	*George*	

From London; 6 June - 14 October, *Prince of Wales* visited York Factory 12 August - 13 September, and reached London on 10 October. *Prince Rupert* visited York Factory 7-31 August, and reached London on 7 October. *Flora* visited York Factory 15-31 August. *George* visited York Factory 24-28 August. *Prince Albert* visited Moose Factory 17 August - 7 September, and reached London on 14 October. *Prince of Wales* carried a second draft of pensioners for the Red River settlement: 20 men and 46 women and children.

HUDSON'S BAY COMPANY (1) and (2).

British naval Franklin search expedition	Henry Kellett	HMS *Herald*	1850

Herald left Honolulu on 24 May for her third and last visit to Bering Strait (see also 1848 and 1849). She met *Investigator* (see 1850-54) off Cape Lisburne, Alaska, returned to Honolulu in September, then sailed back to England, where she arrived in June 1851.

*SEEMANN, 1853, Vol. 2, pp. 175-88; *GREAT BRITAIN, PARLIAMENT, 1851, pp. 19-41.

British Franklin search expedition	Charles Codrington Forsyth	*Prince Albert*	1850
	William Parker Snow		

From Aberdeen; 5 June - 1 October; sponsored by Lady Franklin and by public subscription, and sent to search for Franklin in the region of Boothia Peninsula. They entered Prince Regent Inlet, but ice prevented them from passing beyond Fury Beach. They returned back and entered Wellington Channel. At Cape Riley they discovered a note deposited by Ommanney of Austin's expedition (see 1850-51) describing his discovery of Franklin's winter quarters at Beechey Island. They returned to Britain with this news.

*SNOW, 1851; GREAT BRITAIN, PARLIAMENT, 1851, pp. 68-79.

1850

British depot-laying journey	Robert Dawes Aldrich	sledge

From *Resolute* and *Assistance* off Griffith Island, Barrow Strait (see 1850-51, Austin); 2 - 6 October. They made a depot on Somerville Island, Barrow Strait, in preparation for the sledge journeys of spring 1851 (*qv*).

GREAT BRITAIN, PARLIAMENT, 1852a, pp. 2-4.

1850

British depot-laying journey	Francis Leopold McClintock	3 sledges

From *Resolute* and *Assistance* off Griffith Island, Barrow Strait (see 1850-51, Austin); 2 - 9 October. They made a depot on Cornwallis Island in preparation for the sledge journeys of spring 1851 (*qv*).

GREAT BRITAIN, PARLIAMENT, 1852a, pp. 5-8.

1850

British Franklin search expedition	Sherard Osborn	sledge

From *Resolute* and *Assistance* off Griffith Island, Barrow Strait (see 1850-51, Austin); 10 - 12 October. They searched the shores of Allen Bay, south coast of Cornwallis Island.

GREAT BRITAIN, PARLIAMENT, 1852a, pp. 9-10.

1850

British North-west Passage expedition	Robert John Le Mesurier McClure Stephen Court	sledge

From *Investigator* at Princess Royal Island, Prince of Wales Strait (see below, 1850-54); 21-31 October. They sledged northward along Prince of Wales Strait to Russell Point, Banks Island, where, looking out over Melville Sound, which Parry had discovered from the east (see 1819-20), they confirmed the existence of a Northwest Passage.

OSBORN, 1856, pp. 131-45.

1850-51

British naval Franklin search expedition	Horatio Thomas Austin Erasmus Ommanney John Bertie Cator Sherard Osborn	HMS *Resolute* HMS *Assistance* HMS *Intrepid* HMS *Pioneer*

From London; 3 May 1850 - end of September 1851. They entered Lancaster Sound in mid-August 1850, and Ommanney, whose ships (*Assistance* and *Intrepid*) arrived there before the rest of the fleet, completed the examination of the north shore of the sound between Cape Warrender and Fellfoot Point and, on 23 August, discovered the first traces of Franklin's expedition at Cape Riley and on Beechey Island. The rest of the fleet joined him there on 28 August. They resumed their journey westward, only to be beset in the ice to the north-east of Griffith Island in September. They established winter quarters there, began depot-laying for spring sledge journeys soon after (see below, 1850) and, in spring 1851, undertook many sledge journeys to search for traces of Franklin's expedition on both sides of Barrow Strait (see 1851). The ships were released from the ice on 8 August, and they returned home. *Pioneer* and *Intrepid* made a brief examination of the entrance of Jones Sound during the homeward voyage. Balloons with time fuses were used during the expedition to spread messages for Franklin. Observations of topography, geology, fauna and meteorology were recorded.

*OSBORN, 1852; GREAT BRITAIN, PARLIAMENT, 1850b; 1850c; 1851, pp. 88-96; 1851a; *1852a, pp. 1-298.

1850-51

British Franklin search expedition	Sir John Ross Charles Gerrans Phillips	*Felix* *Mary*

From Loch Ryan, Scotland; 23 May 1850 - 25 September 1851; financed by the Hudson's Bay Company and by public subscription. He sailed in close company with Austin's expedition (see above, 1850-51), and helped to search Beechey Island after the discovery of Franklin's first winter quarters there. He wintered at Assistance Bay, Cornwallis Island, with Penny's expedition (see below, 1850-51). 15 May - 16 June, 1851, Phillips attempted to lead a party northward across Cornwallis Island, but was forced to turn back after he had reached 75°29'N. In summer 1851, Ross made a further search of Beechey Island, and in August he was obliged by shortage of provisions to set off back to Britain. Balloons with time fuses were used during the expedition to spread messages for Franklin; four carrier pigeons were taken to send news home, one of which, released on 5 October 1850, returned to Scotland in five days. The 12-ton yacht *Mary* was left behind at Beechey Island as a depot for future expeditions.

*GREAT BRITAIN, PARLIAMENT, 1851, pp. 83-88; 1852, pp. 4-11; DODGE, 1973, pp. 234-40; SUTHERLAND, 1852, Vol. 2, pp. 67-68, 103-04 [for Phillips's crossing of Cornwallis Island].

1850-51

United States Franklin search expedition (first Grinnell expedition)	Edwin Jesse De Haven Samuel P. Griffin	*Advance* *Rescue*

From New York; 22 May 1850 - 30 September 1851; sponsored by Henry Grinnell and with instructions from the US Navy Department to search Wellington Channel and thence proceed westward towards Cape Walker. They sailed along the south shore of Lancaster Sound as far as Port Leopold, then crossed the sound and helped to search Beechey Island. They joined Austin's and Penny's expeditions (see above and below, 1850-51) at Griffith Island, and soon after were beset in the ice and began to drift northward into Wellington Channel. They drifted as far as 75°24'N, from where they sighted new land (Grinnell Peninsula) to the north. They then drifted southward, and the ice carried them throughout the winter eastwards out of Lancaster Sound and southwards through Baffin Bay. They were finally released in June 1851 in Davis Strait, and returned to the United States.

*KANE, 1834; CORNER, 1972, pp. 85-101; GREAT BRITAIN, PARLIAMENT, 1851, pp. 1-4.

1850-51

British Franklin search expedition	William Penny Alexander Stewart	HMS *Lady Franklin* HMS *Sophia*

From Aberdeen; 13 April 1850 - 21 September 1851; sponsored by the British Government, with instructions from the Admiralty to search Jones Sound, Wellington Channel, Cape Walker, and in the unexplored region beyond. Penny found the entrance to Jones Sound blocked by ice, so they proceeded into Lancaster Sound. In late August, he joined Austin's ships (see above, 1850-51) at Beechey Island, where traces of the Franklin expedition had been found (see above, 1850-51 Austin). He helped to search the island and soon found graves and other evidence to confirm that this was the site of Franklin's first winter quarters. In mid-September, he took up winter quarters in Assistance Bay, Cornwallis Island, next to Sir John Ross's expedition (see above, 1850-51). His sledge journeys of spring 1851 searched the shores of Wellington Channel. Penny wanted to examine the channel more thoroughly with the use of Austin's two steamships *Pioneer* and *Intrepid* but, after consultations between the two captains, they agreed to return home. Penny's ships left Assistance Bay in August 1851, and reached Gravesend on 21 September. During the expedition, they made observations on tides, meteorology, botany, zoology, and geology.

*†SUTHERLAND, 1852; GREAT BRITAIN, PARLIAMENT, 1850a, pp. 133-35; 1850b; 1850c; *1851, pp. 80-83; 1851a; *1852a, pp. 303-67; STEVENSON, ALEX, 1966; †SUTHERLAND, 1853; †SALTER, 1853.

1850-51

Hudson's Bay Company Franklin search expedition	John Rae

Following his failure to reach Victoria Island during Richardson's expedition (see 1847-49), Rae was sent by the company to renew the attempt. He spent the winter of 1850-51 at Fort Confidence, and set out for the mouth of Coppermine River with two men and dog-sledges on 15 April 1851. He crossed to Wollaston Peninsula, explored the coast from Wilbank Bay in the east to near Cape Back in the west, and returned to Kendall River on 10 June. He returned to the coast with boats three days later, crossed Dease Strait from Cape Alexander, and explored the coast of Victoria

Island from Cambridge Bay to Pelly Point on Collinson Peninsula. He returned along the coast of Victoria Island to Cape Lady Franklin, crossed to the mainland, and arrived back at Fort Confidence on 10 September.

*GREAT BRITAIN, PARLIAMENT, 1851, pp. 44-52; 1852c; *RAE, 1852a; 1852b.

1850-54

British naval Franklin search expedition | Robert John Le Mesurier McClure | HMS *Investigator*

From London; 10 January 1850 - 28 September 1854; to search for Franklin by way of Bering Strait. They sailed in company with *Enterprise* (see 1850-55), but the two ships were separated in the Pacific Ocean, and never met again. *Investigator* reached Bering Strait in July 1850, rounded Point Barrow and, on 7 September, sighted south-west Banks Island (which they named Baring Island, not realizing that it was the same island as the one sighted from the north by Parry in 1819 [see 1819-20]). They proceeded up Prince of Wales Strait, and wintered in the middle of the strait at Princess Royal Island.

1851: On 15 August, they continued northward along Prince of Wales Strait, but were stopped by ice within 40 km of Melville Sound. They turned back, sailed around the south and west coasts of Banks Island, and wintered in Mercy Bay on the north coast.

1852: They remained beset and wintered a second time in Mercy Bay.

1853: After members of the crew of *Resolute* (see 1852-54) had found a message deposited by McClure on Melville Island (see 1852), Lieutenant Pim of the *Resolute* was sent to locate *Investigator,* and reached Mercy Bay just as McClure was making plans to abandon ship and march towards the mainland. *Investigator* was abandoned on 3 June, and the crew was transferred to *Resolute.* They wintered again, and finally returned to England on board *North Star* (see 1852-54) and the supply ships *Phoenix* and *Talbot* (see 1854). Records of the expedition include description of physiography, geology, biology, and Eskimo customs.

*OSBORN, 1856; *ARMSTRONG, 1857; *MIERTSCHING, 1855; *NEATBY, 1967; McCLURE, 1853; n.d.; †HOOKER, 1857; *GREAT BRITAIN, PARLIAMENT, 1854, pp. 21-62; 1855, pp. 107-12; NEATBY, 1960; McKENZIE, 1970.

1850-55

British naval Franklin search expedition | Richard Collinson | HMS *Enterprise*

From London; 10 January 1850 - 5 May 1855; to search for Franklin by way of Bering Strait. They sailed in company with *Investigator* (see 1850-54) until the two ships were separated in the Pacific Ocean. *Enterprise* reached Point Barrow in late August 1850, when it was no longer possible to penetrate to the east as *Investigator* had done. She therefore returned south and wintered at Hong Kong.

1851: They returned to Bering Strait, and this time succeeded in passing along the north coast of the continent and through Prince of Wales Strait. However, heavy ice prevented them from entering Melville Sound, and they returned south along Prince of Wales Strait. They tried to sail up the west coast of Banks Island as *Investigator* had done only two weeks before, but were again obliged to turn back, and wintered in Winter Cove, Prince Albert Peninsula, Victoria Island.

1852: They sailed again on 5 August, passed through Dolphin and Union Strait, and wintered at Cambridge Bay, Victoria Island.

1853: On 10 August, they began to return to Point Barrow, and wintered at Camden Bay, Alaska.

1854: They continued westward, met *Plover* (see 1852-54) at Point Barrow, then returned to England.

During the voyage they made observations on meteorology, currents, physiography, Eskimo customs, and collected biological specimens. They also made a number of sledge journeys in search of Franklin (see 1852 and 1853).

COLLINSON, 1855; *1889; GREAT BRITAIN, PARLIAMENT, 1850a, pp. 82-85; 1851, pp. 5-8, 14-18; 1852, pp. 65-78; 1855, pp. 943-52; †BUTLER, 1877.

1851

Hudson's Bay Company supply voyage — David J. Herd, Robert Royal — *Prince of Wales* [II], *Prince Albert*

From London; 4 June - 13 October. *Prince of Wales* visited York Factory 13 August - 9 September, and reached London on 3 October. *Prince Albert* visited Moose Factory 25 August - 10 September, and reached London on 13 October.

HUDSON'S BAY COMPANY (1) and (2).

1851

Hudson's Bay Company exploration of Pelly River — Robert Campbell

From Fort Selkirk (see 1848); 5 June - mid-October; to continue his exploration of Pelly River (see 1842-44). Whilst descending that river, he discovered the mouths of several other rivers, including the Stewart, and met several tribes of Indians who had never before seen white men. The party continued until they reached Fort Yukon (see 1847), thus confirming Campbell's suspicion that Pelly River was a tributary of Yukon River. From there, he continued up Porcupine River to La Pierre's House, travelled on foot to Fort McPherson, and ascended Mackenzie River to Fort Simpson, where he arrived in early August. He then retraced his steps along Mackenzie, Porcupine and Pelly Rivers to Fort Selkirk, where he arrived in mid-October.

HUDSON'S BAY COMPANY, ARCHIVES, B200/b/26/6-7; WILSON, 1970, pp. 111-15; PARNELL AND KIRK, 1942.

1851

Church Missionary Society journey to Moose Factory — John Horden

Horden was the first missionary to visit Moose Factory after the departure of George Barnley in 1848 (see 1840). He arrived from

London on the supply vessel *Prince Albert* (see 1851), and devoted much of the rest of his life to missionary work in the region. In 1852, he was visited by David Anderson, Bishop of Rupert's Land (see 1849 and 1852), who ordained him priest on 24 August. In 1862, he made the first of several visits to Great Whale River, which did not have a permanent missionary until the arrival of Peck in 1876 (see 1876-84). In 1865, he took his first leave, and returned to England for a short time. In 1872, the diocese of Rupert's Land was divided up, and, on 15 December of that year, Horden was consecrated as first bishop of the new diocese of Moosonee. He remained at Moose Factory until his death on 12 January 1893.

BOON, 1962, pp. 76-78; GOULD, 1917, pp. 121-28, 223-25.

British depot-laying journey | George Frederick McDougall | sledge *Endeavour* | 1851

From *Resolute* and *Assistance* off Griffith Island, Barrow Strait (see 1850-51); 4 - 23 April. He examined depots on Somerville and Cornwallis islands made in October 1850 (*qv*), and made fresh depots in preparation for the spring searching expeditions (see below, 1851).

GREAT BRITAIN, PARLIAMENT, 1852a, pp. 13-21.

British Franklin search expedition | Erasmus Ommanney | sledge *Reliance* | 1851
George Frederick Mecham | sledge *Succour*

From *Resolute* and *Assistance* off Griffith Island, Barrow Strait (see 1850-51); 15 April - 14 June. Ommanney sledged to Cape Walker, Russell Island, explored the north coast of Russell Island, then continued along the previously unexplored west coast of Prince of Wales Island, charted Ommanney Bay, then turned back. Mecham's supporting party accompanied Ommanney as far as the north coast of Prince of Wales Island, and returned to the ship on 14 May. The total length of Ommanney's journey was 770 km; Mecham covered 380 km.

GREAT BRITAIN, PARLIAMENT, 1852a, pp. 22-63.

British Franklin search expedition | William Henry James Browne | sledge *Enterprise* | 1851
Charles Ede | sledge *Inflexible*

From *Resolute* and *Assistance* off Griffith Island, Barrow Strait (see 1850-51); 15 April - 28 May. Browne sledged to Cape Walker, Russell Island, then explored the east coast of Prince of Wales Island and offshore islands as far as the north coast of Pandora Island. He covered a total of 600 km. Ede's supporting party accompanied Browne to Cape Walker, searched part of the north coast of Russell Island in company with Ommanney's parties (see above, 1851), and returned to the ships on 5 May. They sledged a total of 280 km.

GREAT BRITAIN, PARLIAMENT, 1852a, pp. 64-82.

1851 British Franklin search expedition — Frederick J. Krabbé — sledge *Success*

From *Resolute* and *Assistance* off Griffith Island, Barrow Strait (see 1850-51); 15 - 28 April. Krabbé's party, an auxiliary to the parties of Ommanney and Browne (see above, 1851), sledged with them to Cape Walker, and covered a total of 185 km.

GREAT BRITAIN, PARLIAMENT, 1852a, pp. 83-87.

1851 British Franklin search expedition — Sherard Osborn — sledge *True Blue*; Richard Vesey Hamilton — sledge *Adventure*

From *Resolute* and *Assistance* off Griffith Island, Barrow Strait (see 1850-51); 15 April - 12 June. Osborn sledged along the west coast of Prince of Wales Island in company with Ommanney (see above, 1851). They parted company on 17 May at about Hawkes Point, on the southern shore of the entrance to Ommanney Bay. From there, Ommanney travelled south-east to explore Ommanney Bay, and Osborn sledged westward, then south-westward, to explore the coast of Prince of Wales Island to about 70°20'N. He made an unsuccessful attempt to march westward across the sea ice, then returned to the ship, having covered a distance of about 837 km. Hamilton's supporting party accompanied Osborn to the north coast of Russell Island, then turned back and examined the east coasts of Young and Lowther islands, Barrow Strait, before reaching the ships. He covered about 320 km.

GREAT BRITAIN, PARLIAMENT, 1852a, pp. 87-112.

1851 British Franklin search expedition — Robert Dawes Aldrich — sledge *Lady Franklin*; Richard Bulkeley Pearse — sledge *Hotspur*

From *Resolute* and *Assistance* off Griffith Island, Barrow Strait (see 1850-51); 15 April - 15 May. Aldrich's party sledged to Bathurst Island and examined the previously unexplored coasts of Bathurst Island between the northern end of Graham Moore Bay and Pell Inlet, and of the western shores of the islands lying north-west of Bathurst Island as far as Cape Aldrich in about 76°11'N. He covered a total distance of about 885 km. Pearse's supporting party accompanied him as far as Cape Cockburn, Bathurst Island, and arrived back at the ship on 9 May, having covered a total distance of 335 km.

GREAT BRITAIN, PARLIAMENT, 1852a, pp. 112-36.

1851 British Franklin search expedition — Francis Leopold McClintock — sledge *Perseverance*; William B. Shellabear — sledge *Dasher*

From *Resolute* and *Assistance* off Griffith Island, Barrow Strait (see 1850-51); 15 April - 4 July. McClintock examined and charted the entrance to McDougall Sound and accompanied Aldrich (see above, 1851) to Cape Cockburn, Bathurst Island. He then continued westward across the sea ice, resurveyed the south coasts of Byam Martin and Melville islands (first discovered by Parry in 1819-20, *qv*), and travelled round Dundas Peninsula into Liddon

Gulf. He then crossed Dundas Peninsula to Winter Harbour and returned to the ship, surveying Graham Moore Bay, Bathurst Island, on the way. He covered a total distance of 1,240 km. Shellabear's supporting party accompanied McClintock to Cape Cockburn, Bathurst Island, and returned to the ships on 9 May, having travelled a distance of 395 km.

GREAT BRITAIN, PARLIAMENT, 1852a, pp. 136-202.

British Franklin search expedition | Abraham Rose Bradford / Walter William May | sledge *Resolute* / sledge *Excellent* | 1851

From *Resolute* and *Assistance* off Griffith Island, Barrow Strait (see 1850-51); 15 April - 4 July. Bradford accompanied McClintock (see above, 1851) to the south-east corner of Melville Island. He then parted from McClintock, and examined the previously unexplored east coast of Melville Island northward to about 76°21'N. Returning to the ships, he made a complete circuit of Byam Martin Island, of which only the south coast had been explored previously. He covered a total of 1,080 km. May's supporting party accompanied him to Cape Gillman, Byam Martin Island, and returned to the ships on 18 May, having covered a distance of about 595 km.

GREAT BRITAIN, PARLIAMENT, 1852a, pp. 202-35

British Franklin search expedition | John P. Cheyne | sledge *Parry* | 1851

From *Resolute* and *Assistance* off Griffith Island, Barrow Strait (see 1850-51); 15-27 April. Cheyne's party sledged to south-east Bathurst Island as an auxiliary to McClintock's and Bradford's parties (see above, 1851).

GREAT BRITAIN, PARLIAMENT, 1852a, pp. 235-41.

British Franklin search expedition | Robert Calder Allen | sledge *Grinnell* | 1851

From *Resolute* and *Assistance* off Griffith Island, Barrow Strait (see 1850-51); 22 April - 12 May. He searched and charted the west coast of Lowther Island, and made a complete circuit of Garrett Island, Barrow Strait.

GREAT BRITAIN, PARLIAMENT, 1852a, pp. 241-50.

British surveying expedition | Frederick J. Krabbé | sledge *Edward Riddle* | 1851

From *Resolute* and *Assistance* off Griffith Island, Barrow Strait (see 1850-51); 20 May - 7 June. He sledged to Cape Walker, Russell Island, to meet the returning sledge expedition led by Browne (see above, 1851), and to survey the cape. He met Browne, who needed no assistance, but he was prevented by bad weather from surveying the cape.

GREAT BRITAIN, PARLIAMENT, 1852a, pp. 270-76.

1851 British surveying expedition — George Frederick McDougall — sledge *Beaufort*

From *Resolute* and *Assistance* off Griffith Island, Barrow Strait (see 1850-51); 22 May - 13 June. Sent to survey "the land connecting Bathurst and Cornwallis Islands," McDougall examined McDougall Sound. However, his examination of the northern shores of the sound was made from a distance, and was insufficient to determine the existence of channels separating the two islands. He also laid down fresh depots for the returning expeditions of McClintock and Bradford (see above, 1851).

GREAT BRITAIN, PARLIAMENT, 1852a, pp. 276-82

1851 British surveying expedition — Robert Calder Allen, John P. Cheyne — sledge *Raper*

From *Resolute* and *Assistance* off Griffith Island, Barrow Strait (see 1850-51); sent to survey the coast of Cornwallis Island between Assistance Harbour and Cape Martyr.

GREAT BRITAIN, PARLIAMENT, 1852a, p. 251.

1851 British Franklin search expedition — Walter William May, Clements Markham — sledge

From *Resolute* and *Assistance* off Griffith Island, Barrow Strait (see 1850-51); 26 June - 2 July. He made a complete circuit of Griffith Island in search of traces of Franklin's expedition.

GREAT BRITAIN, PARLIAMENT, 1852a, pp. 264-70.

1851 British depot laying expedition — George Frederick Mecham — sledge *Russell*

From *Resolute* and *Assistance* off Griffith Island, Barrow Strait (see 1850-51); 27 May - 19 June. He made a complete circuit of Russell Island, examined depots, and made fresh ones for the returning expeditions of Ommanney and Osborn (see above, 1851).

GREAT BRITAIN, PARLIAMENT, 1852a, pp. 251-64.

1851 British Franklin search expedition — John Stuart — sledge

From *Lady Franklin* and *Sophia* in Assistance Bay, Cornwallis Island (see 1850-51); 6 May - 6 June. He crossed Wellington Channel and followed the south coast of Devon Island eastward to Cape Hurd, and found traces of one of Franklin's camps near Caswall Tower in Radstock Bay.

GREAT BRITAIN, PARLIAMENT, 1852a, pp. 360-65.

1851 British Franklin search expedition — Robert Anstruther Goodsir — sledges

From *Lady Franklin* and *Sophia* in Assistance Bay, Cornwallis Island (see 1850-51); 6 May - 17 June. He accompanied Penny (see below, 1851) along the east coast of Cornwallis Island, then explored the north coast as far as Little Cornwallis Island, where open water obliged him to turn back.

GREAT BRITAIN, PARLIAMENT, 1852a, pp. 346-59; SUTHERLAND, 1852, Vol. 2, pp. 105-06.

British Franklin search expedition | Alexander Stewart, Peter Cormack Sutherland | 2 sledges | **1851**

From *Lady Franklin* and *Sophia* in Assistance Bay, Cornwallis Island (see 1850-51); 6 May - 22 June. They crossed Wellington Channel and sledged northward along the west coast of Devon Island. Stewart's party traced the coast as far as Cape Becher, where they found open water and were obliged to turn back to the ships. Sutherland's party turned eastward into Prince Alfred Bay to search for the western outlet of Jones Sound. Failing to find it, they returned to the ships on 21 June.

GREAT BRITAIN, PARLIAMENT, 1852a, pp. 323-39.

British Franklin search expedition | William Penny | sledge | **1851**

From *Lady Franklin* and *Sophia* in Assistance Bay, Cornwallis Island (see 1850-51); 9 - 20 May. He sledged along the east coast of Cornwallis Island. From there, he discovered Queens Channel, which was free of ice, so he returned to the ships for a boat.

SUTHERLAND, 1852, Vol. 2, pp. 122-36.

British Franklin search expedition | William Penny | boat | **1851**

From *Lady Franklin* and *Sophia* in Assistance Bay, Cornwallis Island (see 1850-51); 10 June - 25 July. They returned to Baillie-Hamilton Island and sailed around it; then discovered and landed on Margaret Island. They sighted Dundas Island, but were unable to reach it, and discovered Baring Island. They then crossed Queens Channel to Cape Becher, Devon Island. Returning, they visited Dundas and Baillie-Hamilton islands, abandoned the boat at Abandon Bay, Cornwallis Island, and continued to the ships on foot.

SUTHERLAND, 1852, Vol. 2, pp. 141-86.

British Franklin search expedition | Samuel Gurney Cresswell | sledge | **1851**

From *Investigator* at Princess Royal Island, Prince of Wales Strait (see 1850-54); 18 April - 20 May. They sledged to the northern end of Prince of Wales Strait, then turned north-west and explored about 275 km of the coast of Banks Island.

OSBORN, 1856, pp. 163, 181-82.

1851

British Franklin search expedition | William Henry Haswell | sledge

From *Investigator* at Princess Royal Island, Prince of Wales Strait (see 1850-54); 18 April - 29 May. He sledged southward to the northern shore of Wollaston Peninsula, Victoria Island, 70°45'N, 114°W; a point only 65 km from the furthest reached by Rae (see 1850-51) 10 days later, on 24 May.

OSBORN, 1856, pp. 163, 184-85.

1851

British Franklin search expedition | Robert Wynniatt | sledge

From *Investigator* at Princess Royal Island, Prince of Wales Strait (see 1850-54); 18 April - 7 June. He sledged to the northern end of Princes of Wales Strait, then explored eastward along the north coast of Victoria Island. He turned back on 26 May from a point about 110°W, just east of Wynniatt Bay.

OSBORN, 1856, pp. 163-64, 189.

1851

British Franklin search expedition | Robert John Le Mesurier McClure, Johann August Miertsching

From *Investigator* at Princess Royal Island, Prince of Wales Strait (see 1850-54); 30 May - (?) June. They travelled on foot to Wollaston Peninsula, Victoria Island, to open communication with Eskimos whom Haswell had met during his sledging journey (see above, 1851). These Eskimos assured them that Wollaston Peninsula was a part of Victoria Island.

OSBORN, 1856, pp. 186-88.

1851

British naval supply voyage | George Greville Wellesley | HMS *Daedalus*

Daedalus, which was stationed in the Pacific Ocean, was sent from Valparaiso to Bering Strait to resupply *Plover* (see 1848-52). She met *Plover* at Port Clarence, Alaska, on 30 July, gave her fresh supplies, made repairs to the ship, and replaced most of her crew. *Daedalus* returned south on 1 October.

GREAT BRITAIN, PARLIAMENT, 1852, pp. 79-89.

1851-52

British Franklin search expedition | William Kennedy, Joseph René Bellot | *Prince Albert*

From Aberdeen; 22 May 1851 - 7 October 1852; sponsored by Lady Franklin and by public subscription, they were sent to search Prince Regent Inlet and in the area south-west of Cape Walker. They met De Haven's returning expedition (see 1850-51) in Davis Strait. They tried to enter Port Leopold, Somerset Island, for the winter; Kennedy landed, but was separated from the ship. Bellot took the ship on to Batty Bay, where he established winter quarters, then he returned northward overland to rescue Kennedy.

1852: In early spring, Kennedy and Bellot discovered Bellot Strait during a sledging expedition (see below, 1852). They sailed from Batty Bay on 6 August, visited *North Star* (see 1852-54) at Beechey Island, then returned home. They recorded the physiography of Prince of Wales Island. They took with them six carrier pigeons, which Ross had carried in 1850-51 (*qv*), but apparently did not use them.

*KENNEDY, 1853; *BELLOT, 1854; 1855; HODGSON, 1974.

1851-52

United States whaling voyage	Captain Quayle	*McLellan*

From New London. Following an unsuccessful whaling season, *McLellan* entered Cumberland Sound, where it was decided that 12 members of the crew, led by the first mate, Sidney Buddington, should winter in the sound, begin whaling in the spring, and wait for the return of the *McLellan* in the summer. They built a stone house on the shore of the sound, and remained there until the following summer. *McLellan* sank during her outward voyage in 1852, and the wintering party had to wait until September before they were rescued by Captain Parker in *Truelove* of Hull. This was the first time that whalers had deliberately wintered in the region of Davis Strait and Baffin Bay.

HALL, C.F., 1864, pp. 217-18

1852

Hudson's Bay Company supply voyage	David J. Herd	*Prince of Wales* [II]
	Robert Royal	*Prince Albert*
	William B. Mannock	*Marten*
	Richard Sherris	*Superior*

From London; 2 June - 28 October. *Prince Albert* visited Moose Factory 15 August - 5 September, and reached London on 28 October, *Prince of Wales* visited York Factory 15 August - 16 September, and reached London on 26 October. *Marten* reached York Factory on 19 August, and remained in Hudson Bay. *Superior* visited York Factory 25 August - 1 September, then left for St John's, Newfoundland.

HUDSON'S BAY COMPANY (1) and (2).

1852

Anglican Church journey to James Bay	Bishop David Anderson

On 28 June 1852, Anderson, the first Bishop of Rupert's Land (see 1849), left his seat at Red River to make his first visit to James Bay. On 24 August, at Moose Factory, he ordained John Horden, who had represented the church in that area since 1851 (*qv*). He also met E. A. Watkins (see 1852), whom he sent to establish a mission at Fort George. He then returned to Red River, after an absence of 16 weeks.

ANDERSON, 1854.

1852

Church Missionary Society journey to Fort George — E. A. Watkins

Watkins arrived at Moose Factory with his wife in summer 1852 expecting to represent the Church Missionary Society in that area. However, John Horden had already arrived there the previous year (*qv*), and Bishop David Anderson decided he should remain. Watkins, therefore, went to establish a mission at Fort George, where he remained until 1857.

BOON, 1962, p. 78.

1852

Establishment of Fort Rae

The Hudson's Bay Company built Fort Rae on the northern arm of Great Slave Lake in 1852. It is still in existence today.

MOCCASIN TELEGRAPH, 1970, p. 123.

1852

British Franklin search expedition — William Kennedy, Joseph René Bellot — dog sledge

From *Prince Albert* at Batty Bay, Somerset Island (see above, 1851-52); 25 February - 30 May. They sledged southward to discover Bellot Strait, passed through it, and crossed Prince of Wales Island to Ommanney Bay. They returned to *Prince Albert* by re-crossing Prince of Wales Island to Peel Sound, travelling north to Cape Walker, and then by sledging along the north and east coasts of Somerset Island.

KENNEDY, 1853, pp. 114-52.

1852

British Franklin search expedition — Robert John Le Mesurier McClure — sledge

From *Investigator* in Mercy Bay, Banks Island (see 1850-54); 11 April - 11 May. They sledged across McClure Strait to Winter Harbour, Banks Island, where they left a note describing the position of *Investigator*. Lieutenant Mecham found this note later in the year during an expedition from *Resolute* (see below, 1852-54 and 1852), and this led to the rescue of the crew of *Investigator* in 1853.

OSBORN, 1856, pp. 234-36.

1852

British Franklin search expedition — C. T. Jago — sledge *Victoria*

From *Enterprise* in Winter Cove, Prince Albert Peninsula, Victoria Island (see 1850-55); 12 April - 30 May. They sledged southward across Minto Inlet to the region of Cape Wollaston.

COLLINSON, 1889, pp. 182-183, 204-07.

British Franklin search expedition | Richard Collinson | sledge *Enterprise* | 1852

From *Enterprise* in Winter Cove, Prince Albert Peninsula, Victoria Island (see 1850-55); 16 April - 6 June. They sledged to the northern end of Prince of Wales Sound, then along the north coast of Victoria Island to Reynolds Point.

COLLINSON, 1889, pp. 184-204.

British Franklin search expedition | Murray T. Parkes | sledge *Resolution* | 1852

From *Enterprise* in Winter Cove, Prince Albert Peninsula, Victoria Island (see 1850-55); 16 April - 29 June. He sledged across Melville Sound and landed at Cape Providence, Melville Island. Lack of provisions prevented his reaching Winter Harbour, where he would probably have met McClure's sledge expedition from *Investigator* (see above, 1852).

COLLINSON, 1889, pp. 184-85, 209-18.

British Franklin search expedition | Edward Augustus Inglefield | *Isabel* | 1852

From Peterhead; 10 July - 4 November; sponsored by Lady Franklin and by public subscription, and sent to search Jones Sound and the west coast of Baffin Bay southward to Labrador. They surveyed the west coast of Greenland to Cape Alexander, and penetrated Smith Sound to 78°28'N, the farthest north until then achieved in Baffin Bay. They entered Jones Sound as far as 84°10'W before proceeding through Lancaster Sound to Beechey Island, where they communicated with *North Star* (see 1852-54). They continued the search along the east coast of Baffin Island, and reached Cumberland Sound before being forced by the approach of winter to return home. Observations were made on the botany and physiography.

INGLEFIELD, *†1853; 1853a; †SUTHERLAND, 1856; GREAT BRITAIN, PARLIAMENT, 1852b, pp. 33-41.

British naval supply voyage | Charles Frederick | HMS *Amphitrite* | 1852

Amphitrite was stationed in the Pacific Ocean and was sent from Honolulu to resupply *Plover* (see 1848-52) in Bering Strait. She met *Plover* at Port Clarence, Alaska, on 30 June, gave her fresh supplies, and made replacements in the crew. Captain Moore was replaced by Rochfort Maguire, who continued the search in Bering Strait until 1854 (see 1852-54). *Amphitrite* met *Plover* again at Cape Lisburne and later, in the same place, picked up Maguire's boat expedition from *Plover*, which had been missing. She returned to Port Clarence, then left for Valparaiso on 6 September.

GREAT BRITAIN, PARLIAMENT, 1852b, pp. 60-72.

1852 | British Franklin search expedition | Sir Edward Belcher
George Henry Richards | boats

From *Assistance* and *Pioneer* in Northumberland Sound, Grinnell Peninsula, Devon Island (see below, 1852-54); 23 August - 8 September. They sailed north-eastward to discover Exmouth Island, Table Island, and the south coast of Cornwall Island.

GREAT BRITAIN, PARLIAMENT, 1855, pp. 117-22; BELCHER, 1855, pp. 91-121.

1852 | British depot-laying journey | Francis Leopold McClintock
Robert Charles Scott | 2 carts

From *Resolute* and *Intrepid* at Dealy Island, Melville Sound (see below, 1852-54); 14 September - 2 October. They made a depot in Hecla and Griper Bay, Melville Island, in preparation for the sledge journeys of spring 1853.

GREAT BRITAIN, PARLIAMENT, 1855, pp. 469-78.

1852 | British depot-laying journey | George Frederick Mecham
George Strong Nares | sledge *Discovery*
sledge *Fearless*

From *Resolute* and *Intrepid* at Dealy Island, Melville Sound (see below, 1852-54); 22 September - 14 October. They made a depot in Liddon Gulf, Melville Island, in preparation for the sledge journeys of spring 1853. On the return journey, a note deposited by McClure during a sledge expedition from *Investigator* (see 1852) was found at Winter Harbour. This was the first news of McClure's missing expedition (see 1850-54) since 1850, and it led to the relief of his crew in the following spring (see 1853, Pim).

GREAT BRITAIN, PARLIAMENT, 1855, pp. 489-97, 596-601.

1852 | British depot-laying journey | Bedford Clapperton Trevelyan Pim
Emile Frédéric de Bray | sledge *Marie*

From *Resolute* and *Intrepid* at Dealy Island, Melville Sound (see 1852-54); 22 September - 8 October. They made a depot at Cape Providence, Melville Island, in preparation for the sledge journeys of spring 1853.

GREAT BRITAIN, PARLIAMENT, 1855, pp. 611-17.

1852 | British depot-laying journey | Richard Vesey Hamilton | sledge *Hope*

From *Resolute* and *Intrepid* at Dealy Island, Melville Sound (see 1852-54); 22 September - 7 October. They made a depot at Cape Providence, Melville Island, and reported on the possibility of crossing to Banks Island that year to make contact with *Investigator* (see 1850-54), following the discovery of a note describing the whereabouts of *Investigator* (see above, 1852, Mecham). They said it would be impossible.

GREAT BRITAIN, PARLIAMENT, 1855, pp. 617-23.

British depot-laying journey	Francis Leopold McClintock	sledge *Star of the North*	1852
	Robert Charles Scott	sledge *Hero*	

From *Resolute* and *Intrepid* at Dealy Island, Melville Sound (see 1852-54); 7 - 25 October. They made depots in Hecla and Griper Bay, Melville Island, in preparation for the sledge expeditions of spring 1853.

GREAT BRITAIN, PARLIAMENT, 1855, pp. 587-96.

British Franklin search expedition	Robert McCormick	boat *Forlorn Hope*	1852

From *North Star* at Beechey Island, south-west Devon Island (see below, 1852-54); 19 August - 8 September; to assist the search for Franklin by determining whether there was any water communication between Baring Bay, west Devon Island, and Jones Sound. He followed the west coast of Devon Island to Point Owen, just beyond Baring Bay, where he turned back.

McCORMICK, 1854; 1884, pp. 105-58; GREAT BRITAIN, PARLIAMENT, 1850a, pp. 121-27; 1854, pp. 187-225.

British Franklin search expedition	William John Samuel Pullen	boat	1852

From *North Star* at Beechey Island, south-west Devon Island (see below, 1852-54); 26 August - 2 September; to examine Maxwell Bay, south coast of Devon Island, in search of a passage through to Jones Sound. He was delayed by a storm in Gascoyne Inlet for one day, and reached Maxwell Bay on 29 August. There, he found that there was no water communication between the head of the bay and Jones Sound, so he returned to the ship.

GREAT BRITAIN, PARLIAMENT, 1854, pp. 108-09.

British depot-laying journey	William B. Shellabear	boat	1852

From *North Star* at Beechey Island, south-west Devon Island (see below, 1852-54); 8 - 12 September. He made a depot at Cape Bowden, Devon Island.

GREAT BRITAIN, PARLIAMENT, 1854, pp. 112-13, 133; 1855, pp. 288-90.

British naval Franklin search expedition	Sir Edward Belcher	HMS *Assistance*	1852-54
	Sherard Osborn	HMS *Pioneer*	

From London; 15 April 1852 - 28 September 1854. A fleet of five ships under the general command of Belcher was sent to continue the search for Franklin in the direction of Wellington Channel, and to carry supplies to Melville Island for Collinson's and McClure's searching expeditions (see 1850-54 and 1850-55). *Resolute* and *Intrepid* sailed to Melville Island (see 1852-54, below), *North*

Star remained at Beechey Island as a base (see 1852-54, below) whilst Belcher took *Assistance* and *Pioneer* up Wellington Channel and wintered in Northumberland Sound, newly discovered on north-west Grinnell Peninsula, Devon Island.

1853: Spring sledge parties searched to the north of Grinnell Peninsula and Bathurst Island and on the coasts of Sabine Peninsula, Melville Island (see 1853). The ships sailed south on 14 July, but were beset in Wellington Channel, and wintered in the ice north of Cape Osborn, Devon Island.

1854: The ships were released in August, but could not reach Lancaster Sound. Belcher decided to abandon them, and the crews joined those of *Resolute, Intrepid and Investigator,* which had also been abandoned, on board *North Star.* The supply ships *Phoenix* and *Talbot* (see 1854) arrived soon after, the men were distributed among the three ships, and they returned home.

Scientific results of the expedition included observations of meteorology, geology, tides, and marine biology.

*†BELCHER, 1855; *GREAT BRITAIN, PARLIAMENT, 1854, pp. 62-85; 1855, pp. 12-468; †HOOKER, 1857; OSBORN, 1856a; †1858; NEATBY, 1970, pp. 195-219; †ADAMS, A. L., 1875.

1852-54

British naval Franklin search expedition	Henry Kellett Francis Leopold McClintock	HMS *Resolute* HMS *Intrepid*

From London; 15 April 1852 - 28 September 1854. These two of the five ships under the general command of Belcher (see above, 1852-54) parted from the rest of the fleet at Beechey Island on 15 August 1852 and sailed to Melville Island to search for *Enterprise* (see 1850-55) and *Investigator* (see 1850-54). They wintered at Dealy Island off the south coast of Melville Island. An autumn sledge expedition discovered news of the whereabouts of *Investigator* (see above, 1852, Mecham).

1853: Spring sledge expeditions searched the coasts of Melville, Eglinton, and Prince Patrick islands, and visited Banks and Victoria islands. One party located *Investigator* (see 1853, Pim), which was thereupon abandoned and the crew taken on board *Resolute.* The ships began the return to Beechey Island on 17 August, but were beset off Cape Cockburn, Bathurst Island, and had to winter there.

1854: Belcher prematurely ordered the abandonment of both ships in spring, and the crews set off to join *North Star* (see below, 1852-54) at Beechey Island on 13 May. They later returned home in this ship and in the supply ships *Phoenix* and *Talbot* (see 1854). *Resolute* later drifted out to Davis Strait, where she was discovered in 1855 by Captain Buddington of the US whaler *George Henry* and was later restored to England.

During the expedition, observations were made on physiography, geology, meteorology, and biology. Balloons with time fuses were used to distribute messages for Franklin.

*McDOUGALL, 1857; *GREAT BRITAIN, PARLIAMENT, 1854 pp. 85-98; *1855, pp. 69-106, 469-732; McCLINTOCK, 1858.

British naval Franklin search expedition — William John Samuel Pullen — HSM *North Star*

1852-54

From London; 15 April 1852 - 28 September 1854. One of the five ships in Belcher's squadron (see above, 1852-54), they remained in Erebus and Terror Bay, Beechey Island, as a base for the four other ships. In May 1854, the crews of *Resolute, Intrepid,* and *Investigator* (see above, 1850-54 and 1852-54) arrived on board, and were followed in August by those of *Assistance* and *Pioneer* (see above, 1852-54), all of these ships having been abandoned. The supply vessels *Phoenix* and *Talbot* arrived in late August, took most of these men (263 in all) on board, and the three ships returned to England.

Several sledge and boat expeditions left *North Star* to search for Franklin, to lay depots, and to carry despatches to Belcher at Northumberland Sound, Grinnell Peninsula (see above and below, 1852, 1853, and 1854).

*GREAT BRITAIN, PARLIAMENT, 1854, pp. 98-147; 1855, pp. 733-830; *McCORMICK, 1884, Vol. 2, pp. 1-159; PULLEN, H.F., 1954.

British naval Franklin search expedition — Rochfort Maguire — HMS *Plover*

1852-54

Maguire relieved Moore (see 1848-52) of the command of *Plover* on 30 June 1852 at Port Clarence, Alaska. In July, he sailed to Icy Cape, and from there made a boat journey to Point Barrow to communicate with the Eskimos and to find a winter harbour for *Plover*. When he returned to Icy Cape, *Plover* had already left her rendezvous, and he had to continue by boat to Cape Lisburne, where he was picked up by *Amphitrite* (see 1852). He returned in this ship to Port Clarence, rejoined *Plover,* and sailed back north in her to winter at Point Barrow.

1853: He sailed to Port Clarence to receive supplies from *Rattlesnake* (see 1853-54), and returned at once to Point Barrow.

1854: He sailed to Port Clarence to receive supplies from *Trincomalee* (see 1854), then returned again to Point Barrow, where *Enterprise* appeared after an absence of three years. Both ships sailed to Port Clarence, then continued south – *Plover* to San Francisco

GREAT BRITAIN, PARLIAMENT, 1852b, pp. 60-69; *1854, pp. 160-86; 1854a, pp. 2-3; *1855, pp. 905-42; OSBORN, 1856, pp. 351-405.

Hudson's Bay Company supply voyage	David J. Herd	*Prince of Wales* [II]
	Robert Royal	*Prince Albert*
	William B. Mannock	*Prince Rupert* [VI]

1853

From London; 8 June - 24 October, *Prince of Wales* visited York Factory 16 August - 12 September, and reached London on 6 October. *Prince Albert* visited Moose Factory 22 August - 17 September, and reached London on 24 October. *Prince Rupert* left London on 2 April, visited Moose Factory 16 August - 15 September, and reached London on 20 October.

HUDSON'S BAY COMPANY (1) and (2).

1853

British expedition in search of *Investigator* | Bedford Clapperton Trevelyan Pim | sledge *John Barrow*
William Thomas Domville | sledge *James Fitzjames*

From *Resolute* and *Intrepid* at Dealy Island, Melville Sound; 10 March - 19 April; following the discovery by Mecham (see above, 1852) of a note describing the location of *Investigator* (see 1850-54), which had been missing since 1850, Pim and Domville set out to make contact with that ship. They found her at Mercy Bay, Banks Island, on 6 April, and returned to *Resolute* with Captain McClure to discuss the question of abandoning her.

GREAT BRITAIN, PARLIAMENT, 1855, pp. 646-60, 664-77.

1853

British Franklin search expedition | Francis Leopold McClintock | sledge *Star of the North*
Emile Frédéric de Bray | sledge *Hero*

From *Resolute* and *Intrepid* at Dealy Island, Melville Sound (see 1852-54); 4 April - 18 July. They crossed Melville Island to Hecla and Griper Bay, then explored the coast towards the island's north-western extremity. There, de Bray's supporting party turned back and reached the ship on 18 May. McClintock continued across Fitzwilliam Strait, discovered northern Prince Patrick Island, and explored its north and east coasts between Intrepid Inlet and Satellite Bay. He also explored the northern coasts of Eglinton Island and the west and south coasts of Emerald Island. He covered a total distance of 1,660 km. De Bray covered 615 km.

GREAT BRITAIN, PARLIAMENT, 1855, pp. 540-87, 478-89.

1853

British Franklin search expedition | George Frederick Mecham | sledge *Discovery*
George Strong Nares | sledge *Perseverance*

From *Resolute* and *Intrepid* at Dealy Island, Melville Sound (see 1852-54); 4 April - 6 July. Mecham sledged through Liddon Gulf, Melville Island, discovered the south coast of Eglinton Island, and explored the south and west coasts of Prince Patrick Island. Returning, he crossed the interior of Prince Patrick Island, and sledged around the north coast of Eglinton Island. He covered a total distance of 1,620 km. Nare's supporting party accompanied him to the south coast of Eglinton Island and returned to the ship on 1 June.

GREAT BRITAIN, PARLIAMENT, 1855, pp. 498-540, 601-10.

1853

British depot-laying journey | Richard Vesey Hamilton | sledge *Hope*
Richard Roche | sledge *Beauty*

From *Resolute* and *Intrepid* at Dealy Island, Melville Sound (see 1852-54); 4 - 17 April. They carried depots to Sabine Peninsula, Melville Island.

GREAT BRITAIN, PARLIAMENT, 1855, pp. 624, 643-46.

British Franklin search expedition | Richard Vesey Hamilton | sledge *Hope* | 1853

From *Resolute* and *Intrepid* at Dealy Island, Melville Sound (see 1852-54); 27 April - 21 June. They crossed Melville Island to Sabine Bay, then explored the west coast of Sabine Peninsula to its northernmost tip, and turned back on 3 June at about 76°15'N on the west coast of the peninsula. Returning, they charted Vesey Hamilton Island. During the expedition, they met Richards and Osborn on journeys from *Assistance* and *Pioneer* (see below, 1853). They covered a total of 915 km.

GREAT BRITAIN, PARLIAMENT, 1855, pp. 625-41.

British relief expedition | William Thomas Domville, Robert John Le Mesurier McClure | sledge *John Dyer* | 1853

From *Resolute* and *Intrepid* at Dealy Island, Melville Sound (see 1852-54); 5 May - 9 June. McClure, captain of *Investigator* (see 1850-54), had earlier travelled to *Resolute* to discuss the question of abandoning his ship (see above, 1853). He now returned to *Investigator* to examine the health of his crew, and carried his invalids back to *Resolute.*

GREAT BRITAIN, PARLIAMENT, 1855, pp. 672-77.

British sledging journey | Richard Roche, Samuel Gurney Cresswell | sledge | 1853

From *Resolute* and *Intrepid* at Dealy Island, Melville Sound (see 1852-54). He took Lieutenant Cresswell of the *Investigator* to Beechey Island for a passage home on the supply ship *Phoenix* (see below, 1853) with *Investigator's* despatches. Cresswell, being the first member of the crew of *Investigator* to reach England, thus became the first to travel through the North-west Passage.

GREAT BRITAIN, PARLIAMENT, 1855, pp. 685-89.

British depot-laying journey | Bedford Clapperton Trevelyan Pim | sledge *Murchison* | 1853

From *Resolute* and *Intrepid* at Dealy Island, Melville Sound (see 1852-54); 19 May - 9 June. He carried a depot to Cape Fisher, Melville Island, for the use of McClintock's returning sledge expedition (see above, 1853).

GREAT BRITAIN, PARLIAMENT, 1855, pp. 678-85.

British depot-laying journey | George Strong Nares | sledge | 1853

From *Resolute* and *Intrepid* at Dealy Island, Melville Sound (see 1852-54); 3 - 14 June. He picked up a cart that had been left at Winter Harbour, Melville Island, and took it to Liddon Gulf to lay a depot for the use of Mecham's returning sledge expedition (see above, 1853).

GREAT BRITAIN, PARLIAMENT, 1855, p. 610.

1853

British depot-laying journey | George Henry Richards | 6 sledges

From *Assistance* and *Pioneer* at Northumberland Sound, Devon Island (see 1852-54); 22 - 29 March. He made a depot at Cape Lady Franklin, Cornwallis Island.

GREAT BRITAIN, PARLIAMENT, 1855, pp. 122-34.

1853

British Franklin search expedition | George Henry Richards | sledge *Sir Edward*

From *Assistance* and *Pioneer* at Northumberland Sound, Devon Island (see 1852-54); 10 April - 12 July. They sledged westward along the north coasts of Sherard Osborn, Helena, and Cameron islands to the east coast of Melville Island, then continued around the northern end of Sabine Peninsula and down the west coast of the peninsula to visit HMS *Resolute* (see 1852-54) at Dealy Island. They then returned around the east coast of Melville Island, around the north coast of Cameron Island, and through the strait between Bathurst Island and Helena and Sherard Osborn islands. They covered a total distance of 1,300 km.

GREAT BRITAIN, PARLIAMENT, 1855, pp. 180-83, 304-40.

1853

British Franklin search expedition | Sherard Osborn | sledge *John Barrow*

From *Assistance* and *Pioneer* at Northumberland Sound, Devon Island (see 1852-54); 10 April - 15 July. He accompanied Richards as far as the north-eastern tip of Sabine Peninsula, Melville Island. He then retraced his steps around the northern end of Cameron Island, passed through the strait between Sherard Osborn and Helena islands and Bathurst Island, and explored the east coast of Bathurst Island southward to about 75°40'N before returning to the ship. He covered a total of 1,485 km.

GREAT BRITAIN, PARLIAMENT, 1855, pp. 187-254.

1853

British Franklin search expedition
Francis B. Herbert — sledge *Success*
Walter William May — sledge *Reliance*
James Blair Grove — sledge *Dauntless*
David Lyall — sledge *Lady Franklin*

From *Assistance* and *Pioneer* at Northumberland Sound, Devon Island (see 1852-54); 10 April - 12 June. These were supporting parties to the sledge expeditions of Richards and Osborn (see above, 1853). Grove, Lyall and Herbert accompanied them to north-east Bathurst Island, and returned to the ships on 20 April, 3 May and 13 May respectively. May accompanied them to the coast of Melville Island, and returned to the ships on 12 June.

GREAT BRITAIN, PARLIAMENT, 1855, pp. 145-46, 163-69, 162, 280-88, 137-39, 144-45, 170-79.

British depot-laying journey	John H. Allard J. F. Loney	sledge *Enterprise* sledge *Reward*	1853

From *Assistance* and *Pioneer* at Northumberland Sound, Devon Island (see 1852-54); 12 - 19 April. They made a depot on Princess Royal Island, Belcher Channel, in preparation for Belcher's sledging expedition to the north-east of Northumberland Sound (see below, 1853).

GREAT BRITAIN, PARLIAMENT, 1855, pp. 134-36.

British despatch-carrying journey	John P. Cheyne	sledge *Victory*	1853

From *Assistance* and *Pioneer* at Northumberland Sound, Devon Island; 21 - 28 April. He deposited despatches at Cape Becher, Devon Island, for W.J.S. Pullen in *North Star* (see 1852-54).

GREAT BRITAIN, PARLIAMENT, 1855, pp. 139-43.

British Franklin search expedition	Sir Edward Belcher James Blair Grove John H. Allard	sledge *Londesborough* sledge *Dauntless* sledge *Enterprise*	1853

From *Assistance* and *Pioneer* at Northumberland Sound, Devon Island (see 1852-54); 2 May - 23 June. Belcher sledged along the north coast of Devon Island as far as Cardigan Strait. He discovered Buckingham, North Kent, and Graham islands. Grove's supporting party accompanied him as far as Princess Royal Island, then turned back. Allard's supporting party accompanied him to Princess Royal Island, returned to the ship on 19 May, then went again to Princess Royal Island with a depot for Belcher. He arrived back at the ship on 20 June.

GREAT BRITAIN, PARLIAMENT, 1855, pp. 154-61; *BELCHER, 1855, Vol. 1, pp. 248-328.

British despatch-carrying journey	John P. Cheyne David Lyall	sledge *Victory*	1853

From *Assistance* and *Pioneer* at Northumberland Sound, Devon Island (see 1852-54). They sledged to Cape Becher, Devon Island, where they deposited despatches for Captain Pullen of *North Star* (see 1852-54).

GREAT BRITAIN, PARLIAMENT, 1855, pp. 147-52.

British despatch-carrying journey	John P. Cheyne James Blair Grove	sledge *Dauntless*	1853

From *Assistance* and *Pioneer* in Wellington Channel (see 1852-54); 3 - 20 October. They carried despatches to *North Star* at Beechey Island (see 1852-54).

GREAT BRITAIN, PARLIAMENT, 1855, pp. 294-95.

1853 British despatch-carrying journey — William John Samuel Pullen, Alfred H. Alston — 2 sledges

From *North Star* at Beechey Island; 21 March - 24 April. They deposited despatches for Captain Belcher of *Assistance* and *Pioneer* (see 1852-54) at Cape Becher, Devon Island.

GREAT BRITAIN, PARLIAMENT, 1855, pp. 747-57.

1853 British despatch-carrying journey — William John Samuel Pullen, William B. Shellabear — boat and sledge

From *North Star* at Beechey Island (see 1852-54); 11 July - 11 August. They set out for Cape Becher to deposit despatches for Belcher (see 1852-54), but met *Assistance* and *Pioneer* sailing south down Wellington Channel. They received orders from Belcher and returned to *North Star*.

GREAT BRITAIN, PARLIAMENT, 1855, pp. 766-74.

1853 British despatch-carrying journey — William B. Shellabear — boat

From *North Star* at Beechey Island (see 1852-54); 2-8 September. He tried unsuccessfully to reach *Assistance* and *Pioneer* (see 1852-54) in Wellington Channel.

GREAT BRITAIN, PARLIAMENT, 1855, pp. 291-93.

1853 British despatch-carrying journey — William B. Shellabear — sledge

From *North Star* at Beechey Island (see 1852-54); 23 September - 10 October. He carried despatches to *Assistance* and *Pioneer* in Wellington Channel (see 1852-54).

GREAT BRITAIN, PARLIAMENT, 1855, pp. 296-300.

1853 British despatch-carrying journey — William John Samuel Pullen — sledge

From *North Star* at Beechey Island (see 1852-54); 4 - 22 October. He carried despatches to *Assistance* and *Pioneer* in Wellington Channel.

GREAT BRITAIN, 1855, pp. 300-04, 790-93.

1853 British Franklin search expedition — Richard Collinson, sledge *Enterprise*; C. T. Jago, sledge *Victoria*; George Fowler, sledge *Royal Albert*

From *Enterprise* at Cambridge Bay, Victoria Island (see 1850-55); 12 April - 31 May. They searched northwards along the east coast of Victoria Island and turned back at Gateshead Island, a short distance beyond the farthest point reached by Rae in 1851 (see 1850-51).

COLLINSON, 1889, pp. 263-73; McKENZIE, 1975.

British naval supply voyage	C. Frederick	HMS *Amphitrite*	1853

Amphitrite's second and final visit to Bering Strait (see 1852). She resupplied *Plover* (see 1852-54) off Cape Lisburne, Alaska, returned with her to Port Clarence, then continued to Honolulu.

GREAT BRITAIN, PARLIAMENT, 1854, pp. 156-60.

British naval supply voyage	Edward Augustus Inglefield ? Joseph René Bellot	HMS *Phoenix* HMS *Breadalbane*	1853

Inglefield was sent from England with fresh supplies for the ships under Belcher's command (see 1852-54). On 8 August, he reached Beechey Island, where *North Star*, one of Belcher's ships, was anchored. *Breadalbane* was nipped in the ice and sank off Beechey Island on 21 August, and *Phoenix* set off for England three days later. She carried back Samuel Gurney Cresswell of HMS *Investigator* (see 1850-54). Cresswell was the first of *Investigator's* crew to return to England (the remainder returned in 1854), and was thus the first person to travel through the North-west Passage.

GREAT BRITAIN, PARLIAMENT, 1854, pp. 1-21; 1855, pp. 774-85; McCORMICK, 1884, Vol. 2, pp. 91-98.

British despatch-carrying journey	Joseph René Bellot	sledge	1853

From *Phoenix* at Beechey Island (see above, 1853); 12 - 20 August. Bellot set out up Wellington Channel with despatches for Belcher (see 1852-54) that had been brought from England on board *Phoenix*. On 17 August, he was carried out to sea on an ice floe and drowned. The four men who were with him returned at once to Beechey Island.

GREAT BRITAIN, PARLIAMENT, 1855, pp. 775-82; 1854, pp. 142-47; ANONYMOUS, 1854.

British naval supply voyage	Henry Trollope	HMS *Rattlesnake*	1853-54

From England, 21 February 1853 - (?) 1854. *Rattlesnake* was sent to Bering Strait with supplies for *Plover* (see 1852-54). She sailed by way of Magellan Strait, reached Port Clarence, Alaska, on 22 August 1853, and there met *Plover*. *Rattlesnake* wintered there, but *Plover* returned to winter at Point Barrow.

1854: *Rattlesnake* sailed for Point Barrow in June, but failed to meet *Plover*. She returned to Port Clarence, where *Plover* had already arrived. A few days later, *Enterprise* (see 1850-55) appeared after an absence of three years in the Arctic, and *Rattlesnake* left at once for San Francisco to forward her despatches. She reached San Francisco on 24 September.

GREAT BRITAIN, PARLIAMENT, 1853; *1854, pp. 147-56; 1854a, pp. 1-2; *1855, pp. 861-904.

1853-54

British whaling voyage | William Penny, George Brown | *Lady Franklin*, *Sophia*

From Aberdeen; 9 August 1853 - (?) 1854; to establish a wintering station in Cumberland Sound. This was the first time that whalers had deliberately chosen to winter in the Baffin Bay and Davis Strait region with their ships, although one party had previously wintered in Cumberland Sound without ships (see 1851-52, Quayle). They landed in Cumberland Sound on 16 September, did some whaling during the autumn, boiled down the whales during the winter, and recommenced whaling in spring 1854. They returned to Aberdeen later in 1854. Following the success of this voyage, most of the Davis Strait whalers began to make wintering voyages, and several whaling stations were built on Baffin Island: first in Cumberland Sound, but later also in Pond Inlet and Frobisher Bay.

HOLLAND, 1970.

1853-54

Hudson's Bay Company exploring expedition and Franklin search | John Rae

From York Factory; 25 June 1853 - August 1854; to complete the survey of the north coast of America by exploring the west coast of Boothia Peninsula between Castor and Pollux River and Bellot Strait. He had explored the east coast of Boothia Peninsula during an earlier expedition (see 1846-47). He reached Repulse Bay by boat on 14 August 1853, and wintered there.

1854: He set out northward across Rae Isthmus by sledge on 31 March, and continued to Pelly Bay, where he first heard Eskimo reports of the fate of the Franklin expedition. From these reports and from subsequent enquiries, he learned that about 1850 a party of about 40 white men had been seen north of King William Island walking towards Back River. Later in the same season, about 30 corpses had been seen west of that river. Rae purchased relics taken by the Eskimos from those corpses, including a small silver plate engraved "Sir John Franklin K.C.H." Meanwhile, he continued his survey of the coast, but was able to explore only between Castor and Pollux Bay and Cape Porter before bad weather forced him to turn back. He reached Repulse Bay on 26 May, there made further enquiries among the Eskimos, then returned to York Factory to send to England the first definite news about the fate of the Franklin expedition, last heard of nine years before (see 1845, Martin and Dannett).

RAE, 1855; GREAT BRITAIN, PARLIAMENT, 1855, pp. 831-45.

1853-55

United States Franklin Search expedition (second Grinnell expedition) | Elisha Kent Kane | *Advance*

From New York; 30 May 1853 - 11 August 1855; sponsored by Henry Grinnell, George Peabody, and United States scientific societies, to sail to the most northerly penetrable point in Baffin Bay, and from there to search for Franklin, and to carry out scientific investigations relating to magnetism, meteorology, and the existence of an open polar sea. They followed the west coast of Greenland to Rensselaer Harbour, in lat 78°37'N, where *Advance* was beset for the winter. In autumn, boat and sledge expeditions traced the coast of north-west Greenland to 80°N, making depots for explorations the following spring.

1854: Sledge expeditions explored Kane Basin (see 1854). As the season progressed without any indication that the ship would be freed from the ice, many of the crew became restless, and attempts were made to reach Belcher's squadron at Beechey Island (see 1852-54) and Godhavn in the hope of securing aid. The attempts failed, and they were forced to winter again.

1855: They abandoned *Advance* and made their way by boat and on foot along the Greenland coast to Godhavn, where they met Hartstene's relief expedition (see 1855) on 11 August.

KANE, *†1856; †1858; †1859; †1860; †1860a; *GODFREY, 1857; *VILLAREJO, 1965; CORNER, 1972, pp. 124-224.

1853-58

Roman Catholic missionary journeys in the interior of the Quebec-Labrador peninsula — Charles Arnaud, OMI

Father Arnaud failed to ascend Manicouagan River in his first attempt to reach the interior in June 1853 because of high water. He tried again in August and wintered with Naskapi Indians near Lake Mouchalagane, where he nearly starved. In 1855, he again failed to ascend Manicouagan River, turning back after losing his canoe. In 1858, he could not find suitable guides and, as he did not wish to winter again inland, he gave up the attempt, which was resumed by Father Babel in 1866 (*qv*).

1854

Hudson's Bay Company supply voyage	Robert Royal	*Prince Arthur*
	David J. Herd	*Prince of Wales* [II]

From London; 7 June - 24 October. *Prince Arthur* visited Moose Factory 3 - 21 September *Prince of Wales* visited York Factory 28 August - 20 September. Both ships reached London 24 October.

HUDSON'S BAY COMPANY (1) and (2).

1854

United States Franklin search expedition — Elisha Kent Kane — dog-sledge

From Rensselaer Harbour, north-west Greenland, winter quarters of *Advance* (see 1853-55); 27 April - 14 May. Kane intended to trace the west coast of Greenland to its northernmost extent, but sickness and the destruction by bears of food depots laid the previous autumn forced them to turn back in Dallas Bay at lat 79°05'N.

KANE, 1856, Vol. 1, pp. 219-33.

1854

United States Franklin search expedition | Isaac Israel Hayes, William Godfrey | dog-sledge

From Rensselaer Harbour, north-west Greenland, winter quarters of *Advance* (see 1853-55); 20 May - 1 June. They crossed Smith Sound to Cape Sabine, Ellesmere Island, then explored the east coast of Ellesmere Island north to 79°45'N, 69°12'W.

KANE, 1856, Vol. 1, pp. 237-39; 247-56; Vol. 2, pp. 365-73.

1854

United States Franklin search expedition | James McGary, Amos Bonsall | dog-sledge

From Rensselaer Harbour, north-west Greenland, winter quarters of *Advance* (see 1853-55); 4 - 26 June. They sledged northward along the west coast of Greenland to the northern end of Peabody Bay.

KANE, 1856, Vol. 1, pp. 257-59, 272-79; Vol. 2, pp. 357-65.

1854

United States Franklin search expedition | William Morton, Hans Hendrik | dog-sledge

From Rensselaer Harbour, north-west Greenland, winter quarters of *Advance* (see 1853-55); 4 June - 4 July. They sledged northward through Kane Basin to Franklin Island.

KANE, 1856, Vol. 1, pp. 280-306; Vol. 2, pp. 373-83.

1854

United States relief expedition | Elisha Kent Kane | boat *Forlorn Hope*

From Rensselaer Harbour, north-west Greenland, winter quarters of *Advance* (see 1853-55); July-August. Fearing that *Advance* would not be freed from the ice before the onset of another winter, Kane sailed with five men towards Beechey Island, Barrow Strait, hoping to obtain assistance from Belcher's squadron (see 1852-54). The attempt failed, and Kane turned back before reaching the entrance to Lancaster Sound.

KANE, 1856, Vol. 1, pp. 310-37.

1854

United States relief expedition | Isaac Israel Hayes, Carl Petersen | boat

From Rensselaer Harbour, north-west Greenland, winter quarters of *Advance* (see 1853-55); 28 August - 12 December. Hayes, Petersen, and seven others did not want to spend a second winter in Rensselaer Harbour, where *Advance* was beset. However, they failed in their attempt to reach Upernavik, west Greenland.

HAYES, 1854.

British expedition in search of *Enterprise* | George Frederick Mecham | sledge *Discovery* | **1854**

From the winter quarters of *Resolute* (see 1852-54) in Barrow Strait, off Cape Cockburn, Bathurst Island; 3 April - 12 June. Mecham sledged across Barrow Strait to Prince of Wales Strait in search of some indication of the whereabouts of *Enterprise* (see 1850-55). He found documents deposited from that ship in 1851 and 1852 on Princess Royal Island and Ramsay Island. He returned by way of Dealy Island, where he found orders to proceed direct to *North Star* at Beechey Island, *Resolute* having been abandoned.

British naval supply voyage | Edward Augustus Inglefield, R. Jenkins | HMS *Phoenix*, HMS *Talbot* | **1854**

From England. They were sent to Beechey Island, Barrow Strait, with supplies for Belcher's squadron (see 1852-54). They arrived at Beechey Island on 26 August to find that four of Belcher's ships and McClure's ship *Investigator* (see 1850-54) had all been abandoned, and that their crews were assembled on *North Star*, Belcher's only remaining ship. Most of these men were put on board *Phoenix* and *Talbot*, and the three ships returned to England.

GREAT BRITAIN, PARLIAMENT, 1855, pp. 1-11.

British naval supply voyage | Wallace Houston | HMS *Trincomalee* | **1854**

Trincomalee, then stationed in the Pacific Ocean, was sent to Port Clarence, Alaska, with supplies and orders for *Plover* (see 1852-54). She met *Plover* there on 1 August, and returned southward on 19 August.

GREAT BRITAIN, PARLIAMENT, 1855, pp. 859-60.

British whaling voyage; discovery of Eclipse Sound | John Gray | *Eclipse* | **1854**

From Peterhead. Gray followed the rest of the whaling fleet to Pond Inlet, then sailed into the inlet to discover Eclipse Sound.

LUBBOCK, 1937, p. 359.

Hudson's Bay Company supply voyage | Robert Royal, David J. Herd | *Prince Arthur*, *Prince of Wales* [II] | **1855**

From London; 7 June - 27 October. *Prince Arthur* visited Moose Factory 23 August - 22 September, and reached London 27 October. *Prince of Wales* visited York Factory 4-25 September, and reached London 25 October.

HUDSON'S BAY COMPANY (1) and (2).

1855 Hudson's Bay Company Franklin search expedition — James Anderson, James Green Stewart

From Fort Resolution; 2 June - 16 September; organized by the Hudson's Bay Company on receipt of Rae's report of his discovery of Franklin relics in 1854 (see 1853-54). Anderson and Stewart were to descend Back River to its mouth and to determine finally the fate of the Franklin expedition. They canoed to the mouth of Back River, where they found articles from Franklin's ships in the possession of Eskimos. They searched Montreal Island, where they found more relics, and examined the western shore of Chantrey Inlet, reaching Maconochie Island before turning back. They added little to Rae's findings, for bad weather, the unsoundness of the canoes, and the lack of an interpreter forced them to return to Fort Resolution without having made a thorough examination of the area. They made notes on physiography, flora, and fauna along their route.

*ANDERSON, J., 1856; 1857; 1940-41; GREAT BRITAIN, PARLIAMENT, 1855, pp. 846-58; *1856, pp. 19-29; TYRRELL, J. B., 1909.

1855 United States relief expedition — H. J. Hartstene, C. Simms — *Release*, *Arctic*

From New York; 4 June - (?); sent to search for Kane's expedition (see 1853-55). They reached lat 78°33'N through Smith Sound, but were prevented by ice from progressing farther. They met Eskimos who had seen Kane, and found articles left by him. They moved on to Lancaster Sound, where they suspected Kane might have gone to seek provisions, but were unable to reach Beechey Island. They returned to Greenland, and met Kane and his men at Godhavn on 11 September.

KANE, 1856, Vol. 2, pp. 322-32; LAWS, 1967.

1856 Hudson's Bay Company supply voyage — Robert Royal, David J. Herd — *Prince Arthur*, *Prince of Wales* [II]

From London; 12 June - 31 October. *Prince Arthur* visited Moose Factory 17 August - 7 September, and reached London 16 October. *Prince of Wales* visited York Factory 20 August - 22 September, and reached London 31 October.

HUDSON'S BAY COMPANY (1) and (2).

1857 Hudson's Bay Company supply voyage — D. D. Wishart, David J. Herd, Charles Robert Limstrong — *Prince Arthur*, *Prince of Wales* [II], *Baroness*

From London; 11 June - 19 October. *Prince Arthur* visited Moose Factory 16 August - 18 September, and reached London 17 October. *Prince of Wales* visited York Factory 9 August - 18 September, and reached London 19 October. *Baroness* visited York Factory 29 August - 18 September.

HUDSON'S BAY COMPANY (1) and (2).

Hudson's Bay Company exploration of Anderson River	Roderick MacFarlane	

1857

From Fort Good Hope; 4 June - 14 July; to explore the valley of Anderson River (which was known only from Eskimo reports) with a view to establishing trading relations with the Eskimos of that region. He left Fort Good Hope in company with four Indians and one Métis, travelled by way of Loon and Rorey lakes to Carcajou Lake, descended Iroquois River into Carnwarth River, and, on 13 June, reached the confluence of Carnwarth and Anderson rivers. Soon after, he began to meet Eskimos along Anderson River. On 16 June, when in sight of Wood Bay at the mouth of Anderson River, he turned back on foot to the mouth of Carnwarth River. There, he found an old canoe, which he used to explore the upper Anderson River. It appears that he then travelled by way of Gassend and Niwelin lakes to Colville Lake, where he abandoned the canoe. He continued overland to Hare Indian River, where he found another canoe, and paddled back to Fort Good Hope. MacFarlane revisited Anderson River several times in the years that followed and finally built a trading post there in 1861 (*qv*).

*MACFARLANE, 1890; STAGER, 1967; HOHN, 1963.

British whaling voyage; Moravian missionary journey to Baffin Island	William Penny George Brown Mathias Warmow	*Lady Franklin* *Sophia*

1857-58

From Aberdeen; 30 June 1857 - 22 August 1858. For several years, Captain Penny had been trying to carry a Moravian missionary from Greenland to the Eskimos of Cumberland Sound, but each time he had been unable to land at Lichtenfels to pick up the missionary. On this occasion, he arranged to carry a Moravian, Warmow, from Aberdeen. The ships entered Cumberland Sound on 1 August 1857, and soon after anchored at Kekerten, their winter quarters. Warmow spent much of the winter among the Eskimos, but, when he reported on his journey after his return to Europe in 1858, it was decided not to place a missionary permanently in Cumberland Sound. Warmow was the first missionary to visit Baffin Island; not until 1894 was a second attempt made to establish a mission there (see 1894-96 Peck).

WARMOW, 1859; HOLLAND, 1970.

British Franklin search expedition	Francis Leopold McClintock	*Fox*

1857-59

From Aberdeen; 2 July 1857 - 21 September 1859; a private expedition sponsored by Lady Franklin and by public subscription, sent to search in the area of King William Island for relics of the Franklin expedition. The ship, beset in Melville Bay during the outward voyage, was forced to winter in the ice, drifting southward down Davis Strait.

1858: Returning northward, they entered Lancaster Sound,

stopped at Beechey Island, then entered Peel Sound, but were prevented by ice from penetrating very far. They turned back, intending to reach King William Island by way of Prince Regent Inlet and Bellot Strait. They failed after several attempts to pass through Bellot Strait, and wintered at its eastern end.

1859: They found many relics and two records of the Franklin expedition during spring sledge journeys, and they explored parts of Prince of Wales Island (see 1859). On 10 August, the ship left Bellot Strait and returned home. They made observations on geology and meteorology, and collected fossils and biological specimens.

McCLINTOCK, *†1859; †1862; †HAUGHTON, 1862; †JEFFREYS, 1879; †O'MEARA, 1862; †SABINE, 1863; †WALKER, D., 1862; 1860; MARKHAM, 1909; YOUNG, 1875; NEATBY, 1970, pp. 249-70.

1858

Hudson's Bay Company supply voyage	D. D. Wishart	*Prince Arthur*
	David J. Herd	*Prince of Wales* [II]
	Henry McLaren	*Effort*

From London; 9 June - 9 October. *Prince Arthur* visited Moose Factory 18 August - 8 September, and reached London 9 October. *Prince of Wales* visited York Factory 13 August - 13 September, and reached London 6 October. *Effort* visited York Factory 23 August - 17 September, then left for Montreal. She seems to have been lost on the way to Montreal.

HUDSON'S BAY COMPANY (1) and (2).

1858-59

Anglican missionary journey to Mackenzie River — James Hunter

Archdeacon Hunter travelled from Red River to Fort Simpson to prepare for the establishment of the first Anglican mission in the Mackenzie River district. He spent nearly a year at Fort Simpson and at other posts in the region. On his way back to Red River, Hunter met William West Kirkby travelling north to establish the mission for which he had prepared (see 1859-68).

BOON, 1962, pp. 80-82.

1858-63

Roman Catholic missionary journey to Mackenzie River — Henri Grollier, OMI

Father Grollier was the first Roman Catholic missionary to reach Mackenzie District. He travelled north from Ile-à-la-Crosse in 1858 and established missions at Fort Resolution, Fort Simpson, and at Big Island, at the head of Mackenzie River. The latter mission was moved downstream in 1861 and renamed Fort Providence.

1859: He established missions at Fort Rae, Fort Norman, and at Fort Good Hope, at the last of which he wintered.

1860: He visited the missions at Fort Norman and Fort Simpson and built another at Fort McPherson, where he wintered. During this year, he twice encountered Eskimos.

1861: He returned to Fort Good Hope, where he was joined by Father Séguin and Brother Kearney. He remained here until his death, 4 June 1863.

CARRIÈRE, 1964; BRETON, 1963.

Hudson's Bay Company supply voyage	D. D. Wishart David J. Herd Captain Ellis	*Prince Arthur* *Prince of Wales* [II] *Kitty*	1859

From London; 8 June - 22 October. *Prince Arthur* visited Moose Factory 2-16 September, and reached London 21 October. *Prince of Wales* visited York Factory 25 August - 18 September, and reached London 22 October. *Kitty* visited York Factory, then was wrecked at the entrance to Hudson Strait on 5 September. The chief mate, William Armstrong, and four sailors, reached the Labrador coast by boat, but the captain and ten men in a longboat were never seen again. The ship's entire cargo was lost.

HUDSON'S BAY COMPANY (1) and (2).

British Franklin search expedition	Francis Leopold McClintock	sledge	1859

From Bellot Strait, winter quarters of *Fox* (see 1857-59); 17 February - 14 March. He sledged to Cape Victoria, Boothia Peninsula, where he met Eskimos from whom he purchased relics of the Franklin expedition.

McCLINTOCK, 1859.

British Franklin search expedition	Francis Leopold McClintock Carl Petersen	sledge dog-sledge	1859

From Bellot Strait, winter quarters of *Fox* (see 1857-59); 2 April - 19 June. He sledged to the east coast of King William Island, made a complete circuit of the island in a clockwise direction, and found skeletons and other relics of the Franklin expedition. He also visited Montreal Island, and revisited the Eskimos at Cape Victoria, from whom he purchased more relics.

McCLINTOCK, 1859.

British Franklin search expedition	William Robert Hobson	sledge	1859

From Bellot Strait, winter quarters of *Fox* (see 1857-59). He sledged to the west coast of King William Island, where he found a boat left by Franklin's men on their journey to Back River, and a note which told of Franklin's death on 11 June 1847, of the expedition's establishment of the existence of a North-west Passage, and of the abandonment of *Erebus* and *Terror* in 1848. This was one of only two written records of the expedition's fate that have ever been found. The other, also found during McClintock's expedition, was written in May 1847, and reported all well. Hobson examined the west coast of King William Island southward to near Cape Herschel before returning to the *Fox*.

McCLINTOCK, 1859.

1859

British Franklin search expedition — Allen Young — sledge, dog-sledge

From Bellot Strait, winter quarters of *Fox* (see 1857-59); 7 April - 7 June. He explored the south coast of Prince of Wales Island from Cape McClure in the east to Cape Collinson in the west. He also discovered McClintock Channel.

McCLINTOCK, 1859.

1859-62

United States zoological expedition to Mackenzie River — Robert Kennicott

From Chicago; 28 April 1859 - 17 October 1862; sent by the Smithsonian Institution and the Audubon Club of Chicago to make zoological collections from the trading posts in Mackenzie District and Yukon Territory. He travelled by steamer to Lake Superior, then continued by canoe with a Hudson's Bay Company brigade to Fort Simpson, where he arrived on 14 August. In October, he made a three-week visit to Fort Liard.

1860: Between 9 January and 9 March he again visited Fort Liard. He then spent some time at Fort Simpson and Fort Resolution; in August, he travelled down Mackenzie River and, by way of Peel and Porcupine rivers, to Fort Yukon, where he wintered and spent much of the following year.

1861: He went from Fort Yukon to La Pierre's House, where he wintered.

1862: In spring, he walked to Fort Simpson, where, hearing of his father's illness, he decided to return to Chicago at once.

In addition to large collections of birds and animals of the Mackenzie River region, Kennicott made many valuable observations on the fur trade, but his scientific work did not extend to geographical surveys or exploration.

JAMES, 1942.

1859-68

Anglican missionary journey to Mackenzie River — William West Kirkby

Reverend Kirkby volunteered to travel from Red River to Fort Simpson to establish the mission for which Archdeacon Hunter had prepared in 1858-59 (*qv*). He left Red River on 9 June 1859 and reached Fort Simpson with a brigade of the Hudson's Bay Company on 18 August. With the assistance of the company's officers, he began at once to build his mission house and church. During his first two years in the region, he visited forts Liard and Norman.

1861: He was the first missionary to visit Fort Yukon.

1862: He made a second visit to Fort Yukon, accompanied this time as far as La Pierre's House by Father Séguin.

1866: On 26 August, he opened St David's Church at Fort Simpson, the first church to be erected on the banks of Mackenzie River.

1868: He left Mackenzie District to return to England on furlough.

PEAKE, 1965; BOON, 1962, p. 82; 1965.

c1860

Hudson's Bay Company exploring expedition — C. J. J. Anderson

About 1860, Anderson, officer in charge of Great Whale River post, canoed northward to Mosquito Bay to examine the white whale fisheries of the rivers of northern Quebec. He made a good track survey of the coast along which he sailed.

LOW, 1902, p. 6D.

1860

Hudson's Bay Company supply voyage	D. D. Wishart	*Prince Arthur*
	David J. Herd	*Prince of Wales* [II]

From London; 6 June - 18 October. *Prince Arthur* visited Moose Factory 22 August - 9 September, and reached London 18 October. *Prince of Wales* visited York Factory 16 August - 8 September, and reached London 3 October.

HUDSON'S BAY COMPANY (1) and (2).

1860

British oceanographic voyage — Sir Francis Leopold McClintock — HMS *Bulldog*

From England; 1 July - 20 November; to measure the depth of the North Atlantic Ocean and the nature of the bottom between the Faroe Islands and Labrador in preparation for the laying of a telegraph cable. They began the survey from the Faroe Islands on 6 July and followed a route along the south coast of Iceland and around the southern tip of Greenland to Hamilton Inlet, Labrador, where they arrived in early September. They spent a week surveying the length of Hamilton Inlet, and collected information on climate and ice conditions. From there, *Bulldog* sailed south to Cape Breton, then returned north along the Labrador coast to west Greenland. She then returned by way of Iceland and Ireland to Portsmouth. The cable for which this expedition had prepared was never laid, owing to the impossibility of finding suitable landing sites on the Labrador coast, but another trans-Atlantic cable was laid successfully in 1866 (see 1867, Laberge).

*McCLINTOCK, 1861; WALLICH, †1862; †1860; MARKHAM, C. R., 1909, pp. 252-62.

1860 United States ornithological expedition to James Bay — C. Drexler

Drexler's expedition, the first ornithological expedition to James Bay, appears to have been sponsored by the Smithsonian Institution and Henry Bryant of Boston. He travelled by canoe by way of Lake Temiskaming, Abitibi Lake, and Abitibi and Moose rivers to Moose Factory, where he made collections between 25 May and 3 June. On 4 June he set out with Indians by canoe along the east coast of James Bay to Rupert House, then north to the islands between Cape Hope Islands and Long Point. His farthest north was Fort George, where he deposited his collections to be taken out by ship. On 12 July he set off back along the coast to Moose Factory, then retraced his steps to Fort William, where he arrived on 17 October.

TODD, 1963, pp. 9-10.

1860 United States whalers' boat voyage — John Giles, John Martin, Hiram J. Davis, Williard Hawkins, Thomas Colwell, Joseph Fisher, Samuel J. Fisher, John F. Sullivan, Warren Dutton

These men, of the United States whalers *Ansell Gibbs* and *Daniel Webster,* anchored in Cumberland Sound, complained of ill treatment and bad conditions and decided to sail home in a whale boat. They left the ships on 4 August. On 7 August, they met *George Henry,* Captain S. O. Budington (see 1860-62), who gave them some supplies. After a stop at Resolution Island, they reached Cape Chidley, Labrador, on 16 August. Early in September, they ran short of supplies, and, when Dutton died, they ate him. Samuel Fisher was later killed and eaten. On 29 September, they met Eskimos who took them to Okak mission. From there, they were taken to Nain, Hopedale, and Kibokok, and gradually dispersed. Their conduct at these missions was said to be shameful.

HALL, 1864, Vol. 1, pp. 91-95.

1860-61 United States North Polar expedition — Isaac Israel Hayes — *United States*

From Boston; 7 July 1860 - October 1861; sponsored by several American societies and by public subscription to reach the North Pole by way of Smith Sound and across the Arctic Ocean, which many then believed to be free of ice. After visits to Prøven and Upernavik, west Greenland, in August, they continued to Smith Sound, where they met very heavy ice. In early September, they had reached only Foulke Fjord, where they were obliged to put into winter quarters. There they met Eskimos, who remained with them throughout the winter. In spring 1861, Hayes made a sledge journey northward along the coast of Ellesmere Island (see 1861). He returned to the ship, still convinced that an open polar sea existed, and decided to return to Boston and make a second attempt by steamer the following year. He left Foulke Fjord on 14 July, and reached Boston in October to learn that war had broken out, an event which frustrated his intention to return to the Arctic. Scientific results of the voyage included meteorological and magnetic observations, studies of glacial movement, and collections of plants, rocks, minerals, fossils, and marine invertebrates.

HAYES, 1862; *1867a; †1867b; †DURAND AND OTHERS, 1864; †EMERSON, 1905; †KIRK, 1925; †MEEK, 1865; †STIMPSON, 1864.

1860-62

United States whaling voyage and Franklin search	Charles Francis Hall Sidney O. Budington ?	*George Henry* *Rescue*

From New London; 20 May 1860 - 13 September 1862; Hall intended to visit King William Island in search of relics of the Franklin expedition (see 1845-48) and, having been offered a passage north on the whaler *George Henry,* planned to travel there across Baffin Island and Foxe Basin from Cumberland Sound or Frobisher Bay. After a visit to Holsteinborg, west Greenland, 7-24 July, Budington headed for Baffin Island. He first visited Cornelius Grinnell Bay, which Hall named, then coasted south to Cyrus Field Bay (also named by Hall), where the ships wintered. During a storm on 27 September, *Rescue* and Hall's expedition boat were wrecked. During the winter, he befriended the Eskimo Ebierbing and his wife Tookoolito, who became his companions for the remainder of the expedition.

1861: Between 9 August and 7 September, Hall made an exploratory boat trip around Frobisher Bay, during which he discovered relics of Frobisher's expeditions (see 1576, 1577, 1578) on and around Kodlunarn Island. He also discovered that Frobisher Bay had no western outlet, as had been generally believed. He returned to find *George Henry* already beset and forced to winter again.

1862: Hall visited Kodlunarn Island again in July and collected Frobisher relics. On 9 August, *George Henry* sailed homeward, carrying not only Hall, but also Ebierbing and Tookoolito, who were to accompany him on his next expedition in search of Franklin relics (see 1864-69).

*HALL, 1864; *LOOMIS, 1972, pp. 71-145; †EMERSON, 1879; †STEVENS, R.P., 1863; DCB, Vol. 10, pp. 327-29.

1860-1915

United States whaling in Hudson Bay

In 1860, two American whalers, *Syren Queen,* Captain Christopher Chapel, and *Northern Light,* Captain Edward Chapel, wintered at

Depot Island, near Daly Bay, on the north-west coast of Hudson Bay. Despite several deaths by scurvy, they killed many whales and made a profitable voyage. They were followed during the next decade by 59 ships but, after 1870, the numbers of vessels declined. The last, *A. T. Gifford,* ended in shipwreck in 1915. Whaling was concentrated in Roes Welcome Sound and Repulse Bay. Of the total number of 146 whaling voyages to Hudson Bay, 117 were American and the rest were British. About 105 of the voyages overwintered. Many captains temporarily employed Eskimo labour, and Eskimo trade was often profitable.

ROSS, W. G., 1971.

1861

Hudson's Bay Company supply voyage	D. D. Wishart David J. Herd J. Sennett	*Prince Arthur* *Prince of Wales* [II] *Corea*

From London; 4 June - 5 November. *Prince Arthur* visited Moose Factory 31 August - 19 September, and reached London on 24 October. *Prince of Wales* visited York Factory 22 August - 9 September, and reached London 9 October. *Corea* visited Little Whale River on 7 September, Great Whale River on 5 October, and reached London on 5 November.

HUDSON'S BAY COMPANY (1) and (2).

1861

Establishment of Fort Anderson	Roderick MacFarlane

Since his exploration of Anderson River in 1857 (*qv*), MacFarlane had visited the region several times to trade with the Eskimos, and in 1859 had received permission to build a trading post there for the Hudson's Bay Company. In 1860, he had made a summer journey to the river to search for wood suitable for building a fort. In April 1861, he travelled from Fort Good Hope to Anderson River to cut timber at a place he named Shantyville, a short distance below the mouth of Carnwarth River. In June, he rafted the wood downstream to the chosen site, some 180 km upstream from the mouth of Anderson River, in about lat 68°30'N. The post was completed by winter. MacFarlane had charge of Fort Anderson until 1866 and made annual trips to the shore of Franklin Bay to collect bird and egg specimens. In late 1865, a scarlet fever epidemic killed about 150 of the local Eskimos. This tragedy, together with the difficulty MacFarlane experienced in communicating with Mackenzie River, led to the closure of the post in 1866. Fort Anderson was the first post in the western Arctic built exclusively for Eskimo trade.

*STAGER, 1967; HOHN, 1963; HUDSON'S BAY COMPANY, ARCHIVES, B6/a/1.

1861

United States exploration of Ellesmere Island	Isaac Israel Hayes	2 dog-sledges

From Foulke Fjord, north-west Greenland, winter quarters of *United States* (see 1860-61); 3 April - 3 June. Hayes encountered great difficulty in crossing the ice of Kane Basin and sent back most of his companions, continuing with only three men. He landed on Ellesmere Island at Cape Hawks, having spent 31 days covering only about 120 km. He followed the coast of Ellesmere Island northward, passing Cape Fraser, his farthest north in 1854 (*qv*), and he claimed to have reached 81°35'N on 18 May, before turning back. However, it is probable that he never got so far north; his physiographical descriptions were later proved inaccurate, and members of Hall's *Polaris* expedition (see 1871-73) found his discoveries "quite imaginary."

HAYES, 1867a.

1861

Canadian exploration of the Quebec-Labrador peninsula — Henry Youle Hind, William G. R. Hind

On 10 July, Hind and his brother William (an artist), two Dominion Government surveyors, and seven canoemen, began the ascent of Moisie River. From the confluence of Moisie and Nipissis rivers, they followed the latter to the Siamese Lakes, then took Matinipi River through Mistamoue Lake, and turned back on 2 July just before reaching Chicomo Lake near the height of land: theirs was the first scientific exploration of the interior of the peninsula. Although they penetrated only about 160 km inland, Hind was able to form a good estimate of the elevation of the central interior and described the Indians and geography of the region. His two-volume account was the standard reference to the region for some 30 years.

HIND, *†1863; 1864; COOKE, 1964, p. 151; HARPER, 1967.

1862

Hudson's Bay Company supply voyage — D. D. Wishart, David J. Herd — *Prince Arthur*, *Prince of Wales* [II]

From London; 4 June - 26 October. *Prince Arthur* visited Moose Factory 29 August - 19 September, and reached London on 26 October. *Prince of Wales* visited York Factory 13 August - 9 September, and reached London on 5 October.

HUDSON'S BAY COMPANY (1) and (2).

1862-63

Roman Catholic missionary journey to Fort Yukon — Jean Séguin, OMI

Father Séguin, who had arrived at Fort Good Hope in August 1861, was sent in 1862 to Fort McPherson. He then travelled in company with Reverend Kirkby (see 1859-68) as far as La Pierre's House, which they reached 17 June. On 3 September, having returned to Fort Good Hope, he set out again for Fort Yukon, this time in company with Robert MacDonald, another Anglican clergyman. He spent the winter at Fort Yukon in difficult circumstances, and returned to Fort Good Hope the next summer.

BRETON, 1963.

1862-64

British exploring expedition	Viscount William Fitzwilliam Milton Walter Butler Cheadle	

From Liverpool; 19 June 1862 - 5 March 1864; to discover the shortest route through British territory from the east coast of Canada to the Cariboo gold mines, British Columbia. The expedition travelled by steamer to Quebec, by rail and steamer to Fort Garry, then by horse to Carlton House, near where they wintered. In 1863, they continued west along Saskatchewan River to Edmonton, crossed the Rocky Mountains at Yellowhead Pass, visited the Cariboo District, then continued down Thompson and Fraser rivers to the sea. On 25 November 1863, they left Victoria by steamer, and returned to Liverpool by way of Panama and New York.

MILTON AND CHEADLE, 1865.

1863-64

Hudson's Bay Company supply voyage	J. D. Smythe David J. Herd John Todd (out) John James (home)	*Prince Arthur* *Prince of Wales* [II] *Ocean Nymph*

From London; 3 June 1863 - (?) 1864. *Prince Arthur* visited Moose Factory 28 August - 26 September, and reached London on 6 November 1863. *Prince of Wales* visited York Factory 29 August - 17 September, and reached London 9 October 1863. A labourer on the ship carried typhus, which spread at York Factory. Todd, commanding *Ocean Nymph,* failed to follow his instructions relating to the course on the outward voyage, and was beset in the ice. He turned back, ran ashore on the coast of Labrador, but reached St John's, Newfoundland, on 27 October. When news of his misadventures reached London, the company sent David Herd out by steamer to take charge. On 22 December, after repairs, he sailed the ship to Portland, Maine. There, on 5 January 1864, he turned command of the ship over to John James and returned to England.

HUDSON'S BAY COMPANY (1) and (2).

1864

United States voyage to the Labrador coast	William Bradford Alpheus Spring Packard Captain Brown	*Benjamin S. Wright*

From Boston; 4 June - 4 September. Bradford, a marine artist from New York, organized this voyage "for the purpose of painting icebergs and artistic scenery," but the party also included naturalists and pleasure seekers. Bradford conducted six such voyages between the years 1861 and 1867, but this was the only one to be described in detail. The ship stopped at many points along the

Labrador coast, passed through the Strait of Belle Isle, visited Hopedale 30 July - 4 August, then returned homeward. Members of the expedition made notes on marine invertebrates, distribution of fish and mammals, geology, Eskimos, and the Moravian missions.

PACKARD, 1891, pp. 93-225.

1864-65

Hudson's Bay Company supply voyage	J. D. Smythe	*Prince Arthur*
	John James	*Ocean Nymph*
	James Sennett (out)	*Prince of Wales* [II]
	James Taylor (home)	

From London; 8 June 1864 - 2 November 1865. *Prince Arthur* and *Prince of Wales* both ran aground on Mansel Island, Hudson Bay, on 13 August 1864. *Prince Arthur* was abandoned, *Prince of Wales* was refloated, took on board *Prince Arthur's* crew and some of her cargo, and set out for York Factory on 23 August. She remained at York Factory undergoing repairs 29 August 1864 - 2 October 1865, then returned to London. *Ocean Nymph* visited York Factory 23 August - 20 September, and reached London 23 October 1864.

HUDSON'S BAY COMPANY (1) and (2).

1864-69

United States Franklin search expedition — Charles Francis Hall

From New London; 1 July 1864 - 26 September 1869. Hall, wishing to make a second attempt to search King William Island for Franklin relics (see 1860-62), decided to make his base at Repulse Bay. He was offered a passage out on the whaler *Monticello,* Captain E. A. Chapel, accompanied by her tender *Helen F.,* Captain H. Y. Chapel. He took with him the Eskimos Ebierbing and Tookoolito, who had come home with him after the previous expedition. He was landed from *Helen F.* in Roes Welcome Sound at 64°35'N, 87°33'W, more than 60 km south of Wager Bay, where he had intended to land; the mistake cost him a year. He wintered among the Eskimos a short distance north of his landing site.

1865: Early in April, he continued to Repulse Bay and made his winter quarters at Fort Hope. He remained in touch with the whalers, but now preferred to live among the Eskimos.

1866: On 31 March, he set out across Rae Isthmus for King William Island. At Cape Weynton, on the western shore of Committee Bay, he met Eskimos who showed him Franklin relics. They dissuaded him from continuing to King William Island that year, so he returned to his base and spent the rest of the summer surveying Repulse Bay. He wintered in an igloo on one of the Harbour Islands, Repulse Bay, where several whalers were anchored for the winter.

1867: In February and March, he visited the Eskimos of Igloolik and Tern Island. In May, he inspected a cache of provisions he had left at Cape Weynton. In July, he surveyed Harbour Islands for the benefit of the whalers.

1868: He decided to visit Fury and Hecla Strait, having heard Eskimo reports that some of Franklin's men had reached there. He set out on 23 March, crossed Lyon Inlet, and followed the east coast of Melville Peninsula northward to Hooper Inlet. He then turned inland, followed Crozier River and Grinnell Lake to Cape Englefield, the north-western extremity of Melville Peninsula, then traced the west coast of Melville Peninsula southward to Franklin Bay, where he arrived on 25 April. He returned by way of the south shore of Fury and Hecla Strait and the east coast of Melville Peninsula, and reached Repulse Bay on 26 June. In November, he visited Lyon Inlet to correct Parry's chart (see 1821-23).

1869: On 23 March, he set out again for King William Island. On 12 May, he reached the Todd Islands, off the south-eastern shore of King William Island, where he found the graves of some of Franklin's men. He also searched on the south-eastern shore of King William Island, where he met Eskimos who showed him more Franklin relics and told him of the fate of some of Franklin's men. He then returned to Repulse Bay, where he arrived on 20 June. On 13 August, he left for home on the whaler *Ansell Gibbs,* Captain Fisher, of New Bedford.

*†NOURSE, 1879; *LOOMIS, 1972, pp. 178-227; HALL, 1873; CYRIAX, 1944.

1864-72

Roman Catholic missionary explorations of Mackenzie River district — Emile Petitot, OMI

Father Petitot, who had been based at Fort Resolution since 1862, was ordered to Fort Good Hope to replace Father Grollier (see 1858-63), who had died in 1863. From there, he made many exploring journeys in the region bounded by the Coppermine River and the Rocky Mountains, Great Slave Lake and the Arctic Ocean (see 1865, 1866, 1867-68, 1868, 1869, 1870, 1871, 1872). During these journeys, he made ethno-linguistic, cultural, and distributional studies of the Indians and Eskimos, described the meteorology, topography, geology, routes, trade, rivers, and lakes of the region. He afterwards wrote many books and articles on the native peoples and the geography of the region. In 1864, before travelling to Fort Good Hope, he explored from Fort Resolution across Great Slave Lake. He travelled along Marian River, crossed Faber and Rae lakes to Lac Sainte-Croix, where he turned back. During this journey, he visited Lac La Martre, where he found Indians whom he took back to the mission at Fort Rae. He returned to Fort Resolution in July, then continued to Fort Good Hope.

PETITOT, 1875; 1875a; 1875b; 1886; 1887; SAVOIE, 1970.

Hudson's Bay Company supply voyage | John James, Henry Bishop | *Lady Head*, *Prince Rupert* [VII]

1865

From London; 8 June - 18 November. *Lady Head* visited Moose Factory 13-28 September, and reached London on 18 November. *Prince Rupert* visited York Factory 22 August - 2 October and reached London on 30 October.

HUDSON'S BAY COMPANY (1) and (2).

Establishment of Fort Nelson | W. Cornwallis King

1865

King built a small house for the Hudson's Bay Company on the west bank of Fort Nelson River, about 160 km from its mouth. It was the second Fort Nelson to be built in the region; the first had been operated near the mouth of Fort Nelson River betwen 1806 and 1813 (see 1805, Fort Liard), and, apparently, from about 1818 to 1825. King's post remained in operation on the same site until about 1942, when it was moved to the east bank about 2 km upstream. Between 1951 and 1967, there were two posts on Fort Nelson River, the second one on the Alaska Highway at Fort Nelson Mile 300.

In 1960, the Fort Nelson of 1942 was renamed Fort Nelson River, and Fort Nelson Mile 300 became Fort Nelson. The former was closed in 1967, the latter remains in operation.

POWELL, 1970, p. 127.

Missionary exploration of Anderson River | Emile Petitot, OMI

1865

From Fort Good Hope. Father Petitot travelled to Fort Anderson (see 1861), then explored Anderson River to its mouth with a party of Eskimos. He later made a second descent to the mouth of the river from Fort Anderson. Returning to Fort Good Hope during the winter, he travelled overland from Fort Anderson to Simpson Lake, then southward over an unexplored route (possibly through lakes Tedji and Aubry) to Colville Lake.

PETITOT, 1875; 1887; SAVOIE, 1970, Vol. 1.

British whaling voyage | G. Brown, Edward P. Philpots | *Queen*

1865-66

From Peterhead; 31 March 1865 - (?) 1866. After whaling off Resolution Island in May, and in Lancaster Sound, Navy Board Inlet, and Prince Regent Inlet in August, they searched the east coast of Devon Island for a winter harbour and found a suitable one near Philpots Island (actually a peninsula on the east coast of Devon Island). During winter, they explored Philpots Island and the Devon Island coast south to Cape Sherard. They left on 27 August

1866, made short visits to Lancaster Sound and Cumberland Sound, then returned to Peterhead.

PHILPOTS, 1867; 1869.

1865-66

Scarlet fever epidemic

In 1865 and in winter 1865-66, a scarlet fever epidemic swept through Mackenzie District; deaths from the fever and from measles exceeded 1,000 persons.

MORTON, A.S., 1939.

1866

Hudson's Bay Company supply voyage; re-establishment of Fort Chimo	John James Henry Bishop R. B. Wood	*Lady Head* *Prince Rupert* [VII] *Labrador*

From London; 8 June - 31 October. *Lady Head* visited Moose Factory 24 August - 23 September, and reached London on 25 October. *Prince Rupert* visited York Factory 2 September - 4 October, and reached London on 31 October. *Labrador* sailed from London on 15 June to Quebec, then visited posts along the coast of Labrador on her way to Ungava Bay, where Joseph McPherson reopened Fort Chimo, which had been closed in 1843 (see 1842-43), on 29 September. She left the post on 10 October and returned to London on 10 November.

HUDSON'S BAY COMPANY (1) and (2).

1866

Missionary exploration of Great Bear Lake	Emile Petitot, OMI	

From Fort Good Hope. He followed Hare Indian River to its source and crossed the watershed to Smith Arm, Great Bear Lake. He then travelled southward overland to Keith Arm, and returned westward along Great Bear River to the Mackenzie.

PETITOT, 1875.

1866-67

Hudson's Bay Company whaling and trading voyage	James Taylor	*Ocean Nymph*

From London; 8 June 1866 - 30 September 1867; sent on a whaling voyage along the west coast of Hudson Bay north of Churchill. It was also intended that they should trade with the Eskimos and compete with the American whalers wintering along the coast. *Ocean Nymph* wintered at Marble Island and resumed operations in the spring of 1867. However, it was a bad season, and the expedition was a failure. She visited York Factory 28 August - 5 September, then returned to England.

ROSS, W.G., 1973a.

Missionary journeys in the interior of the Quebec-Labrador peninsula	Louis Babel, OMI	1866-70

Renewing the attempt to reach the Naskapi and Montagnais Indians of the interior (see 1853-58), Father Babel with two Indians began the ascent of St John River on 22 July, then worked across to Romaine River. On 13 August he arrived at the Hudson's Bay Company's post at Lake Winokapau, but found it deserted. They were back at Mingan on 29 August.

1867: Babel travelled from Mingan to North West River on the Hudson's Bay Company's steamer, then went inland with Naskapi Indians to lakes Winokapau and Petitsikapau. He returned to Mingan by canoe following his route of 1866.

1868: He followed the previous summer's course inland, but returned by Moisie River to Sept-Isles, which he found a shorter and easier route.

1869: He followed the previous summer's course inland. Naskapi Indians he met at Lake Petitsikapau tried to persuade him to accompany them to Fort Chimo, but he decided not to go because the post was not to be supplied that year, and he did not wish to winter there.

1870: Owing to the wreck of the Hudson's Bay Company's steamer, he went inland to Lake Petitsikapau by St John River and returned by Moisie River. Babel's unpublished journals of these trips are a valuable account of the geography and ethnography of the region. He drew a map depicting the very difficult rivers he had travelled, which, considering the tools he had, is very good. He seems to have been the first traveller to report the existence of abundant iron ore in the interior of the peninsula.

CARRIÈRE, 1963.

Hudson's Bay Company supply voyage	Henry Bishop John James R. B. Wood	*Prince Rupert* [VII] *Lady Head* *Labrador*	1867

From London; 6 June - 26 October. *Prince Rupert* visited York Factory 12 August - 17 September, and reached London on 15 October. *Lady Head* visited Moose Factory 15 August - 18 September, and reached London on 26 October. *Labrador* sailed on 1 June to posts on the Labrador coast, visited Fort Chimo 4-14 September, and returned to England.

HUDSON'S BAY COMPANY (1) and (2); STEVENSON, 1967.

British naval survey of the Labrador coast	William Chimmo	HMS *Gannet*	1867

From Halifax, NS; 31 July - mid-September; to search the little known north-east coast of Labrador for new fishing grounds and harbours of refuge for the Newfoundland fishing fleet. Chimmo surveyed Indian Harbour, Webeck, Ailik Harbour, where he visited the Hudson's Bay Company's post, and Hopedale harbour. He then returned to Halifax. This was the first official visit by the Royal Navy to these waters. He found the coast marked as much as 15 km too far east on the charts, and its outline very inaccurately surveyed, although the latitudes were reasonably correct.

CHIMMO, 1868.

1867

Western Union Telegraph Company expedition

Mike Laberge
F. E. Ketchum
Michael Byrnes

Laberge and Ketchum were members of a team employed by the Western Union Telegraph Company to explore Alaska in preparation for the laying of a telegraph line from America to Europe by way of Bering Strait. They left the main party at Nulato, Alaska, on 11 March 1867 to explore Yukon River up to Fort Yukon. They sledged to Fort Yukon, then continued up Yukon and Pelly rivers to the ruined site of Fork Selkirk (see 1848). Whilst there, Laberge apparently heard from the Indians of a large lake farther upstream. Although he never visited this lake, he spoke of it to some Indians and traders who, after its discovery, named it Lake Laberge after him. Laberge and Ketchum returned down Yukon River to join the main party at Fort Yukon on 29 June.

Meantime, Byrnes had been exploring the headwaters of Yukon River in British Columbia. In August 1867, he had reached Lake Teslin and was about to descend the river when news reached him that the whole project was to be abandoned immediately because the North Atlantic Telegraph Company's cable had, after three failures between 1857 and 1864, been successfully laid in 1866 and was in use in August 1867.

DALL, 1898, pp. 63, 110; OGILVIE, 1890, p. 21; WHYMPER, 1869, pp. 258-59.

1867

Confederation of Canada

On 29 March 1867, the British North America Act, 1867, passed by the British Parliament, received the Queen's assent. The act, proclaimed 1 July 1867, was based largely on the 72 resolutions drafted at the Quebec Conference in 1864 by the "Fathers of Confederation" and provided for a federal union of the three British North American provinces (Canada, New Brunswick, and Nova Scotia) in one dominion to be named Canada. Provision was made for entry into the dominion of other British possessions in North America: Rupert's Land and the North-Western Territory were transferred to it in 1870 (*qv*), Manitoba became a province 15 July 1870, British Columbia 20 July 1871, Prince Edward Island 1 July 1873, Alberta and Saskatchewan 1 September 1905, and Newfoundland 31 March 1949.

Purchase of Alaska

1867

Although the Hudson's Bay Company had leased the south-east mainland of Alaska from the Russian-American Company since 1839, Great Britain showed no great interest when Russia began to think of withdrawing from the territory. Negotiations between Russia and the United States, interrupted in 1861 by the Civil War of the United States, were resumed when it ended in 1865. Strongly urged by Secretary of State William Seward, the treaty was ratified by the United States Senate on 9 April 1867. The transfer of ownership was effected at Sitka on 3 October 1867, but payment of the agreed $7.2 million was not authorized by the United States House of Representatives until 14 July 1868.

Roman Catholic missionary exploration north of Great Bear Lake — Emile Petitot, OMI

1867-68

From Fort Good Hope. He followed Hare Indian River to its source, then travelled northward by Belot, Colville and Niwelin lakes to Anderson River. Having crossed this river, he travelled north-east and explored the middle courses of Horton and Hornaday rivers. Returning to Ford Good Hope, he visited Keith and McTavish arms of Great Bear Lake.

PETITOT, 1875.

Hudson's Bay Company supply voyage — John James, *Lady Head*; Henry Bishop, *Prince Rupert* [VII]; R. B. Wood, *Labrador*

1868

From London; 3 June - 20 October. *Lady Head* visited Moose Factory 26 August - 14 September, and reached London on 18 October. *Prince Rupert* visited York Factory 24 August - 20 September, and reached London 20 October. After a whaling voyage to Greenland, 16 February - 3 May, *Labrador* sailed on 21 May to the Labrador coast, visited Fort Chimo 19 - 27 September, and returned to England on 29 November.

HUDSON'S BAY COMPANY (1) and (2).

Establishment of Arctic Red River — Jean Séguin, OMI

1868

Father Séguin (see 1862-63) visited Fort McPherson from Fort Good Hope every year between 1862 and 1873, but in 1868, having found that Protestant influence at Fort McPherson made his work there very difficult, he established a new mission just south of Point Separation in 1868. This settlement, a focus of Roman Catholic activity in the first place, eventually became a trading post of the Hudson's Bay Company and still exists.

1868 | Roman Catholic missionary expedition to Mackenzie River delta | Emile Petitot, OMI

From Fort Good Hope. On return from the Anderson River region earlier in the year (see 1867-68), Petitot set out again on 8 June for Fort McPherson, which he reached on 13 June, and then set out to visit islands at the mouth of Mackenzie River. On 1 July, having met Eskimos there, he turned back at about 68°30'N because of illness and because of their threats to his life. At Fort McPherson, he conceived a new plan to visit the East Channel of the delta, but was warned against it. Instead, he took an Eskimo couple, who had been at the fort, back to Fort Good Hope with him to study their language.

PETITOT, 1875.

1868 | Establishment of Hay River post

In this year, the Hudson's Bay Company established the first of several posts to be built on Hay River. This first post was probably closed in 1875. The site remained abandoned until 1901, when the Hudson's Bay Company and other trading companies established new posts there.

USHER, 1971.

1868-69 | United States whaling voyage; discovery of Fisher Strait | Elnathan B. Fisher | *Ansell Gibbs*

From New Bedford; whaling voyage to north-west Hudson Bay. Before this voyage, Southampton and Coats islands in Hudson Bay were represented on maps as being one and the same island. However, on his outward voyage, on 8 August 1868, Fisher sailed into Evans "Inlet" (now Evans Strait) and, two days later, emerged from its western outlet. He thus demonstrated that Southampton and Coats islands were separate. During his passage through the strait, he discovered and named Walrus Island, the existence of which was later confirmed by another whaling captain, George Comer (see 1907-09). Returning home in summer 1869, with the explorer C.F. Hall on board (see 1864-69), Fisher again passed through this strait. Later in the same year, he sent to the British Hydrographic Department a map showing the new strait and naming its western part Fisher Strait. The new straits were first incorporated in Admiralty charts in 1872.

ROSS, W. G., 1971, pp. 43-44.

1869 | Hudson's Bay Company supply voyage | Henry Bishop
John James | *Prince Rupert* [VII]
Lady Head

From London; 5 June - 29 October. *Prince Rupert* visited York Factory 19 August - 26 September, and reached London on 24 October. *Lady Head* visited Moose Factory 8 - 26 September, and reached London on 29 October.

HUDSON'S BAY COMPANY (1) and (2).

Roman Catholic missionary exploration north of Great Bear Lake	Emile Petitot, OMI	1869

From Fort Good Hope; March-May. Petitot revisited the region of Colville Lake, north of Great Bear Lake, and spent several weeks among the Indians of the region. In the same year, he visited the East Channel of the Mackenzie River delta with a party of Eskimos.

PETITOT, 1875.

Establishment of Fort Providence	1869

The Hudson's Bay Company built its first trading post on the present site of Fort Providence, Mackenzie River, in 1869. The company has maintained a post there continuously to the present day.

USHER, 1971, p. 63; *MOCCASIN TELEGRAPH*, 1970, p. 123.

Closure of Fort Yukon; establishment of Rampart House	1869

Following the Purchase of Alaska in 1867 (*qv*), Captain Charles W. Raymond, US Army Corps of Engineers, made a reconnaissance survey of Yukon River, during which he ascertained that Fort Yukon stood in American territory. The Hudson's Bay Company therefore withdrew and opened a new post, Rampart House, Porcupine River. But, in 1890, J. H. Turner, US Coast Survey, found that it, too, was in Alaska, and the post was removed some 30 km upstream to a point just within Yukon Territory.

Hudson's Bay Company supply voyage	John James Henry Bishop George B. Galbraith	*Lady Head* *Prince Rupert* [VII] *Ocean Nymph*	1870

From London; 8 June - 19 October. *Lady Head* visited Moose Factory 27 August - 11 September, and reached London on 14 October. *Prince Rupert* visited York Factory 20 August - 18 September, and reached London on 19 October. *Ocean Nymph* visited York Factory 2-9 September. She then carried cargo and coals to Rigolet, and reached London on 19 October.

HUDSON'S BAY COMPANY (1) and (2).

1870

Transfer to Canada of Rupert's Land and the North-Western Territory

The Hudson's Bay Company had held monopolistic trading, mineral, and fishing rights to the lands draining into Hudson Bay and Hudson Strait since 1670 (*qv*) and to the North-Western Territory since its coalition with the North West Company in 1821 (*qv*). Negotiation went forward between 1864 and 1869, and the company finally agreed to surrender its rights and territories to the Crown for a compensation of £300,000, the retention of specified blocks of land about its posts, and one-twentieth of the fertile belt in southern Canada. The Crown transferred its interest in Rupert's Land and the North-Western Territory to Canada on 15 July 1870. The Canadian Arctic islands were added in 1880 (*qv*), the whole being designated the North-West Territories until 1906, when the present style, Northwest Territories, was adopted.

CANADA YEAR BOOK, 1973, p. 90; CUMMING and MICKENBERG, 1972, pp. 147-53.

1870

Roman Catholic missionary exploration east of Mackenzie River — Emile Petitot, OMI

From Fort Good Hope. Father Petitot attempted to follow a new route from Fort Good Hope to Fort Norman, of which he had learned from an Indian. The route took him behind the mountains east of Mackenzie River, but before he had gone halfway his guides refused to continue, and he was forced to turn back. In the same year, he travelled to Fort Yukon, where he was coolly received as Father Séguin had been (see 1862-63), then returned to Fort Good Hope.

PETITOT, 1875.

1871

Hudson's Bay Company supply voyage — Henry Bishop, *Prince Rupert* [VII]; John James, *Lady Head*; George B. Galbraith, *Ocean Nymph*

From London; 1 June - 22 October. *Prince Rupert* visited York Factory 8 August - 20 September, and reached London 22 October. *Lady Head* visited Moose Factory 27 August - 18 September, and reached London 22 October. *Ocean Nymph* visited York Factory 11 August - 2 September, and was damaged by gales whilst there. She did not, therefore, return direct to London, but sailed to Quebec and wintered at Sorel.

HUDSON'S BAY COMPANY (1) and (2).

1871

Roman Catholic missionary exploration east of Mackenzie River — Emile Petitot, OMI

From Fort Good Hope, Petitot made a second attempt to follow the new route between Fort Good Hope and Fort Norman (see 1870), this time accompanied by Kearney (see 1861) and Charles Gaudet, an officer of the Hudson's Bay Company. It appears that they successfully completed the journey, and that Petitot continued along the inland route as far as Fort Simpson. He found the whole region devoid of Indians, but remarked that, following this journey and two more made in 1872, the Indians began hunting there.

PETITOT, 1875.

1871-73

United States North Polar expedition	Charles Francis Hall Sidney O. Budington George E. Tyson	*Polaris*

From New London; 3 July 1871 - 7 November 1873; to reach the North Pole by way of Smith Sound. The ship entered Smith Sound in late August and made rapid progress northward. On 28 August she entered Kennedy Channel, on 29 August Hall Basin, and on 30 August she reached her highest latitude, 82°11'N, where she was stopped by ice. Hall turned southward to seek a suitable winter harbour and found one at 81°38'N, at the northern end of Polaris Bay, north-west Greenland, which they named Thank God Harbour. Between 12 and 24 October, Hall sledged to the northern side of the entrance to Newman Bay. On 8 November he died on board ship. The circumstances of his death remain a mystery: the official reason given was apoplexy, but it now seems probable that he died of arsenic poisoning. However, whether the poisoning was deliberate or accidental is not known. With Hall's death, the crew's enthusiasm for the North Polar venture evaporated.

1872: A number of spring boat journeys were undertaken, but achieved nothing. On 12 August, the ship began the return southward, but was soon beset. On 15 October, 19 of the party were carried away from the ship on an ice floe. They drifted south throughout the winter, their survival being mainly owing to the efforts of two Eskimos in the party, Joe and Hans, and their senior officer, Captain Tyson. They were finally rescued on 30 April 1873 off Grady Harbour, Labrador, lat 53°10'N, by the Newfoundland sealer *Tigress.* Meanwhile, those still on board *Polaris* spent the winter of 1872-73 at Foulke Fjord, north-west Greenland. They abandoned ship on 3 June 1873, and set off southward in boats. On 23 June, they were picked up by the whaler *Ravenscraig* of Kirkaldy in lat 75°38'N, long 65°35'W. They were carried to Dundee on the whalers *Arctic* and *Eric.* The last of the party arrived back at New York on 7 November 1873.

*DAVIS, 1876; *LOOMIS, 1972, pp. 261-354; BESSELS, 1879; BLAKE, E.V., 1874.

1872

Hudson's Bay Company supply voyage	Henry Bishop John James Henry Tuggey	*Prince Rupert* [VII] *Lady Head* *Walrus*

From London; 6 June - 25 October. *Prince Rupert* visited York Factory 17 August - 22 September, and reached London on 25

October. *Lady Head* visited Moose Factory 20 August - 10 September, and reached London on 11 October. *Walrus* arrived at York Factory on 23 September, and remained in Hudson Bay until 1874, when she returned to London.

HUDSON'S BAY COMPANY (1) and (2).

1872

Formation of the Diocese of Moosonee

When Robert Machray, Bishop of Rupert's Land, visited Moose Factory in 1868, he seems to have formed a strong impression that the Diocese of Rupert's Land was too large, and that the Hudson Bay region needed a bishop of its own. He brought this matter before the Church Missionary Society, and the final decision to create a new diocese was made on 10 June 1872. On 15 December of that year, John Horden (see 1851) was consecrated as the first bishop of the new Diocese of Moosonee. The diocese covered the whole Hudson Bay region and much of northern Quebec, Ontario, and Manitoba, and had missions at Moose Factory, Fort George, Albany, York Factory and Churchill. It was divided in 1899 with the creation of the Diocese of Keewatin (*qv*).

BOON, 1962, pp. 127-29.

1872

Roman Catholic missionary exploration north of Great Slave Lake — Emile Petitot, OMI

From Fort Good Hope. On the last of his exploring journeys in the Mackenzie River region, Petitot visited Lac des Bois and Lac Maunoir, north of Great Bear Lake, and travelled in the region of Anderson River. In November, he visited the headwaters of Anderson River and visited Indian settlements around Smith Arm of Great Bear Lake. In 1873-74 after nine years of almost continuous exploration in Mackenzie District, Petitot was stationed at Lac la Biche. In 1875, he went to Paris, where he publicly presented some of his observations on the culture and history of the northern Indians and Eskimos and arranged for publication of some of his works. He returned to Canada in 1876, but his health was broken. In 1886, he retired to France and, during the next 30 years, published many books and articles on northern Canada.

PETITOT, 1875.

1872

British whaling voyage — William Adams — *Arctic*

According to Captain J. E. Bernier, Captain Adams surveyed Admiralty Inlet during a whaling voyage in 1872, named Richards and Yeoman islands, and entered Adams Sound.

CANADA, DEPARTMENT OF MARINE AND FISHERIES, 1911, p. 34.

1872

Discovery of McDougall Pass — James McDougall

McDougall was sent by the Hudson's Bay Company to find a more suitable line of communication between the Mackenzie River and Yukon River drainage systems than the winter portage route of about 120 km between Fort McPherson on Peel River and La Pierre's House on Bell River. The route McDougall discovered, now known as McDougall Pass, was a portage of only about 55 km between Rat and Bell rivers. It had, in fact, been visited more than 30 years earlier by Bell and Isbister (see 1839 and 1840-41), but their visits had not been followed by a survey, and the portage had been neglected in favour of the McPherson / La Pierre route. McDougall Pass was later used by considerable numbers of prospectors during the Klondike gold rush of 1897-99.

STEFANSSON, 1960, pp. 223-25.

1872-74

Gold prospecting expedition to Yukon Territory — Arthur Harper, Leroy Napoleon McQuesten

In September 1872, Harper and four others set out from Manson River, British Columbia, to prospect for gold along Mackenzie and Yukon rivers. They descended Peace River to Halfway River, which they ascended by canoe to its headwaters. During the winter, they sledged to Sikanni Chief River, a tributary of the Nelson-Liard river system. At the mouth of Nelson River, they met "Jack" McQuesten and two others, who joined them. One of Harper's party, named Wilkinson, left them here to prospect up Liard River, the rest canoed down Liard and Mackenzie rivers to Peel River, continued by the Rat River, McDougall Pass, and Porcupine River route to Yukon River, and arrived at Fort Yukon on 15 July 1873. There, the party divided. McQuesten took one party about 80 km downstream, where they wintered. Harper's party went upstream, discovered Fortymile River, but were dissuaded by Indians from ascending it. They continued up the Yukon to White River, near where they wintered. In spring 1874, the two parties reunited at Fort Yukon, neither having had much success in prospecting. On 4 June they set off down Yukon River to St Michael for new supplies. In the same year, Harper and McQuesten returned to the Klondike region, and formed a new trading company (see 1874).

OGILVIE, 1913, pp. 87-101.

1873

United States relief expedition — D. L. Braine, *Juniata*; James A. Greer, *Tigress*

From New York; 24 June - 10 November; to search for the missing members of the *Polaris* expedition (see 1871-73) after those who had been rescued from an ice floe in April 1873 had indicated that the men still on board *Polaris* might be in trouble. *Juniata* sailed first and followed the west coast of Greenland northward to Tasiussaq, just north of Upernavik. From there, in her steam-launch, *Little Juniata,* Lieutenant G. W. De Long searched Melville Bay between 3 and 11 August, but he found nothing of *Polaris. Tigress* having now arrived, *Juniata* returned homeward, and reached St John's on 10 September. Soon after, news reached her that the crew of *Polaris* had been picked up by British whalers, so she returned to New York, arriving 25 October. Meanwhile, *Tigress,* which had left New York on 14 July, entered Smith Sound in August and, on 14 August, located the spot in Foulke Fjord where *Polaris* had wintered in 1872-73. Commander Greer learned from an Eskimo that the *Polaris* had sunk near there. After visiting Upernavik, he searched the Baffin Island coast for whalers that might have picked up the crew of *Polaris.* Failing to meet any ships, he sailed homeward, and reached New York on 10 November.

DAVIS, 1876, pp. 579-93.

1873

Establishment of the
North-West Mounted Police

Established by the Dominion Government as a civil force to maintain law in what was then the North-West Territories, the prefix Royal was added in 1904 in recognition of this service. In 1918, the force's sphere of operations was expanded to all of Canada west of Thunder Bay, Ontario. In 1920, it absorbed the Dominion Police, moved its headquarters from Regina to Ottawa, and changed its name to the present Royal Canadian Mounted Police. Yukon Territory (see 1894) and the Northwest Territories are policed exclusively by the force.

ZASLOW, 1971.

1873

British whaling voyage — William Adams, Albert Hastings Markham — *Arctic*

From Dundee; 3 May - 19 September; routine whaling voyage to the Davis Strait region. Markham, a commander in the Royal Navy, joined the ship as second mate to gain Arctic experience in preparation for the North Polar expedition of 1875-76 (*qv*). *Arctic* visited the south-west fishing grounds off Frobisher Bay before heading northward into Davis Strait. On 27 June, she entered Lancaster Sound and the next day was off Prince Regent Inlet. On 4 July, they visited Port Leopold, Somerset Island, and found relics of Ross's expedition of 1848-49 (*qv*). They left Lancaster Sound in mid-July, but returned when they found the whaling poor in Baffin Bay. They visited Admiralty Inlet, and Markham corrected the charts of the north coasts of Baffin and Bylot islands. They then entered Prince Regent Inlet, found more relics of past expeditions at Fury Beach, and turned back for home at Cresswell Bay. During landings on the north coast of Baffin Island and the east coast of Somerset Island, Markham collected geological, botanical, and zoological data.

*†MARKHAM, A. H., 1874; MARKHAM AND MARKHAM, 1927, pp. 56-73.

Hudson's Bay Company supply voyage | Henry Bishop, George B. Galbraith | *Prince Rupert* [VII], *Lady Head*

1873-74

From London; 6 June 1873 - 20 October 1874. *Prince Rupert* visited York Factory 10 August - 21 September, and reached London on 17 October 1873. *Lady Head* visited Moose Factory 21 August - 19 September. She returned there from Mansel Island on 13 October, and wintered at Charlton Island. In 1874, she left Moose Factory for London on 12 September, now under the command of Henry Bishop, and reached London on 20 October 1874.

HUDSON'S BAY COMPANY (1) and (2).

Hudson's Bay Company supply voyage | Henry Bishop (out), George B. Galbraith (home) | *Ocean Nymph*
Alexander Main, Gilbert Hackland | *Mink*

1874

From London; 13 June - 22 October. *Ocean Nymph* visited Moose Factory 24 August - 13 September, and reached London on 22 October. *Mink* visited York Factory 10 - 20 September. There, Hackland took command, and sailed her to Moose Factory. She remained in Hudson Bay.

HUDSON'S BAY COMPANY (1) and (2).

Consecration of the Bishop of Athabasca

1874

Reverend William C. Bompas came to Mackenzie River in 1865. He had travelled widely throughout the district before he was consecrated the first Anglican Bishop of Athabasca on 3 May 1874. In 1884 (*qv*) his huge diocese was divided along lat 60°N; he took the northern part of it and a new title, Bishop of Mackenzie River. His diocese was again divided in 1891 (*qv*).

BOON, 1962, pp. 206-16.

Formation of Harper, McQuesten and Company; establishment of Fort Reliance

1874

Soon after their gold prospecting journey to Yukon Territory (see 1872-74), "Jack" McQuesten and two others joined the Alaska Commercial Company, the fur trading company which now owned Fort Yukon (see 1869). They travelled up to Fort Yukon by steamer, arrived there on 20 August, and heard good reports of furs on Klondike (then known as "Trondeg") River. McQuesten therefore continued up Yukon River and, in autumn 1874, established Fort Reliance about 10 km from the mouth of Klondike River. Soon after, he was joined by Harper, and the two men formed a new company, trading on commission from the Alaska Commercial Company. McQuesten's two companions, Frederick Hart and Alfred H. Mayo, also joined the company. They remained at Fort Reliance in 1875 and thereafter traded at several points up and down Yukon River. In 1883, they acquired a steamer, *New Racket.* In 1886, they established a post at Stewart River to trade with the miners who had recently arrived there. In 1887, they followed the miners to Fortymile River, where they established another post, whilst keeping the Stewart River post open for the Indian fur trade. In 1889, the company split up; Harper and McQuesten continued to trade independently of each other, though still on commission from the Alaska Commercial Company.

During their time in Yukon Territory, Harper and McQuesten discovered and named Klondike, Fortymile, Sixtymile, and Fifteenmile rivers. These relate to the rivers' approximate distance from Fort Reliance.

OGILVIE, 1890, pp. 44-45; 1913, pp. 65-67, 97-102; ARMOND, 1973, pp. 114-15.

1874

Establishment of Fort Smith

The Hudson's Bay Company established a trading post at Fort Smith, Slave River, in 1874.

USHER, 1971, p.44.

1875

Hudson's Bay Company supply voyage — Henry Bishop, John McPherson — *Lady Head*, *Ocean Nymph*

From London; 10 June - 7 November. *Lady Head* visited Moose Factory 10-26 September, and reached London on 7 November. *Ocean Nymph* visited York Factory 12-22 September, and reached London on 22 October.

HUDSON'S BAY COMPANY (1) and (2).

1875

British North-west Passage expedition — Allen William Young — *Pandora*

From Southampton; 25 June - 16 October; a private venture with the object of reaching the magnetic pole by way of Baffin Bay and Lancaster Sound, then navigating the North-west Passage in one season. They entered Lancaster Sound on 22 August, and at once ran into heavy ice. On 26 August, they reached Beechey Island, where they examined relics of the Franklin search expeditions. They continued into Peel Sound, where the ice was again heavy, and were finally beset off the De La Roquette Islands in Franklin Strait, with no hope of further progress. They therefore turned homeward, and reached Spithead on 16 October.

*YOUNG, 1876; 1879; *INNES-LILLINGSTON, 1876; *MACGAHAN, 1876; *KOOLEMANS BEYNEN, 1876; †WOODWARD, 1878.

1875

Geological Survey of Canada expedition to Peace River — Alfred R. C. Selwyn, John Macoun

From Victoria, BC; 18 May - 1 November. They travelled by stagecoach and steamer to Quesnel, and continued from there on 5

June with a pack train of 26 horses. They travelled north-westward across West Road, Chilako, Nechako and Stuart rivers to Stuart Lake, then turned eastward past Carp Lake to McLeod Lake trading post. They left there on 3 July, and continued by boat and canoe along Parsnip and Peace rivers, reaching the mouth of Smoky River on 18 August. There, Selwyn turned back, and reached Victoria on 30 October. Macoun, the expedition's botanist, continued down Peace River to Lake Athabasca, then took the route by Methy Portage to Winnipeg, where he arrived on 1 November.

*†SELWYN, 1877.

British naval North Polar expedition	George Strong Nares Henry F. Stephenson	HMS *Alert* HMS *Discovery*	1875-76

From Portsmouth; 29 May 1875 - 2 November 1876; to reach the North Pole by way of Smith Sound. They called at Lievley and Upernavik, west Greenland, to pick up interpreters, and entered Smith Sound on 28 July. They proceeded up the western side of Kane Basin and Kennedy Channel and, on 25 August, reached Discovery Harbour, on the northern side of Lady Franklin Bay, Ellesmere Island. This site was chosen as *Discovery's* winter quarters. *Alert* continued up Robeson Channel to Floeberg Beach, Ellesmere Island, in lat 82°28'N, the highest latitude reached by any ship to that date. There, she put into winter quarters on 1 September. Autumn travelling parties established depots at Cape Joseph Henry.

1876: Spring sledging parties explored towards the North Pole, and along the north coasts of Ellesmere Island and Greenland (see 1876) *Alert* left her winter quarters on 31 July to return home, for scurvy was now afflicting many of the crew. The two ships left Discovery Harbour on 20 August, passed out of Smith Sound on 9 September, and sailed to Portsmouth. The scientific results of the expedition included zoological, geological, botanical, and ethnological observations.

*GREAT BRITAIN, PARLIAMENT, 1877; NARES, 1876; 1876a; 1877; †*1878; MARKHAM, A.H., 1876; *1878; MOSS, *1878; n.d.; †SMITH, A.H., 1919; †BERKELEY, 1878; †BLACK, 1876; †BUSK, 1880; †CAMBRIDGE, 1877; †CARTER, 1877; †DeRANCE, 1878; †DICKIE, 1878; †DUNCAN AND SLADEN, 1877; 1881; †ETHERIDGE, 1878; †FIELDEN, 1877; 1877a; 1877b; 1878; †FIELDEN AND DeRANCE, 1878; †FIELDEN AND JEFFREYS, 1877; †HART, 1880; †HEER, 1878; †JEFFREYS, 1877; †MIERS, 1877; †SMITH, E.A., 1877; †STEPHENSON, 1876; HATTERSLEY-SMITH, 1976; CASWELL, 1977.

Hudson's Bay Company supply voyage	John McPherson Henry Bishop	*Ocean Nymph* *Prince of Wales* [II]	1876

From London; 8 June - 8 November. *Ocean Nymph* visited Churchill 1-8 September, York Factory 12-21 September, and reached London on 8 November. *Prince of Wales* visited Moose Factory 4-20 September, and reached London on 6 November.

HUDSON'S BAY COMPANY (1) and (2).

1876

British North Polar expedition | Albert Hastings Markham, Alfred A.C. Parr | sledge *Marco Polo*, sledge *Victoria*

From Floeberg Beach, Ellesmere Island, winter quarters of *Alert* (see 1875-76); 3 April - 14 June; to attain the North Pole. They followed the coast of Ellesmere Island to Cape Joseph Henry, then set out due north over the pack ice. Signs of scurvy appeared early in the journey, but they managed to achieve a record latitude of 83° 20'26"N before turning back. The scurvy worsened during the return, and Parr had to go on ahead to fetch a relief party from the ship. One man died before they arrived back.

GREAT BRITAIN, PARLIAMENT, 1877, pp. 118-66.

1876

British exploring expedition | Pelham Aldrich, George A. Giffard | sledge *Challenger*, sledge *Poppie*

From Floeberg Beach, Ellesmere Island, winter quarters of *Alert* (see 1875-76); 3 April - 25 June. They accompanied Markham (see above, 1876) to Cape Joseph Henry, where Giffard's party turned back. Aldrich's party continued along the north coast of Ellesmere Island, and turned back at Yelverton Bay, having established capes Aldrich and Columbia, 83°06'N, as the northernmost extremity of the Canadian Arctic Archipelago. The whole party was badly affected with scurvy during the return, but were fortunate in meeting a relief party sent out from the ship.

GREAT BRITAIN, PARLIAMENT, 1877, pp. 167-274.

1876

British exploring expedition | Lewis A. Beaumont | sledge *Sir Edward Parry*

From Discovery Harbour, Ellesmere Island, winter quarters of *Discovery* (see 1875-76); 6 April - 15 August. They travelled first to *Alert* at Floeberg beach, then set out eastward to explore the north coast of Greenland. Sickness soon broke out in the party of 25 men, and Lieutenant Wyatt Rawson was sent back with the sick on 11 May. Beaumont explored the Greenland coast to Osborn Fjord, where he turned back on 22 May. When they reached Polaris Bay, west Greenland, all the men were very sick with scurvy, and they remained there from 25 June until August. They were met there by a relief expedition, which helped them back to the ship. Two men died during the journey.

GREAT BRITAIN, PARLIAMENT, 1877, pp. 338-418.

1876

British relief expedition | Allen William Young | *Pandora*

From Southampton; 17 May - 3 November. Young had intended making a second attempt on the North-west Passage (see 1875), but changed his plans before setting out when he received a request from the Admiralty to rendezvous with Nares's expedition (see 1875-76). Whilst waiting for Nares in the region of Smith Sound, he made observations on ice and tides and landed at Cape Isabella, Ellesmere Island. He returned to Upernavik on 7 September, waited there and at Godhavn, and finally decided to return home. He met with Nares's expedition in the Atlantic Ocean, and returned with them to Portsmouth.

YOUNG, 1879; BECKER, 1878.

Creation of
Keewatin District | **1876**

By SC 1876, c. 21, a separate district to be known as the District of Keewatin was established and provision was made for the local government thereof. The Act, which was to come into force by proclamation, was, in fact, never proclaimed. It provided that portions of the district might be re-annexed to the North-West Territories by proclamation and, in 1886, a portion of the district was re-annexed. In 1905 the entire district was re-annexed (see 1918).

CANADA YEAR BOOK, 1973, p. 91; CUMMING AND MICKENBERG, eds., 1972, pp. 153-54.

Anglican missionary
expedition to Little
Whale River | Edmund James Peck | **1876-84**

Reverend Peck left London on 11 July 1876 on the Hudson's Bay Company's supply ship *Prince of Wales* (*qv*) to carry out missionary work among the Eskimos. On 7 September, the ship reached Moose Factory, where Peck met Bishop Horden (see 1851). Soon after, he set out with an Eskimo guide in a small sailing boat and reached Little Whale River in mid-October; he remained there until autumn 1877.

1877-78: He wintered at Moose Factory, where he translated the Gospels into Eskimo syllabics and prepared to transport a prefabricated iron church to Little Whale River.

1878: He arrived back at Little Whale River in July.

1879: His iron church was erected and completed in October.

1881: In February and April he visited the Eskimos of Great Whale River. In May, he met Cree Indians at Fort George.

1882-83: He made three unsuccessful attempts to cross overland to Ungava Bay.

1884: He finally reached Ungava Bay (see 1884), then went to England on furlough.

LEWIS, A., 1904.

Hudson's Bay Company
supply voyage | Henry Bishop
John McPherson | *Prince of Wales* [II]
Ocean Nymph | **1877**

From London; 9 June - 25 October. *Prince of Wales* visited Moose Factory 20 August - 8 September, and reached London on 13 October. *Ocean Nymph* visited Churchill 23 August - 2 September, York Factory 3 - 22 September, and reached London on 25 October.

HUDSON'S BAY COMPANY (1) and (2).

1877 Establishment of Fort Wrigley

Fort Wrigley was established by the Hudson's Bay Company as an outpost of Fort Simpson to collect provisions. It was first built on the east bank of Mackenzie River, about 50 km upstream from its present site, and was known as Little Rapid. It was soon moved to about 12 km from the present site; in 1887, it was made a separate post and renamed Fort Wrigley after Trade Commissioner Joseph Wrigley. It was moved to its present site in 1966.

HUDSON'S BAY COMPANY, ARCHIVES, B370/e/1; *MOCCASIN TELEGRAPH,* 1970, p. 123; USHER, 1971, p. 70.

1877 Geological Survey of Canada expedition to Hudson Bay — Robert Bell

From Montreal. Bell travelled to Lake Superior, then continued northward by canoe, arriving at Moose Factory on 30 June. He left there on 7 July and sailed along the east coast of James and Hudson bays, making a track-survey of the coast and studying the geology of the coastline and of Nastapoka and Hopewell islands. He also made notes on climate, timber, soils, rivers, fish, and mammals. He turned back at Cape Dufferin on 24 August and returned along the coast to Moose Factory, where he arrived on 22 September. He then returned by canoe to Lake Superior and thence to Montreal, where he arrived on 1 November.

BELL, R., 1879.

1877-78 Howgate Preliminary Arctic Expedition — George E. Tyson — *Florence*

From New London; 3 August 1877 - 26 September 1878; the first stage of an expedition planned by Captain H. W. Howgate, of the US Army, the ultimate objective of which was a permanent scientific colony at Lady Franklin Bay, Ellesmere Island, to be established in 1878. Tyson was to winter at Cumberland Sound, Baffin Island, collect supplies and hire Eskimos for the main expedition, make scientific observations and collections, and support his voyage by whaling. He reached Cumberland Sound in September and put into winter quarters at Ananito Harbour on 7 October.

1878: All the objectives of the expedition were fulfilled by spring, and on 5 July Tyson set out for Disko, where he was to meet the main expedition. On arrival, he learned that the main expedition had been postponed (in fact, it never took place), so he set off homeward on 22 August. He arrived at St John's on 26 September. He had gathered valuable scientific data on meteorology, hydrology, botany, zoology, ethnology, and made geographical notes on Cumberland Sound, Ananito Harbour, American Harbour, Kekerten Harbour, and Kikastan Islands.

*TYSON, 1879; †KUMLIEN, 1879; †SHERMAN, 1883.

1878

Hudson's Bay Company supply voyage | John McPherson, Henry Bishop | *Ocean Nymph*, *Prince of Wales* [II]

From London; 8 June - 12 October. *Ocean Nymph* visited Churchill 14-17 August, York Factory 19-29 August, and reached London on 28 September. *Prince of Wales* visited Moose Factory 11 August - 4 September, and reached London on 12 October.

HUDSON'S BAY COMPANY (1) and (2).

1878

Prospecting expedition to Yukon Territory | George Holt

Holt, a miner, was the first white man to enter Yukon Territory from the Pacific coast. He crossed either Chilkoot Pass or White Pass to the upper Yukon or Lewes River, followed it down to Marsh Lake, then took an old Indian trail eastward to Teslin River. He returned by the same route to the coast and reported having found "coarse gold." Other dates have been given for this journey, including 1872, 1875, and 1884, but G. M. Dawson (see 1887) decided, after conversation with miners who knew Holt, that 1878 was correct.

DAWSON, 1888.

1878-80

United States Franklin search expedition | Frederick Schwatka

From New York; 19 June 1878 - 22 September 1880; sponsored by the American Geographical Society; to follow up recent Eskimo rumours that records and journals of Franklin's expedition (see 1845-48) might still be preserved on King William Island. He left for Hudson Bay on the whaler *Eothen*, whose captain, Thomas Barry, had brought the Eskimo rumour to the United States. On 9 August, he was landed with his party at Camp Daly, opposite Depot Island and just south of Daly Bay, where he wintered. Between 28 December 1878 and 13 January 1879, he made a reconnaissance trip into Winchester Inlet and discovered Lorillard River, which he explored for part of its length in search of a new overland route to the mouth of Back River. In February and March 1879, he visited Marble Island to communicate with the whalers and to obtain dogs. In April, he set out for King William Island. He travelled up Lorillard River, across the head of Wager Bay, and discovered Hayes River, which he followed down to Chantrey Inlet. He reached King William Island on 5 June. He searched the west coast and found some Franklin relics but was assured by Eskimos that the written records had been destroyed. On 1 November, he left to explore Wilmot Bay, Adelaide Peninsula, and discovered Sherman Inlet and Basin. On 5 December, he returned up Back and Meadowbank rivers, then overland back to Daly Bay. He reached Camp Daly on 4 March, and found that *Eothen* had failed to wait for him or to leave provisions. He continued to Marble Island, where he boarded *George and Mary* of New Bedford on 20 March. On 8 August, he sailed home in this ship.

Records of Schwatka's expedition include observations on topography, travelling conditions, Eskimos and their distribution and survival techniques, flora and fauna. His sledge journey to King William Island, covering about 4,500 km in 50 weeks, was a record for any white man at that time.

*GILDER, 1881; *KLUTSCHAK, 1881; *STACKPOLE, 1965; ANONYMOUS, 1899; †SCHWATKA, 1883; 1884; 1884a; AMERICAN GEOGRAPHICAL SOCIETY, 1881; MARKHAM, C.R., 1880.

1879

Hudson's Bay Company supply voyage — Henry Bishop, John McPherson — *Prince of Wales* [II], *Ocean Nymph*

From London; 11 June - 21 October. *Prince of Wales* visited Moose Factory 23 August - 12 September, and reached London on 21 October. *Ocean Nymph* visited Churchill 9-17 August, York Factory 22 August - 11 September, and reached London on 4 October.

HUDSON'S BAY COMPANY (1) and (2).

1879-80

United States whaling voyage — John O. Spicer — *Era*

From New London; 24 June 1879 - (?) 1880. Spicer sailed through Hudson Strait and entered Foxe Channel on 4 August 1879. He then worked northwards through the ice in Foxe Channel and central Foxe Basin. On 7 August, he was beset in central Foxe Basin and, on the same day, he sighted new land ahead: the Spicer Islands. A week later, he had drifted to within 20 km of the islands, but he got no nearer for he then began to drift south-east. On 18 August, he entered open water and he sailed back to Hudson Strait. He wintered in Frobisher Bay, and set off back to New London on 31 October 1880.

The existence of the Spicer Islands remained in doubt for more than 60 years, and they were occasionally removed from maps. It was not until 1946 that their existence was finally confirmed by the RCAF's Northern Survey Expedition.

*CANADA, DEPARTMENT OF MARINE AND FISHERIES, 1898, pp. 59-60; POLUNIN, 1949, pp. 241-47.

1880

Hudson's Bay Company supply voyage — Henry Bishop, John McPherson — *Prince of Wales* [II], *Ocean Nymph*

From London; 10 June - 17 November. *Prince of Wales* visited Moose Factory 23 August - 12 September, and reached London 14 November. *Ocean Nymph* visited Churchill 23-29 August, York Factory 4-13 September, and reached London on 17 November.

HUDSON'S BAY COMPANY (1) and (2).

1880

Transfer to Canada of the Arctic islands

In 1874, the British Government received two requests for grants of land near Cumberland Sound. This raised the question of sovereignty in a region where American explorers and whalers, since about 1850, had become increasingly active. The Government of Canada by an order-in-council of 10 October 1874 declared its interest in including the Arctic islands within the dominion, but it was not until 4 November 1879, and with some reluctance, that the government passed an order-in-council that ". . .all British Territories and Possessions in North America, not already included within the Dominion of Canada, and all Islands adjacent to any of such Territories or Possessions" shall (except for Newfoundland and its dependencies) be annexed to the dominion, and proposed the date 1 September 1880. The British Government approved the order on 31 July and it took effect 9 October. The vagueness of the order's wording and other confusions left the matter still in some doubt until 2 October 1895, when the Government of Canada by another order-in-council constituted the provisional districts of Ungava, Franklin, Mackenzie, and Yukon.

ZASLOW, 1971, pp. 251-55.

1881

Hudson's Bay Company supply voyage	Henry Bishop John McPherson	*Prince of Wales* [II] *Ocean Nymph*

From London; 13 June - 12 October. *Prince of Wales* visited Moose Factory 13 August - 7 September. *Ocean Nymph* visited Churchill 14-27 August, and York Factory 31 August - 10 September. Both ships reached London on 12 October.

HUDSON'S BAY COMPANY (1) and (2).

1881-84

United States Polar Year Expedition	Adolphus Washington Greely Richard Pike	 *Proteus*

From New York; 12 June 1881 - 17 July 1884; part of the United States' contribution to the first International Polar Year (see 1882-83). They were sent to establish a scientific station at Lady Franklin Bay, Ellesmere Island. A similar expedition had been prepared by H. W. Howgate several years previously, but had been only partly completed (see 1877-78). *Proteus* made a good passage through Smith Sound and along the western shore of Kane Basin, and reached Lady Franklin Bay on 4 August 1881. Captain Pike departed for home on 19 August, leaving a shore party of 25 men, who built Fort Conger at Discovery Harbour, on the northern shore of the bay, where they wintered.

In 1882 and 1883, sledging parties carried out explorations in northern Ellesmere Island (*qv*), and smaller parties explored the region of Fort Conger. Relief expeditions were sent out from the United States in each of these years (*qv*), but both failed to reach Greely's party. Their failure left Greely short of essential supplies and, on 9 August 1883, he was obliged to abandon Fort Conger and to retreat southward by boat. On 29 September, they reached Cape Sabine where, unable to continue farther, they wintered. They suffered considerable hardship throughout the winter; many men died from famine and scurvy, at least one killed himself, and another was executed on 6 June 1884 for repeatedly stealing food. The seven survivors (among whom was Greely) were finally rescued on 22 June 1884 by Schley's relief expedition (*qv*). One more man died before they reached St John's on 17 July. The scientific results of the expedition include valuable meteorological, hydrographic, botanical, zoological, geological, and ethnographic observations.

*†GREELY, 1886; 1888; BRAINARD, 1940; LANMAN, 1885; NEATBY, 1961; POWELL, 1961; TODD, A.L., 1961.

1882

Hudson's Bay Company supply voyage	Henry Bishop John Hawes Alexander Gray C. H. D. Ricks	*Prince of Wales* [II] *Ocean Nymph* *Labrador* *Diana*

From London. *Prince of Wales* left London on 13 June, visited Moose Factory 15 August - 13 September, and reached London on 20 October. *Ocean Nymph* left London on 13 June, visited York Factory 15-24 August, Churchill 28 August - 3 September, and reached London on 5 October. *Labrador* left London on 17 May, visited posts on the Labrador coast, and returned to London on 16 October. *Diana* left London on 8 June, visited Fort Chimo 6 August - 13 September, and returned to London on 10 October.

HUDSON'S BAY COMPANY (1) and (2).

1882

United States exploring expedition	Octave Pavy	dog sledge

From Fort Conger, headquarters of the US Polar Year Expedition (see 1881-84); 19 March - 2 May; to determine whether land existed to the north of Cape Joseph Henry, Ellesmere Island, as reported by members of Hall's *Polaris* expedition (see 1871-73). He followed the east coast of Ellesmere Island to Cape Joseph Henry, then set out northward over the pack ice. However, he had travelled only a short distance before he was stopped and forced to turn back by open water.

GREELY, 1888, Vol. 1, pp. 165-78.

1882

United States exploring expedition	James B. Lockwood	dog sledge

From Fort Conger, headquarters of the US Polar Year Expedition (see 1881-84); 4 April - 1 June; to explore the north-west coast of Greenland eastward beyond Cape Britannia. His farthest east was Lockwood Ø, off the north coast of Greenland; there, at 83°24'N, he also established a new record northern latitude, beating the previous record (held by Markham, see 1876) by nearly four minutes.

GREELY, 1888, Vol. 1, pp. 182-232.

First Greely relief expedition | William M. Beebe, William Sopp | *Neptune* | 1882

From St John's; 8 July - 24 September; to transport supplies and a replacement team to the US Polar Year Expedition at Fort Conger (see 1881-84). Shortly after passing Littleton Island in Smith Sound they were stopped by ice. On 5 September, having failed to get north of 79°20'N, they abandoned their attempt to reach Fort Conger, and returned to St John's.

TODD, A.L., 1961, pp. 47-49; UNITED STATES, CONGRESS, SENATE, 1884, pp. 252-53.

First International Polar Year | 1882-83

A plan for international scientific cooperation in the polar regions was first proposed by the Austrian, Karl Weyprecht, in 1875. An International Polar Commission was set up to work out the details of the plan, which was further developed at the International Meteorological Congress in Rome, April 1879, and the first International Polar Conference in Hamburg, October 1879. At the second International Polar Conference, held at Bern in August 1880, the year beginning 1 August 1882 and ending 1 September 1883 was selected for the project, and at the third conference, at St Petersburg in August 1881, the definitive programme was agreed upon. The project called for a continuous series of meteorological and magnetic observations to be recorded simultaneously at stations scattered throughout the Arctic, in the Antarctic, and at many other observatories in lower latitudes. In the Arctic, 12 stations were in operation during the International Polar Year: at Jan Mayen (operated by Austria and Hungary), Godthaab, Greenland (Denmark), Cumberland Sound, Baffin Island (Germany, *qv*), Fort Rae (Great Britain, *qv*), Dickson, Russia (Holland), Alta (Norway), the Lena delta and Novaya Zemlya (Russia), Spitsbergen (Sweden), Point Barrow, Alaska (United States), Lady Franklin Bay, Ellesmere Island (United States, see 1881-84), and Sodankylä (Finland). Several of the Polar Year expeditions made important studies of geology, botany, zoology, ethnology, oceanography, and other subjects, in addition to making intensive meteorological and magnetic observations.

GREELY, 1886, pp. 19-24; STEEN, 1887, p. i.

British Polar Year Expedition | Henry P. Dawson | 1882-83

From Liverpool; 11 May 1882 - 20 November 1883; Britain's contribution to the International Polar Year (see above, 1882-83). Dawson's party was sent to conduct scientific observations at Fort Rae, a Hudson's Bay Company post on the northern shore of Great Slave Lake. They reached Fort Rae by way of Quebec, Winnipeg, and Carlton House on 30 August, then made meteorological and magnetic observations for twelve months, from 1 September 1882 to 31 August 1883. They then returned to England.

DAWSON, H.P., 1886.

1882-83

German Polar Year Expedition — W. Giese, Captain Mahlstede — *Germania*

From Hamburg; 28 June - 17 October; part of Germany's contribution to the first International Polar Year (see above, 1882-83), sent to establish a scientific station on Clearwater Fiord, Cumberland Sound, Baffin Island. They landed at Clearwater Fiord in late August, and commenced their observations on 15 September. On 12 September, *Germania* sailed for home. Their scientific programme included intensive meteorological and magnetic observations, as well as botanical, geological, and ethnological work. They terminated their observations on 15 September; soon after, they left for Hamburg on board *Germania.*

ABBES, *1884; †1884a; †*NEUMAYER, 1890-91; †NEUMAYER AND BÖRGEN, 1886; †PFEFFER, 1886.

1882-83

German supplementary Polar Year Expedition — K. R. Koch

From Stromness; 15 July 1882 - 7 October 1883; sent to make meteorological observations along the Labrador coast, with the aid of the Moravian missionaries. He sailed out in the Moravian supply vessel *Harmony* and reached Hopedale mission on 10 August. For the next few weeks, he distributed meteorological instruments among the missionaries at Zoar, Nain, Okak, Hebron, and Ramah. He wintered at Nain. After a year's observations, he sailed to London on *Harmony.*

*NEUMAYER, 1890-91, Vol. 1, pp. 145-88; †NEUMAYER AND BÖRGEN, 1886.

1882-83

Gold prospecting expedition to Yukon Territory — Joseph Ladue, William Moore

In the summer of 1882, Ladue, Moore, and ten others crossed Chilkoot Pass and descended Yukon River to Fort Reliance, where they wintered. In 1883, they prospected throughout the summer on Fortymile and Sixtymile rivers without any great success, and all but Ladue and one other left the region in autumn. Ladue remained to work for Harper, McQuesten and Company (see 1874), and later took charge of Ogilvie post on Yukon River, opposite the mouth of Sixtymile River.

OGILVIE, 1913, p. 110.

1882-84

United States scientific expedition to Ungava Bay — Lucien McShan Turner

From Quebec; June 1882 - September 1884; the first detailed scientific study of any part of the Quebec - Labrador peninsula. Turner was employed by the Smithsonian Institution, Washington, DC, to make ethnological and other collections in the Ungava Bay region. He did not travel far from his base at the Hudson's Bay Company post, Fort Chimo, because of his programme of meteorological observations, but with native assistance he made large biological and ethnological collections and voluminous notes on the language and culture of the Eskimos and Naskapi Indians.

†TURNER, L. M., 1886; 1888; 1894.

Second Greely relief expedition	Ernest A. Garlington Richard Pike Frank Wildes	 *Proteus* *Yantic*	1883

From St John's; 29 June - 13 September; to transport supplies and a replacement team to the US Polar Year Expedition at Fort Conger (see 1881-84). An earlier relief vessel had failed to make contact with the expedition (see 1882). *Proteus* reached Littleton Island in Smith Sound on 22 July, and men landed at Cape Sabine the same day. Soon after, the ship was nipped by the ice and sank off Cape Albert. The crew retreated in boats, hoping to meet *Yantic,* which was cruising in the vicinity of Cary Islands, but she had already sailed south. They finally met her at Godhavn on 31 August and returned in her to St John's.

*U.S. CONGRESS, SENATE, 1884; *GARLINGTON, 1883; †LAMAR AND ELLIS, 1884; TODD, A. L., 1961, pp. 66-80.

United States exploring expedition	Frederick Schwatka		1883

From Portland, Oregon; 22 May - 5 October; sent by the US Army to study the native peoples of Alaska. He arrived at Chilkoot Inlet by steamer, crossed Chilkoot Pass to Bennett Lake, and set off down Yukon River through a region that was largely unexplored and poorly charted. He discovered and named Marsh Lake, travelled through Lake Laberge, and continued down the Yukon into Alaska. He reached St Michael on Norton Sound in early September and soon after left by ship for San Francisco, where he arrived on 5 October. In addition to providing information on the Indian tribes that he met, Schwatka made a detailed chart of his route.

SCHWATKA, *†1885; *1885a.

Hudson's Bay Company supply voyage	Henry Bishop John Hawes James L. Dunn C. H. D. Ricks	*Prince of Wales* [II] *Ocean Nymph* *Labrador* *Diana*	1883-84

From London. *Prince of Wales* left London on 11 June, visited Moose Factory 20 September - 8 October, and reached London on 16 November. *Ocean Nymph* visited York Factory 13 - 22 September, and arrived at Churchill 8 October. Heavy ice prevented her leaving Churchill, so she wintered there until 24 July 1884, and reached London on 8 September 1884. *Labrador* left London on 19 May 1883, visited the Labrador posts and Quebec, and reached London on 10 November. *Diana* left London on 11 June 1883, visited Fort Chimo 29 July - 10 September, and reached London on 15 October.

HUDSON'S BAY COMPANY (1) and (2).

1883-84

Gold prospecting expedition to Yukon Territory	Richard Poplin Charles McConkey Benjamin Beach C. Marks	

In summer 1883, these men crossed Chilkoot Pass and descended Yukon River, prospecting for gold as they went. They followed Stewart River up to McQuesten River, but there ran short of supplies and had to go back to Fort Reliance, then to Tanana, Alaska, where they wintered. The following spring, they returned to Stewart River and continued prospecting. In autumn, they returned up Yukon River and across Chilkoot Pass to Juneau. On the return journey, they met Boswell and Franklin (see 1884-85) prospecting on upper Yukon River, and recommended Stewart River to them. They appear to have met with little success themselves.

OGILVIE, 1913, pp. 107-08.

1883-84

German anthropological expedition	Franz Boas Captain Mahlstede	*Germania*

From Hamburg; 20 June 1883 - 7 September 1884; Boas sailed to Baffin Island on the relief ship for the German Polar Year Expedition (see 1882-83) with the object of collecting data on the Eskimos of Cumberland Sound. He arrived at Kekerten on 26 August and from there made a number of exploratory trips around Cumberland Sound: 5 September - 4 October he travelled with Eskimos to Clearwater Fiord, later in autumn he explored Pangnirtung and Kingnait fiords by boat, and throughout the winter and the spring of 1884 he made sledge journeys with Eskimos to Kangilo Fiord, Nettilling Lake, and to the southern shore of Cumberland Sound in the vicinity of Kingmiksok. Between 6 May and 4 July 1884, he travelled with Eskimos up Kingnait Fiord and across Cumberland Peninsula to Padloping Island. He then returned along the east coast of Baffin Island to Cumberland Sound. There, he met the whaler *Wolf*, on which he was given a passage to St John's, where he arrived on 7 September. Although Boas was unable to carry out his main object, to cross Baffin Island to Foxe Basin, he recorded a great deal of valuable anthropological and geographical data, as well as some geological material.

BOAS, †1884; *1884a; *†1885; 1885a; 1885b; †1885c; †1885d; †1885e; †1887; *1887a; †1888; †1888a; †1894; †1901 07.

1884

Third Greely relief expedition	W. S. Schley W. H. Emory George W. Coffin Robert Jones	*Thetis* *Bear* *Alert* *Loch Garry*

From St John's; *Bear* sailed on 24 April, *Thetis* and *Loch Garry* on 12 May, and *Alert* on 22 May; sent to relieve the US Polar Year Expedition at Fort Conger (see 1881-84). Two earlier relief ships had failed to make contact with the expedition (see 1882 and 1883). Only *Thetis* and *Bear* succeeded in passing through exceptionally heavy ice in Melville Bay. They reached Littleton Island in Smith Sound on 20 June and, on the same day, a reconnaissance party discovered Greely's records, which he had deposited eight months previously on Brevoort Island, and which informed them that the expedition was permanently encamped at Cape Sabine. On 22 June, Greely and six other survivors were found at Cape Sabine in a starving condition. They were taken back to Godhavn, where one more man died, then to St John's, where they arrived on 17 July.

*SCHLEY AND SOLEY, 1885; *SCHLEY, 1887.

1884

Division of the Diocese of Athabasca

The Anglican Diocese of Athabasca had been created in 1874 (*qv*), when W. C. Bompas was consecrated as its first bishop. In 1884, this huge diocese, which included much of present-day Alberta, Mackenzie District and Yukon Territory, was divided along the 60th parallel of latitude. Bompas retained the northern part, which received the name Diocese of Mackenzie River; the southern part retained the name of Athabasca, and Richard Young was consecrated as its bishop on 18 October 1884.

BOON, 1962, pp. 215-16.

1884

Church Missionary Society expedition to Fort Chimo	Edmund James Peck

From Little Whale River. Peck, who had been at Little Whale River since 1876 (see 1876-84), left there on 17 July to visit Fort Chimo. He canoed by way of Richmond Gulf, Seal Lake, Larch and Koksoak rivers, and reached Fort Chimo on 11 August. He remained for three weeks to preach to the Eskimos, then left on the Hudson's Bay Company's steamer. Later, he sailed from St John's to Liverpool, where he arrived on 15 October. He remained in England on leave until 1885 (see 1885-92).

LEWIS, A., 1904.

1884-85	Hudson's Bay Company supply voyage	Henry Bishop Alexander Main James L. Dunn C. H. D. Ricks	*Prince of Wales* [II] *Cam Owen* *Labrador* *Diana*

From London. *Prince of Wales* left London on 10 June, visited Moose Factory 9 September - 4 October, and left for London. She met impassable ice between Southampton and Mansel islands, and returned to winter at Charlton Island. In 1885, she was repaired at Moose Factory, left there on 22 August, and reached London on 11 September. *Cam Owen* left London on 10 June 1884, visited Churchill 10-17 September, York Factory 21 September - 3 October, and reached London 31 October. *Labrador* left London on 19 May, visited the Labrador posts and Quebec, and reached London on 28 October. *Diana* left London on 11 June, visited Fort Chimo 23 July - 11 September, and reached London on 8 October.

HUDSON'S BAY COMPANY (1) and (2).

1884-85	First Canadian Hudson Bay Expedition	Andrew Robertson Gordon Robert Bell	*Neptune*

From Halifax; 22 July 1884 - 18 October 1885; sent by the Canadian Department of Marine and Fisheries and the Geological Survey of Canada to establish stations for observing ice conditions in Hudson Strait. They were to establish six stations, with a wintering party of three men in each. They sailed through Gray Strait into Ungava Bay and established the first station at Port Burwell on the western end of Killinek Island. They explored this region by ship and named McLean Strait. Leaving Port Burwell on 8 August, they crossed Hudson Strait to Ashe Inlet, Big Island, where they established the second station. They left there on 16 August, recrossed Hudson Strait, and established the third station at Stupart Bay. On 24 August, they reached the southern shore of Nottingham Island, where they established the fourth station at Port de Boucherville. They searched the east shore of Mansel Island for a site for the fifth station, but without success. They then followed the south-east coast of Southampton Island, visited Marble Island, Churchill, and York Factory, and returned to Digges Islands to build the fifth station at Port Laperrière. Then, after paying a final visit to the first four stations, they sailed homeward, establishing the sixth station at Nachvak on the Labrador coast. The ship reached St John's on 11 October. The 18 men at the stations made daily observations on ice conditions until relieved by the second Canadian Hudson Bay Expedition in summer 1885 (see 1885-86). Scientific studies made on board *Neptune* included fisheries, mineral resources, glaciology, geology, flora, fauna, settlement, and trade.

†*GORDON, 1885; †BELL, R., 1884; †STUPART, 1887; *TUTTLE, 1885.

Geological Survey of Canada expedition to Lake Mistassini	John Bignell Albert Peter Low	1884-85

Rumours, based on Indians' reports, of a vast inland sea between Lake St John and James Bay led the Société de géographie de Québec to seek support from the Geological Survey of Canada in a joint expedition to Lake Mistassini, which was not yet surveyed. Bignell had charge of the expedition, but he and Low left Quebec in July with separate parties of canoemen. Low preceded Bignell up Betsiamites River to Lake Pipmuacan, where they met on 10 September. They then followed different routes to Lake Onistagan, where they met again on 4 November. Winter had then come, and they transported their effects by sledges to the Hudson's Bay Company's post on Lake Mistassini, which they reached 23 December.

On 2 February 1885, having finally disagreed with Bignell about their operations, Low set out on snowshoes for Lake St John and then, by more ordinary transport, to Ottawa to obtain new instructions. He reached headquarters on 2 March, received orders to take charge of the party on 23 March, left the next day with J. M. Macoun, a botanist, and was back at Lake Mistassini on 29 April. Low surveyed the lake during the summer, and in autumn, he and Macoun descended Rupert River and returned to Ottawa by way of Moose Factory. Bignell had returned to Quebec in May and had complained to the newspapers. The "battle of the sextants" received much publicity and was a partial cause of a growing public interest in the vast hinterland of the Quebec-Labrador peninsula. Low's description of the peninsula's largest water body was disputed until 1889, when W. James Loudon and G. S. MacDonald from the University of Toronto confirmed his report. Low's and Macoun's observations and collections were a useful complement to those L. M. Turner had already made at Fort Chimo (see 1882-84).

ROUSSEAU, 1948.

Gold prospecting expedition to Yukon Territory	Thomas Boswell Howard Franklin	1884-85

Boswell, Franklin, and some others were prospecting on upper Yukon River when they met Poplin and party (see 1883-84) returning from Stewart River. Poplin recommended Stewart River to them, so they set off in that direction, wintering first at Fort Reliance. In April 1885, they man-hauled a sledge to Stewart River, and prospected along it. About 150 km upstream, at Chapman's Bar, they made the first big gold strike in the Yukon. They worked the bar all summer and stayed to work it again in 1886.

OGILVIE, 1913, pp. 108-09.

1885	Hudson's Bay Company supply voyage	John Hawes W. Barfield James L. Dunn C. H. D. Ricks	*Cam Owen* *Princess Royal* *Labrador* *Diana*

From London. *Cam Owen* left London on 10 June, visited Churchill 30 August - 7 September, York Factory 21-27 September, and reached London on 26 November. *Princess Royal* left London on 10 June and arrived at Moose Factory on 4 September. She was wrecked in Moose River on 3 October. *Labrador* left London on 19 May, visited the Labrador posts and Quebec, and reached London on 29 October. *Diana* left London on 13 June, visited Fort Chimo 10 August - 10 September, and reached London on 7 October.

HUDSON'S BAY COMPANY (1) and (2).

1885-86	Second Canadian Hudson Bay Expedition	Andrew Robertson Gordon Robert Bell	*Alert*

From Halifax; 27 May 1885 - 18 October 1886; to relieve the six scientific stations on Hudson Strait that had been established the previous year (see 1884-85). They called at Nachvak on the Labrador coast on 1 August and picked up the leader of the station, leaving his two assistants to continue their ice observations through the summer. The ship then passed through Gray Strait, visited Port Burwell, then tried without success to visit the station on Big Island. The ice conditions would not permit them to land there, but they went close enough to contact the staff, then continued to Stupart Bay. There they learned that the three observers had run short of provisions during the winter and had gone to Fort Chimo. On 24 August, they reached Port de Boucherville on Nottingham Island and learned that one of the assistants there had died during the winter. On 25 August, they visited Port Laperrière, Digges Islands, and three days later they sailed for Churchill, where they remained about a week. On return, they examined the "North Sleeper" islands, which they renamed Ottawa Islands. Bell landed on one of these islands to study the geology on 11 September, whilst Gordon made a running survey of the west coasts of the most northerly islands in the group. They then made a final visit to all the stations, delivering new crews to man the stations during the following winter, then sailed to Halifax, where they arrived on 18 October.

*†CANADA, DEPARTMENT OF MARINE AND FISHERIES, 1886; †BELL, R., 1886; †PAYNE, 1888.

1885-86	Gold prospecting expedition to Yukon Territory	Richard Poplin Peter Wybourg Francis Morphat Jeremiah Bertrand

Poplin, who had already been prospecting on Stewart River in 1883-84 (*qv*), returned there from Juneau in spring 1885 with another party. He met Boswell and party (see 1884-85) working on their strike at Chapman's Bar and, about 10 km upstream, they made another large strike at Steamboat Bar. After working at it throughout the summer, they wintered about 30 km from the mouth of Stewart River, and, presumably, returned to work at their strike in 1886.

OGILVIE, 1913, p. 109.

1885-92

Church Missionary Society journey to Fort George (Quebec)	Edmund James Peck	

In May 1885, Peck, accompanied by his wife, left Liverpool by ship for Quebec on his second missionary journey to the Eskimos of northern Quebec (see also 1876-84). They reached Moose Factory on 4 July and remained there for the winter. In March 1886, Peck arrived at Little Whale River on a two-month visit, then moved south along the coast to his new headquarters, Fort George. He stayed there for the next five years, making occasional visits to Little Whale River and Moose Factory. In 1891, he conceived a plan for returning to Fort Chimo (see 1884) to compete with the Roman Catholic missionaries, who had plans for establishing themselves in that area. In 1892, W. G. Walton was sent out from England to assist in Peck's venture. In the same year, however, Peck's wife fell ill, and Peck returned with her to England. Thus, the Ungava plan fell through, and Walton succeeded Peck at Fort George. Peck returned to Canada two years later to commence missionary activities on Baffin Island (see 1894-96).

LEWIS, A., 1904.

1886

Hudson's Bay Company supply voyage	John Hawes	*Cam Owen*
	Henry Bishop	*Lady Head* [II]
	W. Neil Shaw	*Labrador*
	C. H. D. Ricks	*Diana*

From London. *Cam Owen* left London on 9 June, and was wrecked off Churchill on 30 August. Her crew was saved, but her cargo lost. *Lady Head* left London on 9 June, visited Moose Factory 22 August - 14 September, and reached London on 24 October. *Labrador* left London on 15 May, visited the Labrador posts and Quebec, and reached London on 16 October. *Diana* left London on 15 June, visited Fort Chimo 5-24 August, George River 25-31 August, then Fort Chimo again 3-20 September, and reached London on 22 October.

HUDSON'S BAY COMPANY (1) and (2).

1886

Third Canadian Hudson Bay Expedition	Andrew Robertson Gordon	*Alert*

From Halifax; 23 June - (?); sent by the Department of Marine and Fisheries to dismantle the sea ice observation stations that had been operating in Hudson Strait since 1884 (see 1884-85 and 1885-86). Passing through Hudson Strait in early July, Gordon visited two of the observatories, at Ashe Inlet and at Digges Islands. He then continued to Churchill, where he arrived on 29 July, and to York Factory. In August, he returned through Hudson Strait, visited all six observatories, embarked the observers, and continued back to Halifax.

*†CANADA, DEPARTMENT OF MARINE AND FISHERIES, 1887; †FEILDEN, 1887; †GELDART, 1887; †LAWSON, 1888.

1886

Gold mining on Stewart River, Yukon Territory

Following the successes of Boswell and party (see 1884-85) and Poplin and party (see 1885-86) on Stewart River in 1885, about 100 gold prospectors arrived there in 1886, forming the first major assembly of gold miners in Yukon Territory. Stewart River was prospected up to the falls, and McQuesten River for an unknown distance, but little was found on the latter. Harper, McQuesten and Company established a supply post at the mouth of Stewart River to meet the miners' needs. However, the discovery of gold on Fortymile River in 1886 (see below) soon attracted attention and, by spring 1887, Stewart River was again practically deserted.

OGILVIE, 1890, pp. 40-41; 1913, pp. 109-12.

1886

Discovery of gold on Fortymile River, Yukon Territory and Alaska

Howard Franklin
Henry Madison

These two men prospected Stewart River early in the year, but decided that there was little gold to be found. They therefore travelled down Yukon River to prospect up Fortymile River, on which they discovered gold just beyond the international boundary. In October, they proclaimed their discovery at the Stewart River trading post, with the result that most of the miners on Stewart River (see above, 1886) left for Fortymile River. From 1887 until the discovery of the Klondike goldfields in 1896, Fortymile River was considered the most important goldfield in Yukon Territory and was mined by about 300 men. This discovery was later followed by the discovery of gold on Sixtymile River, which also remained famous until the Klondike discovery. In September 1887, Harper, McQuesten and Company (see 1874) built a post at Fortymile River to meet the miners' needs. From 1887 to 1890, the company's steamer *New Racket* supplied the post, but it proved inadequate to serve all the miners and, in 1890, the Alaska Commercial Company started sending its steamer *Arctic* on regular supply trips from St Michael, Alaska.

OGILVIE, 1913, pp. 111-12; ARMOND, 1973.

Formation of Hislop and Nagle partnership

c1887

In about this year, J. Hislop and E. Nagle of Edmonton formed a partnership to trade in competition with the Hudson's Bay Company. They provided the Hudson's Bay Company with its first competition in the Northwest Territories since 1821 by opening posts at Fort Resolution and Fort Rae in about 1894. During the next decade, they opened 10 more posts: Fort Smith (c1900?), Hay River (1901), Fort Providence (c1900?), Fort Simpson (c1900?), Fort Liard (c1901?), Fort Norman (c1900?), Wrigley (c1902?), Fort Good Hope (c1900?), Arctic Red River (1901), and Fort McPherson (1902?). In 1911, Hislop and Nagle disposed of their posts and vessels to the Northern Trading Company (later Northern Traders Ltd.), which continued to compete with the Hudson's Bay Company until 1939.

ZASLOW, 1971, pp. 57-58, 240; USHER, 1971, pp. 26, 159.

Hudson's Bay Company supply voyage	Henry Bishop	*Lady Head*
	William Barfield	*Prince Rupert* [VIII]
	W. Neil Shaw	*Labrador*
	C. H. D. Ricks	*Diana*

1887

From London. *Lady Head* left London on 2 June, visited Moose Factory 24 August - 16 September, and reached London 2 November. *Prince Rupert* left London on 2 June, visited Churchill 27 August - 3 September, York Factory 8 - 23 September, and reached London on 3 November. *Labrador* left London on 14 May, visited the Labrador posts and Quebec, and reached London on 14 November. *Diana* left London on 4 May, visited Fort Chimo 26 September - 3 October, and reached London on 5 November.

HUDSON'S BAY COMPANY (1) and (2).

Canadian Yukon Exploring Expedition — George Mercer Dawson

1887

From Ottawa; 22 April - 20 September; sent by the Canadian government to make a geographical and geological survey of Yukon Territory. William Ogilvie (see 1887-89) and Richard George McConnell (see 1887-88) were also sent to survey the region. Dawson left Ottawa with McConnell, and they travelled by steamer from Victoria to Wrangell, up Stikine River to Dease Lake, and from there to Lower Post at the confluence of Dease and Liard rivers. There they parted. Dawson ascended Liard and Frances rivers to Frances Lake, where he arrived on 8 July and which he surveyed. He left a cache of provisions by the lake, crossed the portage to Pelly River, and canoed down this river to its confluence with Lewes River, where he met Ogilvie on 13 August. Having learned of Ogilvie's surveys to that date, he ascended Lewes River to its head, and crossed Chilkoot Pass to Lynn Canal, where he arrived on 20 September. Dawson made surveys of his route and observations on natural history and geology.

DAWSON, 1888; TRIMMER, 1898, pp. 243-381.

1887

British expedition to Churchill Falls — Randle F. Holme

From North West River; 22 August - 19 September; first tourist excursion into the Quebec-Labrador peninsula. Holme was visiting Newfoundland and Labrador with an English friend, H. Duff, who had to return home in mid-summer. He then recruited two local men to assist him to the Grand, now Churchill, Falls, which he believed only two white men, John McLean (see 1839) and William Kennedy, had seen. With great difficulty, they reached Lake Winokapau, having had to portage a heavy and cumbersome boat past many rapids. They were obliged to turn back some 80 km below the falls owing to short rations and bad weather. In 1888, Holme read a paper before the Royal Geographical Society in which he advanced reasons for thinking the falls were some 600 m high. His remarks, reported in the newspapers, were the cause of later expeditions in 1890 and 1891 (*qv*).

HOLME, 1888.

1887-88

Geological Survey of Canada expedition to James Bay — Albert Peter Low

Low's party travelled by rail to Missinaibi, then by boat down Missinaibi and Moose rivers to Moose Factory, where they arrived on 27 June. They crossed to Charlton Island, which they examined for about five days, then surveyed many of the smaller islands in James Bay, including Danby Island, Carey Island, Strutton Islands, Solomon's Temple Island, Weston Island, North Twin and South Twin islands, Spencer Island, Walter Island, Grey Goose Island, and Bear Island. Finally, between 8 and 14 August, they surveyed Akimiski Island. They then returned south for winter.

1888: They returned to James Bay in June, and made a track survey for about 320 km up Fort George River. They then portaged to the head of the south branch of Great Whale River, which they surveyed down to the sea. Continuing northwards to Richmond Gulf, they surveyed Clearwater River up to the outlet of Clearwater Lake. They finished their survey there and returned to Moose Factory, where they arrived on 8 September. In addition to his report on the geology and physical geography of the regions visited, Low also listed the plants collected on the shores and islands of James Bay, and made some zoological and meteorological notes.

LOW, 1888.

1887-88

Canadian Yukon Exploring Expedition — Richard George McConnell

From Ottawa; 22 April 1887 - 1 October 1888. McConnell, one of three men sent by the Canadian government to make a geographical and geological survey of Yukon Territory (see also Dawson, 1887, and Ogilvie, 1887-89), accompanied Dawson to the confluence of Dease and Liard rivers. He left there on 26 June 1887 and descended Liard River to Fort Simpson. He then travelled up Mackenzie River to Fort Providence, from which he visited Slave River, Salt River, Hay River, and the western end of Great Slave Lake. During the winter he visited Fort Rae and Bistcho Lake. On 1 May 1888, he left Fort Providence and travelled by dog-sledge and canoe to Fort McPherson. He met Ogilvie during this part of his journey. He then crossed McDougall Pass to Fort Yukon, where he arrived on 24 July. From there, he travelled up Yukon River, visiting the mining centre at Fortymile River, ascended Lewes River to its head, crossed Chilkoot Pass, and travelled by steamer from Juneau to Victoria, where he arrived on 1 October. In addition to surveying his route, he made observations on the geology, fauna and flora of the region.

*†McCONNELL, 1891; LEECHMAN, 1948.

1887-89

Canadian Yukon Exploring Expedition — William Ogilvie

From Ottawa; 20 April 1887 - 15 January 1889. Ogilvie was the leader of the three-man team sent by the Canadian government to make geographical and geological surveys in Yukon Territory (see also Dawson, 1887, and McConnell, 1887-88). He travelled by steamer from Victoria to Sitka, then crossed Chilkoot Pass to Bennett Lake, and travelled down Lewes River to the mouth of Pelly River, where he met Dawson on 13 August. He continued down Yukon River to White River, which he wished to survey up to the international boundary. However, he was prevented from doing so by its shallowness and the swiftness of the current. He reached Fortymile River on 7 September. He remained there throughout the winter, in the company of about 100 miners, and surveyed the river to the international boundary and made meteorological observations.

1888: He set off down Yukon River on 17 March and surveyed Tatonduk River to its head. He crossed the watershed from there to discover Ogilvie River but, despite assurances from the Indians that Ogilvie River flowed into Peel River, he decided not to navigate that river but to reach the Peel by way of Porcupine River and McDougall Pass. He reached Fort McPherson on 20 June. From there, he surveyed Mackenzie and Slave rivers up to Fort Chipewyan. He left there on 27 November, and reached Ottawa on 15 January 1889. In addition to his geographical surveys and geological studies, Ogilvie made observations on meteorology, fauna and flora, and the economic potential of the regions he visited.

OGILVIE, 1890.

1888

Hudson's Bay Company supply voyage — William Barfield, *Prince Rupert* [VIII]; W. Neil Shaw, *Erik*

From London. *Prince Rupert* left London on 4 June, visited Churchill 20-26 August, York Factory 2-14 September, and reached London on 2 November. *Erik* left London on 19 May, visited the Labrador posts, Fort Chimo, and Quebec, and reached London on 27 October.

HUDSON'S BAY COMPANY (1) and (2).

1888

United States whaling voyage to Herschel Island — Joe Tuckfield

United States whalers had been operating north of Bering Strait since 1848, but Tuckfield was the first commercial whaler to venture far east of Point Barrow. Eskimos at Point Barrow had told him of a large number of whales at Herschel Island, so he set out with some Eskimos in a boat to investigate the region. He saw hundreds of whales near Herschel Island, and killed one. He carried news of his discovery back to Point Barrow and, in the following year, the first whaling ships sailed east of Point Barrow to winter at Herschel Island.

STEVENSON, 1968.

1888-89

British expedition — Lord Lonsdale

From Liverpool; 22 February 1888 - 7 June 1889; a private expedition for the purpose of tourism. Lonsdale travelled on the steamer *Republic* to New York, then by rail to Montreal and to Qu'Appelle, Saskatchewan, where he arrived on 21 March. He then continued by sledge northwards to Athabasca River, Fort Chipewyan, Slave River and Fort Resolution. Ice conditions detained him on the shores of Great Slave Lake throughout June, but he finally crossed the lake in early July and continued his journey by steamer down Mackenzie River to Fort McPherson. He then travelled by boat to Mackenzie River delta, where he met Eskimos, and from where he made a six-day excursion to Cape Kellett, Banks Island. Returning homeward, he visited the gold mining regions of Yukon Territory in October, then travelled by dog sledge down Yukon River to the west coast of Alaska. Early in 1889 he left Kodiak by ship for San Francisco, where he arrived on 13 April. He arrived back in England on 7 June 1889.

DAWSON, LIONEL, 1946, pp. 60-86.

Hudson's Bay Company supply voyage	William Barfield Henry Bishop W. Neil Shaw	*Prince Rupert* [VIII] *Lady Head* *Erik*	1889

From London. *Prince Rupert* left London on 5 June, visited Churchill 22-30 August, York Factory 3-13 September, and reached London on 31 October. *Lady Head* left London on 5 June, visited Moose Factory 30 August - 17 September, and reached London on 1 November. *Erik* left London on 17 May, visited the Labrador posts, Fort Chimo, and Quebec, and reached London on 9 November.

HUDSON'S BAY COMPANY (1) and (2).

United States voyage	Charles H. Stockton	USS *Thetis*	1889

From San Francisco; 20 April - 7 December; to supervise United States whaling and commercial interests in Bering Sea and the Arctic Ocean. Stockton spent the early part of the summer cruising in Bering Sea, then sailed through Bering Strait for Point Barrow. There, having learned that part of the whaling fleet had, for the first time, sailed east of Point Barrow bound for Herschel Island, he decided to follow them. He sailed along the north coast of Alaska, crossed the Alaskan-Canadian boundary on 15 August and, later the same day, anchored off Herschel Island. He remained there for a day with the whaling ships, then set off back to Point Barrow and homeward.

STOCKTON, 1890.

French expedition to Mackenzie River delta	Edouard de Sainville	1889-94

De Sainville travelled by rail from Winnipeg to Edmonton, then continued to Athabasca Landing. He left there by boat with an Indian and Métis crew on 7 May, and followed Athabasca, Slave, and Mackenzie rivers down to Fort McPherson, making notes on trading posts and the fur trade as he went. He arrived at Fort McPherson on 20 July 1889. From there, he made several visits to Mackenzie River delta and, in 1894, he travelled with Frank Russell (see 1893-94) along the Arctic coast to Herschel Island. He also surveyed Peel River up to the mouth of Wind River, and he appears to have surveyed the upper course of Porcupine River. During the expedition, he also made notes on meteorology, Indians, Eskimos, and conditions of travel.

SAINVILLE, 1898.

Exploring expedition to Back River	Warburton Pike James Mackinlay	1890

From Fort Resolution; 7 May - 24 August. The expedition had several purposes; Pike was most interested in shooting musk-oxen for sport and in making a third exploration of Back River (Back, 1833-35, and Anderson and Stewart, 1855, were the only Europeans to visit the river previously); Mackinlay, the Hudson's Bay Company officer in charge of Fort Resolution, wished to examine the possibility of extending the fur trade to the Eskimos of the region. They crossed Great Slave Lake by canoe to Macleod Bay, where they engaged Indians to accompany them. They continued by way of Lac des Morts, Wolverine Lake, Camsell Lake, MacKay Lake, Lockhart River and Aylmer Lake to Back River; they were the first explorers to use this route. Descending the river, on 23 July, they came to Beechey Lake, near where they found a deserted Eskimo encampment. From there, they tried to contact Eskimos who, they suspected, came inland from Bathurst Inlet. Failing in this, they left gifts at the encampment and set off homeward. From Aylmer Lake, they followed Back's route to Great Slave Lake, by way of Clinton-Colden and Artillery lakes. During their return Mackinlay learned that the Hudson's Bay Company had already abandoned its plans to trade with the Eskimos of Back River.

*PIKE, 1892; MACKINLAY, 1957.

1890

Canadian expedition to Churchill Falls — J. G. Alwyn Creighton

Stimulated by R. F. Holme's report (see 1887) of a 600-m waterfall on Churchill River, Creighton tried to reach the location by ascending St Augustine River. He soon turned back, however, because of a forest fire and shortage of provisions.

CREIGHTON, 1891.

1890

North-West Mounted Police patrol — J. V. Begin

From Winnipeg; 2 July - 26 October; to patrol Lake Winnipeg and to visit York Factory. This was the North-West Mounted Police's first northern patrol and the first to visit the shores of Hudson Bay. Inspector Begin set out from Winnipeg in a small sailing vessel, *Keewatin.* After patrolling Lake Winnipeg for a month, he established a detachment at Norway House. There, he gave instructions to his men to patrol the northern end of Lake Winnipeg on board *Keewatin* until the end of August, then to return to Winnipeg. He himself set out by canoe on 13 August to visit York Factory, where he was to appoint a local justice of the peace and control local brewing and other illegal activities. He remained at York Factory 23 August - 3 September, then returned to Norway House, where he arrived about 25 September. There, he learned that *Keewatin* had been wrecked on Lake Winnipeg. Further investigations revealed that two of his men, Corporal H.O. Morphy and Constable G.Q.R. de Beaujeu, had been drowned.

TURNER, J.P., 1950, Vol. 2, pp. 486-88; STEELE, 1936, pp. 23-24; ZASLOW, 1971, p. 97.

Hudson's Bay Company supply voyage	William Barfield Henry Bishop W. Neil Shaw	*Prince Rupert* [VIII] *Lady Head* *Erik*	1890-91

From London. *Prince Rupert* left London on 5 June, visited Churchill 24 August - 3 September, and York Factory 21 September - 11 October. During the return voyage, the ship met heavy ice between Southampton and Mansel islands and sprang a leak. The captain turned back on 23 October and made a difficult voyage to Charlton Island, where he wintered. The ship was repaired at Moose Factory in summer 1891, left on 1 September, and reached London on 7 October. *Lady Head* left London on 5 June 1890, visited Moose Factory 1-24 September, and reached London on 4 November. *Erik* left London on 7 June, visited the Labrador posts, Fort Chimo, and Montreal, and reached London on 1 November.

HUDSON'S BAY COMPANY (1) and (2).

Hudson's Bay Company supply voyage	J. E. Milne John G. Ford Alexander Gray	*Perseverance* *Lady Head* *Erik*	1891

From London and Peterhead. *Perseverance* left Peterhead on 8 June, visited York Factory 25 August - 4 September, then visited Churchill, and reached London on 20 October. *Lady Head* left London on 6 June, visited Moose Factory 23 August - 25 September, and reached London on 15 November. *Erik* left London on 3 June, visited the Labrador posts, Fort Chimo, and Montreal, and reached London on 1 November.

HUDSON'S BAY COMPANY (1) and (2).

United States expeditions to Churchill Falls	Austin Cary D. M. Cole Henry G. Bryant C. A. Kenaston	1891

Stimulated, like J. G. A. Creighton (see 1890) by R. F. Holme's report (see 1887) of a 600-m waterfall on Churchill River, two expeditions, each unknown to the other, set out in 1891. Cary and Cole, members of an expedition of undergraduates at Bowdoin College to Labrador, were the first to arrive. They had set out with Leslie Lee and Jonathan Cilley, but these two turned back after their canoe overturned. Cary and Cole walked the last 15 km to the falls and, on return to their camp, found their canoe and other equipment destroyed by fire, so they had to walk the rest of the way. Bryant and Kenaston, representing the Geographical Society of Philadelphia, arrived at the falls a fortnight later and found Cary and Cole's record. Returning, they passed Cary and Cole without seeing them for the second time. Their two accounts gave a more accurate notion of Churchill Falls' height, about 75m. A. P. Low gave a still better description in 1894 (see 1892-96).

BRYANT, 1892; 1894; CARY, 1891; 1892.

1891

Division of the Diocese of Mackenzie River

The Anglican Diocese of Mackenzie River had been created in 1884 (*qv*) by the division of the Diocese of Athabasca. In 1891, the Mackenzie River diocese was in turn divided. W.C. Bompas became bishop of its western part, the Diocese of Selkirk (present-day Yukon Territory; the name was changed to Diocese of Yukon in 1907). The eastern part retained the name Mackenzie River, and William Day Reeve became its new bishop. The Diocese of Mackenzie River remained excessively large, but it was not divided again until 1933, when its northern portion was absorbed into the new Diocese of the Arctic.

BOON, 1962, pp. 216, 220-21.

1892

Hudson's Bay Company supply voyage	John G. Ford Alexander Gray John Hawes	*Lady Head* *Erik* *Princess*

From London. *Lady Head* left London on 11 June, visited Moose Factory 17 August - 10 September, and reached London on 5 November. *Erik* left London on 12 June, visited the Labrador posts, Fort Chimo, Churchill, and York Factory, and reached London on 7 November. She was the first steam vessel sent by the Hudson's Bay Company to Hudson Bay. *Princess* reached York Factory from London on 22 August, and remained in Hudson Bay.

HUDSON'S BAY COMPANY (1) and (2).

1892

Geological Survey of Canada expedition	Joseph Burr Tyrrell D. B. Dowling

They were sent by the Geological Survey of Canada to survey the unknown territory lying north of Churchill River and south-east of Lake Athabasca. Setting out from Prince Albert on Saskatchewan River, they travelled overland to Green Lake, where they began their canoe trip along Beaver River to Lac Ile-à-la-Crosse, down Churchill River and up Mudjatic River, and across the height of land to Cree Lake, where they arrived on 1 July. From there, they followed Cree River to Black Lake, obtained fresh supplies from Fond du Lac post on Lake Athabasca, then returned to survey Black Lake. They surveyed the western shore of Wollaston Lake, then Dowling went to survey Reindeer Lake, whilst Tyrrell ascended Geikie River to its source, crossed the height of land, and discovered Upper Foster Lake. He returned to Churchill by way of Foster River, arrived at Ile-à-la-Crosse on 20 September, then continued to Prince Albert. During the journey, they surveyed a total distance of about 2,000 km, of which more than 800 km had never before been travelled by white men.

*TYRRELL, J.B., 1894; LOUDON, 1930, pp. 78-88.

National Mission to Deep Sea Fishermen voyage to Labrador | John Trezisse, Wilfred Grenfell | *Albert* | 1892

From Yarmouth; 12 June - 21 November. In response to repeated requests, the National Mission to Deep Sea Fishermen sent a hospital ship, *Albert*, with Grenfell as surgeon, to Newfoundland and Labrador. They reached St John's in late July, just after the city had been destroyed by fire. After giving such assistance as they could, they left on 2 August for Labrador, sailing during the next two months as far north as Hopedale and stopping wherever possible. Grenfell was distressed by the poverty and social state of the fishing population, and he seems immediately to have realized what he might be able to do with and for them. His work, which brought him an international reputation, gradually evolved into the International Grenfell Mission, which brought millions of dollars to spend and hundreds of volunteers to serve in the region. Grenfell died 9 October 1940, but the Grenfell Mission continues.

KERR, 1959, p. 67-85.

United States expedition to Mackenzie District | Elizabeth Taylor | 1892

From Athabasca Landing; a private expedition to make photographs and sketches along Mackenzie River. Miss Taylor, daughter of the United States Consul at Winnipeg, travelled by Hudson's Bay Company steamer down Athabasca, Slave, and Mackenzie rivers to Fort McPherson and back. During the journey, she made collections of birds, mammals, fishes, insects, and plants.

PREBLE, 1908, p. 77; MACOUN, 1892; NUTE, 1948.

British hunting expedition to Yukon Territory | Warburton Pike | 1892-93

From Victoria; early July 1892 - autumn 1893; a private expedition to hunt for moose. He took a steamer to Wrangell, then canoed up Stikine River, crossed the watershed to Dease Lake on 16 August, canoed down Dease River, and reached the Hudson's Bay Company's McDame trading post on 19 August. He continued down to Lower Post near the confluence of Dease and Pelly rivers and, after a short excursion along Hyland River, wintered near there. During the winter, he made a number of moose hunting excursions. At the beginning of February 1893, he began hauling his supplies up Liard and Frances rivers to Frances Lake. After two journeys to Frances Lake and a mining expedition to Hyland River, he finally left Lower Post on 18 April, in company with several Indian and half-breed canoemen and guides. From Frances Lake, he followed Yusezyu River north to McPherson Lake, then portaged to Woodside River, which he explored to its source. He then canoed down Woodside, Pelly, and Yukon rivers to the mouth of the Yukon. He canoed around the Alaskan coast to Bristol Bay, and from there continued by steamer to Vancouver Island. During the expedition, he made a collection of plants and geological specimens.

PIKE, 1896.

1892-96

Geological Survey of Canada explorations of the Quebec-Labrador peninsula

Albert Peter Low

Although officers of the Hudson's Bay Company had explored the major rivers of the Quebec-Labrador peninsula during the first half of the 19th century, the reports they had submitted had not been made public, and the region was virtually unknown to science until the results of A. P. Low's geological reconnaissances, which included important information on every aspect of the natural history and the peoples of the region, reached print. He always travelled with one or two assistants and several experienced Montagnais Indian canoemen.

1892: Low and A. H. D. Ross canoed from Lake St John to Lake Mistassini by way of Ashuapmuchuan River, then descended one of the upper branches of Rupert River about 80 km before portaging across to Eastmain River, which they surveyed to its mouth. They then coasted south to Moose Factory, ascended Moose River to where it is crossed by the Canadian Pacific Railway, and returned to Ottawa, having canoed some 2,000 km.

1893-94: Low and D. I. V. Eaton canoed to Lake Mistassini by substantially the same route Low had followed in 1892, descended another of the upper branches of Rupert River before crossing over to Eastmain River, which they ascended about 160 km, then worked through small rivers and lakes across to the Hudson's Bay Company's post on Lake Nichicun, which they reached on 4 August. From there they crossed to Kaniapiskau Lake to descend Kaniapiskau and Koksoak rivers to Fort Chimo, arriving 27 August. On this journey, Low discovered the vast deposits of iron ore of the Labrador Trough. They had planned to winter at Fort Chimo, but starvation among the Indians decided them to take the company's supply boat to North West River on the Labrador coast and to winter there. During the winter and spring, they carried supplies and canoes to Big Hill, the portage discovered by McLean in 1841 (*qv*) and in May they examined Churchill (Grand) Falls. After break-up, they surveyed Churchill (Hamilton) River to Sandgirt Lake, which they reached on 15 June and, from a base camp there, they spent the rest of the summer exploring the lake-strewn plateau of the central interior of the peninsula. They followed Ashuanipi River to Lake Petitsikapau, crossed to Lake Michikamau and surveyed it, then returned to Sandgirt Lake. They then followed Attikonak River to lakes Ossokmanuan and Attikonak; from the latter, they crossed the height of land to Romaine River, descended it to within 160 km of St Lawrence

River, then crossed over to St John River, descended it, and ended their travels — more than 9,000 km, mainly by canoe — at Mingan on 22 August.

1895: In June, Low and D. I. V. Eaton started up Manicouagan River to Lake Mushalagan, which they reached 12 July, and continued up Mushalagan River, with a side trip to Lake Matonipi and Rivière aux Outardes, until they could cross the height of land to Lake Naococane. They spent some time exploring this region, again discovering massive outcrops of iron formation, then returned by Manicouagan River to St Lawrence River, arriving on 1 September.

1896: Low, G. A. Young, and W. Spreadborough descended Missinabie River to Moose Factory, which they left on 14 June to coast along Hudson Bay in a fishing boat to Richmond Gulf. They then followed Low's route of 1888 (see 1887-88) up Clearwater River to Clearwater Lake, which they reached on 14 July; they continued to Seal Lake, crossed the height of land to Larch River and descended it and Koksoak River to Fort Chimo, arriving on 26 August. They returned on the Hudson's Bay Company's supply vessel, *Erik*. Low reported important extensions of the iron formation he had discovered in 1893 and more iron-bearing strata on upper Larch River. His unified report on his explorations, 1892-95, has with good reason been called the bible of the Quebec-Labrador peninsula.

LOW, 1896.

1893

Hudson's Bay Company supply voyage	John G. Ford Alexander Gray	*Lady Head* *Erik*

From London. *Lady Head* left London on 10 June, visited Moose Factory 21 August - 11 September, and reached London on 12 October. *Erik* left London on 8 June, visited the Labrador posts, Fort Chimo and Churchill, and reached London on 28 October. *Princess* (see 1892) carried York Factory's returns to Churchill, and transported back to York Factory supplies delivered by *Erik*.

HUDSON'S BAY COMPANY (1) and (2).

1893

Geological Survey of Canada expedition	Joseph Burr Tyrrell James Williams Tyrrell

These two men were sent by the Geological Survey of Canada to make an exploratory survey of the unexplored interior of Keewatin District. After travelling by train from Toronto and Ottawa to Edmonton in May, they canoed down Athabasca River and eastwards along Lake Athabasca to Black Lake. From there, on 8 July, they set out into the unexplored country to the north. They followed Chipman River to Selwyn Lake and Wholdaia Lake, then continued down Dubawnt River, discovering Barlow, Carey, and Markham lakes, to Samuel Hearne's Dubawnt Lake (see 1770). They reached there on 7 August and spent the next week exploring it, searching for the outlet of Dubawnt River. Continuing down Dubawnt River, they discovered Grant Lake, the mouth of Chamberlain River, Wharton Lake, Aberdeen Lake, and Schultz Lake. They then continued through Baker Lake and Chesterfield Inlet and, on 13 September, began the voyage down the west coast of Hudson Bay to Churchill. For the last few days of their journey the weather was bitterly cold, and they had to travel a short distance overland to Churchill, where they arrived about 20 October. On 6 November, they set out on snowshoes for Winnipeg, where they arrived on 2 January 1894. In addition to surveying their route, they made geological observations, a collection of plants, and notes on Eskimos.

*†TYRRELL, J.W., 1898; TYRRELL, J.B., 1894; †1894a; †1898.

1893-94

University of Iowa expedition to Mackenzie District — Frank Russell

From Edmonton; 26 April 1893 - 27 October 1894. Russell left Fort Chipewyan by steamer on 20 June, travelled down Slave River to Fort Resolution, then crossed Great Slave Lake to Fort Rae. Between 18 July and 5 August, he canoed up Yellowknife River in company with an Indian, and between 29 August and 29 September, he made a canoe trip with Indians to Fort Resolution. On 7 December, he set out with Indians on a sledge trip around the western end of Great Slave Lake, visiting Fort Providence and Fort Resolution, where he spent a fortnight. He returned to Fort Rae in early February 1894. Between 5 March and 4 May, he accompanied a party of Dogrib Indians on a musk-ox hunt northeastwards from Fort Rae, crossing Coppermine River to the region of Contwoyto Lake, where Russell collected ornithological specimens. On 10 May 1894, he left Fort Rae and travelled alone down Mackenzie River to Fort McPherson. There, he met de Sainville (see 1889-94), and the two men canoed together to Herschel Island to meet the whalers, arriving on 9 July. After making a visit to the mainland to study the Eskimos, Russell departed for home on the whaler *Jeanette*, Captain E. W. Neath. The ship did some whaling off Ostrov Geral'da and Ostrov Vrangelya, reached San Francisco on 27 October, and from there Russell returned to Iowa. During the expedition, Russell made ethnological observations on the Indians and Eskimos, and extensive natural history notes and collections.

*RUSSELL, F., 1898; MOWAT, 1973, pp. 219-37.

1894

Hudson's Bay Company supply voyage	John G. Ford Alexander Gray	*Lady Head* *Erik*

From London; 9 June - 29 October. *Lady Head* visited Moose Factory 9 August - 4 September, and reached London on 18 October. *Erik* visited the Labrador posts, Fort Chimo, and Churchill, and reached London on 29 October. As in 1893 (*qv*), *Princess* carried goods between Churchill and York Factory.

HUDSON'S BAY COMPANY (1) and (2).

Geological Survey of Canada expedition	Joseph Burr Tyrrell Robert Munro Ferguson	**1894**

From Ottawa; 5 July 1894 - 16 January 1895. Tyrrell was sent by the Geological Survey of Canada to continue his exploration of interior Keewatin District (see 1893). From Saskatchewan River, they canoed up Sturgeon Weir River and crossed Frog Portage to Churchill River, then continued up Reindeer River to Reindeer Lake, from where they were to begin their exploration. On 18 July, they reached Brochet trading post at the northern end of the lake. From there, they followed Cochrane River (which they named) to its source, crossed a portage and canoed down Kazan River through Kasba Lake to Ennadai Lake, where they hired Eskimos to guide them to the sea. At Angikuni Lake, they learned that Kazan River did not enter Hudson Bay near Marble Island, as the Eskimos had led them to believe, but at Chesterfield Inlet. They considered that that was too far north for the time of the season, so they left the river at Yathkyed Lake and, guided by the Eskimos, crossed 12 portages to Ferguson Lake and reached Hudson Bay near Bibby Island by way of Ferguson River. They followed the coast to Churchill, arriving on 1 October, left there on snowshoes on 28 November, and travelled to Fort Selkirk, arriving on 5 January 1895.

TYRRELL, J. B., 1895; †1898.

United States Peary Auxiliary Expedition	Henry G. Bryant Henry Bartlett	*Falcon*	**1894**

From St John's; 7 July - 15 September; sent with a relief party to meet R. E. Peary's second North Greenland Expedition of 1893-95, which was wintering at Bowdoin Bay, Inglefield Gulf, northwest Greenland. They had the secondary object of searching for two Swedes, Alfred Björling and Evald Kallstenius, who had disappeared off the Cary Islands in 1892. They had left a message there stating their intention to sail to Clarence Head, Ellesmere Island. Bryant's final object was to explore the north shore of Jones Sound. He landed on the Cary Islands on 24 July and found some relics of the missing expedition, but no indication of the two men's whereabouts. The next day, he reached Inglefield Gulf, where he met Peary's expedition. On 4 August, he set out again to continue the search for the missing men. On 7 and 8 August, he landed at Cape Faraday and at Clarence Head, Ellesmere Island, but, finding nothing, concluded that the two Swedes had never reached the Canadian shore. Bryant then went on to explore Jones Sound. Ice prevented him from penetrating very far into it, but he did manage to make a landing on the northern shore near Cone Island. Coming out of the sound, he landed on Northumberland Island to search for the missing men, but found nothing. He then made a last visit to Inglefield Gulf to pick up the men who were to return home, then set off for St John's. Geological, zoological, and botanical collections were made during the voyage.

BRYANT, 1894; *†1895.

1894

Enforcement of Canadian law in the Klondike District

By 1893, there were so many prospectors for gold in the Klondike that some means of law enforcement was required. The Canadian Government therefore sent Inspector Charles Constantine and Staff-Sergeant Charles Brown of the North-West Mounted Police (see 1873) to the region. They entered by way of Chilkoot Pass and spent July-August inspecting the camps. Brown wintered in the district and Constantine returned by way of St Michael to report to Ottawa. He returned in spring 1895 with a force of 20 men, established his headquarters at Fortymile, and thereafter succeeded in imposing Canadian law on a population made up mainly of United States citizens.

ZASLOW, 1971, pp. 96-100.

1894-95

Hunting expedition to Mackenzie District — Caspar Whitney

From New York; 20 December 1894 - 18 June 1895; a private expedition to hunt wood-bison and musk-oxen in northern Canada. He travelled to Edmonton by rail, to Fort McMurray by sledge and horses, and from there on snowshoes to Fort Chipewyan, where he arrived on 22 January 1895. Five days later, he arrived at Fort Smith, from where he went on a fortnight's trip with Indians to hunt wood-bison. He then continued on snowshoes to Fort Resolution, from where he set out on 8 March with a small party of Indians to hunt for musk-oxen on the Barren Grounds. Walking on snowshoes, he travelled due north from Fort Resolution, crossed Point Lake, and turned back on 9 April at about 67°N, possibly close to Inulik Lake. He arrived back at Fort Resolution on 29 April, having killed a number of musk-oxen. Soon after, he left Fort Resolution by canoe for home.

WHITNEY, 1896.

1894-95

Hunting expedition to Mackenzie District — Henry Toke Munn, Walter Gordon-Cumming

From Edmonton; May 1894 - April (?) 1895; a private expedition to hunt for musk-oxen north-east of Great Slave Lake. They travelled by Hudson's Bay Company steamer, by scow and by canoe down Athabasca and Slave rivers to the eastern end of Great Slave Lake, then by a chain of lakes and portages to Artillery Lake, where they arrived on 21 July. They cached their canoes there and made a ten-day journey on foot in search of musk-oxen, but found none. They then returned to Fort Resolution, where they parted: Gordon-Cumming apparently returned south; Munn went back to the east end of Great Slave Lake to hunt musk-oxen with Indians. They began their hunt on 29 October and continued throughout November, wandering far into the country north-east of Great Slave Lake, though exactly where they went is not recorded. By Christmas, Munn was back at Fort Resolution. At the end of January 1895, he met Caspar Whitney at Fort Smith, and joined him on a short wood-bison hunt. On 5 March, he left Fort Smith and sledged back to Edmonton.

MUNN, 1932, pp. 46-80.

1894-96

Church Missionary Society journey to Baffin Island — Edmund James Peck, J. C. Parker

These two men left Peterhead on the whaler *Alert* on 13 July 1894 to visit the Eskimos of Cumberland Sound. It was only the second time that missionaries had visited Baffin Island (see 1857-58). On 21 August, they reached Blacklead Island, Cumberland Sound, where there was a whaling station owned by Crawford Noble Sr.

1895: On 6 May, Peck left Blacklead Island to visit Noble's whaling station at Kekerten and stayed there for one month. Between 2 July and 27 August, Parker visited a whaling station in Frobisher Bay.

1896: In April and May, the two men made another visit to Kekerten. Parker was drowned in a boating accident on 14 August. On 22 August, C. G. Sampson arrived on *Alert* to relieve Peck. Peck left on 18 September on board *Hope,* Captain John Bartlett, of R. E. Peary's North Greenland Expedition, 1896. He sailed on this ship to Sydney, NS, then continued to England. He returned to Cumberland Sound in 1897 (see 1897-99).

LEWIS, A., 1904.

1894-96

Gold prospecting expedition to Yukon Territory — Robert Henderson

In July 1894, Henderson and two companions arrived at Ogilvie trading post, opposite Sixtymile River. There, Joseph Ladue (see 1882-83) recommended that they should try prospecting on Indian River. Henderson went to Indian River alone a few days later and spent the winter prospecting on Quartz Creek, one of its tributaries. In March 1895, he left Quartz Creek for Australia Creek, where he met with little success. After a visit to Ogilvie for supplies, he returned to Quartz Creek with a companion. The companion later left him, and Henderson spent the winter prospecting alone on Quartz Creek.

1896: After a second brief visit to Australia Creek, Henderson ascended Quartz Creek and crossed the watershed to "Gold Bottom," a tributary of Klondike River. He found better prospects there, went to Ogilvie for supplies, and returned to "Gold Bottom" with four companions. In July, he again travelled to Ogilvie, and this time returned to "Gold Bottom" by way of Klondike River. At the mouth of that river, he met Carmack (see 1896) and told him of his activities. This meeting resulted in Carmack's discovery of the rich Klondike goldfields, though Henderson heard of it too late to make a claim. He also lost his claim on "Gold Bottom" to Andrew Hunker, after whom that creek is now named.

OGILVIE, 1913, pp. 115-32; BERTON, 1958, pp. 34-44.

1895

Hudson's Bay Company supply voyage	John G. Ford Alexander Gray	*Lady Head* *Erik*

From London. *Lady Head* left London on 12 June, visited Moose Factory 19 August - 6 September, and reached London on 5 October. *Erik* left London on 8 June, visited the Labrador posts, Churchill, and Fort Chimo, and reached London on 29 October. As in the previous two years, *Princess* carried goods between Churchill and York Factory.

HUDSON'S BAY COMPANY (1) and (2).

1896

Hudson's Bay Company supply voyage	John G. Ford Alexander Gray	*Lady Head* *Erik*

From London; 9 June - 25 October. *Lady Head* visited Moose Factory 12 August - 1 September, and reached London on 6 October. *Erik* visited the Labrador posts, Churchill, and Fort Chimo, and reached London on 25 October. *Princess,* which was to have carried goods between York Factory and Churchill as in previous years, was severely damaged by floods in May, so open boats were used for transport between the two posts.

HUDSON'S BAY COMPANY (1) and (2).

1896

Discovery of gold on Klondike River	George Washington Carmack Skookum Jim Tagish Charlie

Carmack and his two Indian companions were salmon fishing at the mouth of Klondike River when Robert Henderson (see 1894-96) told them of his discovery of gold on "Gold Bottom," a tributary of Klondike River. About 20 days later, they set off up Klondike River to join Henderson, travelling by way of Bonanza Creek, another tributary of the Klondike. There, they made a good gold strike and decided to return there if "Gold Bottom" was not as good. They stayed a short time at "Gold Bottom" before returning to Bonanza Creek without telling Henderson of their find. On 17 August, they made another find, richer than the previous one, and they set off at once for Fortymile River, where they registered their claims. News of these discoveries soon spread. Edward Monahan staked a claim on Bonanza Creek two days after Carmack, and four more claims were made on 20 August.

OGILVIE, 1913, pp. 125-36; BERTON, 1958, pp. 40-51.

1896

The founding of Dawson

Within a few days of George Carmack's announcement of his discovery of gold on Bonanza Creek, a tributary of Klondike River (see 1896), Joseph Ladue (see 1882-83) went to stake out a town site at the mouth of Klondike River in anticipation of a rush of miners to the area. Soon after, he had floated his sawmill to the site, transported lumber there, built up a rough warehouse and a small cabin, and had named the new encampment "Dawson City" after George Mercer Dawson, the government geologist (see 1887). A steady flow of miners from other sites in the region arrived throughout the rest of 1896 and during the winter; by April 1897 there were 1,500 people there and by August the population had risen to 4,000. As winter approached, it became obvious that the town faced famine, and nearly 1,000 quit the town. Supplies were very low in winter 1897-98 but, because so many had left, there was no starvation. In early June 1898, the main contingent of gold seekers from the outside began to arrive, and by mid-summer a rough estimate by the North West Mounted Police showed the population of Dawson to be 18,000 with another 5,000 working on the creeks. However, there were undoubtedly many thousands more in the region, and certainly more than 30,000 visited Dawson that summer, if only for a few days. For one year, July 1898 - July 1899, Dawson enjoyed a boom. It suddenly became the largest Canadian city west of Winnipeg, and it boasted many of the facilities and services of a modern city, including the well known saloons, dance halls, and gambling houses. Huge fortunes changed hands daily in business deals, gambling, and drinking sprees; many luxury goods were brought into the town and were bought at enormous expense. The price of property in the city was astronomical: lots on the main street were sold for as much as $5,000 a front foot. Large areas of the town were several times destroyed by fire. The largest fire, on 26 April 1899, destroyed a large part of the most affluent section of town and, although it was rapidly rebuilt, Dawson was already on the decline. In spring 1899, news of the discovery of gold at Nome, Alaska, reached the town. The news was confirmed by mid-August, and a huge exodus began; in a single week in August, 8,000 people left.

When the stampeders had gone, a population of about 9,000 remained in Dawson, gold mining continued, but the town declined steadily. Dawson was incorporated as a city in 1902, heavy taxation was imposed, and many residents moved out to smaller settlements on the creeks. The Fairbanks gold rush of 1904-05 caused a further decline in the population which, by 1911, had dropped to 3,000. Even so, Dawson remained the chief population centre of Yukon Territory until 1941. In 1953, Whitehorse replaced Dawson as the territorial capital and, by 1961, the population of Dawson had declined to 881.

BERTON, 1958; LOTZ, n.d.

1897

Hudson's Bay Company supply voyage	John G. Ford Alexander Gray	*Lady Head* *Erik*

From London; *Lady Head* left London on 11 June, visited Moose Factory 10 August - 10 September, and reached London on 16 October. *Erik* left London on 12 June, visited the Labrador posts, Churchill, and Fort Chimo, and reached London on 15 October.

HUDSON'S BAY COMPANY (1) and (2).

1897 Canadian Hudson Strait expedition — William Wakeham — *Diana*

From Halifax; 3 June - 11 November; sent by the Department of Marine and Fisheries to investigate the navigability of Hudson Strait. They reached Button Islands on 22 June and, soon after, entered Hudson Strait and made their way through heavy ice to the northern shore of Mansel Island. There, they turned back to land A.P. Low (see below, 1897) at Douglas Harbour, and Robert Bell (see below, 1897) at Ashe Inlet, Big Island. After that, they sailed to the eastern end of Hudson Strait, made another run to the western end and back, then set out for Cumberland Sound. They anchored at Kekerton on 15 August, visited the whaling station there, then returned to Hudson Strait. They crossed Hudson Bay to Churchill, then returned to pick up Bell at Ashe Inlet and Low at Fort Chimo and to carry them back to St John's, where they arrived on 25 September. The ship then returned to Hudson Strait to investigate end-of-season navigation. They passed through the strait, arriving at Digges Islands on 22 October, then sailed home.

*†CANADA, DEPARTMENT OF MARINE AND FISHERIES, 1898; †OSBURN, 1932.

1897 Geological Survey of Canada expedition — Albert Peter Low, G. A. Young

Low landed from *Diana* (see above, 1897) at Douglas Harbour, northern Quebec, on 16 July. He explored the south-western arm of Douglas Harbour, then, on 22 July, left in a small yacht, *Alle*, and coasted eastwards, making a geological and geographical survey of the coast. He named Wakeham Bay, Doctor Island, Diana Bay, Plover Islands, and Gyrfalcon Islands, and reached Fort Chimo on 24 August. On 27 August, he left Fort Chimo to survey eastwards as far as George River. He arrived back at Fort Chimo on 4 September and, on 19 September, left for home on *Diana*.

LOW, 1899.

1897 Geological Survey of Canada expedition — Robert Bell

Bell landed from *Diana* (see above, 1897) at Ashe Inlet, Big Island, Hudson Strait, on 19 July. On 21 July, he began to make a track survey of the coast westwards to Chorkbak Inlet, Baffin Island, where he arrived on 22 August. During that journey, he made a seven-day trip inland from Amadjuak Bay to Mingo and Amadjuak lakes. Turning back towards Ashe Inlet, he made a track survey of the islands off the south coast of Baffin Island. He then surveyed the south coast of Baffin Island from opposite Big Island eastwards almost to Cape Weymouth. He was picked up by *Diana* at Ashe Inlet on 12 September. During his journey, he made notes on geology, glacial geology, mountains, and glaciers, and collected plants and Lepidoptera.

BELL, R., *†1901; 1901a; †WHITEAVES, 1899.

North-West Mounted Police patrol — A. M. Jarvis, W. Hetherington — 1897

From Fort Saskatchewan; 4 January - 15 April; to patrol Athabasca, Peace, and Slave rivers as far as Great Slave Lake. This was the first police patrol to enter present-day Mackenzie District. Travelling by dog-team, they passed beyond the existing limit of police jurisdiction at Lac la Biche. They then followed Athabasca River to Fort Chipewyan, where they stayed a short time, then continued down Slave River to Fort Resolution, where they stayed 13 - 17 February. Returning, they patrolled Peace River to Fort Vermilion, Lesser Slave Lake, and Athabasca Landing, then continued to Fort Saskatchewan. Jarvis's work in establishing police jurisdiction in the region was followed up later the same year: in December 1897, Inspector A. E. Snyder set out to patrol Lesser Slave Lake and Peace River region, and Inspector W. H. Routledge went to Athabasca, Slave, and Mackenzie rivers. He was the first policeman to patrol Mackenzie River, which he descended as far as Fort Simpson.

STEELE, 1936, pp. 29-30, 37, 43.

United States expedition to Frobisher Bay — Russell Williams Porter, Alfred Victor Shaw — 1897

At the invitation of R. E. Peary, Porter and Shaw organized a party of six students from Massachusetts Institute of Technology to go on a sporting trip to Frobisher Bay. They sailed from Sydney, NS, with Peary's north-west Greenland expedition on board the steamer *Hope,* Captain John Bartlett, and were landed at Cape Haven, near Cyrus Field Bay, Baffin Island. With 12 Eskimos and two whale boats, they sailed into Frobisher Bay by way of Lupton Channel. They explored many locations along the coast and at the head of Frobisher Bay, and ascended Jordan River as far as the first falls. They surveyed and described several geographical features, made notes on glacial phenomena and on geology, collected fossils at Silliman Mount, and surveyed and measured the flow of Boas Glacier, which they named. Porter and Shaw had intended to winter on Baffin Island and explore the interior in 1898, but they were obliged to cut short the expedition and return home, when *Hope* arrived to collect the rest of the party at Cape Haven, because Peary had no dogs to lend them.

PORTER, 1898; WILLARD, 1968; †SHAW, 1945.

United States hunting expedition — Charles Jesse Jones, John R. Rea — 1897-98

From Oklahoma; 12 June 1897 - (?) 1898; "Buffalo" Jones intended to capture musk-oxen on the Barren Grounds and carry them live back to the United States. He travelled by rail and by dog-team by way of Edmonton to Athabasca Landing, and there built a boat in which he descended Athabasca and Slave rivers. He was joined by Rea, a fur trader, at Fort Smith. They crossed Great Slave Lake to its eastern end, built cabins, and wintered close to the ruined site of Fort Reliance (see 1833-35).

1898: They left Great Slave Lake on 28 February and set out north-eastward by sledge. It is not certain which route they took: they crossed Back River to reach about 66°N, then turned south-east to reach Dubawnt River. Apparently they followed this river to the head of Baker Lake, from where they set off back to Great Slave Lake. During the return journey, they captured five musk-ox calves, but all were killed by Indians or Eskimos before they reached Great Slave Lake. From Great Slave Lake, they descended Mackenzie River and crossed McDougall Pass to Yukon River. The two men separated at Fort Yukon: Rea went to Dawson; Jones continued down Yukon River to the sea, and returned from there by ship to the United States.

EASTON AND BROWN, 1964, pp. 84-117.

1897-98

North-West Mounted Police patrol — J. D. Moodie

From Edmonton; 4 September 1897 - 24 October 1898; to obtain information on the best route from Edmonton by way of northern British Columbia to the Yukon goldfields. He set out by way of Lesser Slave Lake, Peace River, and Finlay River, but was held up for some months at Fort Grahame because of shortage of supplies. He continued in spring 1898 by way of Liard River, Frances Lake, and Pelly River to reach Fort Selkirk on Yukon River on 24 October. After his journey, Moodie reported that he considered the route far too difficult and that nobody would use the route in preference to the quicker and easier one by way of Skagway.

STEELE, 1936, pp. 34, 36, 43, 48, 61-62.

1897-99

The Klondike gold rush

Although many miners who were already in Yukon Territory assembled in Dawson (see 1896) soon after the discovery of gold on the Klondike in August 1896 (see 1896, Carmack), and continued to arrive during the rest of the year, the true gold rush did not begin until summer 1897, when news of the Klondike strike had reached the outside world carried by men who had become rich on the Klondike during winter 1896-97. They left for home in early June 1897 when the first steamers of the season, *Alice* and *Porteus B. Weare,* set off down Yukon River from Dawson to St Michael, Alaska, where they boarded the ocean-going steamers *Portland* and *Excelsior. Excelsior* docked at San Francisco on 15 July, *Portland* at Seattle on 17 July, and news of the Klondike spread at once throughout the world. As early as 19 July the first steamer, *Al-ki,* left the Pacific coast laden with stampeders bound for the Klondike; within five weeks, 20 other ships had followed. In a single week in mid-August, 2,800 persons left Seattle for the Klondike; a further 3,000 had already reached Dyea and Skagway. By February 1898, 41 ships were operating regular services to Alaska from

San Francisco alone; their destinations were the major gateways to the Klondike: Skagway, Dyea, Wrangell, and St Michael.

The main routes to the Klondike led over Chilkoot Pass and White Pass from the ports of Dyea and Skagway: 22,000 men crossed Chilkoot Pass in winter 1897-98. Another important route was up Yukon River from St Michael. There were also several overland routes through Canada. At least 1,500 men attempted the "Ashcroft Trail," following the route of the abandoned Western Union Telegraph line (see 1867) from Ashcroft, British Columbia. Very few of the men who attempted this route reached the Klondike; most gave up before they reached Skeena River. Another route, the "Stikine Trail" (following the course of Stikine River from the coast) joined the "Ashcroft Trail" at Glenora. Thousands of men dragged their sledges up the frozen Stikine River in winter 1897-98. A further 2,000 men attempted various routes from Edmonton, but again, few of them ever reached the Klondike: by autumn 1898 fewer than a dozen had arrived, and some who had set off from Edmonton at the very beginning of the gold rush were still trickling into the Klondike in July 1899, when Dawson was already on the decline. There were two main routes from Edmonton: one led by way of Peace, Liard and Pelly rivers, the other led down Mackenzie River, then by way of Liard River, by Gravel (now Keele) River, or by Peel and Rat rivers and McDougall Pass.

In all, during the winter of 1897-98, at least 100,000 people set out for the Klondike, but the effect of this stampede was not felt in Dawson until the spring of 1898. Even most of those who had set out at the very beginning and had crossed White Pass in autumn 1897 had failed to reach Yukon River before freeze-up, and were obliged to wait until break-up in May 1898. But, at the beginning of June 1898, these men flooded into Dawson in a flotilla of more than 7,000 boats. At the same time, steamboats carrying more stampeders began to arrive from Alaska. Altogether, more than 30,000 reached Dawson during summer 1898, when the gold rush reached its climax and, at the same time, began to decline. The outbreak of the Spanish-American War on 24 April 1898 discouraged many prospective American stampeders from setting out and caused many more who were there or on their way to return home. Many prospectors who reached Dawson failed — or did not even attempt — to stake claims, and they left for home after only a few days with the news that all the gold-bearing land had been staked. However, a steady trickle of stampeders still continued to arrive in Dawson through winter 1898-99, then, in summer 1899, the Klondike gold rush ended as suddenly as it had begun when news arrived of a major gold strike at Nome, Alaska, and thousands left Dawson for the new gold strike.

Gold mining continued in the Klondike region, although the richer creeks near Dawson had been worked out by 1900, and gold production in the Klondike region reached its peak in 1900, when about $22,275,000 worth was taken out of the ground, more than twice the estimated $10,000,000 worth that had been mined in

1898, at the height of the gold rush. The first dredge was introduced on Bonanza Creek in 1901, and large-scale dredging began in 1905. From then until the present, an average of about $2,000,-000 worth of gold has been taken annually from the Klondike region.

*BERTON, 1958; LOTZ, n.d.

1897-99

United States zoological expedition — Andrew Jackson Stone

Stone was sent by the American Museum of Natural History to collect mammal and bird specimens in north-western North America. He began his work in British Columbia in 1897, and travelled north the following year. In spring 1898, he travelled down Liard River to Fort Simpson, visiting the Nahanni Mountains on the way. He then descended Mackenzie River to Fort Norman, from where he visited the Rocky Mountains with Indians to collect specimens of mountain sheep *Ovis dalli.* Continuing north to Peel River, he visited Fort McPherson from where, in October, he made another trip into the Rocky Mountains. In November, he travelled to the Mackenzie River delta and along the Arctic coast to Herschel Island. He returned to Fort McPherson in December.

1899: In March-May, he travelled eastwards along the Arctic coast from the Mackenzie delta to Cape Lyon. In July, he set off down Porcupine and Yukon rivers into Alaska.

*†STONE, 1900; †ALLEN, 1899.

1897-99

Church Missionary Society journey to Baffin Island — Edmund James Peck

From Peterhead; early July 1897 - October 1899; Peck sailed on the whaler *Alert* to make his second missionary journey to the Eskimos of Cumberland Sound (see 1894-96). On 22 August, he reached Blacklead Island, where he met another Anglican missionary, G. C. Sampson, who had wintered in Cumberland Sound (see 1894-96). Peck carried out equipment with which to improve the missionary buildings.

1898: On 28 August, Julian William Bilby, another missionary, arrived on *Alert* to assist them.

1899: On 9 October, Peck left on *Alert* for England, leaving behind his two subordinates.

LEWIS, 1904.

1898

Hudson's Bay Company supply voyage — John G. Ford, Alexander Gray — *Lady Head*, *Erik*

From London. *Lady Head* left London on 11 June, visited Moose Factory 18 August - 12 September, and reached London on 23 October. *Erik* left London on 11 June, visited the Labrador posts, Churchill, and Fort Chimo, and reached London on 14 October.

HUDSON'S BAY COMPANY (1) and (2).

The Yukon Act

1898

The provisional district of Yukon, which was established within the North-West Territories (see 1870 and 1880) in 1895, was made a judicial district by proclamation on 16 August 1897.

During winter 1897-98, the governments of Canada and of the North-West Territories became involved in a jurisdictional dispute over the issuing of permits to ship liquor into the Klondike. This dispute proved existing constitutional arrangements for the Yukon region unworkable so, on 13 June 1898, the Canadian Parliament solved the problem with "An Act to Provide for the Government of the Yukon Territory" which created the Yukon a separate territory and provided it with a full complement of executive, legislative, and judicial institutions. Executive powers were conferred upon a commissioner, who was responsible for administering the government of the territory under instructions from the Governor in Council or the Minister of the Interior, and upon a council of six officials. Because of the great distance and the poor communications between Ottawa and Dawson, the commissioner was given authority over all officers of the federal government in the territory. When the provisions of the Act were brought into force, the government appointed William Ogilvie as commissioner, for he had already had considerable experience of the region (see 1887-89). Ogilvie arrived at Dawson to take up his duties as commissioner in September 1898.

MORRISON, 1968, pp. 18-24; CANADA YEAR BOOK, 1973, p. 90.

Geological Survey of Canada expedition — Joseph Burr Tyrrell

1898

From Ottawa; 12 May - 13 October; to make a geological survey of south-western Yukon Territory, between upper Yukon River and the Alaska boundary, and south of the site of Fort Selkirk. He travelled by steamer from Vancouver to Chilkat Inlet, then by horse up Chilkat River and along the "Dalton trail" past Dezadeash Lake to the headwaters of Nordenskiöld River. He surveyed this river to its confluence with Yukon River. Leaving his horses in camp at that point, he descended Yukon River to Dawson, where he met R. G. McConnell (see below, 1898). Together, they surveyed Bonanza, Eldorado, and Dominion creeks in the Klondike region, then Tyrrell returned to his camp on Nordenskiöld River. He found that his horses had become sick, and many of them died during the next part of his journey. He re-ascended Nordenskiöld River for a short distance, then turned westward into a tributary, Rowlinson Creek. He ascended this river for most of its length, then, turning south-west, eventually reached Nisling River, which he followed to its confluence with White River. There, as the summer was ending, they turned homeward. They returned to Nisling River, then followed a trail southward through the deserted Indian village of Aishihik, west of Aishihik Lake, then eastwards across Aishihik River to Hutshi Lakes. There, they met another party, from whom they hired fresh horses to take them back over the "Dalton trail" to Chilkat Inlet, where they arrived on 25 September. They left there by steamer and eventually reached Ottawa by rail on 13 October. During the expedition, he made nearly 500 km of new surveys, and collected 51 species of plants, in addition to making observations on geology and topographical features.

GEOLOGICAL SURVEY OF CANADA, 1899, pp. 36A-46A; LOUDON, 1930, pp. 152-55.

1898

Geological Survey of Canada expedition — Richard George McConnell

From Ottawa; 13 May - (?); to make a geological reconnaissance of the main upper tributaries of Yukon River, to examine the geology of the gold-bearing creeks of the Klondike region, and to make a preliminary geological examination of the route between Teslin Lake and Stikine River. He took a steamer from Vancouver to Chilkoot Inlet, crossed Chilkoot Pass to Bennett Lake and continued by boat from there down Yukon River. On 12 June, he reached the mouth of Big Salmon River. He made a traverse and geological examination up this river to the head of Quiet Lake. He then continued down Yukon River to Dawson, from where, in company with Tyrrell (see above, 1898), he spent two weeks examining part of the Klondike region. He returned by steamer to Big Salmon River and from there surveyed Teslin River, Teslin Lake, Nisutlin River, and the Teslin trail to Telegraph Creek. He then took a small boat down Stikine River to Wrangell and from there returned to Ottawa.

GEOLOGICAL SURVEY OF CANADA, 1899, pp. 46A-62A.

1898-99

Geological Survey of Canada expedition — Albert Peter Low

From Ottawa; 2 June - (?); sent to survey the east coast of Hudson Bay southward from Cape Wolstenholme. Low took a steamer from Quebec to Rigolet, from where, on 19 July, he set out on the Hudson's Bay Company's ship *Erik* for Hudson Strait. He landed at Erik Cove, near Cape Wolstenholme, on 31 July and commenced his survey by yacht. Progressing down the coast, he visited the mouth of Kovik River, landed at Smith Island on 18 August and travelled into its interior, then went a short distance up Sorehead and Povungnituk rivers. On 21 September, he reached Richmond Gulf, of which he surveyed the north shore. Two days later, he reached Great Whale River post, where he wintered.

1899: In February, two of his assistants, G. A. Young and James Schupe, began a survey of the coast southward to Fort George, whilst Low set out by dog-sledge to explore the mouth of Nastapoka River. He then travelled inland and explored part of Lake Minto. Returning to Richmond Gulf, he met Young, who had continued his survey northward to there. They reached Great Whale River together on 13 March. On 30 April, Low and Young travelled up the north branch of Great Whale River and explored part of Coast River. They returned to Great Whale River early in

May. They went north again on 6 June, surveyed Richmond Gulf, then sailed southward, making a geological and geographical survey of the coast and offshore islands of Hudson Bay as far as Rupert River. They ended their survey there and returned to Ottawa.

LOW, 1902.

1898-1900

Construction of
White Pass railway

White Pass, between the Pacific port of Skagway and Bennett Lake, Yukon Territory, was one of the major routes of access to the Klondike during the gold rush (see 1897-99). However, the trail was very difficult, and at the height of the gold rush it became so bad, and so many pack horses died on it, that it became known as "Dead Horse Trail." Realizing the need for easier access into the Yukon, and following the rejection by the Canadian Senate of a plan to build an all-Canadian railway from Glenora on Stikine River to Teslin Lake, a group of British businessmen organized the White Pass and Yukon Railway Company and obtained charters from Washington, DC, Ottawa, and Victoria to build a railway there. Their surveyors reached Skagway in May 1898, and construction of the railway began that summer. The track reached the summit on 22 February 1899, and the first section, from Skagway to Bennett, was completed the following summer; the first train from Skagway reached Bennett on 6 July 1899. Construction of the second section, between Bennett and Whitehorse, began at once, and the last spike was driven on 29 July 1900. By that time, the gold rush was nearly over, and the demand for the railway had diminished. But the company kept it running, and even took over the river service between Whitehorse and Dawson. The company, now a complex of companies known as the White Pass & Yukon Route, has continued to expand and, besides maintaining the railway, runs a wide variety of transport facilities in Yukon Territory and a regular container shipping service between Vancouver and Skagway.

*GRAVES, 1908; HAMILTON, 1964, pp. 87-96; STOBBE, 1969.

1898-1902

United States North Polar expedition	Robert Edwin Peary	*Windward*
	John Bartlett	*Hope*
	Samuel W. Bartlett	*Diana*
	John W. Blakeney	*Erik*

From New York; 4 July 1898 - 17 September 1902; sponsored by the Peary Arctic Club, with the main object of reaching the North Pole. Peary sailed with *Windward* and *Hope* to Etah, north-west Greenland. *Hope* returned home from there on 13 August, whilst *Windward* continued through Smith Sound. She was unable to penetrate farther north than Cape D'Urville, Allman Bay, Ellesmere Island, so Peary landed his stores there and explored the surrounding area. On 2-5 September, he surveyed by sledge the north shore of Princess Marie Bay as far as Copes Bay. On 6-8 September, he surveyed the head of Dobbin Bay. On 18-26 September, he made a more thorough survey of Princess Marie Bay, discovered that Bache "Island" was a peninsula, and that Buchanan "Strait" was a bay. He also explored Sawyer and Woodward bays. During the winter, Peary and his assistant, Matthew Henson, made a number of trips northward to lay depots for their coming journey to Fort Conger (see 1881-84). Peary arrived at Fort Conger on 6 January 1899 and remained there until 18 February, when he set off back to *Windward.* During this journey, several of his toes were frostbitten and later had to be amputated. He set off for Fort Conger again on 19 April, arrived there on 28 April, and made an unsuccessful attempt to cross Robeson Channel. He arrived back at the ship on 29 May. Between 29 June and 28 July, he completed his survey of Princess Marie Bay. He then went to meet *Diana* at Cape Sabine. Late in August, *Diana* and *Windward* departed for home, leaving Peary with a party at Etah, where they wintered.

1900: Peary and his men travelled to Fort Conger in March. Between 11 April and 10 June, they explored the north coast of Greenland. From Kap Morris Jesup, they travelled northward over the pack ice to 83°50'N, which they reached on 16 May. Continuing their exploration of the north coast of Greenland, they reached their farthest east, Wychoff Island, on 20 May. They arrived back at Fort Conger on 10 June and wintered there.

1901: *Windward,* Captain Samuel W. Bartlett, arrived at Payer Harbour in spring, and Peary joined the ship there. *Erik* arrived at Etah early in August. Both ships left for home at the end of August, leaving Peary and party to winter at Payer Harbour.

1902: Peary set out from Payer Harbour on his main attempt on the North Pole on 6 March. He made a brief halt at Fort Conger, then, on 6 April, set out northward over the pack ice from Cape Hecla. He reached his farthest north, 84°17'27"N, on 21 April. He then returned by way of Fort Conger to Payer Harbour, and sailed south in *Windward,* Captain Samuel W. Bartlett, in August. He reached Sydney, Nova Scotia, on 17 September.

*PEARY, 1903; 1903a; 1904; 1904a; 1907, pp. 293-349; †ALLEN, 1901; 1902; †BRITTON, 1909; †WHITFIELD, 1900.

1898-1902

Norwegian Arctic expedition	Otto Sverdrup	*Fram*

From Norway; 24 June 1898 - 9 September 1902; they set out with the intention of sailing through Smith Sound to explore the north coast of Greenland. However, the ship was caught in the pack ice off Cape Sabine, and they were forced to go into winter quarters in mid-August. Soon after, Sverdrup set up an outpost, Fort Juliana, in Hayes Fiord, and from there he explored the immediate neighbourhood, incidentally discovering that Bache Peninsula was not, as had been supposed, an island. Near there, he had a brief meeting with Peary (see above, 1898-1902).

1899: In spring, two sledging parties crossed Ellesmere Island (see 1899). On 24 July, *Fram* sailed again, and they tried to continue northward. They made little progress and decided to abandon the attempt after a month. Instead, they sailed into Jones Sound to explore that area. They entered Harbour Fiord, Jones Sound, on 1 September, and wintered there.

1900: Spring sledging parties (see 1900) discovered Axel Heiberg Island and Amund Ringnes Island. On 9 August, the ship left winter quarters and sailed through Cardigan Strait to Arthur Fiord. She then sailed back to winter at Goose Fiord.

1901: Spring sledging parties extended the discoveries of the previous season (see 1901). During August and September, it became clear that *Fram* had wintered so far inside Goose Fiord that it was impossible to escape again that year. After making only 16 km, they abandoned their attempt to get her out, and wintered there again.

1902: Sledge parties (see 1902) made further discoveries. The ship was able to leave the fiord on 6 August, and they set off for Norway, reaching Stavanger on 9 September. The scientific results of the expedition included observations on geology, botany, zoology, meteorology, and archaeology, in addition to extensive geographical discoveries in the Sverdrup Islands.

SVERDRUP, 1903; 1903a; *†1904; †NORSKE VIDENSKAPS-AKADEMI, 1907-19; †ALEXANDER AND MÜNSTER, 1924; †BRAENDEGAARD, 1936; †GRIEG, 1911; ISACHSEN, 1902; †JUNGERSEN, 1916; †KIEFFER, 1926; NATHORST, 1902; †NATVIG, 1930; †SCHAANNING, 1933; †SCHEI, 1903; †SIMMONS, 1903; †TOLMACHOFF, 1926.

Formation of the Diocese of Keewatin — 1899

In 1899, the Diocese of Moosonee (see 1872) was divided and its western half, covering much of north-western Ontario, northern Manitoba, and the District of Keewatin, became the Diocese of Keewatin. Joseph Lofthouse was consecrated as its first bishop on 17 August 1902.

BOON, 1962, pp. 382-90.

Treaty No. 8 — 1899

On 21 June 1899, Cree, Beaver, Chipewyan, and Slave Indians ceded the greater part of northern Alberta to negotiators representing the Federal Government. On 27 June 1921, Slave, Dogrib, Hare, Loucheux, and other tribes ceded the Mackenzie River region of the Northwest Territories. In addition to cash payments, there was provision for the establishment of reserve lands within the territories ceded. These have yet to be alloted in the Northwest Territories.

CHALMERS, 1972, pp. 52-58; CUMMING AND MICKENBERG, eds., 1972, pp. 118-31.

1899-1900

Geological Survey of Canada expedition — Robert Bell, James Mackintosh Bell

From Ottawa; June 1899 - 12 December 1900; to make a geological and geographical survey of the eastern and north-eastern end of Great Slave Lake, and of the region between that lake and Great Bear Lake. In June-September 1899, they surveyed the eastern end of Great Slave Lake; R. Bell explored around the shoreline of the eastern end of the lake and McLeod Bay, and his nephew, J. M. Bell, explored around the North Arm and Marian Lake. On 13 September, Robert Bell left Fort Resolution for Ottawa, leaving J. M. Bell to continue the survey during the winter and in the following year. Bell wintered at Fort Resolution, making short excursions east and west of Slave River during the winter.

1900: Bell left Fort Resolution by canoe on 11 April and travelled by way of Mackenzie and Bear rivers to Great Bear Lake. He traced the north-western and north-eastern extremities of the lake, Smith and Dease arms, to the head of Dease Arm, where he found the ruined site of Fort Confidence (see 1837-39 and 1847-49) on 31 July. From there he made a trip to Coppermine River and back. He then crossed to the southern shore of Great Bear Lake and ascended Camsell River (which he named) through Clut, Grouard, Hotah, and other lakes to Marian River. He ascended this to Marian Lake, where he reached Fort Rae on 20 September. He canoed from there to Fort Chipewyan, then continued by dogsledge to Edmonton. He reached Ottawa on 12 December.

GEOLOGICAL SURVEY OF CANADA, 1901, pp. 95A-103A; BELL, J. M., 1901.

1899-1901

United States expedition — Robert Stein, Joseph Samuel Warmbath, Leopold Kann

From Sydney, NS; 22 July 1899 - (?) 1901; to explore the interior and the west coast of Ellesmere Island, and to make scientific observations. The party travelled north on Peary's relief ship *Diana* (see 1898-1902) and landed near Cape Sabine, where they built their winter quarters, Fort Magnesia. The following year, a change in Peary's plans led them to abandon their intention of exploring Ellesmere Island, and Kann voted to return home. Leaving Warmbath at Fort Magnesia, Kann and Stein travelled in the company of Eskimos to Cape York, Greenland, where they met the whaler *Eclipse,* of Dundee, in May 1900. Kann travelled to Scotland on that ship, arriving there on 7 November. Stein decided to remain in Greenland among the Eskimos and to return home later on one of Peary's ships. Whether he returned home that year, or in 1901, is not known. Warmbath remained at Fort Magnesia until 1901, when he returned home.

KANN, 1903; GEOGRAPHICAL JOURNAL, 1900; SCOTT POLAR RESEARCH INSTITUTE MANUSCRIPT 1053 (Warmbath collection).

Hudson's Bay Company supply voyage	John G. Ford Alexander Gray	*Lady Head* *Erik*	1899

From London. *Lady Head* left London on 10 June, visited Moose Factory 22 August - 23 September, and reached London on 1 November. *Erik* left London on 10 June, visited the Labrador posts, York Factory, Churchill, and Fort Chimo, and reached London on 25 October.

HUDSON'S BAY COMPANY (1) and (2).

North West Mounted Police patrol	G. M. Skirving	1899

From Dawson; 30 August - 17 December; to search between Dawson and Fort McPherson for three gold seekers who had been lost near Porcupine River on their way to the Klondike. He travelled by canoe and steamer down Yukon River and up Porcupine River to Old Crow River, where he hired a dog team. He pushed on to Bell River, where he found the bodies of two of the men, and satisfied himself that the other had drowned. On 1 October, he continued over the mountains to Fort McPherson. Soon after, he set off back to Dawson. This was the first Mounted Police patrol to link the Yukon and Mackenzie regions.

STEELE, 1936, pp. 70-73.

British exploring expedition to Keewatin District	David Theophilus Hanbury	1899

From Winnipeg; 26 February - October; a private expedition to explore the unknown region between Chesterfield Inlet and Great Slave Lake. Hanbury had first attempted this journey in 1898, but had failed because poor ice conditions in Hudson Bay prevented him from setting out northward from Churchill early enough. He had, however, left a canoe in cache on Marble Island. In 1899, he returned by dog-sledge to Churchill, left there on 12 May, and sledged to Marble Island, where he collected his canoe. On 8 June, he entered Chesterfield Inlet, followed the south shore of Baker Lake to the mouth of Thelon River, and set off up the Thelon by canoe on 19 July. He passed through Schultz and Aberdeen lakes, then, passing the mouth of Dubawnt River, began his exploration of the unknown upper waters of the Thelon. He took its western branch, Hanbury River, crossed from there to Clinton-Colden Lake, then travelled through Artillery Lake to Great Slave Lake, where he reached Fort Resolution on 25 September. He then continued along Slave and Athabasca rivers to Fort McMurray. From there, in October, he returned by dog-sledge to Edmonton. The results of the expedition included observations on geology, natural history, and anthropology.

HANBURY, 1900; *†1904.

1899

Norwegian exploring expedition | Otto Sverdrup, Edvard Bay | dog-sledge

From Hayes Fiord, Ellesmere Island, winter quarters of *Fram* (see 1898-1902); 17 April - mid-May. They travelled west across Ellesmere Island by way of a valley at the head of Flagler Bay. On 29 April, they reached the head of Bay Fiord and from there had a clear view of Axel Heiberg Island, which they were the first Europeans to see. They then returned to *Fram*.

SVERDRUP, 1904, Vol. 1, pp. 120-41.

1899

Norwegian exploring expedition | Gunerius Ingvald Isachsen, Ove Braskerud | dog-sledge

From Hayes Fiord, Ellesmere Island, winter quarters of *Fram* (see 1898-1902); 23 May - 2 July; they set out to cross Ellesmere Island in a more southerly latitude than Sverdrup and Bay had done (see above, 1899). They reached a point at about 83°W, 78°30'N, remained in that area for two weeks, then returned to *Fram*.

SVERDRUP, 1904, Vol. 1, pp. 170-82.

1900

Hudson's Bay Company supply voyage | John G. Ford, Alexander Gray | *Lady Head*, *Erik*

From London. *Lady Head* left London on 12 June, visited Moose Factory 2-29 August, and reached London on 1 October. *Erik* left London on 9 June, visited the Labrador posts, York Factory, Churchill, and Fort Chimo, and reached London on 23 October.

HUDSON'S BAY COMPANY (1) and (2).

1900

Norwegian exploring expedition | Otto Sverdrup, Ivar Fosheim | dog-sledge

From Harbour Fiord, Jones Sound, winter quarters of *Fram* (see 1898-1902); 20 March - 4 July. They explored the south coast of Ellesmere Island westward through Hell Gate, then traced its west coast to Goose Point at the entrance to Baumann Fiord. From there, they crossed Norwegian Bay to Cape Southwest, Axel Heiberg Island, where they arrived on 11 April. They followed the west coast of Axel Heiberg Island north to 80°55'N, where they turned back on 5 May. They then retraced their steps to *Fram*.

SVERDRUP, 1904, Vol. 1, pp. 330-461.

1900

Norwegian exploring expedition | Per Schei, Peder Leonard Hendriksen | dog-sledge

From Harbour Fiord, Jones Sound, winter quarters of *Fram* (see 1898-1902); 20 March - 3 June. They accompanied Sverdrup (see above, 1900) as far as the entrance to Baumann Fiord, Ellesmere Island. There, it was decided that Schei and Hendriksen should return to the ship for fresh supplies, then return to explore the region of Baumann Fiord. Having visited the ship, they reached North Kent Island on 13 April, then went on to explore Buckingham and Graham islands. They crossed Norwegian Bay northwards to land at Hyperite Point, Axel Heiberg Island, then return to *Fram.*

SVERDRUP, 1904, Vol. 1, pp. 330-59, 466-83.

1900

Norwegian exploring expedition — Gunerius Ingvald Isachsen, Sverre Hassel — dog-sledge

From Harbour Fiord, Jones Sound, winter quarters of *Fram* (see 1898-1902); 20 March - 20 June. They accompanied Sverdrup (see above, 1900) until 16 April, when, from the west coast of Axel Heiberg Island, they sighted new land to the west. Isachsen and Hassel set off to explore it. They landed on the new island, Amund Ringnes Island, on 20 April. They remained there only a short time, then returned to Cape Levvel, Axel Heiberg Island, from where they continued back to the ship.

SVERDRUP, 1904, Vol. 1, pp. 330-76; Vol. 2, pp. 1-8.

1900

The founding of Whitehorse

Whitehorse began life during the gold rush (see 1897-99) as a cluster of tents at the terminus of a wooden tramline built to circumvent Miles Canyon and Whitehorse Rapids, but it really came into being as a town in 1900, when it became the terminus of the White Pass railway (see 1898-1900), and a centre for the distribution of goods to nearly all parts of the Yukon. At the beginning of World War II, the population of Whitehorse was about 300, but it suddenly rose to nearly 5,000 with the beginning of the construction of the Alaska Highway in 1942. Whilst Whitehorse was growing, Dawson steadily declined and, in 1953, Whitehorse replaced Dawson as capital of Yukon Territory.

LOTZ, 1961; McCOURT, 1969, pp. 28-30.

1900

Dominion Lands Survey expedition — James Williams Tyrrell, C. C. Fairchild, Joseph Lofthouse

From Edmonton; to explore the region between Great Slave Lake and Chesterfield Inlet. They reached Fort Resolution by winter trail from Edmonton on 1 April. They then sledged to the eastern end of Great Slave Lake and portaged to Artillery Lake. They continued by way of Clinton-Colden Lake to Hanbury River, descended this river to Thelon River, and the Thelon to its confluence with the Dubawnt, where the party divided. Fairchild and Lofthouse carried on to Hudson Bay by way of Aberdeen and Baker Lakes and Chesterfield Inlet; Tyrrell returned to the confluence of Thelon and Hanbury rivers, explored the Thelon for a considerable distance above the point, then continued on foot to Artillery Lake, where Fairchild rejoined him in early September. The party then returned to Fort Resolution, where they arrived on 23 September.

TYRRELL AND FAIRCHILD, 1902.

1900

United States biological expedition — Edward Alexander Preble, Alfred E. Preble

From Winnipeg; 14 June - 22 September; sent by the Division of Biological Survey of the United States Department of Agriculture to make a biological reconnaissance of the region immediately west of Hudson Bay. They travelled by canoe and steamer to Norway House, where their operations began. They made collections there 17-23 June, then continued north by canoe with two Indians, following the established boat route to York Factory and stopping at favourable points to make collections. From York Factory they sailed up the west coast of Hudson Bay to Churchill. Leaving his brother to work there, E. A. Preble set out again on 30 July and continued northward along the coast. He stopped at several points along the coast and remained for several days near the mouth of Thlewiaza River and near Eskimo Point. He turned back at Eskimo Point on 13 August, reached Churchill on 19 August, and the two men then retraced their steps to Winnipeg, where they arrived on 22 September. The report of the expedition contains extensive notes on mammals, birds and batrachians, and notes on life zones of the region and on flora.

PREBLE, 1902.

1900

Brown-Harvard expedition to Labrador — Edmund Burke Delabarre, Reginald Aldworth Daly, Henry Bryant Bigelow, ? — *Brave*

From St John's; 25 June - (?); a scientific exploration of the Labrador coast carried out by five men from Harvard and one from Brown University. Early in the voyage, they were hindered by difficult ice and wind conditions. Making numerous landings along the coast, the expedition reached Nachvak Fiord on 22 August and remained there for 12 days. Continuous geological samplings and topographical observations were made along the entire length of the coast, including the Kaumajet and Torngat mountains. Several coastal features were charted, and a limited study of the Labrador Current was carried out while they were at sea. Meteorological, botanical, and ornithological information was collected, and notes were made on the Eskimos and the medical mission.

*†DELABARRE, 1902; †DALY, 1902; †BIGELOW, 1902; †GRENFELL AND OTHERS, 1909, pp. 81-139, 391-425.

1900-02

Church Missionary Society journey to Baffin Island — Edmund James Peck

From Peterhead; early July 1900 - 5 November 1902. Peck sailed on the whaler *Alert* to make his third missionary journey to the Eskimos of Cumberland Sound (see 1894-96 and 1897-99). He reached Blacklead Island on 20 August to meet Sampson and Bilby, who had wintered there. When *Alert* departed for Britain,

Sampson, who had been in Cumberland Sound since 1896, sailed with her.

1901: Peck visited Kekerten in March. On 18 September, *Alert* arrived at Blacklead Island with another missionary, E. W. Greenshield. Bilby returned home.

1902: Peck visited Kekerten in March and Frobisher Bay in June. On 2 September, *Alert* arrived with Bilby on board. On 17 September, Peck returned to Britain on the whaler *Balaena.*

LEWIS, 1904; PECK, n.d.

Hudson's Bay Company supply voyage	John G. Ford Alexander Gray	*Lady Head* *Pelican*	1901

From London. *Lady Head* left London on 13 June, visited Moose Factory 17 August - 18 September, and reached London on 23 October. *Pelican* left London on 25 June, visited the Labrador posts, Churchill, and Fort Chimo, and reached London on 23 October.

HUDSON'S BAY COMPANY (1).

Norwegian exploring expedition	Otto Sverdrup Per Schei	1901

From Goose Fiord, Jones Sound, winter quarters of *Fram* (see 1898-1902); 8 April - 18 June. They crossed Simmons Peninsula to Norwegian Bay, explored Baumann and Vendom fiords, Ellesmere Island, and reached Blue Man Cape on 4 May. There they parted from the other sledging parties (see below, 1901). They explored northwards along Eureka Sound and reached their farthest north at Butter Porridge Point, Schei Peninsula, Axel Heiberg Island, on 12 May. They then retraced their steps to the ship.

SVERDRUP, 1904, Vol. 2, pp. 149-263.

Norwegian exploring expedition	Ivar Fosheim Oluf Raanes	1901

From Goose Fiord, Jones Sound, winter quarters of *Fram* (see 1898-1902); 8 April - 13 June. They accompanied Sverdrup (see above, 1901) to Blue Man Cape and parted from him on 4 May. They then continued northward along the west coast of Ellesmere Island, passed Slidre Fiord, rounded Iceberg Point, and entered Canyon Fiord. From there, they retraced their steps to *Fram.*

SVERDRUP, 1904, Vol. 2, pp. 273-94.

Norwegian exploring expedition	Gunerius I. Isachsen Sverre Hassel	1901

From Goose Fiord, Jones Sound, winter quarters of *Fram* (see 1898-1902); 8 April - 5 June. They accompanied Sverdrup (see above, 1901) to Norwegian Bay, then travelled by way of the north coast of Graham Island to Cape Southwest, Axel Heiberg Island. They then crossed Hendricksen Strait to south coast of Amund Ringnes Island, and continued westward to Cape Nathorst, Ellef Ringnes Island. They followed the west coast of Ellef Ringnes Island north to Cape Isachsen, returned down the east coast of the island, and crossed Hassel Sound to Cape Sverre, the northern point of Amund Ringnes Island, where they arrived on 24 May. They continued eastward across Massey Sound to Cape Maundy Thursday, Axel Heiberg Island, then returned to Goose Fiord by way of Cape Southwest and the west coast of Graham Island.

SVERDRUP, 1904, Vol. 2, pp. 295-98.

1901

Norwegian exploring expedition — Victor Baumann, Rudolf Stolz

From Goose Fiord, Jones Sound, winter quarters of *Fram* (see 1898-1902); 26 April - 28 May. They charted the region of Baumann Fiord, southwestern Ellesmere Island.

SVERDRUP, 1904, Vol. 2, pp. 264-72.

1901

United States biological expedition — Edward Alexander Preble, Alfred E. Preble

From Washington; 23 April - 15 September; sent by the Bureau of Biological Survey of the United States Department of Agriculture to study the fauna of the region about Athabasca and Great Slave lakes. They travelled by rail to Edmonton and by canoe to Fort Chipewyan, where they remained 17 May - 5 June. Continuing their journey downstream, they stopped frequently to make collections, stayed 10 days at Fort Smith, and reached Fort Resolution on 4 July. A. E. Preble remained there to examine the region thoroughly, whilst E. A. Preble crossed Great Slave Lake to Fort Rae, where he stayed 10 days. He set off back to Fort Resolution on 30 July, and the two men returned up Slave and Athabasca rivers on the steamers *Wrigley* and *Grahame*. They reached Athabasca Landing on 29 August and returned from there to Washington. They went back to Mackenzie District to continue their biological studies in 1903 (see 1903-04).

PREBLE, 1908.

1901

Geological expedition to the Nastapoka Islands — Albert Peter Low

During the summer of 1901, Low made a detailed examination of the geology and physiography of the Nastapoka Islands, Hudson Bay, on behalf of the Dominion Development Company.

LOW, 1903.

1901

United States expedition — Arthur Moore

From New York. Very little is known of this expedition except that, in the summer of 1901, Moore led a party to the Hudson Bay region and obtained caribou on the mainland of Labrador and on Salisbury Island. He remarked on the existence of Eskimos on Salisbury Island and also appears to have visited Resolution Island.

GRANT, 1902, pp. 16-17.

1901

First Carnegie Museum expedition — Walter Edmond Clyde Todd, Daniel A. Atkinson, George E. Mellor

From Pittsburgh; 17 May - 9 October; to collect specimens of mammals, birds, reptiles, insects, and plants along the coast of Labrador. They travelled by rail to Tabusintac, New Brunswick, then by ship to St John's, Newfoundland, where they joined the steamer *Home* on 25 July. The ship called first at Battle Harbour, then made many stops along the coast as far as Nain, where she arrived on 16 August. The expedition remained at Nain until 3 September, and made two trips to Paul Island. They then returned by the same route to Pittsburgh, carrying a large natural history collection.

TODD, 1963, pp. 13-14.

1901-02

British exploring expedition to Keewatin District — David Theophilus Hanbury, Hubert Darrell

From Edmonton; early July 1901 - early December 1902; private expedition to explore interior Keewatin District. Hanbury travelled down Athabasca and Slave rivers to Fort Resolution, where he was joined by Darrell. They left there on 13 July, canoed to the eastern end of Great Slave Lake, and continued by way of Artillery Lake to Campbell Lake at the headwaters of Hanbury River. They then performed in reverse Hanbury's journey of 1899 (*qv*), descending Hanbury and Thelon rivers to Chesterfield Inlet. From there they travelled to Marble Island where they hoped to meet Captain Santos of the New Bedford whaler *Francis Allyn*, who had brought out most of their supplies. However, they found no whalers at all at Marble Island, so they left on 18 September and returned to Baker Lake. There, in an Eskimo camp, they received a letter from Santos, saying that he was wintering at Depot Island. They arrived at Depot Island on 20 October, received their supplies, and returned to winter in an Eskimo camp on Baker Lake. In November and December, Hanbury joined some Eskimos on a musk-ox hunt, travelled northward from Baker Lake along Ketyet River to Tehek Lake, crossed the height of land to the lakes at the head of Meadowbank River (a tributary of Back River), and returned to Baker Lake by way of Tehert and Quoich rivers. At the end of December, he travelled overland by way of Armit Lake to Depot Island, where he spent some time on board *Francis Allyn*.

1902: They left Baker Lake on 9 March and travelled up Thelon River. They turned northwards at Tibielik Lake, and travelled down Consul River to Pelly Lake on Back River. Continuing northwards, they followed Armak River to the sea. They set out westward along the coast on 16 May, crossed Kent Peninsula to Elu Inlet, and continued along the coast to the mouth of Coppermine River. There, Hanbury enquired among the Eskimos about copper deposits along the coast, which was one of the main purposes of his journey. They then ascended Coppermine River and its tributary Kendall River to Dismal Lakes, crossed the height of land to Dease River, and entered Great Bear Lake on 19 August. From there, they reached Fort Norman on 30 August, and returned up Mackenzie, Slave, and Athabasca rivers to Edmonton. During the expedition, they gathered information on geology, natural history, anthropology, and meteorology.

HANBURY, 1903; *†1904.

1902 Hudson's Bay Company supply voyage — John G. Ford, Alexander Gray — *Lady Head*, *Pelican*

From London. *Lady Head* left London on 12 June, visited Charlton Island 3-22 September, and reached London on 28 October. *Pelican* left London on 14 June, visited the Labrador posts, Churchill, and Fort Chimo, and reached London on 21 October.

HUDSON'S BAY COMPANY (1).

1902 Norwegian exploring expedition — Otto Sverdrup, Per Schei

From Goose Fiord, Jones Sound, winter quarters of *Fram* (see 1898-1902); 1 April - 16 June; to complete their surveys in Eureka Sound (see 1901). They followed the east coast of Axel Heiberg Island to Iceberg Point, crossed the mouth of Greely Fiord, and reached the mouth of Hare Fiord on 28 April. They continued along the coast to the western end of Kleybolte Peninsula, and reached their farthest north at Lands Lokk, 81°40'N. They then returned to the ship by way of the western shore of Eureka Sound.

SVERDRUP, 1904, Vol. 2, pp. 355-405.

1902 Norwegian exploring expedition — Victor Baumann, Ivar Fosheim, Oluf Raanes

From Goose Fiord, Jones Sound, winter quarters of *Fram* (see 1898-1902); 23 April - 20 May; to inspect the depots left by British ships at Beechey Island during the 1850's. They travelled around the north side of North Kent Island, followed Arthur Fiord to its head, crossed the isthmus to Wellington Channel, then travelled southward along the west coast of Devon Island to reach Beechey Island on 4 May. There, they found the British depots destroyed and the yacht *Mary* (see 1850-51) wrecked. They then returned to *Fram*.

SVERDRUP, 1904, Vol. 2, pp. 412-15.

1902 Norwegian exploring expedition — Gunerius I. Isachsen, Edvard Bay

From Goose Fiord, Jones Sound, winter quarters of *Fram* (see 1898-1902); 12 April - 22 May. They surveyed Baad Fiord, on the south coast of Ellesmere Island, travelled eastwards to South Cape, then crossed Jones Sound to Cape Sparbo, Devon Island. They explored westward as far as Struis Point, then recrossed Jones Sound to Cape Storm, from where they returned to the ship.

SVERDRUP, 1904, Vol. 2, pp. 415-19.

Geological Survey of Canada expedition — Charles Camsell — 1902

From Edmonton. Camsell, who was not yet on the permanent staff of the Geological Survey, but had accompanied J. M. Bell on his survey of Great Bear Lake region (see 1899-1900), was invited to make an exploratory survey of the country south-west of Fort Smith. He left Athabasca Landing on 17 June and canoed down Athabasca and Slave rivers to Fort Smith. He made two short trips from there before beginning his main survey. On the first, he made a six-day survey of Salt River, and on the second, he travelled overland with horses into the Wood Buffalo country. On his third trip, he surveyed the unexplored country south-west from Fort Smith to Peace River, following a route outlined by an old Indian. He set out on 2 August, travelled a short distance up Salt River, then portaged to Little Buffalo River, which he followed to near its headwaters. He then portaged to Jackfish River, which he followed down to Peace River. He returned to Athabasca Landing by way of Peace River Crossing and Lesser Slave Lake.

CAMSELL, 1954, pp. 148-55.

Geological Survey of Canada expedition — Richard George McConnell, Joseph Keele — 1902

From Ottawa; 7 June - (?); to survey Macmillan River, Yukon Territory. They reached Whitehorse on 17 June and Fort Selkirk, at the mouth of Pelly River, on 22 June. Soon after, they set off up Pelly River by canoe, and they reached the mouth of Macmillan River on 5 July. They surveyed Macmillan River to the confluence of its north and south branches, then the party divided. Keele continued the survey up South Macmillan River and its main tributary, Riddell River. McConnell surveyed North Macmillan River as far as the headwaters of Cache Creek, and turned back on 13 August. Returning, they made a micrometer traverse of Pelly River down to Fort Selkirk.

McCONNELL, 1906.

British exploring expedition — Crawford Noble Jr — 1902

Noble, who worked at the Kekerten and Blacklead whaling stations in Cumberland Sound, decided at the end of the whaling in the summer of 1902 to make a sporting trip to the unexplored Nettilling Lake, Baffin Island. Travelling with two Eskimo families, he portaged a large whale boat from Clearwater Fiord to Nettilling Lake, then sailed around the lake. He studied birds and natural history on the islands of the lake. His sketch map and description were the first attempt to make a geographical survey of the area.

MILLWARD, 1930, pp. 36-39.

1903

Hudson's Bay Company supply voyage	John G. Ford Alexander Gray	*Lady Head* *Pelican*

From London. *Lady Head* left London on 13 June and visited Charlton Island 12 August - 14 September. On 16 September, she was stranded on Gasket Shoals and, ten days later, she was abandoned as a total loss. *Pelican* left London on 16 June, visited the Labrador posts, York Factory, Churchill, and Fort Chimo, and returned to London on 23 October.

HUDSON'S BAY COMPANY (1).

1903

Revillon Frères fur trading voyage	(?) d'Aigneaux William Berry	*Eldorado*

In about 1900, the Revillon Frères fur company decided to enter into direct competition with the Hudson's Bay Company in northern Canada by establishing trading posts on the shores of Hudson Bay. They prepared for this venture in 1902 by setting up five posts on the north shore of St Lawrence River and one at North West River, Labrador.

Eldorado sailed from Quebec on 15 August 1903 with 23 passengers (including d'Aigneaux, the company's Montreal representative) and 24 crew, bound for Fort George (James Bay), Rupert House, Moose Factory, Hannah Bay, and Fort Albany. The ship successfully navigated Hudson Strait but, on about 30 August, was wrecked off Fort George. Finding supplies in the Hudson's Bay Company's store at Fort George inadequate, the passengers and crew had no alternative but to travel by boat to Charlton Island for a passage home on the Hudson's Bay Company supply ship *Lady Head* (see 1903). However, when they reached Charlton Island, *Lady Head* had already departed. The trader at Charlton Island arranged for them to be taken by boat to Moose Factory and from there, on 22 September, they embarked on a canoe journey southward. They canoed by Abitibi River and Lake Abitibi, then walked overland to Ottawa River, where they took a boat to New Liskeard. From there, they continued by steamer and train to Montreal, where they arrived on 18 October.

UPTON, 1968.

1903

Establishment of Fort McPherson police detachment	C. Constantine F. J. Fitzgerald

From Fort Saskatchewan; to establish a police detachment on Mackenzie River. The North-West Mounted Police had patrolled

Mackenzie River region regularly since 1897, but so far had no men stationed there permanently. Constantine, Fitzgerald and four constables left Fort Saskatchewan on 30 May and reached Fort McPherson, the site selected for the new detachment, on 14 July. Constantine rented buildings for the detachment, then, leaving his men to winter there, set off homeward on 16 July. Soon after, Fitzgerald set out from Fort McPherson on his first patrol to Herschel Island (see below, 1903).

Although Constantine considered Fort McPherson an unsuitable site for a detachment and intended keeping it open only one year, it did, in fact, remain open much longer and was an important base for patrols to Herschel Island and across the mountains to Dawson. It appears that there has been a detachment there ever since.

CANADA, ROYAL NORTH-WEST MOUNTED POLICE, 1904.

1903

Establishment of Herschel Island police detachment — F. J. Fitzgerald, F. D. Sutherland

From Fort McPherson, 29 July - (?); to establish a police detachment on Herschel Island to exercise Canadian jurisdiction over the American whalers who wintered there. They reached Herschel Island on 7 August and, between 16 and 19 August, the whaling fleet of six ships arrived. Sergeant Fitzgerald warned the captains of his intention to collect customs duties and forbade them to land liquor or to distribute it among the Eskimos. He had two sod huts built for the police headquarters. Soon after, he set off back to Fort McPherson.

Fitzgerald returned to Herschel Island later the same year and he continued to patrol the island regularly until 1909. The police detachment remained on the island long after the whaling era ended, and it was not finally closed until 1964.

CANADA, ROYAL NORTH-WEST MOUNTED POLICE, 1904; STEVENSON, A., 1969.

1903

Establishment of Cape Fullerton police detachment — J. D. Moodie

Cape Fullerton, the first outpost of the North-West Mounted Police in the eastern Canadian Arctic, was established on 24 September 1903 from CGS *Neptune* (see 1903-04). The main purpose of the detachment was to extend Canadian jurisdiction to the American and British whalers who wintered in north-west Hudson Bay (see 1860-1915). *Neptune* wintered at Cape Fullerton in 1903-04 then departed, leaving three policemen at the post.

The detachment was maintained continuously throughout the following ten years. The police visited whalers in the vicinity and sometimes made longer journeys: in 1906, Constable L. E. Seller travelled north to Lyon Inlet to inspect the Scottish trading vessel *Ernest William*; in 1907-08, Sergeant D. McArthur visited Wager

Bay to prevent Eskimos from killing musk-oxen for the whalers. Following the establishment of the police detachment at Churchill (see 1906, *Adventure*), the coast was patrolled regularly between the two detachments. In about 1914, following the establishment of a Hudson's Bay Company trading post and a Roman Catholic mission at Chesterfield Inlet, a new detachment was opened there and Fullerton was closed. Fullerton was reopened in 1916, but was closed permanently in 1919.

CANADA, ROYAL NORTH-WEST MOUNTED POLICE, *1905; 1907; 1909; STEELE, 1936, pp. 102, 119-20, 127, 167, 180, 211; RIVETT-CARNAC, 1973.

1903

Alaska Boundary Tribunal

The boundaries agreed by the Anglo-Russian Treaty (see 1825) became the subject of dispute between Canada and the United States after the Purchase of Alaska (see 1867) and the gold rush (see 1897-99). The dispute centred on whether, south of 54°40', the boundary ran around the heads of inlets or parallel with the general trend of the coast. The definition of the boundary was effected by an international tribunal of six British and American jurists in London in 1903. The award sustained the main contention of the United States, and gave that nation possession of the inlets of the "Alaska Panhandle," to the lasting resentment of Canadians. The survey began during summer 1904 and continued every summer until 1913.

BAILEY, 1937; TOMPKINS, 1945; NICHOLSON, 1954; BILLMAN, 1971.

1903-04

Canadian Arctic patrol — Albert Peter Low, S. W. Bartlett, J. D. Moodie — CGS *Neptune*

From Halifax; 23 August 1903 - 12 October 1904; sent by the Canadian government to assert Canadian sovereignty over Hudson Bay and the Arctic islands and to establish police and customs stations. They first visited Port Burwell in Hudson Strait and the whaling stations and missions at Kekerten and Blacklead Island, Cumberland Sound, and at Cape Haven near Cyrus Field Bay. They then returned through Hudson Strait to Hudson Bay, surveyed the north shore of Charles Island, landed at Seahorse Point, Southampton Island, sailed through Evans and Fisher straits, then continued to the western shore of Hudson Bay where they anchored at Winchester Inlet on 16 September. On 23 September, they moved northward to near Cape Fullerton, where they established winter quarters in company with the whaler *Era*, Captain Comer. They began at once to build a police post at this place (see 1903).

1904: They undertook several surveying expeditions in spring (*qv*). They left winter quarters on 18 July and sailed to Port Burwell, where they met the Hudson's Bay Company's supply ship *Erik*. They then sailed northward into Baffin Bay and Smith Sound and landed to take possession at Cape Sabine and Cape Herschel, Ellesmere Island. In Lancaster Sound they landed at

Beechey Island and Port Leopold. Returning south, they visited Pond Inlet and Blacklead Island. After a final visit to Cape Fullerton, where they landed supplies for the police post, they returned to Port Burwell. There, they met *Arctic* (see 1904-05), which brought news of their recall. They therefore returned to Halifax. Low's report on the expedition includes observations on physical geography, Eskimos, geology, whaling, navigation, botany, ornithology, and palaeontology.

*LOW, 1906; CANADA, ROYAL NORTH-WEST MOUNTED POLICE, 1905; HALKETT, 1905; †EIFRIG, 1905; BORDEN, 1961; ROSS, W. G., 1976.

1903-04

Revillon Frères fur trading voyage — (?) Draulette, D. Thevenet, Captain Kearney — *Stord*

From Quebec; early July 1903 - (?) 1904; to establish trading posts on Hudson Bay as part of the company's plan to compete with the Hudson's Bay Company in northern Canada. *Stord* ran aground in July in the Gulf of St Lawrence and was replaced for the voyage to Hudson Bay by *Eldorado* (see above, 1903). However, she was refloated in August and was sent north with a modified objective: to establish a trading post on Ungava Bay. After some difficulties, the expedition reached Koksoak River where they built a post a short distance upstream from the Hudson's Bay Company's Fort Chimo, and where they wintered.

With this voyage, Revillon Frères for the first time established themselves successfully in the Hudson Bay region. During the next few years, the company consolidated its position on Hudson Bay. By 1908, it had a depot on Strutton Island, James Bay, trading posts at Fort George, Rupert, Moose, Albany, Great Whale River, Port Harrison, and Attawapiskat, and three more ships had joined the company's fleet serving these posts: *Violet, Adventure,* and a smaller steamer.

SEXÉ, 1923; INNIS, 1956, p. 367.

1903-04

United States biological expedition — Edward Alexander Preble, Alfred E. Preble, Merritt Cary

From Washington, DC; 2 May 1903 - (?) 1904; sent by the Bureau of Biological Survey of the United States Department of Agriculture to continue the biological reconnaissance of the Athabasca and Mackenzie river regions begun in 1901 (*qv*). They travelled by rail to Edmonton, and sailed down Athabasca and Slave rivers, making collections whenever they had the opportunity. They reached Fort Resolution on 19 June. There, the party divided. A. E. Preble and Cary set out for Mackenzie River and worked downstream to Fort Wrigley, where they turned back on 22 July. They returned by steamer to Athabasca Landing, remained there for two weeks to make collections, then continued to Washington. Meanwhile, E. A. Preble made a trip to Great Slave Lake in the company of James MacKinlay, formerly of the Hudson's Bay

Company, and two Indians. He crossed Great Slave Lake to Fort Rae, then canoed to Great Bear Lake, following the chain of lakes and rivers by way of Lac Sainte-Croix. He reached Great Bear Lake on 25 August, and followed its southern shore westward to the site of Fort Franklin, where he remained for ten days to make collections. He then continued down Great Bear River and up Mackenzie River to Fort Simpson. He remained there 20 October 1903 - 1 June 1904, and collected birds and small mammals during the winter. In summer 1904, he travelled down Mackenzie River to Fort McPherson. He set off homeward from there on board the *Wrigley* on 17 July, and reached Edmonton on 4 September. During this trip and the expedition of 1901, Preble and his companions made extensive observations and collections of mammals, birds, reptiles and batrachians, fishes, trees and shrubs, and life zones of the Athabasca-Mackenzie region.

PREBLE, 1908.

1903-04

United States exploring expedition	Leonidas Hubbard Dillon Wallace George Elson

From New York; 20 June 1903 - 27 May 1904; a private expedition to explore the interior of Labrador from North West River post on Lake Melville north-westward by way of Naskaupi River and Michikamau Lake to George River, where Hubbard intended to study the Indians. They travelled by steamer to Indian Harbour, then by canoe and steamer up Hamilton Inlet and Lake Melville to North West River post. They left there by canoe on 15 July and paddled to Grand Lake, searching for Naskaupi River. However, they missed the mouth of this river and mistakenly entered Susan River, believing it to be the Naskaupi. They travelled up Susan River to near its headwaters, then portaged to Mountaineer Lake (which they named) and soon after joined Beaver River. They paddled up Beaver River to discover (and name) Hope Lake and Disappointment Lake. From there, they portaged north-westward by way of Mary Lake to discover Windbound Lake, which they reached on 8 September. There, Hubbard climbed a hill and sighted Michikamau Lake. However, they were held up at Windbound Lake for several days by wind and rain and, as winter was approaching, they turned back on 21 September. They retraced their steps to Beaver River, abandoned the canoe, marched overland to Susan River, and began to follow it downstream, suffering increasingly from starvation and sickness. Before they were halfway down the river, Hubbard broke down, so Wallace and Elson left him and went on to get help. A short distance further on, Wallace was also unable to continue, so Elson went on alone. He found trappers on Grand Lake and four of them set out to rescue Wallace and Hubbard. They found Wallace alive on 30 October but, when they found Hubbard, he was dead. On 6 November, Wallace and Elson reached North West River post, where they wintered.

1904: Wallace had remained at North West River post in the hope of returning for Hubbard's body but, by March 1904, he was still unfit to go. Elson and two trappers undertook the task and set off on 14 March; they arrived back at North West River with Hubbard's body on 27 March. Wallace and Elson then sledged with the body to Battle Harbour, from where they set out by steamer for New York on 27 May.

*WALLACE, D., 1905; HUBBARD, 1908, pp. 239-333; KENNEY, G.I., 1971.

1903-06

Norwegian North-west Passage expedition | Roald Amundsen | *Gjøa*

From Christiania; 16 June 1903 - October 1906; a private expedition to find the exact location of the magnetic pole, to collect magnetic and other scientific information, and to attempt a Northwest Passage. After sailing through Davis Strait, Baffin Bay, and Lancaster Sound, they anchored at Erebus Bay, Beechey Island, on 22 August 1903, and made magnetic observations. A few days later, they entered Peel Sound, sailed through Franklin and James Ross straits, and ran aground near Matty Island. On 12 September, they anchored at Gjoa Haven, south-eastern King William Island, where they wintered.

1904: On spring sledge expeditions (*qv*), they took magnetic observations and discovered Hovgaard Islands. They established a ring of magnetic stations around the ship, which they visited regularly throughout the year. They wintered again at Gjoa Haven.

1905: Spring sledge expeditions visited Victoria Island (*qv*). Eskimos brought them mail from Captains Bernier (see 1904-05) and Comer in Hudson Bay. They left Gjoa Haven on 13 August, sailed through Simpson and Victoria straits, discovered Palander Strait and Nordenskiöld Islands, then continued along the south coast of Victoria Island and through Amundsen Gulf. They were finally beset at King Point, where they wintered. During the winter, Amundsen carried mail to Eagle City (see 1905-06).

1906: They sailed again on 10 June, and became the first men to complete the North-west Passage by sea when they arrived at Nome, Alaska, in late August. They then continued to San Francisco, where they arrived in October. *Gjøa* remained at San Francisco for the next 66 years, and was returned to Norway in 1972.

AMUNDSEN, 1907; 1907a; *1908; 1927, pp. 37-63; HANSSEN, 1936, pp. 9-84; SVERDRUP, H.U., 1959.

1903-10

United States exploring expeditions | William Brooks Cabot

In 1899, Cabot made a mid-winter journey from Lake St John to Lake Mistassini. In the years following 1903, he made a series of journeys to the Labrador coast and inland from it. His published accounts are not easy to follow chronologically, but they are filled with detailed observations on geography, natural history, and Indians. His chapter "Mice" may be the first systematic account of the lemming's population cycle.

1904: With Robert Walcott, Cabot returned to Labrador on *Virginia Lake,* landed at Nain, made his way south to Voisey Bay, and again ascended Kogaluk River, this time meeting Naskapi Indians inland.

1905: With Lewis Quackenbush, Cabot returned to the same region, ascended Kogaluk River in August and, with the help of Naskapi Indians they met along the way, crossed the height of land to Mistinibi Lake. The Indians invited them to accompany them to Indian House Lake on George River, but because the season was late they turned back. Had they gone inland, they would have met Mrs Hubbard's party (see 1905).

1906: Cabot went to Davis Inlet and, with two local men, started up Kogaluk River. The men turned back near the height of land, which event obliged Cabot to return alone. Later in the summer he reascended the river with a young boy and, leaving the canoe part way, they walked to Mistinibi Lake, where again Cabot met Naskapi Indians.

1907 and 1908: Cabot visited the Labrador coast, but did not go inland.

1910: Having satisfied his desire to see Naskapi Indians inland, Cabot now decided to work out the remaining portion of the route inland from Mistinibi Lake to Indian House Lake. He travelled with three friends, Scoville Clark, George P. Howe, and D. B. MacMillan. They used only two canoes, two of the party walking along shore. They achieved their object without great difficulty, thereby becoming the first explorers to reach George River from the coast (they anticipated Prichard's party [see 1910] by only 10 days). On return, they left Kogaluk River to walk to Mistastin Lake, then continued to the coast. Cabot visited Labrador in later years, but does not appear to have gone inland again.

CABOT, 1912; n.d.; ALLEN, E.S., 1963, pp. 131-37; ELTON, 1942, pp. 234-36.

1904

Canadian exploring expedition — G. F. Caldwell

From Cape Fullerton, Keewatin, winter quarters of *Neptune* (see 1903-04); 11 April - 30 May. Travelling with two dog sledges, he surveyed the coast northward to Wager Bay and studied the geology.

LOW, 1906.

1904

Canadian exploring expedition — Albert Peter Low

From Cape Fullerton, Keewatin, winter quarters of *Neptune* (see 1903-04); 4-14 May. He surveyed the coast southward to Chesterfield Inlet "in order to connect the work of Caldwell [see above, 1904] with that of Tyrrell [see 1893]." On the return journey, he explored 65 km inland from Winchester Inlet. He also examined the geology of the region.

LOW, 1906.

Canadian exploring expedition | Albert Peter Low, L. E. Borden

1904

From Cape Fullerton, Keewatin, winter quarters of *Neptune* (see 1903-04); 15 June - 2 July. They accompanied Captain Comer of the whaler *Era* on a boat trip to Cape Kendall, Southampton Island. They spent a few days on Southampton Island, made several trips inland, and collected birds and fossil specimens.

LOW, 1906.

Norwegian exploring expedition | Godfred Hansen, Peder Ristvedt

1904

From Gjoa Haven, King William Island, winter quarters of *Gjøa* (see 1903-06); 17-26 March. They travelled by dog sledge to Point Richardson, Adelaide Peninsula. During the journey, they discovered Hovgaard Islands. In June 1904, the whole crew of *Gjøa* went to chart these islands.

AMUNDSEN, 1908, Vol. 2, supplement.

Norwegian exploring expedition | Roald Amundsen, Peder Ristvedt

1904

From Gjoa Haven, King William Island, winter quarters of *Gjøa* (see 1903-06); 6 April - 27 May; to take magnetic observations in the vicinity of the magnetic pole. They took observations at Cape Hardy, Matty Island, 9-14 April, then crossed James Ross Strait to Cape Christian Frederick, Boothia Peninsula. They travelled north along the coast until they were nearly opposite Tasmania Islands, then turned back on 7 May. They had planned to cross the peninsula to Victory Harbour, but Amundsen injured himself and Eskimos plundered their depot, so they returned directly to the ship.

AMUNDSEN, 1908.

Norwegian exploring expedition | Godfred Hansen, Helmer Hansen

1904

From Gjoa Haven, King William Island, winter quarters of *Gjøa* (see 1903-06); 6 August - mid-September. They travelled by boat westward through Simpson Strait to Cape Crozier with the object of investigating the state of Simpson Strait in preparation for the passage of *Gjøa* in the following summer. They found the skulls and bones of two white men from the Franklin expedition (see 1845-48) at Point Hall, and a grave constructed for them by C. F. Hall (see 1864-69).

AMUNDSEN, 1908.

Hudson's Bay Company supply voyage | John G. Ford, Alexander Gray | *Stork*, *Pelican*

1904-05

From London. *Stork* left London on 20 June 1904 and visited Charlton Island 3-22 September. Returning home, she was beset off Mansel Island and was obliged to sail back to winter at Charlton Island. She left there again on 9 August 1905 and reached London on 4 October. *Pelican* left London on 16 June 1904, visited the Labrador posts, York Factory, Churchill, and Fort Chimo, and reached London on 18 October.

HUDSON'S BAY COMPANY (1).

1904-05

Canadian Arctic patrol — J. D. Moodie, Joseph-Elzéar Bernier — CGS *Arctic*

From Quebec; 17 September 1904 - 6 October 1905; to patrol Hudson Bay and to establish a divisional police headquarters at Churchill. The expedition reached Port Burwell on 1 October and there met *Neptune* (see 1903-04), which thereupon sailed homeward. Moodie and Bernier continued through Hudson Strait and across Hudson Bay to Cape Fullerton, where a police post had been established from *Neptune* in 1903, and where they wintered.

They resumed their voyage on 9 July 1905, but soon after abandoned their intention to visit Churchill on account of difficult ice conditions.They re-crossed Hudson Bay to Erik Cove, where they arrived on 16 July. Soon after, they moved *Arctic* to a new berth nearby, which they named Préfontaine Harbour. On 23 September, *Neptune* arrived (see 1905-06) with orders for *Arctic* to return to Quebec for alterations and repairs. Bernier sailed home with *Arctic;* Moodie joined *Neptune* and returned to Hudson Bay.

*MOODIE, 1907; STEELE, 1936, pp. 113-14; BERNIER, 1939, pp. 305-06.

1905

Hudson's Bay Company supply voyage — A. Gray, G. F. Lovegrove — *Discovery*, *Pelican*

From London. *Discovery* left London on 15 June, visited Charlton Island 27 August - 8 September, and reached London on 3 November. *Pelican* left London on 17 June, visited the Labrador posts, York Factory, Churchill, and Fort Chimo, and returned to London 29 November.

HUDSON'S BAY COMPANY (1); *BEAVER*, 1929.

1905

Norwegian exploring expedition — Godfred Hansen, Peder Ristvedt

From Gjoa Haven, King William Island, winter quarters of *Gjøa* (see 1903-06); 2 April - 25 June; to chart the unexplored part of the east coast of Victoria Island north of Cape Collinson. Travelling by dog sledge, they crossed Victoria Strait to Cape Adelaide, then travelled due north across Edward Albert Bay and Collinson Peninsula to Cape Jensen. They then explored the coast northward to Cape Nansen, in about 72°N, where they arrived on 26 May. They returned south by a similar route to Cape Adelaide, then visited the north shore of Jenny Lind Island, and crossed Victoria Strait to Cape Crozier by way of the south side of the Royal Geographical Society Islands (which they named). They also sighted the Nordenskiöld Islands.

AMUNDSEN, 1908, Vol. 2, pp. 296-364.

Geological Survey of Canada expedition | Charles Camsell | 1905

From Ottawa; mid-March - 17 September; to survey the country between Stewart River, Yukon Territory, and Peel River, by way of Beaver and Wind rivers. This region had never before been surveyed, but it had been visited once before by a party of prospectors, who had travelled by way of Peel and Wind rivers, Bonnet Plume Pass, and Beaver and Stewart rivers to reach the Klondike in 1898-99.

Camsell travelled to Dawson by way of Vancouver and Skagway, and arrived there on 14 April. Five weeks later, he travelled with Keele (see below, 1905) by steamer to Frazer Falls, Stewart River. There, he began his canoe journey on 5 June. He canoed up Stewart and Beaver rivers to Braine Creek, followed Braine Creek up to Braine Pass, and there crossed the height of land to Nash Creek, a tributary of Wind River. He then canoed down Wind and Peel rivers to reach Fort McPherson on 29 July. He spent the next two weeks mapping Husky Channel of Mackenzie River delta, then set off back to Dawson by way of McDougall Pass, reaching Dawson on 17 September. His report on the journey describes the geology and topography of the region.

CAMSELL, 1906; 1954, pp. 179-201.

Geological Survey of Canada expedition | Joseph Keele | 1905

During this year, Keele surveyed upper Stewart River between Frazer Falls and the Tasin Range. Although prospectors had been in this region during the gold rush and it was still visited by trappers, it had never before been surveyed and little was known of it. Leaving Dawson on 22 May, Keele accompanied Camsell (see above, 1905) on a steamer to Frazer Falls, where they parted. Keele then surveyed the greater part of Stewart River above that point, and he also made a reconnaissance of the lower reaches of its major tributaries: Hess, Lansing, Beaver, and Rackla rivers. He reported on the geology and topography of the area, and on its fauna and flora.

KEELE, 1906.

United States exploring expedition | Mina Benson Hubbard, George Elson | 1905

From Halifax; 16 June - 20 November; to complete the work begun by her husband (see 1903-04, Hubbard) by exploring Naskaupi and George rivers, Labrador and northern Quebec. She travelled by steamer to North West River post on Lake Melville, where she arrived on 25 June. From there she set out by canoe with a crew of four, including George Elson, who had accompanied her husband, on 27 June. They travelled through Grand Lake, up Naskaupi River and Wapustan River (which she named), and overland to Seal Lake. There, they rejoined Naskaupi River and ascended it to Michikamau Lake, which they reached on 2 August.

They crossed this lake to the source of George River, which they began to descend on 11 August. They named a number of lakes on the upper course of George River, including Hubbard, Elson, Cabot, and Resolution lakes, and they met several groups of Naskapi Indians during their descent of the river. They reached George River post at the mouth of George River on 24 August, and remained there until 24 October, when they left on board the Hudson's Bay Company's ship *Pelican* for Rigolet. Mrs Hubbard joined another steamer at Rigolet and sailed for Quebec, where she arrived on 20 November.

Mrs Hubbard was probably the first white person to follow the whole length of George River, for there is no reason to suppose that John McLean or any of his men did so (see 1839). And it is clear that there was a strong spirit of rivalry between Mrs Hubbard and Elson's party and that of Dillon Wallace (see 1905-06), although neither leader mentions the other in their respective narratives.

HUBBARD, 1906; 1906a; *1908.

1905

Establishment of Killinek mission

In this year, the Moravians established Killinek mission at Port Burwell, Killinek Island, Labrador. The following year, Bernhard Hantszch stayed at the mission to make ornithological observations (*qv*). In 1920, the Royal Canadian Mounted Police established a post close by. The mission was closed in 1923.

WILLIAMSON, H.A., 1964; HANTSZCH, 1928-29, February 1928, p. 33.

1905-06

Norwegian overland expedition — Roald Amundsen

From King Point, Yukon Territory, winter quarters of *Gjøa* (see 1903-06); 21 October 1905 - 12 March 1906; to carry mail to Eagle City. He travelled in the company of Eskimos by way of Herschel Island, Fort Yukon, and Circle City, and reached Eagle City about the end of November. He remained there until 3 February 1906, when he set off back to the ship.

AMUNDSEN, 1908, Vol. 2, pp. 212-49.

1905-06

United States North Pole expedition — Robert Edwin Peary, Robert A. Bartlett — *Roosevelt*

From New York; 16 July 1905 - 24 December 1906; sponsored by the Peary Arctic Club to attain the North Pole by way of Smith Sound. After stopping at Etah to pick up 40 Eskimos and 200 dogs, they passed quickly through Kane Basin, but encountered very difficult ice conditions in Robeson Channel and Kennedy Channel. They established winter quarters at Cape Sheridan, Ellesmere Island, on 5 September. The expedition was nearly thwarted in early winter when a disease attributed to poisoned whale meat killed 80 of the dogs. Camps were established around the shores of Lake Hazen, where the Eskimos remained throughout the winter, fishing and hunting musk-oxen.

1906: In early spring, Peary made an unsuccessful attempt on the North Pole (*qv*), and in summer he explored the north-west coast of Ellesmere Island (*qv*). The ship left for home on 6 July.

PEARY, 1905; *1907.

United States exploring expedition — Dillon Wallace, Peter Stevens, George M. Richards, Clifford H. Easton, Leigh Stanton — 1905-06

From New York; 30 May 1905 - 30 April 1906; to complete the descent of George River, northern Quebec, which Hubbard and Wallace had failed to accomplish in 1903 (see 1903-04). They travelled by steamer to North West River post and, on 27 June, started up Naskaupi River, which they soon left, on local advice, to follow a lake-and-portage route by Crooked River, Seal and Kasheshibaw lakes to Lake Michikamau. They reached this point after great difficulty and loss of time on 3 September, by which date Mrs Hubbard's party had already completed its descent of George River (see 1905, Hubbard). Here Wallace decided to send back all of his party except Easton, owing to lateness of season and shortage of food. These two continued in freezing weather and, on 29 September, about one-third the way down George River, they capsized and nearly died of exposure. They finally reached George River post on 16 October to find Mrs Hubbard's party already there. Wallace and Easton then decided, against all advice, to go with Eskimos in a small boat to Fort Chimo. They left on 22 October, but were soon stopped by ice and did not reach Whale River post until 8 November and after much suffering. They continued westward on 28 November by dog sled and reached Fort Chimo the next day. They stayed there until 3 January 1906, when they started homeward on a journey by dog sled that took them, with many vicissitudes, from George River post across the Torngat Mountains to Nachvak and, by relays of Eskimo drivers and dog teams, down the Labrador coast to Battle Harbour and then westward to Natashquan on the north shore of St Lawrence River, which point they reached on 17 April. They then took ship to Quebec and returned to New York.

WALLACE, D., *1907; 1932.

Canadian Arctic patrol — J. D. Moodie, S. W. Bartlett — CGS *Neptune* — 1905-06

From Quebec (?); to patrol Hudson Bay and to set up police posts. They met *Arctic* (see 1904-05) off the Labrador coast on 23 September, and Major Moodie left that ship, which was returning home, to continue his police patrol of Hudson Bay on board *Nep-*

tune. *Neptune* sailed in very bad weather conditions through Hudson Bay to the police post at Cape Fullerton. They wintered there, then sailed to Churchill, but were apparently unable to remain there long enough to establish a police headquarters as intended. They then returned to St John's, Newfoundland.

STEELE, 1936, pp. 114-16.

1905-07

British exploring expedition — Alfred H. Harrison

From Liverpool; 15 June 1905 - (?) 1907; a private expedition with the object of discovering a "Polar Continent" or any unknown land in the Arctic Ocean. He travelled by steamer to Quebec and by rail to Edmonton, and was joined *en route* by Hubert Darrell. He continued by wagon to Athabasca Landing, then sailed by scow down Athabasca and Slave rivers, was towed across Great Slave Lake by the steamer *Wrigley,* drifted down Mackenzie River, and reached Arctic Red River post on 4 October 1905. He made several sledge journeys from there to Fort McPherson; whilst he was absent on one journey, Darrell left his service.

1906: He left Fort McPherson for Herschel Island on 13 February. He surveyed Herschel Island, then, on 10 July, left on board the whaler *Jeanette* for Banks Island. On the way, he landed at Pelly and Baillie islands, where he met five whalers that had wintered there. He charted Baillie Island. He then sailed with *Jeanette* to Nelson Head, Banks Island, sailed along the south-west coast of the island to Cape Kellett, then sailed a short distance north-westward before returning to Herschel Island. There, he joined Stefansson (see 1906-07), and set out with him by boat for Shingle Point on 29 August. They parted at Shingle Point, and Harrison continued up Mackenzie River to obtain supplies from Fort McPherson. He returned down the East Channel to Mackenzie River delta and then by way of Campbell and Sitidgi lakes to the southern end of Eskimo Lakes, where he wintered. Stefansson visited him during the late December 1906 and early January 1907.

1907: He began a triangulation survey of Eskimo Lakes on 25 March, and reached the coast at Warren Point on 25 April. From there, he travelled westward to the Mackenzie delta, then returned inland to Eskimo Lakes. On 1 June he set off back for Fort McPherson, then sailed on the steamers *Wrigley, Grahame,* and *Midnight Sun* back to Athabasca Landing. The results of his journey included reports on the Eskimos, and observations on flora, fauna, and topography.

HARRISON, A.H., *1908; 1908a.

1906

Hudson's Bay Company supply voyage	John G. Ford	*Discovery*
	N. E. Freakley	*Stork*
	Alexander Gray	*Pelican*

From London. *Discovery* left London on 14 June, visited Charlton Island 10-23 August, and returned to London on 4 October. *Stork* left London on 12 June, visited Charlton Island 25 August - 15 September, and reached London on 28 October. *Pelican* left London on 2 June, visited the Labrador posts, York Factory, Churchill and Fort Chimo, and reached London on 18 October.

HUDSON'S BAY COMPANY (1).

Canadian Arctic patrol | J. D. Moodie (?) | CGS *Adventure* | 1906

From Halifax; 1 August - (?); to patrol Hudson Bay and to set up a police headquarters at Churchill. They reached Port Burwell on 10 August and there landed coal for future patrol ships. They then stopped at the police post at Cape Fullerton 18-31 August. On 2 September, they reached Churchill and began building about 2 km upstream from the ruined Prince of Wales Fort. *Adventure* sailed for home on 1 October, leaving Moodie and a small party of police to winter at the new post. Churchill was now the divisional headquarters of the RNWMP's "M" Division. At the time, there was only one other detachment in the division: Fullerton (see 1903).

MOODIE, 1907a.

United States North Pole expedition | Robert Edwin Peary | 1906

From Cape Sheridan, Ellesmere Island, winter quarters of *Roosevelt* (see 1905-06); 21 February - late May; to attain the North Pole. At Cape Hecla, Peary met six of his men and 21 Eskimos, who were to carry forward his supplies in five supporting parties. He set out northward over the sea ice from Point Moss on 6 March. At 84°30'N, he was held up by a wide lead of open water. He crossed it, but soon after was driven too far east by a gale. These setbacks made attainment of the pole impossible, but he pressed on to reach a record latitude of 87°06'N on 21 April. Returning, he again had to cross the wide lead, and his party reached Cape Neumayer on the Greenland coast in a starving state. However, they were able to shoot musk-oxen, which saved them.

PEARY, 1907, pp. 97-169.

United States oceanographic expedition | Ross G. Marvin | 1906

From Cape Sheridan, Ellesmere Island, winter quarters of *Roosevelt* (see 1905-06); 2 June - 16 July; instructed by Peary to make ocean soundings from Cape Hecla as far northward as possible. He set out with Peary's western expedition (see below, 1906), and parted from them at Cape Hecla on 4 June. After he had travelled a short distance northward over the sea ice, he was stopped by a big lead of open water, and decided to take his soundings in a westerly direction. He continued westwards as far as Cape Richards, where he turned back on 25 June. On 16 July, he reached the ship, having made 21 soundings in all.

BUSHNELL, 1956.

1906

United States exploring expedition — Robert Edwin Peary

From Cape Sheridan, Ellesmere Island, winter quarters of *Roosevelt* (see 1905-06); 2 June - 30 July; to explore the north-west coast of Ellesmere Island between Aldrich's farthest (see 1876) at Yelverton Bay, and Sverdrup's farthest (see 1902) at Lands Lokk, Kleybolte Peninsula. He followed the north coast of Ellesmere Island westward and passed Yelverton Bay on 16 June. Ten days later, he reached Cape Stallworthy, the northernmost point of Axel Heiberg Island, from where he believed he saw "Crocker Land" to the north. The existence of Crocker Land remained a subject of dispute for several years, but was finally disproved by MacMillan's Crocker Land Expedition, 1913-17 (*qv*).

PEARY, 1907, pp. 173-243.

1906

Second Carnegie Museum expedition — Walter Edmond Clyde Todd

This was a one-man expedition to northern Ontario, with the object of making an avifaunal survey northward to the height of land along the line of the Temiskaming and Northern Ontario Railway. He began his survey at North Bay on 10 June, then made further observations at New Liskeard (11-12 June) and at Boston Creek (13-26 June). He made excursions north-westward from there to the site of present-day Swastika. On 26 June, he continued his journey along the line of the railway to Bourkes. He remained there until 4 July, visiting nearby lakes, then returned to Pittsburgh.

TODD, 1963, pp. 14-15.

1906

German ornithological expedition — Bernhard Adolf Hantzsch

From London; 2 July - 26 November; an expedition to make ornithological observations in the region of Port Burwell, Killinek Island, in preparation for his more extensive ornithological exploration of Baffin Island (see 1909-11). He sailed out on board the Moravian missionary supply vessel *Harmony,* Captain J. E. Jackson, and landed at Killinek (Port Burwell) mission on 4 August. He remained there until 11 October, making brief excursions along the coast and two long excursions into Ungava Bay and down the Atlantic coast to about 59°30'N. *Harmony* returned in October and carried him to St John's, from where he took the steamer *Siberian* to Glasgow, arriving on 26 November. In addition to avifaunal observations, he made notes on topography, flora, fauna, geology, and Eskimos.

HANTZSCH, 1928-29; 1931-32; 1930; STEIN, 1909.

1906

United States ornithological expedition — Charles W. Townsend, Glover M. Allen

From Battle Harbour, Labrador; 11-26 July; to make an ornithological survey of the Labrador coast. They travelled from Boston to Newfoundland, then to Blanc Sablon and reached Battle Harbour on 10 July. From there, they sailed on the mail steamer *Virginia Lake* and, stopping at settlements along the way, went as far up the coast as Nain, then returned to Battle Harbour, where they spent a further week. On the evidence of their field work and the published literature, they compiled the first comprehensive list of Labrador birds.

*TOWNSEND, 1907; †TOWNSEND AND ALLEN, 1907.

Canadian overland expedition	Elihu Stewart	1906

From Edmonton; 2 June - 31 August; Stewart, the Canadian Government's Superintendent of Forestry, undertook this trip for pleasure and to study the timber of the Mackenzie basin. He travelled down Athabasca, Slave, and Mackenzie rivers on the steamers *Midnight Sun, Grahame* and *Wrigley,* on which Stefansson was also travelling north (see 1906-07). He left Fort McPherson on 24 July, portaged to Bell River, then canoed down Bell and Porcupine rivers to Fort Yukon. From there, he took the steamer *Lavelle Young* up Yukon River to Dawson, where he arrived on 17 August. He remained there five days, then continued up Yukon River on the steamer *Dawson* to Whitehorse. He then took the White Pass Railway to Skagway, and on 27 August he boarded the steamer *Princess May* for Vancouver, where he arrived on 31 August. During the journey, he made notes on timber, and brief observations on natural resources, wildlife, traders and missionaries.

STEWART, 1913.

United States expedition	Stephen Tasker Florence A. Tasker George Elson Job Chapies	1906

From Philadelphia; apparently a private expedition undertaken for the sole purpose of adventure. Mr and Mrs Tasker left Philadelphia on 9 June and joined Elson, who had been on both Hubbard expeditions (see 1903-04 and 1905) and Chapies, who had been on the second one, at Missinaibi, Ontario, on 11 June. They left there nine days later and canoed to York Factory. They then took the steamer *Inenew* to Great Whale River, and continued by canoe to Richmond Gulf. Their route from there to Fort Chimo is not entirely clear, but it appears that they followed the route explored by Hendry in 1828 (*qv*) and later surveyed by A. P. Low: that is, by way of Clearwater and Larch rivers. They arrived at Fort Chimo on 3 September and soon after left for Cartwright on the Hudson's Bay Company's supply ship *Pelican.* They then took a steamer to St John's, and from there returned to the United States.

*TASKER, 1908; HUBBARD, 1908, p. 15; COOKE, 1964, p. 160.

1906-07

United States anthropological expedition | Vilhjalmur Stefansson

From New York; April 1906 - (?) 1907. Stefansson's original object was to join an expedition organized by Ernest de Koven Leffingwell and Ejnar Mikkelsen at the Mackenzie River delta, and to study the "blond Eskimos" of Victoria Island. He travelled by rail to Edmonton and by steamer down Athabasca, Slave, and Mackenzie rivers to Fort McPherson. He left there by whaleboat for Herschel Island on 30 June 1906. At Herschel Island, in August, he met Amundsen's expedition (see 1903-06) and the whaling fleet, and he learned that Captain Klengenberg of the whaler *Olga* had recently been at Victoria Island and had seen the "blond Eskimos." Leffingwell's ship *Duchess of Bedford* failed to turn up, so Stefansson sailed with Harrison (see 1905-07) to Shingle Point. In December, he continued eastwards to Tuktoyaktuk to meet more remote Eskimo tribes. Later in the month, he travelled to Eskimo Lakes to spend a few weeks with Harrison. On 1 February 1907, he returned to Herschel Island to meet his ship, and there learned that it was wintering at Flaxman Island on the Alaskan coast. At Flaxman Island, he found *Duchess of Bedford* damaged and sinking. He went alone to Jones Islands to study ancient Eskimo encampments then, on 25 July, joined the whaler *Belvedere* of San Francisco, Captain Stephen F. Cottle, for a trip back to Herschel Island. There, he accepted an assignment to carry to the telegraph station at Eagle City a report that Leffingwell, Mikkelsen, and Storker Storkerson, who had been reported dead, had returned alive. He travelled by boat to Fort McPherson, took a portage route south of McDougall Pass to Bell River, continued by raft and boat to Fort Yukon and by steamer to Eagle City, where he delivered his message. He then travelled up Yukon River and across White Pass to the Pacific. He took a steamer to Seattle and returned by rail to New York. The results of his expedition included extensive notes on physical geography, fauna and flora, and on the Eskimos.

STEFANSSON, †1908; 1908a; 1908b; 1908c; 1910; *1923.

1906-07

Canadian Arctic patrol | Joseph-Elzéar Bernier | CGS *Arctic*

From Quebec; 28 July 1906 - 19 October 1907; sent by the Canadian Department of Marine and Fisheries to take possession of the Arctic islands in the name of Canada, and to collect whaling dues. He sailed through Davis Strait and Baffin Bay into Lancaster Sound, then passed through Navy Board Inlet and Eclipse Sound to visit the Scottish whaling station at Albert Harbour. Returning to Lancaster Sound and Barrow Strait, they made several landings to take possession of Somerset Island (at Port Leopold), Griffith Island, Cornwallis Island (at Sherringham Point), Bathurst Island (at Cockburn Point), and Byam Martin Island (at Cape Gillman). From south-eastern Melville Island, they took possession of Melville, Prince Patrick, and Eglinton islands. Returning east, they landed at Beechey Island, and entered Admiralty Inlet, where they discovered and named Moffet and Levasseur inlets. They then continued to Albert Harbour, where they wintered. From there, one party explored Milne Inlet, and others visited Erik Harbour and Navy Board Inlet.

1907: They left Albert Harbour on 27 July and sailed for Jones Sound. It was blocked by ice so, after they had taken possession of Coburg Island, they returned to Port Leopold to pick up a cache they had left there in 1906. A second attempt to enter Jones Sound also proved fruitless, so they landed at its entrance and took possession of the Sverdrup Islands from King Edward VII Point (which they named). They then returned south along the east coast of Baffin Island and visited whaling stations at Kekerten and Blacklead Island and the Moravian mission at Port Burwell. They then returned to Quebec. The report of the expedition includes data on Arctic whaling and fishing, ice conditions, meteorology, geology, and a list of relics and documents of earlier expeditions which they had collected.

BERNIER, *†1909; 1939, pp. 306-24.

Hudson's Bay Company supply voyage	John G. Ford	*Discovery*	1907
	N. E. Freakley	*Stork*	
	Alexander Gray	*Pelican*	

From London. *Discovery* left London on 20 June, visited Charlton Island 9-27 August, and reached London on 25 September. *Stork* left London on 18 June, visited Charlton Island 22 August - 8 September, and reached London on 20 October. *Pelican* left London on 16 May, visited the Labrador posts, York Factory, Churchill, and Fort Chimo, and reached London on 19 October.

HUDSON'S BAY COMPANY (1).

Canadian prospecting expedition	William Tees Curran	1907
	Horace P. Adams	

This was apparently a private expedition to examine the mineral deposits of the Nastapoka Islands, Hudson Bay. They travelled by canoe down Missinaibi and Moose rivers to Moose Factory, and sailed north from there by yacht on 10 July. During the northward voyage they stopped at Eastmain, and part of the party spent two weeks on a canoe journey up Eastmain River to explore for minerals. They reached the Nastapoka Islands on 9 August and spent the next two weeks examining the islands as far north as Taylor Island; they found encouraging deposits of iron ore. One group visited the mouth of Nastapoka River. On return to James Bay, the party divided: one group returned home by way of Moose and Missinaibi rivers and reached Montreal on 9 October; the rest sailed on the ship *Erik* by way of Hudson Strait and St John's, and reached Montreal on 2 October.

CURRAN AND ADAMS, n.d.

1907

Third Carnegie Museum expedition — Walter Edmond Clyde Todd

From Pittsburgh; to extend the avifaunal survey of northern Ontario he had begun in 1906 (*qv*) from New Liskeard to James Bay. He arrived at New Liskeard on 25 June, but was unable to arrange a canoe trip to James Bay so he returned south to Mattawa. There, a representative of the Hudson's Bay Company advised him to go to Missinaibi, the usual point of departure for James Bay. He remained at Missinaibi 1-8 July making collections and observations, but learned that his funds were insufficient to support a trip to James Bay. He therefore returned home, having made provisional arrangements for a trip the following year (*qv*). He observed 39 species of birds at Mattawa, and 45 at Missinaibi.

TODD, 1963, p. 15.

1907

United States exploring expedition — Ernest Thompson Seton, Edward Alexander Preble

From Winnipeg; 5 May - (?); a private expedition to study the natural history of Mackenzie District, in particular the abundance of caribou, and to survey Aylmer and Clinton-Colden lakes, which had never been properly explored since their discovery by Back (see 1833-35). They travelled by rail to Edmonton, then canoed down Athabasca and Slave rivers to Smith Landing, where they arrived on 7 June. They remained there for a month, and Seton made three short trips into the country to the westward, in the region of Salt, Little Buffalo, and Nyarling rivers, to study the conditions of the herds of wood bison. He was accompanied on these trips by Major A. M. Jarvis of the RNWMP. On the last trip, they travelled down Little Buffalo River to Great Slave Lake and Fort Resolution. Leaving Fort Resolution on 16 July, they sailed to the eastern end of Great Bear Lake, and travelled by way of Pike's Portage, Lockhart River and Artillery Lake to Clinton-Colden and Aylmer lakes. They spent about three weeks in this region, surveying the two lakes, and studying caribou and musk-oxen. They set off on the return journey from Aylmer Lake on 24 August, and canoed back by Great Slave Lake, Slave and Athabasca rivers to Athabasca Landing, which they reached on about 1 November. The scientific results of the expedition include notes on birds, buffalo and other mammals, insects, and plants.

*SETON, 1912; CANADA, ROYAL NORTH-WEST MOUNTED POLICE, 1908.

1907-08

Geological Survey of Canada expedition — Joseph Keele, R. B. Riddell, J. M. Christie

Keele was instructed to examine the country in the vicinity of upper Pelly River, Yukon Territory, and subsequently to carry out an exploration across Mackenzie Mountains to Mackenzie River. Beginning the survey on 1 July 1907, he spent the early part of the summer examining Pelly River for about 220 km above its confluence with Ross River, both to study the topography and geology of that region and to inquire into reports (which Keele found to be untrue) that there was an active volcano in the vicinity. He had not enough time to trace Pelly River to its source as he had hoped.

He then decided to follow Ross River as the route most likely to lead towards Gravel (now Keele) River, along which he expected to find a route to the Mackenzie. He ascended Ross River nearly to its source and found a gap in the mountains (Christie Pass) leading across the height of land to the headwaters of Keele River. They crossed the divide in late winter, using dogs to haul their supplies across in relays, then surveyed Keele River down to Mackenzie River, ending the survey there on 20 July 1908.

This expedition was the first to study the geology and topography of the Mackenzie Mountains. It was also the first to survey upper Pelly and Ross and Keele rivers, of which practically nothing was known. (Adam McBeath had visited Keele River in 1843 [*qv*], and a party had travelled on it to cross Mackenzie Mountains by way of Twitya and Hess rivers to the Klondike during the gold rush, but they had left almost no information on it.) Keele also provided information on fauna (particularly fur-bearing animals), trapping, and forests in the region.

KEELE, 1910.

1907-09

United States North Pole expedition	Frederick Albert Cook Rudolph Francke Moses Bartlett	*John R. Bradley*

From Gloucester, Massachusetts; 3 July 1907 - (?) 1909; sponsored by John R. Bradley and originally organized as a hunting expedition. During the voyage north, the ship visited Eskimos at Wolstenholme Sound, Whale Cove, and Etah. At Annoatok, just north of Etah, Cook decided that conditions were good for an assault on the Pole. He returned to Etah to prepare for the journey, and took Eskimos from there to Annoatok to help him. The ship left for home on 3 September, leaving Cook and Francke behind.

1908: With help from Eskimos, Cook started to transfer supplies to Cape Sabine in mid-January. On 19 February, he left Annoatok on the Pole journey. Accompanied by Francke and a party of Eskimos, he crossed Smith Sound to Cape Sabine, travelled north to Flagler Bay, then crossed Ellesmere Island to Bay Fiord. He travelled through Eureka Sound (incidentally discovering that the so-called Schei Island was a peninsula) to Cape Stallworthy, the northernmost point of Axel Heiberg Island, where he established a camp. Leaving Francke and the main party of Eskimos there, he set off northward over the sea ice with two Eskimos, two sledges, and 26 dogs. He claims to have reached the North Pole on 21 April 1908, and to have remained there two days. Returning south, he missed his depot on Axel Heiberg Island, and made a landfall in Hassel Sound about 15 June. One aspect of his account of the journey that many consider dubious is that, during his return, he apparently passed within 5 km of Meighen Island without seeing it. Continuing through Hassel Sound and Penny Strait, Cook crossed north-western Devon Island to Eidsbotn Fiord, Jones Sound, then travelled by boat and sledge to Cape Sparbo, north coast of Devon Island, where he wintered.

1909: In February, he continued to Annoatok. From there, he travelled by boat and sledge to Upernavik, where he arrived about 20 May. He then sailed to Copenhagen on the ship *Hans Edege.*

Cook's claim to have reached the pole was rejected as a hoax by Peary (see 1908-09) and by many others, and is still in dispute.

COOK, *1911; 1951; GIBBONS, 1956; 1964; 1968; 1970; AMERICAN GEOGRAPHICAL SOCIETY, 1909; MACMILLAN, 1918c; FREEMAN, 1961; WEEMS, 1960; STEFANSSON, 1939; WRIGHT, T., 1970; RAWLINS, 1973.

1907-09

United States whaling voyage; survey of Southampton Island — George Comer — *A.T. Gifford*

From New London; 13 July 1907 - (?) 1909. As Comer had made several previous whaling voyages to the vicinity of Southampton Island (as captain of *Era* in 1895-96, 1897-99, 1900-03 and 1903-05), it is difficult to distinguish the discoveries he made on this voyage from those of earlier years. However, it was in 1907-09 that he undertook a special survey of Southampton Island, and to this voyage may be attributed the bulk of his discoveries, especially along the south coast of the island, which he showed to be quite different from its configuration on earlier maps.

Most important was his discovery, on the south shore, of South Bay, at the head of which he located one of the few good anchorages to be found on Southampton Island, Coral Harbour. On the south-eastern coast, he confirmed that Bell "Island" was a peninsula of Southampton Island, as A.P. Low had conjectured on his visit to Seahorse Point (see 1903-04); to the south-west, he discovered and named Cape Low. In Fisher Strait, he confirmed the existence of Walrus Island (see 1868-69, Fisher). The result of Comer's surveys was the publication of a map of Southampton Island (in 1910) which was much superior to maps that had previously been published.

On this voyage, as on earlier voyages, Comer made extensive observations and collections on the zoology, geology and ethnology of Southampton Island. Franz Boas wrote a number of papers on the Eskimos of Southampton Island on the basis of Comer's ethnological collections, and it is from Comer that we have the only eyewitness accounts by a white man of the Sadlermiut culture before it became extinct in 1903.

COMER, 1906; *1910; 1913; BIRD, 1953, p. 7; SUTTON, 1932, section 2, pp. 19-20; *STACKPOLE, 1969, pp. 34-51.

1908

Hudson's Bay Company supply voyage	John G. Ford	*Discovery*
	N. E. Freakley	*Stork*
	Alexander Gray	*Pelican*

From London. *Discovery* left London on 13 June, visited Charlton Island 19 August - 3 September, and reached London on 4 October. *Stork* left London on 18 June and visited Charlton Island 7 September - 5 October. During the homeward voyage, on 7 October off Mansel Island, Captain Freakley decided to turn back and winter at Charlton Island because of the lateness of the season. Three days later, about 25 km from Charlton Island, the ship ran aground on a reef and was abandoned. There was no loss of life, and the crew wintered at Rupert House. *Pelican* left London on 16 May, visited the Labrador posts, York Factory, Churchill, and Fort Chimo, and reached London on 27 October.

HUDSON'S BAY COMPANY (1); WILLIAMS, J., 1939.

Fourth Carnegie Museum expedition — Walter Edmond Clyde Todd, M. A. Carriker Jr, Will Bain, Jerry Solomon — 1908

From Missinaibi: 4 June - 1 August; to make avifaunal investigations between Missinaibi and Moose Factory by way of Missinaibi and Moose rivers. Todd had made provisional arrangements for this trip during his previous expedition (see 1907). Having made observations and collections along Missinaibi and Moose rivers, they reached Moose Factory on 26 June. Whilst Carriker made collections in the vicinity of the factory, Todd made a short boat journey along the south shore of James Bay, during which he spent three days at the mouth of Partridge River, which he found to be a very good collecting ground for water birds. They left Moose Factory on 11 July, and retraced their steps to Missinaibi.

TODD, 1963, pp. 15-16.

United States overland expedition — Agnes Deans Cameron — 1908

From Chicago; mid-May - (?); a private venture undertaken for pleasure. She travelled by rail to Edmonton, left there for Athabasca Landing on 1 June, then continued by scow and by Hudson's Bay Company steamer down Athabasca, Slave and Mackenzie rivers to Mackenzie River delta. She then returned south by Mackenzie, Slave and Peace rivers and Lesser Slave Lake to arrive back at Edmonton about the end of October. She made notes on flora, fauna, native peoples, and on the settlements she visited.

CAMERON, A.D., *1910; 1910a.

United States expedition to Labrador — Francis W. Peabody, John Bryant, Owen Bryant, Edward S. Bryant, Henry S. Forbes — 1908

From Boston. The party chartered a schooner and cruised along the Labrador coast north to McLelan Strait. At Komaktorvik Fiord, E. S. Bryant carried a transit to the top of Mt Tetragona (named by Coleman in 1915-16, *qv*) and triangulated the summits of the Four Peaks, which proved to be less than 1,500 m high, not the 2,500 m or more that had been supposed. Bryant's unpublished notes were later used by O. M. Miller and Alexander Forbes in their aerial survey of northern Labrador, 1931-35. The only published records of the expedition appear to be papers based on O. Bryant's collections of marine invertebrates.

FORBES, 1938, pp. 10, 212; †BALCH, 1911; †BIGELOW, 1910; †MOORE, J.P., 1910.

1908-09

Canadian Arctic patrol — Joseph-Elzéar Bernier — CGS *Arctic*

From Quebec; 28 July 1908 - 5 October 1909; to continue the work of annexing the islands of the Canadian Arctic to the Dominion Government and of issuing whaling licences. Sailing northward along the west coast of Greenland, Bernier anchored at Etah, where he left stores for Frederick Cook (see 1907-09), and met *Erik,* Captain Samuel Bartlett, relief ship for Peary's 1908-09 expedition (*qv*). Bernier then sailed into Lancaster Sound and Barrow Strait, landing at Beechey Island. He continued westward through Melville Sound into McClure Strait, where he found ice conditions so good that he believed he could have accomplished a North-west Passage by that route if he had wished. He then returned to Winter Harbour, Melville Island, where he wintered. Whilst there, Bernier organized a search of the surrounding area and of Dealy Island for relics of Parry's (see 1819-20) and Kellett's (see 1852-54) expeditions. Numerous relics of these and other expeditions were found and carried home in *Arctic.*

1909: Spring sledge parties visited Banks and Victoria islands to establish sovereignty (see 1909). On 3 August, Bernier left winter quarters, returned to Baffin Bay, and visited whaling stations in Pond Inlet and Cumberland Sound to issue licences. On 15 September, he visited the mission at Port Burwell, Hudson Strait. The scientific results of the expedition include reports on ice conditions, tides and meteorology at Winter Harbour, collections of flora, fauna, and rock specimens from all the places visited.

BERNIER, *†1910; 1939, pp. 325-46.

1908-09

Royal North-West Mounted Police patrol — E. A. Pelletier, M. A. Joyce, R. H. Walker, P. R. Conway

From Fort Saskatchewan; 1 June 1908 - 18 March 1909; to make an overland patrol from Great Slave Lake to Hudson Bay. They travelled by scow, by steamer and by canoe from Athabasca Landing to Fort Resolution, Great Slave Lake. They left there by canoe on 1 July and travelled by way of Pike Portage, Artillery Lake, Hanbury and Thelon rivers to Chesterfield Inlet, which they reached on 31 August. There, they met the police launch *MacTavish,* which had been sent to carry them to Churchill. However,

MacTavish was wrecked on the journey to Churchill, and the party was obliged to travel north to the police post at Cape Fullerton and to wait for freeze-up. On 29 November, they set out by dog-sledge for Churchill, which they reached on 11 January 1909. They remained there until 7 February then continued southward, and ended their patrol at Gimli, on the west shore of Lake Winnipeg, on 18 March. This patrol, covering a total of some 5,300 km, was one of the longest that had ever been made by the North-West Mounted Police.

CANADA, ROYAL NORTH-WEST MOUNTED POLICE, 1910.

1908-09

United States North Pole expedition — Robert Edwin Peary, Robert A. Bartlett, Samuel Bartlett — *Roosevelt*, *Erik*

From New York; 6 July 1908 - 21 September 1909; to attain the North Pole. In early August, the ships stopped at Cape York, north-west Greenland, to pick up a party of Eskimo helpers that included 22 men and 17 women with 246 dogs. On 18 August at Etah, *Roosevelt* parted from the supply ship *Erik* and steamed northward into Kane Basin. She reached her winter quarters at Cape Sheridan, north-eastern Ellesmere Island, on 5 September. During the autumn, Peary began the work of transferring stores to Cape Columbia, which he had selected as starting point for the Pole. In the first week of October, Peary and three Eskimos explored Clements Markham Inlet to its head. As in the previous expedition (see 1905-06), the Eskimos were set to hunt and fish in the vicinity of Lake Hazen during the winter. During December 1908 and January 1909, Ross G. Marvin and Captain Bartlett made tidal observations at Cape Brevoort, Greenland.

During February - April 1909, Peary made his successful assault on the Pole (see 1909). On 18 July, *Roosevelt* left Cape Sheridan. After having landed the Eskimos at Cape York, she continued to Sydney, NS, where she arrived on 21 September.

PEARY, *1910; 1910a; UNITED STATES, CONGRESS, HOUSE, 1911; LEWIN, 1911; MITCHELL, 1959; GIBBONS, 1956; RAWLINS, 1970; 1970a; 1973; MILLER, 1963; WEEMS, 1960; BARTLETT, 1940; HENSON, 1969; MACMILLAN, 1934.

1908-09

United States hunting expedition — Harry Whitney

From Sydney, NS; 17 July 1908 - 28 September 1909; a private expedition to hunt for sport in the region of Etah and Annoatok, north-west Greenland. He travelled north as a guest on board Peary's exploring ship *Erik* (see above, 1908-09). On 21 August, he landed at Etah, and *Erik* departed southward, leaving him alone among the Eskimos. For the remainder of the year and throughout the winter, he lived and hunted with these Eskimos.

1909: On 9 May he set out from Annoatok with a party of Eskimos to take part in a musk-ox hunt on Ellesmere Island. They landed at Flagler Bay, Ellesmere Island, and spent some time hunting to the west of there. They returned to Annoatok about the end of May, and Whitney spent the remainder of the summer hunting on the coast of Greenland. On 18 August, Peary's ship *Roosevelt* arrived from the north. Five days later, *Jeanie,* Captain Samuel Bartlett, arrived to collect Whitney. He returned with her to St John's, where he arrived on 28 September.

WHITNEY, H., 1910.

1908-11

Expedition to Great Bear Lake	James Cosmo Dobrée Melvill John Hornby James Mackinlay Peter McCallum

From Athabasca Landing; early May 1908 - (?) 1911; a private venture for hunting and adventure with the secondary object of trading. They travelled by boat down Athabasca, Slave and Mackenzie rivers to Fort Norman, where McCallum joined the party. On his advice, Melvill decided to winter on the north-eastern corner of Great Bear Lake. With help from Indians, he took his boats up Bear River and across Great Bear Lake to "Caribou Bay" (now named Hornby Bay) where, in mid-September, they began to build a house. During the winter, McCallum took the Indians on a musk-ox hunt to the north-west whilst Melvill, Mackinlay and Hornby remained alone in their house, living by hunting and trapping.

1909: From the end of March to early May, the party travelled to Coppermine River to hunt musk-oxen and to trade with Eskimos. However, they met no Eskimos and saw few musk-oxen. On 8 May, Melvill, Mackinlay and McCallum travelled by dog-sledge to open trade at Fort Franklin whilst Hornby made another trip to Coppermine River. He probably joined the rest of the party for winter at Fort Franklin.

1910: After a visit to Fort Norman in July, they set off back to Great Bear Lake. Mackinlay probably remained to continue trade at Fort Franklin, the rest sailed to Dease Bay to build a new post, having abandoned the Hornby Bay house. They took with them Joe Hodgson, a retired fur trader, and his family, who built a house close to Melvill's on Dease Bay. Melvill's house was close to the ruins of Fort Confidence. Stefansson (see 1908-12) joined them there for a few weeks in autumn.

1911: Finding that the trading venture had not been as successful as they had hoped, Melvill and Mackinlay returned south in summer. Hornby remained in the north with the intention of returning to Great Bear Lake.

WHALLEY, 1962, pp. 22-50.

1908-12

Stefansson-Anderson Arctic Expedition	Vilhjalmur Stefansson Rudolph Martin Anderson

From New York; 22 April 1908 - 18 August 1912; sponsored by the American Museum of Natural History and the Geological Survey of Canada, to study the Eskimos and the biology of Coronation

Gulf, Victoria Island and Banks Island. From New York, they travelled by rail to Edmonton and by boat down Athabasca, Slave and Mackenzie rivers to Arctic Red River post, where they arrived on 5 July. Soon after, they went by whale boat to Herschel Island, where they separated. They spent the winter travelling in the company of Eskimos on the north coast of Alaska. The two men reunited in spring 1909 at Flaxman Island, and later travelled to Herschel Island, where they arrived on 18 August. They then sailed by whaler to Cape Parry, made a journey westward along the shores of Franklin Bay and, soon after, discovered the mouth of Horton River. They spent the winter of 1909-10 on the lower reaches of Horton River and on the shore of Franklin and Langton bays.

1910: Whilst Anderson went to obtain supplies from Herschel Island, Stefansson, on 21 April, set out eastward along the coast in search of unknown Eskimo tribes. He located a tribe on the shore of Dolphin and Union Strait on 13 May. Two days later he crossed this strait and met the "Blond Eskimos" of Victoria Island. He then returned to the mainland and travelled inland to the Eskimos' hunting grounds near Dismal Lakes. On Great Bear Lake, at the mouth of Dease River, he met Melvill and Hornby (see 1908-11). In November, he located the middle course of Horton River and explored it downstream to the part he had explored in 1909. He then went to Langton Bay to rejoin Anderson, who had successfully completed his journey to Herschel Island. They returned together to the Eskimo winter camp at the head of Dease River.

1911: In March, they travelled to Coronation Gulf. Stefansson crossed the gulf to Victoria Island on 30 April, whilst Anderson remained to make zoological collections. Stefansson crossed Victoria Island from Simpson Bay to Prince Albert Sound (the first crossing of any part of Victoria Island by a white man) where he met another tribe of unknown Eskimos. He later rejoined Anderson at Langton Bay, near where they wintered.

During 20 March - 13 June, Stefansson sledged along the mainland coast to Point Barrow, Alaska, leaving Anderson to follow later by whaler with their collections. Stefansson reached Nome, Alaska, on the revenue cutter *Bear*, on 18 August.

This expedition resulted in a wealth of new knowledge relating to the Eskimo, physical geography, travelling and survival techniques, and natural history.

STEFANSSON, 1912; 1912a; 1912-13; *1913; 1913a; †1919; †ANDERSON, R.M., 1913.

1909

Hudson's Bay Company supply voyage	John G. Ford A. Cleveland Smith	*Discovery* *Pelican*

From London. *Discovery* left London on 19 June, visited Charlton Island 12-25 August, and reached London on 30 September. *Pelican* left London on 15 May and visited the Labrador posts, Churchill and Fort Chimo. She also called twice at Wolstenholme, 12-23 August and 5-6 October, to establish the new post there (see below, 1909). She reached London 3 January 1910.

HUDSON'S BAY COMPANY (1); *BEAVER*, 1929.

1909

Attainment of the North Pole

Robert Edwin Peary
Matthew Henson
Robert A. Bartlett
George Borup
Ross G. Marvin
Donald B. Macmillan
J. W. Goodsell

From Cape Sheridan, Ellesmere Island, winter quarters of *Roosevelt* (see 1908-09); 15 February - 25 April. Supporting parties led by Bartlett, Borup, Marvin, Macmillan and Goodsell were to carry provisions, blaze a trail, and build igloos for Peary and Henson, who were to make the final assault on the pole. Bartlett left with the first advance party on 15 February; Peary was last to leave the ship, on 22 February. He set out northward over the sea ice from Cape Columbia on 1 March. At the fourth camp, he was halted by a big lead of open water and by accidents to supporting sledges, and he remained there until 11 March. Goodsell's party turned back on 14 March, Macmillan's on 15 March, and Borup's on 20 March. Marvin, who had taken ocean soundings during the outward journey, turned back on 26 March at 86°38'N. On 10 April, he was drowned crossing the big lead. Bartlett turned back on 1 April at 87°47'N, leaving Peary, Henson and four Eskimos (Ootah, E-ging-wah, See-gloo and Oo-ke-ah) to continue to the Pole. They reached the Pole (or its close vicinity) on 6 April and remained there 30 hours. They reached land again at Cape Columbia on 23 April, and two days later arrived at the ship. Although Peary's claim to have reached the North Pole has been generally accepted, some have doubted it. Among other reasons, the speed of his return from the Pole was certainly remarkable, and his detractors have insisted that he must have turned back soon after the last support party left him and without having continued northward to the Pole.

SEE BIBLIOGRAPHY AT 1908-09.

1909

Canadian expedition

O. Jules Morin

From Winter Harbour, Melville Island, winter quarters of *Arctic* (see 1908-09); 6 April - 9 May; to make landings on Banks and Victoria islands to establish Canadian sovereignty, and to search for records of McClure's and Collinson's expeditions (see 1850-54 and 1850-55). They crossed M'Clure Strait from Cape Providence, Melville Island, to Russell Point, Banks Island. From there, two of Morin's men, Reuben Pike and Napoléon Chassé, crossed Prince of Wales Strait to Peel Point, Victoria Island, and took possession of the island, but left no record of having done so. No records of earlier expeditions were found.

BERNIER, 1910, pp. 126-37.

1909

Canadian expedition

Charles W. Green

From Winter Harbour, Melville Island, winter quarters of *Arctic* (see 1908-09); 1 May - 10 June. They crossed M'Clure Strait to

Mercy Bay, Banks Island, to search for records of McClure's expedition (see 1850-54) which had wintered there in 1851-53. They found no records.

BERNIER, 1910, pp. 147-61.

1909

Canadian expedition — O. Jules Morin

From Winter Harbour, Melville Island, winter quarters of *Arctic* (see 1908-09); to leave a record on Banks Island indicating that they had taken possession of both Banks and Victoria islands on their previous journey (see above, 1909), and to search for coal deposits reported by earlier expeditions. They sledged across M'Clure Strait to Mercy Bay, Banks Island, and found small deposits of coal on the coast of Banks Island. They left their record at Mercy Bay and, at the same time, searched unsuccessfully for relics of the McClure expedition (see 1850-54).

BERNIER, 1910, pp. 167-77.

1909

Establishment of Wolstenholme post — Ralph Parsons

In this year, Parsons established for the Hudson's Bay Company Wolstenholme trading post at Erik Cove, Quebec, on the south shore of Hudson Strait.

ANDERSON, J. W., 1939; *MOCCASIN TELEGRAPH,* 1970, pp. 86-87.

1909-10

United States geological expedition — Charles Kenneth Leith, Arthur T. Leith

C. K. Leith, of the University of Wisconsin, led a four-man geological expedition to the east coast of Hudson Bay. They canoed down Moose River to Moose Factory, then sailed along the east coast of James and Hudson Bays to Richmond Gulf, where they studied the iron formations. Two of the party returned home from there during summer; A. T. Leith and one companion remained on the shores of Hudson Bay until December, then returned home by dog team and snowshoes early in 1910.

LEITH, C.K., 1910; LEITH, C.K. AND LEITH, A.T., 1912.

1909-11

Establishment of Lake Harbour mission — Julian William Bilby, Archibald Lang Fleming

From St John's; 30 July 1909 - (?) 1911. Bilby and Fleming sailed north on board *Lorna Doone* to continue Anglican missionary activity on Baffin Island. This was Fleming's first visit to the Arctic, though Bilby had had nine years' experience at the missions in Cumberland Sound since his first visit in 1898 (see 1897-99, Peck). On their arrival off south-east Baffin Island, they learned that Robert Kinnes had opened a mica mining depot at Lake Harbour and that many Eskimos were living there, so they decided they should establish a mission there. The ship reached Lake Harbour towards the end of August and, on 30 August, Bilby and Fleming began to build a mission house. The ship departed for home soon after. For the next two years, the two men lived and worked among the Eskimos in and around Lake Harbour. Early in 1911, Fleming made a long journey north-eastward across Meta Incognita Peninsula and Frobisher Bay to visit the Eskimos at O. C. Forsyth-Grant's whaling station at Cape Haven. After a short stay with Forsyth-Grant, he retraced his steps to Lake Harbour. In mid-September, *Stella Maris* arrived with two other missionaries to relieve Bilby and Fleming. At the same time, the Hudson's Bay Company's ship *Pelican* arrived to set up a trading post at Lake Harbour under the leadership of Ralph Parsons. A few days later, Bilby and Fleming departed for home.

FLEMING, A. L., 1957, pp. 40-135.

1909-11

German exploring expedition — Bernhard Adolf Hantzsch

From Dresden; 4 July 1909 - June 1911; sponsored by the Geographical Society of Dresden to make the first crossing of Baffin Island and to make ornithological and ethnological observations. He sailed from Dresden to Dundee, then joined the Dutch ship *Janthina Agatha,* bound for the whaling stations in Cumberland Sound. The ship was wrecked at the entrance to Cumberland Sound on 26 September, and most of Hantzsch's supplies were lost. On 1 October, Hantzsch reached the whaling station on Blacklead Island.

1910: In early January, he sledged to the Kekerten whaling station, where he obtained sledges and dogs. On 23 April, he set off on his journey, accompanied by a party of Eskimos. He travelled along the south shore of Cumberland Sound and up Nettilling Fiord to Nettilling Lake, on the south shore of which he established a summer camp. He resumed his journey on 15 August and travelled down Koukdjouak River to reach the shores of Foxe Basin on 20 September, thus completing the first crossing of Baffin Island. He explored the coast about 160 km northward from Koukdjouak River before making a winter camp. He suffered hardship and near starvation during the winter.

1911: On 16 April, he continued his northward march, reaching his farthest north at 68°45'N on 8 May. During the return journey, he died in early June of sickness, probably trichinosis, and his Eskimo companions buried him near the mouth of Hantzsch River. The Eskimos returned to Cumberland Sound with his journal and his notes on birds and mammals.

MILLWARD, 1930, pp. 103-30; ROSENMÜLLER, 1913; NEATBY, 1975; ROWLEY, 1977.

1910

Hudson's Bay Company supply voyage	John G. Ford A. Cleveland Smith	*Discovery* *Pelican*

From London. *Discovery* left London on 19 June, visited Charlton Island 4-28 August and York Factory 6-24 September, and reached London on 5 November. *Pelican* left London on 25 June, visited the Labrador posts, Wolstenholme, Churchill, York Factory, and Fort Chimo, and reached London on 3 November.

HUDSON'S BAY COMPANY (1).

British exploring expedition	H. Hesketh Prichard G.M. Gathorne-Hardy Robert Porter Boaz Obed		1910

From London; 22 June - (?); a private hunting expedition to Indian House Lake on George River, northern Quebec. Prichard had visited the Hopedale district of Labrador in 1903 on a hunting trip and he decided then to make a journey inland. He and Gathorne-Hardy sailed on *Harmony* to Labrador and reached Nain on 20 July. They started up Nain Bay on 23 July and found that Fraser River discharges into it, not into a bay farther north as indicated on the charts of the time. Passing through Tasisuak Lake, they left their canoe on 3 August and on the 5th began to walk inland carrying their outfits, hoping to kill caribou for food on their way. The Eskimo, Obed, had left them on 31 July, having been refused higher wages. They reached Indian House Lake on 19 August. Here, Prichard sprained an ankle and was obliged to rest it for 10 days. They saw abundant signs and some stragglers of the immense migration of caribou that had taken place earlier in the season, and eked out their rations with caribou and fish. On 31 August, they set out on a more southerly route back to their canoe on Fraser River, which they reached on 4 September. They were back at Nain on the 9th.

PRICHARD, 1910; 1911.

Governor-General's tour of Hudson Bay	Earl Grey J.D. Moodie (?) Bartlett	*Earl Grey*	1910

From Norway House; 8 August - (?); organized by the RNWMP and the Hudson's Bay Company to select a site for a terminus of the planned Canadian Northern Railway. The party travelled by canoe along Echimamish and Hayes rivers and arrived at York Factory on 19 August. The next day, they joined the steamer *Earl Grey* for the second part of the journey. They sailed to Churchill to examine the port facilities there, then set off homeward. During the voyage back to Quebec, they called at Port Burwell, Okak, Indian Harbour and St Anthony, inspecting the mission stations.

HINDS, 1971-72.

Canadian Arctic patrol	Joseph-Elzéar Bernier	DGS *Arctic*	1910-11

From Quebec; 7 July 1910 - 25 September 1911; to issue whaling licences and to patrol Davis Strait, Baffin Bay, Lancaster Sound,

Barrow Strait, Melville Sound and M'Clure Strait and, if possible, to sail through the North-west Passage. Bernier sailed through Pond and Navy Board inlets and entered Lancaster Sound on 15 August. They met heavy ice conditions in Barrow Strait and Melville Sound and, on reaching Cape Dundas, south-west Melville Island, they found M'Clure Strait impenetrable. They therefore retreated eastwards and made their winter quarters in Arctic Bay, Admiralty Inlet. During autumn, sledging parties surveyed the surrounding region. A longer sledge expedition, led by J.T.E. Lavoie, set out on 10 October to explore the southern end of Admiralty Inlet and the west coast of Brodeur Peninsula between Fury and Hecla Strait and Cape Kater. The party travelled southward through Admiralty Inlet, then crossed to the Gulf of Boothia by way of Bell Bay, Saputing Lake, and Agu Bay. There, Lavoie spent some time in an Eskimo encampment. He then surveyed the unexplored coast northward to Cape Landry, about halfway to Cape Kater, before turning back and retracing his steps to the ship.

1911: Landry made another sledge journey to finish his surveys of 1910 (see below, 1911). *Arctic* was freed from her winter quarters about 20 July, and Bernier took her into Pond Inlet. From there, in August, Lavoie surveyed by launch the southern shores of Pond Inlet and Eclipse Sound. Bernier then sailed into Prince Regent Inlet and resurveyed the west coast of Brodeur Peninsula down to about the latitude of Bernier Bay. He had intended to sail through Fury and Hecla Strait into Hudson Bay, but heavy ice conditions prevented him. On 18 August, he turned back. He returned to Baffin Bay, visited whaling stations at Kekerten and Blacklead Island, Cumberland Sound, then continued back to Quebec.

*†CANADA, DEPARTMENT OF MARINE AND FISHERIES, 1911; BERNIER, 1939, pp. 347-69.

1910-11

Canadian expedition to Hudson Bay — Robert Joseph Flaherty

Flaherty was sent by Sir William Mackenzie, President of the Canadian Northern Railway, to examine the Nastapoka Islands, Hudson Bay, and to report on the commercial value of iron-ore deposits that had been found there by Robert Bell (see 1877). He travelled by rail to northern Ontario, then set out by canoe in late August 1910 down Mattagami and Moose rivers to Moose Factory. He then sailed across James Bay to Fort George and remained there until mid-December, when he set out by dog-sledge northwards along the coast to the Nastapoka Islands. He reached these islands in early January 1911 and spent five days there examining the iron ore deposits, but concluded that they were of no immediate commercial interest. As he was about to return south, one of his Eskimo dog-drivers mentioned the existence of a group of large islands to the west of the Nastapokas. He was referring to the Belcher Islands, a group that was known to exist and which had been described and roughly charted by Captain Coats of the Hudson's Bay Company in the 18th century. However, they had never been properly mapped, and even as late as 1910 they appeared on the Admiralty chart as a small cluster of tiny islands. Flaherty's interest was aroused when his dog-driver described them as being much larger, and when he gave an account of a tribe of Eskimos who inhabited them. He made further enquiries about the Belcher Islands among the Eskimos at Great Whale River, then returned homeward, having resolved to return to Hudson Bay in the summer of 1911 to explore the islands (see 1911-12).

*FLAHERTY, 1918; 1924; CALDER-MARSHALL, 1963, pp. 26-36.

Hudson's Bay Company supply voyage	John G. Ford A. Cleveland Smith E. Falk	*Discovery* *Pelican* *Beothen*	1911

From London and St John's. *Discovery* left London on 21 June, visited Charlton Island 4-28 August, York Factory 6-24 September, and reached London on 5 November. *Pelican* left London on 21 May, visited the Labrador posts, Churchill, Wolstenholme and Fort Chimo. She also visited Chesterfield Inlet, on 3-9 September, and Lake Harbour, on 17-23 September, to establish new posts for the company. *Beothen* sailed from St John's to deliver supplies at York Factory 1-5 August and at Charlton Island 8-22 August, and reached St John's on 30 August.

HUDSON'S BAY COMPANY (1); *BEAVER*, 1929.

Canadian exploring expedition	J.T.E. Lavoie	1911

From Arctic Bay, Admiralty Inlet, winter quarters of *Arctic* (see 1910-11); 15 March - 11 May; to complete the exploration of the unknown west coast of Brodeur Peninsula between Cape Kater and Fury and Hecla Strait. He crossed Brodeur Peninsula to Port Bowen, then followed the coast southward. South of Cape Kater, he discovered and named Bourassa Bay, Morin Point, Bernier Bay, Van Koenig Point, Thiboult Bay, and Easter Cape. He continued along the coast to Agu Bay, then crossed Brodeur Peninsula to the southern end of Admiralty Inlet, and returned from there to the ship.

CANADA, DEPARTMENT OF MARINE AND FISHERIES, 1911, pp. 89-106.

Establishment of Chesterfield Inlet trading post	1911

The Hudson's Bay Company established a trading post at Chesterfield Inlet in 1911.

USHER, 1971, p. 141.

United States exploring expedition	Donald Baxter MacMillan	1911

This was a private expedition to explore the Labrador coast by canoe. MacMillan travelled by rail and by steamer to Davis Inlet, then canoed along the coast to Hebron, visiting the Moravian mission stations on the way. At Hebron, he joined the Moravian supply vessel *Harmony* for a voyage to Killinek. From there, he made a short canoe trip through McLelan Strait before returning on board *Harmony* to St John's.

During the journey, he made a close examination of the coastline and made notes on birds, mammals, and the flora of the coast.

ALLEN, E.S., 1963, pp. 137-38.

1911-12

Canadian geological expedition

George Mellis Douglas
Lionel Dale Douglas
August Sandberg

From Edmonton; 23 May 1911 - October 1912; a private expedition to investigate the copper-bearing rocks of the Coppermine Mountains. They travelled by scow and by Hudson's Bay Company steamer down Athabasca, Slave and Mackenzie rivers to Fort Norman. From there, they continued by York boat up Bear River and across Great Bear Lake to the mouth of Dease River, where they built their winter quarters. On 28 July, G.M. Douglas and Sandberg set off up Dease River by canoe on a preliminary journey to Dismal Lakes and the Coppermine Mountains. The existing charts of the route proved to be deficient and they had considerable difficulty in finding their way from Dease River to Dismal Lakes. However, they finally arrived and they made their way through the lakes and down Kendall River to Coppermine River. They spent a week making a preliminary geological examination of the mountains, then retraced their steps to their winter quarters, where they arrived on 11 September. During winter, they were joined by Father Rouvière and John Hornby (see 1911-13).

1912: During April, Douglas's party sledged their supplies to Lac Rouvière (which Douglas named after the missionary); on 30 April, they made their final start for Coppermine Mountains. Hornby joined the party for this main journey. They soon reached Coppermine River, and Sandberg began his survey of the mountains. Douglas followed Coppermine River down to the sea, where he met Eskimos. In mid-June, Sandberg finished his work, and they returned to their camp on Dease River. On 26 June, they set off homeward across Great Bear Lake (leaving Hornby behind), and reached Fort Norman at the end of July. They left Fort Norman by steamer on 18 August, and returned to Edmonton.

†*DOUGLAS, G.M., 1914; WHALLEY, 1962, pp. 51-80.

1911-12

Canadian exploring expedition

Robert Joseph Flaherty

On his previous expedition (see 1910-11), Flaherty had gathered Eskimo descriptions of the Belcher Islands and had made plans to visit them. In 1911, under the sponsorship of Sir William Mackenzie, he set out again for Hudson Bay to explore the Belcher Islands, to search them for economic minerals, and to study the Eskimos who inhabited them. He canoed down Missinaibi River to Moose Factory, where he obtained a sailing boat. However, the boat proved inadequate and, when he arrived at Great Whale River post, the season was too far advanced for him to sail to the Belcher Islands. He therefore wintered at Fort George, waiting to sledge across the sea ice to the islands in February and March 1912. He was obliged to change his plans, though, for when he was about to make the journey, a heavy gale broke up the sea ice and made the crossing impossible. Flaherty's alternative plan was to cross the Quebec-Labrador peninsula to Ungava Bay to examine an ore-bearing rock series discovered by Low in 1893 (see 1892-96). He set out on this journey in March 1912. He travelled northward along the coast of Hudson Bay to about 57°N, then set off inland to Lake Minto, which Low had discovered in 1899 (see 1898-99). From then on, he was in unexplored territory as he surveyed Leaf River down to Ungava Bay. He reached Fort Chimo on 14 April. He left Fort Chimo in June and commenced his return to Hudson Bay. He travelled by sloop to Payne Bay, canoed a short distance up Payne River, then followed a circuitous lake-and-river route to the headwaters of North Payne River. From there, he located Povungnituk River and followed it down to Hudson Bay, which he reached on 1 August. He then made his way to Wolstenholme trading post and returned to James Bay on the Hudson's Bay Company ship *Nascopie*. He arrived home in southern Ontario in October 1912.

FLAHERTY, 1918, p. 442; *1918a; 1924; BÉDARD, 1919; CALDER-MARSHALL, 1963, pp. 37-53.

1911-12

United States zoological expedition — Harry V. Radford, Thomas George Street

This was a private expedition to study fauna and flora along the north coast of Canada and to collect musk-oxen and wood bison for American museums. Radford engaged Street to accompany him at Smith Landing in early 1911. The two men left Fort Smith on 27 June 1911 and travelled by way of Great Slave Lake, Thelon River, and Beverley, Aberdeen, Schultz and Baker lakes to Chesterfield Inlet. There, they tried to engage two Eskimo guides but, failing to do so, retraced their steps to Schultz Lake, where they wintered. Early in 1912, having engaged two Eskimo guides at Schultz Lake, they travelled to Bathurst Inlet, intending to follow the coast westward to the Mackenzie River. However, both men were killed by Eskimos at Bathurst Inlet in June. One of their Eskimo guides carried the news of their deaths back to Chesterfield Inlet and, in May 1913, news of the incident reached the police post at Cape Fullerton. The police there reported the matter to headquarters at Regina and, in July 1914, a police patrol led by Inspector W. J. Beyts sailed from Halifax to investigate the killings. Beyts established a base camp at Baker Lake, but failed in his attempts to reach Bathurst Inlet. It was not until 1917 that Inspector F. H. French, Beyts's replacement, finally reached the north coast and contacted the Eskimos who had killed the explorers (see 1916-18). French learned that Radford had provoked the Eskimos into the killings, and he made no arrests.

ADAM, 1950; ANONYMOUS, 1953; ANDERSON, 1972.

1911-13

Roman Catholic missionary expedition — Jean-Baptiste Rouvière OMI, Guillaume LeRoux OMI

In 1910, John Hornby (see 1908-11) had written from Great Bear Lake to the Roman Catholic mission at Fort Norman informing them of the presence of Eskimos north-east of Great Bear Lake during summer, and suggesting that they should establish a mission there. Consequently, Father Rouvière was sent from Fort Good Hope to Fort Norman to prepare for the journey. Hornby joined him there and, early in July 1911, they set out for Hornby's camp on Dease Arm, Great Bear Lake. They arrived there on 29 July. On 12 August, Rouvière and Hornby set out up Dease River in search of Eskimos. Rouvière met Eskimos three days later, spent several days among them and, encouraged by this meeting, decided to winter at a place that was accessible to the Eskimos, on Dismal Lakes. However, neither he nor Hornby was able to locate Dismal Lakes so, instead, Rouvière decided to build his house on the small lake they had found at the head of Dease River, now named Lac Rouvière. Whilst Rouvière was building the house, Hornby returned to Great Bear Lake, and there met Douglas's expedition (see 1911-12). Later, he returned to Lac Rouvière and persuaded Rouvière to abandon his new house and to winter with Douglas on Dease Arm.

1912: In March, Rouvière sledged back to Fort Norman. He set off back to Lac Rouvière in July, accompanied by another missionary, Father LeRoux. They met Eskimos in early September then, later in the year, returned to winter in Douglas's cabin (Douglas having now departed) on Dease Arm.

1913: Rouvière travelled to Fort Norman with Hornby in January, and returned in February. In the meantime, LeRoux had built a new cabin for the missionaries on Ritch Island, Dease Arm. The movements of the missionaries during summer 1913 are not clear, though it is known that they did travel to Lac Rouvière. Hornby left them there in autumn. On 8 October, Rouvière and LeRoux set off from Lac Rouvière for the mouth of Coppermine River, where they intended to establish a mission. They were never seen again by white men, and it was later established by the RNWMP (see 1915-17, LaNauze) that they were killed by two Eskimos at Bloody Falls later in the same year.

WHALLEY, 1962, pp. 55-97, 118-23, 349-54.

1912

Hudson's Bay Company supply voyage — A. C. Smith — *Nascopie*

From St John's; 30 June - 31 October. *Nascopie* called at Montreal 3-24 July, then at the Labrador posts 1-9 August; Lake Harbour

11-17 August; Wolstenholme 19-22 August; Churchill 25-31 August; Chesterfield Inlet 3-5 September; York Factory 10-18 September; Charlton Island 22 September - 2 October; Wolstenholme 6 October; Fort Chimo 10-16 October; then again at the Labrador posts 19-29 October, and returned to St John's.

HUDSON'S BAY COMPANY (1); MACK, 1938.

Gold prospecting expedition to Baffin Island	"Lucky" Scott Robert S. Janes	*Neptune*	1912

From St John's; July - 24 September. Janes persuaded Scott that he had found gold on Salmon River, near Eclipse Sound, during the cruise of *Arctic* (see 1910-11), so Scott organized this expedition to search for the gold. On their arrival in Eclipse Sound, Scott sent prospectors ashore to examine the reported gold placers, but they found nothing. They met there the shipwrecked crew of *Algerine* (see above, 1912), and sailed back to St John's with these men on board.

TREMBLAY, 1921, pp. x-xi, 421; MUNN, 1932, pp. 161, 178-83.

Canadian prospecting expedition	William Tees Curran Harold A. Calkins	1912

From Montreal; 20 May - 31 December; to investigate mineral deposits on the Nastapoka Islands, especially on Clarke Island, where Curran's previous expedition had found valuable iron ore deposits. They travelled by rail to Missinaibi, then by canoe and motor boat down Missinaibi and Moose rivers, examining mineral deposits on the way, and reached Moose Factory on 25 June. They set out over James Bay with a motor boat and a York boat on 1 July, and reached Clarke Island on 12 August. Whilst the prospectors were examining that island, Curran took a party to visit the mouth of Nastapoka River. The prospectors finished their work soon after his return and they set off homeward on 2 September. They spent seven weeks at Moose Factory, 26 October - 18 December, then continued by sledge to Cochrane, where they arrived on 29 December. They reached Montreal by rail two days later. The expedition confirmed the existence of very large deposits of iron ore on the Nastapoka Islands.

CURRAN AND CALKINS, 1917.

Gold prospecting expedition to Baffin Island	Henry Toke Munn George Bartlett John Bartlett	*Algerine*	1912

From St John's; 26 June - 24 September; to search for gold on northern Baffin Island with the aid of a map drawn some 50 years earlier by one of the crew of a Dundee whaler who reported having found gold there. The ship reached Pond Inlet on 10 July. A week later, off Cape McCulloch, just south of Pond Inlet, *Algerine* was nipped in the ice and sank whilst Munn and the prospectors were on shore searching (in vain) for the gold. There was no loss of life, and the whole party was rescued by the *Neptune* of St John's (see above, 1912), which arrived at Pond Inlet (also in search of the gold) soon after. After a short examination of the Pond Inlet region, *Neptune* sailed back to St John's with both crews on board.

MUNN, 1932, pp. 158-84.

1912

United States exploring expedition — Donald Baxter MacMillan, Jonathan Cook Small

From Boston; a private expedition to examine the Labrador coast from the 21-foot power boat *George Borup,* continuing the work that MacMillan had begun in 1911 (*qv*). They visited Hopedale, Davis Inlet, Nain and Hebron. They turned south again at Hebron, left the boat at Battle Harbour, where it was to be collected the following year by the Crocker Land Expedition (see 1913-17), and returned by steamer to the United States.

ALLEN, E.S., 1963, pp. 148-64.

1912

Fifth Carnegie Museum Expedition — Walter Edmond Clyde Todd, Alfred E. Preble, Charles H.M. Barrett

Todd set out to continue the series of avifaunal surveys in the James Bay region that he had begun in 1906 (*qv*). He began his survey at Cochrane, northern Ontario, on 19 May and, after working on Mattagami and Moose rivers, reached Moose Factory on 9 June and remained in its vicinity for two weeks. They then continued along the southern and eastern shores of James Bay, with trips to Charlton, Strutton, and other islands, and reached their farthest north at Fort George on 9 August. Continuing the survey, they returned along the shore of James Bay to Moose Factory, travelled up Abitibi River, and reached Cochrane, where they ended their survey, on 5 November.

TODD, W.E.C., 1963, pp. 17-20.

1912

Survey of Button Islands — F. Anderson

In this year, Anderson surveyed the Button Islands. He named the individual islands after his assistants, J.H. Knight, H.H. Lawson, A.M. Lacy and E.B. MacColl.

LEECHMAN, 1950, p. 152.

1912-13

Canadian gold prospecting expedition — Joseph-Elzéar Bernier, Alfred Tremblay — *Minnie Maud*

From Quebec; 29 July 1912 - 25 September 1913; a private expedition to search for gold on Salmon River, northern Baffin Island. Early in 1912, it had been reported in the press that Robert S. Janes had discovered gold near Eclipse Sound during the cruise of the *Arctic* (see 1910-11). As a result, three expeditions set out in 1912 to search for the gold, led by Munn (see above, 1912), Scott (see above, 1912), and Bernier. Bernier's expedition reached Pond Inlet on 29 August 1912, and the ship anchored at Albert Harbour, where she wintered. They traded with the Eskimos of Pond Inlet and prospected for gold, but without success. However, Tremblay made a long journey from Pond Inlet to Igloolik, during which he remapped some 4,800 km of coastline and discovered and named many new bays, islands, lakes, rivers and sounds. He set out from Albert Harbour on 25 October 1912 and travelled through Pond and Navy Board inlets and down Admiralty Inlet to Arctic Bay. He remained there from 19 November 1912 to 29 January 1913, when he set out southwards for Igloolik. He travelled to the head of Admiralty Inlet, then crossed overland to Foxe Basin by way of Gifford River and Gifford Fiord, and reached Igloolik on 20 March. He was the first white man to reach this island since Parry (see 1821-23). He made two excursions from Igloolik: 26 March - 5 April he travelled to Agu and back by way of Fury and Hecla Strait; 8-12 April, he made a trip to Hooper Inlet, Foster Bay, Hall and Sarcpa lakes, Melville Peninsula. He finally left Igloolik on 15 April and travelled by way of Murray Maxwell Bay and overland to Milne Inlet, and arrived back at the ship on 11 May. On a second journey, 15 June - 11 July, Tremblay examined the coasts of Eclipse Sound and Navy Board Inlet. During this trip, he explored Tremblay Sound. *Minnie Maud* left Albert Harbour and sailed for home on 25 August 1913. The ship arrived back at Quebec with a valuable cargo of furs and skins, ivory, and seal oil.

TREMBLAY, 1921.

Hudson's Bay Company supply voyage	J. Meikle A. Cleveland Smith	*Nascopie* *Pelican*	1913

From London and St John's (?). *Nascopie* visited Montreal 3-17 July; the Labrador posts 22 July - 10 August; Churchill 16-29 August; Charlton Island 3-19 September; Churchill again 27-29 September, and returned to St John's 8 October. *Pelican* left London on 12 June and called at the Labrador posts, Lake Harbour, at the new post at Cape Dorset 30 August - 2 September, Wolstenholme, Churchill, Chesterfield Inlet, Stupart Bay, and returned to London on 9 December.

HUDSON'S BAY COMPANY (1).

Canadian anthropological expedition	John Alden Mason	1913

From Athabasca Landing; 26 May - 23 October; sent at the instigation of the Anthropological Division of the Geological Survey of Canada to make a preliminary reconnaissance of the Indians of the Great Slave Lake region. A longer expedition that was supposed to follow this preliminary survey never took place.

Mason travelled by scow and by steamer down Athabasca and Slave rivers to Fort Resolution. He remained there 1-22 July, studying the Chipewyan, Yellowknife, Slave and Dogrib Indians, recording their songs and folklore and making ethnological notes and collections. When the Indians left the fort, he crossed Great Slave Lake to Fort Rae, where he continued his work for a further month. He arrived back at Fort Resolution on 16 September and, having missed the southbound steamer, canoed back to Athabasca Landing.

MASON, 1946.

1913

Establishment of Cape Dorset trading post — Hayward Parsons

In this year, the Hudson's Bay Company established a trading Post at Cape Dorset, south-west of Baffin Island. The post has been in operation ever since.

USHER, 1971, p. 127; FLEMING, A.L., 1957, p. 156.

1913-15

Anglican missionary expedition — Archibald Lang Fleming

From Montreal; 17 July 1913 - 19 October 1915; to continue the missionary work at Lake Harbour, Baffin Island, that he had begun in 1909 (see 1909-11). He sailed north on the Hudson's Bay Company's ships *Nascopie* and *Pelican* and reached Lake Harbour on 21 August 1913. On 21 December, he opened the first church at Lake Harbour. Early in 1914, he made a journey westward along the south coast of Baffin Island to King Charles Cape, visiting each Eskimo encampment along the way and also making a visit to the new Hudson's Bay Company post at Cape Dorset (see 1913). In 1915, he made another journey westwards along the south coast, this time to determine whether any Eskimos lived on the coast of Foxe Basin. He left Lake Harbour on 25 January and, at the beginning of March, reached Cape Dorset, where he met Eskimos from Foxe Basin who were willing to guide him there. His guides took him across Foxe Peninsula to the shores of Foxe Basin, close to Cape Weston, and there he met Eskimos from whom he obtained a map of the Foxe Basin coastline which later proved to be more accurate than the official maps of the time. After a short stay there, he returned by way of Cape Dorset to Lake Harbour.

In mid-August 1915, *Nascopie* arrived to take him home, though it brought no missionary to replace him. He sailed with her on her visit to the settlements of Hudson Bay, then returned with her to St John's, where they arrived on 19 October. It was not until 1920 that Fleming was able to return to Lake Harbour and reopen his mission.

FLEMING, A.L., 1957, pp. 139-94.

Canadian exploring expedition | Robert Joseph Flaherty, H. Bartlett | *Laddie* | 1913-16

From St John's; 17 August 1913 - (?) 1916; Flaherty's second attempt to reach the Belcher Islands, under the sponsorship of Sir William Mackenzie, for the purpose of exploring, prospecting for minerals, and studying the Eskimos (see 1911-12 for his first expedition). The ship reached Hudson Strait very late in the season and was unable to pass through into Hudson Bay. Flaherty therefore built his winter quarters at Amadjuak Bay, on the south coast of Baffin Island, and sent *Laddie* back to Newfoundland with orders to return the following year. During winter, he carried out some exploration in southern Baffin Island. He made an attempt to explore the coast of Foxe Basin, but was forced to turn back near Cape Dorset. He then explored northward from Amadjuak Bay to Amadjuak Lake, and there made an ethnological collection and a moving picture film of Eskimo life. *Laddie* arrived back on 19 August 1914 and they sailed for Hudson Bay four days later. On 9 September, they reached the eastern coastline of the Belcher Islands. They spent several days on the east coast, making short trips inland and studying the iron ore series Flaherty had found in their harbour. However, as the ship had been badly damaged on the voyage through Hudson Bay and, as their provisions were low, they were unable to remain long on the Belcher Islands, and they set off on 13 September to winter at Moose Factory. Whilst *Laddie* and her crew remained there, Flaherty travelled by canoe to Toronto to report to Sir William Mackenzie.

1915: Flaherty returned to Moose Factory in summer and sailed back to the Belcher Islands, where he arrived in September to spend a year, exploring, prospecting, mapping the islands, and filming the Eskimos. In February 1916, he sent two of his crew back to the mainland to report to Sir William Mackenzie and to ask for geologists to examine his ore finds. In response to his request, E.S. Moore, a geologist, W.H. Howard of the Dominion Lands Survey, and Robert H. Flaherty, his father, arrived in August. In September 1916, the whole party left for Moose Factory on the Hudson's Bay Company's boat *Nastapoka* (*Laddie* having been abandoned on the Belcher Islands); they then returned home by canoe.

*FLAHERTY, 1918; 1924; 1923; LE VASSEUR, 1915; CALDER-MARSHALL, 1963, pp. 54-75.

Norwegian anthropological expedition | Christian Leden | 1913-16

From Montreal; 17 July 1913 - 10 October 1916; sponsored by the University Museum in Kristiania and by King Haakon and Queen Maud of Norway to study the Eskimos west and north-west of Hudson Bay. He sailed to Hudson Bay on an unnamed ship (probably the Hudson's Bay Company supply vessel *Nascopie*, *qv*), and reached Churchill on 16 August. On 30 September, he set out northwards along the coast with a party of Eskimos. For the next three years, he remained in the company of the Eskimos of the west coast of Hudson Bay, travelling inland with them in the region of Chesterfield Inlet, Rankin Inlet, and Ferguson River. He also made a journey northward from Chesterfield Inlet to Roes Welcome Sound, and he made a brief excursion into north-western Southampton Island. On 16 September 1916, *Nascopie* carried him from Chesterfield Inlet south to Port Nelson. From there, he travelled by canoe up Nelson River and by rail to Winnipeg, where he ended his journey on 10 October.

LEDEN, 1927.

1913-17

United States "Crocker Land" Expedition	Donald Baxter MacMillan Walter Elmer Ekblaw Maurice Cole Tanquary FitzHugh Green Jerome Lee Allen Jonathan Cook Small

From New York; 2 July 1913 - 24 August 1917; sponsored by the American Museum of Natural History, the American Geographical Society, and the University of Illinois, to search for "Crocker Land," which Peary believed he had sighted north of Axel Heiberg Island in 1906 (*qv*). They also intended to carry out an extensive range of scientific observations in the Sverdrup Islands. They left New York on the sealer *Diana*, Captain Waite, but this ship ran aground in the Strait of Belle Isle and they were obliged to transfer to *Erik*, Captain Kehoe, at St John's. This ship was unable to penetrate through Smith Sound so, on 26 August, MacMillan abandoned his intention to winter at Flagler Bay, Ellesmere Island, and unloaded his stores at Etah, north-west Greenland, where he built his winter quarters. *Erik* returned at once to St John's.

1914: After several months of sledging advance depots over to Ellesmere Island, MacMillan and Green started out for "Crocker Land" on 10 March. They crossed Ellesmere Island from the head of Beitstad Fiord to Bay Fiord, and travelled northward through Eureka and Nansen sounds to Cape Thomas Hubbard, northern Axel Heiberg Island. From there, they set out north-westward over the pack ice. They reached their farthest north-west at 82°30'N, 108°22'W, 240 km from Cape Hubbard and well into the region where "Crocker Land" was supposed to exist. Realizing that there was no hope of finding new land there, they turned back and retraced their steps to Etah, where they arrived on 19 May. For the rest of the year, the party carried out its scientific programme in north-west Greenland. Between 24 December 1914 and 7 February 1915, MacMillan and Tanquary sledged to Upernavik to send out instructions for a relief ship to be sent in 1915.

1915: On 24 March, Ekblaw set out from Etah to make a thorough exploration of Greely Fiord, north-west Ellesmere Island, and its tributary fiords. He crossed Ellesmere Island from Flagler Bay to Bay Fiord, then travelled northwards along Eureka Sound to Greely Fiord. Whilst examining the north shore of Greely Fiord, he discovered, explored, and named Borup and Tanquary fiords. He also sighted and named Osborn Mountains. He then crossed Ellesmere Island by way of Lake Hazen to Lady Franklin Bay, and returned through Kennedy Channel and Kane Basin to Etah, arriving on 16 June. In mid-September, the relief ship *George B. Cluett* arrived at North Star Bay, north-west Greenland, to take home Ekblaw, Tanquary, Green and Allen. However, she was beset at Parker Snow Bay, near Cape York, and was obliged to winter there.

1916: On 22 March, MacMillan set out to explore the Findlay Group and King Christian Island (sighted by Sherard Osborn in 1853, *qv*, and by Isachsen and Hassel in 1901, *qv*), south-west of Ellef Ringnes Island. He crossed Ellesmere Island by the Flagler Bay - Bay Fiord route, travelled west along the south shores of Axel Heiberg and Ellef Ringnes islands, and landed on King Christian Island on 17 April. However, his plans to survey the island were thwarted by bad weather and shortage of food so, on 22 April, he set off back to Etah. Returning, he visited Cornwall Island, landing on its north shore at McLeod Head (which he named). He arrived back at Etah on 16 May. *George B. Cluett* sailed home on 29 July, but MacMillan, Hunt and Ekblaw remained at Etah. With them remained two men brought out on the *Cluett*: E.O. Hovey and George Comer. Another relief ship expected in 1916 failed to reach them, so the party wintered at Etah.

1917: Hovey returned home in March on the second relief ship, *Danmark*. From 3-25 May, MacMillan surveyed the coast of Ellesmere Island between Cape Sabine and Clarence Head, making many corrections to existing charts and naming several features along the coast. The relief ship *Neptune*, Captain Robert Bartlett, reached Etah on 31 July and carried the whole party home. They reached Sydney, NS, on 24 August.

MACMILLAN, *†1918; †1918a, 1918b; †1928; †EKBLAW, 1919; 1927-28; GREEN, 1928.

Canadian Arctic expedition — Vilhjalmur Stefansson, Rudolph Martin Anderson — 1913-18

This expedition was sponsored by the Canadian government to carry out a wide variety of geographical exploration and scientific work in the western Canadian Arctic. Stefansson divided the expedition into two parties. One (the Northern Division), under his own command, was to have geographical discovery as its main objective. The other party (the Southern Division), led by Anderson, was to carry out the scientific work in the region of Coronation Gulf. The expedition started with three ships (though several others were used from time to time during the expedition): *Karluk*, Captain Robert Bartlett, was to be used by the Northern Division; *Alaska* was to be used by the Southern Division; and *Mary Sachs*, Captain Peter Bernard, was to be a tender to both parties and was incidentally to carry out oceanographic work.

Northern Division

Karluk, with Stefansson on board, sailed from Victoria on 20 July 1913. She became icebound in the Beaufort Sea and was unable to reach Banks Island, where the party was to have wintered. In September 1913, Stefansson took a hunting party ashore near Jones Islands, Alaska. During a gale which followed soon after, *Karluk* was carried westwards in the ice and drifted for three months, until January 1914, when she sank. Eight members of the party were lost in the attempt to reach land; the remainder reached Wrangel Island, where they were rescued, and were taken back to Victoria. Meanwhile, Stefansson's hunting party joined *Alaska* and *Mary Sachs,* which were wintering at Collinson Point, Alaska.

1914: On 22 March, Stefansson, Ole Andreasen, and Storker Storkerson set out from Martin Point, Alaska, on a journey over the ice of the Beaufort Sea. They landed on Norway Island, off the west coast of Banks Island, on 25 June. They spent the rest of the summer surveying the coasts in the vicinity of Norway Island, and discovered Bernard Island and Bernard River. They then travelled to Cape Kellett, Banks Island, where they met *Mary Sachs* carrying supplies for them. They wintered at Cape Kellett.

1915: In mid-February, Stefansson and three others set out over the sea ice to explore the northern Beaufort Sea. They travelled north-west over the ice until 6 May, then turned eastwards towards Lands End, Prince Patrick Island, where they landed on 4 June. They then travelled north-east along the west coast of Prince Patrick Island, completing the survey that Mecham had begun in 1853 (*qv*). From there, they discovered Brock Island, but Stefansson decided against exploring it and they turned back on 22 June. Returning south, they touched at Mackenzie King Island, passed between Prince Patrick and Melville islands, crossed M'Clure Strait, then travelled across Banks Island from near Mercy Bay to Cape Kellett, where they arrived on 9 August. Soon after, Stefansson purchased the trading ship *Polar Bear* and sailed on her to the mainland. At Baillie Islands, he purchased another ship, *Gladiator,* for use between Cape Kellett and a more northerly base that he hoped to establish. He then returned northwards on *Polar Bear,* hoping to reach Prince Patrick Island. However, the ship was unable to penetrate through Prince of Wales Strait and wintered at Princess Royal Island. In the meantime, *North Star*, which Stefansson had sent for the previous year, attempted to pass northwards along the west coast of Banks Island, but reached only Robillard Island, where she wintered.

1916: Stefansson set out from Cape Kellett on 26 January to continue the exploration of the new land to the north. He reached Mercy Bay on 5 April, crossed M'Clure Strait to Dundas Peninsula, Melville Island, then travelled through Liddon Gulf and Hecla and Griper Bay to Cape Grassy, northern Melville Island. From there, he travelled to Brock Island. He crossed Wilkins Strait to explore Borden Island, then set out for Cape Isachsen, Ellef Ringnes Island, which he reached on 29 May. From there, he travelled north-east to discover and survey Meighen Island. Returning south, he surveyed the west coast of Amund Ringnes Island

and corrected the existing sketchy surveys of King Christian Island and the Findlay Group, discovering that the Findlay Group consisted of three main islands. He then travelled back by way of Brock Island to Cape Grassy, near which he wintered.

1917: Stefansson left Melville Island on 30 March and set out northwards over the Beaufort Sea from Borden Island. Part of his party reached 80°30'N before turning back. They landed again at Cape Isachsen, Ellef Ringnes Island, then travelled by way of Lougheed Island, Byam Martin Channel, the south coast of Melville Island, across M'Clure Strait and across Banks Island to Cape Kellett, where they arrived on 17 August. Ten days later, *Challenge* arrived at Cape Kellett on a trading voyage. Stefansson purchased this ship and sailed with her to Cape Bathurst. The party then sailed west to Barter Island, Alaska, where they wintered. There, Stefansson made plans for the expedition's last journey, an ice-drift over the Beaufort Sea. However, he fell ill and was taken to Fort Yukon, where he spent most of the winter before returning south. Storkerson took over command of the ice-drift party, which camped on the ice on 15 March 1918 and drifted for nine months over a large area of the Beaufort Sea, taking depth soundings and current measurements. They landed again on 8 November. Storkerson then spent the winter at Herschel Island before returning south by way of Mackenzie River in spring 1919.

Southern Division

Alaska left Victoria on 19 July 1913 and wintered at Collinson Point, Alaska, in company with *Mary Sachs.*

1914: *Alaska* left Collinson Point on 25 July and continued eastwards along the coast to discover an uncharted harbour, Bernard Harbour, where the Southern Division established its headquarters. In September, *Alaska*, with Anderson on board, sailed to Herschel Island for supplies. The ship was frozen in at Baillie Islands during the return voyage and Anderson continued on foot back to Bernard Harbour. During winter, J. R. Cox surveyed the coast eastwards from Bernard Harbour to Lockyer Point, and D. Jenness began studies of the Eskimos of the region.

1915: Jenness set out for Victoria Island on 18 April, and he studied the Eskimos there until November, when he returned to Bernard Harbour. On 17 March, K. G. Chipman and J. J. O'Neill began a survey and geological examination of the coast westwards from Bernard Harbour to Darnley Bay. From 16 April to 25 May, Cox and J. Sullivan surveyed the coast of Coronation Gulf between Lockyer Point and the mouth of Rae River, and surveyed up Rae River for about 110 km. During summer, Cox led a survey party eastwards along the coast to Bathurst Inlet. They made a detailed geographical and geological survey of the coast and explored part of Tree River.

1916: During March-May, Cox again travelled eastwards towards Bathurst Inlet to finish his survey of that region. On 13 July, the Southern Division having finished its work, *Alaska* left Bernard Harbour with all members of the party on board. She reached Nome on 15 August. The scientific results of the division's work include extensive studies and collections in the fields of geology and mineralogy, ethnology and archaeology, terrestrial and marine biology, and botany.

STEFANSSON, 1918; 1919a; 1921; *1921a; ANDERSON, R.M., 1915-17; †1917; 1917a; †1919; BARTLETT AND HALE, 1916; †CANADIAN ARCTIC EXPEDITION, 1919-46; †JENNESS, 1917; 1928; McDIARMID, 1923; MONTGOMERY, 1932; NOICE, 1924; †O'NEILL, 1916-17; McKINLAY, 1976.

1914

Hudson's Bay Company supply voyage	A. Cleveland Smith	*Nascopie*

From St John's; 22 June - 16 October. *Nascopie* called at Montreal 25 June - 6 July; at the Labrador posts 13-30 July; Wolstenholme 13-16 August; Churchill 20-25 August; York Factory 26 August - 11 September; Charlton Island 17-26 September, and at Fort Chimo 7-11 October. She then returned to St John's.

HUDSON'S BAY COMPANY (1).

1914

United States expedition to Hudson Bay	W. J. Peters D. W. Berky ?	*George B. Cluett*

From Battle Harbour, Labrador; 31 July - 3 October. The Carnegie Institution of Washington chartered *George B. Cluett* with its captain and crew from the International Grenfell Association to make a magnetic survey of the Labrador coast and of Hudson Bay, with Peters in charge and Berky as observer. E. W. Hawkes (*qv*) also took part in the latter part of the expedition. Because of heavy ice conditions, the ship could not leave Battle Harbour until 31 July. Stations were made there and at 11 other locations along the coast, including Domino, Green Island, Hopedale and Port Burwell. Continuing through Hudson Strait, they made stations at Ashe Inlet and Erik Cove; at Smith Island, off the north-west coast of Quebec; at Coats Island in northern Hudson Bay; and at Eskimo Point and Mistake Bay, Keewatin District.

BAUER, L. A. AND OTHERS, pp. 307-08; CARNEGIE INSTITUTION OF WASHINGTON YEAR BOOK, Vol. 13, 1914, pp. 299-300, 317.

1914

Geological Survey of Canada anthropological expedition	Ernest W. Hawkes	

Hawkes, who had previously lived among Alaskan Eskimos, was commissioned to make a comprehensive survey of the Eskimo culture of the whole Quebec-Labrador peninsula. He spent the early part of the summer in the Sandwich Bay - Hamilton Inlet section of the coast of Labrador to ascertain the southern limit of the Eskimo population, then spent the rest of the summer with the Carnegie Museum Magnetic Expedition (*qv*), which continued up the coast of Labrador to Cape Chidley, visited both sides of Hudson Strait, the east coast of Hudson Bay as far south as Cape Dufferin, the west coast of Hudson Bay between Churchill and Chesterfield Inlet and some of the islands in the bay. He made important archaeological and ethnological collections.

HAWKES, 1916.

1914

Establishment of
Frobisher Bay
trading post

The Hudson's Bay Company established its first trading post on Frobisher Bay, Baffin Island, in 1914, as an outpost of Lake Harbour (see 1909-11, Lake Harbour mission). The post was situated on Charles Francis Hall Bay, on the southern shore of Frobisher Bay. The post was moved to Hamlen Bay, on the north shore of Frobisher Bay, in 1920. The present Hudson's Bay Company post, at the head of Frobisher Bay, was built in 1948.

USHER, 1971, p. 127.

1914

Establishment of
Stupart Bay post

In this year, the Hudson's Bay Company established a trading post at Stupart Bay on the south side of Hudson Strait.

ANDERSON, J. W., 1939.

1914

Canadian fisheries research expedition (Hudson Bay party) — Napoleon Alexander Comeau — *Burleigh*

From Halifax; 8 July - 19 October; sent by the Department of the Naval Service to investigate the potential value of commercial fisheries in Hudson Bay. Two other parties, led by A.R.M. Lower and C.D. Melvill, set out in the same year to conduct similar investigations in James Bay (see below, 1914). Comeau sailed directly across Hudson Bay to Port Nelson, where he began his work by investigating the fisheries of Nelson River. He continued the study along the coast and on rivers northward as far as Owl River and eastward as far as Cape Tatnam. *Burleigh* set off home on 8 September and reached Halifax on 19 October. Comeau himself continued his investigations until 24 September and returned home on board SS *Bonaventure,* reaching Sydney on 6 October. He reported on the numbers of fishes seen during his study, on the value of fisheries in south-western Hudson Bay, and listed the birds sighted during the expedition.

COMEAU, 1915.

1914

Canadian fisheries research expedition (James Bay parties) — James Cosmo Dobrée Melvill, Arthur Reginald Marsden Lower

From Ottawa; 4 June - mid-October. Melvill and Lower were sent by the Department of the Naval Service to investigate the potential value of commercial fisheries in James Bay. They travelled together by rail to Cochrane, where they split into two parties. Lower's party canoed down Albany River and investigated the western shoreline of James Bay northwards to Lakitusaki River. They also visited Akimiski Island. Melvill's party canoed down Moose River to Moose Factory, and investigated the eastern shoreline of James Bay northwards to Cape Jones. Returning south, they visited North Twin Island. On their return to Moose Factory, they were joined by Lower's party, who accompanied them back up Moose and Mattagami rivers to Cochrane. The reports of these two men include annotated lists of fishes suitable for food, geological and geographical notes, and remarks on tides.

MELVILL, 1915; LOWER, 1915; 1944.

1914

Geological Survey of Canada expedition — Charles Camsell, Francis Harper, A.J.C. Nettell

From Ottawa; 5 May - 18 October; to make a geographical, geological, topographic and natural history survey of the unexplored region between Lake Athabasca and Great Slave Lake and east of Slave River. The only white man who might have visited this region previously was Samuel Hearne, who could have been there briefly during his return from Coppermine River (see 1770-72). Camsell's party travelled by rail to Edmonton, then by boat down Athabasca River and north-eastward across Lake Athabasca to Camsell Portage, about mid-way along the lake's northern shore, where their survey began at the end of June. They crossed the portage to Tazin Lake, then descended Tazin River to join Taltson River, which they followed down to Great Slave Lake, where they arrived on 15 August. They then returned to Ottawa by way of Fort Resolution, Slave and Athabasca rivers and Edmonton.

Camsell wrote a detailed report on the geology and physical geography of the region, with notes on climate, agricultural possibilities, fauna, and Indians. Harper wrote reports on the fauna and flora.

*†CAMSELL, 1916; 1954, pp. 219-30; †HARPER, F., 1915; 1931; 1931a; 1932.

1914-15

Sixth and seventh Carnegie Museum expeditions — Walter Edmond Clyde Todd, Olaus J. Murie

This expedition continued Todd's avifaunal surveys of the James Bay region that had been taking place since 1906 (*qv*). They began their survey at Cochrane, northern Ontario, on 23 May and, after working in north-eastern Ontario and in the region of Nottaway and Rupert rivers, Quebec, reached James Bay at Rupert House on 27 June. They then surveyed northwards along the east coasts of James Bay and Hudson Bay to Great Whale River, where they turned back. Returning, they visited Charlton Island 30 August - 20 September, and worked in the region of Moose Factory 23 September - 6 October. Todd then returned to Cochrane, whilst Murie wintered on James Bay. He spent the winter at Moose Factory, Rupert House, Eastmain, and Great Whale River. In April, he travelled north to Richmond Gulf, and during May - July he examined the coast and off-lying islands as far north as Port Harrison. On 14 August, he sailed on the Hudson's Bay Company's steamer *Inenew* to the Belcher Islands, where he remained for two weeks. He made further surveys in James Bay until mid-October, then returned up Abitibi River to Cochrane.

TODD, W.E.C., 1963, pp. 20-26.

1914-15

British trading expedition | Henry Toke Munn, Captain Milne, Captain Adams | *Albert*

From Peterhead; 7 July 1914 - 18 October 1915; a private venture to establish trading posts on Baffin Island. They reached Pond Inlet on 2 September, and there met Captain J.-E. Bernier who had also just arrived as a free trader. On 24 September, after Munn had built his trading post at the entrance to Pond Inlet, the ship departed for home, leaving Munn and one seaman to winter there alone. On 30 November, Munn left the station on a journey of about 150 km southwards along the Baffin Island coast to trade with Eskimos. On 15 March 1915, he travelled from his station to Admiralty Inlet to trade at another Eskimo encampment. In July, he again travelled southward along the Baffin Island coast, this time to make a thorough search for the gold that he had looked for on his previous voyage (see 1912). He concluded that no gold existed. On 13 August, *Albert* arrived back under the command of Captain John Murray, and Munn sailed in her back to Peterhead.

MUNN, 1932, pp. 184-206.

1915

Hudson's Bay Company supply voyage | G. Edmund Mack | *Nascopie*

From Montreal; 2 August - 18 October. *Nascopie* called at Port Burwell 9 August; Lake Harbour 11 - 14 August; Wakeham Bay 15 - 16 August; Cape Dorset 19 - 21 August; Wolstenholme 21 - 23 August; Churchill 25 - 30 August; Chesterfield Inlet 31 August - 2 September; York Factory 5 - 19 September; Charlton Island 25 - 30 September; Wolstenholme 4 October; Lake Harbour 6 - 7 October; Fort Chimo 10 - 13 October, and reached St John's on 18 October.

HUDSON'S BAY COMPANY (1); WATT, 1930.

1915-16

Canadian expedition to northern Labrador | Arthur P. Coleman

Coleman, professor of geology in the University of Toronto, spent 24 July - 31 August 1915 with two Eskimo assistants and 22 July - 5 September 1916 with two Newfoundland fishermen as assistants surveying the coast of Labrador from Cape Mugford north to Komaktorvik Fiord. He was especially interested in the geology and evidence of glaciation in the region, and he made important corrections to the Admiralty charts of the time.

COLEMAN, 1921.

1915-17

Royal North-West Mounted Police patrol — Charles D. LaNauze

From Edmonton; to investigate the fate of two Belgian Oblate missionaries, Fathers Rouvière and LeRoux, who had left their mission at Lac Rouvière, north-east of Great Bear Lake, in autumn 1913 to visit Eskimos at the mouth of Coppermine River, and who had not been seen since. LaNauze's party travelled by steamer down Slave and Mackenzie rivers to Fort Norman. They set out from Fort Norman by York boat on 23 July and travelled up Bear River and across Great Bear Lake to Dease River, where they made their base camp. At the end of September, LaNauze visited the priests' cabin at Lac Rouvière, then returned to winter at his base camp.

1916: The party set out for the mouth of Coppermine River on 29 March, and arrived there on 30 April. Soon after, he met Eskimos, members of the Canadian Arctic Expedition (see 1913-18), and RNWMP Corporal Bruce, who had been sent from Herschel Island to join LaNauze's party. LaNauze then began to interview Eskimos in the region of Dolphin and Union Strait, and soon pieced together the story of how Rouvière and LeRoux had been killed by two Eskimos in the autumn of 1913. He arrested the two killers, one on Victoria Island and the other on Coppermine River, and took them to Bernard Harbour. From there, the Canadian Arctic Expedition's ship *Alaska* carried them to Herschel Island, where they wintered.

1917: LaNauze escorted his prisoners by way of Mackenzie River to Calgary, where they arrived in August, and where the Eskimos were tried and convicted of murder. They were sentenced to life imprisonment to be served at Fort Resolution, but were released in 1919, after it had become clear that they had committed the killings under provocation.

LaNAUZE, 1918; MOWAT, 1973, pp. 276-94, 310; STEELE, 1936, pp. 178-79, 181, 187-89, 190-92.

1916

Hudson's Bay Company supply voyage — G. Edmund Mack — *Nascopie*

From Cardiff; 25 July - 13 November. *Nascopie* visited St John's 31 July - 2 August; Montreal 5 - 11 August; Port Burwell 18 August; Lake Harbour 19 - 22 August; Wakeham Bay 23 - 26 August; Cape Dorset 27 - 30 August; Wolstenholme 31 August - 1 September; Churchill 4 - 8 September; Chesterfield Inlet 9 - 16 September; York Factory 18 September - 4 October; Charlton Island 10 - 15 October; Wolstenholme 21 October; Lake Harbour 22 - 24 October; Wakeham Bay 24 - 27 October; Fort Chimo 31 October - 6 November, and reached St John's on 13 November.

HUDSON'S BAY COMPANY (1).

1916

Establishment of Bernard Harbour trading post — W.G. Phillips

On 28 July 1916, the Hudson's Bay Company's schooner *Fort Mac-*

pherson left Herschel Island to establish a trading post at Bernard Harbour, Dolphin and Union Strait. The post remained in operation until 1932.

STEFANSSON, 1921a, p. 757; USHER, 1971, p. 110.

Royal North West Mounted Police patrol	Francis Henry French Thomas Benjamin Caulkin	1916-18

French arrived at Baker Lake in September 1916 to investigate the killing by Eskimos of the explorers Radford and Street at Bathurst Inlet in 1912 (see 1911-12). He replaced Inspector W. J. Beyts, who had been trying unsuccessfully to reach Bathurst Inlet since 1914. After wintering at Baker Lake, French and Caulkin set out for Bathurst Inlet on 21 March 1917. They travelled by dog sledge by way of Schultz and Aberdeen lakes on Thelon River and Garry and Pelly lakes on Back River. They then struck out north-westward to Ellice River, which they followed to Queen Maud Gulf, arriving there on 7 May. From there, they followed the coast westward to Bathurst Inlet, where they met a band of Eskimos. They took statements from some of those Eskimos, then continued westward along the coast, meeting many Eskimos and taking many statements from those who were present at the killing of the explorers. These statements all led French to the conclusion that the Eskimos had committed the killings under severe provocation from Radford, and he therefore made no arrests. On 13 June, they reached the Hudson's Bay Company post at Bernard Harbour on Dolphin and Union Strait, where they rested for ten weeks. They set off homeward on 1 September, travelled back along the coast to Bathurst Inlet, continued overland south-eastward to join Thelon River at Beverley Lake, then followed Thelon River to Baker Lake, where they arrived back at their detachment on 29 January 1918, having completed a patrol of more than 8,000 km.

ADAM, 1950; ANDERSON, I.S., 1972; *FRENCH, 1919.

British trading expedition	Henry Toke Munn John Murray William Duval	*Albert*	1916-18

From Peterhead; 22 June 1916 - 2 October 1918; to continue his trading activities on Baffin Island and to set up new trading posts in north-west Hudson Bay. Munn reached his trading post at the entrance to Pond Inlet (see 1914-15) on 12 August. He unloaded his cargo and left a man to winter there, then sailed southward to open up his trade in north-west Hudson Bay. At Durban Island, on the coast of Baffin Island, he picked up Duval, who had been trading there, and a party of Eskimos to help him to run his new posts in Hudson Bay. The ship reached South Bay, Southampton Island, at the end of September; Munn set up his trading post there and sent the ship home.

1917: During the summer, Munn made several journeys along the coast of South Bay and inland to the north of South Bay, extending the survey of Southampton Island that Captain Comer had begun (see 1907-09). *Albert* failed to return to Southampton Island this year, as arranged, and Munn was forced to winter there again.

1918: In February, he crossed Southampton Island from South Bay to Roes Welcome (he is the first white man known to have crossed the island) in the hope of crossing Roes Welcome and continuing overland to southern Canada to get a ship to relieve his post. On this journey, he explored Kirchoffer River (which he named) up to near its headwaters, then turned westward to reach Roes Welcome at about 64°50'N. He found that ice conditions in the Welcome made a crossing on foot to the mainland impossible, so he returned to South Bay. On 10 August, *Albert* finally reached South Bay from Scotland. Returning home on board her, Munn first called at Cumberland Sound to land the Baffin Island Eskimos who had accompanied him, and to set up a new trading post under Duval's leadership. He then attempted to reach his post at Pond Inlet. However, the ship was unable to find a passage northward through the ice, and the man in charge at Pond Inlet was left unrelieved for a further year. *Albert* finally reached Halifax, NS, on 2 October 1918.

MUNN, 1919; 1932, pp. 207-20.

1916-20

Canadian fur trading expedition — Robert S. Janes

Janes sailed north on board the Newfoundland whaler *Kite* in 1916 with the intention of establishing a trading post on Pond Inlet. He reached Pond Inlet in mid-August, built a trading post on Eclipse Sound, then sent *Kite* back to Newfoundland and wintered alone. Whilst Janes was in the Arctic, his backer went insane and sold *Kite,* leaving Janes stranded. So, when Munn arrived at his own trading post on Pond Inlet in 1919 (*qv*), he offered to take Janes home. Janes turned down the offer, saying that he intended to return home overland during the winter of 1919-20, by way of Foxe Basin and the west coast of Hudson Bay. He set off on this journey early in 1920, but was murdered by Eskimos at Cape Crauford, at the mouth of Admiralty Inlet, on 15 March 1920.

In 1921, Sergeant Alfred H. Joy of the Mounted Police sailed to Pond Inlet to investigate the murder, and he arrested three Eskimos. In 1923, a court was sent to Pond Inlet to try the three men. The murderer was sentenced to 10 years imprisonment and his two companions to three years each.

MUNN, 1932, pp. 209; 230-31; 243-44; LEE, H.P., 1928, pp. 15-25; STEVENSON, 1973.

1917

Hudson's Bay Company supply voyage — G. Edmund Mack — *Nascopie*

From St John's; 3 July - 17 October. *Nascopie* visited Montreal 8-26 July; Port Burwell 7-8 August; Wakeham Bay 10 August; Lake Harbour 11-13 August; Wolstenholme 14 August; Charlton Island 18-30 August; York Factory 3-6 September; Churchill 7-9 September; Chesterfield Inlet 12-15 September; Wolstenholme 18-19 September; Cape Dorset 20-21 September; Lake Harbour 25-27 September; Wakeham Bay 27-29 September; Fort Chimo 2-6 October; Port Burwell 7-8 October; Cartwright 11-15 October, and reached St John's on 17 October.

HUDSON'S BAY COMPANY (1).

1917

Eighth Carnegie Museum expedition — Walter Edmond Clyde Todd, Olaus J. Murie, Alfred Marshall

From Quebec; 5 May - late October; to make an avifaunal survey of the interior of the Quebec-Labrador peninsula. They sailed by steamer to Sept-Iles, where they decided to penetrate into the interior by way of Rivière Ste-Marguerite. They set off up this river on 26 May, canoed to near its headwaters, then portaged to Rivière aux Pékans, a tributary of Moisie River, which they reached on 12 July. They crossed the height of land on 22 July and, soon after, reached Shibogamo Lake. From there, they followed Ashuanipi River north to Menihek Lake, from where they portaged by way of Petitsikapau Lake to the head of Swampy Bay River. They then followed Swampy Bay and Kaniapiskau rivers down to Fort Chimo, which they reached on 22 August. They continued making collections in the region of Fort Chimo until the beginning of October. The Hudson's Bay Company's steamer *Nascopie* arrived on 2 October and carried them to Montreal.

TODD, 1963, pp. 26-31.

1918

Creation of districts in the Northwest Territories

Keewatin District had been created in 1876 (*qv*), and by Order in Council on 2 October 1895, the provisional districts of Ungava, Franklin, Mackenzie, and Yukon were created. The boundaries of these districts were redefined in 1897. Subsequently Yukon Territory was formed (see 1898) and other portions annexed to the provinces. By Order in Council of 16 March 1918 (effective 1 January 1920) the remaining portions of the Northwest Territories were divided into three provisional districts: Mackenzie, Keewatin, and Franklin.

CANADA YEAR BOOK, 1973, p. 90.

1918

Hudson's Bay Company supply voyage — Thomas F. Smellie, G. H. Mead — *Nascopie*, *Discovery*

From Montreal; 20 July - 15 October. *Nascopie* called at St John's 24-27 July; Port Burwell 1-3 August; Lake Harbour 5-8 August; Wolstenholme 11-14 August; Churchill 17-21 August; Chesterfield Inlet 23-30 August; Churchill 31 August - 1 September; Charlton Island 5-13 September; Wolstenholme 17-18 September; Cape Dorset 18-20 September; Lake Harbour 21-22 September; Wakeham Bay 23-24 September; Fort Chimo 26-30 September; Port Burwell 1-2 October; Cartwright 4-6 October, and reached Montreal on 15 October. *Discovery* left Montreal on 13 July, and was beset in the ice at the eastern end of Hudson Strait on 24 July. She drifted in the ice through Hudson Strait and met with considerable difficulty in penetrating southwards through Hudson Bay, but finally entered open water near the Belcher Islands. She then, presumably, visited the James Bay posts and returned to Montreal.

HUDSON'S BAY COMPANY (1); GEOGRAPHICAL JOURNAL, 1920.

1919

Hudson's Bay Company supply voyage — G. H. Mead — *Nascopie*

From Hull; 26 June - 31 October. *Nascopie* called at St John's 7-9 July; Montreal 13-19 July; Port Burwell 29-31 July; Lake Harbour 1-5 August; Wolstenholme 7-8 August; Charlton Island 11-25 August; Churchill 29 August - 1 September; Chesterfield Inlet 4-6 September; Wolstenholme 9-11 September; Lake Harbour 12-14 September; Fort Chimo 15-18 September; Port Burwell 19-22 September; St John's 26 September - 1 October, and reached London on 31 October.

HUDSON'S BAY COMPANY (1).

1919

British trading expedition — Henry Toke Munn, John Murray — *Albert*

From Halifax, NS; 28 June - 15 November; to continue his trading activities on Baffin Island (see 1914-15, 1916-18). Munn had also bought out Captain J.-E. Bernier's trading post on Pond Inlet. The ship arrived at Munn's post in Cumberland Sound on 17 August. They landed their cargo there then, a week later, set off for Pond Inlet. Munn reached Pond Inlet in mid-September, relieved his own station and Bernier's, and set off back for Peterhead on 20 September.

MUNN, 1932, pp. 225-39; 1920.

1920

Hudson's Bay Company supply voyage — G. H. Mead, R. Bergner — *Nascopie*, *Pelican*

From Hull and London. *Nascopie* left Hull on 19 June, called at St John's 28-29 June; Montreal 3-23 July; Port Burwell 3-7 August; Lake Harbour 8-12 August; Wolstenholme 14-16 August; Charlton Island 21 August - 1 September; Churchill 7-14 September; Chesterfield Inlet 16-19 September; Wolstenholme 24-26 September; Port Harrison 28 September - 4 October; Lake Harbour 7-9 October; Port Burwell 11-13 October; St John's 18-21 October, and arrived at Glasgow on 24 November. *Pelican* left London on 8 April, called at Montreal 24 June - 9 July; Port Burwell 26 July - 1 August; Lake Harbour 6-23 August; Stupart Bay 24-25 August; Wolstenholme 28 August - 1 September; Lake Harbour 10-15 September; Port Burwell 17-21 September, and reached St John's on 28 September. This was *Pelican's* last voyage to Hudson Bay.

HUDSON'S BAY COMPANY (1); *BEAVER*, 1929.

Geological expedition to the Belcher Islands — Dwight E. Woodbridge, George H. Rupp — **1920**

During the summer of 1920, Woodbridge and Rupp travelled by canoe and motor boat to the Belcher Islands, where they carried out an examination of the iron formations. Woodbridge's report gives no further details of their journey, but concentrates on presenting an encouraging account of the iron formations and discusses the possibility of commercial mining.

WOODBRIDGE, 1921.

British archaeological expedition to Labrador — G.M. Gathorne-Hardy, Robert Gathorne-Hardy — **1920**

Geoffrey Gathorne-Hardy had visited Labrador with Hesketh Prichard (see 1910); in 1920, he returned with his cousin Robert to Nain to examine some nearby house ruins that he thought might be Norse. He considered that the ruins were not typically Eskimo, but thought they might be the work of Eskimos, alien to Labrador, who had lived among European dwellings in Greenland.

GATHORNE-HARDY, G.M., 1922; GATHORNE-HARDY, R., 1963, pp. 23-144; 1972, p. 447.

Establishment of Port Harrison trading post — L.A. Learmonth, H.J. Pardy, Johnny Hayward — **1920**

Revillon Frères had established a trading post at Port Harrison before 1920 and were cutting off much of the trade that had previously gone to the Hudson's Bay Company's posts at Great Whale River and Wolstenholme. The Hudson's Bay Company therefore felt it necessary to establish their own post at Port Harrison and an outpost at Povungnituk. They sent ss *Pelican*, Captain Bergner, from St John's with the outfit, building materials and staff for the post. *Pelican* called first at Lake Harbour, Baffin Island, to pick up Learmonth and Pardy, who had spent the previous winter there. She was damaged by ice at Lake Harbour so, having landed the staff and materials at Wolstenholme, returned to St John's. MV *Nascopie* picked up the materials left at Wolstenholme and carried them to Port Harrison. There, Learmonth built his post about 1 km east of the Revillon Frères post, on the north bank of Inuksuak River. The outpost at Povungnituk was built in 1921.

LEARMONTH, 1970.

Ninth Carnegie Museum expedition — Walter Edmond Clyde Todd, George M. Sutton — **1920**

From Pittsburgh; to study avifaunal conditions along the Labrador coast, especially along its northern part, of which little was known. They reached Battle Harbour by rail and steamer on 7 July, and made collections there for the next ten days. On 17 July, they sailed northwards on the hired motor yawl *Northern Messenger*. During the northward voyage, they stopped at many places along the coast, with prolonged stops at Indian Harbour (20-25 July) and Grenfell Sound (5-13 August). On 14 August, they sailed around Cape Chidley, where they remained until 18 August. From there, Sutton made his way back southward on the motor yawl, again making numerous stops along the coast, and reached Battle Harbour on 13 September. Meanwhile, Todd continued into Ungava Bay on board the Revillon Frères power schooner *Dorothy G. Snow*. He examined the coast of the bay around to Fort Chimo, where he remained 26 August - 1 September. He then sailed back to Fort Chimo and returned south along the Labrador coast, reaching Battle Harbour on 5 October. He remained there a further 20 days, making one excursion into St Lewis Inlet, then made his way by steamer and rail back to Pittsburgh.

TODD, 1963, pp. 31-34.

1920

United States expedition — Donald Baxter MacMillan — *Thetis*

A Boston business company engaged MacMillan to visit Hudson Bay and to report on the possibility of opening up trade in the area in competition with the Hudson's Bay Company. The company acquired *Thetis*, on which MacMillan travelled as second mate. The ship made several stops in Hudson Bay, including Chesterfield Inlet and Churchill, but MacMillan returned home to report that the Hudson's Bay Company could not seriously be challenged. The Boston company later attempted to trade in Hudson Bay in spite of MacMillan's advice, but their venture was shortlived.

ALLEN, E.S., 1963, pp. 220-21.

Roster of Men's Names

In this index, comprising some 5,000 entries, we have attempted a compilation of the names of all leaders and senior members of all the expeditions listed. In addition, we have included the names of the lesser members (seamen, voyageurs, guides, etc.) of many of the major expeditions. Thus, the names of all those known to have taken part in the North-west Passage, Franklin search, North Polar, and numerous other important maritime and overland expeditions should appear in this index, although there are bound to be many omissions.

Each entry includes (wherever possible) the man's full name, his rank or status on each expedition, and reference by date and by ship (or some other suitable indicator) to the expedition(s) in which he took part. We also include, when known, dates of birth and death and selected biographical references. The biographical references, in square brackets at the end of an entry, refer either in the usual way to the main bibliography or, in an abbreviated form, to one of the biographical dictionaries listed below. For the sake of brevity, some journal titles are abbreviated also (*eg*, GJ 1915 = *Geographical Journal* 1915). A key to these also appears below but full references will be found, under the titles of the journals, in the main bibliography.

Abbreviations

BOA BOASE, Frederick. *Modern English Biography. Vols. 1-6.* London, Frank Cass, 1965.

DAB *Dictionary of American Biography.* Vols. 1-. New York, Charles Scribner's Sons, 1928-.

DCB *Dictionary of Canadian Biography,* Vols. 1-. Toronto, University of Toronto Press, 1966-.

DNB *Dictionary of National Biography,* Vols. 1-63, and supplements. London, Smith, Elder, & Co., 1885-.

EA *Encyclopedia Arctica.* Typescripts held in the Stefansson Collection, Dartmouth College Library, Hanover, New Hampshire, USA. Published on microfilm by Xerox University Microfilms, Ann Arbor, 1975.

GJ *Geographical Journal.* (See main bibliography).

GR *Geographical Review.* (See main bibliography).

MAR MARKHAM, Clements Robert. *The Arctic navy list; or, a century of Arctic & Antarctic officers, 1773-1873.* London and Portsmouth, Griffin and Co., 1875.

MDC WALLACE, W. Stewart, *ed. The Macmillan Dictionary of Canadian Biography. 3rd ed.* London and Toronto, Macmillan, 1963.

NE *A naval encyclopaedia; comprising a dictionary of nautical words and phrases, biographical notices, and records of naval officers.* Philadelphia, L. R. Hamersly, 1881.

O'B (1) O'BYRNE, William. *A naval biographical dictionary. Comprising the life and service of every living officer in Her Majesty's navy, from the rank of Admiral of the Fleet to that of Lieutenant, inclusive.* London, John Murray, 1849.

O'B (2) O'BYRNE, William. *O'Byrne's naval biography; a new and enlarged edition, comprising a record of the services of all living officers, naval and marine.* London, O'Byrne Brothers, 1858-62.

PR *Polar Record.* (See main bibliography).

RGS *Royal Geographical Society.* (See main bibliography).

RNB MARSHALL, John. *Royal Naval Biography. . . .* 8 Vols. in 12. London, Longman, Hurst, Rees, 1823-35.

WWW *Who Was Who.* Vols. 1-. London, A. and C. Black Ltd., 1920-.

ABBES, H. Assistant, German Polar Year Expedition, 1882-83.
ABBOT, John. AB, *Isabella,* 1818.
ABBOTT, John. AB, *Assistance,* 1850-51.
ABBOTT, Joseph. Captain of the Forecastle, *Pioneer,* 1850-51; Ice Quartermaster, *Pioneer,* 1852-54.
ABERNETHY, John. Gunner's Mate, *Plover,* 1848-52.
ABERNETHY, Thomas (-1860). AB, *Hecla,* 1824-25; Mate, *Victory,* 1829-33; (?), *Enterprise* and *Investigator,* 1848-49; Master, *Felix,* 1850-51; Ice Master, *Isabel,* 1852. [MAR; GB, Parliament, 1851a, p. 97].
ABLE, James. Cook, *Fury,* 1824-25.
ABRAHAM, John. Captain, *George,* 1683; Second-in-command, *Hayes,* 1686; Captain, *Mary,* 1688. [Rich, 1948, pp. 370-73; DCB Vol. 1].
ABRAHAM, Mahomet. Seaman, *Resolution,* 1776-80.
ABRAHAM, Mark. AB, *Herald,* 1848; 1849; 1850.
ABRAHAMSEN, Ismael. Seaman, *Enhiörningen,* 1619-20. Died at Churchill, 29 March 1620.
ACKERMAN, Albert A. Ensign, *Alert,* 1884.
ADAM, Jean-Baptiste. Interpreter, Franklin's expedition, 1819-22.
ADAMS, Andrew. AB, *Sophia,* 1850-51.
ADAMS, Edward. Assistant Surgeon, *Plover,* 1848-52.
ADAMS, Edward (1824-c1890). Assistant Surgeon, *Investigator,* 1848-49; *Enterprise,* 1850-55.
ADAMS, George. (?), *Nancy Dawson,* 1849.
ADAMS, Horace P. Joint leader, geological expedition, 1907.
ADAMS, James. Private RM, *Enterprise,* 1850-55.
ADAMS, John. AB, *Phoenix,* 1853.
ADAMS, Luke. Sailor (?), *Sunneshine,* 1585.
ADAMS, *Sir* Thomas (-1770). Captain, *Niger,* 1776. [Lysaght, 1971, pp. 63-64; Charnock, 1794-98, Vol. 6].
ADAMS, William (-c1890). Captain, *Arctic,* 1872; 1873.
ADAMS, William (1869-1942). (Son of William Adams, above). Mate, *Albert,* 1914-15 [1914 only]. [Wordie, 1944].
ADAMSON, William. AB, *Prince Albert,* 1850; 1850-51.
ADAMSON, William. Middleman, HBC expedition, 1846-47. [Rich, 1953, p. 352].
ADLER, Thomas. Leader, HBC voyage, 1804.
AFFERTON, Henry. Sailor, *Ayde,* 1578.
AFSUNEQUNTOR, Jean-B. Voyageur (?), HBC Franklin search, 1855.
AGAR, Robert. AB, *Alexander,* 1818.
AGGLETON, William. AB, *Phoenix,* 1853.
AGSTON, Robert. AB, *Breadalbane,* 1853.
AHER, Richard. AB, *Herald,* 1848; 1849; 1850.
AH-WEH-LAH. (Eskimo). Accompanied Cook to North Pole, 1907-09.
AIRD, Gustavus. Bowman, *Lion,* Arctic Land Expedition, 1825-27. Drowned during expedition.
AIRES, Christopher. Seaman, *Resolution,* 1776-80.
AITKEN, Aneas. Boatswain, *Resolution,* 1776-80.
AKAITCHO (Copper Indian chief). Accompanied Arctic Land Expedition, 1819-22. [Weekes and King, 1939].
ALBANEL, Charles (1616 [or 1613]-1696). Leader, overland expeditions, 1671-72; 1674. [DCB Vol. 1].
ALBION, William A. Seaman, *Florence,* 1877-78.
ALDRED, W. Cook, *Neptune,* 1903-04.
ALDRICH, George. Private RM, *Rattlesnake,* 1853-54.
ALDRICH, Pelham (1844-1930). Senior Lieutenant, *Alert,* 1875-76. [PR 1931; EA].
ALDRICH, Robert Dawes (1808-1891). Lieutenant, *Resolute,* 1850-51. [O'B(1)-(2); BOA].
ALDRIDGE, Benjamin. AB, *Phoenix,* 1853.
ALDRIDGE, William. AB, *Blossom,* 1825-28.
ALFONSE [FONTENEAU], Jean (1484-1544). Pilot, French expedition, 1542-43; Leader, North-west Passage expedition, c1543. [DCB Vol. 1].
ALHELY (?), Joseph. AB, *Phoenix,* 1854.
ALIECTE, Antoine. (?), North-west Passage expedition, 1535-36.
ALIECTE, Guillaume. (?), North-west Pas-

sage expedition, 1535-36.

ALIRYD, Thomas. Sailor, *Ayde,* 1578.

ALLARD, John Hillary. Second Master, *Pioneer,* 1850-51; Master, *Pioneer,* 1852-54. [O'B(2)].

ALLCOCK, James. Corporal RM, *Herald,* 1848; 1849; 1850.

ALLDRIDGE, William. Carpenter's crew, *Blossom,* 1825-28.

ALLEN, Christopher. AB, *Pioneer,* 1852-54.

ALLEN, G.H. Master's Assistant, *Herald,* 1848; 1849; 1850.

ALLEN, George. Sailmaker, *Hecla,* 1819-20.

ALLEN, Glover Morrill (1879-1942). Joint leader, U.S. ornithological expedition, 1906. [Tyler, 1943].

ALLEN, James. HBC overland expeditions, 1766-67; 1767-68; 1768-69; 1769-70.

ALLEN, Jerome Lee (1891-). Electrician and wireless operator, "Crocker Land" Expedition, 1913-17. [Macmillan, 1918, p. 5].

ALLEN, John. Private of Marines, *Resolution,* 1776-80. Killed 14 February 1779.

ALLEN, John. AB, *Discovery,* 1791-95.

ALLEN, Robert Calder (1812-1903). Master, *Resolute,* 1850-51. [O'B(2); DNB].

ALLEN, W.G. Quartermaster, *Herald,* 1848; 1849; 1850.

ALLEN, William. Captain, *Bonaventure,* 1696.

ALLEN, William. AB, *Hecla,* 1824-25.

ALLISON, John. Master and Greenland Pilot, *Alexander,* 1818; Greenland Master, *Hecla,* 1819-20; *Fury,* 1821-23; *Hecla,* 1824-25.

ALLISON, John. AB, *Phoenix,* 1854.

ALSING, Iffuer. Seaman, *Enhiörningen,* 1619-20. Died at Churchill 3 April 1620.

ALSTON, Alfred Henry (1830-1874). Mate, *North Star,* 1852-54. [O'B(2); BOA].

AMBRONN, L. Deputy commander, German Polar Year Expedition, 1882-83.

AMERY, L.S. (London *Times* journalist). Passenger, Governor-General's tour, 1910.

AMES, Howard E. Passed Assistant Surgeon, *Bear,* 1884.

AMUNDSEN, Peder. Seaman, *Enhiörningen,* 1619-20. Died at Churchill 19 April 1620.

AMUNDSEN, Roald Engelbregt Gravning (1872-1928). Commander of expedition, *Gjøa,* 1903-06. [Amundsen, 1927; Partridge, 1953; Sverdrup, H.U., 1959].

ANARIZE, Joseph. (Iroquois Indian). HBC Franklin search, 1855.

ANDERSEN, Karsten. (?), Canadian Arctic Expedition, 1913-18.

ANDERSEN, Olluff. Captain's Servant, *Enhiörningen,* 1619-20. Died at Churchill, 17 April 1620.

ANDERSON, Alexander. AB, *Terror,* 1836-37.

ANDERSON, Alexander. AB, *Prince Albert,* 1850.

ANDERSON, Alexander. First Officer, *Karluk,* Canadian Arctic Expedition, 1913-18.

ANDERSON, C.J.J. Leader, HBC expedition, 1860.

ANDERSON, C.W. AB, *Herald,* 1848; 1849; 1850.

ANDERSON, *Bishop* David (1814-1885). Anglican Church journey, 1852.

ANDERSON, F. Surveyed Button Islands, 1912.

ANDERSON, Henry. AB (or Mate?), *Prince Albert,* 1850; Chief Mate, *Prince Albert,* 1851-52.

ANDERSON, James (-1867). Leader, HBC Franklin search, 1855. [DCB Vol. 9; EA].

ANDERSON, John. Captain of the Forecastle, *Herald,* 1848; 1849; 1850.

ANDERSON, John Bielt. Midshipman, *Herald,* 1848; 1849; 1850.

ANDERSON, Joseph. Private RM, *Assistance,* 1852-54.

ANDERSON, M. AB, *Enterprise,* 1848-49.

ANDERSON, R. AB (?), *Enterprise,* 1848-49; 1850-55.

ANDERSON, Robert. Gunner, *Resolution,* 1776-80.

ANDERSON, Robert. Surgeon, *Enterprise,* 1850-55; Surgeon, *Investigator,* 1848-49.

ANDERSON, Rudolph Martin (1876-1961). Joint leader, Stefansson-Anderson expedition, 1908-12; Second-in-command and zoologist, Canadian Arctic Expedition, 1913-18. [Jenness, 1961; Russell, L.S., 1962].

ANDERSON, Swen. Seaman, *Arctic,* 1908-09.

ANDERSON, Thomas. Cook, *Michael,* 1578.

ANDERSON, Thomas. Fisherman, Back's expedition, 1833-35.

ANDERSON, William (c1748-1778). Surgeon, *Resolution,* 1776-80. Died at sea, 3 August 1778. [DNB Vol. 1; Iredale, 1938; Keevil, 1933].

ANDREASEN, Ole (1883-1948). (?), Canadian Arctic Expedition, 1913-18. [PR 1949; EA].

ANDREW, Edward. AB, *Rattlesnake,* 1853-54.

ANDREWES, William. Captain, *Squirrel,* 1583.

ANDREWS, . Captain, *Dee,* 1815-16.

ANDREWS, Henry. Quartermaster, *Pandora,* 1875; 1876.

ANDREWS, Joseph. Captain of the Hold, *Erebus,* 1845-48.

ANDRIES, George Peter. Established Harrison's House, 1819. [Rich, 1938, p. 428].

ANGLIN, William. Boatswain's Mate, *Herald,* 1848; 1849; 1850.

ANTHOINE, *dom.* Chaplain (?), North-west Passage expedition, 1535-36.

ANTONIO, J. Seaman, *George Henry,* 1860-62.

APPLEGARTH, Robert. AB, *Terror,* 1836-37.

ARBUTHNOT, Charles R. Lieutenant, *Pandora,* 1876.

ARBUTHNOT, George. Ice Mate, *Enterprise,* 1850-55.

ARCHER, John. AB, *Hecla,* 1824-25.

ARCHER, Robert H. Lieutenant, *Discovery,* 1875-76. [MAR].

ARDEN, D'Arcy. Guide and Special Constable, RNWMP patrol, 1915-17.

ARDINGTON, John. Boy, *Ayde,* 1578.

ARFUEDDSEN, Suend. Carpenter, *Enhiörningen,* 1619-20. Died at Churchill 30 March 1620.

ARMITAGE, Thomas. Gun-Room Steward, *Terror,* 1845-48.

ARMSHOW, William. Gentleman, *Ayde,* 1577.

ARMSTRONG, Alexander (1818-1899). Surgeon, *Investigator,* 1850-54. [DNB; Keevil, 1936].

ARMSTRONG, Robert. AB, *Discovery,* 1776-80.

ARMSTRONG, Samuel. Purser's Servant, *Terror,* 1836-37.

ARMSTRONG, Thomas. Gunner's Mate, *Assistance,* 1850-51.

ARMSTRONG, W.J. Captain's Steward, *Fury,* 1824-25.

ARNAUD, Charles (1826-1914). RC missionary journey, 1853-58. [Carrière, 1958].

ARNOLD, Thomas. Private RM, *Rattlesnake,* 1853-54.

ARNOT, . Surgeon, *Rosamond,* 1814.

ARTEAGA, Ignacio. Captain, *Princesa,* 1779.

ARTHUR, Daniel. Quartermaster, *Erebus,* 1845-48.

ARTHUR, Walter. Sailor, *Sunneshine,* 1585.

ASASELA, James [Jim Fiji]. (?), *Polar Bear,* Canadian Arctic Expedition, 1913-18.

ASH, Arthur. Prospector, Curran's expedition, 1912.

ASH, John. Armourer, *Discovery,* 1791-95.

ASHDOWNE, Clark. AB, *Hecla,* 1821-23.

ASHE, W.A. Officer in charge, Ashe Inlet Station, Canadian Hudson Bay Expedition, 1884-85.

ASHE, William. Seaman, *Sunneshine,* 1586.

ASHMORE, William. Seaman (?), *Nancy Dawson,* 1849.

ASHWORTH, James. Corporal RM, *Hecla,* 1821-23.

ASSINIJUNTON, Bte. (Iroquois Indian). HBC Franklin search, 1855.

ATEASTA, Pierre. (?), Back's expedition, 1833-35.

ATKINS, Henry. Captain, *Whale,* 1729; Captain, New England voyages, 1753; 1758.

ATKINSON, Daniel A. General collector and assistant, Carnegie Museum expedition, 1901.

ATKINSON, Edmund. Master's Mate, *Chatham,* 1791-95.

ATKINSON, George (1777-1830). Established Great River post (Fort George), 1803; established Great Whale River post, 1813; leader, HBC expeditions, 1816; 1818; 1828. [Davies, 1963, pp. 329-41].

ATKINSON, Jacob. Established Great River post (Fort George), 1803.

ATKINSON, John. Ice Mate, *Enterprise,* 1850-55.

ATKINSON, Robert. AB, *Phoenix,* 1853.

ATKINSON, Thomas. Established Wapiscogamy House, 1777.

AUDIEPVRE, Michel. (?), North-west Passage expedition, 1535-36.

AUGUSTINA. (Eskimo). On board *Polaris,* 1871-73.

AUGUSTUS. *See* TATTANNOEUCK.

AULUK. (Eskimo). Pilot, *George,* 1774.

AUNTING, Frederick. Seaman, *Polaris,* 1871-73.

AUSTIN [AUSTEN], Frauncys. (?), *Ayde,* 1578.

AUSTIN, Horatio Thomas (1801-1865). Lieutenant, *Fury,* 1824-25; Captain, *Resolute,* and commander of expedition, 1850-51. [O'B(1)-(2); DCB Vol. 9].

AWLE, Robert. Sailor, *Judith,* 1578.

AXWORTHY, William. Carpenter's crew, *Herald,* 1848; 1849; 1850.

AYLEN, John Franklin Robins (1832-). Acting Master's Assistant, *North Star,* 1849-50. [O'B(2)].

AYLES, Adam. Second Captain of the Foretop, *Alert,* 1875-76.

AYLMER, William. Private RM, *Hecla,* 1821-23.

AYLMORE [AYLMER], Richard. Gun-Room Steward, *Erebus,* 1845-48.

AYRE [AYER], Henry. *See* EYRE, Henry.

BABEL, Louis-François (1826-1912). Missionary journey, 1866-70. [Carrière, 1963].

BABSTOCK, W. Seaman, *Neptune,* 1903-04.

BACHELOR, Robert [William]. Boatswain's Mate, *Pioneer,* 1852-54.

BACK, George (1796-1878). Midshipman, Arctic Land Expedition, 1819-22; Lieutenant, Arctic Land Expedition, 1825-27; Commander RN and leader, British expedition, 1833-35; Captain, *Terror,* and leader of expedition, 1836-37. [DNB Vol. 2; DCB Vol. 10].

BACON, Joseph. AB, *Pioneer,* 1850-51; Cook, *Resolute,* 1852-54.

BADCOCK, George. Carpenter, *Ayde,* 1578.

BADGER, Charles Johnston (1853-1932). Lieutenant USN, Executive and Navigating Officer, *Alert,* 1884. [DAB].

BADLINGTON, Jacob. AB, *Hecla,* 1824-25.

BAFFIN, William (c1584-1622). Mate and Pilot, *Discovery,* 1615; 1616. [DNB Vol. 2; DCB Vol. 1].

BAGGE, John. Sailor (?), *Sunneshine,* 1585.

BAGGS, Rufus C. Seaman, *Rescue,* 1850-51.

BAGLEY, William. Prospector, Curran's expedition, 1912.

BAILEY, A. Boat-steerer, *George Henry,* 1860-62.

BAILEY, James. AB, *Discovery,* 1791-95.

BAILEY, James. Private RM, *Blossom,* 1825-28.

BAILEY, John [alias HOLLAND, John]. AB, *Terror,* 1845-48.

BAILEY, John [William]. Private RM, *Assistance,* 1850-51.

BAILEY, Joseph [Henry P.]. AB, *Griper,* 1819-20.

BAILEY, Richard. AB, *Phoenix,* 1854.

BAILIE, Alexander. Carpenter's crew, *Assistance,* 1852-54.

BAILLIE, . Navigating sub-Lieutenant, *Gannet,* 1867.

BAIN, Alexander. Sailmaker, *Lady Franklin,* 1850-51.

BAIN, Will. Guide, Carnegie Museum expedition, 1908.

BAINARD, John J. *See* BARNARD, John J.

BAINBRIDGE, Thomas. Bombardier RMA, *Resolute,* 1852-54.

BAKER, Charles. Private RM, *Phoenix,* 1854.

BAKER, George. Boatswain's Mate, *Plover,* 1848-52.

BAKER, George. Captain of the Forecastle, *Herald,* 1848; 1849; 1850.

BAKER, Henry. Captain, *Prince of Wales* [I], 1833-34; 1836-37; 1837.

BAKER, Henry. AB (or Boatswain?), *North Star,* 1849-50.

BAKER, James. Steward, *Breadalbane,* 1853.

BAKER, Jefferson. Seaman, *Advance,* 1853-55. Died 8 April 1854.

BAKER, John. AB, *Phoenix,* 1854.

BAKER, Joseph. Third Lieutenant, *Discov-*

ery, 1791-95.

BAKER, Richard. AB, *North Star,* 1849-50. Died 7 April 1850.

BAKER, Richard. AB, *Rattlesnake,* 1853-54.

BAKER, Robert. Sailmaker's crew, *Plover,* 1852-54.

BAKER, Samuel. AB, *Terror,* 1836-37.

BAKER, Thomas. Boatswain's Mate, *Phoenix,* 1853.

BAKER, Thomas. Cook, *Phoenix,* 1854.

BALBOA, Vasco Nunez de (c1475-1517). Overland expedition, 1513.

BALDOCK, William. AB, *Rattlesnake,* 1853-54.

BALDWIN, M. Fireman, *Neptune,* 1903-04.

BALE, James. AB, *North Star,* 1849-50.

BALEY, David. Captain of the Foretop, *Release,* 1855.

BALEY [BAYLY, BAYLEY], Henry (- 1701), Captain, *Prosperous,* 1692-93; *Dering* [III], 1694; 1696; *Pery,* 1698-1700. [Davies, 1965, pp. 353-55; DCB Vol. 2].

BALL, Alfred. AB, *Enterprise,* 1850-55.

BALL, Benjamin. First Engineer, *Pandora,* 1875; 1876.

BALLANTYNE, Robert Michael (1825-1894). Apprenticeship to HBC, 1841-47. [DNB, 1st Supp. Vol. 1; Quayle, 1967].

BALLARD, V.V. AB, *Discovery,* 1791-95.

BANCE, Henry Prescott. Mate, *Assistance,* 1850-51.

BANCROFT, Thomas. Marine, *Investigator,* 1850-54.

BANFIELD, W.H. AB, *Phoenix,* 1853.

BANKS, John. AB, *Phoenix,* 1854.

BANKS, Joseph (1743-1820). Naturalist, *Niger,* 1766. [DNB Vol. 1].

BANNERMAN, James. Mate, *Arctic,* 1873.

BANNISTER, John. Private RM, *Herald,* 1848; 1849; 1850.

BARBER, D. Private RM, *Investigator,* 1849-50.

BARBER, George. Carpenter's Mate, *Resolution,* 1776-80.

BARBER, James. AB, *Hecla,* 1824-25.

BARBER, John. Private RM, *Herald,* 1848; 1849; 1850.

BARBER, Thomas. Sailmaker, *Assistance,* 1852-54.

BARDIN, Richard. Engineer, *Isabel,* 1852.

BARFF, Richard. AB, *Phoenix,* 1853.

BARFIELD, William. Captain, HBC supply voyages. *Princess Royal,* 1885; *Prince Rupert* [VIII], 1887; 1888; 1889; 1890-91.

BARFLEUR, Henry. AB, *Chatham,* 1791-95.

BARGERY, David. AB, *Rattlesnake,* 1853-54.

BARK, George. Private RM, *North Star,* 1849-50.

BARKER, Charles. Second Mate, *Karluk,* Canadian Arctic Expedition, 1913-18. Died off Wrangel Island after sinking of *Karluk,* 1914.

BARKER, George. Captain's Cook, *Terror,* 1836-37.

BARKER, George. AB, *Phoenix,* 1854.

BARKER, James [Thomas]. Sergeant RM, *Assistance,* 1850-51.

BARKER, John. AB, *Phoenix,* 1854.

BARLOW, James. Quartermaster, *Hecla,* 1824-25.

BARLOW, John Clark. Second Class Volunteer, *Blossom,* 1825-28. [O'B(2)].

BARLOW, Richard. Armourer's Mate, *Fury,* 1824-25.

BARLOW, Robert. AB, *Fury,* 1821-23.

BARNARD, John James. Lieutenant, *Investigator,* 1848-49; Lieutenant, *Enterprise,* 1850-55. Killed by Indians at Nulato, February 1851. [BOA].

BARNARDA, Pedro de (fictitious character). Commander, *Rosario,* 1640.

BARNES, Benjamin. Blacksmith, *Investigator,* 1848-49.

BARNES, John. Drummer, Marines, *Discovery,* 1791-95.

BARNES, John. Seaman, *Roosevelt,* 1908-09.

BARNES, Joseph. Captain of the Foretop, *Assistance,* 1852-54.

BARNES, Samuel. AB, *Phoenix,* 1854.

BARNETT, George. Cook, *Herald,* 1848; 1849; 1850.

BARNETT, Isaac. *See* BURNETT, Isaac.

BARNETT, William. Leading Seaman, *Blossom,* 1825-28.

BARNLEY, George. Methodist missionary expedition, 1840.

BARNOTT, Isaac. *See* BURNETT, Isaac.

BARNS, Henry. AB, *Phoenix,* 1854.

BARNUM, Thomas. Seaman, *United States,* 1860-61.

BARRETT, Charles H.M. Scientific assistant, Carnegie Museum expedition, 1912.

BARRETT, Edward. AB, *Resolution,* 1776-80.

BARRETT, William. Carpenter's crew, *Blossom,* 1825-28.

BARRETT, James. Sailor, *Ayde,* 1578.

BARRIE, John James. First Officer, *Alert,* 1885.

BARRIE, Robert. Midshipman, *Discovery,* 1791-95.

BARRIEU, François. Canadian voyageur, Mackenzie's expedition, 1789.

BARRON, Richard. Seaman, *Arctic,* 1906-07.

BARRY, David. AB, *Herald,* 1848; 1849; 1850.

BARRY, J.J. Passed Assistant Engineer, *Juniata,* 1873.

BARRY, James. Leading Seaman, *Hecla,* 1824-25.

BARRYMORE, John. AB, *Discovery,* 1791-95.

BARTLETT, George. (?), *Algerine,* 1912.

BARTLETT, H. Captain, *Laddie,* 1913-16.

BARTLETT, Henry. Captain and Ice Master, *Falcon,* 1894.

BARTLETT, James H. AB, *Hecla,* 1819-20.

BARTLETT, John. Private RM, *Resolute,* 1850-51.

BARTLETT, John. Captain, *Algerine,* 1912.

BARTLETT, John W. Captain, *Windward,* U.S. North Polar Expedition, 1898-1902 [1898-99 only].

BARTLETT, Moses. Second Mate, *Neptune,* 1903-04; Mate, *Roosevelt,* 1905-06; Captain, *John R. Bradley,* 1907-09.

BARTLETT, Robert Abram (1875-1946). Captain, *Hope* [1898], *Diana* [1899], *Windward* [1901, 1902], U.S. North Polar expedition, 1898-1902; Captain, *Roosevelt,* 1905-06; 1908-09; *Karluk,* Canadian Arctic Expedition, 1913-18 [1913-14 only]; *Neptune,* "Crocker Land" Expedition, 1913-17 [1917 only]. [Bartlett, 1931; Gould, A.T., 1946; Putnam, 1947; EA].

BARTLETT, Samuel W. Captain, *Neptune,* 1903-04; 1905-06; *Erik,* 1908-09.

BARTON, John. Gunner's Mate, *Henrietta Maria,* 1631-32. Drowned 17 October 1631.

BASSELE, George. AB, *Investigator,* 1848-49.

BASTILLE, Guillaume le Breton. Captain, *Émérillon,* 1535-36.

BASTIN, Isaac. Corporal RM, *Rattlesnake,* 1853-54.

BATCHELOR, William Robert. *See* BACHELOR, Robert [William].

BATERBY, Harry. Sailor, *Ayde,* 1578.

BATES, John. AB, *Terror,* 1845-48.

BATES, Robert. AB, *Plover,* 1848-52.

BATEY, David. Captain of the Foretop, *Release,* 1855.

BATT, Isaac. HBC overland expeditions, 1757-58; 1759-60; 1763-64; 1765-66 1766-67; 1767-68; 1768-69; 1771-72.

BATTAN, William. AB, *Pandora,* 1876.

BATTEN, William. AB, *Investigator,* 1850-54.

BATTERBY, Thomas. (?), Frobisher's expedition, 1578. Died 6 September 1578.

BATTERICK, Thomas. (?), *Ayde,* 1578.

BAUGHMAN, George E. Assistant Paymaster, *Tigress,* 1873. [NE].

BAUMANN, H.A. Victor (1870-1932). Second-in-command, *Fram,* 1898-1902. [EA].

BAUR, William J. Steward, Canadian Arctic Expedition, 1913-18.

BAXTER, George. Stoker, *Victory,* 1829-33.

BAY, Edvard (1867-). Zoologist, *Fram,* 1898-1902.

BAYLEY, Richard. Captain of the Forecastle, *North Star,* 1849-50.

BAYLEY, Richard. Boatswain's Mate, *Assistance,* 1852-54.

BAYLY, Charles (-1680). (First overseas governor of HBC). HBC voyages, 1670-71; 1672-75; 1674; retired 1679. [Rich, 1942, pp. 209-13; DCB Vol. 1].

BAYLY, William. Astronomer, supernumary, *Discovery,* 1776-80. [DNB Vol. 3].

BAYNE, Peter. Assistant to C.F. Hall, 1864-69 [1868 only].

BEACH, Benjamin. Gold prospecting expedition, 1883-84.

BEACH, Mathew. AB, *Resolution,* 1776-80.

BEADLING, Robert. AB, *Phoenix,* 1854.

BEAL, Charles [or BEALL, James?]. Private RM, *Assistance,* 1850-51.

BEAMS [BEANS], James [Joseph]. Boatswain's Mate, *Resolute,* 1850-51; *Assistance,* 1852-54.

BEAN, John. Captain, *Churchill,* 1755; 1756; 1757. [DCB Vol. 3].

BEAN, Thomas. AB, *Discovery,* 1776-80.

BEARD, Edward. Private RM, *Herald,* 1848; *Plover,* 1848-52; *Rattlesnake,* 1853-54.

BEARE, James. Master, *Michael,* 1577; *Anne Francis,* 1578.

BEATON, D.G. (Editor of Winnipeg *Times*). Supernumary, *Alert,* 1885-86.

BEAUCHAMP, Jacques. Voyageur, Mackenzie's expedition, 1792-93.

BEAUCHAMP, James. Private RM, *Resolute,* 1850-51.

BEAUCROFT, Thomas. Private RM, *Investigator,* 1850-54.

BEAUJEU, G.Q.R. de. Constable, NWMP, Begin's patrol, 1890.

BEAULIEU, François (1771-1872). Canoeman, Mackenzie's expedition, 1792-93; Guide and interpreter, Franklin's expedition, 1825-27. [DCB Vol. 10].

BEAULIEU, Louis. Second Cook, *Arctic,* 1906-07; 1910-11.

BEAUMONT, Lewis Anthony (1847-1922). Senior Lieutenant, *Discovery,* 1875-76. [GJ 1922; WWW Vol. 2; EA].

BEAUPARLANT, Gabriel. Voyageur, Franklin's expedition, 1819-22. Died 17 October 1821.

BEAUPRÉ, *Vicomte de.* (?), Cartier's expedition, 1541-42.

BECK, Adam. (Eskimo). Interpreter, *Felix,* 1850-51.

BECK, Martin. AB, *Blossom,* 1825-28.

BECKER, Alois *Ritter von.* Lieutenant, *Pandora,* 1876.

BECKETT, James. Carpenter's Mate, *Chatham,* 1791-95.

BECKWITH, George. Seaman, *George Henry,* 1860-62.

BEDFORD, E.I. Volunteer, *Griper,* 1824.

BEDGOOD, Joseph. AB, *Phoenix,* 1854.

BEDUTZMAN, James. AB, *Griper,* 1824.

BEEBE, William M. Private in General Service and leader of expedition, *Neptune,* 1882.

BEECH, Samuel. Private RM, *Phoenix,* 1854.

BEECHEY, Frederick William (1796-1856). Lieutenant, *Hecla,* 1819-20; Captain, *Blossom,* 1825-28. [DNB Vol. 4; Brown, 1961; O'B(1)].

BEECHEY, Richard Brydges (1808-1895). Midshipman, *Blossom,* 1825-28. [O'B(1); BOA].

BEECKER, Jan Arno de. AB, *Resolution,* 1776-80.

BEEDLING, William. AB, *Assistance,* 1850-51.

BEER, James. Corporal RMA, *Pioneer,* 1850-51.

BEGG, Charles. Middleman, Dease and Simpson's expedition, 1837-39.

BÉGIN, I. Cook, *Arctic,* 1908-09.

BEGIN, J.V. Inspector, NWMP, patrol, 1890.

BÉGIN, Ulrich. Fireman, *Arctic,* 1906-07.

BELANGER, Jean-Baptiste. Voyageur, Franklin's expedition, 1819-22.

BELANGER, Solomon. Voyageur, Franklin's expedition, 1819-20.

BELCHER, *Sir* Edward (1799-1877). Lieutenant and Assistant Surveyor, *Blossom,* 1825-28; Captain and leader of expedition, *Assistance,* 1852-54. [DCB Vol. 10; DNB Vol. 4; O'B(1)-(2)].

BELCHER, James. Captain, HBC supply voyages. *Port Nelson,* 1714; 1715-16; 1717; *Mary* [I], 1718; 1719; 1720; 1721; 1722; 1723; 1724.

BELITHER, Samuel. AB, *Enterprise,* 1850-55.

BELL, Alexander. AB, *Discovery,* 1791-95.

BELL, Benjamin. Captain, HBC supply voyages. *Emerald,* 1816-17; *Eddystone,* 1818; 1819-20; 1820; 1821; 1822; 1823; *Camden,* 1824; 1825; 1826; *Prince Rupert* [V], 1827; 1828; 1829; 1830-31; 1831; 1832; 1833-34.

BELL, David. Corporal RM, *Plover,* 1848-52 [1851-52 only]; 1852-54.

BELL, Edward. Clerk, *Chatham,* 1791-95.

BELL, George. AB, *Resolute,* 1852-54.

BELL, James Mackintosh (1877-1934). Second-in-command, GSC expedition, 1899-1900. [Canadian Geographical Journal, 1934; GJ 1934].

BELL, John (1799-1868). Leader, HBC expeditions, 1839; 1840-41; 1842; 1845; Assistant, Richardson's expedition, 1847-49. [Fleming, 1940, pp. 427-28; EA].

BELL, John. (?), *Investigator,* 1848-49.

BELL, Robert (1841-1917). Leader, GSC expedition, 1877; Geologist and naturalist, *Neptune,* 1884-85; *Alert,* 1885-86; Leader, GSC expeditions, 1897; 1899-1900.

[Scottish Geographical Magazine, 1917; GR 1917; EA].

BELL, Thomas. Assistant Surgeon, *Fury,* 1824-25.

BELL, Thomas. Stoker, *Phoenix,* 1854.

BELL, William. Quartermaster, *Erebus,* 1845-48.

BELLEAU, Jean-Baptiste. Canadian voyageur, Arctic Land Expedition, 1819-22. Discharged, December 1820.

BELLINGS, W.J. Assistant Surgeon, *Herald,* 1848; 1849; 1850.

BELLOT, Joseph-René (1826-1853). Lieutenant and second-in-command, *Prince Albert,* 1851-52; *Phoenix,* 1853. Drowned, 17 August 1853. [Woodward, 1950; Rouch, 1945; BOA].

BENDALL, Harry. Sailor, *Ayde,* 1578.

BENDALL, John. AB, *Phoenix,* 1853.

BENDER, Jacob. Private, 9th Cavalry, U.S. Polar Year Expedition, 1881-84. Died 6 June 1884.

BENDTSEN, Hans. Seaman, *Enhiörningen,* 1619-20. Died at Churchill, 16 April 1820.

BENNES, William. Master's Mate, *Michael,* 1578.

BENNETT, Peter. AB, *Resolution,* 1776-80.

BENNETT, Robert. Coxswain, *Herald,* 1848; 1849; 1850.

BENNETT, Samuel. AB, *Phoenix,* 1853.

BENNETT, Spencer. Assistant, Curran's expedition, 1912.

BENNETT, Thomas. Private RM, *Rattlesnake,* 1853-54.

BENNETT, W.H. AB, *Phoenix,* 1854.

BENNETT, William. AB, *Chatham,* 1791-95.

BENNOIT, Joseph. Canadian voyageur, Arctic Land Expedition, 1819-22.

BENSON, William. Seaman, *Advance,* 1850-51.

BENTHAM, Gregory. AB, *Discovery,* 1776-80.

BENTLY, James. Fireman, *Roosevelt,* 1908-09.

BENTLY, Robert. AB, *Rattlesnake,* 1853-54.

BERCHERVAISE, John. Master-at-Arms, *Blossom,* 1825-28.

BERE, Samvell. Sailor and Gunner, *Ayde,* 1578.

BERGEN, Laurids. Seaman, *Enhiörningen,* 1619-20. Died at Churchill 5 February 1620.

BERGNER, R. Captain, *Pelican,* HBC supply voyage and establishment of Port Harrison, 1920.

BERKY, D.W. Magnetic Observer, U.S. expedition to Hudson Bay, 1914.

BERLEY, George. Captain, *Albany,* 1716; 1717; 1718; 1719-21. [DCB Vol. 2].

BERNARD, Peter. Captain, *Mary Sachs,* Canadian Arctic Expedition, 1913-18.

BERNIER, Joseph-Elzéar (1852-1934). Captain, *Arctic,* 1904-05; 1906-07; 1908-09; 1910-11; *Minnie Maud,* 1912-13. [Bernier, 1939; Fairley and Israel, 1957; EA].

BERNIER, Louis. Fireman, *Arctic,* 1906-07; AB, *Arctic,* 1910-11.

BERRIE, James. Ice Quartermaster, *Alert,* 1875-76.

BERRY, Alexander. AB, *Terror,* 1845-48.

BERRY, Andrew. AB, *Hecla,* 1821-23; Greenland Mate, *Griper,* 1824.

BERRY, Charles. Seaman, *Advance,* 1850-51.

BERRY, Edward. Gunner's Mate, *Discovery,* 1791-95.

BERRY, John. AB, *Plover,* 1848-52 [1851-52 only]; 1852-54.

BERRY, Robert Mallary (1846-1929). Lieutenant USN, *Tigress,* 1873. [NE; EA].

BERRY, William. Orderly, *Plover,* 1848-52. Discharged 1851.

BERRY, William. Quartermaster, *Pandora,* 1876.

BERRY, William. Captain, *Eldorado,* 1903.

BERRYMAN, James. Private RM, *Investigator,* 1848-49.

BERTRAM, Thomas. Assistant Surgeon, *Phoenix,* 1854.

BERTRAND, Jeremiah. Gold prospecting expedition, 1885-86.

BESSELS, Emil (1847-1888). Surgeon, *Polaris,* 1871-73.

BEST, Charles. AB, *Erebus,* 1845-48.

BEST [BESTE, BEAST], George. Lieutenant, *Ayde,* 1577; Captain, *Anne Francis,* 1578. [DCB Vol. 1].

BEST, John. Cook, *Chatham,* 1791-95.

BESWICK, James. Corporal RM, *Alexander,* 1818.

BETTISON, John. Carpenter's Mate, *Resolute,* 1850-51.

BETTON, Roderick. AB, *Discovery*, 1791-95.

BEUCHAT, Henri (1878-1914). Anthropologist, Canadian Arctic Expedition, 1913-18. Died off Wrangel Island after sinking of *Karluk*, 1914.

BEVAN, John. AB, *North Star*, 1849-50.

BEVERLEY, Charles James (1788-1868). Assistant Surgeon, *Isabella*, 1818; *Griper*, 1819-20. [DNB Vol. 4; BOA].

BEWERS, Thomas. Private RM, *Herald*, 1848; 1849; 1850.

BEX, Richard. Private RM, *Assistance*, 1852-54.

BEYNEN, L.R. Koolemans. *See* KOOLEMANS BEYNEN.

BICKFORD, Henry. Carpenter's crew, *Herald*, 1848; 1849; 1850.

BIDDLE, Robert. Ordinary Seaman, *Herald*, 1848; 1849; 1850.

BIDGOOD [BIGWOOD], Emanuel. AB, *Resolute*, 1852-54.

BIDWOLD, George. Boatswain's Mate, *Arctic*, 1855.

BIEDERBICK, Henry (1848-). Private, 17th Infantry, U.S. Polar Year Expedition, 1881-84. [EA].

BIGELOW, Henry Bryant (1879-). Scientist, Brown-Harvard Expedition, 1900.

BIGGS, James (1818-). Acting Paymaster, *Enterprise*, 1848-49. [O'B(2)].

BIGGS, James. Private RM, *Investigator*, 1850-54.

BIGGS, John. AB, *Herald*, 1848; 1849; 1850.

BIGNELL, John. Joint leader, GSC expedition, 1884-85.

BIGWOOD, Emanuel. *See* BIDGOOD.

BILBY, Julian William (1871-1932). Anglican missionary expeditions, 1897-99; 1900-02; established Lake Harbour mission, 1909-11. [EA].

BILLETT, Henry. AB, *Assistance*, 1852-54.

BILLING, Edwin. Sailmaker, *Herald*, 1848; 1849; 1850.

BILLINGS, Joseph. AB, *Discovery*, 1776-80.

BILLINGS, William Thomas. Assistant Surgeon, *Herald*, 1848; 1849; 1850. [O'B(2)].

BILSON, Nathaniel. Boatswain, *Henrietta Maria*, 1631-32.

BINDER, Otto. (?), *Challenge*, Canadian Arctic Expedition, 1913-18 [1917 only].

BINSKIN, Edward. AB, *Enterprise*, 1848-49. Died during expedition.

BINSTEAD, Richard. Captain of the Hold, *Resolute*, 1850-51.

BIRCH, Daniel. Boatswain's Mate, *Phoenix*, 1854.

BIRCH, John. AB, *Phoenix*, 1854.

BIRD, Edward Joseph (-1881). Midshipman, *Hecla*, 1821-23; *Fury*, 1824-25; Captain, *Investigator*, 1848-49. [O'B(1)-(2); BOA].

BISHOP, Henry. Captain, HBC supply voyages. *Prince Rupert* [VII], 1865; 1866; 1867; 1868; 1869; 1870; 1871; 1872; 1873-74; *Ocean Nymph*, 1874; *Lady Head*, 1875; *Prince of Wales* [II], 1876; 1877; 1878; 1879; 1880; 1881; 1882; 1883-84; 1884-85; *Lady Head* [II], 1886; *Lady Head*, 1887, 1889; 1890-91.

BISHOP, Samuel. AB, *Resolution*, 1776-80.

BISHOP, William. Private RM, *Plover*, 1848-52.

BISSETTE, A.R. Assistant, Digges Islands station, Canadian Hudson Bay Expedition, 1885-86.

BISSON, Baptiste. Canadian voyageur, Mackenzie's expedition, 1792-93.

BISSON, Philip. Midshipman, *Alexander*, 1818.

BISSOT, François. *See* BYSSOT.

BITTON, Henry. AB, *Isabella*, 1818.

BJARNI. *See* HERJULFSSON, Bjarni.

BLACK, Samuel (1780-1841). Leader, HBC expedition, 1824. [Rich, 1938, pp. 429-30; 1955, pp. xxvii-c].

BLACKBURN, James. AB, *Hecla*, 1821-23; *Fury*, 1824-25.

BLACKWELL, Thomas. Paymaster's Steward, *Assistance*, 1850-51; *Resolute*, 1852-54; Ship's Steward, *Fox*, 1857-59. Died of scurvy, 14 June 1859.

BLADES, Samuel. Seaman (?), *Charles*, 1631.

BLAIR, Alexander. AB, *Fury*, 1824-25.

BLAIZE, Paul. (?), *Nancy Dawson*, 1849.

BLAKE, Charles. (?), HBC Franklin search, 1855.

BLAKE, George. AB, *Phoenix*, 1853.
BLAKE, Gilbert. (Halfbreed Eskimo). Assistant, Hubbard expedition, 1905.
BLAKE, John. AB, *Discovery*, 1791-95.
BLAKE, John [HUSSEY, William]. Seaman, *Advance*, 1853-55.
BLAKENEY, John W. Captain, *Erik*, U.S. North Polar Expedition, 1898-1902 [1901 only].
BLAKEY, David. AB, *Phoenix*, 1854.
BLAKEY, William. AB, *Phoenix*, 1854.
BLANCHARD, John. Carpenter's Mate, *Terror*, 1836-37.
BLANCHARD, Thomas. Captain of the Forecastle, *Blossom*, 1825-28.
BLAND, Richard. AB, *Resolute*, 1850-51.
BLAND, William. AB, *Phoenix*, 1853.
BLANKEY, Thomas. AB, *Griper*, 1824.
BLANKY [BLANKEY], Thomas. First Mate, *Victory*, 1829-33; Acting Icemaster, *Terror*, 1845-48.
BLIGH, William (1754-1817). Master, *Resolution*, 1776-80. [DNB Vol. 4].
BLINN, John. Seaman, *Advance*, 1850-51; Boatswain's Mate, *Release*, 1855.
BLOOM, William. AB, *Discovery*, 1776-80.
BLOOMFIELD, D. Sergeant RM, *Rattlesnake*, 1853-54.
BLUE, Daniel Wallace. Engineer, *Mary Sachs*, Canadian Arctic Expedition, 1913-18. Died at Baillie Islands, 2 May 1915.
BLUFF, Henry. Boatswain's Mate, *Investigator*, 1850-54.
BLUNDELL, William. Armourer's Mate, *Fury*, 1821-23.
BLUNDEN, William. Armourer's Mate, *Griper*, 1819-20.
BOAKE, A.F. Constable, NWMP, Skirving's patrol, 1899.
BOAKE, E.P. Constable, NWMP, Skirving's patrol, 1899.
BOAKES, John. Ordinary Seaman, *Herald*, 1848; 1849; 1850.
BOAS, Franz (1858-1942). Leader, anthropological expedition, 1883-84. [Kroeber, 1943].
BÖDEKER, George A. Seaman, *Arctic*, 1906-07.
BÖDEKER, Joseph. Seaman, *Arctic*, 1908-09.
BODEGA Y QUADRA MOLLINEDO, Juan Francisco de la (c1742-1794). Lieutenant commanding, *Felicidad*, 1775; *Favorita*, 1779.
BOECKLEN, C. Assistant, German Polar Year Expedition, 1882-83.
BOG, Jacques du. Passenger, North-west Passage expedition, 1535-36.
BOGAR, Roger. Sailor, *Ayde*, 1578.
BOILEAU, E. Acting Mate, *Rattlesnake*, 1853-54.
BOLDUC, Emile. Second Engineer, *Arctic*, 1906-07; 1908-09; 1910-11.
BOLDUC, Joseph-Etienne. Medical Officer, *Arctic*, 1908-09; 1910-11.
BOLT, Ikey. (Eskimo). Canadian Arctic Expedition, 1913-18.
BOMPAS, William Carpenter (1834-1906). Consecrated Bishop of Athabasca, 1874. [Cody, 1908; EA].
BONA, Gregory. Goldfinder, *Ayde*, 1578.
BONCHIN, William. Private of Marines, *Discovery*, 1791-95.
BOND, James. Cook, *Herald*, 1848; 1849; 1850.
BOND, James. AB, *Phoenix*, 1853.
BOND, Silvanus. Cooper, *Discovery*, 1610-11.
BOND, Thomas. Ice Quartermaster, *Resolute*, 1850-51; *Assistance*, 1852-54.
BOND, William. Captain, HBC supply voyages. *Craven*, 1682-83; *George*, 1683; *Happy Return*, 1684; 1685; 1686; *Dering* [I], 1687-88; *Churchill*, 1688-89. [Rich, 1948, pp. 373-77; DCB Vol. 1].
BONE, Eli. AB, *Resolute*, 1850-51.
BONNETT, Samuel. Private RM, *Enterprise*, 1848-49.
BONNSALL, Samuel. AB, *Investigator*, 1850-51.
BONSALL, Amos. Officer, *Advance*, 1853-55.
BOOTH, J.W. Fireman, *Polaris*, 1871-73.
BORDEN, Lorris Elijah (1877-1963). Surgeon and botanist, *Neptune*, 1903-04. [*Arctic*, 1963].
BORDEN, Richard. Seaman, *Sunneshine*, 1586.
BORGLE, H. Steward, *Neptune*, 1903-04.

BRAITHWAITE, George. First Officer, *Arctic*, 1908-09.
BRAMS, James. Boatswain's Mate, *Resolute*, 1850-51.
BRAND, Alexander. Captain, *Moose*, 1786.
BRANDS, George. Engineer, *Fox*, 1857-59. Died 6 November 1858.
BRANDS, George. AB, *Lady Franklin*, 1850-51.
BRASEN, Christopher. *See* BRAASEN.
BRASKERUD, Ove. Stoker, *Fram*, 1898-1902. Died October 1899.
BRATT, Thomas. Private RM, *Enterprise*, 1850-55.
BRAY, Émile-Frédéric de (1829-1879). Mate, *Resolute*, 1852-54. [Rouch, 1938; 1945; DCB Vol. 10].
BRAY, James. Carpenter, *Chatham*, 1791-95.
BREDDY, G. Fireman, *Karluk*, Canadian Arctic Expedition, 1913-18. Committed suicide on Wrangel Island, 1914.
BRENDAN, *Saint*. Irish voyage, c500. [DCB Vol. 1, p. 675; DNB Vol. 6].
BRENNAN, John. Sailing Mate, *Advance*, 1850-51.
BRETH, William. Captain of the Forecastle, *Investigator*, 1848-49.
BRIAN, John. AB, *Phoenix*, 1853.
BRIANT, Henry. Musician, *Assistance* (or *Resolute*?), 1852-54.
BRIDGE, James. Cook, *Isabella*, 1818.
BRIDGENS, John. Subordinate Officers' Steward, *Erebus*, 1845-48.
BRIDGMAN, H.L. Historian, *Falcon*, 1894.
BRIGGS, John. AB, *Herald*, 1848; 1849; 1850.
BRIGHT, Charles. AB, *Herald*, 1848; 1849; 1850.
BRIGHT, John. AB, *Plover*, 1852-54.
BRISLEY, William. Boatswain's Mate, *North Star*, 1849-50. Died 30 January 1850.
BROCK, Hans. Second Mate, *Enhiörningen*, 1619-20. Died at Churchill 23 January 1620.
BROCK, Reginald Walter (1874-1935). (Director, Geological Survey of Canada). Passenger, Governor-General's tour, 1910.
BROCKENHENSER, Frederick. Oiler, *Arctic*, 1906-07. Died 11 February 1907.
BROCKLEY, James. Gunner's Mate, *Hecla*, 1821-23.
BRODIE, David. Private, Sappers and Miners, Richardson's expedition, 1847-49.
BROKEN, Hugh. Seaman, *Sunneshine*, 1586.
BROOK [BROOKE, BROOKS], Frederick [James]. Coxswain, *Assistance*, 1850-51; Captain's Coxswain, *Resolute*, 1852-54.
BROOKE, M. Gentleman, Frobisher's expedition, 1578.
BROOKER, Thomas. Private RM, *Herald*, 1848; 1849; 1850; *Plover*, 1848-52 [1850-52 only].
BROOKS, Charles. Steward, *Alaska*, Canadian Arctic Expedition, 1913-18 [1913 only].
BROOKS, Henry. Boatswain and Second Officer, *Rescue*, 1850-51; First Officer, *Advance*, 1853-55.
BROOKS, John. Cook, *Assistance*, 1852-54.
BROOM, William. Private of Marines, *Discovery*, 1776-80.
BROOMAN, John Egmont (-1858). Paymaster and Purser, *Resolute*, 1850-51. [O'B(2)].
BROTHERS, John. AB, *Alexander*, 1818; Gunner, *Hecla*, 1824-25.
BROUGHTON, William Robert (1762-1821). Lieutenant commanding *Chatham*, 1791-95. [DNB Vol. 6].
BROWN, . Captain, *Benjamin S. Wright*, 1864.
BROWN, Adam. AB, *Chatham*, 1791-95.
BROWN, Alfred. Carpenter's crew, *Phoenix*, 1853.
BROWN, E.T. Captain of the Maintop, *Herald*, 1848; 1849; 1850.
BROWN, Fraser. AB, *Discovery*, 1791-95.
BROWN, G. Captain, *Queen*, 1865-66.
BROWN, George. Armourer's Mate, *Fury*, 1824-25.
BROWN, George. AB, *North Star*, 1849-50.
BROWN, George. AB (later Quartermaster), *Investigator*, 1850-54.
BROWN, George. Captain, *Sophia*, 1853-54; 1857-58.
BROWN, Henry. AB, *Isabella*, 1818.
BROWN, Henry. Private RM, *Terror*, 1836-

BORRINGHOLM, Jens. Seaman, *Enhiörningen*, 1619-20. Died at Churchill 1 March 1620.

BORUP, George (1884-1912). Assistant, *Roosevelt*, 1908-09. [GJ 1912; EA].

BOSQUET, G.M.T. AB, *Enterprise*, 1850-55.

BOSS, David. AB, *Resolute*, 1852-54.

BOSTOCK, John. Seaman, *Resolution*, 1776-80.

BOSWELL, Thomas. Gold prospecting expedition, 1884-85.

BOTSFORD, James. Seaman, *Arctic*, 1855.

BOTTING, Arthur. Private of Marines, *Chatham*, 1791-95.

BOUCHARD, Alphé. Seaman, *Arctic*, 1908-09; Quartermaster, *Arctic*, 1910-11.

BOUCHÉ [BOUCHER], Jean-Marie. Voyageur, middleman, Black's expedition 1824. [Rich, 1955, pp. 215-18].

BOUCHER, François. Half-breed lad, HBC expedition, 1837-39.

BOUCHER, Joseph. Middleman, HBC Franklin search, 1855.

BOUCHERVILLE, C. de. Officer in charge of Port De Boucherville station, Canadian Hudson Bay Expedition, 1884-85.

BOUCHIER, Thomas. *See* BOURCHIER.

BOULANGER, Charles. Canoeman (?), Back's expedition, 1833-35.

BOULTON, Henry. Captain, *Prince Albert*, 1841.

BOURCHIER [BOUCHIER], Thomas (-1866). Second Master (?), *Herald*, 1848; 1849; Acting Second Master, *Plover*, 1848-52 [1849-52 only]; Acting Master, *Rattlesnake*, 1853-54. [O'B(2); Dawson, L.S., 1885, pt. 2, p. 85; MAR].

BOURDON, Jean (c1601-1668). Leader, French expedition, 1657. [DCB Vol. 1].

BOURGET, A. Oiler, *Arctic*, 1908-09.

BOURQUE, E. Boy, *Neptune*, 1903-04.

BOUTHLAND, Thomas. Private RM, Buchan's expedition, 1811.

BOUTILLIER, Albert. Assistant, Stupart Bay station, Canadian Hudson Bay Expedition, 1885-86.

BOW, Elias. Private RM, *Investigator*, 1850-54.

BOWDEN, . Mate, *California*, 1746-47.

BOWER, Joseph. Private RM, *Blossom*, 1825-28.

BOWGLE, William. Sailor, *Ayde*, 1578.

BOWMAN, Charles. Leading Seaman, *Fury*, 1824-25.

BOWN, Richard. AB, *Discovery*, 1791-95.

BOWREY, William. Carpenter, *Ayde*, 1578.

BOYD, John. Seaman, *Resolution*, 1776-80.

BOYE, Oluf. Seaman, *Enhiörningen*, 1619-20. Died at Churchill 8 March 1620.

BOYER, Charles. Established Old Establishment, Peace River, 1788.

BOYLE, John. AB, *Investigator*, 1850-54. Died at Mercy Bay, 6 April 1853.

BOYS, Steven. Sailor, *Ayde*, 1578.

BRAASEN, Christopher. Moravian missionary, missionary expeditions, 1771; 1774. Drowned, 14 September 1774.

BRACE, James. Seaman, *Arctic*, 1908-09; 1910-11.

BRACKENBURYE, Frauncis. Gentleman, *Ayde*, 1577; *Gabriel*, 1578.

BRADBURY, Mark. Steward, *Enterprise*, 1848-49; AB, *Investigator*, 1850-54.

BRADFORD, Abraham Rose. Surgeon, *Resolute*, 1850-51. [O'B(2)].

BRADFORD, William (1823-1892). Marine artist and leader of expedition, *Benjamin S. Wright*, 1864.

BRADLEY, A.S. Seaman, *George Henry*, 1860-62.

BRADLEY, Josiah. AB, *Phoenix*, 1854.

BRADLEY, Mark. AB, *Investigator*, 1850-54.

BRADLEY, William. Seaman, *Resolution*, 1776-80.

BRADSHAW, Benjamin. Boatswain's Mate, *Resolute*, 1850-51.

BRADY, John. Sailor, *Karluk*, Canadian Arctic Expedition, 1913-18. Died off Wrangel Island after sinking of *Karluk*, 1914.

BRADY, Robert. AB, *Phoenix*, 1854.

BRAINARD, David Legge (1856-1946). Sergeant, 2nd Cavalry, U.S. Polar Year Expedition, 1881-84. [GR 1946; Hurd 1950; EA].

BRAINE, D.L. Commander USN, commanding *Juniata*, 1873. [EA].

BRAINE, William. Private RM, *Erebus*, 1845-48. Died on Beechey Island, 3 April 1846.

37.

BROWN, James W. Caulker, *Erebus,* 1845-48.

BROWN, John. AB (later Quartermaster), *Resolution,* 1776-80.

BROWN, John. Quartermaster, *Discovery,* 1791-95.

BROWN, John. Captain of the Foretop, *Blossom,* 1825-28.

BROWN, John. Quartermaster, *Rattlesnake,* 1853-54.

BROWN, John. Seaman, *Arctic,* 1855.

BROWN, John. Seaman, *George Henry,* 1860-62. Died 17 March 1861.

BROWN, Joseph. Cook, *Arctic,* 1855.

BROWN, Mathew. AB, *Phoenix,* 1854.

BROWN, Richard. Private of Marines, *Resolution,* 1776-80.

BROWN, Robert. AB, *Prince Albert,* 1850.

BROWN, Samuel [alias HARDY, W.]. Boatswain's Mate, *Erebus,* 1845-48.

BROWN, Thomas. AB, *Discovery,* 1791-95.

BROWN, Thomas. Blacksmith, *Resolute,* 1850-51.

BROWN, Thomas. AB, *Phoenix,* 1854.

BROWN, Tom. Seaman (?), *Charles,* 1831.

BROWN, William. Seaman, *Resolution,* 1776-80.

BROWN, William. AB, *Isabella,* 1818.

BROWN, William. Carpenter, *Alexander,* 1818.

BROWN, William. Quartermaster, *Griper,* 1819-20.

BROWN, William. AB, *Plover,* 1848-52; 1852-54.

BROWN, William. Steward, *North Star,* 1849-50.

BROWNE, John Aisley. AB, *Discovery,* 1791-95.

BROWNE, Maurice. Captain, *Swallow,* 1583.

BROWNE, William Henry James (-1871). Lieutenant, *Enterprise,* 1848-49; *Resolute,* 1850-51. [O'B(2); BOA].

BROWNEL [BROWNE, BRUNEL], Oliver. (?), *Hopewell,* 1606.

BRUCE, Bruce. (?), Richardson's expedition, 1847-49.

BRUCE, H.P. Midshipman, *Assistance,* 1850-51.

BRUCE, Henry. AB, *Plover,* 1848-52; 1852-54.

BRUCE, J. Seaman, *George Henry,* 1860-62.

BRUCE, John. AB, *Alexander,* 1818.

BRUCE, Robert. Seaman (or Armourer?), *Rescue,* 1850-51.

BRUCE, Robert. Boatswain's Mate, *Arctic,* 1855.

BRUCE, W.V. Corporal, RNWMP, LaNauze's patrol, 1915-17.

BRUCE, William. AB, *Lady Franklin,* 1850-51.

BRUCHETT, S. Third Engineer, *Neptune,* 1903-04.

BRUNTON, Alexander. First Engineer, *Victory,* 1829-33.

BRUNTON, John (1794-). Midshipman, *Hecla,* 1824-25. [O'B(1)].

BRUSKOME, John. Seaman. *Sunneshine,* 1586.

BRUTON, William. Captain, *Mooneshine,* 1585; Master, *Elizabeth,* 1587.

BRYAN, R.W.D. Astronomer, *Polaris,* 1871-73. [EA].

BRYAN, William. AB, *Plover,* 1848-52.

BRYANT, David. Sergeant RM, *Erebus,* 1845-48.

BRYANT, Edward S. (Brother of John and Owen Bryant). Joint leader, U.S. expedition, 1908.

BRYANT, George. Captain of the Maintop, *Discovery,* 1875-76.

BRYANT, Henry Grier (1859-1932). Joint leader, U.S. expedition, 1891; leader, Peary Auxiliary Expedition, 1894. [GR 1933; EA].

BRYANT, John. (Brother of O. and E.S. Bryant). Joint leader, U.S. expedition, 1908.

BRYANT, Owen. (Brother of J. and E.S. Bryant). Joint leader, U.S. expedition, 1908.

BUCHAN, David (-c1838). Lieutenant commanding *Adonis,* 1811. [RNB; MAR].

BUCK, Anthony. *See* BURK.

BUCK, Charles. AB, *Rattlesnake,* 1853-54.

BUCKET, [BUCKETT], Joshua [Joseph]. AB, *Phoenix,* 1854.

BUCKINGHAM, George. (?), *Blossom,* 1825-28.

BUCKLEY, J. Seaman, *George Henry,* 1860-

62.

BUDINGTON [BUDDINGTON], Sidney O. Captain, *George Henry,* 1860-62; Sailing Master, *Polaris,* 1871-73. [EA].

BUFF, Henry. *See* BLUFF.

BUGBEE, Thomas. Private, Sappers and Miners, Richardson's expedition, 1847-49.

BUGHEE [BUGBEE], Thomas. Blacksmith, *Enterprise,* 1850-55.

BULKELEY, Vincent. Captain's Steward, *Enterprise,* 1850-55.

BULKLEY, W. F. Ensign, *Juniata,* 1873.

BULL, Bartholomew. Master, *Michael,* 1578.

BULL, George. Private of Marines, *Discovery,* 1791-95.

BULLAN, Christopher. AB, *Fury,* 1824-25.

BULLEN, George. Mate, *Breadalbane,* 1853-54.

BULLEY, Samuel. Stoker, *Discovery,* 1875-76.

BUMBRELE, Edward. AB, *Hecla,* 1824-25.

BUNN, Charles. Canoeman, GSC expedition, 1899-1900.

BUNYAN, George. Ropemaker, *Discovery,* 1875-76.

BURGESS, William James. Sergeant RM, *Herald,* 1848; 1849; 1850.

BURGOYNE, D. Sailing Master and Carpenter, Low's GSC expedition, 1897.

BURK, Anthony, Harpooner, *Victory,* 1829-33.

BURK, T.W. Seaman, *Arctic,* 1908-09.

BURNETT, . Master, *Bulldog,* 1860.

BURNETT, [BARNETT, BARNOTT], Isaac. Captain of the Maintop, *Pioneer,* 1852-54. Died of scurvy, 28 January 1854.

BURNEY, James (1750-1821). First Lieutenant, *Discovery,* 1776-80. [DNB Vol. 7; Mainwaring, 1931].

BURNIT, Rychard. Gunner, *Judith,* 1578.

BURNS, George. AB, *Enterprise,* 1850-55.

BURNS, George. AB (or Steward?), *North Star,* 1852-54.

BURNS, John. AB, *Hecla,* 1824-25.

BURNS, John. AB, *Investigator,* 1848-49.

BURROUGHS, George I. Ship's Steward, *Alert,* 1875-76.

BURROWS, Denison. Second Officer, *Florence,* 1877-78.

BURROWS, William. Cook, *Enterprise,* 1848-49.

BURTON, Thomas. AB, *Griper,* 1824.

BURWELL, W.H. Officer in charge of Port Burwell station, Canadian Hudson Bay Expedition, 1884-85.

BUSBRIDGE [BUSHBRIDGE], John Henry. Quartermaster, *Plover,* 1848-52.

BUSBRIDGE [BUSHBRIDGE], Joseph. Cook, *Plover,* 1848-52.

BUSHMAN, I. AB, *Isabella,* 1818.

BUSHNAN, John (- 1824). Midshipman and Clerk, *Isabella,* 1818; Midshipman, *Hecla,* 1819-20; Midshipman and Assistant Surveyor, *Fury,* 1821-23.

BUSTAMENTE, José de. Captain, *Atrevida,* 1791.

BUTCHER, Joseph. Private RM, *Rattlesnake,* 1853-54.

BUTCHER, Philip. AB, *Discovery,* 1791-95.

BUTCHER, Thomas. Seaman, *Resolution,* 1776-80.

BUTE, George H. Corporal RM, *Phoenix,* 1853.

BUTERMAN, James. *See* BEDUTZMAN.

BUTEUX, Jacques (1599-1652). (Jesuit missionary). Leader, French expeditions, 1651; 1652. Killed by Indians, 10 May 1652. [DCB Vol. 1].

BUTLER, . Captain, *Raleigh,* 1583.

BUTLER, George. AB, *North Star,* 1849-50.

BUTLER, James. Corporal RM, Buchan's expedition, 1811.

BUTLER, James. AB, *Resolute,* 1852-54.

BUTLER, Joel B. Seaman, *Florence,* 1877-78.

BUTLER, John. Workman, second founding of Churchill, 1717.

BUTLER, William. Seaman, *Resolution,* 1776-80.

BUTS, Thomas. (?), *Trinity* and *William,* 1536.

BUTT, George H. Sergeant RM, *Resolute,* 1852-54.

BUTT, Michael. Seaman, *Discovery,* 1610-11. Set adrift with Hudson by mutineers.

BUTTER, James. AB, *Resolute,* 1852-54.

BUTTER, John. AB, *Herald,* 1848; 1849; 1850.

BUTTERFIELD, Henry. AB, *Hecla,* 1824-25.
BUTTERS, James. Gunner's Mate, *Discovery,* 1791-95.
BUTTERWICK, William Taylor. Captain, *Esquimaux,* 1836-37.
BUTTERY, John. Private RM, *Phoenix,* 1854.
BUTTON, Thomas (-1634). Commander, *Resolution,* 1612-13. [DNB Vol. 8; DCB Vol. 1].
BUTTS, Thomas. Passenger, *William,* 1536.
BYLOT, Robert. Mate, *Discovery,* 1610-11; (?), *Resolution* and *Discovery,* 1612-13; (?), *Discovery,* 1614; Commander, *Discovery,* 1615; 1616. [DNB Vol. 8; DCB Vol. 1].
BYRNES, Michael. Joint leader, Western Union Telegraph expedition, 1867.
BYSSOT [BISSOT] DE LA RIVIÈRE, François (c1612-1673). Received Seigneurie de Mingan, 1661. [DCB Vol. 1].

CAAMAÑO, Jacinto. Captain, *Aranzazu,* 1792.
CABLETON, John. Roper, *Phoenix,* 1853.
CABOT, John. Leader, North-west Passage expeditions, 1497; 1498. [DCB Vol. 1].
CABOT, Sebastian. (Son of John Cabot). Possibly member of John Cabot's expedition, 1497; leader, North-west Passage expedition, 1508-09. [DCB Vol. 1].
CABOT, William Brooks (1858 -). Leader, U.S. expeditions, 1903-10.
CADE, . Master, *Squirrel,* 1583.
CADOTTE, J.-B. Montreal pedlars' expedition, 1775-76.
CAIN, Sylvester. Seaman, *Arctic,* 1906-07.
CALDER, John. AB, *Investigator,* 1848-49; Captain of the Forecastle, *Investigator,* 1850-54.
CALDWELL, G.F. Photographer, *Neptune,* 1903-04.
CALEY, Edward. Government Surveyor, Hind's expedition, 1861.
CALKINS, Harold A. Assistant, Curran's geological expedition, 1912.
CAMERON, Agnes Deans (1863-1912). Leader, U.S. expedition, 1908.
CAMERON, Alexander. Captain of the Foretop, *Plover,* 1848-51.
CAMPBELL, C. Captain, *Rosamond,* 1811-12.
CAMPBELL, Charles. Cooper, *Intrepid,* 1850-51.
CAMPBELL, John. AB, *Discovery,* 1791-95.
CAMPBELL, Robert. AB, *Blossom,* 1825-28.
CAMPBELL, Robert (1808-1894). Established Dease Lake post, 1836-37; leader, HBC expeditions, 1840; 1842-44; established Fort Selkirk, 1848; leader, HBC expedition, 1851. [Parnell and Kirk, 1942; EA].
CAMPBELL, Walter F. Fireman, *Polaris,* 1871-73.
CAMSELL, Charles (1876-1958). Canoeman, GSC expedition, 1899-1900; leader, GSC expeditions, 1902; 1905; 1914. [Camsell, 1954; Bostock and Rowley, 1959; Shaw, M.M., 1958].
CANE, James Frederick. Armourer, *Alert,* 1875-76.
CANN, George J. AB, *Terror,* 1845-48.
CANNON, Thomas. AB, *Hecla,* 1824-25.
CANON, Hugh. Artilleryman, Back's expedition, 1833-35.
CANOT, Auguste. Seaman, *Rescue,* 1850-51.
CAPATO, Spero. Captain's Steward. *Alert,* 1875-76.
CARDINAL, Louis. Canoeman, Camsell's GSC expedition, 1905.
CAREW, Henry. Gentleman, *Ayde,* 1577; Captain, *Hopewell,* 1578.
CAREY, George. AB, *Fox,* 1857-59.
CAREY, William. Gunner's Mate, *Release,* 1855.
CARLER, Robert C. Acting Master, *Rescue,* 1850-51.
CARLEY, Isaac. Private of Marines, *Resolution,* 1776-80.
CARLYLE, Henry. AB, *Terror,* 1845-48.
CARLYLE, Joseph. AB, *Terror,* 1845-48.
CARMACK, George Washington. Discovered gold on Klondike River, 1896.
CARMICHAEL, Thomas S. AB, *Investigator,* 1850-54.
CARNE, William. AB, *Investigator,* 1848-49.
CARNES, Charles. Leading Seaman, *Hecla,* 1824-25.
CARON, Wilfred C. Seaman, *Arctic,* 1910-11; (?), *Minnie Maud,* 1912-13.
CARPENTER, William. Private RM, *Alexan-*

der, 1818.

CARR, . (*Griper,* 1824). *See* KERR.

CARR, Henry. AB, *Plover,* 1848-52 [1850-52 only].

CARRICK, John. Sailmaker, *Plover,* 1848-52; 1852-54.

CARRIKER, M.A. Jr. Assistant, Carnegie Museum expedition, 1908.

CARROLL, Lewis, Captain of the Forecastle, *Enterprise,* 1850-55.

CARROLL, William. AB, *Investigator,* 1848-49; 1850-54.

CARRON, Hugh. *See* CANON.

CARRUTHERS, John. Workman (?), second founding of Churchill, 1717.

CARSTAIRS, John. Boatswain's Mate, *Enterprise,* 1850-55.

CARTER, Edward. Captain of the Maintop, *Rattlesnake,* 1853-54.

CARTER, John. AB, *Discovery,* 1791-95.

CARTER, Mark. Mate, *Sunneshine,* 1586.

CARTER, Robert R. Passed Midshipman, Acting Master and First Officer, *Rescue,* 1850-51. [EA].

CARTIER, Jacques (1491-1557). Commander, French expeditions, 1534; 1535-36; 1541-42. [DCB Vol. 1].

CARTIER, Laurent. Bowman, Dease and Simpson expedition, 1837-39.

CARTMEL, Daniel. Engineer, *Discovery,* 1875-76.

CARTWRIGHT, Henry. Leading Stoker, *Phoenix,* 1853.

CARTWRIGHT, John. Preacher, *Discovery,* 1602.

CARUTHERS, Gibson. Seaman (or Carpenter's Mate?), *Advance,* 1850-51; Boatswain and Carpenter, *United States,* 1860-61. Died at Upernavik, August 1860.

CARY, Allwyn. Agent and supercargo, *Discovery,* 1616.

CARY, Austin (1865-1936). Joint leader, U.S. expedition, 1891.

CARY, Merritt (1880-1918). Assistant, U.S. biological expedition, 1903-04.

CASAR, David. Private RM, *Blossom,* 1825-28.

CASPERSEN, Casper. Surgeon, *Enhiörningen,* 1619-20. Died at Churchill 21 March 1620.

CASTEL, Aarnout. Master, *North Star* and *Challenge,* Canadian Arctic Expedition, 1913-18.

CATER, Thomas. Carpenter's Mate, *Discovery,* 1776-80.

CATLEY, Josh. Private RM, *Hecla,* 1819-20.

CATOR, John Bertie (1820-). Lieutenant commanding *Intrepid,* 1850-51. [O'B(1)-(2)].

CATTO, J. AB, *Felix,* 1850-51.

CAULKER, John. AB, *Fury,* 1821-23.

CAULKIN, Thomas Benjamin. Sergeant-Major RNWMP, patrol, 1916-18.

CAVE, John. Seaman, *Resolution,* 1776-80.

CAWN, Edward. Seaman, *Resolution,* 1776-80.

CHAFE, Ernest F. Messroom boy, *Karluk,* Canadian Arctic Expedition, 1913-18.

CHAKE [CHACKE, CHAQUE], Martin. North-west Passage expedition, 1555.

CHALKLEY, Thomas. AB, *Discovery,* 1875-76.

CHALLIS, Abraham. Sergeant RM, *Plover,* 1848-52; 1852-54.

CHAMBERES, Rychard. Soldier, Frobisher's expedition, 1578.

CHAMBERLAINE, Thomas. Gentleman, *Michael,* 1577; *Gabriel,* 1578.

CHAMBERLIN, J. Quartermaster, *Herald,* 1848; 1849; 1850.

CHAMBERLIN, James. AB, *Hecla,* 1821-23.

CHAMBERLIN, Thomas Chrowder (1843-1928). Geologist, *Falcon,* 1894. [EA].

CHAMBERS, George. Boy, *Erebus,* 1845-48.

CHAMBERS, Washington I. Ensign, *Thetis,* 1884.

CHAMBERS, William. Boatswain's Mate, *Plover,* 1852-54.

CHAMPION, George. Greenland Mate *Hecla,* 1824-25.

CHAMPLAIN, Samuel de (c1570-1635). Established Quebec, 1608. [DCB Vol. 1].

CHANDLER, Thomas. Boatswain, *Herald,* 1848; 1849; 1850; *Intrepid,* 1852-54.

CHAPIES, Job. (Cree Indian). Guide, Hubbard expedition, 1905; Tasker expedition, 1906.

CHAPMAN, George. AB, *Herald,* 1848; 1849; 1850.

CHAPMAN, J.H. AB, *Phoenix,* 1854.

CHAPPELL, Edward (1792-). Lieutenant, *Rosamund,* 1814. [O'B (1)-(2)].

CHAPPLE, George. AB, *Plover,* 1848-52 [1851-52 only].

CHAQUE, Martin. *See* CHAKE.

CHARLES, Samuel. Cook, *Herald,* 1848; 1849; 1850.

CHARLESWORTH, J. AB, *Rattlesnake,* 1853-54.

CHARLOIT, Antoine de. (?), Back's expedition, 1833-35. Discharged 1833.

CHARLOW, William. Ordinary Seaman, *Herald,* 1848; 1849; 1850.

CHARLTON, John Fitzgerald. Surgeon, *Phoenix,* 1853. [O'B (2)].

CHARLTON [CHARLETON], William. Seaman. *Resolution,* 1776-80.

CHARRON, Harry. Waiter, *Arctic,* 1906-07.

CHARTERS, John. AB, *Terror,* 1836-37; *Enterprise,* 1848-49; 1850-55.

CHARTIER, Joseph (c1817-). Middleman, Richardson's expedition, 1847-49. [Rich, 1953, p. 354].

CHARTRAND, Benoid. Waiter and portager, *Arctic,* 1910-11.

CHASSÉ, Napoléon. Quartermaster, *Arctic,* 1906-07; 1908-09; Boatsman, *Arctic,* 1910-11; (?), *Minnie Maud,* 1912-13.

CHATEL, Frank. Captain of the Forecastle, *Discovery,* 1875-76.

CHAUNCELLOR [CHAUNSLER], Nicholas. Purser, *Gabriel,* 1576; Frobisher's expeditions, 1577; 1578.

CHAUVIN DE TONNETUIT, Pierre de (-1603). Commander, French expedition, 1600-02. [DCB Vol. 1].

CHAWCHINAHAW, Captain. (Cree Indian). Guide to Hearne, Coppermine River expedition, 1769.

CHEADLE, Walter Butler (1835-1910). Joint leader, British expedition, 1862-64. [GJ 1910; WWW Vol. 1].

CHEATHAM, William. Private RM, *Isabella,* 1818.

CHEESEMAN, William. AB, *Enterprise,* 1850-55.

CHERRY, William T. Caulker, *Assistance,* 1850-51; *Phoenix,* 1854.

CHESNAYE, Pierre Duquet de la. *See* DUQUET DE LA CHESNAYE.

CHESSELL, John. AB, *Enterprise,* 1848-49.

CHESTER, H.C. Mate, *Polaris,* 1871-73; First Mate, *Monticello,* 1864-65 [see 1864-69 Hall].

CHEYNE, John Powles (1826-1902). Midshipman, *Enterprise,* 1848-49; *Resolute,* 1850-51; Lieutenant, *Assistance,* 1852-54. [O'B(2); Gould, R.T., 1935].

CHEYNE, Lawrence. AB, *Terror,* 1836-37.

CHILTON, Robert. Accompanied Clouston, 1819-20.

CHIMHAM, Thomas. Carpenter, *Victory,* 1829-33.

CHIMMO, William (1828-1891). Mate, *Herald,* 1848; 1849; 1850; Commander RN, commanding *Gannet,* 1867. [O'B(2); BOA; Dawson, L.S., 1885].

CHINCH, James, AB, *Phoenix,* 1854.

CHIPMAN, E.J. Acting Master and Assistant Ice-master, *Tigress,* 1873.

CHIPMAN, Kenneth Gordon (1884-). Topographer, Canadian Arctic Expedition, 1913-18.

CHIPP, Charles W. (-1881). Lieutenant USN, *Juniata,* 1873. [NE; EA].

CHOUART DES GROSEILLIERS, Médard (c1618-c1696). French expedition, 1659-60; New England voyage, 1663; Supercargo, *Eaglet,* 1668-69; *Prince Rupert,* 1670-71; *Ste Anne,* 1682-83. [Rich, 1942, pp. 231-36; DCB, Vol. 1].

CHRICHTON, George. *See* CRICHTON.

CHRISTIAN, Anton. (Greenland Eskimo). Interpreter, *Fox,* 1857-59.

CHRISTIANSEN, Thorlip Frederick. (Greenland Eskimo). Dog-driver, U.S. Polar Year Expedition, 1881-84. Died 5 April 1884.

CHRISTIE, J.M. Assistant to Keele, GSC expedition, 1907-08.

CHRISTOPHER, William. Captain, HBC voyages. *Churchill,* 1761; 1762; 1763; *Seahorse* [II], 1770; 1771; 1772; 1773; 1774; 1775; 1776; 1777; 1778; 1779; 1780; 1781; *Prince Rupert* [IV], 1782; *King George* [III], 1783; 1784; 1785; 1786; 1787; 1788.

CHURCHYARD, John. Pilot, *Ellen,* 1587.

CILLEY, Jonathan Prince (1835-1920). Member of U.S. expedition to Churchill

Falls, 1891.
CLARK [CLARKE], Daniel. AB, Richardson's expedition, 1847-49.
CLARK, E.J. Private RM, *Plover,* 1848-52; 1852-54.
CLARK, Eleazer John. Private RM, *North Star,* 1852-54.
CLARK, George. Workman, second founding of Churchill, 1717.
CLARK, J. Seaman, *Neptune,* 1903-04.
CLARK, James. AB, *Isabella,* 1818.
CLARK, John. AB, *Rattlesnake,* 1853-54.
CLARK, John A. Cook, *North Star,* 1849-50.
CLARK, Joseph. Steward, *North Star,* 1852-54.
CLARK, Scoville. Joint leader, Cabot's expedition, 1910 [see 1903-10].
CLARK, William. Private of Marines, *Chatham,* 1791-95.
CLARKE, . Leader, exploration of Mackenzie River, 1809.
CLARKE, Charles. Fireman, *Roosevelt,* 1905-06.
CLARKE, Edward. Sergeant RM, *Plover,* 1848-52; 1852-54.
CLARKE, James. Captain of the Maintop, *Blossom,* 1825-28.
CLARKE, James. (?), *Nancy Dawson,* 1849.
CLARKE, John. Established Neoskweskau post, 1793.
CLARKE, John (1781-). Leader, HBC expeditions, 1815-16; 1816-17; second-in-command, HBC expedition, 1818-19. [Rich, 1938, pp. 433-35].
CLARKE, John. AB, *Plover,* 1848-52.
CLARKE, Richard M. Surgeon's Steward, *Release,* 1855.
CLARKE, Thomas. Midshipman, *Discovery,* 1791-95.
CLARKE, William. Captain of the Foretop, *Herald,* 1848; 1849; 1850.
CLARKE, William. Stoker, *Phoenix,* 1853.
CLARKSON, Alexander. Private RM, *North Star,* 1849-50.
CLARKSON, James [Joseph]. Boatswain, *Blossom,* 1825-28.
CLARRY, John. Private RM, *Rattlesnake,* 1853.
CLAY, Henry. "Military employé," U.S. Polar Year Expedition, 1881-84 [1881 only].
CLAY, Job. Seaman, *Resolution,* 1776-80.
CLEARKE, Richard. Master, *Delight,* 1583.
CLEMENDSEN, Rasmus. Gunner's Mate, *Enhiörningen,* 1619-20. Died at Churchill, 5 April 1620.
CLEMENS, Francis. Boatswain, *Discovery,* 1610-11.
CLEMENTS, Peter. Captain, HBC supply voyages. *Prosperous,* 1713-14; *Diligence,* 1717.
CLEMENTS, William. Lieutenant, *Henrietta Maria,* 1631-32.
CLENCH, Joseph. Private RM, *Enterprise,* 1848-49.
CLERKE, Charles (1743-1779). Commander RN, commanding *Discovery,* 1776-80. Died at sea, 22 August 1779. [DNB Vol. 11].
CLEVELY, James. Carpenter, *Resolution,* 1776-80.
CLEVERLY [CLEVERLEY], John. Gunner's Mate, *Intrepid,* 1852-54.
CLINCH, James. AB, *Phoenix,* 1854.
CLOSSON [CLOSSAM], William. AB, *Erebus,* 1845-48.
CLOUSTON, James. (?), Rae's expedition, 1853-54. [Rich, 1953, p. 354].
CLOUSTON, James. Leader, HBC expedition, 1819-20. [Davies, 1963, pp. 341-52].
CLYSDALE, Hugh. Cook, *Hecla,* 1819-20.
COAD, James. Cook, *North Star,* 1849-50.
COADY, John. Seaman, *Roosevelt,* 1908-09.
COADY, William. Seaman, *Arctic,* 1906-07.
COATESWORTH, John. Boatswain's Mate, *Charles,* 1631.
COATS, William (-1751). Captain, HBC supply voyages. *Mary* [II], 1727; *Hannah,* 1728; 1729; 1730; 1731; 1732; 1733; *Hudson's Bay* [IV], 1734; 1735; 1736; *Mary* [III], 1737; *Mary* [IV], 1738; 1739; 1740; 1741; 1742; 1743; 1744; 1745; 1746; 1747; 1748; 1749; *King George* [I], 1750; 1751. [DCB Vol. 3; Briffault, 1950].
COBBIE, Walsall. Captain, *Colleton,* 1680-81. [Rich, 1945, pp. 335-36; DCB Vol. 1].
COBHAM, John. Private RM, *Plover,* 1848-52.
COBRETH [COBREATH, COLBERT], William. Master, *Discovery,* 1602.

COCKING, Matthew (-1799). Leader, HBC expedition, 1772-73. [DCB Vol. 4].

COCKRAM, Edward. Corporal RM, *Plover,* 1848-52; 1852-54.

CODE, James. AB, *Phoenix,* 1854.

COFFIN, George W. Commander USN, commanding *Alert,* 1884. [NE].

COFFIN, N.J. Carpenter, *Polaris,* 1871-73.

COGHLAN, John. AB, *Resolute,* 1852-54.

COGLIN, Jeremiah. AB, *Resolute,* 1852-54.

COLAN, Thomas. Fleet Surgeon, *Alert,* 1875-76. [MAR].

COLE, Benjamin. AB, *Hecla,* 1821-23.

COLE, D.M. Joint leader, U.S. expedition, 1891.

COLE, James. Musician, *Sunneshine,* 1585.

COLE, James. Stoker, *Pandora,* 1875.

COLE, John. HBC expedition, 1772-73.

COLE, Samuel. Captain, *Employ,* 1672-75.

COLE, William. Carpenter, *Henrietta Maria,* 1631-32. Died 18 May 1632.

COLE, William Sea. Intending colonist, Frobisher's expedition, 1578.

COLEMAN, Arthur P. Leader, Canadian expedition, 1915-16.

COLEMAN, James. Quartermaster, *Discovery,* 1776-80.

COLEMAN, Joseph. Captain's Coxswain, *Enterprise,* 1848-49.

COLEMAN, Patrick. Seaman, Hall's expedition, 1864-69 [1867-68 only]. Died 14 August 1868.

COLEMAN, William. Seaman, *Isabella,* 1818.

COLES, William [Tom]. Gunner's Mate, *Resolute,* 1850-51.

COLLEGE, Paul. AB, *Hecla,* 1824-25.

COLLETT, Joseph. Seaman and Gunroom Servant, Cook's expedition, 1776-80.

COLLETT, Richard. AB, *Discovery,* 1776-80; Gunner, *Discovery,* 1791-95.

COLLETT, William. Seaman, *Resolution,* 1776-80.

COLLIE, Alexander (1793-1835). Surgeon, *Blossom,* 1825-28.

COLLINS, . Special Constable RNWMP, Governor-General's tour, 1910.

COLLINS, George. Ordinary Seaman, *Herald,* 1848; 1849; 1850.

COLLINS, Henry Foster. Second Master, *Erebus,* 1845-48.

COLLINS, Michael. Ice Quartermaster, *Assistance,* 1850-51.

COLLINS, Thomas. Paymaster's Steward, *Herald,* 1848; 1849; 1850.

COLLINSON, Richard (1811-1883). Captain and commander of expedition, *Enterprise,* 1850-55. [DNB Vol. 11; RGS 1883; O'B (1)-(2); Dawson, L.S., 1885 DCB Vol. 3].

COLLINSON, Thatwell Benjamin. (Nephew of Richard Collinson). Acting Mate, *Herald,* 1848; 1849; 1850. [O'B(2)].

COLMAN, Patrick. AB, *Rattlesnake,* 1853-54.

COLOMB, Philip Howard (1831-1899). Mate, *Phoenix,* 1854. [DNB; O'B(2); BOA].

COLVIL, William. *See* COLWILL.

COLWELL, John C. Lieutenant USN, *Yantic,* 1883; *Bear,* 1884.

COLWELL, Thomas. Seaman, whalers' boat voyage, 1860.

COLWILL, William. Blacksmith, *Assistance,* 1850-51; *Resolute,* 1852-54.

COMEAU, Napoleon Alexander (1848-1923). Leader, fisheries research expedition, 1914.

COMER, George. Captain, *A.T. Gifford,* 1907-09; Ice Pilot, *George B. Cluett,* "Crocker Land" Expedition, 1913-17 [1915-17 only].

COMLY [COMLEY], Samuel P. Ensign, *Juniata,* 1873. [NE].

COMMANDA, Paul. (Ojibwa Indian). Guide and canoeman, Carnegie Museum expeditions, 1912; 1914-15; 1917.

COMMYNGES, John. Purser, *Ayde,* 1578.

COMPTE, Eustache le. Assistant to Cuthbert Grant, 1785-88.

COMSTOCK, R.A. Seaman, *George Henry,* 1860-62.

CONE, Eleazor. Steward, *Florence,* 1877-78.

CONGER, Nycholas. Soldier, Frobisher's expedition, 1578.

CONLEY, W. Seaman, *George Henry,* 1860-62.

CONNE-E-QUESE. (Chipewyan Indian). Guide to Hearne, 1770.

CONNEL, John. Armourer, *Isabella,* 1818.

CONNELL, Maurice. Private, 3rd Cavalry, U.S. Polar Year Expedition, 1881-84.

CONNELLY, . Constable NWMP, *Nep-*

tune, 1903-04.

CONNELLY, John. Seaman, *Resolution,* 1776-80.

CONNELLY, Michael. Seaman, *North Star,* 1849-50.

CONNOLLY, J. AB, *Rattlesnake,* 1853-54.

CONNORS, John. Seaman, *Roosevelt,* 1908-09.

CONNOLLY, John. Ordinary Seaman, *Blossom,* 1825-28.

CONSTANTINE, Charles. Superintendent NWMP, established Fort McPherson detachment, 1903.

CONWAY, P.R. Constable RNWMP, Pelletier's patrol, 1908-09.

CONYBEARE, Crawford John Markland (1854-1937). Sub-Lieutenant, *Discovery,* 1875-76. [Conybeare, 1930; www].

CONYER, Nicholas, Servant, Frobisher's expedition, 1577.

COOK, Frederick Albert (1865-1940). Leader, U.S. North Polar Expedition, 1907-09. [DAB: Gibbons, 1964].

COOK, James (1728-1779). Commander, *Resolution,* 1776-80. Killed at Kealakekua Bay, Hawaii, 14 February 1779. [DNB Vol. 12; Kitson, 1907; Campbell G., 1936].

COOK, John. AB, *Discovery,* 1791-95.

COOKE, Rychard. Tailor, Frobisher's expedition, 1578.

COOKSLEY, F. Sergeant, Royal Horse Artillery, and Observer, British Polar Year Expedition, 1882-83.

COOMBES, Henry. Ordinary Seaman, *North Star,* 1849-50.

COOMBES, John. Stoker, *Intrepid,* 1852-54.

COOMBES, Thomas [William]. Carpenter's crew, *Investigator,* 1848-49. Died 27 October 1848.

COOMBS, Charles. AB, *Erebus,* 1845-48.

COOMBS, John. *See* COOMBES.

COOMBS, Thomas. *See* COOMBES.

COONING, John. AB, *Isabella,* 1818.

COOPER, . Captain, *Earl Fauconberg,* 1820.

COOPER, . Captain, *Aurora,* 1837.

COOPER, Edward James Lloyd (1814-1852). Lieutenant, *Herald,* 1848; 1849; *Plover,* 1848-52 [1849-51 only]. [O'B(1); MAR].

COOPER, James. Second Captain of the Maintop, *Discovery,* 1875-76.

COOPER, Valentine, Boatswain's Mate. *Fury,* 1824-25.

COOPER, William. AB, *Enterprise,* 1850-55.

COOPER, William. AB, *North Star,* 1852-54.

COOPER, William. AB, *Phoenix,* 1853.

COOTE, James. AB, *Chatham,* 1791-95.

COPELAND [COOPLAND, COPLAND, COPLANDS], Thomas. AB, *Assistance,* 1850-51; *Pioneer,* 1852-54.

COPPINGER, Richard William (1847-1910). Surgeon, *Discovery,* 1875-76. [DNB 2nd Supp. Vol. 1; British Medical Journal, 1910].

CORBET, Owen. Sailor, *Ayde,* 1578.

CORK, John. Private RM, *Plover,* 1852-54.

CORMACK, William Epps (1796-1868). Leader, first crossing of Newfoundland, 1822.

CORNELIUS, James. Boatswain's Mate, *Herald,* 1848; 1849; 1850; *Resolute* 1852-54.

CORNISH, Robert. Musician. *Sunneshine,* 1585.

CORNISHE, John. Sailor, *Ayde,* 1578.

CORRIGAL, James. Canoeman, Tyrrell's expedition, 1893.

CORRIGAL, John. Steersman, Rae's expedition, 1846-47. [Rich, 1953, p. 355].

CORTE-REAL, Gaspar. Commander, Portuguese expeditions, 1500; 1501. [DCB Vol. 1].

CORTE-REAL, Miguel. (Brother of Gaspar Corte-Real). Commander, Portuguese expedition, 1502. [DCB Vol. 1].

CORVIC [CORVIL], John [alias SMITH, John]. (?), *Erebus,* 1845-48.

COSTA [COSTER], Louis. Seaman (or Armourer?), *Advance,* 1850-51.

COTESWORTH, Charles. Captain, *Pery,* 1692-93.

COTTELL, John. AB, *Resolute,* 1850-51.

COTTERELL, John. AB, *Fury,* 1824-25. Drowned at Port Bowen, 6 July 1825.

COTTMAN, John. AB, *Nancy Dawson,* 1849.

COTTON, John. Sailor, *Ayde,* 1578.

COUCH, Edward. Mate. *Erebus,* 1845-48.

COUCHAI, Jacko. (Ojibwa Indian). Guide and canoeman, Carnegie Museum expeditions, 1912; 1914-15.

COULDERY, W.S. Acting Master, *Enterprise,* 1848-49.

COUNZER, Nicholas. (?), Frobisher's expedition, 1578.

COURNOYÉ [COURNOYER, CORNOYER], Joseph. Middleman, Black's expedition, 1824. [Rich, 1955, pp. 218-19].

COURNOYEE [COURNOYER, COURNORGEE], Emanuel. Canadian voyageur, Franklin's expedition, 1819-22.

COURT, Stephen (1826-1861). Second Master, *Investigator,* 1850-54; *Enterprise,* 1848-49. [O'B(2); MAR].

COURTNEY, . Captain, *Emmanuel,* 1578.

COURTOIS, François. Canadian voyageur, Mackenzie's expedition, 1792-93.

COUSINS, George. Stoker, Pioneer, 1852-54.

COUSINS, Thomas. (?), Richardson's expedition, 1847-49.

COUSINS, Thomas. AB, *Enterprise,* 1850-55.

COUSINS, Thomas, AB, *Plover,* 1848-52; 1852-54.

COUTURE, Guillaume (c1616-1701). Leader, French expedition, 1663.

COVEY, . Navigating sub-Lieutenant, *Gannet,* 1867.

COWIE, John. Stoker, *Erebus,* 1845-48.

COWIE, Robert. Medical Officer, *Prince Albert,* 1851-52.

COWLEY, Rychard, Intending colonist, *Gabriel,* 1578.

COX, John Raffles (1887-). Assistant Topographer, Canadian Arctic Expedition, 1913-18.

COX, Joseph. Quartermaster, *Discovery,* 1776-80.

COXE, Richard. Master Gunner, *Ayde,* 1577; Pilot, *Thomas,* 1578.

COXE, William. Master, *Golden Hind,* 1583.

COXWORTHIE, Robert. Sailor (?), *Sunneshine,* 1585.

COYER, Joe. Prospector, Curran's expedition, 1912.

COYLE, Neil. Private of Marines, *Discovery,* 1791-95.

CRACKNELL, Lyonell. Labourer, Frobisher's expedition, 1578.

CRAIG, Peter. AB, *Discovery,* 1875-76.

CRAIG, Walter. AB, *Lady Franklin,* 1850-51.

CRANDELL, Cecil. Acting Surgeon, *Daedalus,* 1851. [O'B(2)].

CRANE, George. AB, *Rattlesnake,* 1853-54.

CRANSTON, George. AB, *Alert,* 1875-76.

CRANSTOUN, A.P. Surgeon, *Discovery,* 1791-95.

CRAWFORD, George. Greenland Mate, *Hecla,* 1819-20; *Fury,* 1821-23; 1824-25.

CRAWFORD, George. AB, *Fury,* 1824-25.

CRAWFORD, J. R. (?), Canadian Arctic Expedition, 1913-18.

CRAWLEY, John. First Class Volunteer, *Blossom,* 1825-28.

CREAKE, Alexsander, Sailor, *Ayde,* 1578.

CREBER, Richard J. Quartermaster, *Plover,* 1848-52; 1852-54.

CREELMAN, D. Assistant, Ashe Inlet station, Canadian Hudson Bay Expedition, 1885-86.

CREESE, Charles. AB, *Phoenix,* 1853.

CREIGHTON, J. G. Alwyn. Leader, Canadian expedition, 1890.

CRESPI, *Father.* (?), Spanish expedition, 1774.

CRESSWELL, Daylas. Captain's Steward, *Rattlesnake,* 1853-54.

CRESSWELL, Samuel Gurney (-1867). Mate, *Investigator,* 1848-49. Lieutenant, *Investigator,* 1850-54. [O'B(2); BOA: DCB Vol. 9].

CREW, George. Stoker, *Phoenix,* 1853.

CREWS, Arthur. Midshipman, *Discovery,* 1791-95.

CRICHTON [CHRICHTON], George. Assistant Surgeon, *Enterprise,* 1848-49.

CRISPE, Samuel. AB, *Terror,* 1845-48.

CROCKER [CROKER], George. AB, *Plover,* 1848-52; 1852-54.

CROCKFORD, James. AB, *Alexander,* 1818.

CROGGAN, Samuel. Carpenter's crew, *Herald,* 1848; 1849; 1850.

CROOK, Edwin. Master, *Phoenix,* 1854.

CROOKSHANKS, George. AB, *Fury,* 1824-25.

CROPP, John. Gunner, RM, *Discovery,* 1875-76.

CROSBY, Freeman H. Lieutenant USN, executive and navigating officer, *Bear,* 1884.

CROSS, William H. Sergeant in General Service, U.S. Army, U.S. Polar Year Expedition, 1881-1884. Died of scurvy 18 January 1884.
CROSSE, William. Boatswain, *Sunneshine,* 1585.
CROSSLEY, Thomas. Private RM, *Phoenix,* 1854.
CROSSMAN, J. Second Engineer, *Neptune,* 1903-04.
CROSSMAN, J. A. AB, *Phoenix,* 1853.
CROSSMAN, W. First Engineer, *Neptune,* 1903-04.
CROTCH, William. Seaman, *Resolution,* 1776-80.
CROTEAU, Ernest. Electrician, *Arctic,* 1906-07.
CROW, Robert. Captain, *Musquash,* 1736-37.
CROZIER, Francis Rawdon Moira (1800-1848). Midshipman, *Fury,* 1821-23; *Hecla,* 1824-25; Captain, *Terror,* 1845-48. [DNB Vol. 13; O'B(1); Fluhmann, 1976].
CULL, William. Guide. Buchan's expedition, 1811.
CULLIMORE, John. Private RM, *North Star,* 1849-50.
CULVER, William. AB, *Pioneer,* 1850-51; *Resolute,* 1853-54.
CUMBER, Henry. Officers' Cook, *Pioneer,* 1850-51.
CUMMINGS, John AB, *Discovery,* 1791-95.
CUNDY, William. Captain of the Hold, *Investigator,* 1848-49. Died 12 May 1849.
CUNNINGHAM, John. AB, *Investigator,* 1848-49; *Assistance,* 1850-51.
CUNNYNGHAM [CONINGHAM], Thomas. Sailor, *Ayde,* 1578. Died 25 September 1578.
CURRAN, James. AB, *Alexander,* 1818. Discharged 23 July 1818.
CURRAN, William Tees. Leader, geological expeditions, 1907; 1912.
CURRIE, Alexander. Middleman, Franklin's expedition, 1825-27.
CURTIS, James [Joseph]. Harpooner, *Victory,* 1829-33.
CURTIS, John. AB, *Isabella,* 1818.
CURTIS, John. Private RM, *Hecla,* 1824-25.
CUSTANCE, George. Stoker, *Intrepid,* 1850-51; *Pioneer,* 1852-54.
CUTBUSH, William. Private RM, *Assistance,* 1852-54. Died on board, 27 February 1853.
CUTHBERT, . Midshipman, *Gannet,* 1867.
CUTTER, John. Cooper, *Ayde,* 1578.
CUTTILL, John. AB, *Resolute* (or *Assistance*), 1850-51.

DABLON, Claude (c1619-1697). (Jesuit missionary). Leader, overland expedition, 1661. [DCB Vol. 1].
DADD, John. Cox of Pinnacle, *Rattlesnake,* 1853. Died 29 October 1853.
D'AIGNEAUX, . Montreal representative of Revillon Frères, passenger, *Eldorado,* 1903.
DAILEY, Mathew. Seaman, *Resolution,* 1776-80.
DALL, Robert. Private, Sappers and Miners, Richardson's expedition, 1847-49.
DALY, Alexander. Seaman, *Rescue,* 1850-51.
DALY, James. Private RM, *Terror,* 1845-48.
DALY, Reginald Aldworth (1871-1957). Joint leader, Brown-Harvard expedition, 1900. [Geological Society of London, 1959].
DAMARESQ, James. AB, *Enterprise,* 1850-55.
DANIEL, George. Middleman, HBC Franklin search, 1855.
DANNETT, . Captain, *Prince of Wales,* 1845.
DARDES, Roger. Labourer, Frobisher's expedition, 1578.
DARKE, John. AB, *Hecla,* 1824-25.
DARKE, Thomas. Private RM, *Discovery,* 1875-76.
DARLINGTON, Thomas. Caulker, *Terror,* 1845-48.
DARRELL, Hubert. Assistant, Hanbury's expedition, 1901-02; Harrison's expedition, 1905-07 (1905 only). [Neufeld, 1973].
DAVID, Peter. AB, *Griper,* 1819-20.
DAVIDGE, James. AB, *Enterprise,* 1850-55.
DAVIDSON, . Carpenter, *Isabel,* 1852.
DAVIDSON, Edward. (?), *Nancy Dawson,* 1849.

DAVIDSON, J. AB, *Lady Franklin*, 1850-51.
DAVIDSON, J.E. AB, *Enterprise*, 1848-49; 1850-55.
DAVIDSON, James. Carpenter's mate, *Enterprise*, 1848-49.
DAVIDSON, James. Carpenter's mate, *Phoenix*, 1853.
DAVIDSON, John. AB, *Discovery*, 1791-95.
DAVIE, Henry. Gunner, *Sunneshine*, 1585.
DAVIES, Harry. Sailor, *Ayde*, 1578.
DAVIES, John. AB, *Discovery*, 1791-95.
DAVIES, John. Bombardier, RMA, *Resolute*, 1850-51.
DAVIES, John. AB, *Investigator*, 1850-54.
DAVIES, Robert. Surgeon's mate, *Resolution*, 1776-80.
DAVIES, W.H.A. Leader, HBC expedition, 1840.
DAVIES, W.W. AB, *Enterprise*, 1850-55.
DAVIES, William. Sailor, *Ayde*, 1578.
DAVIES, William. AB, *Hecla*, 1824-25.
DAVIS, David. Seaman, *Rescue*, 1850-51.
DAVIS, George. AB, *Assistance*, 1850-51.
DAVIS, Hiram J. Seaman, whalers' boat voyage, 1860.
DAVIS [DAVYS], John (c1552-1605). Captain, *Sunneshine*, 1585; *Mermaid*, 1586; *Elizabeth*, 1587. [Markham, C.R., 1889; DCB Vol. 1; DNB Vol. 14].
DAVIS, John. Seaman, *Resolution*, 1776-80. Died at sea, 20 September 1780.
DAVIS, John. AB, *Herald*, 1848; 1849; 1850.
DAVIS, John. AB, *Plover*, 1848-52; 1852-54.
DAVIS, Robert. Master, *Gabriel*, 1578.
DAVIS, Robert. Master, *Raleigh*, 1583.
DAVIS, Robert. AB, *Phoenix*, 1853.
DAVIS, Thomas. AB, *Herald*, 1848; 1849; 1850.
DAVIS, W.W. *See* DAVIES, W.W.
DAVIS, William. Quartermaster, *North Star*, 1849-50.
DAVIS, William. Quartermaster, *Herald*, 1848; 1849; 1850.
DAVIS, William. AB, *Pandora*, 1875.
DAVISON, John. Captain, HBC supply voyages. *Mainwaring*, 1807; *Robert Taylor*, 1812; *Hadlow*, 1815-16; *Eddystone*, 1817-18; *Prince of Wales* [I], 1818; 1819-20; 1820; 1821; 1822; 1823; 1824; 1825; 1826; 1827.
DAW, Walter. AB, *Plover*, 1852-54.
DAWBERRY [DAWBENY], Oliver. Passenger, *William*, 1536.
DAWSON, George Mercer (1849-1901). Leader, Yukon exploring expedition, 1887. [EA; DNB 2nd Supp. Vol. 1].
DAWSON, Henry P. Captain RA and leader, British Polar Year Expedition, 1882-83.
DAWSON, James. AB, *Assistance*, 1850-51.
DAWSON, Robert. AB, *Fury*, 1824-25.
DEACON, Thomas. AB, *Chatham*, 1791-95.
DEALY [DEALEY], William Justin. Midshipman, *Hecla*, 1819-20. [O'B(1)-(2)].
DEAN, Alfred. (?), *Nancy Dawson*, 1849.
DEAN, John. Gunner's Mate, *Herald*, 1848; 1849; 1850.
DEAN, John. Quartermaster, *Plover*, 1848-52.
DEAN, Samuel J. Carpenter's Mate, *Intrepid*, 1852-54.
DEAN [DEANS], William. Carpenter, *Investigator*, 1848-49; *Assistance*, 1850-51; *Resolute*, 1852-54. [MAR].
DEAN, William. AB, *Breadalbane*, 1853.
DEANE, W.L. Gunner's Mate, *Investigator*, 1848-49.
DEARING, James. Leader, HBC expeditions, 1766-67; 1767-68; 1768-69; 1769-70.
DEASE, Peter Warren (1788-1863). Chief Factor, HBC and joint leader, HBC expedition, 1837-39. [Fleming, 1940, pp. 434-36].
DECOIGNE, François. (?), HBC expedition, 1816-17. [Rich, 1939, p. 210].
DEE, C.H. Staff Sergeant, NWMP, *Neptune*, 1903-04.
DEE, John. Scientist, *Gabriel*, 1576. [Deacon, 1968].
DE HAVEN, Edwin Jesse (1816-1865). Lieutenant commanding, *Advance*, 1850-51. [EA; DAB].
DELABARRE, Edmund Burke (1863-). Joint leader, Brown-Harvard expedition, 1900.
DELANO [DELANA], Edward C. Seaman, *Advance*, 1850-51.
DELLER [DELLAR], Henry. AB, *Resolute*, 1850-51; Cook, *Assistance*, 1852-54.
DE LONG, George Washington (1844-

1881). Lieutenant USN, *Juniata,* 1873. [EA; DAB].

DENHAM, Robert. Goldsmith, *Ayde,* 1578.

DENNIS, Henry. AB and Stoker, *Pandora,* 1876.

DENNIS, William. AB, *North Star,* 1852-54.

DENYS, Jean [Jehan]. Captain, French fishing voyage, 1506. [DCB Vol. 1].

DENYSE, John. Sailor, *Ayde,* 1578.

DEOWER, George. *See* DROVER.

DERMOT, James. Seaman, *Resolution,* 1776-80.

DE ROCQUE [DEROCKLE], Henry. Seaman (or Cook?), *Advance* 1850-51.

DESCHESNEAU, L. Boy, *Neptune,* 1903-04.

DESJARDIN, Arthur. Quartermaster, *Arctic,* 1908-09.

DESLAURIERS, Jack. Canoeman and Cook, Camsell's expedition, 1905.

DES VOEUX, Charles Frederick. Mate, *Erebus,* 1845-48. [O'B (1)].

DEUCHARS, David. Ice Quartermaster, *Alert,* 1875-76.

DEVERELL, George. AB, *North Star,* 1849-50. Died on board, 17 May 1850.

DEVERELL, George. Private RM, *Enterprise,* 1850-55.

DEVYS, George. Seaman, *Release,* 1855.

DEWAR, Alexander. Seaman, *Resolution,* 1776-80.

DEWCHERS, Davy. Harpooner, *Arctic,* 1873.

DE WILDE, George R. Artist, *Pandora,* 1875.

DIAMOND, Robert. Corporal RM, *Plover,* 1852-54.

DICK, William. AB, *Hecla,* 1819-20; AB, *Isabella,* 1818.

DICKE, William. Sailor (?), *Sunneshine,* 1585.

DICKENS, Arthur. Captain of the Forecastle, *Pioneer,* 1852-54.

DICKENS, Arthur. Boatswain's Mate, *Enterprise,* 1850-55.

DIEBITSCH, Emil (1867-1897). (Civil engineer and brother-in-law of R.E. Peary). (?), *Falcon,* 1894. [EA].

DIGGLE, John. Cook, *Terror,* 1845-48.

DILLON, Walter. AB, *Discovery,* 1791-95.

DINES, John. AB, *Blossom,* 1825-28.

DINKHAM, Thomas. Gunner's Mate, *Phoenix,* 1854.

DIX, Simon. Private RM, *Assistance,* 1850-51; *Pioneer,* 1852-54.

DIXON, George. Armourer, *Discovery,* 1776-80.

DIXON, James. Stoker, *Victory,* 1829-33. Died during expedition.

DIXON, Thomas. AB, *Herald,* 1848; 1849; 1850.

DOBING, Wilson. Gunner RM, *Discovery,* 1875-76.

DOD, William. Sailor, *Ayde,* 1578.

DODD, Edward [Edmund]. Private, Sappers and Miners, Richardson's expedition, 1847-49.

DODD, John. AB, *Herald,* 1848; 1849; 1850.

DODGE, Henry W. Mate, *United States,* 1860-61. [EA].

DOIDGE, James. Captain of the Foretop, *Alert,* 1875-76.

DOMINICS, Vincent. Ship's Cook, *Alert,* 1875-76.

DOMVILLE, William Thomas (-1879). Surgeon, *Resolute,* 1852-54. [O'B(2); BOA].

DONALDSON, R. M. L. Constable, NWMP, *Neptune,* 1903-04.

DONALDSON, Thomas. Leading Seaman, *Griper,* 1819-20.

DONALDSON, Thomas. Greenland Mate, *Fury,* 1824-25.

DONALDSON, Thomas. Gunner, *Terror,* 1836-37. Died 4 February 1837.

DONALDSON, W. J. Prospector, Curran's expedition, 1912.

DONALDSON, Wilfred. Prospector, Curran's expedition, 1912.

DONNET, James John Louis (1816-1905). Surgeon, *Assistance,* 1850-51. [GJ 1905; DNB, O'B(2)].

DONNETT, William. AB, *Terror,* 1836-37.

DONOVAN, James. Surgeon, *Terror,* 1836-37. [O'B(2)].

DORE, William. AB, Richardson's expedition, 1847-49.

DORE, William. Captain of the Maintop, *Assistance,* 1850-51.

DORMAN, David. AB, *Chatham,* 1791-95.

DOUGALL, William. Ice Quartermaster, *Discovery,* 1875-76.

DOUGHTREE, Thomas. Boatswain, *Blossom,* 1825-28.
DOUGHTY, Thomas. AB, *Phoenix,* 1854.
DOUGLAS, George Mellis (1875-1963). Leader, geological expedition, 1911-12. [Finnie, R. S., 1963; *Arctic,* 1963a].
DOUGLAS, Lionel Dale. (Lieutenant RNR; brother of G. M. Douglas). Assistant, geological expedition, 1911-12.
DOUGLAS, Thomas *5th Earl of Selkirk* (1771-1820). Received Grant of Assiniboia, 1811. [DNB Vol. 15].
DOVE, Benjamin. Fifer, *Herald,* 1848; 1849; 1850.
DOW, Francis. AB, *Assistance,* 1850-51.
DOW, Ledger. AB, *Terror,* 1836-37.
DOWLING, Donaldson Bogart (1858-1925). Assistant, GSC expedition, 1892. [EA].
DOWMAN, Thomas. AB, *Enterprise,* 1848-49.
DOWN, E. H. *See* DUNN.
DOWN, John. AB, *Plover,* 1848-52; 1852-54.
DOWNEY, Cornelius. Carpenter's Crew, *Discovery,* 1791-95.
DOWNING, John. Quartermaster, *Erebus,* 1845-48.
DOYLE, Thomas. Seaman, *Arctic,* 1906-07.
DOYLE, William. Seaman, *Resolution,* 1776-80.
DOYLE, William. Seaman, *Arctic,* 1908-09; 1910-11.
DRACHARDT, Christian (1711-1778). Moravian missionary. Missionary voyages, 1765; 1770; 1771. [Olsen, 1961].
DRAGE, John. Captain's Steward, *Investigator,* 1848-49.
DRAGE, Theodore Swaine. Clerk, *California,* 1746-47. [Eavenson, 1949; EA].
DRAGFORD, Thomas. Sailor, *Ayde,* 1578.
DRAPER, Esdras. Sailor, *Ayde,* 1578.
DRAPER, Thomas. Captain, *Albemarle,* 1680-81. [Rich, 1945, pp. 336-37; DCB Vol. 1].
DRAULETTE, . Leader, Revillon Frères voyage, 1903-04.
DRAWARD, George. Seaman, *Sunneshine,* 1586.
DRAY, John. Ordinary Seaman, *Blossom,* 1825-28.
DREW, John. AB, *Intrepid,* 1852-54.
DREW, Richard. AB, *Hecla,* 1819-20.
DREWE, John. Master, *Godspeed,* 1602.
DREXLER, C. Leader, ornithological expedition, 1860.
DRIVER, William. Cook, *Enterprise,* 1850-55. Died 14 May 1852.
DROVER [DEOWER], George. Captain of the Forecastle, *Assistance,* 1850-51; *Intrepid,* 1852-54. Died 12 December 1852.
DROVER, Richard. Captain of the Maintop, *Blossom,* 1825-28.
DROWLEY, Samuel. Blacksmith, *Rattlesnake,* 1853-54.
DRUILLETTES, Gabriel (1610-1681). (Jesuit priest and missionary). Accompanied Dablon, 1661. [DCB Vol. 1].
DRUMMOND, James. AB, *Discovery,* 1791-95.
DRUMMOND, Robert. Assistant Engineer, *Phoenix,* 1853. [O'B(2)].
DRUMMOND, Thomas (-1835). Assistant naturalist, Franklin's expedition, 1825-27. [DNB Vol. 16].
DUBRILL [LEBRULE], Louis. (?), Richardson's expedition, 1847-49. [Rich, 1953, p. 363].
DUCETTE, Charles. Canadian voyageur, Mackenzie's expeditions, 1789; 1792-93.
DUGUID, William. AB, *Prince Albert,* 1850.
DUMAS, Michel (c1826-). Middleman, Rae's expedition, 1850-51. [Rich, 1953, p. 355.].
DUMAS, Pierre. Canadian voyageur, Franklin's expedition, 1819-22.
DUMURESQ, Amos. Captain, *Albemarle,* 1689.
DUN, Charles. AB, *Griper,* 1824.
DUNBAR, John. AB, *Sophia,* 1850-51.
DUNBAR, John. AB, *North Star,* 1852-54.
DUNCAN, Charles. Captain RN and leader, North-west Passage expeditions, 1790; 1791-92.
DUNCAN, J. Captain, *Esquimaux,* 1835.
DUNCAN, James. Customs House operator, *Arctic,* 1906-07.
DUNCAN, William. Coxswain, *Lion,* Franklin's expedition, 1825-27.
DUNE, . Master, *Charles,* 1631.
DUNN [DOWN], E.H. Chief Officer, *Nancy*

Dawson, 1849.
DUNN, Francis. Caulker's Mate, *Erebus*, 1845-48.
DUNN, George. AB, *Hecla*, 1821-23.
DUNN, James L. Captain, *Labrador*, 1883-84; 1884-85; 1885.
DUNN, T. Captain, *Marten*, 1844.
DUNNET, William. Private RM, *Terror*, 1836-37.
DUNNING [DUNNIN], Thomas. Seaman, *Advance*, 1850-51.
DUNSTALL, Thomas. Quartermaster, *Plover*, 1852-54.
DUQUET DE LA CHESNAYE, Pierre (1643-1687). Accompanied Couture, 1663. [DCB Vol. 1].
DURTNOW, J. Caulker's Mate, *Herald*, 1848; 1849; 1850.
DUTTON, Warren. Seaman, whalers' boat voyage, 1860. Died in Labrador, 3 September 1860.
DUVAL, William. Free trader, *Albert*, 1916-18.
DWYER, John. AB, *Phoenix*, 1853; Stoker, *Phoenix*, 1854.
DYER, Andrew. Master, *Gabriel*, 1578; Pilot, *Ayde*, 1577; (?), *Gabriel* and *Michael*, 1576.
DYMOND, Joseph. Astronomer, British scientific expedition, 1768-69.

EAMES, John. AB, *Investigator*, 1850-54. Died at Mercy Bay, 11 April 1853.
EASTON, Clifford Hamilton (-1958). Assistant, U.S. exploring expedition, 1905-06. [Schwarz, 1958].
EATON, D.I.V. Assistant, GSC expedition, 1892-96.
EBIERBING [JOE]. (Eskimo). Accompanied Hall, 1860-62; 1864-69; 1871-73; Interpreter, *Pandora*, 1875; Interpreter to Schwatka, 1878-80. [Neatby, 1959].
ECCLESTONE [ECCLESTON], Peter Thomas. Cook, *Felix*, 1850-51.
EDDIE, John. Carpenter's Mate, *Sophia*, 1850-51.
EDDY, E. C. Chief Captain's Mate, *Discovery*, 1875-76.
EDE, Charles. Assistant Surgeon, *Assistance*, 1850-51. [MAR].
EDEY, George. Bombardier RMA, *Pioneer*, 1852-54.
EDGAR, Thomas. Master, *Discovery*, 1776-80.
EDGCOMBE, Leonard (-1696). Captain, HBC supply voyages. *John and Thomas*, 1684; 1687-88; 1688-89; *Royal Hudson's Bay*, 1690-91; 1691; *Dering* [III], 1692-93; 1693. [DCB Vol. 1].
EDMAN, John. Captain, *Britannia*, 1817-18.
EDMUNDS, James. AB, *Enterprise*, 1850-55.
EDNEY, John. Quartermaster, *Terror*, 1836-37.
EDWARD, Jens. (Eskimo). Dog-driver, U.S. Polar Year Expedition, 1881-84. Drowned 29 April 1884.
EDWARDS, Germon. AB, *Plover*, 1852-54.
EDWARDS, George. Carpenter's Mate, *Assistance*, 1852-54.
EDWARDS, George. Carpenter's Mate, *Fox*, 1857-59.
EDWARDS, George. Cook, *Assistance*, 1850-51.
EDWARDS, Henry W. AB, *Discovery*, 1875-76.
EDWARDS, John. Surgeon, *Isabella*, 1818; *Hecla*, 1819-20; *Fury*, 1821-23.
EDWARDS, Richard. Gunner, *Henrietta Maria*, 1631-32. Died 22 November 1631.
EDWARDS, Samuel. Captain, *Friendship*, 1682-83.
EDWARDS, William. Ship's steward, *Pandora*, 1875; 1876.
EENOOLOOAPIK (-1847). (Eskimo). Penny's expeditions, 1839; 1840. [McDonald, 1841; Tillotson, 1869, pp. 277-84].
EGERTON, George Le Clerc (c1853-1940). Sub-Lieutenant, *Alert*, 1875-76.
EGGLETON, William. AB, *Phoenix*, 1853.
E-GING-WAH. (Eskimo.) Accompanied Peary to North Pole, 1908-09.
EIRIKSSON, Leif. Vinland voyage, c1000. [DCB Vol. 1, pp. 448-49].
EIRIKSSON, Thorstein. Vinland voyage, c1000.
EIRIKSSON, Thorvald. Vinland voyage, c1000.
EKBLAW, Walter Elmer (1882-1949).

Geologist and botanist, "Crocker Land" Expedition, 1913-17. [Schorger, 1952; Schlaikjer and Schwarz, 1949; GR 1949].

ELDER, Alexander. Leading Seaman, *Alexander,* 1818; Greenland Mate, *Griper,* 1819-20; *Hecla,* 1821-23. Died 15 April 1823.

ELISON, Joseph. Corporal, 10th Cavalry, U.S. Polar Year Expedition, 1881-84. Died 8 July 1884.

ELIZA, Francisco. Leader, Spanish Northwest Coast Survey, 1791.

ELLARD, William. Seaman, *George Henry,* 1860-62.

ELLARD, William. Private RM, *Alert,* 1875-76.

ELLIOT, John. AB, *Enterprise,* 1850-55.

ELLIOTT, James Edward (1813-1865). Lieutenant, *Assistance,* 1850-51; *Phoenix,* 1853; 1854. [O'B(2); MAR].

ELLIOTT, John. AB, *Griper,* 1824.

ELLIOTT, John. Bombardier RMA, *Assistance,* 1850-51.

ELLIOTT, Thomas H. Acting Master's Apprentice, *Rattlesnake,* 1853-54.

ELLIOTT, William. Carpenter's Mate, *Pioneer,* 1850-51.

ELLIOTT, William. Clerk in Charge, *North Star,* 1852-54.

ELLIS, . Captain, *Kitty,* 1859.

ELLIS, Christopher. AB, *Blossom,* 1825-28.

ELLIS, Henry. Agent, *Dobbs,* 1746-47.

ELLIS, John. Master, *Mooneshine,* 1585.

ELLIS, John. Sailor (?), *Sunneshine,* 1585.

ELLIS, Richard. Stoker, *Pioneer,* 1850-51.

ELLIS, Samuel J. Carpenter's Mate, *Phoenix,* 1854.

ELLIS, Simon. Seaman, *Sunneshine,* 1586.

ELLIS, Thomas. Sailor, Frobisher's expedition, 1578.

ELLIS, William. Surgeon's Second Mate, *Discovery,* 1776-80.

ELLIS, William. Private, 2nd Cavalry, U.S. Polar Year Expedition, 1881-84. Died of starvation, 19 May 1884.

ELLYS, John. Sailor, *Ayde,* 1578.

ELMORE, Charles. Captain of the Maintop, *Enterprise,* 1850-55.

ELSEY, John. Boatswain's Mate, *Enterprise,* 1850-55.

ELSON, George. (Half-breed Cree Indian). Guide, Hubbard expeditions, 1903-04; 1905; Tasker expedition, 1906.

ELSON, Thomas. Master, *Blossom,* 1825-28.

ELSTON, J. W. Acting Assistant Surgeon, *Tigress,* 1873.

ELYOT, Richard. Master, *William,* 1536.

EMANUEL, Samuel. (Greenland Eskimo). *Fox,* 1857-59.

EMELIN [EMLIN]. Baptiste. See HAMLIN

EMMERSON, George W. Chief Boatswain's Mate, *Discovery,* 1875-76.

EMORY, William H. Lieutenant USN, commanding *Bear,* 1884. [NF].

ENDERSON, George. AB, *Hecla,* 1824-25.

ENGLAND, John. AB, *Discovery,* 1776-80.

ENGLEHART, James. Sailmaker, *Discovery,* 1791-95.

ENGLISH, Arthur. Prospector and taxidermist, *Arctic,* 1910-11.

ENGLISH, J. Sergeant, Royal Horse Artillery, and Observer, British Polar Year Expedition, 1882-83.

ENGLISH, William. Sailor, *Ayde,* 1578.

ENGLISH CHIEF, The. (Chipewyan Indian Chief). Guide, Mackenzie's expedition, 1789.

ENTWISTLE, Henry. Private RM, *Hecla,* 1824-25.

EOWAN, William. AB, *Fury,* 1824-25.

ERELEASE, Thomas. Sailor, *Ayde,* 1578.

ERHARDT, John Christian. (Moravian missionary). Interpreter, *Hope,* 1752. Died (probably killed by Eskimos) September 1752.

ERIC THE RED. *See* THORVALDSSON, Eirik.

ERLANDSON, Erland (c1790-1875). Established Fort Chimo, 1830; South River House, 1832-33; Fort Nascopie, 1838; leader, HBC exploring expedition, 1834. [Fleming, 1940, pp. 437-38; Davies, 1963, pp. 352-67; DCB Vol. 10].

ESDAILE, W.F. Station Assistant at Port de Boucherville, Canadian Hudson Bay expedition, 1884-85.

ESKIMO JOE. *See* EBIERBING.

ESSEX, Cipio. Gentleman, Frobisher's expedition, 1578.

ESTON, William. Master, *Sunneshine,* 1585;

Mermaid, 1586.
ETCHINSON, James. AB, *Chatham,* 1791-95.
E-TUK-I-SHUK. (Eskimo). Accompanied Cook to North Pole, 1907-09.
EVANS, David. AB, *Phoenix,* 1854.
EVANS, Davy. Baker, Frobisher's expedition, 1578.
EVANS, Evan. Seaman, *Resolution,* 1776-80.
EVANS, G.B. Ordinary Seaman, *Breadalbane,* 1853.
EVANS, George. AB, *Discovery,* 1791-95.
EVANS, Hugh. AB, *Plover,* 1848-52.
EVANS, Humphrey. Quartermaster, *Discovery,* 1791-95.
EVANS, James. Caulker, *Enterprise,* 1848-49; *Investigator,* 1850-54.
EVANS, John. AB, *Discovery,* 1791-95.
EVANS, John. Clerk, *Blossom,* 1825-28.
EVANS, Thomas. Purser, *Griper,* 1824.
EVANS, Thomas. Boy, *Terror,* 1845-48.
EVANS, Thomas. AB, *Investigator,* 1848-49.
EVANS, Thomas. AB, *Phoenix,* 1854.
EVANS, William. Carpenter's crew, *Discovery,* 1776-80.
EVANS, William. AB, *Assistance,* 1852-54.
EWIN, William. Boatswain, *Resolution,* 1776-80.
EWINS, John. Botanist's Lieutenant, *Discovery,* 1791-95.
EYRE, Henry. Cook, *Victory,* 1829-33.
EYRE, John. Marine, *Enterprise,* 1850-55.

FACEY, Joseph, Sailmaker, *Investigator,* 1850-54.
FAGUNDES, João Alvares. Leader, Portuguese expedition, 1520. [DCB Vol. 1].
FAIGAN, J. AB, *Plover,* 1848-52; 1852-54.
FAIRBRASS, William. Private RM, *Alexander,* 1818.
FAIRCHILD, C.C. Assistant to Tyrrell, 1900.
FAIRHOLME, James Walter (1821-c1848). Lieutenant, *Erebus,* 1845-48. [O'B (1)].
FAITHFUL [FAITHFULL], James. Gunroom Steward, *Plover,* 1852-54.
FALCONER, William. Captain, *Severn,* 1783.
FALCUSS, J. AB, *Phoenix,* 1854.
FALK, E. Captain, *Beothen,* 1911.
FANNER, Syracke. Seaman, *Discovery,* 1610-11. Set adrift with Hudson by mutineers, 1611.
FARCE [FARCIE], George. AB, *Lady Franklin,* 1850-51.
FARIBAULT, G.B. Assistant Surgeon, *Neptune,* 1903-04. Died 27 April 1904.
FARIS, Peter. Private of Marines, *Discovery,* 1791-95.
FARQUHARSON, J.B. Corporal RM, *Investigator,* 1850-54.
FARR, Thomas R. Captain of the Maintop, *Terror,* 1845-48.
FARRELL, I. James. AB, *Griper,* 1824.
FARRINGDON, James. AB, *Breadalbane,* 1853.
FARRIS, Peter. Private of Marines, *Chatham,* 1791-95.
FATCHETT, Thomas. Private of Marines, *Resolution,* 1776-80.
FAWCETT, Edward. Boatswain's Mate, *Investigator,* 1848-49; 1850-54.
FAWCKNER, William Henry. Second master RN and Government Agent, *Breadalbane,* 1853. [O'B(2)].
FEARNLEY, George. Gunroom Steward, *Herald,* 1848; 1849; 1850.
FEDDIS, George. Carpenter, *Isabella,* 1818; *Fury,* 1821-23.
FEILDEN, Henry Wemyss (1838-1921). Naturalist, *Alert,* 1875-76. [GJ 1921].
FELIX, François. Canadian voyageur, Franklin's expedition, 1825-27; HBC expedition, 1837-39.
FELLBROOK, John. Sergeant of Marines, *Chatham,* 1791-95.
FENESE, Peter. Ice Quartermaster, *North Star,* 1852-54.
FENNING, William. Officers' Steward, *Resolute,* 1850-51; *Assistance,* 1852-54.
FENTON, Edward. Captain, *Gabriel,* 1577; *Judith,* 1578.
FERBRACHE, William. AB, *Alert,* 1875-76.
FERDINAND, George. AB, *Hecla,* 1824-25.
FERGUSON, Mordoch. (?), McLean's expedition, 1838.
FERGUSON, Robert Munro. Second-in-command, GSC expedition, 1894.
FERLIE, Robert. AB, *Phoenix,* 1854.
FERRIER [FERRIN], Robert. AB, *Erebus,* 1845-48.

FERRIER, William. AB, *Breadalbane,* 1853.
FIDDIS, George. Carpenter, *Fury,* 1821-23; *Hecla,* 1824-25.
FIDLER, Henry. Middleman, Rae's expedition, 1853-54; HBC Franklin search, 1855. [Rich, 1953, p. 356].
FIDLER, John. Middleman, Rae's expedition, 1850-51; Steersman, HBC Franklin search, 1855. [Rich, 1953, p. 356].
FIDLER, Peter (1769-1822). Surveyor, HBC expeditions, 1790-92; 1793; 1802-03; 1803-04; 1804-05; 1805-06; 1807. [Wallace, W.S., 1952; MacGregor, 1966].
FIELDER [TIELDER], Cornelius. AB, *Assistance,* 1852-54.
FIFE, George. Greenland Master, *Griper,* 1819-20; *Hecla,* 1821-23. Died 6 September 1823.
FILE, James. Stoker, *Phoenix,* 1853.
FILPE, John. Seaman, *Sunneshine,* 1586.
FINCH, George. Prospector, gold prospecting expedition, 1872-74.
FINDLAY, George. AB, *Lady Franklin,* 1850-51.
FINLAY, James. Travelled to Saskatchewan River, 1768 [see 1768-69, Pink].
FINLAY, John. Established Fork Fort, 1792; explored Finlay River, 1797.
FINLAYSON, Nicol (1795-1877). Established Fort Chimo, 1830. [Fleming, 1940, pp. 438-40; Davies, 1963, pp. 367-78; DCB Vol. 10].
FINNBOGI. Vinland voyage, c1000.
FINNECY, Peter. Captain of the Maintop, *Resolute,* 1850-51.
FINNREY, Peter. Ice Quartermaster, *North Star,* 1852-54.
FISHBURNE, Rychard. Sailor, *Judith,* 1578.
FISHER, Alexander. Assistant Surgeon, *Alexander,* 1818; *Hecla,* 1819-20; Surgeon, *Hecla,* 1821-23.
FISHER, Anthonie. Trumpeter, *Ayde,* 1578.
FISHER, Elnathan B. Captain, *Ansell Gibbs,* 1868-69.
FISHER, George. Chaplain and Astronomer, *Fury,* 1821-23. [DNB Vol. 19; BOA Vol. 1].
FISHER, John. Seaman, *Resolution,* 1776-80.
FISHER, Joseph. Seaman, whalers' boat voyage, 1860.
FISHER, Peter. AB, *Isabella,* 1818; *Griper,* 1819-20.
FISHER, Peter. Master, *Terror,* 1836-37.
FISHER, Samuel J. Seaman, whalers' boat voyage, 1860. Killed by his companions, September 1860.
FITCH, John. Blacksmith, *Investigator,* 1848-49.
FITZGERALD, F.J. (-1911). Sergeant, NWMP, establishment of Fort McPherson detachment, 1903; established Herschel Island detachment, 1903.
FITZGERALD, Patrick. AB, *Herald,* 1848; 1849; 1850.
FITZJAMES, James. Commander RN (Captain 31 December 1845), *Erebus,* 1845-48. [O'B (1)].
FITZPATRICK, James. AB, *Phoenix,* 1854.
FLAHERTY, Robert Henry. (Father of R.J. Flaherty). Geologist, Canadian exploring expedition, 1913-16 [1916 only].
FLAHERTY, Robert Joseph (1884-1951). Leader, Canadian exploring expeditions, 1910-11; 1911-12; 1913-16. [Calder-Marshall, 1963; Griffith, 1953; Holman, 1951].
FLATMAN, John. Seaman, *Resolution,* 1776-80.
FLAY, James. AB, *Plover,* 1848-52; *Rattlesnake,* 1853-54.
FLEMING, Archibald Lang (1883-1953). Anglican missionary; established Lake Harbour mission, 1909-11; Anglican missionary expedition, 1913-15. [Fleming, A.L., 1957; PR 1954].
FLEMING, Maurice. Assistant, Nottingham Island station, Canadian Hudson Bay expedition, 1885-86.
FLETCHER, John, (-1697). Captain, *Hampshire,* 1697. [DCB Vol. 1].
FLETT, George (-1849). Middleman, HBC expedition, 1837-39; Steersman, Rae's expedition, 1846-47. [Rich, 1953, pp. 356-57].
FLETT, James. Carpenter, *Breadalbane,* 1853.
FLETT, John. (?), Tyrrell's expedition,

1893.

FLOOD, James. AB, *Discovery,* 1776-80.

FLOOD, Peter. Ordinary Seaman, *North Star,* 1849-50.

FLORANCE [FLORENCE], Thomas. Stoker, *North Star,* 1852-54; *Fox,* 1857-59; Captain of the Hold, *Pandora,* 1875; 1876.

FLOWEY, Christopher. Fishmonger, Frobisher's expedition, 1578.

FLYNN, Edward. Sergeant of Marines, *Discovery,* 1791-95.

FLYNN, Michael. AB, *Enterprise,* 1848-49.

FLYNN, Michael. Quartermaster, *Investigator,* 1850-54.

FLYNN, Patrick. Established Waswanipi post, 1819; accompanied Clouston, 1819-20.

FODDER, William R. Gunner's Mate, *Phoenix,* 1853.

FOGGET, James. HBC carpenter, established Nemiscau, 1794.

FOKES [FOOKES], William. Carpenter's Mate, *Assistance,* 1850-51.

FOLDS, Thomas. Ordinary Seaman, *Blossom,* 1825-28.

FOLLY [FOLEY], James. Cook, *Resolute,* 1850-51.

FOLSTER, John [George]. Middleman, Rae's expedition, 1846-47. [Rich, 1953, p. 357].

FOLSTER, Robert. North West Company agent, invasion of James Bay, 1803-06.

FONTANO, Vincenza. Canadian voyageur, Franklin's expedition, 1819-22.

FONTARABIE, Pierre. Soldier, accompanied Buteux, 1652. Killed by Iroquois, 10 May 1652.

FONTE, Bartholomew de. (Fictitious character). Commander, *St Spiritus,* 1640. [DCB Vol. 1, p. 677].

FONTENEAU, Jean. *See* ALFONSE, Jean.

FOOKES, William. *See* FOKES.

FOORD, Alex. Carpenter's crew, *Discovery,* 1791-95.

FORBES, Henry S. Joint leader, U.S. expedition, 1908.

FORCIER, Baptiste. Canoeman, Camsell's expedition, 1914.

FORCIER, Joseph. Canadian voyageur, Franklin's expedition, 1819-22.

FORD, George J. Carpenter, *Investigator,* 1850-54.

FORD, Henry. Sailor and Eskimo interpreter, GSC expedition, 1898-99.

FORD, John G. Captain, HBC supply voyages. *Lady Head,* 1891; 1892; 1894; 1895; 1896; 1897; 1898; 1899; 1900; 1901; 1902; 1903; *Stork,* 1904-05; *Discovery,* 1906; 1907; 1908; 1909; 1910; 1911.

FORD, Thomas. Seaman, *Release,* 1855.

FORD, William J. Gunner, *Herald,* 1848; 1849; 1850.

FORDER, Frauncis. Corporal of the Shot, *Ayde,* 1577.

FORDER, Henry. Captain of the Hold, *Enterprise,* 1850-55.

FORDER, William R. Assistant Gunner, *Phoenix,* 1854.

FORDYCE, William. AB, *Phoenix,* 1854.

FOREMAN, Thomas. AB, *Griper,* 1819-20.

FORESTER, Henry. AB, *Discovery,* 1776-80.

FORSTER, Thomas Balle. Assistant Surgeon, *Plover,* 1848-52 [1851-52 only]; *Rattlesnake,* 1853-54. [O'B(2)].

FORSTER, William, Sailmaker, *Herald,* 1848; 1849; 1850.

FORSYTH, Charles Codrington. Commander RN, commanding *Prince Albert,* 1850. [O'B(1)-(2)].

FOSHEIM, Ivar. (?), *Fram,* 1898-1902.

FOSTER, Henry (1796-1831). Lieutenant, *Hecla,* 1824-25. [DNB Vol. 20].

FOSTER, Robert. AB, *Investigator,* 1848-49.

FOSTER, Thomas Balle. *See* FORSTER.

FOSTER, William. AB, *Enterprise,* 1848-49.

FOUNDLING, Thomas. AB, *Phoenix,* 1853.

FOWLER, George. Captain's Coxswain, *Investigator,* 1848-49; Quartermaster, *Enterprise,* 1850-55.

FOWLER, John. Captain, HBC voyages. *Seahorse* [I], 1744-45; 1746; 1747; 1748; 1749; 1750.

FOWLER, Jonathan, Sr. Captain, HBC supply voyages. *Hudson's Bay* [V], 1744; 1745; 1746; 1747; 1748; 1749; 1750; *Seahorse* [I], 1751; 1752; 1753; 1754; 1755; *Prince Rupert* [II], 1756; *Prince Rupert* [III], 1757; 1758; 1759.

FOWLER, Jonathan, Jr. Captain, HBC supply voyages. *Hudson's Bay* [V], 1757; 1758;

1759; *Prince Rupert* [III], 1760; *Seahorse* [I], 1761; 1762; 1763; *King George* [II], 1764; 1765; 1766; 1767; 1768; 1769; 1770; 1771; 1772; 1773; 1774; 1775; 1776; 1777; 1778; 1779; 1780; *King George* [III], 1781; 1782.

FOWLER, William [WILLIAM, Samuel]. Paymaster and Purser's Clerk, *Erebus*, 1845-48.

FOWLER, William. Cook, *North Star*, 1849-50.

FOX, Bartolemew. Commander's Steward, *Plover*, 1848-52; 1852-54.

FOX, Charles. Carpenter's crew, *Terror*, 1836-37.

FOX, George. AB, *Discovery*, 1791-95.

FOX, James. AB, *Prince Albert*, 1850; *Resolute*, 1850-51.

FOX, John. Second Class Fireman, *Arctic*, 1855.

FOXE [FOX], Luke (1586-1636). Captain and Pilot, *Charles*, 1631. [DCB Vol. 1; DNB Vol. 20].

FRAIZE, Thomas. Seaman, *Arctic*, 1906-07.

FRANCIS, George. Private RM, *Assistance*, 1850-51.

FRANCKE [FRANKE], Rudolph. Accompanied Cook, 1907-09.

FRANCOMBE, Reuben. AB, *Alert*, 1875-76.

FRANKLIN, Howard. Gold prospecting expedition, 1884-85; discovered gold on Fortymile River, 1886.

FRANKLIN, John (1786-1847). Lieutenant RN and leader, overland expedition, 1819-22; Captain RN and leader, overland expedition, 1825-27; Captain, *Erebus*, 1845-48. Died 11 June 1847. [DNB Vol. 20; Traill, 1896; Lamb, G.F., 1956].

FRANKLIN, Thomas. Purser's Steward, *Release*, 1855.

FRASER, Alexander. North West Company agent, invasion of James Bay, 1803-06.

FRASER, James. Captain, Moravian missionary voyages, 1782- . [see 1771].

FRASER, Simon (1776-1862). Established Rocky Mountain Portage House and Fort McLeod, 1805-06; leader, North West Company expeditions, 1806-07; 1808. [Lamb, W.K., 1960; McKeever, 1968; DCB Vol. 9].

FRAZER, Donald. Private, Sappers and Miners, Richardson's expedition, 1847-49.

FRAZER, James. AB, *Felix*, 1850-51.

FREAKLEY, N.E. Captain, *Stork*, HBC supply voyages, 1906; 1907; 1908.

FREDERICK, Julius. Private, 2nd Cavalry, U.S. Polar Year Expedition, 1881-84.

FREDERIQUE, Baptiste (dit Paul). Canoeman, Richardson's expedition, 1847-49. [Rich, 1953, p. 357].

FREELAND, John. Private RM, *Isabella*, 1818.

FRENCH, . Pilot, *Benjamin S. Wright*, 1864.

FRENCH, Francis Henry. Inspector RNWMP, patrol, 1916-18.

FRENCH, K. Seaman, *Laddie*, 1913-16.

FRENCH, Louis. Canoeman, Tyrrell's expedition, 1893.

FRENCH, Michel. (?), Tyrrell's expedition, 1893.

FRENCH, Pierre. Canoeman, Tyrrell's expedition, 1893.

FRENCH, William. AB, *Enterprise*, 1850-55.

FREW, Christian. AB, *Alexander*, 1818.

FREYDIS. Vinland voyage, c1000.

FRICKER, Charles. Paymaster's Steward, *Herald*, 1848; 1849; 1850.

FRIE, John. Sailor, *Ayde*, 1578.

FRIEND, Daniel. Captain, *Prince George*, HBC supply voyages, 1834; 1837.

FRIMECY (?), Peter. Captain of the Maintop, *Resolute*, 1850-51.

FROBISHER, Joseph (1740-1810). Leader, Montreal pedlars' expeditions, 1773-74; 1774-75; 1775-76. [DCB Vol. 5].

FROBISHER, Martin (1539-1594). Leader, British expeditions, 1576; 1577; 1578. [DNB Vol. 20; DCB Vol. 1; McFee, 1928].

FROBISHER, Thomas. Montreal pedlar, fur trading expeditions, 1775-76; 1776-77.

FROMONT, Thomas. Master, *Grande Hermine*, 1535-36.

FROST, John. AB, *North Star*, 1852-54.

FROST, Thomas. Private RM, *Investigator*, 1848-49.

FROUDE, F. Fireman, *Neptune*, 1903-04.

FRY, Richard. AB, *Herald*, 1848; 1849; 1850.

FRY, Thomas. Captain of the Hold, *North*

Star, 1849-50.

FUCA, Juan de (1536-1602). Leader, North-west Passage expedition, 1592. [DCB Vol. 1, p. 677].

FULBISTER, James. Captain, *Lord Wellington,* 1821.

FULFORD, Reginald B. Lieutenant, *Discovery,* 1875-76. [MAR, p. 61].

FULLARTON, James. *See* FULLERTON.

FULLER, Charles H. "Green hand," *Florence,* 1877-78.

FULLER, Thomas. Carpenter, *Dolphin,* 1825-27.

FULLERTON [FULLARTON], James. Gunner's Mate, *Assistance,* 1850-51.

FULLOCK, James. AB, *Plover,* 1848-52; 1852-54.

FUSSELL, Henry. Private RM, *Enterprise,* 1848-49; *Resolute,* 1850-51.

FYFFE, Joseph P. (1832-). Passed Midshipman, *Release,* 1855. [NE].

GAGNÉ, Gedeon. Carpenter, *Arctic,* 1908-09.

GAGNÉ, Joseph. Canadian voyageur, Franklin's expedition, 1819-22.

GALAUP, Jean-François, *Comte de la Pérouse* (1741-1788). Commander, *Sceptre,* 1782; *Boussole,* 1785-88. [NE].

GALAVAN, Jeremiah [SALAVAN, J.; *cf* SULLIVAN, M.]. Captain of the Forecastle, *Assistance,* 1852-54.

GALBRAITH, George B. Captain, HBC supply voyages. *Ocean Nymph,* 1870; 1871; *Lady Head,* 1873-74; *Ocean Nymph,* 1874.

GALIANO, Dionisio Alcala. Captain, *Sutil,* 1792.

GALLICHAND, Philip. AB, *Phoenix,* 1854.

GALLIE, David. AB, *Herald,* 1848; 1849; 1850.

GALT, George. AB, *Phoenix,* 1854.

GALWAY, Alexander. Merchant Master, *Isabella,* 1818.

GAMAIGE, John. Fisher and sailor, *Judith,* 1578.

GAMARD, . Pilot, French voyage, 1506.

GAMBLE, William. AB, *Chatham,* 1791-95.

GANDER, Charles. AB, *Plover,* 1848-52; 1852-54.

GANNICLIFT [GANNICLEFD], Robert. AB, *Intrepid,* 1852-54.

GARANT, Napoléon. Fireman, *Arctic,* 1910-11.

GARBUT, John. Captain, *Success,* 1766.

GARDINER, A.J. Second Officer, *George Henry,* 1860-62.

GARDINER, Hampden Sydney. Sergeant, Signal Corps, U.S. Army, U.S. Polar Year Expedition, 1881-84. Died of starvation, 12 June 1884.

GARDINER, John. AB, *Phoenix,* 1854.

GARDINER, Charles. AB, *North Star,* 1849-50.

GARDNER, William. Carpenter's Mate, *Blossom,* 1825-28.

GARLAND, William. AB, *Fury,* 1824-25.

GARLINGTON, Ernest Albert (1853-). Lieutenant, 7th Cavalry, and leader of expedition, *Proteus,* 1883.

GARNIER, John Miller. Midshipman, *Chatham,* 1791-95.

GARRETT, Thomas. Carpenter, *Blossom,* 1825-28.

GATENBY, John. Boatswain's Mate, *Discovery,* 1776-80.

GATHORNE-HARDY, Geoffrey Malcolm (1878-1972). Joint leader, hunting expedition, 1910. [Gathorne-Hardy, R. 1972].

GAUDET, J.F. Government Surveyor, Hind's expedition, 1861.

GAUEN [GAVEN], Henry. Carpenter's Mate, *Investigator,* 1850-54.

GAUL, George. Stoker, *Phoenix,* 1853; 1854.

GAULTIER DE VARENNES, Pierre, *Sieur de la Vérendrye.* See 1750, Fort Paskoyac.

GAWLER, Henry Barnes (1821-). Second Master, *North Star,* 1849-50. [O'B(2)].

GEATER, Josephus [GEALEO, Joseph]. AB, *Erebus,* 1845-48.

GEDDES, Hugh. Private, Sappers and Miners, Richardson's expedition, 1847-49.

GENGE, Edward. Subordinate Officers' Steward, *Terror,* 1845-48.

GEORGE, William. AB, *Pioneer,* 1850-51.

GERMAIN [JERMAIN], John. Purser, *Hecla,* 1821-23.

GERMAN, Robert D. Quartermaster, *Plover,* 1848-52; 1852-54.

GERRARD, William. AB, *Phoenix,* 1854.
GEYER, George. Supercargo, *Lucy,* and founder of York Factory, 1684. [DCB Vol. 1].
GIBBES, M. Master, *Thomas Allen,* 1578.
GIBBONS, William. (Cousin of Thomas Button). Volunteer, British expedition, 1612-13; Captain, *Discovery,* 1614. [DCB Vol. 1].
GIBBS, George. AB, *Investigator,* 1850-54.
GIBBS, James. Captain's Steward, *Terror,* 1836-37.
GIBLIN, Patrick. Master-at-Arms, *Blossom,* 1825-28.
GIBSON, Andrew. Second Master, *Discovery,* 1791-95.
GIBSON, John [alias HIGGINS, John]. Private RM, *Blossom,* 1825-28.
GIBSON, John [GIBSON, Joseph; JOSEPH, Josh]. Carpenter's Mate, *Resolute,* 1852-54.
GIBSON, Richard. AB, *Phoenix,* 1853.
GIBSON, Robert. AB, *Fury,* 1821-23.
GIBSON, Samuel. Sergeant of Marines, *Resolution,* 1776-80.
GIBSON, William. Subordinate Officers' Steward, *Terror,* 1845-48.
GIBSON, William J. Carpenter's Mate, *Investigator,* 1848-49.
GIDDARD, William. Captain of the Hold, *Terror,* 1845-48.
GIDDY, Henry. Acting Sailmaker, *Enterprise,* 1848-49; Boatswain's Mate, *Resolute,* 1852-54.
GIESE, W. Leader, German Polar Year Expedition, 1882-83.
GIFFARD, George Augustus (1849-1925). Lieutenant, *Alert,* 1875-76. [WWW Vol. 2; EA].
GIFFORD, William. Gunner, *Chatham,* 1791-95.
GILBERT, George. AB, *Resolution,* 1776-80.
GILBERT, *Sir* Humphrey (c1537-1583). Commander, British expedition, 1583. [DCB Vol. 1; DNB Vol. 2].
GILBERT, John. Private RM, *Plover,* 1848-52 [1851-52 only].
GILBERT, William. AB, *North Star,* 1852-54.
GILBERT, William. AB, *Phoenix,* 1853.
GILDER, William Henry (1838-1900). Second-in-command, Schwatka's expedition, 1878-80. [EA].
GILES, John. Boat steersman, whalers' boat voyage, 1860.
GILES, Peter. AB (?), *Nancy Dawson,* 1849.
GILES, William. Quartermaster, *Hecla,* 1821-23.
GILL, Charles. AB, *Phoenix,* 1854.
GILLAM, Benjamin (c1662-1706). (Son of Zachariah Gillam). Captain, *Bachelor's Delight,* 1682-83. [DCB Vol. 2].
GILLAM, Zachariah (1636-82). Captain, *Nonsuch,* 1668-69; *Prince Rupert,* 1670-71; 1672-75; 1674-75; 1682-83. Drowned off Nelson River, 21 October 1682. [Rich, 1942, pp. 226-30; DCB Vol. 1].
GILLET, Thomas. Coxswain, Franklin's expedition, 1825-27.
GILLETT, Thomas. AB, *Hecla,* 1824-25.
GILLIE, William. AB, *Hecla,* 1824-25.
GILLIES, Alec. Gold prospector (?), *Neptune,* 1912.
GILLIES, Allan. Harpooner, *Pandora,* 1875; AB, *Pandora,* 1876.
GILLIES, John. Carpenter's Mate, *Griper,* 1824.
GILLIOTT, Charles. Corporal RM, *Hecla,* 1824-25.
GILMOUR, William. AB, *Isabella,* 1818.
GILPIN, Henry. Clerk, *Rattlesnake,* 1853-54.
GILPIN, James D. Clerk in Charge, *Investigator,* 1848-49.
GIRARD, Daniel. AB, *Discovery,* 1875-76.
GIRIDO, Lucke. Vice-Admiral, *Gabriel,* 1578.
GIRLEY, Thomas. Private of Marines, *Resolution,* 1776-80.
GLADSTONE, George. AB, *Herald,* 1848; 1849; 1850.
GLANVILLE, Edward. Officer (?), British expedition, 1612-13.
GLASS, William. AB, *North Star,* 1852-54.
GLASSPOLE, John. Private of Marines, *Discovery,* 1791-95.
GLENNIE, James. Cook, *Prince Albert,* 1850; 1851-52.
GLOSTER, Charles. Sergeant, First Cavalry, Schwatka's expedition, 1883.
GNUPSSON, *Bishop* Eirik. Vinland voyage, 1121-? [DCB Vol. 1, pp. 675-76].

GOBOURN, John. Carpenter's crew, *Discovery,* 1791-95.

GODDARD, William. Captain of the Hold, *Terror,* 1845-48.

GODEFROY DE NORMANVILLE, Thomas. *See* NORMANVILLE.

GODFREY, George T. Boatswain, *Blossom,* 1825-28.

GODFREY, William. Seaman, *Advance,* 1853-55.

GOFF, Elijah. Mate, *Hope,* 1752; Captain, *Hope,* 1753.

GOFORD, William. Officers' Clerk, *Resolute,* 1852-54.

GOFTON, Ingram. Captain, *Hannah,* 1720; 1721; 1722; 1723.

GOGGIN, Patrick. AB, *Rattlesnake,* 1853-54.

GOLDING, Robert [alias ROBERT, Absolomo]. Boy, *Terror,* 1845-48.

GOLDSMITH, Edward. AB, *North Star,* 1849-50.

GOLDTHORPE, George. Corporal RM, *Alexander,* 1818.

GOLKOWSKY, . Moravian missionary, *Hope,* 1752.

GOMES, Estevão [GOMEZ, Esteban] (c1483-1538). Commander and Pilot, *La Anunciada,* 1524-25. [DCB Vol. 1].

GONZALES, Henry. Master, *Polar Bear,* Canadian Arctic Expedition, 1913-18 [1915-17 only].

GOOD, Joseph. Chief Boatswain's Mate, *Alert,* 1875-76.

GOODCHILD, Thomas. Cook, *Rattlesnake,* 1853-54.

GOODEY, William [GOOCHY, John]. Carpenter's Mate, *Resolute,* 1852-54.

GOODFELLOW, Henry. Seaman, *Advance,* 1853-55.

GOODMAN, Thomas. AB, *Discovery,* 1776-80.

GOODRIDGE, John O. (-1865). Surgeon, *Herald,* 1848; 1849; 1850.

GOODSELL, John W. (-1949). Surgeon, *Roosevelt,* 1908-09. [EA; Schwarz, 1950].

GOODSIR, Harry D.S. Acting Assistant Surgeon, *Erebus,* 1845-48.

GOODSIR, Robert Anstruther. (Brother of Harry Goodsir). Passenger, *Advice,* 1849; Surgeon, *Lady Franklin,* 1850-51.

GOODWILLIE, J. Captain's Cook, *Blossom,* 1825-28.

GOOJE, George. Second Master, *Rattlesnake,* 1853-54.

GOOLEY, G.P. Assistant, Nottingham Island station, Canadian Hudson Bay Expedition, 1885-86.

GORDON, Alexander. Leading Seaman, *Griper,* 1819-20.

GORDON, Andrew Robertson. Lieutenant commanding, *Neptune,* 1884-85; *Alert,* 1885-86; 1886. [EA].

GORDON, Daniel. AB, *Phoenix,* 1854.

GORDON, George F. Acting Mate, *Plover,* 1848-52 [1851-52 only]; 1852-54.

GORDON, John. AB, *Alexander,* 1818.

GORDON, John. Captain of the Hold, *Sophia,* 1850-51.

GORDON [GOURDON], William. Mate, *Enhiörningen,* 1619-20. Died at Churchill, 8 April 1620.

GORDON-CUMMING, Walter. Joint leader, Whitney's expedition, 1894-95.

GORE, Graham. Mate, *Terror,* 1836-37; Lieutenant, *Erebus,* 1845-48. [O'B(1)].

GORE, James. Gunroom Steward, *Enterprise,* 1848-49.

GORE, James. Musician, *Resolute,* 1850-51.

GORE, James. Captain's Steward, *Assistance,* 1852-54.

GORE, John (c1730-1790). First Lieutenant, *Resolution,* 1776-80.

GORE, William I. Stoker, *Alert,* 1875-76.

GORNEY, Christopher. Boy, *Sunneshine,* 1585.

GORRELL, Edward. Mate, *Hopewell,* 1606.

GORST, Thomas. Supercargo, fur-trading voyage, 1668-69; *Wivenhoe,* 1669; *Prince Rupert,* 1670-71; fur trading voyage, 1672-75. [Rich, 1942, p. 230; DCB Vol. 1].

GOSFRAY, John. Private of Marines, *Discovery,* 1791-95.

GOSSAGE, James. Ordinary Seaman, *Herald,* 1848; 1849; 1850.

GOSSELIN, George. Laundryman, *Arctic,* 1910-11.

GOUCH, Alexander. Carpenter's crew, *Terror,* 1836-37.

GOUGH, Daniel. Sergeant RM, *Resolute,* 1850-51.

GOUIN, J. Boy, *Neptune,* 1903-04.
GOULD, Samuel. Captain's Cook, *Plover,* 1851-54.
GOULDING, Robert. Cook, *Discovery,* 1776-80.
GOULET, Joseph. Laundryman, *Arctic,* 1906-07; 1908-09.
GOULSTON, William. AB, *Discovery,* 1776-80.
GOW, James. Captain's Steward, *Assistance,* 1852-54.
GOWAN, J. Private RM, *Assistance,* 1850-51.
GOWANS, George. AB, *Terror,* 1836-37.
GOWEN, William. Captain of the Foretop, *Enterprise,* 1850-55.
GRACE, Edward. AB, *Pandora,* 1875.
GRAHAM, Andrew. Doctor, *Arctic,* 1873.
GRAHAM, James [Joseph]. AB, *Pioneer* (?), 1852-54.
GRAHAM, Robert. Private, Sappers and Miners, Richardson's expedition, 1847-49.
GRAHAM, Robert. AB, *Resolution,* 1850-51.
GRAHAM, William. Private RM, *Herald,* 1848; 1849; 1850.
GRANCHES, Antoine des. (Relative of Jacques Cartier). Passenger, Cartier's expedition, 1535-36.
GRANCHES, Guyon des. (Relative of Jacques Cartier). Passenger, Cartier's expedition, 1541.
GRANDY, Samuel. AB, *Hecla,* 1824-25.
GRANGE, *Captain. See* WATKINS, *Captain.*
GRANSELL, James. Cook, *Discovery,* 1791-95.
GRANSELL, Robert H. Master's Apprentice, *Enterprise,* 1848-49.
GRANT, Augustus Boyd. Midshipman, *Chatham,* 1791-95.
GRANT, Charles. Possibly established Fort Alexander, c1817.
GRANT, Cuthbert (-1799). Second-in-command, Pond's expedition, 1785-88; established Great Slave Lake post, 1786. [DCB Vol. 4].
GRANT, John. AB, *Resolution,* 1776-80.
GRANT, Robert D. AB, *Enterprise,* 1850-55.
GRANT, W. Photographer, *Pandora,* 1876.
GRANT, William. Private RM, *Investigator,* 1848-49.
GRATE, Robert. AB, *Prince Albert,* 1850; Boatswain, *Prince Albert,* 1851-52.
GRAVE, John Costellow. Captain, HBC supply voyages. *Prince Rupert* [V], 1833-34; 1835; 1836-37; 1838. [Rich, 1941, pp. 344-45; O'B (1)].
GRAVES, Thomas. Sailor, *Judith,* 1578.
GRAVILL, John. Captain, *William Ward,* 1849.
GRAY, Alexander. Ice Quartermaster, *Discovery,* 1875-76.
GRAY, Alexander. Captain, HBC supply voyages. *Labrador,* 1882; *Erik,* 1891; 1892; 1893; 1894; 1895; 1896; 1897; 1898; 1899; 1900; *Pelican,* 1901; 1902; 1903; 1904-05; *Discovery,* 1905; *Pelican,* 1906; 1907; 1908.
GRAY, J. Seaman, *George Henry,* 1860-62.
GRAY, James. AB, *Enterprise,* 1848-49. Died at Port Leopold, 16 March 1849.
GRAY, Jemmy. Harpooner, *Arctic,* 1873.
GRAY, John. Captain, *Eclipse,* 1854.
GRAY, William. AB, *Plover,* 1848-52; 1852-54.
GRAY, William. AB, *Herald,* 1848; 1849; 1850.
GRAY, William. Sailmaker, *Investigator,* 1848-49.
GREAVES, Henry. AB, *Phoenix,* 1854.
GREELY, Adolphus Washington (1844-1935). First Lieutenant, 5th Cavalry, commanding U.S. Polar Year Expedition, 1881-84. [GJ 1935; GR 1936; DAB].
GREELY, John Pitt. Surveyed route to Waswanipi, 1817 [see 1819].
GREEN, Charles W. Third Officer, *Arctic,* 1906-07; 1908-09.
GREEN, Edward H. Passed Assistant Surgeon, *Thetis,* 1884.
GREEN, FitzHugh (1888-1947). Ensign, USN, and Physicist, "Crocker Land" Expedition, 1913-17. [EA].
GREEN, George. Captain, *Moose River* [II], 1732.
GREEN, George. AB, *Terror,* 1836-37.
GREEN, George. Ice Mate, *Terror,* 1836-37.
GREEN, George. Boatswain's Mate, *Enterprise,* 1848-49.
GREEN, George. Caulker, *Resolute,* 1850-51.
GREEN, George. Ice Quartermaster, *Resolute,* 1852-54.

GREEN, George. Private RM, *Assistance,* 1850-51; *Pioneer,* 1852-54.

GREEN, James. Carpenter's Mate, *Discovery,* 1791-95.

GREEN, James. AB, *Phoenix,* 1853.

GREEN, John. AB, *Hecla,* 1821-23.

GREEN, John. Private RM, *Resolute,* 1850-51; *Assistance,* 1852-54.

GREEN, Thomas. Private of Marines, *Discovery,* 1791-95.

GREEN, Thomas. AB, *Alexander,* 1818.

GREENAWAY [GREENWAY], William. AB, *Enterprise,* 1850-55. Died 4 November 1853.

GREENE, Henry. Protegé of Henry Hudson, *Discovery,* 1610-11. Killed by Eskimos, 29 July 1611.

GREENE, Rychard. Soldier, Frobisher's expedition, 1578.

GREENFIELD, William. Sailmaker, *Pandora,* 1876.

GREENLAND, J. Sailor, Low's expedition, 1897.

GREENLEAF, George. AB, *Plover,* 1848-52; 1852-54.

GREENSHIELD, Edgar William Tyler (1877-1938). CMS Missionary, journey to Baffin Bay, 1901 [see 1900-02, Peck]. [EA; Fleming, A.L., 1932].

GREENSKY, George. (Ojibwa Indian). Canoeman, Camsell's expedition, 1914.

GREENSTED [GRIMSTEAD], William. Sailmaker's Mate, *North Star,* 1852-54.

GREENWAY, Richard. Captain, *Prudent Mary,* 1680-81.

GREENWAY, William. See GREENAWAY.

GREER, James Augustin (1833-1904). Commander USN, commanding *Tigress,* 1873. [DAB; NE; EA].

GREG, John. AB, *Hecla,* 1821-23.

GREGERSEN, Christen. Seaman, *Enhiörningen,* 1619-20. Died at Churchill, 29 March 1620.

GREGORY, John. Engineer, *Erebus,* 1845-48.

GREILL, Joseph. Established Berens House, 1819. [Rich, 1938, p. 440].

GRENFELL, Wilfred Thomason (1865-1940). Surgeon, National Mission to Deep Sea Fishermen expedition, 1892. [DNB].

GREY, . Lieutenant, *Bulldog,* 1860.

GREY, *Rt Hon* Albert Henry George, *4th Earl Grey* (1851-1917). Governor-General of Canada, toured Hudson Bay, 1910. [DNB].

GREY, George (Cousin of Earl Grey). Passenger, Governor-General's tour, 1910.

GREY, Henry R.E. Midshipman, *Plover,* 1848-52; 1852-54.

GREY, Thomas. Stoker and Blacksmith, *Isabel,* 1852.

GRIEG, George. Stoker, *Phoenix,* 1854.

GRIFFEY, Edwin. Blacksmith and Stoker, *Pandora,* 1875; AB and Stoker, *Pandora,* 1876.

GRIFFIES, Richard. Captain, *Moose River,* 1730.

GRIFFIN, Fraser. AB, *Discovery,* 1791-95.

GRIFFIN, Samuel P. Acting Master, commanding *Rescue,* 1850-51. [EA].

GRIFFIN, William (1755-1839). Cooper and Ship's Corporal, *Resolution,* 1776-80.

GRIFFITHS, Ellis. AB, *Investigator,* 1850-54.

GRIFFITHS, Mark. AB, *Investigator,* 1850-54.

GRIFFITHS, William. AB, *Assistance,* 1850-51; *Resolute,* 1852-54.

GRIFFITHS, William Nelson. Midshipman, *Griper,* 1819-20; *Hecla,* 1821-23. [O'B (1)].

GRIMARD, Ekie. Assistant, first founding of Churchill, 1689.

GRIMINGTON, Michael, Sr. Captain, HBC voyages. *Hayes,* 1686; *Prosperous,* 1690-91; *Royal Hudson's Bay,* 1692-93; 1694; 1696; *Dering* [III], 1697; 1698-1700; 1699; *Pery,* 1701; *Hudson's Bay* [II], 1705-06; 1708; 1710-11. Died at Harwich, 15 June 1710. [Davies, 1965, pp. 368-76; DCB Vol. 2].

GRIMINGTON, Michael, Jr. Captain, *Prosperous,* 1717. [DCB Vol. 2].

GRIMOLFSSON, Bjarni. Karlsefni's Vinland voyage, c1000.

GRIMSHER, William. Private RM, *Hecla,* 1821-23.

GRINSTEAD, Thomas. Sailmaker's Mate, *North Star,* 1852-54; Sailmaker, *Fox,* 1857-59.

GROLLIER, Pierre Henri (1826-1864). RC Missionary journey, 1858-63. [Carrière,

1964; Choque, 1964].

GROOMBRIDGE, Richard. Quartermaster, *Phoenix,* 1853.

GROSEILLIERS, Médard Chouart Sieur des. *See* CHOUART.

GROSELIN, G. Fireman, *Arctic,* 1908-09.

GROVE, James Blair. Mate, *Assistance,* 1852-54.

GROVER, William. Quartermaster, *Arctic,* 1855.

GRUCHY, P. de. AB, *Pandora,* 1876.

GRUNDY, William. AB, *Investigator,* 1848-49. Died at Port Leopold.

GRYFFIN, Manus. Passenger, *Gabriel,* 1578.

GUAY, L. Third Mate, *Neptune,* 1903-04.

GUDLAUGSSON, Gudlief. Voyage to America, 1025.

GUDRID. (Wife of Karlsefni). Karlsefni's Vinland voyage, c1000.

GULL, Thomas. Captain, *Edward and Ann,* 1811-12.

GUMAER, G.G. (?), Canadian Arctic Expedition, 1913-18.

GUSHUE [GUSHIE], S. Mate, *Laddie,* 1913-16.

GUSHUE, Thomas. Mate, *Roosevelt,* 1908-09.

GUTHRIE, Charles. Boatswain's Mate, *Chatham,* 1791-95.

GUY, William. AB, *Discovery,* 1791-95.

HACKLAND, Gilbert. Captain, *Mink,* 1873.

HACKLER, Thomas. AB, *Phoenix,* 1854.

HADDLE, Thomas. HBC expedition, 1869-70.

HADLEY, John R. (-1918). Petty Officer, *Thetis,* 1889; Carpenter, *Karluk* [1913-14], First Officer [1914] and Master [1915-17], *Polar Bear,* Canadian Arctic Expedition, 1913-18.

HADMAN, Benjamin. AB, *Alexander,* 1818.

HAGUE, James. Leading Seaman, *Isabella,* 1818.

HAGUE, John. Leading Seaman, *Isabella,* 1818.

HAINES, Samuel. AB, *Pandora,* 1876.

HALES, John. Stoker, *Pioneer,* 1852-54.

HALES, Richard M. Leading Stoker, *Pioneer,* 1850-51; 1852-54.

HALEY, John. Gunner's Mate, *Release,* 1855.

HALKETT, A. Naturalist, *Neptune,* 1903-04.

HALL, Charles Francis (1821-1871). Leader, Franklin search expeditions, 1860-62; 1864-69; North Polar expedition, 1871-73. Died 8 November 1871. [DCB Vol. 10; Loomis, 1972].

HALL, Christopher. Master, *Gabriel,* 1576; *Ayde,* 1577; Chiefe Pylot of the Fleete, *Ayde,* 1578.

HALL, John. Boy, *Ayde,* 1578.

HALL, Robert. Carpenter, *Resolute,* 1850-51; *Enterprise,* 1848-49.

HALL, Thomas. Ice Master, *Resolute,* 1850-51.

HALL, Thomas. Gunner's Mate, *Pioneer,* 1852-54.

HALL, Thomas A. Second Master, *Plover,* 1848-52; 1852-54.

HALL, Van Rensselaer. Boatswain, *Release,* 1855.

HALLOM, Robert. Corporal RM, Franklin's expedition, 1825-27.

HALLORAN, John. AB, *Resolute,* 1852-54.

HALLOWELL, Thomas. Private RM, *Blossom,* 1825-28.

HALSE, James. Clerk, *Alexander,* 1818; *Hecla,* 1819-20; *Fury,* 1821-23; Purser, *Fury,* 1824-25.

HAMILTON, Fred. Prospector, Curran's expedition, 1912.

HAMILTON, Henry. Captain of the Maintop, *Enterprise,* 1848-49.

HAMILTON, John. AB, *Blossom,* 1825-28.

HAMILTON, Richard Vesey (1829-1912). Mate, *Assistance,* 1850-51; Lieutenant, *Resolute,* 1852-54. [DNB; GJ 1912a].

HAMLIN, [EMELIN, EMLIN], Baptiste. Bowman and fisherman, Richardson's expedition, 1847-49. [Rich, 1953, pp. 355-56].

HAMMON, David. Seaman (?), *Henrietta Maria,* 1631-32.

HAMMOND, James. Corporal RM, *Fury,* 1821-23.

HAMMOND, John. Private RM, *Terror,* 1845-48.

HAMPTON, Robert. AB, *North Star,* 1852-

54; *Fox*, 1857-59.

HANBURY, David Theophilus. Leader, exploring expeditions, 1899; 1901-02.

HANCOCK, John. Captain, *Good Success*, 1719; *Prosperous*, 1720.

HAND, James J. AB, *Discovery*, 1875-76. Died 3 June 1876.

HANDCOCK, Richard. AB, *Blossom*, 1825-28.

HANDFIELD, F.O. Lieutenant, *Rattlesnake*, 1853-54.

HANDFORD, John. AB, *Terror*, 1845-48.

HANDY, Ichabod. Pilot, *Benjamin S. Wright*, 1864.

HANKINS, Henry. AB, *Discovery*, 1791-95.

HANNA, James. Captain, *Harmon* (or *Sea Otter*), 1785.

HANNAH. *See* TOOKOOLITO.

HANNAN [HANNON], William. Private RM, *Resolute*, 1852-54.

HANS. (Eskimo). *See* HENDRIK, Hans Christian.

HANSEN, Godfred (1876-1937). First Lieutenant, *Gjøa*, 1903-06. [Isachsen, 1938; EA].

HANSOM, Joseph. HBC expedition, 1773-74.

HANSON, Charles. AB, *Phoenix*, 1853.

HANSON, James. Second Lieutenant, *Chatham*, 1791-95.

HANSSEN, Helmer Julius (1870-1956). Second Mate, *Gjøa*, 1903-06. [PR 1957; EA].

HANTZSCH, Bernhard Adolf (1875-1911). Leader, ornithological expedition, 1906; exploring expedition, 1909-11. Died June 1911. [Jacobi, 1913; EA].

HANWELL, Henry, Sr. Captain, HBC supply voyages. *Seahorse* [III], 1790, 1791; 1792; *Prince of Wales* [I], 1793; 1794; 1795; 1796; 1797; 1798; 1799; 1800; *King George* [III], 1801; *Prince of Wales* [I], 1802; 1803; 1804; 1805; 1806; 1807; 1808; 1809; 1810; 1811-12; 1814; 1816-17.

HANWELL, Henry, Jr. (-1834). Captain, *Prince of Wales* [I], 1828; 1829; 1830-31; 1832; 1833-34. [Rich, 1941, p. 345].

HARBERD, Richard. Mariner, *Judith*, 1578.

HARBOURN, John. Private RM, *Pioneer*, 1850-51; *Resolute*, 1852-54.

HARDCASTLE, William. Cooper, *Herald*, 1848; 1849; 1850.

HARDIE, Alexander. *See* HARDY.

HARDING, Francis (1799-1875). Lieutenant, *Griper*, 1824. [O'B(1); BOA Vol. 1].

HARDING, J. Second Cook, *Neptune*, 1903-04.

HARDING, James. AB, *Resolution*, 1776-80.

HARDING, William. Private RM, *Rattlesnake*, 1853-54.

HARDY [HARDIE], Alexander. Steward, *Sophia*, 1850-51.

HARDY, W. *See* BROWN, Samuel.

HARDY, William. Cook, *Rattlesnake*, 1853-54.

HARFORD, Thomas. Private RM, *Resolution*, 1776-80.

HARKNESS, George. Boatman, *Union*, Franklin's expedition, 1825-27.

HARLE, Richard. Captain, *Union*, HBC supply voyage, 1714.

HARLEY, Daniel W. Captain of the Foretop, *Alert*, 1875-76.

HARLOW, Charles H. Ensign, *Thetis*, 1884.

HARMAH, Thomas. Carpenter's Mate, *Hecla*, 1824-25.

HARMITAGE, Thomas. *See* ARMITAGE.

HARO, Gonzalo Lopez de. Commander, *San Carlos*, 1788.

HARPER, Arthur. Prospecting expedition, 1872-74; formed Harper, McQuesten & Co., 1874.

HARPER, Francis. Naturalist, Geological Survey of Canada expedition, 1914.

HARPER, William. AB, *North Star*, 1852-54.

HARPER, William. AB, *Phoenix*, 1853.

HARRINGTON, Frederick. Ordinary Seaman, *Rattlesnake*, 1853-54.

HARRIS, Charles. AB, *Investigator*, 1848-49; *Enterprise*, 1850-55.

HARRIS, David. AB, *Investigator*, 1850-54.

HARRIS, E.C. AB, *Discovery*, 1791-95.

HARRIS, Francis L. Volunteer, *United States*, 1860-61.

HARRIS, George. AB, *Pioneer*, 1852-54. Died 10 January 1854.

HARRIS, George K. Captain's steward, *Plover*, 1848-52 [1851-52 only].

HARRIS, James. AB, *Discovery*, 1791-95.

HARRIS, S.A. Assistant Engineer, *Phoenix*, 1853.

HARRISON, Alfred Henry (1865-1933). Leader, exploring expedition, 1905-07. [Stonyhurst Magazine, 1933].

HARRISON, Charles. (?), Rae's expedition, 1853-54. [Rich, 1953, p. 357].

HARRISON, Edward N. Clerk in charge, *Assistance*, 1850-51.

HARRISON, I. Silvester. AB, *Fury*, 1824-25.

HARRISON, James. Clerk, *Hecla*, 1824-25.

HARRISON, John. Private RM, *Resolution*, 1776-80.

HARRISON, John. AB, *Fury*, 1824-25.

HARRISON, John S. Acting Assistant Surgeon, *Yantic*, 1883.

HARRISON, Thomas. AB, *Fury*, 1824-25.

HARRISON, William. AB, *Hecla*, 1821-23.

HARROW, James. AB, *Phoenix*, 1853; Boatswain's Mate, *Phoenix*, 1854.

HART, Frederick. Prospecting expedition, 1872-73; joined Harper, McQuesten & Co., 1874.

HART, George. Quartermaster, *Discovery*, 1791-95.

HART, Henry Chichester. Naturalist, *Discovery*, 1875-76.

HART, James. Leading Stoker, *Erebus*, 1845-48.

HARTGILL, John. Gentleman, Frobisher's expedition, 1578.

HARTLEY, Richard. Captain of the Hold, *Arctic*, 1855.

HARTNELL, John. AB, *Erebus*, 1845-48. Died on Beechey Island, 4 January 1846.

HARTNELL, Thomas. AB, *Erebus*, 1845-48.

HARTNELL, Thomas. AB, *Intrepid*, 1852-54.

HARTSTENE, Henry J. (-1868). Lieutenant commanding, *Release*, 1855. [NE].

HARTY, John. Ordinary Seaman, *Herald*, 1848; 1849; 1850.

HARVEY, George. Sailmaker, *Plover*, 1848-52.

HARVEY, J. AB, *Phoenix*, 1853.

HARVEY, James. Private RM, *Phoenix*, 1854.

HARVEY, William. Master's Mate, *Resolution*, 1776-80.

HARVEY, William. Boatswain's Mate, *Resolute*, 1850-51; *North Star*, 1852-54.

HARVEY, William. Chief Quartermaster, *Fox*, 1857-59.

HARVIE [HARVEY], Edwarde. Gentleman, *Ayde*, 1577; Captain, *Gabriel*, 1578.

HARWOOD, J. Apprentice Engineer, *Pioneer*, 1850-51; Engineer, *Pioneer*, 1852-54.

HARWOOD, John. Surgeon, *Ayde*, 1578.

HASHAM, Richard. AB, *Phoenix*, 1854.

HASLETON, John A. *See* HAZLETON.

HASSEL, Sverre Helge (1876-1928). (?), *Fram*, 1898-1902. [EA].

HASSEL, Thomas. Interpreter, Back's expedition, 1833-35.

HASSELL, James. AB, *Phoenix*, 1854.

HASWELL, William Henry (1818-1900). Lieutenant, *Investigator*, 1850-54. [O'B(1); BOA Vol. 5].

HATCHER, James. AB, *Plover*, 1852-54.

HATLEY, John. AB, *Resolution*, 1776-80.

HAVEN, Jens. Moravian missionary. Missionary voyages, 1764; 1765; 1770; 1771; 1774. Established Okak, 1775.

HAWES, John. Captain, HBC supply voyages. *Ocean Nymph*, 1882; 1883-84; *Cam Owen*, 1885; 1886; *Princess*, 1892.

HAWKERIDGE [HAWKRIDGE], William. Volunteer (?), *Resolution*, 1612-13; Captain and commander of expedition, *Lion's Whelp*, 1625. [DCB Vol. 1].

HAWKES, Ernest William (1883-). Leader, GSC anthropological expedition, 1914.

HAWKINS, Charles. AB (?), *Nancy Dawson*, 1849.

HAWKINS, J. Private RM, *Phoenix*, 1854.

HAWKINS, John. Cooper, *Alert*, 1875-76.

HAWKINS, Thomas. AB, *North Star*, 1852-54.

HAWKINS, Williard. Seaman, whalers' boat voyage, 1860.

HAWKRIDGE, William. *See* HAWKERIDGE.

HAWLEY, Henry. Paymaster, *Phoenix*, 1853.

HAY, Henry. (?), McLean's expedition, 1838.

HAY, James. AB, *Griper*, 1824.

HAYDON, George. Boatswain's Mate, *Plover*, 1848-52; 1852-54.

HAYES, Edward. Captain, *Golden Hind*, 1583.

HAYES, George. Chief Officer, *Arctic*, 1906-

07.

HAYES, Isaac Israel (1832-1881). Surgeon, *Advance*, 1853-55; commander of expedition, *United States*, 1860-61. [DAB; EA].

HAYES, Noah. AB, *Polaris*, 1871-73.

HAYSON, Robert. Gunner, *Judith*, 1578.

HAYWARD, Johnny. Established Port Harrison, 1920.

HAZLEHURST, Thomas. Marine, *Blossom*, 1825-28.

HAZLETON [HASLETON], John A. AB, *Fox*, 1857-59.

HAZLITWOOD, Thomas. Private RM, *Blossom*, 1825-28.

HEAD, Horatio Nelson. Midshipman, *Hecla*, 1824-25.

HEAD, James. Chief Mate, *Phoenix*, 1854.

HEAD, John. Private RM, *Plover*, 1848-52; 1852-54.

HEAD, Josiah. AB, *Phoenix*, 1854.

HEALEY, Joseph. Private RM, *Erebus*, 1845-48.

HEARD, William. Corporal RM, *Plover*, 1852-54.

HEARN, J. First Mate, *Neptune*, 1903-04.

HEARN, James. Seaman, *Arctic*, 1906-07.

HEARNE, Samuel (1745-1792). Mate, *Churchill*, 1767; *Charlotte*, 1769; leader, overland expeditions, 1769; 1770; 1770-72; established Cumberland House, 1774. [DCB Vol. 4; Glover, 1947].

HEATHER, William. Private RM, *Terror*, 1845-48.

HEAYNES [HEAYNS, RAYNES], Samuel. Cooper, *Assistance*, 1852-54.

HEBERT, John (*dit* Fabien) (c1828-1856). Interpreter, Rae's expedition, 1850-51. [Rich, 1953, p. 358].

HECETA, Bruno de. Lieutenant commanding, *Santiago*, 1775.

HEDDINGTON, Thomas. Midshipman, *Chatham*, 1791-95.

HEDGES, William. Corporal RM, *Terror*, 1845-48.

HEDLEY, Matthew. AB, *Griper*, 1824.

HEELS [HIEELS, HEELES], John. Private RM, *Investigator*, 1848-49; *Resolute*, 1850-51; 1852-54.

HEIDEMAN, William. Seaman, *Resolution*, 1776-80.

HEINDRICK, Hans Christian. *See* HENDRIK.

HELGI. Leader, Vinland voyage, c1000.

HELMAN, Edward. Sailor (?), *Sunneshine*, 1585.

HELLMICH, A. Cook, German Polar Year Expedition, 1882-83.

HELPMAN, Edwin James Howard. Clerk in charge, *Terror*, 1845-48.

HELSING, Jens. Seaman, *Enhiörningen*, 1619-20. Died at Churchill, 27 January 1620.

HEMINSTER, George. Workman, second founding of Churchill, 1717.

HEMMELT, John. Sailmaker, *Plover*, 1848-52; 1852-54.

HEMPSON, Hugh. Seaman, *Sunneshine*, 1586.

HENDAY, Anthony. Leader, HBC expeditions, 1754-55; 1759-60. [DCB Vol. 3].

HENDERSON, Donald. (?), McLean's expedition, 1838.

HENDERSON, John. Midshipman, *Fury*, 1821-23. [O'B(1)].

HENDERSON, Robert. Prospecting expedition, 1894-96.

HENDRICHSEN, Jens. Master, *Lamprenen*, 1619-20. Died at Churchill, 12 May 1620.

HENDRIK [HEINDRICK], Hans (1834-1889). (Greenlander). Hunter, *Advance*, 1853-55; *United States*, 1860-61; *Polaris*, 1871-73; Nares expedition, 1875-76. [Rink, 1877; Lidegaard, 1968; Laursen, 1956; Ryder, 1889-90].

HENDRIKSEN, Peder Leonard (1859-). Seaman (?), *Fram*, 1898-1902. [EA].

HENDRY, Anthony. *See* HENDAY.

HENDRY, Daniel. Carpenter, *Lady Franklin*, 1850-51.

HENDRY, William. Leader, HBC expedition, 1828. [Davies, 1963, pp. 378-83].

HENLEY, Richard. AB, *Discovery*, 1791-95.

HENNESSEY, Frank. Assistant naturalist, *Arctic*, 1908-09.

HENRY, Alexander. Fur trading expeditions, 1775-76; 1776-77.

HENRY, Charles B. Private, 5th Cavalry, U.S. Polar Year Expedition, 1881-84. Executed 6 June 1884.

HENRY, Daniel. *See* HENDRY.

HENRY, W. Corporal, Lefroy's expedition, 1843-44.
HENRY, William. Ship's cook, *Release*, 1855.
HENSON, Matthew Alexander (1866-1955). (American Negro). Assistant, Peary's expeditions, 1898-1902; 1905-06; 1908-09. [Miller, 1963; MacMillan, 1955; Robinson, B., 1947].
HENSON, Phillip. Bombardier RMA, *North Star*, 1852-54.
HEPBURN, John (1794-1864). AB, Franklin's expedition, 1819-22; supercargo, *Prince Albert*, 1851-52. [DCB Vol. 9].
HEPBURN [HEPBURNE], William. Boatman, Pullen's expedition, 1849-51 [1850-51 only]. [Rich, 1953, pp. 358-59].
HERBERT, Francis B. Mate, *Assistance*, 1852-54.
HERD, David J. Captain, HBC supply voyages. *Prince Rupert* [V], 1839; 1840; *Prince Rupert* [VI], 1841; 1842; 1843; 1844; 1845; 1846; 1847; 1848; 1849; *Prince of Wales* [II], 1850; 1851; 1852; 1853; 1854; 1855; 1856; 1857; 1858; 1859; 1860; 1861; 1862; 1863-64.
HERD, John. Private RM, *Plover*, 1848-52; *North Star*, 1852-54.
HERGEST, Richard. Midshipman, Cook's expedition, 1776-80.
HERJULFSSON, Bjarni. Vinland voyage, c1000.
HERNANDEZ, Juan Josef Perez. *See* PEREZ HERNANDEZ.
HEROLD, William. AB, *Resolution*, 1776-80.
HERON, F. Canoeman, Camsell's expedition, 1905.
HERRIOTT, John. Private RM, *Discovery*, 1776-80.
HERRON, John. Steward, *Polaris*, 1871-73.
HESKET, John. AB, *Fury*, 1824-25.
HESTER, Henry. AB, *Investigator*, 1848-49; Captain's Coxswain, *Enterprise*, 1850-55.
HESTER, Jack. (Indian). Employee of NWC, invasion of James Bay, 1803-06.
HETHERINGTON, W. Staff-sergeant, NWMP Patrol, 1897.
HETHERTON, Augustin. AB, *Resolution*, 1776-80.
HEVICKE, P. Sailmaker, German Polar Year Expedition, 1882-83.
HEWISON, John. Sailmaker's Mate, *Phoenix*, 1854.
HEWITT, G.C. Surgeon's First Mate, *Discovery*, 1791-95.
HEWS, Anthonie. Soldier, Frobisher's expedition, 1578.
HEWSE, Tege. Sailor, *Gabriel*, 1578.
HEYDON, Gabriel. Captain of the Maintop, *Plover*, 1851-54.
HEYDON [HEYDEN], John. AB, *Investigator*, 1848-49; *Assistance* 1850-51.
HEYWARD, Charles. AB, *Herald*, 1848; 1849; 1850.
HEYWODD, John. Cook and miner, Frobisher's expedition, 1578.
HEYWOOD, Charles. AB, *Breadalbane*, 1853.
HEYWOOD, Harvey. Volunteer, *United States*, 1860-61.
HICCLES, John. Marine, *Intrepid*, 1852-54.
HICKEY, Cornelius. Caulker's Mate, *Terror*, 1845-48.
HICKEY, Thomas. Seaman, *Advance*, 1853-55.
HICKLEY, James. Private RM, *Plover*, 1852-54.
HIEELS, John. *See* HEELS.
HIGGINS, John. *See* GIBSON, John.
HILCHEN, George. AB, *Phoenix*, 1854.
HILL, . Boatswain's Mate, Frobisher's expedition, 1578.
HILL, Edward. Ropemaker, *Rattlesnake*, 1853-54.
HILL, Elias, Gunner, RM, *Alert*, 1875-76.
HILL, James S. Master, *Herald*, 1848; 1849; 1850.
HILL, John. Moravian missionary, missionary expeditions, 1765; 1774.
HILL, John. Carpenter, *Herald*, 1848; 1849; 1850.
HILL, Richard. Corporal RM, *Enterprise*, 1850-55.
HILL, Thomas. Sailor (?), *Sunneshine*, 1585.
HILLIER, Thomas. Cook, GSC expedition, 1875.
HILLS, Edward H. Second Master, *Phoenix*, 1853.
HILLS, Samuel. AB, *Alexander*, 1818.
HILLS, William. AB, *North Star*, 1852-54.
HILLS, William. AB, *Phoenix*, 1853.

HILLSEY, William. Sailmaker, *Discovery*, 1776-80.
HILPE, John. Sailor boy, *Ayde*, 1578.
HIND, Henry Youle (1823-1908). Leader, Canadian expedition, 1861.
HIND, Robert. Surgeon, *Michael*, 1578.
HIND, William G.R. (1833-1888). (Brother of H.Y. Hind). Artist, Canadian expedition, 1861. [Harper, J.R., 1967].
HINDLE, Alfred. AB, *Discovery*, 1875-76.
HINGE, Thomas. AB, *Phoenix*, 1854.
HINKS, Theophilus. Private RM, *Resolution*, 1776-80.
HISLOP, J. Formed Hislop and Nagle partnership, c1887.
HITCHCOCK, James. AB, *Discovery*, 1791-95.
HITCHCOCK, Robert W. AB, *Discovery*, 1875-76.
HITCHCOCKE, John. Sailor, *Ayde*, 1578.
HOAR, Edmund [Edward] Charles. Captain's Steward, *Erebus*, 1845-48.
HOBBS, Jacob. Private, Sappers and Miners, Richardson's expedition, 1847-49.
HOBBS, Richard. Sergeant RM, *Resolute*, 1852-54.
HOBBY, Henry (1842-1874). Seaman, *Polaris*, 1871-73.
HOBDAY, George. Captain of the Hold, *Fox*, 1857-59.
HOBSON, William Robert (1831-1880). Mate, *Rattlesnake*, 1853-54 [1853 only]; *Plover*, 1852-54 [1853-54 only]; Lieutenant, *Fox*, 1857-59. [BOA Vol. 1].
HOCKADAY, William. Cooper, *Rattlesnake*, 1853-54.
HOCKLEY, John Baker. First Class Volunteer, *Blossom*, 1825-28. [O'B(1)].
HODGE, Edwin. AB, *North Star*, 1853-54.
HODGE, William. Quartermaster, *Fury*, 1824-25.
HODGES, Henry. Quartermaster, *Phoenix*, 1853.
HODGES, John. AB, *Discovery*, 1875-76.
HODGSON, Charles E. Chaplain, *Discovery*, 1875-76.
HODGSON, George Henry. Lieutenant, *Terror*, 1845-48. [O'B(1)].
HODGSON, Joseph. Accompanied Melvill to Great Bear Lake, 1910 [see 1908-11].
HODGSTON, James. AB, *Lady Franklin*, 1850-51.
HODGYS, John. Smith, Frobisher's expedition, 1578.
HOEOOTOEROCK, Junius. (Eskimo). Interpreter, Franklin's expedition, 1819-22.
HOFF, J.E. Engineer, *Alaska*, Canadian Arctic Expedition, 1913-18 [1916 only].
HOGG, Alexander. AB, *Resolution*, 1776-80.
HOGGIT, William R. *See* HUGGETT.
HOILE [HOILL, HOYLE], James. AB, *Assistance*, 1850-51.
HOILE [HOILL], Robert. Sailmaker, *Resolute*, 1850-51; 1852-54.
HOLDEN, Thomas. Seaman, *Arctic*, 1908-09; Boatswain, *Arctic*, 1910-11.
HOLDING, . Lieutenant, *California*, 1746-47.
HOLLAMBY, William. Quartermaster, *Discovery*, 1776-80.
HOLLAND, Daniel. Captain of the Hold, *Investigator*, 1848-49.
HOLLAND, Ernest. Gold prospector, *Neptune*, 1912.
HOLLAND, Harry. Gold prospector, *Neptune*, 1912.
HOLLAND, John. *See* BAILEY, John.
HOLLIDAY, William. AB, *Isabella*, 1818.
HOLLINS, John. Private RM, *Alert*, 1875-76.
HOLLOWAY, Jeremiah. Drummer, Marines, *Discovery*, 1776-80.
HOLLY, Robert [HOLLEY, James]. AB, *Resolute*, 1850-51.
HOLMAN, John R. Assistant Surgeon, *Phoenix*, 1853; 1854.
HOLMAN, William. AB, *Phoenix*, 1853.
HOLME, Randle F. Leader, British expedition, 1887.
HOLMES, Quartermaster, *Ayde*, 1578.
HOLMES, Charles. Stoker, *Intrepid*, 1850-51.
HOLMES, Israel. AB, *North Star*, 1852-54.
HOLMES, James T. Caulker, *Herald*, 1848; 1849; 1850.
HOLMES, John. AB, *Fury*, 1824-25.
HOLMES, Richard. Ordinary Seaman, *Rattlesnake*, 1853-54.
HOLMES, William. Seaman, *Advance*, 1850-51.

HOLT, Adam. Shipwright, *Judith*, 1578.

HOLT, George. Miner and prospector, 1878.

HOLTON, Nicholas. Stoker, *Pioneer*, 1850-51.

HOME, Alexander (-1823). Master's Mate, *Discovery*, 1776-80.

HOMES, Walter. Sailor, *Ayde*, 1578.

HONEY, Samuel. Blacksmith, *Terror*, 1845-48.

HONEY, Thomas. Carpenter, *Terror*, 1845-48.

HONMAN, Charles A. Topographical Assistant, U.S. Army; topographer, Schwatka's expedition, 1883.

HONTING, John. Seaman, *Charles*, 1631.

HOOD, Robert (1797-1821). Midshipman, Franklin's expedition, 1819-22. Murdered 20 October 1821. [Houston, 1974; 1974a; 1974b; Jones, A.G.E., 1975].

HOOD, Thomas. Private RM, *Investigator*, 1848-49; *Assistance*, 1850-51; *Intrepid*, 1852-54. Died 2 January 1854.

HOOK, David. AB, *North Star*, 1852-54.

HOOLE, François. (?), Back's expedition, 1833-35.

HOOPER, William Harvey (-1833). Purser, *Alexander*, 1818; *Hecla*, 1819-20; *Fury*, 1821-23; *Hecla*, 1824-25. [MAR].

HOOPER, William Hulme (1827-1854). Acting mate, *Plover*, 1848-52; second-in-command, Pullen's expedition, 1849-51. [DNB Vol. 27; BOA Vol. 1].

HOPCRAFT, Robert. Private RM, *Erebus*, 1845-48.

HOPE, Alfred. Ordinary Seaman, *North Star*, 1849-50.

HOPE, James. (Cree Indian). Bowman, Dease and Simpson's expedition, 1837-39; Steersman, Richardson's expedition, 1847-49. [Rich, 1953, p. 359].

HOPE, Thomas. (Cree Indian). Richardson's expedition, 1847-49. [Rich, 1953, p. 360].

HOPKINS, William. Cook, *Fury*, 1821-23.

HOPPER, Layton. AB, *Terror*, 1836-37.

HOPPNER, Henry Parkyns (-1833). Lieutenant, *Alexander*, 1818; *Griper*, 1819-20; *Hecla*, 1821-23; Commander, *Fury*, 1824-25. [RNB Vol. 5; MAR].

HORDEN, John (1828-1892). Church Missionary Society journey, 1851; made Bishop of Moosonee, 1872. [EA; Buckland, n.d.].

HORE, Richard. Commander, British expedition, 1536. [DCB Vol. 1].

HORN, Thomas. Captain of the Forecastle, *North Star*, 1849-50.

HORN, Timothy. AB, *Enterprise*, 1850-55.

HORNBY, Frederick John (1819-1848). Mate, *Terror*, 1845-48. [Jones, A.G.E., 1955].

HORNBY, John (1880-1927). Accompanied Melvill's expedition, 1908-11; part-time member of Douglas's expedition, 1911-12; assistant to Rouvière and Le Roux, 1911-13. [Whalley, 1962].

HORNER, Arthur C. Surgeon and Naturalist, *Pandora*, 1875; 1876.

HORNER, John. Captain, HBC supply voyages. *Prince Rupert* [III], 1764; *Seahorse* [II], 1765; 1766; 1767; 1768; 1769.

HORSEY, Edmond. Labourer, Frobisher's expedition, 1578.

HORTON, William. AB, *Terror*, 1836-37.

HOULSTON, Robert. Stoker, *Phoenix*, 1853.

HOUNSELL, James M. Private RM, *Rattlesnake*, 1853-54.

HOUSE, William. Boatswain, *Chatham*, 1791-95.

HOUSTON [HOUSTOUN], Wallace (1811-1891). Captain, *Trincomalee*, 1854. [O'B(1); BOA].

HOVEY, Edmund Otis (1862-1924). Representative of American Museum of Natural History, "Crocker Land" Expedition, 1913-17 [1915-17 only].

HOWARD, John. AB, *Griper*, 1824.

HOWARD, W. H. Dominion Land Surveyor, Flaherty's expedition, 1913-16 [1916 only].

HOWARD, William. Gunner's mate, *Chatham*, 1791-95.

HOWARD, William. Captain, *Graham*, HBC supply voyage, 1849.

HOWE, George P. Joint leader, Cabot's expedition, 1910 [see 1903-10].

HOWE, Stephen. AB, *Assistance*, 1850-51.

HOWELL, John. Ordinary Seaman, *Herald*,

1848; 1849; 1850.

HOWELLS, Samuel. Private RM, *Herald*, 1848; 1849; 1850.

HOWES, Robert. Ordinary Seaman, *Rattlesnake*, 1853-54.

HOWES, Thomas. Private RM, *Plover*, 1848-52.

HOY, John. AB, *North Star*, 1849-50.

HOYLE, James. See HOILE, James.

HUBART, Josias. Pilot, *Resolution*, 1612-13; possibly also with Bylot and Baffin, 1616.

HUBBARD, Leonidas (1872-1903). Leader, U.S. exploring expedition, 1903-04. [Hubbard, M.B., 1908, pp. 33-48].

HUBBARD, Mina Benson. Leader, U.S. exploring expedition, 1905. [EA].

HUDSON, Henry (-1611). Captain, *Discovery*, 1610-11. [DNB Vol. 28; DCB Vol. 1].

HUDSON, J.R. Steward, *George Henry*, 1860-62.

HUDSON, John (-1611). (Son of Henry Hudson). Boy, *Discovery*, 1610-11. [DCB Vol. 1].

HUGGETT [HOGGIT], William R. AB, *Resolute*, 1850-51; 1852-54.

HUGHES, Joseph. AB, *Investigator*, 1848-49.

HUGHES, Timothy. AB, *North Star*, 1849-50.

HUGHES, W. AB, *North Star*, 1849-50.

HUGHSTER, Matthew. Guide, Buchan's expedition, 1811.

HULL, Thomas A. Master's Assistant, *Herald*, 1848; 1849; 1850; Second Master, *Plover*, 1852-54.

HULL, William. Corporal RM, *Plover*, 1848-51.

HULOTT, Cornelius (*alias* HULLOTT, George). Captain's Coxswain, *Investigator*, 1850-54.

HUME, Andrew. Quartermaster, *Charles*, 1631.

HUME, Edward. Captain, *Perpetuana Merchant*, 1685.

HUMFREY, William. "Goldfyner," *Ayde*, 1578.

HUMPHREY, Halcro. Orkney boatman, Richardson's expedition, 1847-49. [Rich, 1953, p. 360].

HUMPHREYS, Charles. Captain, *Eagle*, 1836-37. [Rich, 1943, pp. 391-92].

HUMPHREYS, Henry. Midshipman, *Discovery*, 1791-95.

HUMPHREYS, Richard W. Captain, *Isabella*, 1833.

HUMPHRIES, Edward. Sailmaker's mate, *Assistance*, 1852-54.

HUMPHRIES [HUMPHREY], William. Private RM, *Resolute*, 1852-54.

HUNKER, John J. Ensign, *Juniata*, 1873. [NE].

HUNT, Harrison J. Surgeon, "Crocker Land" Expedition, 1913-17.

HUNT, Henry J. Lieutenant, *Alert*, 1884.

HUNT, William. Wardroom Cook, *Alert*, 1875-76.

HUNTER, Harky. Harpooner, *Arctic*, 1873.

HUNTER, James (1817-). Archdeacon, Anglican missionary journey, 1858-59.

HUNTER, John. AB, *Hecla*, 1821-23.

HURDITCH [HURDILCH], William. Sergeant RM, *Enterprise*, 1848-49.

HURIN. *See* URIN.

HURST, Captain, *Ariel*, 1820.

HURST, John. Private RM, *Isabella*, 1818; Corporal RM, *Hecla*, 1824-25.

HURST, Thomas. Corporal RM, *Griper*, 1819-20.

HUSSEY, James. Blacksmith, *Herald*, 1848; 1849; 1850.

HUSSEY, William. *See* BLAKE, John.

HUTCHENSON, William. AB, *Enterprise*, 1850-55.

HUTCHINS, William. Prospector, Curran's expedition, 1912.

HUTCHINSON, Edward. Orkney middleman, Rae's expedition, 1846-47. [Rich, 1953, p. 360].

HUTCHINSON, James. AB, *Phoenix* , 1853.

HUTCHINSON [HUTCHISON], John. Mate, *Herald*, 1848; 1849; 1850.

HUTCHISON, John (c1798-). Established Fort Halkett, 1829. [Rich, 1939, pp. 223-24].

HUTTON, George. Quartermaster, *Phoenix*, 1854.

HYDE, Joshua. Stoker, *Phoenix*, 1854.

IBBETT [IBBETS], William Joshua. Engineer, *Intrepid*, 1852-54. [MAR].

IBESTER, William. Sailor, Hearne's expedition, 1769.

IDE, George E. Lieutenant, *Juniata*, 1873. [NE].

INCENT, John. Intending colonist, *Gabriel*, 1578.

INGLEDON, Alfred. *See* INGLETON.

INGLEFIELD, Edward Augustus (1820-1894). Commander, RN, commanding *Isabel*, 1852; *Phoenix*, 1853; Captain, *Phoenix*, 1854. [Woodward, 1948; DNB 1st Supp. Vol. 3; O'B(1); GJ 1894].

INGLETON [INGLEDON], Alfred. Steward, *Assistance*, 1852-54.

INGLIS, A.D. Station assistant at Port de Boucherville, Canadian Hudson Bay Expedition, 1884-85. Died of scurvy during winter 1884-85.

INGRAM, John. Captain and Master, *Discovery*, 1612-13.

INGRAM, William. Private RM, *North Star*, 1849-50.

INNES-LILLINGSTON, Frederick George. Lieutenant, *Pandora*, 1875.

IRVIN, Henry. Private RM, *Hecla*, 1824-25.

IRVIN, Richard. AB, *Resolution*, 1776-80.

IRVINE, Andrew. AB, *Prince Albert*, 1851-52.

IRVING, John (1815-1847). Lieutenant, *Terror*, 1845-48. [O'B(1); Bell, B., 1881].

ISAACS, Thomas. Gunner's Mate, *Assistance*, 1852-54.

ISACHSEN, Gunnerius (Gunnar) Ingvald (1868-1939). Cartographer, *Fram*, 1898-1902. [Hoel, 1940; GR 1940; PR 1940].

ISBISTER, Alexander Kennedy (1822-1883). Second-in-command, HBC expedition, 1840-41. [Bryce, 1883; Knox, 1957].

ISBISTER, John. Established Nichicun post, 1816.

ISBISTER, Thomas. Established Nichicun post, 1816.

ISERHOFF, Joseph. (Russian half-breed). (?), Hubbard's expedition, 1905.

ISRAEL, Edward. Sergeant, Signal Corps, U.S. Army, U.S. Polar Year Expedition, 1881-84. Died of starvation, 27 May 1884.

IVES, William. Blacksmith, *North Star*, 1849-50.

JACK, Alexander. AB, *Hecla*, 1819-20.

JACKLIN, James. Sailor, *Ayde*, 1578.

JACKLIN, James. Sailor, *Gabriel*, 1578.

JACKMAN, Charles. Mate, *Ayde*, 1577; Master and chief pilot for discovery, *Thomas Allen*, 1578.

JACKMAN, George. Established Little Whale River post, 1793.

JACKSON, Christopher. Trumpeter, *Ayde*, 1578.

JACKSON, J. Private RM, *Phoenix*, 1853.

JACKSON, John. Private of Marines, *Resolution*, 1776-80.

JACKSON, John. AB, *Isabella*, 1818.

JACKSON, John. AB, *North Star*, 1852-54.

JACKSON, W.E. Meteorologist, *Arctic*, 1908-09.

JACKSON, William. Cook, *Polaris*, 1871-73.

JACOBSON, Chrn. [JAKOBSON, Charles]. AB, *Enterprise*, 1850-55.

JAGO, C.J. Lieutenant, *Enterprise*, 1850-55.

JAGO, Edwin. Clerk, *Herald*, 1848; 1849; 1850; Clerk, *Plover*, 1852-54.

JAKOBSON, Charles. *See* JACOBSON.

JALOBERT, Macé. Commander, *Petite Hermine*, 1535-36; Pilot, Cartier's expedition, 1541-42.

JAMES, James. Seaman, *Resolution*, 1776-80.

JAMES, John. Private of Marines, *Resolution*, 1776-80.

JAMES, John. Captain, HBC supply voyages. *Ocean Nymph*, 1863-64; 1864-65; *Lady Head*, 1865; 1866; 1867; 1868; 1869; 1870; 1871; 1872.

JAMES, John. Leading Seaman, *Isabella*, 1818.

JAMES, Robert. Carpenter, *Pandora*, 1875; 1876.

JAMES, Thomas (c1593-c1635). Captain, *Henrietta Maria*, 1631-32. [DCB Vol. 1].

JAMES, Thomas. AB, *Fury*, 1824-25.

JAMKA, Frederick. Seaman, *Polaris*, 1871-73.

JANE [JANES], John. Merchant, *Sunneshine*, 1585; *Elizabeth*, 1587.

JANE, John. Revenue Officer, *Discovery*, 1791-95.

JANE, Philip. Seaman, *Sunneshine*, 1586.

JANES, Robert S. (-1920). Second Officer, *Arctic*, 1910-11; joint leader of expedition, *Neptune*, 1912; leader, fur trading expedition, 1916-20. Killed by Eskimos, 15 March 1920. [Stevenson, 1973].

JANTZEN, A. Labourer and seaman, German Polar Year Expedition, 1882-83.

JARVIS, A.M. Inspector, NWMP patrol, 1897.

JARVIS, F.A. Constable, NWMP, *Neptune*, 1903-04.

JARVIS, George. Captain's Steward, *Resolute*, 1850-51.

JARVIS, Samuel. Ordinary Seaman, *Blossom*, 1825-28.

JEFFERIES, George. Sergeant RM, *Assistance*, 1852-54.

JEFFERSON, William. AB, *Enterprise*, 1850-55.

JEFFREY [JEFFEREY], Isaac. Sergeant RM, *Enterprise*, 1850-55.

JEFFREY, Louis. Boy, *Arctic*, 1906-07.

JEFFREYS [JEFFREY], James. AB, *Resolute*, 1852-54.

JEMMINGS, Cornelius. Steward, *Phoenix*, 1853.

JENKINS, David. AB, *Enterprise*, 1848-49. Died at Port Leopold, 30 March 1850.

JENKINS, Robert. Commander RN, commanding *Talbot*, 1854.

JENKINS, Robert P. Mate, *North Star*, 1852-54.

JENKINS, Thomas. Pumpmaker, *Ayde*, 1578.

JENNESS, Diamond (1886-1969). Anthropologist, Canadian Arctic Expedition, 1913-18. [PR 1970a; *Arctic*, 1970].

JENNYNGES, Thomas. Founder, intending colonist, Frobisher's expedition, 1578.

JENSEN, Peter. Interpreter and dog manager, *United States*, 1860-61.

JENSEN, Rasmus. Chaplain, *Enhiörningen*, 1619-20. Died at Churchill, 20 February 1620.

JENSEN, Stephen. Moravian missionary, missionary expeditions, 1770; 1771; establishment of Okak, 1775.

JERMAIN, John. *See* GERMAIN.

JERRY, William. AB, *Terror*, 1845-48.

JEWELL, Winfield Scott (-1884). Sergeant, Signal Corps, U.S. Army, U.S. Polar Year Expedition, 1881-84. Died of starvation, 12 April 1884.

JIM. *See* SKOOKUM JIM.

JOAS, Thomas. *See* JOYCE, Thomas.

JOE (Eskimo). *See* EBIERBING.

JOHANSEN, Frits. Botanist and marine biologist, Canadian Arctic Expedition, 1913-18.

JOHNS, John. Carpenter, *Plover*, 1848-52; 1852-54.

JOHNS, Philip. Second Mate, *Phoenix*, 1854.

JOHNS, Robert. AB, *Erebus*, 1845-48.

JOHNS, S.W. Sailmaker, *Phoenix*, 1854.

JOHNSON, . Captain, *Cumbrian*, 1820.

JOHNSON, . AB, *North Star*, 1852-54.

JOHNSON, A.B. Private RM, *Hecla*, 1824-25.

JOHNSON, Alexander. Gunroom Steward, *Intrepid*, 1852-54.

JOHNSON, Charles. AB, *Terror*, 1845-48.

JOHNSON, Charles. Captain of the Maintop, *Release*, 1855.

JOHNSON, D. Boatman (?), Dawson's expedition, 1887.

JOHNSON, George W. Caulker, *Enterprise*, 1850-55.

JOHNSON, Godfrey. Shoemaker, Frobisher's expedition, 1578.

JOHNSON, Henry. AB, *Fury*, 1821-23.

JOHNSON, James. Seaman, *Rescue*, 1850-51.

JOHNSON, John. Soldier, Frobisher's expedition, 1578.

JOHNSON, Peter. Seaman, *Polaris*, 1871-73.

JOHNSON, Robert. AB, *Griper*, 1824.

JOHNSON, Thomas. Sergeant RM, *Plover*, 1852-54.

JOHNSON, Thomas. Boatswain's Mate, *Terror*, 1845-48.

JOHNSON, Thomas. Second Master, *Investigator*, 1848-49.

JOHNSON, Thomas. AB, *Rattlesnake*, 1853-54.

JOHNSON, William. Stoker, *Terror*, 1845-48.

JOHNSON, William. Boatswain's Mate, *Herald*, 1848; 1849; 1850.

JOHNSON, William. Captain of the Forecastle, *Resolute*, 1852-54.

JOHNSON, William. Seaman, *George Henry*,

1860-62.

JOHNSON, William. Boatman, *Arctic*, 1908-09.

JOHNSTON, Jerry. Middleman, HBC Franklin search expedition, 1855.

JOHNSTON, Magnus. Captain, HBC expeditions, *Churchill*, 1764; 1765; 1766; 1767; 1768.

JOHNSTON, Magnus. AB, *Breadalbane*, 1853.

JOHNSTON, Phil. Engineman and cook, Curran's expedition, 1912.

JOHNSTON, William. Acting Third Assistant Engineer, *Arctic*, 1855.

JOHNSTONE, Alexander. *See* JOHNSON.

JOHNSTONE, James. Master, *Chatham*, 1791-95.

JOHNSTONE, James. Middleman, labourer and fisherman, Rae's expedition, 1853-54. [Rich, 1953, p. 361].

JOHNSTONE, Thomas. Private RM, *Assistance*, 1850-51.

JOHNSTONE, William. Stoker, *Phoenix*, 1853.

JOINER, Robert. Leading Stoker, *Alert*, 1875-76.

JOLLIET, Louis (1645-1700). Leader, overland expedition, 1679; French voyage, 1694. [DCB Vol. 1].

JOLLIET, Zacharie. (Brother of Louis Jolliet). French overland expedition, 1679.

JOLLIFFE, Thomas. Captain of the Maintop, *Alert*, 1875-76.

JON. Leader, Vinland voyage, 1059.

JONAS, Pawll. Sailor, *Ayde*, 1578.

JONES, Charles Jesse ("Buffalo") (1844-1919). Leader, U.S. hunting expedition, 1897-98. [Easton and Brown, 1964].

JONES, David. AB, *Terror*, 1836-37.

JONES, George. Ordinary Seaman, *Herald*, 1848; 1849; 1850.

JONES, Frank. Stoker, *Discovery*, 1875-76.

JONES, Henry. AB, *Assistance*, 1852-54.

JONES, Hugh. AB, *North Star*, 1849-50.

JONES, Hughe. Sailor and gunner, *Ayde*, 1578.

JONES, J. AB, *Hecla*, 1824-25.

JONES, J.J. Engineer, *Polar Bear*, Canadian Arctic Expedition, 1913-18.

JONES, J.W. Steward, *Investigator*, 1848-49.

JONES, John. AB, *Fury*, 1821-23.

JONES, John. Steward, *Felix*, 1850-51.

JONES, Lewis. AB, *Discovery*, 1791-95.

JONES, Peter. Captain's Master, *Blossom*, 1825-28.

JONES, Robert. Captain, *Loch Garry*, 1884.

JONES, Stephen. AB, *Herald*, 1848; 1849; 1850.

JONES, Thomas. Carpenter's Mate, *Hecla*, 1821-23.

JONES, William. AB, *Herald*, 1848; 1849; 1850.

JONES, William. Cook, *Investigator*, 1848-49.

JONES, William. AB, *Phoenix*, 1854.

JONES, William. Dog-driver, *Fox*, 1857-59.

JOPSON, Thomas. Captain's Steward, *Terror*, 1845-48.

JORDAN, A. Station hand at Nachvak, Canadian Hudson Bay Expedition, 1884-85.

JÖRGENSEN, Jens. Carpenter, *Enhiörningen*, 1619-20. Died at Churchill, 12 May 1620.

JOY, Alfred H. Sergeant RNWMP. Investigated murder of R.S. Janes, 1921-22 [*see* 1916-20, Janes].

JOY, Thomas. AB, *Investigator*, 1848-49.

JOY, Thomas. Quartermaster, *Resolute*, 1852-54.

JOYCE, M.A. Corporal RNWMP, Pelletier's patrol, 1908-09.

JOYCE, Thomas. AB, *Enterprise*, 1848-49.

JOYCE, Patrick. Fireman, *Roosevelt*, 1908-09.

JOYNER, William. Miner, Frobisher's expedition, 1578.

JUET, Robert. Mate, *Discovery*, 1610-11. Died August (?) 1611.

JUNIPER, Charles. Stoker, *Phoenix*, 1854.

JUNIUS. *See* HOEOOTOEROCK.

KALLIHIRUA [YORK, Erasmus] (1856). Eskimo guide, *Assistance*, 1850-51. Returned to England with the expedition. [Murray, 1857; Bugge, 1965; 1966].

KANAQUASSE, Pierre. Boatman (?), Back's expedition, 1833-35.

KANE, Elisha Kent (1820-1857). Passed Assistant Surgeon, *Advance*, 1850-51; commander of expedition, *Advance*, 1853-55. [Corner, 1972; Kendall, 1963; DAB].

KANE, John Kintzing Jr. (1833-). (Brother of E.K. Kane). Acting Assistant Surgeon, *Arctic*, 1855.

KANN, Leopold. Joint leader (?), Ellesmere Island expedition, 1899-1900.

KANSELLAR, Andrew. First mining expedition to Yukon, 1872-73.

KARLSEFNI. *See* THORDARSSON.

KATELY, Joseph. Private RM, *Hecla*, 1819-20.

KAY, William. Chief Mate, *Prince Albert*, 1850.

KEARNEY, . Captain, *Stord*, 1903-04.

KEARNEY, Joseph Mary Patrick (1834-1918). RC missionary journey, 1861. [Breton, 1963].

KEARNY, F. Seaman, *Neptune*, 1903-04.

KEATH, Daniel. AB, *Breadalbane*, 1853.

KEEFE, John. AB, *Investigator*, 1850-54.

KEELE, Joseph. Joint leader, GSC expedition, 1902; leader, GSC expedition, 1905; leader, GSC expedition, 1907-08. [EA].

KEELER, J.D. Ensign, *Juniata*, 1873.

KEENE, George R. Mate, *Assistance*, 1850-51.

KEENEY, C. Boat-steerer, *George Henry*, 1860-62.

KEEVILLO, Robert. Corporal RM, *Hecla*, 1821-23.

KEHOE, . Captain, *Erik*, "Crocker Land" Expedition, 1913-17 [1913 only].

KELD, Thomas. Quartermaster, *Discovery*, 1791-95.

KELD, Thomas. Boatswain, *Discovery*, 1791-95.

KELLETT, Henry (1806-1875). Captain, *Herald*, 1848; 1849; 1850; Captain, *Resolute*, 1852-54. [DNB Vol. 30; DCB Vol. 10; O'B(1)].

KELLEY, Walter. Carpenter, *Ayde*, 1578.

KELLOCK, James. Re-opened Nichicun post, 1834.

KELLY, George. Captain of the Foretop, *Resolute*, 1852-54.

KELLY, John. Sailor, *Sunneshine*, 1585.

KELSEY, Henry (c1667-1724). First founding of Churchill, 1689; leader, HBC expeditions, 1689; 1690-92; Captain, *Prosperous*, 1719; 1721. [DCB Vol. 2; Davies, 1965, pp. 376-94].

KELSEY, William. AB, *Terror*, 1836-37.

KELTY, Hugh. Private of Marines, *Discovery*, 1791-95.

KEMBILL, Jonathan. (?), *Nancy Dawson*, 1849.

KEMISH, George. Wardroom Steward, *Alert*, 1875-76.

KENARD, John. AB, *Phoenix*, 1854.

KENASTON, C.A. Joint leader, U.S. expedition, 1891.

KENDALL, . Captain, *Dennis*, 1578.

KENDALL, Edward Nicholas (c1800-1845). Surveyor, *Griper*, 1824; Franklin's expedition, 1825-27. [Dawson, L.S., 1885, Part 1, p. 105; O'B(1)].

KENDELL, Abraham W. Surgeon's Steward, *Arctic*, 1855.

KENDELL, J. Midshipman, *Blossom*, 1825-28.

KENLEY, John. Quartermaster, *Terror*, 1845-48.

KENNEDY, Charles. AB, *Herald*, 1848; 1849; 1850. Drowned 26 September 1850.

KENNEDY, Charles. (?), Rae's expedition, 1850-51. [Rich, 1953, p. 361].

KENNEDY, George. Captain, *Beaver*, 1723.

KENNEDY, George. Acting Boatswain, *Investigator*, 1850-54.

KENNEDY, William (1814-1890). Commander of expedition, *Prince Albert*, 1851-52. [BOA Vol. 2; Shaw, E.C., 1970].

KENNICOTT, Robert (1835-1866). Leader, U.S. expedition, 1859-62. [EA].

KENT, Allan. AB, *Hecla*, 1824-25.

KENT, George. Ordinary Seaman, *Rattlesnake*, 1853-54.

KERR [CARR], . Gunner, *Griper*, 1824.

KERR, Alexander. AB, *Plover*, 1852-54.

KERR, John. Gunner's Mate, *Investigator*, 1850-54. Died at Mercy Bay, 13 April 1853.

KERR, William. Carpenter, *Assistance*, 1852-54.

KERWIN [KERVIN], Christopher. Private of Marines, *Discovery*, 1776-80.

KETCHUM, F.E. (?), Western Union Telegraph expedition, 1867.

KICH, James. Sergeant of Marines, *Discov-*

ery, 1776-80.
KIDNEY, William. AB, *Herald*, 1848; 1849; 1850.
KILBERRY, Thomas. Quartermaster, *Plover*, 1848-52 [1851-52 only].
KILGOAR, John. AB, *Fury*, 1824-25.
KILIAN, Herman. (?), Canadian Arctic Expedition, 1913-18.
KILIAN, Martin. (?), Canadian Arctic Expedition, 1913-18.
KILLEY, J. Fireman, *Neptune*, 1903-04.
KILSON, Richard. Stoker, *Intrepid*, 1852-54.
KIMBLE, Richard. AB, *Enterprise*, 1850-55.
KING, A. Sailor, *Karluk*, Canadian Arctic Expedition, 1913-18. Died off Wrangel Island, 1914.
KING, C. Boatswain, *Neptune*, 1903-04.
KING, C.F. Topographer and meteorologist, *Neptune*, 1903-04.
KING, Edward. Stoker, *Phoenix*, 1853.
KING, James (1750-1784). Second Lieutenant, *Resolution*, 1776-80.
KING, James. Seaman, *Resolution*, 1776-80.
KING, James. Gunroom Steward, *Herald*, 1848; 1849; 1850.
KING, John. Quartermaster, *Discovery*, 1610-11.
KING, John. AB, *Discovery*, 1791-95.
KING, John. AB, *Plover*, 1848-52.
KING, Richard (c1810-1876). Surgeon and naturalist, Back's expedition, 1833-35. [DCB Vol. 10; DNB Vol. 31].
KING, Richard. Assistant Surgeon, *Resolute*, 1850-51.
KING, Thomas. Private RM, *Investigator*, 1850-54.
KING, William Cornwallis (c1845-1940). Established Fort Nelson, 1865. [*Beaver*, 1941].
KINGSMAN, William. Captain of the Forecastle, *Phoenix*, 1853.
KINNAIRD, George. AB, *Terror*, 1845-48.
KINNERSLEY, Captain. *See* KYNERSLEY.
KINNES, Robert. Opened mica mining depot at Lake Harbour, c1909 [*see* 1909-11 Lake Harbour mission].
KINSEY, Joseph. Sergeant RM, *Investigator*, 1848-49.
KINSMAN, William. Paymaster's steward, *Resolute*, 1850-51.
KIÖBENHAUFFIN, Rasmus. Seaman, *Enhiörningen*, 1619-20. Died at Churchill, 17 February 1620.
KIPLING, Edward. Middleman, HBC Franklin search, 1855.
KIRK, John. Private RM, *Blossom*, 1825-28.
KIRKBY, William West (1827-1907). Anglican missionary journey, 1859-68. [Peake, 1965; Boon, 1965; Morgan and others, 1907].
KIRKCALDY [KIRKALDY], James. Boatswain's Mate, *Plover*, 1848-52 [1850-52 only]; *Herald*, 1848; 1849.
KIRKMAN, Henrie. Gentleman, *Gabriel*, 1578.
KIRKNESS, George. Boatbuilder, Rae's expedition, 1850-51. [Rich, 1953, p. 361].
KISLINGBURY [KISLINBURY], Frederick F. Second Lieutenant, 23rd Infantry, and acting Signal Officer, U.S. Polar Year Expedition, 1881-84. Died of starvation, 1 June 1884. [EA].
KITSON, Richard. Captain of the Hold, *Lady Franklin*, 1850-51; AB, *Intrepid*, 1852-54.
KLUTH, William. Coxswain, *Resolute*, 1852-54.
KLUTSCHAK, Heinrich W. Scientist, Schwatka's expedition, 1878-80. [EA].
KMOCH, Georg. Moravian missionary expedition, 1811.
KNAUSS, Jean A. Cook, *Resolute*, 1850-51.
KNIGHT, Errol Lorne (1893-1923). (?), Canadian Arctic Expedition, 1913-18.
KNIGHT, Gabriel. (?), *Hopewell*, 1606.
KNIGHT, James (c1640-c1719). Leader, HBC supply voyage, 1692-93; overseas governor of HBC, supply voyage, 1698-1700; passenger, HBC supply voyage, 1714; leader, second founding of Churchill, 1717; leader, HBC expedition, 1719-21. [DCB Vol. 2; Davies, 1965, pp. 394-410].
KNIGHT, John. Captain, *Hopewell*, 1606. Disappeared (probably killed by Eskimos) 26 June 1606. [DCB Vol. 1].
KNIGHT, John. AB, *Plover*, 1848-52.
KNORR, George F. (1842-). Commander's secretary, *United States*, 1860-61. [EA].

KNOTT, Jesse. Private RM, *Rattlesnake,* 1853-54.

KNOWLES, George. AB, *Sophia*, 1850-51.

KNOX, James [Thomas]. Cook, *Sophia,* 1850-51.

KNOX, Robert. Captain, *Luna*, 1820.

KOCH, K.R. Leader, German supplementary Polar Year Expedition, 1882-83.

KOENIG, John Van. *See* VAN KOENIG.

KOHLMEISTER, Benjamin Gottlieb. Moravian missionary expedition, 1811.

KOLNO, John. *See* SCOLVUS.

KOOLEMANS BEYNEN, Laurens Rijnhart (1852-1879). Lieutenant, Dutch RN, *Pandora*, 1875; 1876. [Boissevain, 1885].

KOZDAW. (Chipewyan Indian). Guide, Thompson's expedition, 1796.

KRABBÉ, Frederick J. (-1868). Second Master, *Assistance,* 1850-51; Master, *Intrepid*, 1852-54. [MAR; EA].

KRELLE, Walter. (?), Frobisher's expedition, 1578. Died 24 September 1578.

KRUGER, J.W.C. Seaman, *Polaris*, 1871-73.

KRUMM, . Moravian missionary, *Hope*, 1752.

KUDLAGO. (Eskimo). Guide, Hall's expedition, 1860-62. Died 1 July 1860.

KUMLIEN [KUMLEIN], Ludwig (1853-1902). Scientific staff, *Florence,* 1877-78.

KUNZ, . Moravian missionary, *Hope*, 1752.

KURNER [KUNER], William J. Seaman, *Rescue*, 1850-51.

KYNERSLEY [KYNDERSLEY, KINNERSLEY], Mathew. Gentleman, *Ayde,* 1577; Captain (?), *Michael,* 1578.

KYNERSLEY [KYNDERSLEY, KINNERSLEY], Robert. Gentleman, *Ayde*, 1577; Captain (?), *Michael,* 1578.

LABELLE, William. Quartermaster, *Arctic,* 1910-11.

LABERGE [LABARGE], Mike. (?), Western Union Telegraph Expedition, 1867.

LABORNE, William. Sailor, *Michael,* 1578.

LACEY, John. AB, *Phoenix,* 1854.

LA CHARITÉ, . (?), Back's expedition, 1833-35. Discharged, December 1833.

LACHEY, Barney. *See* LAUGHY, Bernard.

LADRILLO, . Sailor, North-west Passage expedition, 1574.

LADUE, Joseph. Gold prospecting expedition, 1882-83; founded Dawson, 1896.

LADUKE, E.E. (?), *Laddie,* 1913-16.

LA GUARDE [LA GARDE, LAGARD], Joseph. Bowman, Black's expedition, 1824. [Rich, 1955, pp. 219-21].

LAHAIE, Ernest. Fireman, *Arctic,* 1906-07; Oiler, *Arctic,* 1908-09.

LAILOR, Frank. Seaman, Hall's expedition, 1864-69 [1868 only].

LAITHWOOD, Thomas. Carpenter's mate, *Discovery,* 1791-95.

LAJAILLE, *Marquis de.* Lieutenant commanding *Engageante,* 1782.

LAKE, William. AB, *Fury,* 1821-23.

LAKES, John. Master, *Moone,* 1578.

LALIBERTÉ, Alexis (*dit* Lachouette). Bowman, Rae's expedition, 1850-51. [Rich, 1953, p. 361].

LA MALICE, . (?), establishment of Rocky Mountain Portage House, 1805-06.

LAMB, John. Gunner's Mate, *Resolute,* 1850-51.

LAMB, Reuben. Third Officer, *George Henry,* 1860-62.

LAMBELL, John. (?), *Ayde,* 1578.

LAMBERT, P.G. Apprentice Engineer, *Phoenix,* 1854.

LAMBERT, Thomas. AB, *Breadalbane,* 1853.

LAMBERT, William. AB, *Isabella,* 1818; *Hecla,* 1824-25.

LAMPTON, Thomas. *See* LANGSTER.

LANAUZE, Charles Deering ("Denny"). Inspector, RNWMP patrol, 1915-17.

LANCASTER, William. AB, *Hecla,* 1819-20.

LANCEFIELD, George, R. Photographer, *Arctic,* 1906-07.

LANDAM, Titus. Labourer, Frobisher's expedition, 1578.

LANDON, James. Private of marines, *Chatham,* 1791-95.

LANDRY, Joseph. Voyageur, Mackenzie's expeditions, 1789; 1792-93.

LANE, Daniel. Seaman, *Arctic,* 1908-09.

LANE, John. Mate, *Godspeed,* 1602.

LANE, John. Boatswain, *Terror,* 1845-48.
LANE, Michael. Master and surveyor, *Lyon,* 1776. [Taylor, 1966, p. 237].
LANGLE, P.-A. de. Lieutenant commanding *Astrée,* 1782; commander, *Astrolabe,* 1785-88.
LANGLEY, Edward. Boatswain, *Resolute,* 1850-51.
LANGLEY, John. Private of Marines, *Chatham,* 1791-95.
LANGLOIS, Jean. (?), Couture's expedition, 1663.
LANGSTER [LAMPTON], Thomas. AB, *Lady Franklin,* 1850-51.
LANTZ, James. Sailor and carpenter, Low's expedition, 1898-99; cook, Low's expedition, 1897.
LANYON, William Tregoney. Seaman, *Resolution,* 1776-80.
LAPÉROUSE, Jean François, *Comte de. See* GALAUP.
LAPERRIÈRE, A.N. Officer in charge of Port Laperrière station, Canadian Hudson Bay Expedition, 1884-85.
LAPLANTE, Xavier [PLANTE, Antoine]. Canoeman, Richardson's expedition, 1847. [Rich, 1853, p. 362].
LA PRISE, Baptiste. (Chipewyan Indian). Hunter and interpreter, Black's expedition, 1824.
LARCEN, Andrew. Seaman, *Release,* 1855.
LARCOMBE, Edward. AB, *Investigator,* 1848-49.
LARGE, George. Ordinary seaman, *North Star,* 1849-50.
LARGE, George. Ordinary Seaman, *North Star,* 1849-50.
LARKIN, Raphe. Sailor, *Judith,* 1578.
LARMAN, George. Sailor, *Judith,* 1578.
LARN, John. AB, *Hecla,* 1824-25.
LA ROCQUE DE ROBERVAL, Jean François de (c1500-1560). Commander, *Valentine,* 1542-43. [DCB Vol. 1].
LARONDE, Louison. Canoeman, Richardson's expedition, 1847-49. [Rich, 1953, pp. 362-63].
LARRANCE, Nicholas. Labourer, Frobisher's expedition, 1578.
LARTIMS, James. Stoker, *Phoenix,* 1854.
LAUGHY, Bernard. Stoker, *Victory,* 1829-33.
LAURIE, James. *See* LOWRIE, William.
LAVERICK, John. AB, *Hecla,* 1821-23.
LAVOIE, J.T.E. Meteorologist and geologist, *Arctic,* 1910-11.
LAW, John. Surgeon, *Discovery,* 1776-80.
LAWES, Phillip. (?), *Nancy Dawson,* 1849.
LAWES, William. Clerk in Charge, *Terror,* 1836-37.
LAWLER, E. Quartermaster, *Phoenix,* 1854.
LAWRENCE, Benjamin. Carpenter's Mate, *Fury,* 1824-25.
LAWRENCE, Edwin. AB, *Terror,* 1845-48.
LAWRENCE, Edwin. Gunner's Mate, *Alert,* 1875-76.
LAWRENCE, Joseph. Wardroom steward, *Pandora,* 1875.
LAWRENCE, R.C. Private RM, *Phoenix,* 1854.
LAWS, James. Assistant Surgeon, *Release,* 1855.
LAWSON, G. (?), *Minnie Maud,* 1912-13.
LAWSON, John. Sailor, *Michael,* 1578.
LAWSON, John. AB, *Sophia,* 1850-51.
LAWSON, John H. AB, *Plover,* 1848-52 [1851-52 only].
LAY, George Thomas [Tradescant]. Naturalist, *Blossom,* 1825-28.
LAYTON, George. AB, *Herald,* 1848; 1849; 1850.
LEARMONTH, L.A. Established Port Harrison, 1920.
LEASK, John. Ice Master, *North Star,* 1849-50; Master, *Prince Albert,* 1851-52.
LeBEL, William. Seaman, *Arctic,* 1908-09.
LE BRETON-BASTILLE, Guillaume. *See* BASTILLE.
LEBRULE, Louis. *See* DUBRILL.
LECLAIR, Jos. Fireman, *Arctic,* 1908-09.
LEDEN, Christian (1882-1957). Leader, Norwegian expedition, 1913-16. [Isachsen, F., 1961].
LEDYARD, John (1751-1789). Corporal of Marines, *Resolution,* 1776-80. [DNB Vol. 32; Halliday, 1961].
LEE, Francis. Acting second Ice Master, *Plover,* 1848-52.
LEE, James. Cook, *Herald,* 1848; 1849; 1850.
LEE, James W. Seaman, *Florence,* 1877-78.
LEE, John. Gentleman, *Ayde,* 1577; *Gabriel,*

1578.

LEE, Leslie. Member of U.S. expedition to Churchill Falls, 1891.

LEE, Richard. Seaman, *Resolution,* 1776-80.

LEE, Thomas. Captain, *Prince of Wales,* 1848.

LEE, William, AB, *Rattlesnake,* 1853-54.

LEE, William Alfred Rose. Lieutenant, *Plover,* 1848-52 [1848-49 only]. [O'B(1)].

LEECH, James. Master, *Emanuel (Busse),* 1578.

LEFROY, John Henry (1817-1890). Leader, Magnetic Survey, 1843-44. [DNB Vol. 32].

LEGENDRE, Aurelien. Oiler, *Arctic,* 1910-11.

LEGG, R.T. Mate, *Enterprise,* 1850-55.

LEGGATT, Charles. Boatswain's mate, *Enterprise,* 1850-55.

LEGGATT, George. AB, *Discovery,* 1875-76.

LEHMAN, Gottfried. Passenger, *George,* 1774. Drowned, 14 September 1774.

LEIGH, Thomas. Carpenter's crew, *Phoenix,* 1854.

LEIPER, Alexander. Carpenter's mate, *Lady Franklin,* 1850-51.

LEIPER, John (-1852). Second mate, *Lady Franklin,* 1850-51.

LEITH, Arthur T. Second-in-command, geological expedition, 1909-10.

LEITH, Charles Kenneth (1875-). Leader, geological expedition, 1909-10.

LE MESWIER, William. AB, *Chatham,* 1791-95.

LEMLY, Samuel C. Lieutenant, *Thetis,* 1884.

LEO, Frederick. AB, *Plover,* 1848-52; 1852-54.

LEONARD, Dominick. (?), *Michael,* 1578.

LE PITOURAL [LE PATOURAL], James. Captain of the Hold, *Resolute,* 1852-54.

LEROUX, Guillaume. RC missionary expedition, 1911-13. Murdered by Eskimos, 1913.

LEROUX, Laurent. Gregory, McLeod and Co. expedition, 1785-87; established Great Slave Lake post, 1786; established Fort Providence, 1789.

LESLIE, James. Armourer, *Lady Franklin,* 1850-51.

LESSARD, Etienne. Accompanied Jolliet and Silvy, 1679.

LESSARD, George. Quartermaster, *Arctic,* 1908-09.

LESSARD, Joseph. Quartermaster, *Arctic,* 1906-07; 1910-11.

LETT, John. Servant to William Bailey, *Discovery,* 1776-80.

LEVASSEUR, Paul. Chief Steward, *Arctic,* 1906-07.

LEVER, Charles. Captain's clerk, *Release,* 1855.

LE VESCONTE, Henry Thomas Dundas. Lieutenant, *Erebus,* 1845-48. [O'B(1)].

LEVINGSTON, Duncan. *See* LIVINGSTON.

LEWCOCK, George. Captain of the Forecastle, *Rattlesnake,* 1853-54.

LEWES, John Lee (1791-). Second-in-command, HBC expedition, 1843. [Rich, 1938, pp. 446-47].

LEWES, Peter. Gunroom Cook, *Blossom,* 1825-28.

LEWIS, Benjamin. Master and Greenland Pilot, *Isabella,* 1818.

LEWIS, Charles. Second Class Volunteer, *Blossom,* 1825-28.

LEWIS, David. AB, *Phoenix,* 1854.

LEWIS, James. Clerk in Charge, *Assistance,* 1852-54; Clerk, *Resolute,* 1850-51.

LEWIS, John. AB, *Herald,* 1848; 1849; 1850.

LEWIS, L. Boatman (?), Dawson's expedition, 1887.

LEWIS, Lawrence. Seaman, *Release,* 1855.

LEWIS, Levi. Ordinary Seaman, *Phoenix,* 1854.

LEWIS, Michael. AB, *Phoenix,* 1854.

LEWIS, Tege. Sailor, *Ayde,* 1578.

LEWIS, William. Gunroom Steward, *Assistance,* 1850-51.

LEYS, David. AB, *Terror,* 1845-48.

LEYSON, William. Assistant Surgeon, *Griper,* 1824.

LI, Erich Hansen. Seaman, *Enhiörningen,* 1619-20. Died at Churchill, 19 May 1620.

LIBBEY, William (1855-1927). Geographer, *Falcon,* 1894.

LIDDLE, George. Boatswain, *Rattlesnake,* 1853-54.

LIDDON, Matthew (-1869). Lieutenant commanding *Griper,* 1819-20. [RNB Vol. 8;

MAR; EA; O'B(1)].

LIGHT, William. AB, *Hecla,* 1824-25.

LIGHT, William. Steward, *Victory,* 1829-33.

LILLEY [LILLY], Joseph. Boatswain, *Hecla,* 1821-23.

LILLINGSTON, F.G. Innes. *See* INNES-LILLINGSTON.

LILSTER, Thomas. AB, *Terror,* 1836-37.

LIMSTRONG, Charles Robert. Captain, *Baroness,* 1857.

LINCOLN, William. Seaman, *Rescue,* 1850-51.

LINDSAY, E. AB, *Plover,* 1848-52.

LINDSAY, John J. Clerk in Charge, *Plover,* 1848-52.

LINDSAY, Robert [alias ROBERTSON, Andrew (or Edward)]. AB, *Herald,* 1848; 1849; 1850; *Plover,* 1852-54.

LINDSAY, Thomas. Cook, *Phoenix,* 1854.

LINDSTRÖM, Adolf Erik [Henrik] (1866-1939). Steward, *Fram,* 1898-1902; Cook, *Gjøa,* 1903-06. [EA].

LINKLATER, Andrew. AB, *Prince Albert,* 1851-52.

LINKLATER, Peter (c1830-c1882). (?), Rae's expedition, 1850-51. [Rich, 1953, p. 363].

LINQUIST, Gustavus W. Seaman, *Polaris,* 1871-73.

LION, Henry. AB, *Blossom,* 1825-28.

LISTER, Christian. Passenger, *George,* 1774.

LITTLE, Daniel F. Clerk in Charge, *Plover,* 1848-52; 1852-54.

LITTLE, Edward. Lieutenant, *Terror,* 1845-48. [O'B(1)].

LITTLEFIELD, William. AB, *Enterprise,* 1850-55.

LITTLESTONNE, Roger. General's servant, Frobisher's expedition, 1578. Died 28 August 1578.

LIVINGSTON [LIVINGSTONE], Duncan. Established Livingston's Fort, 1796; leader, North West Company expedition, 1799. Killed by Eskimos, 1799. [DCB Vol. 4].

LLOYD, . Captain, *Owner's Love,* 1697.

LLOYD, Edwin. AB, *North Star,* 1852-54.

LLOYD, Frederick. AB, *Enterprise,* 1848-49.

LLOYD, Hawkins. AB, *Chatham,* 1791-95.

LLOYD, Henry. AB, *Erebus,* 1845-48.

LLOYD, John. AB, *Fury,* 1821-23.

LLOYD, Richard. AB, *Fury,* 1821-23.

LLOYD, Thomas. AB, *Blossom,* 1825-28.

LOCKWOOD, Francis. AB, *Isabella,* 1818; *Fury,* 1821-23.

LOCKWOOD, James Booth (1852-1884). Second Lieutenant, 23rd Infantry, and acting Signal Officer, U.S. Polar Year Expedition, 1881-84. Died of starvation, 9 April 1884. [DAB; EA].

LODLO, Arnold. Seaman, *Discovery,* 1610-11.

LOE, John. Cooper, Frobisher's expedition, 1578.

LOFTHOUSE, Joseph *Archdeacon of Keewatin* (1855-1933). Assistant, Tyrrell's expedition, 1900.

LOHMANN, Bartholemi. *See* LOWMAN, Bartholomew.

LONEY, John Felix. Master, *Assistance,* 1852-54.

LONG, Francis. Private, Company F, 9th Cavalry, U.S. Polar Year Expedition, 1881-84.

LONG, George. AB, *Enterprise,* 1848-49.

LONG, Thomas. Cook, *Alexander,* 1818.

LONGDON, Thomas. Private RM, *Hecla,* 1824-25.

LONGLAND, John. Captain, *Phoenix,* 1744.

LONGMAN, William. AB, *Investigator,* 1848-49.

LONGMOOR, Robert. Established Hudson House, 1778-79.

LONSDALE, *Lord. See* LOWTHER, Hugh Cecil.

LOPEZ, Peter. (?), Canadian Arctic Expedition, 1913-18.

LORIMER, William. AB, *Alert,* 1875-76.

LORME, Pierre de. Voyageur, Mackenzie's expedition, 1789.

LOVEGROVE, G.F. Captain, *Pelican,* 1905.

LOVELL, William I. Midshipman and second officer, *Advance,* 1850-51.

LOVELL, William S. Acting Master, *Release,* 1855.

LOW, Albert Peter (1861-1952). Leader, GSC expeditions, 1884-85; 1887-88; 1892-96; 1897; 1898-99; leader, geological expedition, 1901; leader of expedition, *Neptune,*

1903-04. [Alcock, 1954; PR 1955].
LOW, Fulke. AB, *Discovery,* 1776-80.
LOWE, . Captain, *Crocodile,* 1846-48.
LOWE, John. Chief Engineer, *Bear,* 1884.
LOWER, Arthur Reginald Marsden (1889-). Joint leader, fisheries research expedition (James Bay parties), 1914.
LOWER, George. Officers' Steward, *Resolute,* 1850-51.
LOWERY, George. AB, *Phoenix,* 1854.
LOWETH, Benjamin. Ice Quartermaster, *Resolute,* 1850-51.
LOWMAN, Bartholomew [LOHMAN, Bartholemi]. AB, *Discovery,* 1776-80.
LOWRIE, William [LAURIE, James]. AB, *Resolute,* 1850-51.
LOWRING, John. Sailor, *Judith,* 1578.
LOWTHER, Hugh Cecil, *5th Earl of Lonsdale.* Leader, British expedition, 1888-89. [Dawson, L., 1946].
LOYE, John. AB, *Discovery,* 1776-80.
LUCAS, John. AB, *Discovery,* 1791-95.
LUCAS, John P. AB, *Lady Franklin,* 1850-51.
LUCAS, Richard (c1656-). Captain, *Owner's Goodwill,* 1685. [Rich, 1946, pp. 329-31; DCB Vol. 1].
LUCAS, Thomas. Ordinary seaman, *Plover,* 1848-52; 1852-54.
LUCAS, Thomas. AB, *Phoenix,* 1853.
LUDLOW, John. Boatswain's Mate, *Griper,* 1819-20.
LUFFINGHAM, Henry. Boatswain's Mate, *Blossom,* 1825-28.
LUNA, Pedro de. Controller, *Anunciada,* 1524-25.
LUND, Anton. First Mate, *Gjøa,* 1903-06.
LUNT, John. Master's Mate, *Gabriel,* 1578.
LUTIT, Edward. Leader, HBC expeditions, 1766-67; 1767-68; 1768-69.
LUXFORD, William. Quartermaster, *Enterprise,* 1850-55.
LYALL, David. Surgeon, *Assistance,* 1852-54.
LYDALL, William (c1625-). Second Overseas Governor of HBC; supercargo, *Prince Rupert,* 1674-75. [DCB Vol. 1].
LYDGER, Androwe. Sailor, *Ayde,* 1578.
LYDGER, George. Shipwright, *Judith,* 1578.
LYDIET, William. Soldier, Frobisher's expedition, 1578.
LYNN, David. Sergeant, Company C, 2nd Cavalry, U.S. Polar Year Expedition, 1881-84. Died of starvation, 6 April 1884.
LYNS [LINCHE], Abraham. Gentleman, *Ayde,* 1577; *Gabriel,* 1578.
LYON, Benjamin. AB, *Resolution,* 1776-80.
LYON, George Francis (1795-1832). Commander, RN, commanding *Hecla,* 1821-23; *Griper,* 1824. [DNB Vol. 34].
LYONN, William. AB, *Enterprise,* 1848-49.
LYONS, Edward. Blacksmith, *Assistance,* 1852-54.
McAULAY, Aulay [McAULEY, Auley]. Member of Clarke's expedition, 1815-16. [Rich, 1938, pp. 447-48].
McAIRES [McCURES], George. AB, *Breadalbane,* 1853.
McALPINE, John. AB, *Discovery,* 1791-95.
McARTHUR, D. Sergeant RNWMP, patrol to Wager Bay, 1907-08 [see 1903, Moodie].
McARTHUR, William. AB, *Resolute,* 1850-51; *Pioneer,* 1852-54.
McAULIS, D. AB, *Hecla,* 1824-25.
MacBEAN [MACLEAN], Gillies Alexander. Second Master, *Terror,* 1845-48.
McBEATH, Adam. Leader, HBC expedition, 1843.
McBRIDE, . Captain, *Friendship,* 1820.
McCALLUM, Peter. Trapper, accompanied Melvill's expedition, 1908-11.
McCARDELL, John. Private of Marines, *Discovery,* 1791-95.
McCARTHY, John. Captain of the Foretop, *Plover,* 1848-52; 1852-54.
McCARTHY, John. AB, *Phoenix,* 1853.
McCARTHY, Simon. Carpenter's Mate, *Assistance,* 1850-51.
McCARTHY, William. Ordinary Seaman, *Herald,* 1848; 1849; 1850.
McCARTHY, William. AB, *Plover,* 1848-52; 1853-54.
MACARTNEY, James (-1913). AB, *Assistance,* 1852-54.
McCLAIN, Charles S. Ensign, *Alert,* 1884.
McCLAREN, James. Second Corporal, Richardson's expedition, 1847-49.
McCLEAN, Robert. Cook, *Intrepid,* 1850-51.

McCLELAND, J. Private RM, *Hecla*, 1824-25.

McCLELLAN, Edward P. Lieutenant, *Juniata*, 1873. [NE].

McCLENNAN, John. Foreman, GSC expedition, 1875.

McCLINTOCK, Francis Leopold (1819-1907). Lieutenant, *Enterprise*, 1848-49; *Assistance*, 1850-51; Commander, RN, commanding *Intrepid*, 1852-54; Captain, *Fox*, 1857-59; *Bulldog*, 1860. [Markham, C.R., 1909; GJ 1908; DNB 2nd Supp. Vol. 2].

McCLURE, Robert John le Mesurier (1807-1873). Mate, *Terror*, 1836-37; First Lieutenant, *Enterprise*, 1848-49; Commander, RN, commanding *Investigator*, 1850-54. [DNB Vol. 35; DCB Vol. 10].

McCONKEY, Charles. Gold prospecting expedition, 1883-84.

McCONNELL, Burt Morton (1888-1960). Meteorologist, Canadian Arctic Expedition, 1913-18. [PR 1962; EA; Science 1960].

McCONNELL, Richard George (1857-1942). Leader, Yukon exploring expedition, 1887-88; GSC expeditions, 1898; 1902. [EA].

McCONVEY, Thomas. AB, *Erebus*, 1845-48.

McCORMACK, Robert. Marine, *Pioneer*, 1852-54.

McCORMICK, Charles. Seaman, *United States*, 1860-61.

McCORMICK, Francis [John]. Private RM, *Assistance*, 1852-54.

McCORMICK, Robert (1800-1890). Surgeon, *North Star*, 1852-54. [DNB Vol. 35; Keevil, 1943].

McCORMICK, S.J. Sailing Master, *United States*, 1860-61. [EA].

McCORMICK, Thomas. Captain of the Foretop, *Rattlesnake*, 1853-54.

McCRAE, John. (?), Governor-general's tour, 1910.

McCREADIE, James. AB, *Phoenix*, 1853.

McCULLUM, A. AB, *Prince Albert*, 1850.

McCURES, George. *See* McAIRES.

McCURRUS, Magnus. AB, *Prince Albert*, 1851-52.

McDIARMID, . Constable RNWMP, Governor-General's tour, 1910.

MACDIARMID, George. Surgeon, *Victory*, 1829-33.

MACDONALD, Alexander. Surgeon, *Bon Accord*, 1840; Assistant Surgeon, *Terror*, 1845-48.

McDONALD, Daniel. Captain of the Foretop, *Herald*, 1848; 1849; 1850.

MACDONALD, David. Quartermaster, *Terror*, 1845-48.

McDONALD, Donald. (?), Back's expedition, 1833-35.

MacDONALD, Edward. Third Officer, *Arctic*, 1910-11.

McDONALD, Hugh. Cook, *Plover*, 1848-52; 1852-54.

McDONALD, J(?). H. Mate, *Phoenix*, 1853.

McDONALD, James. Captain of the Hold, *Herald*, 1848; 1849; 1850.

McDONALD, James. AB, *Investigator*, 1850-54.

MACDONALD, Joe. Gold prospector, *Algerine*, 1912.

McDONALD, John. Private of Marines, *Resolution*, 1776-80.

MACDONALD, John. Leading Seaman, *Alexander*, 1818.

McDONALD, John. AB, *Griper*, 1824.

McDONALD, John. Cook, *North Star*, 1852-54.

McDONALD, John. Middleman and labourer, Rae's expedition, 1853-54. [Rich, 1953, p. 363].

McDONALD, John. Seaman, *United States*, 1860-61.

McDONALD [McDONNELL], Murdoch. Middleman and labourer, Rae's expedition, 1853-54. [Rich, 1953, p. 364].

MacDONALD, Neil. Voyageur, Franklin's expedition, 1825-27.

MacDONALD, Robert. Anglican missionary. Travelled with Jean Séguin to Fort Yukon, 1862-63.

McDONALD, William. Middleman, HBC expedition, 1837-39.

McDONALD, William. AB, *Phoenix*, 1854.

McDOUGALD [McDOUGALL], John George. Middleman and labourer, Rae's expedition, 1853-54. [Rich, 1953, p. 364].

McDOUGALL, Duncan. North West Company agent, invasion of James Bay, 1803-06.

McDOUGALL [M'DOUGAL], George Frederick (c1825-1871). Second Master, *Resolute,* 1850-51; Master, *Resolute,* 1852-54. [Dawson, L.S., 1885, Part 2, p. 165].

McDOUGALL, James. Established Rocky Mountain Portage House, 1805-06.

McDOUGALL, James. Discovered McDougall Pass, 1872.

McDOUGALL, Lester. Ice Quartermaster, *Assistance,* 1850-51.

McDOWALL, Hugh. Cook, *Plover,* 1852-54.

McDUFFEY, John. Boatman, Franklin's expedition, 1825-27.

McEVOY, J. Assistant, Yukon Exploring Expedition, 1887.

McFARLAN [McFARLANE], Robert. AB, *Plover,* 1852-54.

McFARLANE, L.C. Motor-boat engineer, Curran's expedition, 1912.

MACFARLANE, Roderick Ross (1833-1920). Explored Anderson River, 1857; established Fort Anderson, 1861. [Hohn, 1963; EA].

MacGAHAN, Januarius Aloysius (1844-1878). *New York Herald* correspondent, *Pandora,* 1875.

McGARY, James. Second Officer, *Advance,* 1853-55.

McGILLIVRAY, Simon (1790-1840). Established Fort Smith, 1836. [Rich, 1938, p. 451].

McGRA [McGRAS], John. AB, *Plover,* 1852-54.

MACINNES, Allan. Second Engineer, *Victory,* 1829-33.

McINTOSH, Alexander. AB, *Resolution,* 1776-80. Died at Petropavlovsk, 16 May 1779.

McINTOSH, J.B. (?), Schwatka's expedition, 1883.

McINTOSH, John. AB, *Discovery,* 1776-80. Died 28 October 1778.

McINTOSH, William. AB, *Hecla,* 1824-25.

McINTYRE, Neal. AB, *Enterprise,* 1848-49.

MACK, G. Edmund. Captain, *Nascopie,* 1915; 1916; 1917.

MACK, William. Ordinary Seaman, *Blossom,* 1825-28.

McKARNALL, B. AB, *Herald,* 1848; 1849; 1850.

MACKAY, Alexander. Second-in-command, Mackenzie's expedition, 1792-93.

MACKAY, Alister Forbes. Surgeon, Canadian Arctic Expedition, 1913-18. Died off Wrangel Island, 1914.

McKAY, James. Steersman, Back's expedition, 1833-35; HBC expedition, 1837-39.

MACKAY, Michael. Ordinary Seaman, *Herald,* 1848; 1849; 1850.

MACKAY [MACKIE], Robert. Midshipman, *Resolution,* 1776-80.

McKEE, William. Steward, *Investigator,* 1848-49.

MACKENZIE, Alexander (1764-1820). Joint leader, Gregory, McLeod and Co. expedition, 1785-87; leader, North West Company expeditions, 1789; 1792-93; assisted in establishment of Fork Fort, 1792. [DCB Vol. 5; Lamb, W.K., 1970].

MACKENZIE, Charles. Bowman, Franklin's expedition, 1825-27.

MACKENZIE, Daniel. AB, *Resolution,* 1776-80.

McKENZIE, Foster. AB, *Intrepid,* 1850-51.

McKENZIE, George. Midshipman, *Discovery,* 1791-95.

MACKENZIE, Hector Aeneas. Assistant, Rae's expedition, 1850-51. [Rich, 1953, pp. 364-65].

McKENZIE, James. AB, *Sophia,* 1850-51.

McKENZIE, John. Stoker, *Phoenix,* 1853.

McKENZIE, John. Observer in charge of Nottingham Island station, Canadian Hudson Bay Expedition, 1885-86.

MACKENZIE, Roderick. Joint leader, Gregory, McLeod and Co. expedition, 1785-87; established Fort Chipewyan, 1788.

MACKENZIE, Samuel. AB, *Investigator,* 1850-54.

McKEY, John. Middleman, HBC expedition, 1837-39.

MACKIE, Robert. Lance Corporal, Sappers and Miners, Richardson's expedition, 1847-49.

MacKINLAY, James. Assistant, Pike's expedition, 1890; Preble's expedition, 1903-04; Melvill's expedition, 1908-11.

McKINLAY, William Laird. Magnetician, Canadian Arctic Expedition, 1913-18. [1913-14 only].

McKINLEY, John. AB, *Discovery,* 1791-95.

McKIRDY, John. Captain of the Hold, *Enterprise,* 1848-49.

MACKLIN, Joseph. Gunner, *Hecla,* 1821-23.

McKNIPP, James. Gold prospecting expedition, 1872-74.

MACKRELL, William. AB, *Resolution,* 1776-80.

McLAREN, Allan. Assistant Surgeon, *Hecla,* 1821-23; Surgeon, *Fury,* 1824-25.

McLAREN, Henry. Captain, *Effort,* 1858.

McLASKEY, John. AB, *Enterprise,* 1850-55.

McLEAN, E.J. (?), *Blossom,* 1825-28.

MACLEAN, Gillies Alexander. *See* MACBEAN.

McLEAN, John (1799-1890). Leader, HBC expeditions, 1838; 1839; 1840; 1841. [Wallace, W.S., 1932, pp. xi-xxiii].

McLEAN, John. AB, *Herald,* 1848; 1849; 1850.

MACLEAN, John. AB, *Resolute,* 1852-54.

McLEAN, Robert. Cook, *Assistance,* 1850-51.

McLEARY, . Engineer, *Laddie,* 1913-16.

McLEAY, . Coxswain, Franklin's expedition, 1825-27.

McLELLAN, John. Bowman, Franklin's expedition, 1825-27.

McLELLAN [McLENNAN], Murdoch (c1826-). Labourer, Rae's expedition, 1853-54; Middleman, HBC Franklin search, 1855. [Rich, 1953, pp. 365-66].

McLEOD, Alexander. Established McLeod's Fort, 1790; Fork Fort, 1792.

McLEOD, Alexander Roderick (c1782-1840). Leader, HBC expedition, 1823; HBC representative attached to Back's expedition, 1833-35. [Fleming, 1940, pp. 448-50].

McLEOD, Archibald Norman. Established Fort Dunvegan, 1805. [Rich, 1938, pp. 454-55].

McLEOD, Donald. Middleman, HBC Franklin search, 1855.

McLEOD, John. Private of Marines, *Resolution,* 1776-80.

McLEOD, John. Leader, HBC expedition, 1834.

McLEOD, John M. Leader, HBC expeditions, 1823; 1824.

McLEOD, Neil. Boatman, Richardson's expedition, 1847-49; Pullen's expedition, 1849-51 [1850 only]. [Rich, 1953, p. 366].

McMAHON, D(?). Sergeant RM, *Hecla,* 1819-20.

MacMILLAN, Donald Baxter (1874-1970). Joint leader, Cabot's expedition, 1910 [see 1903-10]; leader, U.S. expeditions, 1911; 1912; Assistant, *Roosevelt,* 1908-09; leader, "Crocker Land" Expedition, 1913-17; Second Mate, *Thetis,* 1920. [Allen, E.S., 1963; Baird, P.D., 1971].

McMILLAN, J.G. Geologist, *Arctic,* 1908-09.

McMILLAN, Stan L. Canoeman, Camsell's expedition, 1914.

McMURDO, Archibald (1812-). Third Lieutenant, *Terror,* 1836-37.

McMURDO, H.R.D. Midshipman, *Herald,* 1848; 1849; 1850.

McNAUGHTON, . (Professor at McGill University). Passenger, Governor-general's tour, 1910.

McNAUGHTON, James. Assistant Geologist, *Alert,* 1885-86.

McNEAL, Donald. AB, *Discovery,* 1791-95.

MACNUTT, C.H. Assistant, Low's expedition, 1887-88.

MACOUN, James Melville (1862-1920). (Son of John Macoun). Assistant and botanical collector, Low's expedition, 1887-88. [Malte, 1920].

MACOUN, John (1831-1920). Botanist, GSC expedition, 1875. [Macoun, 1922; *Canadian Field-Naturalist,* 1920].

McPARTLAND, John. Seaman, *Florence,* 1877-78.

McPHERSON, John. Captain, *Ocean Nymph,* 1875; 1876; 1877; 1878; 1879; 1880; 1881.

McPHERSON, Murdoch, Leader, HBC expedition, 1824.

McQUESTEN, Leroy Napoleon ("Jack"). Gold prospecting expedition, 1872-74; formed Harper, McQuesten and Co., 1874.

MACREADY, Andrew. AB, *Discovery,* 1791-95.

McTAVISH, John George. Leader, invasion of James Bay, 1803-06. [Rich, 1938, pp. 456-57].

McVICAR, Robert (c1794-). Established Fort Resolution, 1819. [Rich, 1938, p. 457].

MADDEN, William. AB, *North Star,* 1852-54.

MADDOCKE [MADOCKE], Andrew. Sailor, *Sunneshine,* 1585; 1586.

MADGESON, . Captain, *Hope,* 1752. Killed by Eskimos during the voyage.

MADISON, Henry. Discovered gold on Fortymile River, 1886.

MAGUIRE, Rochfort (-1867). Lieutenant, *Herald,* 1848; 1849; 1850; Commander, RN, commanding *Plover,* 1852-54. [O'B(1); BOA Vol. 4].

MAHLSTEDE, . Captain, *Germania,* 1882-83; 1883-84.

MAILEY, James. *See* MAYLEY.

MAIN, Alexander. Captain, *Mink,* 1874; *Cam Owen,* 1884-85.

MAKER, Dennis. Cooper, *Rattlesnake,* 1853-54.

MALASPINA, Alejandro. Captain, *Descubierta,* 1791.

MALCOLM, George. Captain of the Hold, *Resolute,* 1850-51.

MALDONADO, Lorenzo Ferrer. Claimed to have navigated the North-west Passage in 1588. [DCB Vol. 1, p. 676].

MALE, Reuben. Captain of the Forecastle, *Terror,* 1845-48.

MALLETT(?), J.C. Captain, *Marten,* 1843.

MALLEY, William. Artilleryman, Back's expedition, 1833-35.

MALLEY, William. AB, *Alert,* 1875-76.

MALLINSON, . Captain, *Blenheim,* 1846-48.

MALLOCH, George S. Geologist, Canadian Arctic Expedition, 1913-18. Died off Wrangel Island, spring 1914.

MALONE, Murtaugh J. Assistant Engineer, *Roosevelt,* 1905-06.

MALPAR, Thomas. Ordinary Seaman, *Blossom,* 1826.

MAMEN, Bjarne. Assistant geologist, Canadian Arctic Expedition, 1913-18. Died off Wrangel Island, 1914.

MAN, Stephen. AB, *Discovery,* 1791-95.

MANDIE, John. Seaman, *Sunneshine,* 1586.

MANGER, Thomas. (?), *Nancy Dawson,* 1849.

MANGER, William. *See* MAUGER.

MANICO, Peter Smith. Lieutenant, *Griper,* 1824. [O'B(1)].

MANLEY, Thomas. Master's Mate, *Discovery,* 1791-95.

MANN, Henry. Shipwright, *Alert,* 1875-76.

MANNEO, Prefecto. Packer, Selwyn's GSC expedition, 1875.

MANNERYNG, William. Sailor, *Ayde,* 1578.

MANNING, Samuel. Boatswain's Mate, *Discovery,* 1791-95.

MANNOCK, William B. Captain, *Marten,* 1852.

MANSON, Donald (1798-1880). Second-in-command, Black's expedition, 1824. [Rich, 1938, p. 458; 1955, pp. 222-41].

MANSON, Donald. Mate, *Sophia,* 1850-51; Ice Master, *Isabel,* 1852; Mate, *Phoenix,* 1853; Ice Master, *Phoenix,* 1854.

MANSON, Magnus. AB, *Terror,* 1845-48.

MANSON, Thomas [William]. AB (or Cooper?), *Resolute,* 1852-54.

MANTE, Andrew. (?), *Nancy Dawson,* 1849.

MARCELLAIS [MARSELLOIS], Baptiste (c1818-). Guide, Rae's expedition, 1850-51. [Rich, 1953, pp. 366-67].

MARCUARD, Charles Richard. Extra Mate, *Terror,* 1836-37. [O'B(1)].

MARCUS. Eskimo. Hunter and dog-driver, *United States,* 1860-61.

MARK, William. AB, *Erebus,* 1845-48.

MARK, William. Cook, *Lady Franklin,* 1850-51.

MARKHAM, Albert Hastings (1841-1918). Second Mate, *Arctic,* 1873; Commander, *Alert,* 1875-76. [Markham, M.E., and Markham, F.A., 1927; DNB; GJ 1919; FitzGerald, W.G., 1895].

MARKHAM, Clements Robert (1830-1916). Midshipman, *Assistance,* 1850-51. [Scottish Geographical Magazine, 1916; Markham, A.H., 1917; DNB; GJ 1916].

MARKHAM, David. Sailmaker's Mate, *Discovery,* 1776-80.

MARKHAM, John. AB, *Isabella,* 1818.

MARKS, C. Prospecting expedition, 1883-

84.

MARRYAT, Joseph Henry (1830-1881). Lieutenant, *Phoenix,* 1853. [BOA Vol. 2].

MARSELLOIS, Baptiste. *See* MARCELLAIS.

MARSH, George. Purser, *Blossom,* 1825-28.

MARSH, Thomas. Private RM, *Herald,* 1848; 1849; 1850.

MARSH, William. Cook, *Blossom,* 1826-28.

MARSHAL, Thomas. Captain of the Forecastle, *Assistance,* 1852-54.

MARSHALL, Alfred. Scientific assistant, Carnegie Museum expedition, 1917.

MARSHALL, James. AB, *Discovery,* 1776-80.

MARSHALL, John. Mate, *Lady Franklin,* 1850-51.

MARSHALL, Thomas. AB, *Resolute,* 1850-51.

MARSHALL, Walter. AB, *Assistance,* 1852-54.

MARSHALL, William. AB, *Alexander,* 1818.

MARSHALL, William. Carpenter's Crew, *Investigator,* 1848-49.

MARSHALL, William. AB, *Sophia,* 1850-51.

MARSHALL, William. AB, *Enterprise,* 1850-55.

MARSLIN, James. Armourer, *Victory,* 1829-33. Died of consumption, 20 January 1830.

MARSTRAND, Anders. Seaman, *Enhiörningen,* 1619-20. Died 4 May 1620.

MARSTRAND, Morten. Boatswain's Mate, *Enhiörningen,* 1619-20. Died 4 May 1620.

MARSTRAND, Suend. Seaman, *Enhiörningen,* 1619-20. Died 12 May 1620.

MARTIN, . Artillery Sergeant, *Hecla,* 1819-20.

MARTIN, George. Private RM, *Resolute,* 1850-51.

MARTIN, George. AB, *Phoenix,* 1854.

MARTIN, Henry. AB, *Isabella,* 1818.

MARTIN, Henry (-1853). Second Master, *Plover,* 1848-52.

MARTIN, James L. Private RM, *Fury,* 1821-23.

MARTIN, John. Sergeant RM, *Isabella,* 1818.

MARTIN, John. Greenland Mate, *Fury,* 1824-25.

MARTIN, John. Seaman, whalers' boat voyage, 1860.

MARTIN, John Henry (-1823). Master's Mate, *Discovery,* 1776-80.

MARTIN, Joseph. (?), *Nancy Dawson,* 1849.

MARTIN, Robert. Captain, *Enterprise,* 1845. [Jones, A.G.E., 1969].

MARTIN, Sandford. Carpenter, *Chatham,* 1791-95.

MARTIN, William. (?), *Blossom,* 1825-28.

MARTIN, William. AB, *Plover,* 1852-54.

MARTINEZ, Esteban José. Pilot, *Santiago,* 1774; Captain, *Princesa,* 1788; 1789.

MARVIN, Ross Gilmore (1875-1909). Secretary and assistant, Peary's expeditions, 1905-06; 1908-09. Drowned 10 April 1909. [EA; Stowell, 1954].

MARVITI, J. Private RM, *Hecla,* 1821-23; Corporal RM, *Griper,* 1824.

MASIK, August (1888-). (?), Canadian Arctic Expedition, 1913-18. [Masik and Hutchison, 1935].

MASKELL, William (1852-1942). AB, *Alert,* 1875-76.

MASKILL, Charles. AB, *Chatham,* 1791-95.

MASLIN, John. Corporal RM, *Terror,* 1836-37.

MASON, Benjamin. AB, *Phoenix,* 1854.

MASON, Charles. AB, *Discovery,* 1791-95.

MASON, George. Private RM, *Herald,* 1848; 1849; 1850.

MASON, John Alden (1885-). Leader, anthropological expedition, 1913.

MASON, Peter. AB, *Resolution,* 1776-80.

MASON, Robert. Quartermaster, *Plover,* 1848-52; 1852-54.

MASON, Tom. Harpooner, *Arctic,* 1873.

MASSIE, George. AB, *Prince Albert,* 1850.

MASTEGON. (?), Richardson's expedition, 1847-49.

MATHÉ, Joseph Eugene. Assistant Steward, *Arctic,* 1910-11.

MATHESON, Peter. Middleman, *Magnet,* 1846-47. [Rich, 1953, p. 367].

MATHEW, Edward. Sailor, *Michael,* 1578.

MATHEWS, James. AB, *Blossom,* 1825-28.

MATHEWS, John. AB, *Blossom,* 1825-28.

MATHIAS [MATTHIAS], Henry (c1822-1849). Assistant Surgeon, *Enterprise,* 1848-49. Died at Port Leopold, 15 June 1849.

MATHIESON, Alexander. AB, *Prince Albert*, 1850; 1851-52.

MATHUES, Bennett. Landsman, *Discovery*, 1610-11.

MATONABBEE (-1782). Chipewyan Indian. Guide, Hearne's expedition, 1770-72.

MATTHEWS, James. Ropemaker, *Herald*, 1848; 1849; 1850.

MATTHEWS, John. Gunner's Mate, *Plover*, 1850-54.

MATTHEWS, Thomas. Carpenter, *Lion*, Franklin's expedition, 1825-27; Carpenter, Back's expedition, 1833-35.

MATTHEWS, W.H. AB, *Phoenix*, 1853.

MATTHEWS, William. Carpenter, Back's expedition, 1833-35.

MATTHIAS, Henry. *See* MATHIAS.

MAUCH, J.B. Seaman, *Polaris*, 1871-73.

MAUGER [MANGER], William. Ice Quartermaster, *Assistance*, 1850-51.

MAURELLE, Francisco Antonio (1755-1820). Quartermaster, *Felicidad*, 1775; Second Officer, *Favorita*, 1779.

MAURER, Frederick W. Fireman, Canadian Arctic Expedition, 1913-18.

MAURICE, François. Canoeman, Tyrrell's expedition, 1893.

MAY, Henry. Quartermaster, *Investigator*, 1850-54; *Enterprise*, 1848-49.

MAY, Isaac. AB, *Terror*, 1836-37.

MAY, J. AB, *Hecla*, 1824-25.

MAY, Sidney H. Ensign, *Juniata*, 1873. [NE].

MAY, Walter William (1830-1896). Mate, *Resolute*, 1850-51; Lieutenant, *Assistance*, 1852-54. [BOA, Vol. 4].

MAY, William Henry (1849-1930). Lieutenant, *Alert*, 1875-76. [DNB].

MAYERS [MAYO], James. Private RM, *Enterprise*, 1850-55.

MAYLEY, James. AB, *Resolution*, 1776-80.

MAYO, Alfred H. Prospecting expedition, 1872-74; joined Harper, McQuesten and Co., 1874.

MEAD, G.H. Captain, *Nascopie*, 1919; 1920.

MEAD, John. Stoker, *Phoenix*, 1853.

MEAGER, George. Armourer, *Terror*, 1836-37.

MEARA, Edward S. [L.?]. Lieutenant, *Phoenix*, 1854.

MEARS, Joseph. Surgeon's second mate, *Discovery*, 1791-95.

MEAULEY, Richard. Ordinary Seaman, *Blossom*, 1825-28.

MECHAM, George Frederick (1828-1858). Lieutenant, *Assistance*, 1850-51; *Resolute*, 1852-54. [Murchison, 1859; Frazer, 1947a].

MEDD, William P. *See* MIDD.

MEDDES, Anthony [Arthur] P. AB, *Phoenix*, 1854.

MEDDES, Robert. AB, *Phoenix*, 1854.

MEE, William. *See* BLUNDEN, William.

MEIKLE, J. Captain, *Nascopie*, 1913.

MELLIRK, Thomas. AB, *Plover*, 1848-52; 1852-54.

MELLISH, Thomas. AB, *Assistance*, 1852-54.

MELLOR, George E. Volunteer Assistant, Todd's expedition, 1901.

MELMS, Frank E. (?), Schwatka's expedition, 1878-80. [EA].

MELUISH, William. Private RM, *Rattlesnake*, 1853-54.

MELVILL, James Cosmo Dobrée (1877-). Leader, Great Bear Lake expedition, 1908-11; joint leader, fisheries research expedition (James Bay parties), 1914.

MELVILLE, George Wallace (1841-1912). Engineer, *Tigress*, 1873; Chief Engineer, *Thetis*, 1884. [DAB; EA].

MENAUX [MINAUX], Hilard. Middleman, Rae's expedition, 1846-47. [Rich, 1953, p. 367].

MENZIES, Archibald. Botanist, *Discovery*, 1791-95.

MENZIES, William. AB, *Herald*, 1848; 1849; 1850.

MERCAO, Basileo. Assistant, GSC expedition, 1875.

MERCER, Thomas. Coxswain's mate, *Rattlesnake*, 1853-54.

MERCIER, John. Assistant, Port Burwell station, Canadian Hudson Bay Expedition, 1885-86.

MERCIER, Paul. Oiler, *Arctic*, 1906-07; 1910-11.

MERCIER, Telesphore. Assistant, Port Burwell station, Canadian Hudson Bay Expedition, 1885-86.

MERIDA, Edward. Sailmaker's Mate,

Rattlesnake, 1853-54.

MERKUT [Mrs Hans]. Eskimo, wife of Hans Hendrik. On board *United States*, 1860-61; *Polaris*, 1871-73.

MERRICK, William. Cook, *Pandora*, 1875-76.

MERRIEFIELD, John. Cook, *Herald*, 1848; 1849; 1850.

MERRIMAN, Edgar C. Lieutenant-Commander, *Juniata*, 1873. [NE].

MERRIMAN, Thomas. Landsman, Hearne's expedition, 1769.

MESSINGHAM, John. AB, *Chatham*, 1791-95.

METCALF, . Second Mate, *Dobbs*, 1746-47.

MEYER, Friedrich. Meteorologist, *Polaris*, 1871-73.

MICHAUX, André. Naturalist and leader, overland expedition, 1792.

MICHEL (The Iroquois). *See* TEROAHAUTÉ.

MICHEL, Auguste. AB, *Arctic*, 1910-11.

MICHIE, Forbes. Captain, *Westminster*, 1847.

MIDD [MEDD], William Plaistred. AB, *Resolution*, 1776-80.

MIDDLETON, Alexander. Second Mate, *Felix*, 1850-51.

MIDDLETON, Christopher (-1770). Captain, HBC supply voyages. *Hannah*, 1725; 1726; 1727; *Hudson's Bay* [IV], 1728; 1729; 1730; 1731; 1732; 1733; *Seahorse* [I], 1734; 1735; 1736; *Hudson's Bay* [V], 1737; 1738; 1739; 1740. Captain, *Furnace*, 1741-42. [Rich, 1949, pp. 325-34; DCB Vol. 3; DNB Vol. 37].

MIDDLETON, Nicholas. Corporal RM, *Assistance*, 1852-54.

MIERTSCHING, Johann August (1817-1875). Interpreter, *Investigator*, 1850-54. [DCB Vol. 10].

MIHILI, Henry. Sailmaker, *Plover*, 1848-52; 1852-54.

MIHILL, Henry. Quartermaster, *Pandora*, 1875.

MILES, James. Leading Stoker, *Intrepid*, 1852-54.

MILL, Adam. Surgeon's First Mate, *Discovery*, 1791-95.

MILLAR, John. AB, *Enterprise*, 1848-49.

MILLAR, William. AB, *Prince Albert*, 1851-52.

MILLARD, George. Caulker, *Rattlesnake*, 1853-54.

MILLER, J.P. Lieutenant, *Rattlesnake*, 1853-54.

MILLER, James. Workman, second founding of Churchill, 1717.

MILLER, Matthew R. Engineer, *Discovery*, 1875-76.

MILLER, Thomas. Quartermaster, *Chatham*, 1791-95.

MILLER, William. Seaman, *United States*, 1860-61.

MILLETT, James. AB, *Resolution*, 1776-80.

MILLIKIN [MILLIKEN], Edward. AB, *North Star*, 1852-54.

MILLION, William. Carpenter's Mate, *Herald*, 1848; 1849; 1850.

MILLS, Edward. AB, *Phoenix*, 1854.

MILLS, J. AB, *Enterprise*, 1848-49.

MILLS, William. Assistant, Ashe Inlet station, Canadian Hudson Bay Expedition, 1885-86.

MILLWARD, Thomas. Private of Marines, *Discovery*, 1791-95.

MILNE, . Captain, *Albert*, 1914-15 [1914 only].

MILNE, J.E. Captain, *Perseverance*, 1891.

MILNE, William. AB, *Discovery*, 1791-95.

MILNER, David. AB, *Phoenix*, 1854.

MILNER, George L. Gunroom Steward, *Investigator*, 1850-54.

MILTON, *Viscount* William Wentworth-Fitzwilliam. *See* WENTWORTH-FITZWILLIAM.

MINAUX, Hilard. *See* MENAUX.

MINE, John. Gunner, *Fury*, 1824-25.

MINTZER, W.A. Engineer, *Tigress*, 1873.

MISKIN, William. AB, *Phoenix*, 1853.

MISTEAGUN [MISTAGAN, MISTEGAN], Thomas. (Indian). Steersman, Richardson's expedition, 1847-49; Middleman and hunter, Rae's expedition, 1853-54; steersman, HBC Franklin search, 1855. [Rich, 1953, pp. 367-68].

MITCHELL, Alexander. AB, *Isabella*, 1818.

MITCHELL, David. (?), *Nancy Dawson*, 1849.

MITCHELL, David. AB, *Alert*, 1875-76.
MITCHELL, Henry. Sailmaker, *Plover*, 1848-52 [1851-52 only].
MITCHELL, Henry. Boatswain, *Pandora*, 1875; Gunner, *Pandora*, 1876.
MITCHELL, James. Lance Corporal, Sappers and Miners, Richardson's expedition, 1847-49.
MITCHELL, John. AB, *Discovery*, 1791-95.
MITCHELL, John E. Crewman, Todd's expedition, 1920.
MITCHELL, Peter. Carpenter, *Prince Albert*, 1850.
MITCHELL, Thomas. Captain, *Eastmain*, 1744; *Success*, 1749; *Eastmain*, 1750. [DCB Vol. 3].
MITCHELL, Thomas (-1924). Assistant Paymaster, *Discovery*, 1875-76. [Bell, M., 1972].
MITCHELL, William. AB, *Assistance*, 1850-51.
MOAR, Andrew. Established Cheashquacheston post, 1799.
MOBLEY, Thomas. Private RM, *Resolute*, 1852-54. Died 18 October 1852.
MOGG, William (1796-1875). Clerk, *Hecla*, 1821-23; *Fury*, 1824-25. [Tavener, 1953].
MONE, Robert. AB, *Blossom*, 1825-28.
MONEY, William. Private RM, Franklin's expedition, 1825-27.
MONLAN, Henry. Private RM, *Phoenix*, 1854.
MONNERAYE, de la. Officer, French attack on HBC posts, 1782.
MONNTIFIELD, William. Cook, *Blossom*, 1826.
MONROE, John. Quartermaster, *Discovery*, 1791-95.
MONTFORT, Eugene. AB, *Arctic*, 1910-11.
MONTOUR, Ignace. (Iroquois Indian). (?), HBC Franklin search, 1855.
MOODIE, J.D. Inspector, NWMP patrol, 1897-98; Superintendent, establishment of Fullerton detachment, 1903; Superintendent, *Neptune*, 1903-04; *Arctic*, 1904-05; *Neptune*, 1905-06; *Adventure*, 1906; Major, Governor-general's tour, 1910.
MOODY, George. AB, *Phoenix*, 1854.
MOON, Robert. *See* MOORE, Robert.
MOONEY, David. Chief Engineer, *Alert*, 1885-86.
MOOR, George. AB, *Griper*, 1819-20.
MOOR, William. Captain, *Discovery*, 1741-42; *Dobbs*, 1746-47. [Rich, 1949, pp. 334-36; DCB Vol. 3].
MOORE, Adame. Seaman, *Discovery*, 1610-11.
MOORE, Arthur. Leader, U.S. expedition, 1901.
MOORE, Elwood S. (1878-). Geologist, Flaherty's expedition, 1913-16 [1916 only].
MOORE, James. Gunner, *Fury*, 1824-25.
MOORE, John. Master's Assistant, *Plover*, 1848-52; 1852-54.
MOORE, Lewis J. Mate, *Investigator*, 1848-49.
MOORE [MOON], Robert. (?), *Prince of Wales*, 1848; Ice Quartermaster, *Pioneer*, 1850-51.
MOORE, Thomas Edward Laws (1819-1872). Commander RN, commanding *Plover*, 1848-52. [O'B(1); BOA Vol. 2].
MOORE, William. Gold prospecting expedition, 1882-83.
MOORE, William George. AB, *Investigator*, 1848-49.
MOORE, William H. Master's Assistant, *Plover*, 1848-52.
MORFIN, John. AB, *Erebus*, 1845-48.
MORGAN, E.W. Boat-steerer, *George Henry*, 1860-62.
MORGAN, Eli. AB, *Phoenix*, 1853.
MORGAN, George. Quartermaster, *Rattlesnake*, 1853-54.
MORGAN, Henry. Purser, *Sunneshine*, 1586.
MORGAN, Henry. Private RM, *Assistance*, 1850-51.
MORGAN, Joseph. AB, *Discovery*, 1791-95.
MORGAN, Patrick. Caulker's crew, *Plover*, 1852-54.
MORGAN, Thomas. Corporal RMA, *Intrepid*, 1850-51.
MORGAN, Thomas. AB, *Investigator*, 1850-54. Died at Beechey Island, 22 May 1854.
MORIN, O. Jules. Second Officer, *Arctic*, 1906-07; 1908-09; 1910-11.
MORIN, Pierre. Bowman, HBC expedition, 1837-39.
MORIN, William. AB, *Arctic*, 1910-11.

MORPHAT, Francis. Gold prospecting expedition, 1885-86.

MORPHY, H.O. Corporal NWMP, Begin's patrol, 1890. Drowned on Lake Winnipeg during patrol.

MORRIS, Joseph. Seaman, *Release*, 1855.

MORRIS, Robert. Captain, *Messenger*, 1672-75.

MORRIS, Robert. Cook, *Resolution*, 1776-80.

MORRIS, S. Stanley. Sailor, *Karluk*, Canadian Arctic Expedition, 1913-18. Died off Wrangel Island, 1914.

MORRIS, Thomas. Private of Marines, *Resolution*, 1776-80.

MORRIS, William. AB, *North Star*, 1852-54.

MORRISON [MORISON], Hector (1814-1902). Middleman, HBC expedition, 1837-39; (?), Richardson's expedition, 1847-49. [Rich, 1953, p. 368].

MORRISON, John. Sergeant RM, *Fury*, 1824-25.

MORRISON, Joseph. AB, *Terror*, 1836-37.

MORRISON, Morrison. (?), Back's expedition, 1833-35.

MORRISON, Robert. AB, *Resolution*, 1776-80.

MORRISON, William. Guide, Todd's expedition, 1914-15.

MORTIMER, James. AB, *Phoenix*, 1853.

MORTON, James. AB, *Hecla*, 1824-25.

MORTON, William. Seaman, *Advance*, 1850-51; Steward, *Advance*, 1853-55; Second Mate, *Polaris*, 1871-73.

MORWICK, James. Assistant, Taylor's HBC expedition, 1831.

MOSELEY, Jacob. Steward, *Phoenix*, 1854.

MOSS, Edward Lawton (1843-1880). Surgeon, *Alert*, 1875-76.

MOSS, James. Bombardier, *Enterprise*, 1850-55.

MOTT, John. Sailmaker, *Pandora*, 1875.

MOTTER, Adrian. Seaman, *Discovery*, 1610-11.

MOUAT, Alexander. Midshipman, *Discovery*, 1776-80.

MOULD, James [John] A. Assistant Surgeon, *Terror*, 1836-37. [MAR].

MOURELLE, Francisco Antonio. *See* Maurelle.

MOYLE, Thomas. Boy(?), *Herald*, 1848; 1849; 1850.

MOYLES, Thomas. Captain, *Francis*, 1578.

MOYNE, Alayne. Pilot, *William*, 1536.

MUDGE, Zachary (1770-1852). First Lieutenant, *Discovery*, 1791-95. [DNB Vol. 39].

MUGFORD, Francis. Captain, *Jersey Packet*, 1770; *Amity*, 1771.

MÜHLEISEN, A. Navigator and Assistant, German Polar Year Expedition, 1882-83.

MUIRHEAD, . Captain, *Larkins*, 1817.

MULCOCK, John. AB, *Phoenix*, 1853.

MUMFORD, William. Carpenter's crew, *Resolute*, 1852-54.

MUNCHIN, Edward. Coxswain, *Rattlesnake*, 1853-54.

MUNDEN, Gwylm. Bombardier RMA, *Assistance*, 1852-54.

MUNDEN, James. AB, *Rattlesnake*, 1853-54.

MUNK, Erich. (Nephew of Jens Munk). (?), *Enhiörningen*, 1619-20. Died at Churchill, 1 April 1620.

MUNK, Jens Eriksen (1579-1628). Captain, *Enhiörningen*, 1619-20. [DCB Vol. 1; Hansen, 1970].

MUNN, Henry Toke (1864-). Joint leader, hunting expedition, 1894-95; leader of expedition, *Algerine*, 1912; *Albert*, 1914-15; 1916-18; 1919. [Munn, 1932; EA].

MUNRO, . (Eskimo). Interpreter, Rae's expedition, 1853-54. [Rich, 1953, p. 369].

MUNRO, David. AB, *Chatham*, 1791-95.

MUNRO, Duncan. AB, *Phoenix*, 1854.

MUNRO, John. Chief Engineer, *Karluk*, 1913-18.

MUNROE, George. Bowman, Franklin's expedition, 1825-27.

MUNRO-FERGUSON, Robert. *See* FERGUSON.

MURDAUGH, William H. Passed Midshipman, Acting Master and First Officer, *Advance*, 1850-51. [EA].

MURGATROYD, Joseph. Carpenter's Mate, *Discovery*, 1791-95.

MURIE, Olaus Johan (1889-1963). Scientific Assistant, Todd's expeditions, 1914-15; 1917. [Casebeer, 1964].

MURPHY, Dennis. Seaman, *Roosevelt*,

1908-09.
MURPHY, J. Seaman, *Neptune*, 1903-04.
MURPHY, John. Carpenter's Mate, *Enterprise*, 1850-55.
MURPHY, John. Boatswain, *Roosevelt*, 1905-06; 1908-09.
MURRAY, Alexander. AB, *Felix*, 1850-51.
MURRAY, Alexander Hunter (c1818-1874). Established Fort Yukon, 1847. [DCB Vol. 10; Andrews, C.L., 1947].
MURRAY, George. Ice Quartermaster, *Intrepid*, 1850-51; *Resolute*, 1852-54.
MURRAY, James. Oceanographer, Canadian Arctic Expedition, 1913-18. Died after sinking of *Karluk*, 1914.
MURRAY, John. Sailmaker, *Erebus*, 1845-48.
MURRAY, John. Private RM, *Discovery*, 1875-76.
MURRAY, John. Captain, *Albert*, 1914-15 [1915 only]; 1916-18; 1919.
MURRAY, Thomas. Quartermaster, *North Star*, 1849-50.
MURRAY, William. AB, *Investigator*, 1848-49.
MURRAY, William. AB, *Investigator*, 1848-49.
MURRAY, William. AB, *Enterprise*, 1850-55.
MUSH, William. Captain's Steward, *Blossom*, 1825-28.
MUSHIR, J. Sergeant RM, *Terror*, 1836-37.
MYATT, Joseph (-1730). Established Eastmain post, 1723-24. [DCB Vol. 2].
MYCHAELL, Harry. Sailor, *Ayde*, 1578.
MYERS, John. AB, *Herald*, 1848; 1849; 1850.

NADEAU, François. Assistant to Grant, North West Company expedition, 1785-88.
NAGLE, E. Formed Hislop and Nagle partnership, c1887.
NAHMENS, Otto. Master, *Alaska*, Canadian Arctic Expedition, 1913-18 [1913-14 only].
NANCARNE, Steven. Sailor, *Judith*, 1578.
NAPPER, James (-1737). Captain, *Good Success*, 1720; *Churchill*, 1736-37. Died at Whale Cove, 7 August 1737. [DCB Vol. 2].
NARES, George Strong (1831-1915). Mate, *Resolute*, 1852-54; Captain, *Alert*, 1875-76. [DNB; GJ 1915; Royal Society 1915; Deacon and Savours, 1976].
NARVAEZ, José Maria. Second in command, Spanish expedition, 1791.
NASH, Francis S. Passed Assistant Surgeon, *Alert*, 1884.
NASH, Percy. Canoeman, Camsell's expedition, 1905.
NASH, William. AB, *Resolution*, 1776-80.
NAUMAN, William H. Passed Assistant Engineer, *Alert*, 1884.
NAYLOR, Thomas. Captain of the Forecastle, *Enterprise*, 1850-55.
NEIL, J.B. Seaman, *George Henry*, 1860-62.
NEILL, Samuel. Surgeon, *Hecla*, 1824-25.
NEILSON, Thomas. Assistant Surgeon, *Blossom*, 1825-28.
NELSON, David (-1789). Servant of William Bailey, *Discovery*, 1776-80.
NELSON, Jacob. AB, *Terror*, 1836-37.
NELSON, James. AB, *Investigator*, 1850-54. [McKenzie, W.G., 1970].
NELSON, Robert. Master, *Resolution*, 1612-13. [Died off Nelson River].
NEPITABO [NEPETABO]. *See* NIBITABO.
NESFIELD, Peter. Quartermaster, *Charles*, 1631.
NESS, Alexander. AB, *Griper*, 1824.
NETTELL, A.J.C. Geological and topographical surveyor, GSC expedition, 1914.
NEVEN, Pierre. (?), McLean's expedition, 1838.
NEVILL, Matthews. Cooper, *Resolute*, 1850-51.
NEWELL [NEWALL], Harman. First Assistant Engineer, *Arctic*, 1855.
NEWLAND, Robert. Captain, *Wivenhoe*, 1670-71. Died at Nelson River, 1670.
NEWMAN, Francis. AB, *Hecla*, 1824-25.
NEWMAN, J. Captain of the Forecastle, *Plover*, 1848-52; 1852-54.
NEWMAN, Joseph. Gunner, *Rattlesnake*, 1853-54.
NEWMAN, Michael. Private of Marines, *Discovery*, 1776-80.
NEWRING, Charles. AB, *North Star*, 1849-50.
NEWTON, . Captain, *Emanuell (Buss)*, 1578.
NEWTON, William. Ice Mate, *Investigator*,

1850-54.

NIAS, Joseph (1793-1879). Admiralty Midshipman, *Alexander,* 1818; Midshipman, *Hecla,* 1819-20; Lieutenant, *Fury,* 1821-23. [DNB Vol. 40; BOA Vol. 2; O'B(1)].

NIBITABO [NEPITABO; NEPETABO]. (Cree Indian). Middleman and hunter, Rae's expedition, 1846-47. [Rich, 1953, p. 369].

NICHOLAS, John. Midshipman, *Discovery,* 1791-95.

NICHOLL, William. Surgeon's Mate, *Chatham,* 1791-95.

NICHOLLS, . Sergeant RNWMP, Governor-General's tour, 1910.

NICHOLSON, John. Captain of the Maintop, *Enterprise,* 1848-49.

NICK, James. Surgeon, *Hecla,* 1824-25.

NIELSEN, Morten. Butler, *Enhiörningen,* 1619-20. Died at Churchill, 28 April 1620.

NIGHTINGALE, James. Ordinary Seaman, *Rattlesnake,* 1853-54.

NIGUELS, James T. AB, *North Star,* 1852-54.

NINDEMANN, William. Seaman, *Polaris,* 1871-73.

NINNIS, Belgrave. Staff Surgeon, *Discovery,* 1875-76.

NISBET [NISBETT], Charles. AB, *Resolute,* 1852-54.

NIVERVILLE, *Chevalier* de. Established Fort Paskoyac, 1750.

NIXON, George. Acting Second Mate, *Rattlesnake,* 1853-54.

NIXON, John (c1623-1692). Overseas Governor of HBC; supercargo, *John and Alexander,* 1679. [Rich, 1946, pp. 331-33; DCB Vol. 1].

NOBLE, Crawford, Jr. (1880-1933). Leader, British expedition, 1902. [EA].

NOBLE, John. Cooper, *Lady Franklin,* 1850-51.

NÖDTVEDT, Jacob. Second Engineer, *Fram,* 1898-1902.

NOICE, Harold. (?), Canadian Arctic Expedition, 1913-18.

NOLAN, Joe. (Ojibwa Indian). Cook and canoeman, Camsell's expedition, 1914.

NOLET, Albert. Oiler, *Arctic,* 1910-11.

NONES, Henry B. Chief Engineer, *Juniata,* 1873. [NE].

NOOT, John. Boatswain's Mate, *Discovery,* 1791-95.

NOREM, Andrew. Steward, *Mary Sachs,* Canadian Arctic Expedition, 1913-18. Committed suicide, 16 April 1914.

NORKETT, Henry. Quartermaster, *Plover,* 1852-54.

NORMAN, Masters. Second Master, *North Star,* 1849-50.

NORMAN [NORMAND], Napoléon. Carpenter, *Arctic,* 1910-11.

NORMANVILLE, Thomas Godefroy de (1610-1652). Accompanied Buteux, 1651. [DCB Vol. 1, p. 341].

NORQUAY, John. Middleman, HBC expedition, 1837-39.

NORTHHOUSE, Thomas. AB, *Assistance,* 1850-51; *Resolute,* 1852-54.

NORTON, John. Boy, *Michael,* 1578.

NORTON, Moses (-c1773). (?), HBC expeditions, 1755; 1762. [DCB Vol. 4].

NORTON, Richard. (-1741). (?), second founding of Churchill, 1717; leader, HBC expedition, 1717-18; (?), *Prosperous,* 1721; *Whalebone,* 1721-22. [DCB Vol. 3].

NORTON, William. Captain, HBC supply voyages. *Hudson's Bay* [V], 1752; 1753; 1754; 1755; 1756; *Seahorse* [I], 1757; 1758; 1759; 1760; *Prince Rupert* [III], 1761; 1762; 1763.

NORVAL, Alexander. Cook's mate, *Discovery,* 1791-95.

NOWALL, Thomas. Private RM, *Hecla,* 1821-23.

NUDDS, Robert. Boatswain, *Blossom,* 1825-28.

NUGENT, James. AB, *North Star,* 1852-54.

NUTCHER, John. AB, *Enterprise,* 1850-55.

NYBORG, Peder. Carpenter, *Enhiörningen,* 1619-20. Died at Churchill, May or June 1620.

OAKLEY, Thomas. Gunner RM, *Alert,* 1875-76.

OBED, Boaz. (Eskimo). Guide, Prichard's expedition, 1910.

O'BRYAN [O'BRIAN], Thomas. AB, *Enterprise,* 1850-55.

O'BRYANT, Edward. AB, *Herald,* 1848; 1849; 1850.

O'CONNOR, William. Ordinary Seaman, *North Star,* 1849-50.

O'CONNELL, F. Boy, *Neptune,* 1903-04. Drowned, 11 December 1903.

ODELL, A.A. Assistant Engineer, *Polaris,* 1871-73.

ODJICK, Joseph. Guide and canoeman, Todd's expedition, 1917.

ODJICK, Moses. Guide and canoeman, Todd's expedition, 1917.

OGILVIE, William (1846-1912). Leader, Yukon exploring expedition, 1887-89; first commissioner of the Yukon (see 1898 Yukon Act). [EA; Thomson, 1966].

OGIVIE, George. AB, *Herald,* 1848; 1849; 1850.

OGSTON, Alexander. *See* OYSTON.

OHLIN, Axel Gabriel (1867-1903). Zoologist, *Falcon,* 1894.

OHLSEN, Christian. Carpenter, *Advance,* 1853-55. Died 12 June 1855.

O'LEARY, R. Seaman, *Laddie,* 1913-16.

OLESEN, Louis. (?), Canadian Arctic Expedition, 1913-18.

OLIVER, James. Carpenter, *Rattlesnake,* 1853-54.

OLIVER, William. AB, *Blossom,* 1825-28.

OLIVIER, Louis (c1819-c1850). Voyageur, Richardson's expedition, 1847-49. [Rich, 1953, p. 369].

OLLEY, George. AB, *Hecla,* 1824-25.

OLLEY, George. AB, *Investigator,* 1848-49; 1850-54.

OLLEY, William. AB, *Griper,* 1824-25.

OLLUFFSEN, Jan. Navigating Officer, *Enhiörningen,* 1619-20. Died at Churchill, 25 March 1620.

OLSEN, Christian. *See* OHLSEN.

OLSEN, Karl. Chief Engineer, *Fram,* 1898-1902.

OLSWIG, Carl Emil. Sailor, *United States,* 1860-61.

OMMANNEY, Erasmus (1819-1904). Captain, *Assistance,* 1850-51. [O'B(1); GJ 1904; DNB 2nd Supp. Vol. 3].

ONE-EYE, Albert (c1824-1849). (Eskimo). Interpreter, Richardson's expedition, 1847-49. Drowned at Bloody Fall, 24 August 1849. [Rich, 1953, p. 370].

O'NEILL, John J. Geologist, Canadian Arctic Expedition, 1913-18.

OO-KE-AH. (Eskimo). Reached North Pole with Peary, 1909.

OOLIGBUCK [OULIGBUCK, OULLIBUCK, OULYBUCK, ULLEBUCK] (-1852). (Eskimo). Guide, Franklin's expedition, 1825-27; establishment of Fort Chimo, 1830; exploration of George River, 1831; Interpreter, HBC expedition, 1837-39; Rae's expedition, 1846-47. [Rich, 1953, pp. 370-73].

OOLIGBUCK, William (-c1895). (Son of Ooligbuck). Interpreter, Rae's expeditions, 1846-47; 1853-54. [Rich, 1953, pp. 373-74].

OO-TAH. (Eskimo). Reached North Pole with Peary, 1909.

OPSLOE, Christoffer. Chief Gunner, *Enhiörningen,* 1619-20. Died at Churchill, 5 April 1620.

ORCHARD, H.M. Clerk, *Discovery,* 1791-95.

ORGAN, Joseph. Quartermaster, *Enterprise,* 1848-49; Ice Quartermaster, *Pioneer,* 1850-51; *Assistance,* 1852-54.

ORMESHAWE, William. Soldier, Frobisher's expedition, 1578.

OROUST, Anders. Seaman, *Enhiörningen,* 1619-20. Died at Churchill, 16 April 1620.

ORREN, William. AB, *Erebus,* 1845-48.

OSBORN, Noel (-1875). Mate, *North Star,* 1852-54. [MAR].

OSBORN, Sherard (1822-1875). Lieutenant commanding *Pioneer,* 1850-51; 1852-54. [DCB Vol. 10; DNB Vol. 42; O'B(1); *Geographical Magazine,* 1875].

OSBORNE, . Ice Master, *Talbot,* 1854.

OSBORNE, Henry. Boatswain, *Investigator,* 1848-49; *Assistance,* 1850-51.

OSMER, Charles Hamilton. Clerk, *Blossom,* 1825-28; Paymaster and Purser, *Erebus,* 1845-48.

OSSIN, Louis (c1800-). Middleman, Black's expedition, 1824. [Rich, 1955, pp. 241-43].

OULIGBUCK [OULLIBUCK, OULYBUCK]. *See* OOLIGBUCK.

OUTLAW, John. Captain, HBC supply voyages. *Lucy,* 1684; *Success,* 1685; second-in-command, *Mary,* 1688. [Rich, 1948, pp.

PARSONS, William. Captain of the Forecastle, *Plover*, 1848-52.
PASCOE, John. AB, *Phoenix*, 1853.
PASHE, John. AB, *Hecla*, 1821-23.
PASSMORE, William. AB, *Discovery*, 1776-80.
PATENAUDE, Alexandre. Waiter, *Arctic*, 1906-07.
PATERSON, Alexander. Corporal RM, *Erebus*, 1845-48.
PATERSON, John. AB, *Fury*, 1821-23.
PATON, John. AB, *Griper*, 1824.
PATTEN [PATTIN], John. (?), *Argo*, 1753; 1754. Killed by Eskimos, 1854. [Eavenson, 1949, pp. 1-70].
PATTENDEN, J.I. Quartermaster, *Rattlesnake*, 1853-54.
PATTERSON, Charles. Montreal pedlars' expedition, 1775-76.
PATTERSON, James. Quartermaster, *Rattlesnake*, 1853-54.
PATTERSON, James. Captain of the Foretop, *North Star*, 1849-50.
PATTERSON, John. Gunner's Mate, *Hecla*, 1824-25.
PATTERSON, William. Sailmaker's Mate, *Discovery*, 1791-95.
PAUL, Baptiste. Voyageur, Richardson's expedition, 1847-49.
PAUL, Charles. AB, *Discovery*, 1875-76. Died 29 June 1876.
PAUL, D. AB, *Hecla*, 1824-25.
PAUL, Frank. Assistant, Stupart Bay station, Canadian Hudson Bay Expedition, 1885-86.
PAUL, Mathew. AB, *Resolution*, 1776-80.
PAVIE, Edward. Boy, *Ayde*, 1578.
PAVY, Octave (1844-1884). Surgeon, U.S. Polar Year Expedition, 1881-84. Died of starvation, 6 June 1884. [EA].
PAYNE, Frank F. Observer in charge of Stupart Bay station, Canadian Hudson Bay Expedition, 1885-86.
PAYNE, William. Carpenter, *Ayde*, 1578.
PEABODY, Francis W. Joint leader, U.S. expedition, 1908.
PEAD, John. AB, *Fury*, 1821-23.
PEARCE, Alfred R. AB, *Alert*, 1875-76.
PEARCE, George, Captain of the Maintop, *Pandora*, 1876.
PEARCE, William. AB, *Rattlesnake*, 1853-54.
PEARCH, George. Steward, *Rattlesnake*, 1853-54.
PEARD, George (1783-1837). Lieutenant, *Blossom*, 1825-28. [Gough, 1973, pp. 50-52; RNB].
PEARSE, Richard Bulkeley (1830-1895). Mate, *Resolute*, 1850-51. [BOA Vol. 2].
PEARSON, John. Private RM, *Isabella*, 1818; *Griper*, 1819-20.
PEARSON, John. AB, *Alert*, 1875-76.
PEARSON, William. Captain of the Foretop, *Plover*, 1848-52; 1852-54.
PEARY, Robert Edwin (1856-1920). Leader, U.S. North Pole expeditions, 1898-1902; 1905-06; 1908-09. Reached North Pole 6 April 1909. [DAB; Hobbs, 1936; Stafford, 1959; Weems, 1967].
PECK, Edmund James (1850-1924). Leader, Anglican missionary expeditions, 1876-84; 1884; 1885-92; 1894-96; 1897-99; 1900-02. [EA; Lewis, A., 1904].
PECKOVER, William. Gunner, *Discovery*, 1776-80.
PECOCKE, John. (?), *Ayde*, 1578.
PEDDAR, M.J. Stoker, *Phoenix*, 1853.
PEDDIE, John Smart. Acting Surgeon, *Terror*, 1845-48.
PEDERSEN, Povel. Seaman, *Enhiörningen*, 1619-20. Died 21 March 1620.
PEEBLES, Thomas. AB, *Griper*, 1819-20.
PEERSON, . Carpenter, *Elizabeth*, 1587.
PEGLAR, Henry Peter. Captain of the Maintop, *Terror*, 1845-48. [Cyriax and Jones, 1954].
PELLETIER. E.A. Inspector RNWMP, patrol, 1908-09.
PELONQUIN (*dit* Credit), Mathew. Voyageur, Franklin's expedition, 1819-22.
PELTIER, Baptiste. Middleman, Rae's expedition, 1850-51. [Rich, 1953, p. 374].
PELTIER, Joseph. Voyageur, Franklin's expedition, 1819-22.
PENDER, William. Private of Marines, *Discovery*, 1791-95.
PENNARA [PENNASCA, PROVENCHER], Sébastien. Accompanied Albanel, 1671-72.
PENNELOSSA, Diego. (Fictitious character). Commander, *St Lucia*, 1640.

383-85; DCB Vol. 1].

OWEN, Gryffyn. Master, *Michael,* 1576.

OWEN, Henry. Private RM, *Phoenix,* 1853.

OWEN, John. AB, *Alexander,* 1818.

OWEN, Joseph. AB, *Phoenix,* 1854.

OWEN, Roger. Sailor, *Ayde,* 1578.

OWEN, William. Apprentice Engineer, *Phoenix,* 1853.

OWEN, William. AB, *Pandora,* 1876.

OWENS, David. Captain of the Hold, *Herald,* 1848; 1849; 1850.

OYSTON [OGSTON], Alexander. Mate, *Isabel,* 1852.

PACKARD, Alpheus Spring (1839-1905). Passenger, *Benjamin S. Wright,* 1864. [American Geographical Society, 1905].

PACKENHAM [PAKENHAM], Robert E. Midshipman, *Herald,* 1848; 1849; 1850.

PACKER, Joseph. AB, *Phoenix,* 1853.

PACKHAM, John. AB, *Phoenix,* 1853.

PADDY. (Chipewyan Indian). Accompanied Thompson, 1796.

PAGE, J. AB, *North Star,* 1849-50.

PAGE, John. Miner, Frobisher's expedition, 1578.

PAGE, John. AB, *Fury,* 1824-25. Died 27 August 1825.

PAGE, Joshua [Joseph]. Gunner's Mate, *Enterprise,* 1850-55.

PAGE, Thomas. Private RM, *Enterprise,* 1848-49.

PAGE, William. Ice Quartermaster, *North Star,* 1852-54.

PAIGAN, Joseph. AB, *Plover,* 1852-54.

PAINE, Joseph C. Clerk in Charge, *Investigator,* 1850-54.

PAINTER, Charles. AB, *Phoenix,* 1854.

PAKENHAM, Robert E. *See* PACKENHAM.

PALL, Mathew. *See* PAUL.

PALMER, Charles. Midshipman, *Hecla,* 1819-20; Lieutenant, *Hecla,* 1821-23. [O'B(1); BOA Vol. 6].

PALMER, George W. Gunner's Mate, *Enterprise,* 1848-49.

PALMER, Henry. Carpenter, *Phoenix,* 1853.

PALMER, John. Boatswain, *Henrietta Maria,* 1631-32.

PALMER, John H. Clerk, *Breadalbane,* 1853.

PAMPHLET, John. AB, *Investigator,* 1848-49.

PANGMAN, Peter (c1744-1819). Joined Gregory, McLeod and Co., 1784; leader, Gregory, McLeod and Co. expedition, 1785-87. [DCB Vol. 5].

PAPANAKIES, Paulette. Steersman, HBC Franklin search, 1855.

PARADICE, John. Surgeon, *Judith,* 1578.

PARDY, H.J. Established Port Harrison, 1920.

PARENT, Jean Baptiste. Voyageur, Franklin's expedition, 1819-22.

PARFITT, George. Private RM, *Investigator,* 1850-54.

PARK, John. AB, *Victory,* 1829-33.

PARKER, George. AB, *Griper,* 1819-20.

PARKER, J.C. CMS missionary, missionary journey, 1894-96. Drowned in Cumberland Sound, 14 August 1896.

PARKER, James. AB, *North Star,* 1824-25.

PARKES, Murray T. Mate, *Enterprise,* 1850-55.

PARKINSON, William. AB, *Alexander,* 1818.

PARR, Alfred Arthur Chase (1849-1926). Lieutenant, *Alert,* 1875-76. [EA].

PARR, James. Captain of the Maintop, *North Star,* 1849-50.

PARR, Joseph. Gunner's Mate, *Resolute,* 1852-54.

PARRY, Charles (1833-1868). (Son of W.E. Parry). First Lieutenant, *Bulldog,* 1860. [Parry, E., 1870].

PARRY, William Edward (1790-1855). Lieutenant commanding *Alexander,* 1818; Lieutenant commanding *Hecla,* 1819-20; Commander RN, commanding *Fury,* 1821-23; Captain, *Hecla,* 1824-25. [Parry, A., 1963; Parry, E., 1859; Beechey, 1857; O'B(1); DNB Vol. 43].

PARSONS, James. Stoker, *Phoenix,* 1854.

PARSONS, Edward S. Crewman, Todd's expedition, 1920.

PARSONS, Hayward. (Brother of Ralph Parsons). Established Cape Dorset post, 1913.

PARSONS, Joseph. AB, *Rattlesnake,* 1853-54.

PARSONS, Ralph. Established Wolstenholme post, 1909; established Lake Harbour post, 1911 [see 1909-11, Lake Harbour mission].

PENNINGTON, James. AB, *Pandora*, 1875.

PENNY, William (1809-1892). Mate, *Traveller*, 1833; Captain, *Neptune*, 1839; *Bon Accord*, 1840; *St Andrew*, 1847; *Advice*, 1849; *Lady Franklin*, 1850-51; 1853-54; 1857-58. [Holland, 1970].

PEPIN, Joseph R. Doctor, *Arctic*, 1906-07.

PERCIVAL, Joseph. Corporal RM, *Investigator*, 1848-49.

PERCY, Charles. Steward, *Roosevelt*, 1905-06; 1908-09.

PERCY, George. Seaman, *Roosevelt*, 1908-09.

PEREZ HERNANDEZ, Juan Josef. Captain, *Santiago*, 1774; Quartermaster, *Santiago*, 1775.

PERKINS, John. Private of Marines, *Resolution*, 1776-80.

PERKINS, Samuel. Carpenter's Mate, *Enterprise*, 1850-55.

PERNE, Edward. AB, *Phoenix*, 1853.

PEROUSE, Jean François Galaup, *Comte de la*. *See* GALAUP.

PERRAULT, Ignace. Voyageur, Franklin's expedition, 1819-22. Died October 1821.

PERREAULT [PERRAULT], Antoine (-1828). Bowman, Black's expedition, 1824. [Rich, 1955, pp. 243-45].

PERRON, Hervé. Waiter, *Arctic*, 1906-07.

PERRY, John. AB, *Rattlesnake*, 1853-54.

PERRYMAN, Frederick. (?), *Nancy Dawson*, 1849.

PERSE, Michael. Seaman, *Discovery*, 1610-11. Died 31 July 1611.

PETERS, William John (1863-1942). Leader, U.S. expedition, 1914. [Harradon, 1942].

PETERSEN, Johan Carl Christian (1813-1880). Interpreter, *Lady Franklin*, 1850-51; *Advance*, 1853-55; *Fox*, 1857-59. [RGS 1880; Irminger, 1880; Sorensen, 1941].

PETERSEN, N.C. Eskimo interpreter, *Alert*, 1875-76.

PETITOT, Emile Fortuné Stanislas Joseph (1838-1917). Leader, RC missionary explorations, 1864-72. [EA; Savoie, 1970].

PETRIE, David. AB, *Terror*, 1836-37.

PETTERSEN, Jan. Second Mate, *Lamprenen*, 1619-20. Died at Churchill, 31 March 1620.

PETTY, Henry. Private RM, *Discovery*, 1875-76.

PHAYRE, G.A. Lieutenant, *Enterprise*, 1850-55.

PHILBRECK, Benjamin. Boatswain's Mate, *North Star*, 1849-50.

PHILIPS [PHILLIPS], Joseph. Mate and Greenland Pilot, *Alexander*, 1818.

PHILLIPPES, Thomas. Intending colonist, *Gabriel*, 1578.

PHILLIPS [PIPPEN], Benjamin. AB, *Blossom*, 1825-28.

PHILLIPS, Charles Gerrans. Commander, RN, and second-in-command, *Felix*, 1850-51. [O'B(1)].

PHILLIPS, Henry. Carpenter, *Discovery*, 1791-95.

PHILLIPS, James. Wardroom Cook, *Discovery*, 1875-76.

PHILLIPS, John. AB, *Investigator*, 1848-49.

PHILLIPS, John. Ordinary Seaman, *Plover*, 1848-52.

PHILLIPS, Molesworth (1755-1832). Second Lieutenant RM, *Resolution*, 1776-80. [DNB 1st Supp. Vol. 3].

PHILLIPS. W.G. Established Bernard Harbour trading post, 1916.

PHILLISKIRK, George. Boatswain's Mate, *Discovery*, 1791-95.

PHILP, Robert K. Gunroom Steward, *North Star*, 1852-54.

PHILPOT [FILPOT], Richard. Ensign, *Ayde*, 1577; Captain, *Beare*, 1578.

PHILPOTS, Edward P. Surgeon, *Queen*, 1865-66.

PHINNEY, William. Quartermaster, *Release*, 1855.

PHIPPS, Constantine John (1744-1792). Lieutenant, *Niger*, 1766. [DNB Vol. 45].

PICKERSGILL, Richard. Lieutenant commanding *Lion*, 1776.

PICKETI, John. AB, *Blossom*, 1825-28.

PICKTHORN [PICTHORN], Thomas R. Assistant Surgeon, *Pioneer*, 1850-51.

PIERS, Henry. Assistant Surgeon, *Investigator*, 1850-54.

PIGOT, Robert. Midshipman, *Discovery*, 1791-95.

PIKE, Richard (1834-1893). Captain, *Proteus*, 1881 [see 1881-84]; 1883. [DNB Vol.

45].

PIKE, Reuben. Seaman, *Arctic*, 1906-07; Waiter, *Arctic*, 1908-09.

PIKE, Warburton Mayer (1861-1915). Leader, overland expeditions, 1890; 1892-93. [GJ 1915a; MDC p. 595].

PILCER, James. Stoker, *Phoenix*, 1853.

PILCHER [PITCHER], George. AB, *North Star*, 1852-54.

PILKINTON, William. Private RM, *Erebus*, 1845-48.

PILLAR, James. Sailmaker, *Isabella*, 1818.

PIM, Bedford Clapperton [Capperton; Chapperton] Trevelyan (1826-1886). Midshipman, *Herald*, 1848; 1849; *Plover*, 1849-52; Lieutenant, *Resolute*, 1852-54. [DNB Vol. 45; EA; Anonymous, n.d.].

PIM, Frederick W. *See* PYM.

PINK, William. HBC fur trading expeditions, 1766-67; 1767-68; 1768-69; 1769-70.

PIPPEN, Benjamin. *See* PHILLIPS.

PIRIE, George. Lieutenant, *Pandora*, 1875; Navigating sub-Lieutenant, *Pandora*, 1876.

PITCHER, George. *See* PILCHER.

PITCHER, James. Stoker, *Fox*, 1857-59.

PITT, *Hon.* Thomas. AB, *Discovery*, 1791-95.

PITTUCAN, George. Leading Stoker, *Phoenix*, 1853.

PLANTE, Antoine [Josephe]. *See* LAPLANTE, Xavier.

PLATER, Thomas. Stoker, *Erebus*, 1845-48.

PLATT, Henry. AB, *Alexander*, 1818.

PLUMSTEAD, Henry. Corporal RM, *Terror*, 1836-37.

PLUNKETT, William. Private RM, *Phoenix*, 1854.

POCOCK, Francis. AB, *Erebus*, 1845-48.

POMEROY, C. Seaman, *Neptune*, 1903-04.

POND, Christopher. Carpenter's crew, *Assistance*, 1852-54.

POND, Peter (1740-c1807). Leader, Montreal pedlars' expeditions, 1775-76; 1778-80; 1781-82; 1785-88. [Innis, H.A., 1930; DCB Vol. 5].

POOLE, George. AB, *Blossom*, 1825-28.

POOLE, James. Private RM, *Discovery*, 1776-80.

POPE, Richard. Master's Mate, *Sunneshine*, 1585; Captain, *Sunneshine*, 1586.

POPLIN, Richard. Gold prospecting expeditions, 1883-84; 1885-86.

PORTEN, Robert. Captain, *Abraham and Robert*, 1686.

PORTEOUS, Archibald. Second Engineer, *Pandora*, 1875; 1876.

PORTEOUS, David. Surgeon, *Felix*, 1850-51.

PORTER, Frank. Miner, Curran's expedition, 1912.

PORTER, George. Gunner RMA, *Alert*, 1875-76. Died on Markham's sledge journey, 8 June 1876.

PORTER, James. Established Fort of the Forks, 1803.

PORTER, Robert. (?), hunting expedition, 1910.

PORTER, Russell Williams (1871-1949). Joint leader, U.S. expedition, 1897. [Willard, 1968; EA].

PORTLOCK, Nathaniel (c1748-1817). Master's Mate, *Discovery*, 1776-80. [DNB Vol. 46].

PORTSMOUTH, Michael. Drummer, Marines, *Resolution*, 1776-80.

POST, . Moravian missionary, *Hope*, 1752.

POTT, William. AB, *Hecla*, 1824-25.

POTTEN, Josh. Private RM, *Chatham*, 1791-95.

POTTER, Byron. Seaman, *Release*, 1855.

POTTS, George. HBC expedition, 1757-58.

POTTS, John. Established Richmond Fort, 1750.

POTVINE, J. Purser's Steward, *North Star*, 1849-50.

POULTER, William. AB, *Discovery*, 1776-80.

POUND, Benjamin. AB, *Fox*, 1857-59.

POWER, Richard (-1681). Captain, *Prince Rupert*, 1678. [Rich, 1945, pp. 340-41; DCB Vol. 1].

POWER, William. AB, *Resolute*, 1852-54.

POWER, William. AB, *Herald*, 1848; 1849; 1850.

POYER, James. Carpenter's crew, *Assistance*, 1852-54.

PRADER, John. Cook, *Herald*, 1848; 1849; 1850.

PRATT, John. Private RM, *Rattlesnake*, 1853-54.

PREBBLE, . Captain, [?], New England voyage, 1753.

PREBLE, Alfred E. Assistant, U.S. biological expeditions, 1900; 1901; 1903-04; Scientific assistant, Carnegie Museum expedition, 1912.

PREBLE, Edward Alexander (1871-). Leader, U.S. biological expeditions, 1900; 1901; 1903-04; joint leader, U.S. exploring expedition, 1907.

PRESLEY, John. AB, *Plover*, 1848-52.

PRESSICK, Henry. Leader, HBC expedition, 1761-62.

PRESSLEY, F. AB, *Pandora*, 1876.

PRESTON, James. (?), *Nancy Dawson*, 1849.

PRESTON, John. Private RM, *Blossom*, 1825-28.

PRICE, Arthur. Master, *Henrietta Maria*, 1631-32.

PRICE, Francis. AB, *Plover*, 1848-52; 1852-54; *Herald*, 1848.

PRICE, George. Seaman, *Release*, 1855.

PRICE, John. Baker, Frobisher's expedition, 1578.

PRICE, Thomas. Sailor, *Ayde*, 1578.

PRICE, Thomas. Master, *Gabriel*, 1578.

PRICE, Thomas. AB, *Resolution*, 1776-80.

PRICHARD, Hesketh Vernon Hesketh (1876-1922). Leader, hunting expedition, 1910. [Parker, E., n.d.; WWW, Vol. 2].

PRICKETT, Abacuck. Landsman, *Discovery*, 1610-11; *Resolution*, 1612-13.

PRIDDLE, Jonah. Private RM, *Plover*, 1848-52.

PRIMEAU [PRIMO], Louis. HBC overland expeditions, 1765-66; 1766-67; 1767-68; 1769-70; 1770-71; 1771-72; 1772-73; Montreal pedlars' expeditions, 1773-74; 1774-75; 1775-76; 1776-77. [DCB Vol. 4].

PRINGLE, James. AB, *Hecla*, 1821-23. Died 15 May 1822.

PRINGLE, Lawrence. AB, *Fury*, 1824-25.

PRINPHILOR, Frederick. Private RM, *Herald*, 1848, 1849; 1850.

PRIOR, William. Private RM, *Enterprise*, 1848-49.

PRITCHARD, William. Cabin boy, *Roosevelt*, 1908-09.

PRIVETT, Edward. AB, *Assistance*, 1850-51.

PROVENCHER, Sébastien. *See* PENNARA, Sébastien.

PRUDEN, James. Leader, HBC expedition, 1843.

PUGET, Peter. Second Lieutenant, *Discovery*, 1791-95.

PULLEN, Thomas C. Master, *North Star*, 1852-54.

PULLEN, W.H. Chaplain, *Alert*, 1875-76.

PULLEN, William John Samuel (1813-1887). Lieutenant, *Plover*, 1848-52; leader, boat journey in search of Franklin, 1849-51; Commander RN commanding *North Star*, 1852-54. [DNB Vol. 47; AB].

PUNNY, Eskimo, *Polaris*, 1871-73.

PURCHASE, Thomas. Engineer, *Intrepid*, 1852-54.

PURFER [PURFUR], Charles. Carpenter, *Hecla*, 1821-23; *Fury*, 1824-25.

PURVER, Richard. Captain of the Foretop, *Phoenix*, 1853.

PYM [PIM], Frederick W. Mate, *Assistance*, 1852-54.

QUACKENBUSH, Lewis. Joint leader, Cabot's expedition, 1905 [*See* 1903-10].

QUADRA, Juan Francisco de la Bodega y. *See* BODEGA Y QUADRA.

QUAIL, Edward. Private RM, *Terror*, 1836-37.

QUAYLE, . Captain, *McLellan*, 1851-52.

QUEN, Jean de (1603-1659). Discovered Lake St John, 1647. [DCB Vol. 1].

QUESNEL, Jules Maurice (1786-1842). Fraser's expedition, 1808.

QUIDDINGTON, George. Carpenter's Mate, *Assistance*, 1852-54.

QUIMPER, Manuel. Captain, *Princess Royal*, 1790.

QUIN, Thomas. AB, *Resolution*, 1776-80.

QUIROS, Fernando. Officer (?), *Favorita*, 1779.

RAANES, Oluf. Mate, *Fram*, 1898-1902. [EA].

RADCLIFFE, Henry G. Assistant Astronomer, *United States*, 1860-61.

RADFORD, Harry V. Leader, zoological expedition, 1911-12. Killed by Eskimos, June 1912.

RADISSON, Pierre Esprit (c1640-1710). French expedition, 1659-60; New England voyage, 1663; Supercargo, *Eaglet,* 1668-69; *Wivenhoe,* 1669; 1670-71; *St Pierre,* 1682-83; *Lucy,* 1684. [Rich, 1942, pp. 243-49; DCB Vol. 2].

RADMORE, John N. Chief Carpenter's Mate, *Alert,* 1875-76.

RAE, Charles. Boatswain, *Prince Albert,* 1850.

RAE, James. Assistant Surgeon, *North Star,* 1849-50.

RAE, John (1813-1893). Leader, HBC expedition, 1846-47; assistant, Richardson's expedition, 1847-49; leader, Franklin search expedition, 1850-51; leader, HBC expedition, 1853-54. [DNB Vol. 47; Royal Society, 1897a; GJ 1893; EA; *Canadian Record of Science,* 1893].

RAINSFORD, . Station hand at Nachvak, Canadian Hudson Bay Expedition, 1884-85.

RALEIGH, Walter (1552-1618). Captain, *Raleigh,* 1583. [DNB Vol. 47].

RALPH, Henry John. Private, Sappers and Miners, Richardson's expedition, 1847-49.

RALSTON, David C. Sergeant, Signal Corps, U.S. Army, U.S. Polar Year Expedition, 1881-84. Died 23 May 1884.

RAMSAY, John. AB, *Resolution,* 1776-80.

RAMSAY, John. AB, *Investigator,* 1850-54.

RAMSEY, Thomas. Captain, HBC supply voyages. *Ceres,* 1802; 1803; 1804; *Eddystone,* 1807; 1808; 1810; 1811-12; 1812; 1813.

RANDAL, Hugh. Captain, *Solomon,* 1578.

RANDALE, Joseph. Carpenter's crew, *Herald,* 1848; 1849; 1850.

RANDALL, William. Private of Marines, *Discovery,* 1776-80.

RANDERSON, William. Quartermaster, *Pandora,* 1875; 1876.

RANDS, Thomas. Ice Quartermaster, *North Star,* 1852-54.

RANKIN, John. Lieutenant, *Furnace,* 1741-42.

RAMSOM, Thomas. AB, *Resolute,* 1850-51; *Enterprise,* 1848-49.

RASTELL, John. Passenger, Hore's expedition, 1536. [DCB Vol. 1].

RAWLINS, Thomas. Steward, *Herald,* 1848; 1849; 1850.

RAWLINS, Thomas. Captain of the Forecastle, *Alert,* 1875-76.

RAWSON, Wyatt (1853-1882). Lieutenant, *Discovery,* 1875-76. [RGS 1882; EA].

RAY, Nicholas. AB, *Hecla,* 1824-25.

RAYMOND, Charles Walker (1842-1913). Captain, U.S. Army Corps of Engineers; surveyed Yukon River (see 1869, closure of Fort Yukon). [DAB].

RAYNER, Elijah. Gunner, RM, *Discovery,* 1875-76.

RAYNES, Samuel. *See* HEAYNES.

REA, John R. (1857-). Fur trader and assistant, Jones's expedition, 1897-98.

READ, Charles. Private RM, (?), Austin's expedition, 1850-51.

READ, James. *See* REID.

READ, Louis. AB, *Assistance,* 1852-54.

READ, Roger. Captain of the Forecastle, *Enterprise,* 1848-49.

READER, A.B. (?), *Minnie Maud,* 1912-13.

REARDON, J.W. AB, *Assistance,* 1852-53.

RECORD, Thomas. Musician, *Resolute,* 1850-51.

REDDINGTON, Philip. Captain of the Forecastle, *Erebus,* 1845-48.

REDDON, James. Ordinary Seaman, *Blossom,* 1825-28.

REED, Henry. AB, *Plover,* 1848-52; 1852-54.

REED, John. Carpenter's Mate, *Fury,* 1821-23.

REED, John William. Master and Assistant Surveyor, *Bulldog,* 1860. [Dawson, L.S., 1885, Part 2, pp. 167-68].

REED, William. Marine, *Erebus,* 1845-48.

REES, Richard. AB, *Griper,* 1824.

REESE, David T. Seaman, *Florence,* 1877-78.

REEVE, Ben. AB, *Discovery,* 1791-95.

REFAD (?), Frederick. Ordinary Seaman, *Rattlesnake,* 1853-54.

REGAN, Michael. AB, *Discovery,* 1875-76.

REID, Andrew (1788-). Midshipman, *Griper,* 1819-20; Lieutenant, *Fury,* 1821-23. [O'B(1)].

REID [READ], James. Ice Quartermaster, *Assistance,* 1852-54.

REID, James. Acting Ice Master, *Erebus,* 1845-48.

REID, James. Second Mate, *Sophia,* 1850-51.

REID, Roger. Quartermaster, *Enterprise,* 1850-55.

REID, William. Middleman, HBC Franklin search, 1855.

REILLY, William. AB, *Phoenix,* 1853.

RELFE, Samuel. AB, *Investigator,* 1850-54.

RENDELL [RENDALL], John (1797-). Midshipman, *Blossom,* 1825-28. [O'B(1)].

RENNIE, . Captain, *Levant,* 1818.

RENWICK, Charles K. Chief Engineer, *Phoenix,* 1853.

RENWICK, Thomas, AB, *Hecla,* 1824-25.

REYMER, Nicholas. Captain, *Lucy,* 1682-83.

REYNERS, John. Private of Marines, *Chatham,* 1791-95.

REYNOLDS, Edward. Ordinary Seaman, *Rattlesnake,* 1853-54.

REYNOLDS, Henry. Sailmaker, *Rattlesnake,* 1853-54.

REYNOLDS, J. AB, *Hecla,* 1821-23.

REYNOLDS, John. Stoker, *Phoenix,* 1854.

REYNOLDS, Lovell K. Ensign, *Bear,* 1884.

REYNOLDS, Peter. Carpenter, *Discovery,* 1776-80.

REYNOLDS, Philip C.F. Cook, *Arctic,* 1910-11.

RHOADES, H.E. Assistant Engineer, *Juniata,* 1873.

RHODES, William. Quartermaster, *Terror,* 1845-48.

RICARD, Edward. AB, *Rattlesnake,* 1853-54.

RICARDS, J.B. Assistant Surgeon, *Pioneer,* 1852-54.

RICE, George W. Sergeant, Signal Corps, U.S. Army, U.S. Polar Year Expedition, 1881-84. Died 9 April 1884.

RICE, John. Boatswain, *North Star,* 1849-50; *Rattlesnake,* 1853-54.

RICE, Thomas. AB, *Resolution,* 1776-80.

RICH, James. AB, *Enterprise,* 1850-55.

RICHARD, Richard. Boatswain, *Discovery,* 1791-95.

RICHARDS, . Captain, *Eddystone,* 1803-06.

RICHARDS, Charles. Midshipman, *Hecla,* 1821-23; 1824-25.

RICHARDS, Charles. Clerk, *Assistance,* 1850-51.

RICHARDS, George Henry (1820-1896). Commander RN, *Assistance,* 1852-54. [Dawson, L.S., 1885, Part 2, pp. 134-39; GJ 1897; Royal Society 1897; EA].

RICHARDS, George M. Geologist, U.S. exploring expedition, 1905-06.

RICHARDS, Henry. AB, *Resolute,* 1852-54.

RICHARDS, John. Captain, HBC supply voyages. *Seahorse* [II], 1783; *Seahorse* [III], 1784; 1785; 1786; 1787; *Beaver,* 1788; *Seahorse* [III], 1789; *King George* [III], 1790; 1791; 1792; 1793; 1795; 1796; 1797; 1798; 1799; 1800.

RICHARDS, John. Gunroom Steward, *Plover,* 1848-52; 1852-54.

RICHARDS, Joseph (-1782). Captain, HBC supply voyages. *Seahorse* [II], 1764; *Prince Rupert* [III], 1765; 1766; 1767; 1768; *Prince Rupert* [IV], 1769; 1770; 1771; 1772; 1773; 1774; 1775; 1776; 1777; 1778; 1779; 1780; 1781; *Seahorse* [III], 1782. (He died during this last voyage).

RICHARDS, William. AB, *Assistance,* 1850-51.

RICHARDS, William H. Clerk in Charge, *Resolute,* 1852-54.

RICHARDSON, James. AB, *Hecla,* 1819-20; Leading Seaman, *Fury,* 1821-23.

RICHARDSON, James (or John?). Gunner, *Blossom,* 1825-28.

RICHARDSON, John. Workman, second founding of Churchill, 1717.

RICHARDSON, John. Boatswain's Mate, *Discovery,* 1776-80.

RICHARDSON, John. Carpenter's Crew, *North Star,* 1849-50.

RICHARDSON, *Sir* John (1787-1865). Surgeon, Franklin's expeditions, 1819-22; 1825-27; leader, Franklin search expedition, 1847-49. [Murchison, 1861a; DNB Vol. 48; DCB Vol. 9; McIlraith, 1868; Johnson, 1976].

RICHARDSON, Thomas. AB, *Blossom,* 1825-28.

RICHARDSON, William. Acting Carpenter, *Arctic,* 1855.

RICHE, Corneyles. (?), Frobisher's expedi-

tion, 1578. Died 27 September 1578.
RICKETTS, William. AB, *Pandora,* 1876.
RICKMAN, John. Second Lieutenant, *Discovery,* 1776-80.
RICKMAN, William. AB, *Plover,* 1852-54.
RICKS, C.H.D. Captain, *Diana,* HBC supply voyages, 1882; 1883-84; 1884-85; 1885; 1886; 1887.
RIDDELL, R.B. Assistant, Keele's GSC expedition, 1907-08.
RIDLEY, Francis. Musician, *Sunneshine,* 1585.
RIDLEY, Nathaniel. Cook, *Discovery,* 1791-95.
RIGDEN, James. Captain's Coxswain, *Erebus,* 1845-48.
RIGG, William. Blacksmith, *North Star,* 1852-54.
RIGGS, James Halcan. AB, *Herald,* 1848; 1849; 1850.
RIGGS, William. Blacksmith, *Plover,* 1848-52.
RILEY, George. Assistant Steward, *Advance,* 1853-55.
RIOU, Edward (c1758-1801). Midshipman, *Discovery,* 1776-80. [DNB Vol. 48].
RISTVEDT, Peder (1873-). Meteorologist and First Engineer, *Gjøa,* 1903-06. [EA].
ROBERT, Absolmo. *See* GOLDING, Robert.
ROBERTS, David. Ordinary Seaman, *Blossom,* 1825-28.
ROBERTS, Edward. AB, *Discovery,* 1791-95.
ROBERTS, Henry (c1747-1796). Master's Mate, *Resolution,* 1776-80.
ROBERTS, James. Private RM, *Fury,* 1821-23.
ROBERTS, John. AB, *Blossom,* 1825-28.
ROBERTS, Thomas. AB, *Resolution,* 1776-80. Died at sea 27 January 1778.
ROBERTS, W.F. Boat-steerer, *George Henry,* 1860-62.
ROBERTSON, Alexander. Steward, *Lady Franklin,* 1850-51.
ROBERTSON, Andrew. *See* LINDSAY, Robert.
ROBERTSON, Colin (c1779-1842). Leader, HBC expedition, 1818-19. [Rich, 1938, pp. 461-63].
ROBERTSON, Edward. *See* LINDSAY, Robert.
ROBERTSON, Frederick. Boatswain's Mate, *Investigator,* 1848-49.
ROBERTSON, J. Seaman, *Laddie,* 1913-16.
ROBERTSON, James. Carpenter, *Felix,* 1850-51.
ROBERTSON, John. Surgeon, *Enterprise,* 1848-49.
ROBERTSON, W. Cook, *Laddie,* 1913-16.
ROBERTSON, William. Lieutenant, *Isabella,* 1818. [O'B(1)].
ROBERVAL, Jean François de la Rocque de. *See* LA ROCQUE DE ROBERVAL.
ROBINSON, Edward. (?), *Ayde,* 1578.
ROBINSON, Frederick. Lieutenant, *Investigator,* 1848-49. [O'B(1); Dawson, L.S., 1885, Part 2, pp. 84-85].
ROBINSON, Hercules (1789-1864). Captain, *Favorite,* 1820. [DNB Vol. 49; O'B(1)].
ROBINSON, James. AB, *Chatham,* 1791-95.
ROBINSON, James. Captain's Coxswain, *Resolute,* 1850-51; Captain of the Forecastle, *Assistance,* 1852-54.
ROBINSON, Moses. Boatswain, *Lady Franklin,* 1850-51.
ROBINSON, Peter. Sailor, *Judith,* 1578.
ROBINSON, Rehobeth. AB, *Pioneer,* 1852-54.
ROBINSON, Robert. Leader, HBC trading voyage, 1792.
ROBINSON, Thomas. Captain, *Charlotte,* 1771; 1772.
ROBINSON, William. Leader, HBC trading voyage, 1791.
ROBISON, Andrew. AB, *Sophia,* 1850-51.
ROBITAILLE, A. Second Steward, *Arctic,* 1908-09.
ROBSON, D. Fireman, *Arctic,* 1908-09.
ROCHE, Richard. Midshipman, *Herald,* 1848; 1849; 1850; Mate, *Resolute,* 1852-54.
ROCHER, Joseph. Assistant, Taylor's HBC expedition, 1831.
RODD, William. AB, *Herald,* 1848; 1849; 1850.
RODES, R. AB, *Hecla,* 1824-25.
RODGERS, James. AB, *Assistance,* 1850-51.
RODGERS, John. AB, *Plover,* 1848-52.
ROGER, Joseph. Boatswain's Mate, *Phoenix,* 1854.

ROGERS, . Midshipman, *Gannet,* 1867.
ROGERS, B.F. Assistant Surgeon, *Juniata,* 1873.
ROGERS, Frank. First Officer, *George Henry,* 1860-62.
ROGERS, Henry. AB, *Herald,* 1848; 1849; 1850.
ROGERS, John. Quartermaster's Mate, *Chatham,* 1791-95.
ROGERS, John. AB, *Discovery,* 1791-95.
ROGERS, Richard. AB, *Herald,* 1848; 1849; 1850.
ROGERS, Samuel. Private RM, *Resolute,* 1850-51; 1852-54.
ROI, Joseph. *See* ROY.
ROLFE, Samuel. *See* RELFE.
ROLLINGS, James W. Private RM, *Phoenix,* 1853.
ROMAINE, A. Fireman, *Neptune,* 1903-04.
RONALD, John. Stoker, *Phoenix,* 1853.
RONQUILLO, Philip de (Apocryphal character). Commander, *King Philip,* 1640.
ROOK, T.J. AB, *Plover,* 1848-52; 1852-54.
ROOME, John. AB, *Discovery,* 1791-95.
ROOT, John. AB, *Plover,* 1848-52.
ROPER, George. Soldier, Frobisher's expedition, 1578.
ROSE, J.W. AB, *Phoenix,* 1853.
ROSE, John. Private of Marines, *Chatham,* 1791-95.
ROSE, Joseph. (?), *Nancy Dawson,* 1849.
ROSEWELL, George. AB, *Chatham,* 1791-95.
ROSS, Alexander H.D. Assistant, Low's GSC expedition, 1892-96.
ROSS, David. Bombardier, *Resolute,* 1852-54.
ROSS, James Clark (1800-1862). Midshipman, *Isabella,* 1818; *Hecla,* 1819-20; *Fury,* 1821-23; Lieutenant, *Fury,* 1824-25; Commander, RN, *Victory,* 1829-33; Captain, *Enterprise,* 1848-49. [DNB Vol. 49; DCB Vol. 9; Dodge, 1973].
ROSS, John. Joint leader, Gregory, McLeod and Co. expedition, 1785-87.
ROSS, John (1777-1856). Commander, RN, and leader of expedition, *Isabella,* 1818; Captain, *Victory,* 1829-33; Captain, *Felix,* 1850-51. [DNB Vol. 49; Dodge, 1973].
ROSS, John. Artilleryman, Back's expedition, 1833-35.
ROSS, M. Purser, *Neptune,* 1903-04.
ROSS, Malcolm. Assistant, HBC expedition, 1790-92; established Fairford House, 1795.
ROSS, Melville George Hope Warrender. Lieutenant, *Investigator,* 1848-49. [O'B(1)].
ROSS, Richard A. Quartermaster, *Investigator,* 1850-54.
ROSS, William. Boatswain, *Arctic,* 1906-07.
ROSTAING, de. (?), *Sceptre,* 1782.
ROTH, John. Private, Company 1, 21st Infantry, Schwatka's expedition, 1883.
ROTHERY, David. AB, *Hecla,* 1824-25.
ROUFFIGNOE (?), A. AB, *Rattlesnake,* 1853-54.
ROURKE, Jeremiah. Leading Stoker, *Discovery,* 1875-76.
ROUSSEAU, J. Boy, *Neptune,* 1903-04.
ROUTLEDGE, W.H. Inspector NWMP, patrol, 1897 (see 1897 Jarvis).
ROUVIÈRE, Jean-Baptiste, OMI. RC missionary expedition, 1911-13.
ROWE, William. AB, *Hecla,* 1821-23.
ROWLAND, Thomas. Cook, *Phoenix,* 1853.
ROWLAND, William. Fisherman, Back's expedition, 1833-35.
ROWLAND, William. Assistant Surgeon, *Hecla,* 1824-25.
ROWLANDS, Thomas. Cook, *Rattlesnake,* 1853-54.
ROWLEY, William. AB, *Fury,* 1821-23.
ROXBOROUGH, Robert. Ordinary Seaman, *Herald,* 1848; 1849; 1850; *Plover,* 1852-54.
ROY [ROI], Joseph (c1807-). (?), Rae's expedition, 1850-51. [Rich, 1953, p. 374].
ROYAL, Robert. Captain, HBC supply voyages. *Montcalm,* 1829; 1830-31; 1831; *Ganymede,* 1834; *Prince of Wales* [I], 1835; 1838; 1839; 1840; 1841; *Prince Albert,* 1842; 1843; 1844; 1845; 1846; 1847; 1848; 1849; 1850; 1851; 1852; 1853; *Prince Arthur,* 1854; 1855; 1856.
RUDALL, James T. Assistant Surgeon, *Phoenix,* 1854.
RUDDOCK, John. AB, *Hecla,* 1821-23; Leading Seaman, *Fury,* 1824-25.
RUDOLPH, Charles. Seaman, Hall's expedition, 1864-69 [1864 only].

RUMBLE, Thomas. AB, *Assistance,* 1850-51.

RUPP, George H. Assistant, Geological expedition, 1920.

RUSSEL, John. Musician, *Sunneshine,* 1585.

RUSSEL, William. Sailor (?), *Sunneshine,* 1585.

RUSSELL, . "Exposer," *Charles,* 1631.

RUSSELL, Frank (1868-1903). Leader, University of Iowa expedition, 1893-94. [Nutting, 1903].

RUSSELL, W.B. Seaman, *George Henry,* 1860-62.

RUT, John. Leader, British voyage, 1527. [DCB Vol. 1].

RUTTER, James J. Clerk in Charge, *North Star,* 1849-50.

RUTTER, William. AB, *Isabella,* 1818.

RYAN, James. Private, Company H, 2nd Cavalry, U.S. Polar Year Expedition, 1881-84 [1881 only].

RYAN, James. Seaman, *Arctic,* 1906-07.

RYAN, Michael. Carpenter, *Neptune,* 1903-04; *Arctic,* 1906-07.

RYCRAFT, John. AB, *Chatham,* 1791-95.

RYDER, J.N. (-1864). First Engineer, *Intrepid,* 1850-51. [MAR].

RYDER, John. AB, *Phoenix,* 1853.

SABESTON [SABISTON, SABESTER], George. Acting Second Master and Icemaster, *North Star,* 1849-50; Icemaster, *Breadalbane,* 1853.

SABINE, Edward (1788-1883). Captain, RA and astronomer, *Isabella,* 1818; *Hecla,* 1819-20. [DNB Vol. 50; Georgi, 1958; EA].

SABISTON, William. Labourer, Richardson's expedition, 1847-49 [1847-48 only]. [Rich, 1953, p. 375].

SACHEUSE, John. (Eskimo). Interpreter, *Isabella,* 1818. Died in Edinburgh, 14 February 1819.

SAGGERS, John S. AB, *Discovery,* 1875-76.

SAIL, Josiah. Sailmaker, *Assistance,* 1850-51.

SAINSBURY, H.H. Mate, *Investigator,* 1850-54. Died 14 November 1853.

SAINSBURY, S. (?), *Laddie,* 1913-16.

ST CROIX, Thomas. Caulker, *Resolute,* 1852-54.

ST GEORGE, Jerome (*dit* Laporte) (c1820-1890). (?), Pullen's expedition, 1849-51 [1850 only]. [Rich, 1953, p. 375].

ST GERMAIN, Jacques (c1819-). Middleman, Rae's expedition, 1846-47. [Rich, 1953, pp. 375-76].

ST GERMAIN, Pierre. Interpreter, Franklin's expedition, 1819-20. [Rich, 1938, p. 464].

ST MICHEL, Louis (c1829-). (?), Rae's expedition, 1853-54. [Rich, 1953, p. 376].

ST ONGE, Philip. Guide, Carnegie Museum expedition, 1917.

ST SIMON, Paul Denis Sieur de. Accompanied Albanel, 1671-72.

SAINVILLE, Edouard de. Leader, French expedition, 1889-94.

SAIT, Henry. AB, *Terror,* 1845-48.

SALAVAN, Jeremiah. *See* GALAVAN.

SALMON, John. AB, *Enterprise,* 1848-49; *Intrepid,* 1850-51; 1852-54.

SALMON, William. Icemaster, *North Star,* 1852-54.

SALMON, William. AB, *Resolution,* 1776-80.

SALMONS, William. Captain of the Maintop, *Plover,* 1848-52; 1852-54.

SALOR, Nicholas. Corporal, Company H, 2nd Cavalry, U.S. Polar Year Expedition, 1881-84. Died 3 June 1884.

SALT, Richard. Miner, Frobisher's expedition, 1578.

SALTER, J.W. Ordinary Seaman, *Herald,* 1848; 1849; 1850.

SAMANDRE, François. Voyageur, Franklin's expedition, 1819-22.

SAMPSON, C.G. Missionary journey, 1896-1900 [see Peck, 1894-96, 1897-99, 1900-02].

SAMPSON, John. Sailor, *Ayde,* 1578.

SAMPSON, William. Sailmaker, *Blossom,* 1825-28.

SAMUEL, Alexander. Boatswain, *Sophia,* 1850-51.

SAMWELL, David (1751-1798). Surgeon's 2nd Mate, *Resolution,* 1776-80. [DNB Vol. 50].

SANDBERG, August. Geologist, Geological expedition, 1911-12.

SANDERS, John. Captain of the Forecastle, *Plover,* 1852-54.

SANDFORD, Ezbon (c1645-1682). Captain, *Albemarle,* 1682-83. [Rich, 1946, pp.

333-34; DCB, Vol. 1].
SANDLE, Osmond. AB, *Isabella*, 1818.
SANE, Anthonie. Sailor, *Ayde*, 1578.
SANFORD, Ezbon. *See* SANDFORD.
SANGSTER, Thomas. AB, *Lady Franklin*, 1850-51.
SARAH, George R. Ship's Steward, *Discovery*, 1875-76.
SARGENT, Henry. AB, *Enterprise*, 1850-55.
SARGENT, Robert Orme. Mate, *Erebus*, 1845-48. [O'B(1)].
SAUNDERES [SAUNDERS], William. Sailor, *Ayde*, 1578.
SAUNDERS, Edward. AB, *North Star*, 1852-54.
SAUNDERS, George. AB, *Rattlesnake*, 1853-54.
SAUNDERS, James. Acting Master, *Terror*, 1836-37; Master commanding *North Star*, 1849-50.
SAUNDERS, James. Private RM, *Investigator*, 1850-54.
SAUNDERS, John. Boatswain's Mate, *Plover*, 1852-54.
SAUNDERS, Thomas. Corporal RM, *Enterprise*, 1848-49.
SAUNDERSON, John. Canoeman, Bell's expedition, 1899-1900.
SAVAGE, Thomas. (?), First founding of Churchill, 1689.
SAVAGE, William. AB, *Resolute*, 1852-54.
SAWNDERS, John. Sailor, *Ayde*, 1578.
SAWYER, Henry. Blacksmith, *Plover*, 1848-52.
SAXON, Thomas. Corporal RM, *Blossom*, 1825-28.
SCALLON, James. Gunner, *Hecla*, 1819-20; *Fury*, 1821-23.
SCARLETT, John. AB, *Intrepid*, 1850-51.
SCHEI, Per Elisaeus (1875-1905). Geologist, *Fram*, 1898-1902. [EA].
SCHLEY, Winfield Scott (1839-1911). Commander, USN, commanding *Thetis*, 1884. [EA].
SCHLEIPHAKE, H. Surgeon and naturalist, German Polar Year Expedition, 1882-83.
SCHLOEZER [SCHOLZER], Andrew. Missionary, Moravian expedition, 1765.
SCHNEIDER, John. Missionary, Moravian expedition, 1771.
SCHNEIDER, Roderick Robert. Private, 1st Artillery, U.S. Polar Year Expedition, 1881-84. Died 18 June 1884.
SCHUBERT, Peter. Cook, *Advance*, 1853-55. Died about 12 May 1854.
SCHUMANN, Emil. Chief Engineer, *Polaris*, 1871-73. [EA].
SCHUPE, James. Sailor and Cook, Low's expedition, 1898-99.
SCHWATKA, Frederick (1849-1892). Lieutenant commanding U.S. expeditions 1878-80; 1883. [DAB; EA].
SCOLVUS, John. Voyage to America, c1476. [DCB Vol. 1].
SCOONER, George. Ordinary Seaman, *Herald*, 1848; 1849; 1850.
SCOTT, ("Lucky"). Captain, *Neptune*, 1912.
SCOTT, Banks. Second Engineer, *Roosevelt*, 1908-09.
SCOTT, James Wood. Midshipman, *Chatham*, 1791-95.
SCOTT, Robert. AB, *Terror*, 1836-37.
SCOTT, Robert. Leading Stoker, *Fox*, 1857-59. Died 4 December 1857.
SCOTT, Robert Charles. Assistant Surgeon, *Intrepid*, 1852-54.
SCOTT, Samuel. Private of Marines, *Discovery*, 1791-95.
SCOTT, William. Boatswain's Mate, *Hecla*, 1819-20. Died 30 June 1820.
SCOTT, William. Officers' Cook, *Enterprise*, 1850-55.
SCRATTI [SRATTI], John. Carpenter, *Plover*, 1848-52.
SCRIVEN, Alfred. Ordinary Seaman, *Herald*, 1848; 1849; 1850.
SCRIVENER [SCRIBNER], Benjamin. AB, *Griper*, 1819-20.
SCROGGS, John. Captain, *Whalebone*, 1721-22. [DCB Vol. 1].
SCRUSE, William. Private of Marines, *Resolution*, 1776-80.
SEARLE, Samuel. Private RM, *Plover*, 1848-52; 1852-54.
SEARLE, William. AB, *Phoenix*, 1853; Assistant Master, *Phoenix*, 1854.
SEATH, John. Acting Carpenter, *Plover*, 1848-52 [1850-52 only]; 1852-54.

SEBREE, Uriel S. Lieutenant USN, *Tigress*, 1873; *Thetis*, 1884. [NE].

SEE-GLOO. (Eskimo). Reached North Pole with Peary, 1909. [EA].

SEELEY, Abraham. AB, *Erebus*, 1845-48.

SEELY, William. (?), Frobisher's expedition, 1578.

SEEMAN, Berthold Carl (1825-1871). Naturalist, *Herald*, 1848; 1849; 1850. [Rawlinson, 1872; DNB Vol. 51].

SEEMAN, C. Mechanic and assistant, German Polar Year Expedition, 1882-83.

SELF, James. AB, *Alert*, 1875-76.

SEGUIN, Jean (1833-1902). RC missionary journeys, 1862-63; 1868.

SEGUIN, Olivier. (?), Back's expedition, 1833-35.

SELBY, Henry. AB, *Enterprise*, 1850-55.

SELLER, L.E. Constable RNWMP, patrol to Lyon Inlet, 1906 [see 1903, Moodie].

SELLMAN, Edward. Notary, *Ayde*, 1578.

SELWYN, Alfred Richard Cecil (1824-1902). Leader, GSC expedition, 1875. [*Ottawa Naturalist*, 1902].

SENIOR, John. Captain of the Foretop, *Plover*, 1848-52; 1852-54.

SENNETT, James. Captain, *Corea*, 1861; *Prince of Wales* [II], 1864-65.

SERGEANT, J. AB, *Phoenix*, 1854.

SETON, Ernest Thompson (1860-1946). Joint leader, U.S. expedition, 1907. [EA].

SETTELL, Jamees. Gentleman, Frobisher's expedition, 1578.

SEVERIGHT, Alexander. *See* SIVEWRIGHT.

SEYMOUR, William. AB, *Plover*, 1848-52; 1852-54.

SEYMOUR, William. Second Officer, *Polar Bear*, Canadian Arctic Expedition, 1913-18.

SHANKS, William. AB, *Terror*, 1845-48.

SHANNON, Anthony. (?), *Polar Bear*, Canadian Arctic Expedition, 1913-18.

SHANNON, J.L. (?), *Nancy Dawson*, 1849.

SHARP, Peter. AB, *Fury*, 1821-23.

SHARP, William. AB, *North Star*, 1849-50. Died 1 November 1849.

SHARPE, Philip Roffle. Mate, *Rattlesnake*, 1853-54.

SHAW, Alfred Victor. Joint leader, U.S. expedition, 1897.

SHAW, Angus. Agent, North West Company invasion, 1803-06.

SHAW, Gilbert R. Observer in charge of Port Burwell station, Hudson Bay Expedition, 1885-86.

SHAW, Jeremiah. Private RM, *Intrepid*, 1852-54.

SHAW, Thomas. Gunner's Mate, *Discovery*, 1776-80.

SHAW, W. Neil. Captain, HBC supply voyages. *Labrador*, 1886; 1887; *Erik*, 1888; 1889; 1890-91.

SHAW, William. AB, *North Star*, 1852-54.

SHEA, John. AB, *Isabella*, 1818.

SHEA, John. AB, *Herald*, 1848; 1849; 1850.

SHEATH, Thomas. AB, *Discovery*, 1776-80.

SHEDDEN, Robert (c1819-1849). Captain, *Nancy Dawson*, 1849. Died at Matlazan, November 1849. [Jones, A.G.E., 1958].

SHEEN, Thomas. Carpenter's Assistant, *Phoenix*, 1854.

SHEENRY (?), J. AB, *Hecla*, 1824-25.

SHEERING, J. AB, *Fury*, 1821-23.

SHELLABEAR [SHELLABEER], William B. Master's assistant, *Enterprise*, 1848-49; Second Master, *Intrepid*, 1850-51; Master, *North Star*, 1852-54.

SHELTON, Joseph. Cook, *Pandora*, 1875.

SHEPARD [SHEPHEARD], Thomas. Captain, *Shaftesbury*, 1674-75. [DCB Vol. 1].

SHEPHERD, Robert. Captain of the Forecastle, *Plover*, 1848-52; 1852-54.

SHERER, Joseph (1798-). Midshipman, *Hecla*, 1821-23; Lieutenant, *Hecla*, 1824-25. [O'B(1); MAR].

SHERIFF, John. Master's Mate, *Chatham*, 1791-95.

SHERLOCK, George. Stoker, *Phoenix*, 1854.

SHERMAN, Orray Taft (1856-1921). Scientific Staff, *Florence*, 1877-78.

SHERRIS, Richard. Captain, *Superior*, 1852.

SHIELDS, George. AB, *Blossom*. 1825-28.

SHIELLS [SHIELS], Matthew. Carpenter, *Sophia*, 1850-51.

SHINGLETON, James. AB, *Resolute*, 1850-51.

SHINGLETON, Richard. Gunroom Steward, *Investigator*, 1848-49; *Enterprise*, 1850-55; Officers' Steward, *Fox*, 1857-59.

SHIPLEY, Conway Mordaunt (1824-). Captain, RN, supernumary, *Bulldog*, 1860. [O'B(1)].

SHIRCLIFF, William H. Corporal, Company G, Second Infantry, Schwatka's expedition, 1883.

SHIRDY, Robert. Ordinary Seaman, *Blossom*, 1825-28.

SHIRLEY, John. Stoker, *Alert*, 1875-76.

SHOOD, Robert. AB, *Phoenix*, 1853.

SHORT, Thomas. Ordinary Seaman, *Herald*, 1848; 1849; 1850.

SHOVE, John. Carpenter's crew, *Blossom*, 1825-28.

SHREEVE, Robert. Carpenter's Mate, *Victory*, 1829-33.

SHUTTLEWORTH, William. Midshipman, *Resolution*, 1886-80.

SIEMENS, Hermann. Seaman, *Polaris*, 1871-73.

SIGGERS, Henry. AB, *Fury*, 1821-23.

SILK, William. AB, *Blossom*, 1825-28.

SILVA, F. Seaman, *George Henry*, 1860-62.

SILVA, M. Seaman, *George Henry*, 1860-62.

SILVERLOCK, William. AB, *Breadalbane*, 1853.

SILVEY, John [TILVEY, William]. Ice Quartermaster, *Resolute*, 1852-54.

SILVY, Antoine (1638-1711). Chaplain, Jolliet's expedition, 1679; de la Martinière's expedition, 1684-85; de Troyes' expedition, 1686. [DCB Vol. 2].

SIMMONDS [SIMMETT, SINNETT], John. AB, *Pioneer*, 1852-54.

SIMMONDS, John. Boatswain's Mate, *Fox*, 1857-59.

SIMMONDS, William. Cook, *Assistance*, 1850-51.

SIMMONS, Herman Georg (1866-1943). Botanist, *Fram*, 1898-1902. [EA].

SIMMONS, John. Second Captain of the Maintop, *Alert*, 1875-76.

SIMMONS, Thomas. Captain of the Forecastle, *Discovery*, 1875-76.

SIMMS, C. Lieutenant commanding *Arctic*, 1855.

SIMMS, John. Seaman, *Arctic*, 1908-09.

SIMMS, Robert. AB, *Terror*, 1836-37.

SIMONS, Joseph. Captain of the Maintop, *Herald*, 1848; 1849; 1850.

SIMPKINS, Daniel. Sailmaker's crew, *Herald*, 1848; 1849; 1850.

SIMPKINS, William. AB, *Enterprise*, 1850-55.

SIMPSON, Daniel. Captain of the Hold, *Herald*, 1848; 1849; 1850.

SIMPSON, George. Corporal, *Discovery*, 1791-95.

SIMPSON, George (1792-1860). Arrival at Lake Athabasca, 1820-21. [Rich, 1938, pp. 466-67; Murchison, 1861; Morton, A.S., 1944; Chalmers, 1960; Galbraith, 1976].

SIMPSON, George. Captain, *Traveller*, 1833.

SIMPSON, J.A. Customs House Clerk, *Arctic*, 1906-07.

SIMPSON, John. Captain, *Yonge*, 1688-89.

SIMPSON, John. Private of Marines, *Discovery*, 1791-95.

SIMPSON, John (-1858). Assistant Surgeon, *Plover*, 1848-52. [RGS 1860; MAR].

SIMPSON, Peter. Boatswain's Mate, *Assistance*. 1850-51.

SIMPSON, Thomas (1808-1840). Joint leader, HBC expedition, 1837-39. [Fleming, 1940, pp. 455-56; Simpson, A., 1845; McKay and Lamb, 1938].

SIMPSON, Thomas H. AB, *Alert*, 1875-76.

SIMPSON, William. Sergeant RM, *Plover*, 1848-52.

SIMS, David. AB, *Terror*, 1845-48.

SINCLAIR, George. Steersman, Back's expedition, 1833-35; Dease and Simpson's expedition, 1837-39.

SINCLAIR, J. AB, *Griper*, 1819-20.

SINCLAIR, Robert. Captain of the Foretop, *Erebus*, 1845-48.

SINCLAIR, Samuel. (?), Rae's expedition, 1850-51. [Rich, 1953, p. 376].

SINCLAIR, William. AB, *Terror*, 1845-48.

SINGLETON, John. AB, *Alexander*, 1818.

SINNETT, James (John). *See* SIMMONDS, John.

SISSON, William. First Officer, *Florence*, 1877-78.

SIVEWRIGHT [SEVERIGHT], Alexander (c1795-c1860). First Mate, *Felix*, 1850-

51. [Great Britain, Parliament, 1850a, p. 114].

SIVEWRIGHT, James. Carpenter's Mate, *Felix*, 1850-51.

SIVEWRIGHT, Thomas. AB, *Felix*, 1850-51.

SIVEWRIGHT, William. AB, *Felix*, 1850-51.

SKEAD, Francis. Second Master, *Enterprise*, 1850-55. [Heaton, 1975].

SKEANS, Patrick. Fireman, *Roosevelt*, 1908-09.

SKENE, Alexander Motz. Midshipman, *Isabella*, 1818; *Griper*, 1819-20. [O'B(1)].

SKEOCH, James. Assistant Surgeon, *Fury*, 1821-23.

SKINN, Robert. Ordinary Seaman, *Blossom*, 1825-28.

SKINNER, J. Captain of the Maintop, *Rattlesnake*, 1853-54.

SKIRVING, G.M. Acting Corporal, NWMP, patrol, 1899.

SKOLVSSON, John. *See* SCOLVUS.

SKOOKUM JIM. (Indian). Discovered Klondike gold with Carmack, 1896.

SKUDENES, Hans. Seaman, *Enhiörningen*, 1619-20. Died at Churchill, 8 March 1620.

SKUDENES, Knud Lauritzsen. Seaman, *Enhiörningen*, 1619-20. Died at Churchill May or June 1620.

SKYNNER, . Observer in charge of Nachvak station, Hudson Bay Expedition, 1884-85.

SKYPWITH, Lionel. Intending colonist, Frobisher's expedition, 1578.

SLAKER, George. Quartermaster, *Plover*, 1848 52; 1852-54.

SLATER, J. Private RM, *Phoenix*, 1853.

SLAVE WOMAN, The (-1717). (Chipewyan Indian). Guide and interpreter to Stewart, 1715-16. [Davies, 1965, pp. 410-13; DCB Vol. 2].

SLESSAR [SLESSOR], Robert. AB, *Assistance*, 1850-51.

SMALL, Jonathan Cook (1876-). Second-in-command, MacMillan's expedition, 1912; mechanic and cook, "Crocker Land" Expedition, 1913-17.

SMALLPIECE, John. AB, *Discovery*, 1776-80.

SMELLIE, Thomas Farrar (1881-1963). Captain, *Nascopie*, 1918. [Wild, 1955; *Beaver*, 1945].

SMITH, A. Cleveland. Captain, HBC supply voyages. *Pelican*, 1909; 1910; 1911; *Nascopie*, 1912; 1913; 1914.

SMITH, Alexander. AB, *Sophia*, 1850-51.

SMITH, Benjamin. Seaman, *Rescue*, 1850-51.

SMITH, Boreas A. AB, *Lady Franklin*, 1850-51.

SMITH, Charles. AB, *Assistance*, 1852-54.

SMITH, Davey. Speksioneer, *Arctic*, 1873.

SMITH, Edward. Carpenter's crew, *Blossom*, 1825-28.

SMITH, Francis. Captain, *Churchill*, 1738; 1739; 1740; 1742; 1743; 1744; *California*, 1746-47. [Rich, 1949, pp. 336-37; DCB Vol. 3].

SMITH, Gideon. Blacksmith, *Prince Albert*, 1851-52.

SMITH, H. Seaman, *George Henry*, 1860-62.

SMITH, Henry. Bowman and fisherman, Richardson's expedition, 1847-49. [Rich, 1953, p. 376].

SMITH, Henry. AB, *Resolute*, 1852-54.

SMITH, Henry. AB, *Phoenix*, 1854.

SMITH, J. AB, *Hecla*, 1819-20.

SMITH, J. Private RM, *Hecla*, 1819-20.

SMITH, J.F. AB, *Phoenix*, 1854.

SMITH, J.W. Boatswain, *North Star*, 1849-50.

SMITH, James. Seaman, *Advance*, 1850-51.

SMITH, John. AB, *Isabella,* 1818.

SMITH, John. Artilleryman, *Hecla*, 1819-20.

SMITH, John. AB, *Blossom*, 1825-28.

SMITH, John. Carpenter, *Terror*, 1836-37.

SMITH, John. Ordinary Seaman, *Herald*, 1848; 1849; 1850.

SMITH, John. Steward, *Prince Albert*, 1850; Clerk in Charge, *Prince Albert*, 1851-52.

SMITH, John. Quartermaster, *Release*, 1855.

SMITH, John E. Sailmaker, *Discovery*, 1875-76.

SMITH, Joseph. Joint leader, HBC expeditions, 1756-57; 1757-58; 1759-60; 1763-64.

SMITH, Joseph. Seaman, *Resolution*, 1776-

80.

SMITH, Luke. Stoker, *Terror,* 1845-48.

SMITH, Malcolm. (?), Back's expedition, 1833-35.

SMITH, Robert. AB, *Isabella,* 1818.

SMITH, Robert. Steward, *Terror,* 1836-37.

SMITH, Robert. (?), *Nancy Dawson,* 1849.

SMITH, Robert. Fourth Officer, *George Henry,* 1860-62.

SMITH, Thomas. (?), *Griper,* 1824.

SMITH, Thomas. AB, *Alert,* 1875-76.

SMITH, Watson. Acting Master, *Arctic,* 1855.

SMITH, William. Boatswain, *Fury,* 1821-23; *Hecla,* 1824-25.

SMITH, William. Sailmaker, *Hecla,* 1821-23; *Fury,* 1824-25.

SMITH, William. Blacksmith, *Erebus,* 1845-48.

SMITH, William. Purser's Steward, *Enterprise,* 1850-55.

SMITH, William. Quartermaster, *Enterprise,* 1850-55.

SMITH, William. AB, *Phoenix,* 1854.

SMITH, William. Boatswain's Mate, *Release,* 1855.

SMITH, William. Blacksmith, Curran's expedition, 1912.

SMITHERS, George. Captain's Coxswain, *Pandora,* 1876.

SMITHERS, Joseph. Stoker, *Intrepid,* 1852-54.

SMITHSEND, Nicholas. Captain, *Knight,* 1696; *Royal Hudson's Bay,* 1697. [DCB Vol. 1].

SMITHSEND, Richard. Captain, *Huband,* 1687-88. [DCB Vol. 1].

SMITHSON, William. AB, *Phoenix,* 1854.

SMITHTON, Edward. Intending colonist, Frobisher's expedition, 1578.

SMYTH, William. Master, *Gabriel,* 1577.

SMYTH, William (1800-1877). Mate, *Blossom,* 1825-28; Lieutenant, *Terror,* 1836-37. [O'B(1); Alcock, R., 1878; Aurousseau, 1946; BOA].

SMYTHE, J.D. Captain, HBC supply voyages. *Prince Arthur,* 1863-64; 1864-65.

SNAGG, James. Surgeon's First Mate, *Discovery,* 1776-80.

SNOOK, John. Boatswain, *Enterprise,* 1848-49.

SNOW, George. Private RM, *Enterprise,* 1848-49.

SNOW, Thomas. AB, *Alexander,* 1818.

SNOW, William Parker (1817-1895). Passenger, *Prince Albert,* 1850. [DNB Vol. 53].

SNYDER, A.E. Inspector, NWMP, patrol, 1897 [see 1897, Jarvis].

SOBY, William. Seaman, *Resolution,* 1776-80.

SODENS, Anders. Seaman, *Enhiörningen,* 1619-20. Died at Churchill, 8 April 1620.

SOLOMON, Jerry. Indian guide, Todd's expedition, 1908.

SOLON, John. Armourer, *Blossom,* 1825-28.

SONNTAG, August (1832-1860). Astronomer, *Advance,* 1853-55; Astronomer and second-in-command, *United States,* 1860-61. Died December 1860. [EA].

SOPP, William. Captain, *Neptune,* 1882.

SORRY, Alexander. AB, *Griper,* 1824.

SOTLE, Dennis. Gentleman, *Gabriel,* 1578.

SOULIERE [SOULIER], John. (Ojibwa Indian). Canoeman, Camsell's expedition, 1914.

SOULTER, Alexander. Boatswain, *Felix,* 1850-51.

SOUTER, William. AB, *Isabella,* 1818.

SOUTER, William. Quartermaster, *Fury,* 1821-23. Died 25 June 1822.

SPACKMAN, John. Sergeant RM, *Fury,* 1821-23.

SPARK, Sidney J. Paymaster, *Rattlesnake,* 1853-54.

SPARKS, John. AB, *Isabella,* 1818.

SPARROW, Anthonie. Quartermaster, *Ayde,* 1578. Died 31 July 1578.

SPARROW, Robert. AB, *Terror,* 1836-37.

SPEARER, Thomas. AB, *Discovery,* 1791-95.

SPEARMIN, John S. Base camp superintendent, Hall's expedition, 1864-69 [1868 only].

SPEED, J. AB, *Phoenix,* 1854.

SPEER, Denton H. Second Master, *Talbot,* 1854.

SPENCE, J. AB, *Blossom,* 1825-28.

SPENCE, James. (?), Back's expedition, 1833-35.

SPENCE, Robert. Middleman, Franklin's expedition, 1825-27.

SPENCER, Edward. AB, *Pioneer*, 1850-51.

SPENCER, Michael. Seaman, *Resolution*, 1776-80.

SPENCER, William. Private RM, *Blossom*, 1825-28.

SPEURR, Ralph. AB, *Blossom*, 1825-28.

SPICER, John O. Captain, *Era*, 1879-80.

SPILSBURY, William. Seaman, *Resolution*, 1776-80.

SPINKS, Robert. Coxswain, Franklin's expedition, 1825-27.

SPRACKLIN, H. Boatswain, *Laddie*, 1913-16.

SPRAGE, [SPRAGGE], Harry. Shipwright, *Michael*, 1578.

SPREAD, George. AB, *North Star*, 1852-54.

SPREADBOROUGH, W. Assistant, Low's expedition, 1892-96.

SPUREL, B. AB, *Hecla*, 1824-25.

SPURRELL, George. Captain, HBC supply voyages. *Hudson's Bay* [IV], 1722; 1723; 1724; 1725; 1726; 1727; *Mary* [III], 1728; 1729; 1730; 1731; 1732; 1733; 1734; 1735; 1736; *Seahorse* [I], 1737; 1738; 1739; 1740; 1741; 1742; 1743; *Prince Rupert* [II], 1744; 1745; 1746; 1747; 1748; 1749; 1750; 1751; 1752; 1753; 1754; 1755; 1756. [DCB Vol. 3].

SPURRELL, Joseph. Captain, HBC supply voyages. *Hudson's Bay* [V], 1751; *King George* [I], 1752; 1753; 1754; 1755; *Seahorse* [I], 1756; *Prince Rupert* [III], 1757; 1758; 1759; 1760; *King George* [II], 1761; 1762; 1763.

SRATTI, John. *See* SCRATTI.

STAFFE, Philip. Carpenter, *Discovery*, 1610-11.

STAFFORD, Edmund. Gentleman, *Ayde*, 1577; *Gabriel*, 1578.

STAFFORD, Henry. AB, *Phoenix*, 1854.

STAIRS, . Seaman, Richardson's expedition, 1847-49.

STAKER, George. AB, *Plover*, 1848-52 [1850-52 only]; 1852-54.

STALLARD, Isaac. Private RM, *Assistance*, 1852-54.

STANLEY, J. Carpenter's Mate, *Rattlesnake*, 1853-54.

STANLEY, James. Carpenter's crew, *Blossom*, 1825-28.

STANLEY, John. Soldier, Frobisher's expedition, 1578.

STANLEY, John. Seaman, *Resolution*, 1776-80.

STANLEY, Owen (1811-1850). Second Lieutenant, *Terror*, 1836-37. [Lubbock, A., 1968; O'B(1)].

STANLEY, Stephen Samuel. Surgeon. *Erebus*, 1845-48.

STANNARD, William. Captain, *Eaglet*, 1668-69; *Wivenhoe*, 1669. [EA].

STANTON, Leigh. Assistant, Wallace's expedition, 1905-06.

STANTON, William. Master, *Phoenix*, 1853.

STAPLES, William. AB, *Phoenix*, 1853.

STARES, George. Captain's Coxswain, *Assistance*, 1852-54.

STARR, Collin C. Master's Mate, *United States*, 1860-61.

STARR, Daniel C. Corporal, Company F, 2nd Cavalry, U.S. Polar Year Expedition, 1881-84 [1881 only].

STAUNTON, William. Gentleman, Frobisher's expedition, 1578.

STAWKER, George. House carpenter, Frobisher's expedition, 1578.

STEAL, George. AB, *North Star*, 1849-50.

STEEBING, R.H. Master-at-Arms, *Herald*, 1848; 1849; 1850.

STEEL, Charles. AB, *Investigator*, 1850-54.

STEELE, Daniel. Captain of the Maintop, *North Star*, 1852-54.

STEER, Samuel. Musician, *Herald*, 1848; 1849; 1850.

STEFANSSON, Vilhjalmur (1879-1962). Leader, anthropological expeditions, 1906-07; 1908-12; anthropologist and leader, Canadian Arctic Expedition, 1913-18. [Lebourdais, 1963; Barry, 1966; Stefansson, 1964; Sullivan, 1963; Hanson, 1941].

STEIN, Robert (1857-1917). Leader, Ellesmere Island expedition, 1899-1900.

STEINBRUCK, John. Voyageur, Mackenzie's expedition, 1789.

STEPHENS, Edward. Boatswain, *Blossom*, 1825-28.

STEPHENS, George. Captain of the Maintop, *Plover*, 1848-52 [1851-52 only].

STEPHENS, J.M. AB, *Phoenix,* 1853.
STEPHENS, Richard. Private of Marines, *Chatham,* 1791-95.
STEPHENS, Robert. Midshipman, *Discovery,* 1791-95.
STEPHENSON, Charles W. Master's assistant, *Plover,* 1848-52.
STEPHENSON, George. Seaman, *Advance,* 1853-55.
STEPHENSON, Henry Frederick (1842-1919). Captain, *Discovery,* 1875-76. [WWW Vol. 2; EA; MAR].
STERRY, William R. Blacksmith and cooper, *George Henry,* 1860-62.
STEVENS, Joseph. Captain, *Success,* 1767; 1768; *Charlotte,* 1769; 1770.
STEVENS, Peter. (Ojibwa Indian). Hunter and camp servant, Wallace's expedition, 1905-06.
STEVENS, T. Captain of the Foretop, *Enterprise,* 1848-49.
STEVENS, William. AB, *Discovery,* 1776-80.
STEVENSON, Edmund. Steersman, Richardson's expedition, 1847-49. [Rich, 1953, p. 376].
STEVENSON, W.G. Master's assistant, *Rattlesnake,* 1853-54.
STEWARD, Henry. AB, *Plover,* 1848-52; 1852-54.
STEWARD, Peter. Boatswain's Mate, *Fury,* 1824-25.
STEWART, Alexander. AB, *Hecla,* 1821-23.
STEWART, Alexander. Captain, *Sophia,* 1850-51.
STEWART, Archibald. Soldier, Franklin's expedition, 1825-27. Died 1827.
STEWART, David. Captain of the Foretop, *Discovery,* 1875-76.
STEWART, Elihu. Leader, overland expedition, 1906.
STEWART, George. Seaman, *Resolution,* 1776-80.
STEWART, Henry J. AB, *Plover,* 1852-54.
STEWART, J. (?), *Nancy Dawson,* 1849.
STEWART, James. Seaman, *Rescue,* 1850-51.
STEWART, James. Prospector, Curran's expedition, 1912.
STEWART, James Green (-1881). Second-in-command, HBC Franklin search, 1855. [MDC].
STEWART, John. AB, *Discovery,* 1791-95.
STEWART, John. Cooper at Eastmain, accompanied Atkinson, 1816.
STEWART [STUART], William (c1678-1719). Leader, HBC expedition, 1715-16; took part in second founding of Churchill, 1717. [Davies, 1965, pp. 413-17; DCB Vol. 2].
STICKLAND, John. AB, *Erebus,* 1845-48.
STIRLING, James. Commander, RN, commanding *Brazen,* 1813. [DNB Vol. 54; O'B(1)].
STIRLING, William. Seaman, *Resolution,* 1776-80.
STOBERN, . (?), Frobisher's expedition, 1578.
STOCKTON, Charles Herbert (1845-1924). Lieutenant Commander, USN, commanding *Thetis,* 1889. [NE].
STOEKS, James. AB, *Fury,* 1824-25.
STOKES, W. Seaman, *George Henry,* 1860-62.
STOLZ, Rudolf (1872-). Stoker, *Fram,* 1898-1902.
STONE, Andrew Jackson. (1859-). Leader, U.S. expedition, 1897-99.
STONE, George. Second Captain of the Foretop, *Discovery,* 1875-76.
STONE, Henry (-1912). Blacksmith, *Investigator,* 1850-54. [Scott Polar Research Institute, MS 692].
STOPFORD, Edward. Captain, *Rosamund,* 1814. [RNB].
STORER, J. AB, *Hecla,* 1821-23.
STOREY, Thomas. AB, *Phoenix,* 1854.
STORKERSON, Storker (1883-1940). (?), Canadian Arctic Expedition, 1913-18. [*Norsk Geografisk Tidsskrift,* 1946].
STOTT, William. Leading Stoker, *Phoenix,* 1853.
STREATE, Walter. Sailor, *Ayde,* 1578.
STREET, Thomas George. Assistant to Radford, 1911-12. Killed by Eskimos, June 1912.
STREMENSKI, Napoléon. Assistant Steward, *Arctic,* 1906-07.
STRICKLER, J. Leading Seaman, *Hecla,* 1821-23.
STRONG, G.H. Mate, *Plover,* 1848-52;

1852-54.

STRONG, Horace F. Mining engineer, Curran's expedition, 1912.

STRONG, William. AB, *Terror,* 1845-48.

STROUD, George. Stoker, *Phoenix,* 1854.

STROUD, J. AB, *Griper,* 1824.

STRURNS, Charles. AB, *Blossom,* 1825-28.

STRUTT, Benjamin. AB, *Resolute,* 1850-51.

STUART, A. AB, *Phoenix,* 1853.

STUART, *Hon.* C. AB, *Discovery,* 1791-95.

STUART, John (1779-1847). Established Rocky Mountain Portage House, 1805-06; joint leader, overland expedition, 1806-07; assistant, exploration of Fraser River, 1808. [Rich, 1938, p. 469].

STUART, John. Third Mate, *Lady Franklin,* 1850-51.

STUART, William. *See* STEWART.

STUBBERFIELD, Isaac. Cook, *Investigator,* 1850-54.

STUBBLE, Thomas. Sailor, *Michael,* 1578.

STUBBS, Thomas. AB, *Alert,* 1875-76.

STUBLEY, J.H. Stoker, *Phoenix,* 1853.

STUCKBERRY, Thomas. Captain of the Maintop, *Alert,* 1875-76.

STUPART, Robert Frederick (1857-1940). Observer in charge of Stupart Bay station, Hudson Bay Expedition, 1884-85.

STYGGE, Mauritz. Lieutenant, *Enhiörningen,* 1619-20. Died at Churchill, 10 April 1620.

SUCCI. Eskimo, *Polaris,* 1871-73.

SUGDEN, Henry. Sub-Officers' Steward, *Investigator,* 1850-54.

SUGRAE [SUGRA], Joseph. AB, *Plover,* 1848-52.

SULLIVAN, John. Captain of the Maintop, *Erebus,* 1845-48.

SULLIVAN, John F. Seaman, Boat voyage, 1860.

SULLIVAN, M. [*cf* GALAVAN, Jeremiah]. Captain of the Forecastle, *Resolute,* 1850-51.

SULLOCK, Robert. Private RM, *Plover,* 1848-52.

SULLOCK, Robert. (?), *North Star,* 1852-54.

SULLY, . AB, Richardson's expedition, 1847-49.

SULLY, Abraham. *See* SEELEY.

SULTER, John. Private, Sappers and Miners, Richardson's expedition, 1847-49.

SUMPTION (?), Robert. Cooper, *Blossom,* 1825-28.

SUNDMÖER, Olluff. Captain of the Hold's Mate, *Enhiörningen,* 1619-20. Died at Churchill, 24 April 1620.

SURREY [SURRY], Abraham. Cooper, *Resolute,* 1852-54.

SUTHERLAND, Donald. Sailmaker, *Sophia,* 1850-51.

SUTHERLAND, F.D. Constable, NWMP, establishment of Herschel Island detachment, 1903.

SUTHERLAND, James. Assistant, Livingston's expedition, 1799. Killed by Eskimos during the journey.

SUTHERLAND, Kenneth. Carpenter, *Prince Albert,* 1851-52.

SUTHERLAND, Peter Cormack (1822-1900). Surgeon, *Sophia,* 1850-51; *Isabel,* 1852. [Rosenthal, 1966, p. 369; Kock, 1968, Vol. 1, pp. 781-82].

SUTHERLAND, Thomas. AB, *Phoenix,* 1854.

SUTTON, Edward. AB, *Plover,* 1852-54.

SUTTON, George Miksch (1898-). Assistant, Carnegie Museum expedition, 1920.

SUTTON, Henry. Captain of the Maintop, *Rattlesnake,* 1853-54.

SVENDSEN, Johan (1866-1899). Doctor, *Fram,* 1898-1902. Died June 1899. [EA].

SVERDRUP, Otto Neumann (1854-1930). Commander of expedition, *Fram,* 1898-1902. [Isachsen, 1931; Dietrichsen, 1931; Ahlmann, 1931; GJ 1931].

SWAIN, Thomas (c1779-). Assistant, Fidler's expeditions, 1802-03; 1803-04; 1804-05. [Rich, 1938, p. 470].

SWAINE, Charles [*cf* DRAGE, Theodore Swaine]. Clerk, *California,* 1746-47; Captain, *Argo,* 1753; 1754. [Eavenson, 1949; EA].

SWAINE, Spelman. Master's Mate, *Discovery,* 1791-95.

SWAINSON, . Private RM, *Isabella,* 1818.

SWALLOW, Frederick. AB, *Plover,* 1848-52; 1852-54.

SWANEY, William. AB, *Resolute,* 1850-51.

SWANSEA, Jacob. Boatswain, *Hecla,* 1819-

20.

SWEENEY, Daniel. Master, *Alaska,* Canadian Arctic Expedition, 1913-18 [1914-16 only].

SWEENEY, Patrick. AB, *Herald,* 1848; 1849; 1850.

SWEENEY, Peter. Carpenter's Crew, *Herald,* 1848; 1849; 1850.

SWEENEY, William. AB, *Phoenix,* 1853.

SWEENY, John. AB, *North Star,* 1849-50.

SWEET, William R. Stoker, *Discovery,* 1875-76.

SWIFT, William. AB, *Phoenix,* 1854.

SYKES, John. Master's Mate, *Discovery,* 1791-95.

SYLLEBIN, Giles. Boatswain, *Michael,* 1578.

SYMONDES, Sebastian. Cooper, Frobisher's expedition, 1578.

SYMONDS, Robert D. AB, *Alert,* 1875-76.

SYMS, Nycholas. Boy, *Discovery,* 1610-11.

SZKOLNY, John. *See* SCOLVUS.

TADMAN, Thomas. AB, *Erebus,* 1845-48.

TAGISH CHARLIE. (Indian). Discovered Klondike gold with Carmack, 1896.

TAILOR, Rychard. Miner, Frobisher's expedition, 1578.

TAILOR, William. Shoemaker, Frobisher's expedition, 1578.

TAIT, Robert. AB, *Griper,* 1824.

TAMFIELD [TANFIELD, TANFILDE], William. Gentleman, *Gabriel,* 1577; Captain, *Thomas,* 1578.

TANQUARY, Maurice Cole. Zoologist, "Crocker Land" Expedition, 1913-17.

TARRANGEAU [TAUANGEAU, TOURANGEAU], Jean-Baptiste. Middleman, Black's expedition, 1824. [Rich, 1955, pp. 245-46].

TASKER, Florence. (Wife of Stephen Tasker). U.S. expedition, 1906.

TASKER, Stephen. Leader, U.S. expedition, 1906.

TATHAM, William. Ice Master, *Investigator,* 1848-49.

TATNAM, James. Captain, *Colleton,* 1679.

TATTANNOEUCK, Augustus. (Eskimo). Guide, Franklin's expeditions, 1819-22; 1825-27; Back's expedition, 1833-35. Died 1834.

TAUNT, Emory H. Lieutenant, *Thetis,* 1884. [NE].

TAVERNER, Benjamin. Boatswain's Mate, *Arctic,* 1906-07.

TAWS, Edward. Ice Quartermaster, *Discovery,* 1875-76.

TAYLOR, Elizabeth. Leader, U.S. expedition, 1892.

TAYLOR, Francis. Captain of the Hold, *Release,* 1855.

TAYLOR, Frederick. AB, *Investigator,* 1850-54.

TAYLOR, George. Mate, *Beaver,* 1791-92.

TAYLOR, George. Captain, *Beaver,* 1793.

TAYLOR, George. Third Mate, *Victory,* 1829-33.

TAYLOR, Henry. Private RM, *Hecla,* 1821-23.

TAYLOR, J. AB, *Isabella,* 1818.

TAYLOR, James. Captain, *Prince of Wales* [II], 1864-65; *Ocean Nymph,* 1866-67.

TAYLOR, John. Captain (?) North-west Passage expedition, 1753.

TAYLOR, John. Sailor, *Ayde,* 1578.

TAYLOR, John. Sailmaker, *Terror,* 1836-37.

TAYLOR, Peter. (?), Black's expedition, 1833-35; Middleman, Dease and Simpson's expedition, 1837-39. Died at Great Bear Lake, late 1837.

TAYLOR, Samuel. Stoker, *Pioneer,* 1850-51.

TAYLOR, Thomas. AB, *Discovery,* 1791-95.

TAYLOR, Thomas. Guide, Buchan's expedition, 1811.

TAYLOR, Walter. AB, *Phoenix,* 1854.

TAYLOR, William. Seaman, *Resolution,* 1776-80.

TAYLOR, William. Leader, HBC expedition, 1831.

TAYLOR, William. Boatswain, *Assistance,* 1852-54.

TAYLOR, William. Boatswain, *Pandora,* 1876.

TEDDER, Robert. Smith, Frobisher's expedition, 1578.

TELFER, Isaac. Stoker, *Phoenix,* 1854.

TEGAN, Thomas. AB, *Rattlesnake,* 1853-54.

TEMPLEMAN, Robert. Steward, *Karluk,* Canadian Arctic Expedition, 1913-18.

TEROAHAUTÉ AU TROQUOIS. (Iroquois). Voyageur, Franklin's expedition, 1819-22. Executed for murder of Hood, 23 October 1821.

TERRY, Thomas. Boatswain, *Erebus,* 1845-48.

TETCHNEN, Frederick. AB, *Phoenix,* 1854.

TEW, Moses. AB, *Resolution,* 1850-51.

TEWRY, Michael. Revenue Officer, *Discovery,* 1791-95.

THEVENET, D. Joint leader, Revillon Frères voyage, 1903-04.

THIBAULT, Joseph. Chief Cook, *Arctic,* 1906-07; Chief Steward, *Arctic,* 1908-09; Purser and Steward, *Arctic,* 1910-11.

THOM, William. Purser, *Isabella,* 1818; *Victory,* 1829-33.

THOMAS, Chimham. Carpenter, *Victory,* 1829-33. Died 22 February 1833.

THOMAS, Edward. AB, *Resolute,* 1850-51.

THOMAS, James. Corporal of Marines, *Resolution,* 1776-80. Killed at Kealakekua Bay, 14 February 1779.

THOMAS, John. Seaman, *Discovery,* 1610-11. Killed by Eskimos, 29 July 1611.

THOMAS, John. AB, *Discovery,* 1791-95.

THOMAS, John. AB, *Alexander,* 1818.

THOMAS, John. Private RM, *Plover,* 1852-54.

THOMAS, Philip. Captain of the Forecastle, *Resolute,* 1852-54.

THOMAS, Richard. Cook, *Enterprise,* 1850-55.

THOMAS, Robert. Mate, *Terror,* 1845-48. [O'B(1)].

THOMPSON, Alexander. AB, *Resolute,* 1850-51; 1852-54; Quartermaster, *Fox,* 1857-59.

THOMPSON, Daniel. Carpenter, *North Star,* 1849-50.

THOMPSON, David (1770-1857). Leader, HBC expedition, 1796. [Hopwood, 1968; 1971; Glover, 1962].

THOMPSON, Edward. Surgeon, *Furnace,* 1741-42; *Dobbs,* 1746-47. [Rich, 1949; pp. 337-38].

THOMPSON, George. AB, *Erebus,* 1845-48.

THOMPSON, George T. AB, *Sophia,* 1850-51.

THOMPSON, Hamlet. Private of Marines, *Discovery,* 1776-80.

THOMPSON, J. AB, *Hecla,* 1821-23.

THOMPSON, James. Engineer, *Terror,* 1845-48.

THOMPSON, John. Seaman, *Resolution,* 1776-80.

THOMPSON, John (-1828). Established Rocky Mountain House, 1800. [Rich, 1938, p. 471].

THOMPSON, John. Sailmaker, *Alexander,* 1818.

THOMPSON, John. Carpenter's Mate, *Arctic,* 1855.

THOMPSON, Joseph. Captain, *Shaftesbury,* 1676; 1677; 1678. [Rich, 1945, pp. 341-42; DCB Vol. 1].

THOMPSON, Peter. AB, *Investigator,* 1848-49.

THOMPSON, Peter. Captain of the Foretop, *Investigator,* 1850-54.

THOMPSON, T.S. Passed Assistant Paymaster, *Juniata,* 1873.

THOMPSON, Thomas. Captain, *Wear,* 1819-20.

THOMPSON, Thomas. AB, *Blossom,* 1825-28.

THOMPSON, William. Greenland Mate, *Hecla,* 1821-23.

THOMPSON, William. AB, *Enterprise,* 1848-49.

THOMPSON, William. AB, *Phoenix,* 1854.

THOMSEN, Charles [Karl]. (?), Canadian Arctic Expedition, 1913-18. Died on Banks Island, winter 1916-17.

THÖNSBERG, Thoer. Seaman, *Enhiörningen,* 1619-20. Died at Churchill, 28 April 1620.

THORDARSSON, Thorfinn [KARLSEFNI]. Vinland voyage, c1000. [DCB Vol. 1].

THORES, John. Ice Quartermaster. *Alert,* 1875-76.

THORN, George. AB, *North Star,* 1852-54.

THORNBACK, James. AB, *Discovery,* 1875-76.

THORNE, George W. AB, *Pandora,* 1875; Captain of the Maintop, *Pandora,* 1876.

THORNE, John. Boy, *Ayde,* 1578.

THORNETON, Thomas. Purser, *Ayde,* 1578.

THORNLEY, Henry. AB, *Plover,* 1848-52 [1850-52 only].

THORTE, Thomas. Sailor, *Ayde,* 1578.

THORVALDSSON, Eirik [Eric the Red]. Discovery of Baffin Island, 982. [DCB Vol. 1].

THWAITES, John. Leading Seaman, *Fury,* 1824-25.

TICKLE, James. AB, *Resolution,* 1776-80.

TIELDER, Cornelius. *See* FIELDER.

TIERNEY, D. Second Steward, *Neptune,* 1903-04.

TIFFENY, Robert. Captain of the Maintop, *Investigator,* 1850-54.

TILBROOK, Charles. Cook, *Phoenix,* 1853.

TILE, James. Stoker, *Phoenix,* 1854.

TILVEY, William [John]. *See* SILVEY, John.

TIMPSON, James. Quartermaster, *Pandora,* 1875.

TINTOR, Arthur. AB, *Plover,* 1848-52; 1852-54.

TIRPITT, Samuel. AB, *North Star,* 1852-54.

TIZZARD, Charles. Boatswain's Mate, *Pandora,* 1875; 1876.

TOBIAS. Eskimo, *Polaris,* 1871-73.

TOBIN, Thomas, AB, *Plover,* 1848-52.

TODD, John. Captain, *Ocean Nymph,* 1863-64.

TODD, Robert. Assistant, Selwyn's expedition, 1875.

TODD, Walter Edmond Clyde (1874-1969). Leader, Carnegie Museum expeditions, 1901; 1906; 1907; 1908; 1912; 1914-15; 1917; 1920. [Parkes, 1970; Woods, L.C., 1969; PR 1970].

TOM, John (1800-). Midshipman, *Griper,* 1824. [O'B(1)].

TOMISON, William (1739-). Joint leader, HBC expeditions, 1767-68; 1769-70; established Hudson House, 1778-79. [DCB Vol. 5].

TOMLINSON, W.I. Caulker, *Investigator,* 1848-49.

TOMS, Frank Y. Assistant Surgeon, *North Star,* 1852-54.

TOMS, Henry. AB, *Rattlesnake,* 1853-54.

TOMS, Henry. Quartermaster, *Fox,* 1857-59; Gunner, *Pandora,* 1875.

TONNETUIT, Pierre de Chauvin de. *See* CHAUVIN DE TONNETUIT.

TOOKOOLITO [HANNAH]. (Eskimo woman). Accompanied Hall, 1860-62; 1864-69; 1871-73; Schwatka, 1878-80. [Neatby, 1959; Burgess, H., 1958].

TORRINGTON, John. Leading Stoker, *Terror,* 1845-48. Died on Beechey Island, 1 January 1846.

TORT, Thomas. (?), Frobisher's expedition, 1578. Died 24 September 1578.

TOWN, George. Private RM, *Herald,* 1848; 1849; 1850.

TOWNSEND, Charles Wendell (1859-1934). Leader, U.S. expedition, 1906.

TOWNSEND, Thomas. Quartermaster, *Chatham,* 1791-95.

TOWNSHEND, Thomas. AB, *Discovery,* 1791-95.

TOY, Thomas. AB. *Investigator,* 1850-54.

TOZER, Solomon. Sergeant RM, *Terror,* 1845-48.

TRACEY, Robert L. Master's Assistant, *Investigator,* 1848-49.

TRANTER, Henry. AB, *Assistance,* 1852-54.

TREGONING, Hugh. Carpenter's Mate, *Plover,* 1848-52.

TREGUIMO, William. Cook, *North Star,* 1852-54.

TREMAINE, C.E.G. Constable, NWMP, *Neptune,* 1903-04.

TREMBLAY, Alfred. AB, *Arctic,* 1910-11.

TREMBLAY, Alfred. (?), *Minnie Maud,* 1912-13.

TREMBLAY, Louis. Canoeman, Bell's expedition, 1899-1900.

TREMBLAY, P. Seaman, *Arctic,* 1908-09; Waiter, *Arctic,* 1910-11.

TREMBLIER, du. Marine Officer, *Engageante,* 1782.

TREMBLY, Narcisse (c1825-). Canoeman, Richardson's expedition, 1847-49. [Rich, 1953, p. 377].

TREMEER, Henry. Ordinary Seaman, *Rattlesnake,* 1853-54.

TRENAMAN, M. Carpenter's Mate, *Phoenix,* 1854.

TRETCHER, Thomas. AB, *Discovery,* 1776-80.

TREVENEN, James (1760-1790). Midshipman, *Resolution,* 1776-80. [DNB Vol. 57].

TREVILLER, James. Sailor, *Ayde,* 1578.

TREVOR, Charles. AB, *Phoenix,* 1854.

TREW, Christopher. AB, *Alexander,* 1818.

TREZISSE, John. Captain, *Albert,* 1892.

TRIGGS, Samuel. Steward, *Enterprise,* 1848-49.

TROLLOPE, Henry. Lieutenant, *Herald,* 1848; 1849; 1850; Commander, RN, commanding *Rattlesnake,* 1853-54.

TROTTER, . Earl Grey's aide-de-camp, Governor-general's tour, 1910.

TROW, Thomas. AB, *Isabella,* 1818.

TRUSCOTT, John. AB, *Enterprise,* 1848-49.

TRYBE, Robert. Sailor, *Judith,* 1578.

TSONDOUTANNEN, Thomas. (Huron Indian). Accompanied Buteux, 1652.

TUCKFIELD, Joe. Leader, U.S. whaling voyage, 1888.

TUDOR, Arthur. AB, *Plover,* 1848-52.

TUGGEY, Henry, Captain, *Walrus,* 1872.

TULLETT, James [William]. Captain of the Foretop, *Assistance,* 1850-51; Boatswain's Mate, *Resolute,* 1852-54.

TULLOCK, Thomas. AB, *Griper,* 1824.

TUNSTALL, Joseph. Captain, *Charlotte,* 1778.

TUNSTALL, Joshua. Captain, HBC supply voyages. *Prince Rupert* [IV], 1783; 1784; 1785; 1786; *Seahorse* [III], 1787; 1788; *King George* [III], 1789.

TURNBULL, M. Stoker, *Phoenix,* 1853.

TURNER, J.H. U.S. Coast Survey expedition, 1890 [see 1869 closure of Fort Yukon].

TURNER, J.T. Clerk in Charge, *Enterprise,* 1850-55. Died 28 November 1854.

TURNER, John. Captain, HBC supply voyages. *Queen Charlotte,* 1790; 1791; 1792; 1793; 1794; 1795; 1796; 1797; 1798; 1799; 1800; *Prince of Wales* [I], 1801; *King George* [III], 1802; 1803; 1804; 1805; 1806; 1807; 1808; 1809; 1812; *Prince of Wales* [I], 1813; *Eddystone,* 1814; 1815-16.

TURNER, Lucien McShan. Leader, U.S. expedition, 1882-84.

TURNER [TURNOR], Richard. Middleman, Rae's expedition, 1846-47. [Rich, 1953, p. 377].

TURNER, Sandy. Accompanied Hanbury, 1901-02 [1901 only].

TURNOR, Philip. Established Frederick House, 1785; leader, HBC expedition, 1790-92.

TURPITT, Samuel. *See* TIRPITT.

TWINE, Jonathan G. Gunner, *North Star,* 1849-50.

TWYMAN, Edward. Boatswain, *North Star,* 1849-50.

TYLER, George. Second Class Fireman, *Arctic,* 1855.

TYLER, Robert. Armourer's Mate, *Hecla,* 1819-20.

TYRRELL, James Williams (1863-1945). Observer in charge of Ashe Inlet station, Hudson Bay Expedition, 1885-86; second in command, J. B. Tyrrell's expedition, 1893; leader, Dominion Lands Survey expedition, 1900. [PR 1945; Mills, E., 1946].

TYRRELL, Joseph Burr [1858-1957]. Leader, GSC expeditions, 1892; 1893; 1894; 1898. [Eagan, 1971; Shaw, M.M., 1958; PR 1958; Loudon, 1930; Alcock, 1953].

TYSOE, Shadrach. Marine, Franklin's expedition, 1825-27.

TYSON, Benjamin. AB, *Herald,* 1848; 1849; 1850.

TYSON, George Emory (1829-1906). Assistant Navigator, *Polaris,* 1871-73; Captain, *Florence,* 1877-78.

UDELL, William. Revenue Officer, *Discovery,* 1791-95.

UGGANES, George. Seaman (?), *Henrietta Maria,* 1631-32.

ULLEBUCK. *See* OOLIGBUCK.

UNDERWOOD, William. AB, *Discovery,* 1791-95.

UNTHANK, Ransom. Ice Quartermaster, *Assistance,* 1852-54.

UPCOT, . Captain, *Moone,* 1578.

UPTON, F.E. Master, *Juniata,* 1873.

UREN, Thomas. AB, *Blossom,* 1825-28. Drowned 26 August 1827.

URIN, [HURIN, YOURIN], . Master's Mate, *Charles,* 1631.

URQUHART, Robert. Private RM, *Assistance,* 1850-51; 1852-54.

USHER, Nathaniel Reilly (1855-1931). Lieutenant, *Bear,* 1884. [DAB].

USHER, Robert. AB, *Phoenix,* 1854.

1751; 1752; 1753; 1754.
WALKER, James. AB, *Fury*, 1824-25.
WALKER, James. AB, *Terror*, 1845-48.
WALKER, Nehemiah. Captain, *John and Alexander*, 1679; *Diligence*, 1681-82; 1683. [Rich, 1946, pp. 334-37; DCB Vol. 1].
WALKER, R.H. Constable, RNWMP, Pelletier's patrol, 1908-09.
WALKER, Samuel. Carpenter's Mate, *Pioneer*, 1852-54.
WALKER, Thomas. AB, *Fury*, 1824-25.
WALKER, W.M. AB, *Discovery*, 1791-95.
WALKER, William. Carpenter's crew, *Discovery*, 1776-80.
WALKER, William. Surgeon, *Chatham*, 1791-95.
WALKER, William. AB, *Intrepid*, 1852-54.
WALL, Richard. Cook, *Erebus*, 1845-48.
WALL, Richard. Harpooner, *Victory*, 1829-33.
WALLACE, Dillon (1863-1939). Second-in-command, U.S. expedition, 1903-04; leader, U.S. expedition, 1905-06. [Mauro, 1975; EA].
WALLACE, George. AB, *Fury*, 1824-25.
WALLACE, Thomas. Armourer's Mate, *Terror*, 1836-37.
WALLACE, William. AB, *Phoenix*, 1853.
WALLER [WELLER], Charles Crump. Midshipman, *Fury*, 1824-25.
WALLER, William. Private RM, *Discovery*, 1875-76.
WALLICH, George Charles (1815-1899). Naturalist, *Bulldog*, 1860. [DNB Vol. 59].
WALLIS, James. (?), Frobisher's expedition, 1578.
WALLIS, William. Carpenter, *Hecla*, 1819.
WALTER, James. Sailor, *Ayde*, 1578.
WALTERS, William. Carpenter's crew, *Fox*, 1857-59.
WALTON, T.C. Surgeon, *Juniata*, 1873.
WALTON, Thomas. AB, *Plover*, 1852-54.
WALTON, W.G. Relieved E.J. Peck at Fort George mission, 1892 [see 1885-92].
WARD, James. AB, *Resolution*, 1776-80.
WARD, John. Private RM, *Plover*, 1848-52; 1852-54.
WARD, John. Assistant Surgeon, *Intrepid*, 1850-51.
WARD, Richard. Captain, *Pery*, 1711-12; *Hudson's Bay* [III], 1716; 1717; 1718; 1719. [DCB Vol. 2].
WARD, Thomas. AB, *Assistance*, 1850-51.
WARD, William. Master's Mate, *Judith*, 1578.
WARD, William. AB, *Resolute*, 1850-51.
WARD, William. Armourer, *Discovery*, 1875-76.
WARDEN, John. Master's Chief Mate, *Henrietta Maria*, 1631-32. Died 6 May 1632.
WARDENER, Robert. Second Mate, *Breadalbane*, 1853.
WARDWELL, George A. Chief Engineer, *Roosevelt*, 1905-06; 1908-09.
WARINGS, Peter. AB, *Herald*, 1848; 1849; 1850.
WARK, Thomas. *See* WORK.
WARLEY, James. AB, *Plover*, 1848-52; 1852-54.
WARMBATH, Joseph Samuel. (?), Ellesmere Island Expedition, 1899-1900.
WARMOW, Mathias. Moravian missionary, expedition to Baffin Island, 1857-58.
WARNE, Charles. AB, *North Star*, 1852-54.
WARNE, Richard. AB, *Intrepid*, 1852-54.
WARRIN, Nycolas. Carpenter, *Ayde*, 1578.
WATERS, Joseph. Ordinary Seaman, *Herald*, 1848; 1849; 1850.
WATKINS,   . Captain, *Seaford*, 1696.
WATKINS. E.A. Church Missionary Society journey, 1852.
WATKINS, William. AB, *Phoenix*, 1854.
WATMAN, William. AB, *Resolution*, 1776-80. Died at Hawaii, 1 February 1779.
WATS, Robert. Carpenter, *Sunneshine*, 1585.
WATSON, John. AB, *Terror*, 1836-37.
WATSON, John. Gunroom Steward, *Plover*, 1848-52; 1852-54.
WATSON, John. Mate, *Enhiörningen* (or *Lamprenen*) 1619-20. Died at Churchill, 6 May 1620.
WATSON, Rychard. Sailor, *Ayde*, 1578.
WATSON, Thomas. Carpenter's Mate, *Erebus*, 1845-48.
WATSON, William. Seaman, *Resolution*, 1776-80.
WATT, J. AB, *Hecla*, 1824-25.
WATT, James. AB, *Prince Albert*, 1850.
WATTS, Edward. Second Officer, *Alert*,

VAILLANCOURT. W. Second Cook, *Arctic*, 1908-09.

VAILLANT, Registre. Voyageur, Franklin's expedition, 1819-22.

VALDÉS, Flores Bazan y Peon Cayetano. Captain, *Mexicana*, 1792. [Michaud, 1843-65, Vol. 42].

VAN, John. AB, *North Star*, 1852-54.

VANASSE, Fabien. Historian, *Arctic*, 1906-07; 1908-09; 1910-11.

VANCOUVER, George (1757-1798). Midshipman, *Discovery*, 1776-80; Captain, *Discovery*, 1791-95. [DNB Vol. 58].

VAN DYKE, John. Purser's Steward, *Arctic*, 1855.

VAN HEEMSKERK, . (?), *Wivenhoe*, 1669.

VAN KOENIG, John. Chief Engineer, *Arctic*, 1906-07; 1908-09; 1910-11.

VANT, Isaac. AB, *Phoenix*, 1853.

VAUGHAN, David (-c1720). Captain, *Good Success*, 1717; *Discovery*, 1719-21. [DCB Vol. 2].

VAUGHAN, John. Boatswain, *Terror*, 1836-37.

VAUGHAN, William. Leading Stoker, *Phoenix*, 1853.

VENABLES, George Edward. Third Engineer, *Eldorado*, 1903.

VENABLES, James. Captain's Steward, *Enterprise*, 1848-49.

VERNON, Charles E.H. (1827-1873). Lieutenant, *Plover*, 1852-54. [MAR].

VÉRENDRYE, Pierre Gaultier de Varennes *Sieur de la*. *See* GAULTIER DE VARENNES.

VERRAZZANO, Giovanni da (c1485-c1528). Captain, *Dauphine*, 1524. [DCB Vol. 1].

VEZINA, Auguste. Oiler, *Arctic*, 1910-11.

VIDAMOUR, Peter. AB, *Herald*, 1848; 1849; 1850.

VIGNEAU, Claude. Quartermaster, *Arctic*, 1908-09.

VINCENT, Peter. Gentleman, Frobisher's expedition, 1578.

VINE, Charles. Captain's Coxswain, *Pandora*, 1875; Wardroom Steward, *Pandora*, 1876.

VIVIER, Alexis. Voyageur, Franklin's expedition, 1825-27.

VOLKI, Fred. (?), Canadian Arctic Expedition, 1913-18.

VOLLANT, Charles. Guide, Todd's expedition, 1917.

VOLSKE, David. Surgeon, *Lamprenen*, 1619-20. Died at Churchill, 12 December 1619.

VREELAND, Benjamin. Assistant Surgeon, *Rescue*, 1850-51. [EA].

VTEY, James. Soldier, Frobisher's expedition, 1578.

WADDELL, James. *See* WEDDELL.

WADDELL, William. Engineer, *Phoenix*, 1854.

WADE, Whitehouse. AB, *Isabella*, 1818.

WADEN, Jean-Etienne (-1782). North West Company expedition, 1779-82.

WAGER, John. AB, *Plover*, 1848-52 [1851-52 only].

WAGER, William. Quartermaster, *Chatham*, 1791-95.

WAGGE, Robert. Carpenter, *Sunneshine*, 1586.

WAGGONER, Joseph. Joint leader, HBC expeditions, 1756-57; 1757-58; 1760-61.

WAGGONER, Rowland. Workman, second founding of Churchill, 1717.

WAINWRIGHT, John. Lieutenant, *Blossom*, 1825-28. [O'B(1)].

WAKEHAM, Cyrus. Clerk, *Griper*, 1819-20.

WAKEHAM, Henry. Seaman, *Arctic*, 1908-09.

WAKEHAM, William (1845-). Commander, Hudson Strait expedition, 1897. [EA].

WALCOTT, Robert. Joint leader, Cabot's expedition, 1904 [see 1903-10].

WALDRON, William. Carpenter, *Enterprise*, 1850-55.

WALES, William (1734-1798). Astronomer, British expedition, 1768-69. [DNB Vol. 59].

WALKER, David (1837-). Surgeon and Naturalist, *Fox*, 1857-59.

WALKER, Donald. Assistant, Selwyn's expedition, 1875.

WALKER, Graham. AB, *Terror*, 1836-37. Died 13 January 1837.

WALKER, James. Captain, *Churchill*, 1750;

1885.

WATTS, John. Midshipman, *Resolution*, 1776-80.

WATTS, Peter. AB, *Rattlesnake*, 1853-54.

WAY, Henry. Private RM, *Rattlesnake*, 1853-54.

WAY, John. Senior Second Master, *North Star*, 1849-50.

WAY, Robert. AB, *Herald*, 1848; 1849; 1850.

WAYMOUTH, George. *See* WEYMOUTH.

WAYNAM, Richard. Cook, *Discovery*, 1616. Died 26 July 1616.

WEAKAM, William. Revenue Officer, *Discovery*, 1791-95.

WEATHERALL, John. AB, *Resolute*, 1852-54.

WEBB, Henry P. Assistant Engineer, *Pioneer*, 1850-51; 1852-54.

WEBB, John. AB, *Griper*, 1819-20.

WEBB, John. Private RM, *North Star*, 1852-54.

WEBB, Richard. Private, Sappers and Miners, Richardson's expedition, 1847-49.

WEBB, Richard. Steward, *Prince Albert*, 1851-52.

WEBBER, Francis. AB, *Assistance*, 1850-51.

WEBBER, James. AB, *Intrepid*, 1850-51.

WEBBER, John (1752-1793). Landscape painter, *Resolution*, 1776-80. [DNB Vol. 60].

WEBSTER, . Captain, *Lady Fitzherbert*, 1846-48.

WEBSTER, Arthur. Surveyor, Selwyn's expedition, 1875.

WEBSTER, James. Boatswain's Mate, *Chatham*, 1791-95.

WEBSTER, Thomas. AB, *Terror*, 1836-37.

WEBSTER, Tom. Second Mate, *Arctic*, 1873.

WEDDELL [WADDELL], James. Private, Sappers and Miners, Richardson's expedition, 1847-49.

WEDDERBURN, Joseph. Private RM, *Isabella*, 1818.

WADENBY, C. Gunner, RA, Artificer, British Polar Year Expedition, 1882-83.

WEEKES, John. Carpenter, *Erebus*, 1845-48.

WEEKS, Emanuel. Private RM, *Plover*, 1848-52 [1851-52 only].

WEEKS, W.H. Purser, *Arctic*, 1906-07; 1908-09.

WEILAND, Charles. Quartermaster, *North Star*, 1849-50.

WEIR, Charles. Cooper, *North Star*, 1849-50.

WEISE, R. Cook. German Polar Year Expedition, 1882-83.

WELCH, William. AB, *Plover*, 1848-52; 1852-54.

WELLER, Charles Crump. *See* WALLER.

WELLINGTON, William C. Sergeant RM, *Discovery*, 1875-76.

WELLS, F. Seaman, *Neptune*, 1903-04.

WELLS, Henry. Coxswain, *Blossom*, 1825-28.

WENMAN, John. Carpenter's crew, *Enterprise*, 1850-55.

WENTWORTH, William. Boatswain's Mate, *Hecla*, 1821-23; Boatswain, *Fury*, 1824-25.

WENTWORTH-FITZWILLIAM, William *Viscount Milton* (1839-1877). Joint leader, British expedition, 1862-64. [Alcock, 1877; DCB Vol. 10].

WENTZALL, William. AB, *Terror*, 1845-48.

WENTZEL, Ferdinand. Half-breed lad, Dease and Simpson's expedition, 1837-39.

WENTZEL, Willard Ferdinand (c1777-1832). North West Company clerk and assistant, Franklin's expedition, 1819-22. [Rich, 1938, p. 472; Holand, 1930].

WESSON, William. Private RM, *Phoenix*, 1854.

WEST, Alexander. AB, *Phoenix*, 1853.

WEST, John (-1845). Missionary, CMS expedition, 1820-23 [EA].

WEST, Joseph. Private RM, *Phoenix*, 1853.

WEST, Richard. Boatswain, *Phoenix*, 1854.

WEST, Thomas. Captain of the Maintop, *Resolute*, 1852-54.

WEST, William. Seaman, *Advance*, 1850-51.

WESTAL [WESTOL], . Second Mate, *California*, 1746-47.

WESTROPP, Berkley. Midshipman, *Fury*, 1825-28. [O'B (1)].

WETHERHILL, H. Emerson. Surgeon, *Falcon*, 1894.

WEYMOUTH [WAYMOUTH], George. Captain, *Discovery*, 1602. [DCB Vol. 1].

WHARTON, William James Lloyd (1843-

1905). Second Lieutenant and assistant surveyor, *Gannet*, 1867. [DNB 2nd Supp. Vol. 3].
WHELAN, Patrick. Seaman, *Resolution*, 1776-80.
WHETLEYE, Rychard. Sailor, *Ayde*, 1578.
WHITBEY, Joseph. Master, *Discovery*, 1791-95.
WHIFFIN, J.G. Clerk, *Herald*, 1848; 1849; 1850.
WHIPPLE, George. Seaman, *Advance*, 1853-55.
WHISLER, William. Private, Company F, 9th Infantry, U.S. Polar Year Expedition, 1881-84. Died 24 May 1884.
WHIT, Guy. Sailor, *Judith*, 1578.
WHITAKER, Robert. AB, *Phoenix*, 1854.
WHITE, Adam. AB, *Hecla*, 1819-20.
WHITE, Benjamin. AB, *Hecla*, 1821-23.
WHITE, Charles. AB, *Phoenix*, 1854.
WHITE, George. Engineer, *Alert*, 1875-76.
WHITE, H. J. Seaman, *Rescue*, 1850-51.
WHITE, Henry C. Lieutenant-Commander and executive officer, *Tigress*, 1873. [NE].
WHITE, John. AB, *Isabella*, 1818.
WHITE, John. AB, *Phoenix*, 1854.
WHITE, Robert. AB, *Isabel*, 1852.
WHITE, Thomas. Seaman, *Arctic*, 1908-09.
WHITE, Thomas. Captain's Coxswain, *Phoenix*, 1853.
WHITE. Thomas. AB, *Phoenix*, 1854.
WHITE, Watson. AB, *Hecla*, 1821-23.
WHITE, William. AB, *Assistance*, 1850-51.
WHITEFIELD, William. Carpenter's Crew, *Investigator*, 1850-54.
WHITEHEAD, Edward. Clerk, *Enterprise*, 1848-49; Clerk in Charge, *Enterprise*, 1850-55. Died 19 June 1851.
WHITEHEAD, Richard. AB, *North Star*, 1849-50.
WHITEMAN, Paul. AB, *Terror*, 1836-37.
WHITING, Samuel. Acting Boatswain, *Arctic*, 1855.
WHITNEY, Caspar (1862-1929). Leader, hunting expedition, 1894-95. [DAB].
WHITNEY, Harry Payne (1873-1936). Leader, hunting expedition, 1908-09. [DAB; EA].
WHITTON, Benjamin. Seaman, *Resolution*, 1776-80.
WHITTERED, John. Master's second mate, *Henrietta Maria*, 1631-32.
WIARS, Thomas. Passenger, *Emanuell (Buss)*, 1578.
WICKETTS, George. AB, *Pioneer*, 1852-54.
WIDDALL, William. Sailmaker, *Resolution*, 1776-80.
WIEGAND, Thomas. Visited Belcher Islands for HBC, 1847.
WIGGINS, [WIGGANS], Joseph. AB, *Enterprise*, 1850-55.
WIGHT, J.E.F. Constable, RNWMP, LaNauze's patrol, 1915-17.
WIGMORE, Thomas R. AB, *Hecla*, 1819-20.
WIIK, Gustav Juel. Second Engineer, *Gjøa*, 1903-06. Died 31 March 1906.
WILCOX, John. Paymaster and Paymaster's Steward, *Investigator*, 1850-54.
WILCOX, Thomas. Mate and Greenland Pilot, *Isabella*, 1818.
WILDES, Frank. Commander, USN, commanding *Yantic*, 1883. [NE].
WILKES, Henry. Private RM, *Terror*, 1845-48.
WILKES, Samuel. Armourer's Mate, *Hecla*, 1821-23.
WILKIE [WILKEY], James. Captain of the Foretop, *Investigator*, 1848-49; Captain of the Forecastle, *Assistance*, 1850-51; Ice Quartermaster, *Intrepid*, 1852-54. Died on board, 2 February 1854.
WILKINS, George. Lieutenant and Navigator, *Tigress*, 1873.
WILKINS, George Hubert (1888-1958). Photographer, Canadian Arctic Expedition, 1913-18. [Wood, 1959; Grierson, 1960; Thomas, 1961].
WILKINSON, C.R.B. Second Master, *Rattlesnake*, 1853-54.
WILKINSON, James. AB, *Discovery*, 1791-95.
WILKINSON, John. Quartermaster's Mate, *Chatham*, 1791-95.
WILKINSON, Robert. Boatswain's Mate, *North Star*, 1852-54.
WILKINSON, Samuel. Prospecting expedition, 1872-74.
WILKINSON, Walter. Quartermaster, *Arctic*, 1855.
WILLCOCKS, John. Corporal Master-at-

Arms, *Discovery,* 1791-95.
WILLIAM, Samuel. *See* FOWLER, William.
WILLIAMS, . Corporal of Marines, *Discovery,* 1791-95.
WILLIAMS, Charles. Carpenter's Mate, *Release,* 1855.
WILLIAMS, Edward. AB, *Chatham,* 1791-95.
WILLIAMS, George. AB, *Hecla,* 1824-25.
WILLIAMS, George. AB, *Erebus,* 1845-48.
WILLIAMS, H. Sailor, *Karluk,* Canadian Arctic Expedition, 1913-18.
WILLIAMS, J. AB, *Terror,* 1836-37.
WILLIAMS, James. Captain of the Hold, *Investigator,* 1850-54.
WILLIAMS, John. Master's Mate, *Ayde,* 1578.
WILLIAMS, John. Gunner, *Discovery,* 1610-11. Died mid-November 1610.
WILLIAMS, John. Captain of the Maintop, *Investigator,* 1848-49.
WILLIAMS, John. Seaman, *Rescue,* 1850-51.
WILLIAMS, John. Seaman, *United States,* 1860-61.
WILLIAMS, Martin. (?), *Judith,* 1578.
WILLIAMS, Robert. Seaman, *Discovery,* 1776-80.
WILLIAMS, William. Carpenter's crew, *Discovery,* 1776-80.
WILLIAMS, William. AB, *Fury,* 1824-25.
WILLIAMS, William. Captain, *Nonpareil,* 1835.
WILLIAMSON, David. Artilleryman, Back's expedition, 1833-35. Died 1834.
WILLIAMSON, David. Carpenter's Mate, *Plover,* 1848-52; 1852-54.
WILLIAMSON, Edward. AB, *Discovery,* 1791-95.
WILLIAMSON, John (-1798). Lieutenant, *Resolution,* 1776-80.
WILLIAMSON, Robert. AB, *Phoenix,* 1854.
WILLIAMSON, Robert J. Second Engineer, *Karluk,* Canadian Arctic Expedition, 1913-18.
WILLIAMSON, Thomas. Captain, HBC supply voyages. *Nimble,* 1792; 1793.
WILLIS, G. Fireman, *Neptune,* 1903-04.
WILLIS, John. AB, *Discovery,* 1791-95.
WILLOUGHBY, Joseph. AB, *Fury,* 1824-25.
WILLIS, John. Sailmaker, *Herald,* 1848; 1849; 1850.
WILLSON, William. Quartermaster's Mate, *Chatham,* 1791-95.
WILMOT [WILMET], John. Sailor, *Ayde,* 1578. Died 28 September 1578.
WILNATER, Robert. Sailor, *Ayde,* 1578.
WILSON, Alexander. Master, *George,* 1774.
WILSON, Alexander. Carpenter's Mate, *Terror,* 1845-48.
WILSON, Andrew. Private RM, *Terror,* 1836-37.
WILSON, Daniel (c1847-). (?), Richardson's expedition, 1847-49. [Rich, 1953, pp. 377-78].
WILSON, David [Robert]. Ice Quartermaster, *Assistance,* 1850-51.
WILSON, Edward. Surgeon, *Discovery,* 1610-11.
WILSON, Edward. AB, *Advance,* 1850-51.
WILSON, G. (?), *Minnie Maud,* 1912-13.
WILSON, George. Private RM, Franklin's expedition, 1825-27.
WILSON, George F. Assistant Surgeon, U.S. Army, Surgeon, Schwatka's expedition, 1883.
WILSON, Henry. Voyageur, Richardson's expedition, 1847-49.
WILSON, J. Gunroom Steward, *Rattlesnake,* 1853-54.
WILSON, John. Sailor, *Judith,* 1578.
WILSON, John. Captain's Coxswain, *Terror,* 1845-48.
WILSON, John Wall. Sailing Master, *Advance,* 1853-55.
WILSON, Joseph. AB, *Phoenix,* 1854.
WILSON, Robert. Master, *Furnace,* 1741-42.
WILSON, Sam. Seaman, *George Henry,* 1860-62.
WILSON, Thomas. Leading Seaman, *Alexander,* 1818.
WILSON, Thomas. AB, *Assistance,* 1850-51; Sailmaker, *Resolute,* 1852-54.
WILSON, Watson, AB, *Enterprise,* 1850-55.
WILSON, William. Seaman, *Discovery,* 1610-11. Killed by Eskimos, 29 July 1611.
WILSON, William. AB, *Blossom,* 1825-28.
WILSON, William. Second Mate, *Prince Albert,* 1850.
WINDILL, Richard. AB, *Terror,* 1836-37.

WINDSER, Henry. Carpenter's crew, *Discovery,* 1875-76.

WINSTONE, George. AB, *Alert,* 1875-76.

WINTER, William. Captain, *Delight,* 1583. [DNB Vol. 62].

WISE, Charles. Sergeant RM, *Isabella,* 1818; *Hecla,* 1821-23; 1824-25.

WISE, W.N. Acting Mate, *Enterprise,* 1850-55.

WISEMAN, John. Fireman, *Roosevelt,* 1908-09.

WISEMAN, Stephen. AB, *Phoenix,* 1854.

WISHART, D. D. Captain, *Prince Arthur,* 1857; 1858; 1859; 1860; 1861; 1862.

WISTLE, Louis. Seaman, *Arctic,* 1908-09.

WITHERS, . Constable, RNWMP, Governor-General's tour, 1910; LaNauze's patrol, 1915-17.

WITHERS, John. Private RM, *Discovery,* 1791-95.

WOLCOME, William. Seaman, *Sunneshine,* 1586.

WOLF, Louie J. Surgeon, *Roosevelt,* 1905-06.

WOLFALL [WOOLFALL], . Chaplain, *Ayde,* 1578.

WOLFE, George. AB, *Phoenix,* 1853.

WOLFE, James (1800-1849). Mate, *Blossom,* 1825-28. [O'B(1); Dawson, L.S., 1885].

WOOD, David. AB, *Victory,* 1829-33.

WOOD, George. Boatswain's Mate, *Assistance,* 1850-51; Captain of the Hold, *Assistance,* 1852-54.

WOOD, James. Captain, *Churchill,* 1758; 1760.

WOOD, John. AB, *Victory,* 1829-33.

WOOD, Joseph. Sailmaker, *Enterprise,* 1850-55.

WOOD, R. B. Captain, *Labrador,* 1866; 1867; 1868.

WOOD, William. AB, *Hecla,* 1819-20.

WOOD, William. Cook, *Pioneer,* 1852-54.

WOOD, William. Colour Sergeant RM, *Alert,* 1875-76.

WOODBRIDGE, Dwight Edwards (1864-). Leader, geological expedition, 1920.

WOODEN, Isaac. Seaman, *Discovery,* 1791-95.

WOODERSON, William. Seaman, *Discovery,* 1791-95.

WOODFIELD, Philip. Carpenter's Mate, *Discovery,* 1776-80.

WOODHEAD, William. Private RM, *Enterprise,* 1850-55.

WOODLEY, William. Gold prospector, *Algerine,* 1912.

WOODRUFF, Simon [Simeon]. Gunner's Mate, *Discovery,* 1776-80.

WOODS, John. AB, *Isabella,* 1818.

WOODWARD, James. Boatswain, *Enterprise,* 1850-55.

WOODWARD, Thomas. Purser, *Herald,* 1845-51. Died 26 January 1851.

WOODWORTH, Percy. Observer in charge of Digges Islands station, Hudson Bay Expedition, 1885-86.

WOOLFORD, James. Private RM, *Blossom,* 1825-28.

WOOLLEY, William P. AB, *Alert,* 1875-76.

WOOLNER, James. Carpenter's Mate (?), *North Star,* 1849-50.

WOON, John. Sergeant RM, *Investigator,* 1850-54.

WOOTTON, James. Engineer, *Alert,* 1875-76.

WORCHURST, William. Carpenter's crew, *Enterprise,* 1850-55.

WORK [WARK], Thomas. AB, *Erebus,* 1845-48.

WORTH, Josiah. AB, *Herald,* 1848; 1849; 1850.

WRADON, I. G. AB, *Plover,* 1848-52; 1852-54.

WREN, Thomas. AB, *Blossom,* 1825-28.

WRIGHT, Arthur R. Lieutenant, *Phoenix,* 1854.

WRIGHT, Edward. AB, *Enterprise,* 1850-55.

WRIGHT, Henry. AB, *Alexander,* 1818.

WRIGHT, J. Cook, *Rattlesnake,* 1853-54.

WRIGHT, John. Private RM, *Resolute,* 1850-51.

WRIGHT, Thomas D. Midshipman, *Plover,* 1848-52; 1852-54.

WRIGHT, William. AB, *Alexander,* 1818.

WRIGHT, William. AB, *Hecla,* 1819-20.

WYATT, Benjamin. AB, *Discovery,* 1875-76.

WYATT, William. Ordinary Seaman, *Herald,* 1848; 1849; 1850.

WYBOURG, Peter. Gold prospecting expedition, 1885-86.

WYDOWSE [WOODHOUSE], Thomas. Mathematician, *Discovery,* 1610-11.

WYLIE, Gilbert. Ordinary Seaman, *Phoenix,* 1854.

WYNN, John Land. Lieutenant, *Hecla,* 1824-25. [RNB].

WYNNIATT, Robert. Mate, *Investigator,* 1850-54.

WYRELL [WYVILL], Thomas. AB, *Alexander,* 1818.

YEADON, W. F. Carpenter, *Alert,* 1885.

YONG, George. Miner, Frobisher's expedition, 1578. Died 17 September 1578.

YORK, Erasmus. *See* KALLIHIRUA.

YORK, Richard B. Seaman, *Florence,* 1877-78.

YORKE, Gilbert. Captain, *Michael,* 1577; *Thomas Allen,* 1578.

YOUNG, Alexander. Private RM, *Terror,* 1836-37. Died 25 April 1837.

YOUNG, Allen William (1827-1915). Sailing Master, *Fox,* 1857-59; Captain, *Pandora,* 1875; 1876. [DNB; Markham, C. R., 1916; EA].

YOUNG, Benjamin. Captain of the Foretop, *Resolute,* 1850-51; Ice Quartermaster, *Pioneer,* 1852-54.

YOUNG, David. Boy, *Erebus,* 1845-48.

YOUNG, F.N.W. Mate, *Rattlesnake,* 1853-54.

YOUNG, George Albert (1878-1943). Assistant, Low's expeditions, 1892-96 [1896 only]; 1897; 1898-99.

YOUNG, Henry. (?), *Nancy Dawson,* 1849.

YOUNG, James. Captain, HBC supply voyages. *Dering* [I], 1688-89; *Hopewell,* 1689; *Dering* [II], 1690-91; *Loyalty,* 1691; *Supply,* 1693. [DCB Vol. 1].

YOUNG, John. AB, *Alexander,* 1818.

YOUNG, Richard. AB, *Alexander,* 1818.

YOUNG, Richard. Seaman, *Resolution,* 1776-80.

YOUNG, Richard. AB, *Enterprise,* 1850-55.

YOUNG, Thomas. AB, *Chatham,* 1791-95.

YOUNG, Walter. Lieutenant commanding, *Lion,* 1777.

YOUNGSON, George. Officers' Cook, *Assistance,* 1852-54.

ZENO, Antonio. Voyage to America, c1380. [DCB Vol. 1].

ZENO, Nicolo. Voyage to America, c1380. [DCB Vol. 1].

ZIMMERMANN, Heinrich. AB, *Discovery,* 1776-80.

Bibliography

ABBES, H. (1884). "Die deutsche Nordpolar-Expedition nach dem Cumberland-Sunde." *Globus,* Bd. 46, 1884, pp. 294-98, 312-15, 328-31, 343-45, 365-68.

ABBES, H. (1884a). "Die Eskimos des Cumberland-Sundes." *Globus,* Bd. 46, 1884, pp. 198-201, 213-18.

ADAM, C.I. (1950). "The Bathurst Inlet patrol." *Royal Canadian Mounted Police Quarterly,* Vol. 16, No. 1, July 1950, pp. 12-25.

AHLMANN, Hans W:son (1931). "Otto Sverdrup," *Ymer,* Arg. 51, Häft 1, 1931, pp. 85-87.

ALCOCK, F.J. (1953). "Joseph Burr Tyrrell." *Canadian Geographical Journal,* Vol. 47, No. 6, 1953, pp. 253-55.

ALCOCK, F.J. (1954). "Albert Peter Low." *Canadian Geographical Journal,* Vol. 48, No. 4, 1954, pp. 160-63.

ALCOCK, *Sir* Rutherford (1877). "Lord Milton." *Royal Geographical Society. Proceedings,* Vol. 21, No. 5, 1876-77 (pub. 1877), pp. 416-18.

ALCOCK, *Sir* Rutherford (1878). "Rear-Admiral William Smyth." *Royal Geographical Society. Proceedings,* Vol. 22, No. 5, 1877-78 (pub. 1878), pp. 321-22.

ALEXANDER, Charles Paul *and* MÜNSTER, T.G. (1924). "Report of the second Norwegian Arctic Expedition in the *Fram,* 1898-1902. The crane-flies. Coleoptera, supplement." *Norsk Entomologisk Tidsskrift,* Bd. 1, 1923 (pub. 1924), pp. 296-97.

ALLEN, Everett S. (1963). *Arctic odyssey; the life of Rear-Admiral Donald B. MacMillan.* New York, Dodd, Mead and Co., 1963.

ALLEN, J.A. (1899). "On mammals from the Northwest Territory collected by Mr A.J. Stone." *American Museum of Natural History. Bulletin.* Vol. 12, 1899, pp. 1-9.

ALLEN, J.A. (1901). "The musk-oxen of Arctic America and Greenland." *American Museum of Natural History. Bulletin,* Vol. 14, 1901, pp. 69-86.

ALLEN, J.A. (1902). "A new caribou from Ellesmere Land." *American Museum of Natural History. Bulletin,* Vol. 16, 1902, pp. 409-12.

AMERICAN GEOGRAPHICAL SOCIETY (1881). "Arctic meeting at Chickering Hall, October 28th, 1880. Reception of Lieut. Frederick Schwatka and his associates of the Franklin search party of 1878, 1879 and 1880." *American Geographical Society. Bulletin,* Vol. 12, No. 4, 1880 (pub. 1881), pp. 237-96.

AMERICAN GEOGRAPHICAL SOCIETY (1905). "Dr. Alpheus S. Packard." *American Geographical Society. Bulletin,* Vol. 37, No. 3, 1905, p. 172.

AMERICAN GEOGRAPHICAL SOCIETY (1909). "Discovery of the North Pole." *American Geographical Society of New York. Bulletin,* Vol. 41, 1909, pp. 575-78.

AMUNDSEN, Roald (1907). "The Northwest Passage." *Harper's Magazine,* Vol. 114, 1907, pp. 659-74.

AMUNDSEN, Roald (1907a). "To the North Magnetic Pole and through the Northwest Passage." *Geographical Journal,* Vol. 29, 1907, pp. 485-518.

AMUNDSEN, Roald (1908). *The North West Passage; being the record of a voyage of exploration of the ship "Gjøa" 1903-1907, by Roald Amundsen, with a supplement by First Lieutenant Hansen, vice-commander of the expedition.* London, Archibald Constable, 1908. 2 vols.

AMUNDSEN, Roald (1927). *My life as an explorer.* London, Heinemann, 1927.

[ANDERSON, David] (1854). *The net in the bay; or, journal of a visit to Moose and Albany.* London, Thomas Hatchard, 1854. (Reprinted by S.R. Publishers Ltd., and Johnson Reprint Corporation, 1967).

ANDERSON, Ian Stuart (1972). "Bathurst Inlet patrol." *Beaver,* Outfit 302, No. 4,

Spring 1972, pp. 20-25.

ANDERSON, J.W. (1939). "Trading north of Hudson's Bay." *Beaver,* Outfit 270, No. 3, December 1939, p. 43.

ANDERSON, James (1856). "Letter from Chief Factor James Anderson, to Sir George Simpson, F.R.G.S., Governor in Chief of Rupert Land. Communicated by the Hudson's Bay Company." *Royal Geographical Society. Journal,* Vol. 26, 1856, pp. 18-25.

ANDERSON, James (1857). "Extracts from Chief Factor James Anderson's Arctic journal. Communicated by Sir John Richardson." *Royal Geographical Society. Journal,* Vol. 27, 1857, pp. 321-28.

ANDERSON, James (1940-41). "Chief Factor James Anderson's Back River journal of 1855." *Canadian Field-Naturalist,* Vol. 54, 1940, pp. 63-67, 84-89, 107-09, 125-26, 134-36; Vol. 55, 1941, pp. 9-11, 21-26, 38-44.

ANDERSON, Rudolph Martin (1913). "Arctic game notes. Distribution of large game animals in the far north; extinction of the musk ox; the chances for survival of moose and caribou; mountain sheep; polar bear and grizzly." *Natural History,* Vol. 13, January 1913, pp. 5-21.

ANDERSON, Rudolph Martin (1915-17). "Canadian Arctic Expedition, 1913-14, 1915, 1916." *Geological Survey of Canada. Summary Report,* 1914 (pub. 1915), pp. 163-66; 1915 (pub. 1916), pp. 220-36; 1916 (pub. 1917), pp. 314-30.

ANDERSON, Rudolph Martin (1917). "Canadian Arctic Expedition, 1916. Zoology." *Geological Survey of Canada. Summary Report,* 1916 (pub. 1917), pp. 374-84.

ANDERSON, Rudolph Martin (1917a). "Recent explorations on the Canadian Arctic coast." *Geographical Review,* Vol. 4, October 1917, pp. 241-66.

ANDERSON, Rudolph Martin (1919). "Recent zoological explorations in the western Arctic." *Washington Academy of Science. Journal,* Vol. 9, 1919, pp. 312-14.

ANDREWS, C.L. (1947). "Alexander Hunter Murray founder of Fort Yukon." *Alaska Life,* Vol. 10, No. 4, April 1947, pp. 4-6, 21.

ANONYMOUS (1821). *Letters written during the late voyage of discovery in the western Arctic Sea. By an officer of the expedition.* London, Sir Richard Phillips and Co., 1821.

ANONYMOUS (1854). *Memoir of the late Lieutenant Bellot . . . to accompany his engraved portrait: from the original picture, painted expressly for Lady Franklin, by Stephen Pearce, and engraved by James Scott.* London, Thomas Boys, 1854.

ANONYMOUS (1899). *The search for Franklin. A narrative of the American expedition under Lieutenant Schwatka, 1878 to 1880.* London, T. Nelson and Sons, 1899.

ANONYMOUS (1953). "The Radford and Street murders." *Beaver,* Outfit 284, September 1953, pp. 28-29.

ANONYMOUS (n.d.). *Commander Bedford Pim, R.N., F.R.G.S., Assoc. Inst. C.E.* No publisher, n.d., 8 pp. [Pamphlet in Scott Polar Research Institute library].

ARCTIC (1963). "Dr L.E. Borden." *Arctic,* Vol. 16, No. 4, 1963, p. 279.

ARCTIC (1963a). "George M. Douglas (1875-1963)." *Arctic,* Vol. 16, No. 4, 1963, pp. 278-79.

ARCTIC (1970). "Diamond Jenness (1886-1969)." *Arctic,* Vol. 23, No. 2, June 1970, pp. 71-81.

ARMOND, R.N. de (1973). "A letter to Jack McQuesten: 'gold on the Fortymile'." *Alaska Journal - History and Arts of the North - Quarterly,* Vol. 3, No. 2, Spring 1973, pp. 114-21.

ARMSTRONG, Alexander (1857). *A personal narrative of the discovery of the North-west Passage; with numerous incidents of travel and adventure during nearly five years' continuous service in the Arctic regions while in search of the expedition under Sir John Franklin.* London, Hurst and Blackett, 1857.

ARTHUR, Jean (1974). "La famille Le Vesconte." *Société Jersiaise. Annual Bulletin*, 99th, Vol. 21, Pt. 2, 1974, pp. 264-68.

ASHER, G.M. (1860). *Henry Hudson the navigator. The original documents in which his career is recorded, collected, partly translated, and annotated, with an introduction.* London, Hakluyt Society, 1860. (Works issued by the Hakluyt Society, First Series, No. 27).

AUROUSSEAU, M. (1946). "A correction to Arctic history." *Geographical Journal*, Vol. 107, Nos. 5-6, 1946, pp. 273-74.

BACK, George (1836). *Narrative of the Arctic land expedition to the mouth of the Great Fish River, and along the shores of the Arctic Ocean, in the years 1833, 1834, and 1835.* London, John Murray, 1836. (Reprinted by M.G. Hurtig Ltd., Edmonton, 1970).

BACK, George (1837). "On the north-eastern shore of Southampton Island." *Royal Geographical Society. Journal*, Vol. 7, 1837, pp. 460-66.

BACK, George (1838). *Narrative of an expedition in H.M.S. Terror, undertaken with a view to geographical discovery on the Arctic shores, in the years 1836-7.* London, John Murray, 1838.

BAILEY, T.A. (1937). "Theodore Roosevelt and the Alaska boundary settlement." *Canadian Historical Review*, Vol. 18, June 1937, pp. 123-30.

BAIRD, P.D. (1971). "D.B. MacMillan (1874-1970)." *Arctic*, Vol. 24, No. 1, March 1971, p. 80.

BALCH, Francis Noyes (1911). "On a new Labradorean species of *Onchidiopsis*, a genus of mollusks new to eastern North America; with remarks on its relationships." *U.S. National Museum. Proceedings*, Vol. 38, 1911, pp. 469-84.

BANCROFT, Hubert Howe (1890). *History of Alaska 1730-1885.* San Francisco, The History Co., 1890. (Reprinted by the Antiquarian Press, New York, 1960).

BARRATT, Glynn (1976). "Rae's Fort Hope." *Beaver*, Outfit 304, No. 3, Winter 1973, pp. 9-15.

BARROW, John (1818). *A chronological history of voyages into the Arctic regions; undertaken chiefly for the purpose of discovering a North-east, North-west, or polar passage between the Atlantic and Pacific: from the earliest periods of Scandinavian navigation, to the departure of the recent expeditions, under the orders of Captains Ross and Buchan.* London, John Murray, 1818. (Reprinted by David & Charles Reprints, 1971).

BARTLETT, Robert Abram (1931). *The log of Bob Bartlett; the true story of forty years of seafaring and exploration.* New York, Blue Ribbon Books, 1931.

BARTLETT, Robert Abram (1940). "Peary's extended explorations of Arctic lands culminating in the attainment of the North Pole." *American Philosophical Society. Proceedings*, Vol. 82, 1940, pp. 935-47.

BARTLETT, Robert A., *and* HALE, Ralph T. (1916). *Northward ho! The last voyage of the Karluk.* Boston, Small, Maynard & Co., 1916.

BAUER, L.A. *and others.* (n.d.). "Land magnetic observations, 1914-20. By L.A. Bauer, J.A. Fleming, H.W. Fisk, and W.J. Peters." *Carnegie Institution of Washington Publication*, No. 175, Vol. 4.

BEAGLEHOLE, J.C. *ed* (1967). *The journals of Captain James Cook on his voyages of discovery. Four volumes and a portfolio. Volume 3. The voyage of the Resolution and Discovery 1776-1780. Parts 1 and 2* (2 vols). Cambridge, Cambridge University Press for the Hakluyt Society, 1967. (Hakluyt Society Extra Series, No. 36).

BEAVER (1929). "S.S. Pelican." *Beaver*, Outfit 260, No. 1, June 1929, p. 215.

BEAVER, (1941). [Obituary of William Cornwallis King]. *Beaver,* Outfit 271, March 1941, p. 58.

BEAVER (1945). "Arctic mariner." *Beaver,* Outfit 276, December 1945, p. 46.

BECKER, Alois *Ritter von* (1878). *Arktische Reise der englischen Yacht Pandora im Jahre 1876. Unter Commando des Capitain Sir Allen Young.* Pola, Carl Gerold's Sohn in Wien, 1878.

BÉDARD, A. (1919). "Un voyage dans l'Ungava." *Société de Géographie de Québec,* Vol. 12, 1919, pp. 20-23; Vol. 13, 1919, pp. 139-41.

BEECHEY, Frederick William (1831). *Narrative of a voyage to the Pacific and Beering's Strait, to co-operate with the polar expeditions: performed in His Majesty's Ship Blossom, under the command of Captain F. W. Beechey . . . in the years 1825, 26, 27, 28.* 2 vols. London, Henry Colburn and Richard Bentley, 1831.

BEECHEY, Frederick William (1857). "Parry, Sir William Edward, Kt., D.C.L., F.R.S." *Royal Geographical Society. Proceedings,* Vol. 1, No. 5, 1855-57 (pub. 1857), pp. 127-30.

BELCHER, *Sir* Edward (1855). *The last of the Arctic voyages; being a narrative of the expedition in H.M.S. Assistance, under the command of Captain Sir Edward Belcher, C.B., in search of Sir John Franklin, during the years 1852-53-54. With notes on the natural history, by Sir John Richardson, Professor Owen, Thomas Bell, J. W. Salter, and Lovell Reeve. 2 vols.* London, Lovell Reeve, 1855.

BELL, Benjamin (1881). *Lieut. John Irving, R.N., of H.M.S. "Terror" in Sir John Franklin's last expedition to the Arctic regions; a memorial sketch with letters.* Edinburgh, D. Douglas, 1881.

BELL, James Mackintosh (1901). "Explorations in the Great Bear Lake region." *Geographical Journal,* Vol. 18, No. 3, September 1901, pp. 249-58.

BELL, Michael (1972). "Thomas Mitchell, photographer and artist in the high Arctic, 1875-76." *Image* (Journal of Photography and Motion Pictures at the International Museum of Photography at George Eastman House), Vol. 15, No. 3, September 1972, pp. 12-20.

BELL, Robert (1879). "Report on an exploration of the east coast of Hudson's Bay 1877." *Geological Survey of Canada. Report of Progress for 1877-78,* pub. 1879, Part C.

BELL, Robert (1884). *Observations on the geology, minerals, zoology and botany of the Labrador coast, Hudson's Strait and Bay.* Montreal, Dawson Bros., 1884. (Geological and Natural History Survey of Canada. Report of Progress, 1882-83-84, Section DD).

BELL, Robert (1886). "Observations on the geology, zoology and botany of Hudson's Strait and Bay, made in 1885." *Geological and Natural History Survey of Canada. Annual Report,* 1885 (pub. 1886), Part DD.

BELL, Robert (1901). *Report of an exploration on the northern side of Hudson Strait.* Ottawa, Geological Survey of Canada, 1901. (Geological Survey of Canada Annual Report, NS Vol. 11, 1898 (pub. 1901), Section M).

BELL, Robert (1901a). "A survey in Baffinland, with a short description of the country." *Geographical Journal,* Vol. 18, July 1901, pp. 25-46.

BELLOT, Joseph-René (1854). *Journal d'un voyage aux mers polaires exécuté à la recherche de Sir John Franklin, en 1851 et 1852.* Paris, Perrotin, 1854.

BELLOT, Joseph-René (1855). *Memoirs of Lieutenant Joseph René Bellot. . .with his journal of a voyage in the polar seas, in search of Sir John Franklin.* 2 vols. London, Hurst and Blackett, 1855.

BERKELEY, M. J. (1878). "Enumeration of the fungi collected during the Arctic expedition, 1875-76." *Linnean Society, London. Journal. Botany,* Vol. 17, 1878, pp. 13-17.

BERNARD, *Sir* Francis (1792). "Account of the coast of Labrador." *Massachusetts Historical Society Collection,* Series 1, Vol. 1, 1792, pp. 233-37.

BERNIER, Joseph-Elzéar (1909). *Report on the Dominion Government expedition to Arctic Islands and the Hudson Strait on board the* C.G.S. *"Arctic" 1906-1907* Ottawa, King's Printer, 1909.

BERNIER, Joseph-Elzéar (1910). *Report on the Dominion of Canada Government expedition to the Arctic islands and Hudson Strait on board the* D.G.S. *'Arctic.'* Ottawa, Government Printing Bureau, 1910.

BERNIER, Joseph-Elzéar (1939). *Master mariner and Arctic explorer. A narrative of sixty years at sea from the logs and yarns of Captain J. E. Bernier . . . .* Ottawa, Printed by Le Droit, 1939.

BERRY, Erick (1966). *Mr. Arctic; an account of Vilhjalmur Stefansson.* New York, David McKay Co., 1966.

BERTON, Pierre (1958). *The Klondike fever; the life and death of the last great gold rush.* New York, Alfred A. Knopf, 1958.

BERTON, Pierre (1972). *Klondike; the last great gold rush, 1896-1899.* Revised ed. Toronto, McClelland and Stewart, 1972.

BESSELS, Emil (1879). *Die amerikanische Nordpol-Expedition.* Leipzig, Verlag von Wilhelm Engelmann, 1879.

BIGELOW, Henry Bryant (1902). "Birds of the northeastern coast of Labrador." *Auk,* Vol. 19, January 1902, pp. 24-31.

BIGELOW, Henry Bryant (1910). "Coelenterates from Labrador and Newfoundland, collected by Mr. Owen Bryant from July to October, 1908." U.S. *National Museum. Proceedings,* Vol. 37, 1910, pp. 301-20.

BIGGAR, Henry Percival (1901). *The early trading companies of New France. A contribution to the history of commerce and discovery in North America.* Toronto, University of Toronto Library, 1901. (Reprinted by Argonaut Press Ltd., New York, 1965).

BIGGAR, Henry Percival (1903). *The voyages of the Cabots and the Corte-Reals to North America and Greenland, 1497-1503.* Paris, Macon, Protat Frères, 1903.

BIGGAR, Henry Percival (1924). *The voyages of Jacques Cartier. Published from the originals with translations, notes and appendices.* Ottawa, Public Archives of Canada, 1924. (PAC Publications, No. 11).

BIGGAR, Henry Percival (1930). *A collection of documents relating to Jacques Cartier and the Sieur de Roberval.* Ottawa, Public Archives of Canada, 1930. (PAC Publications, No. 14).

BILLMAN, Christine W. (1971). "Jack Craig and the Alaska boundary survey." *Beaver,* Outfit 302, No. 2, Autumn 1971, pp. 44-49.

BIRD, C.D. (1967). "The mosses collected by Thomas Drummond in western Canada, 1825-1827." *Bryologist,* Vol. 70, No. 2, 1967, pp. 262-66.

BIRD, J. Brian (1953). "Southampton Island." *Canada. Department of Mines and Technical Surveys. Geographical Branch. Memoir,* No. 1, 1953.

BLACK, Patrick (1876). *Scurvy in high latitudes; an attempt to explain the cause of the "medical failure" of the Arctic expedition of 1875-6.* London, Smith, Elder & Co., 1876.

BLACKADAR, R. G. (1967). "Kodlunarn Island and Frobisher's 'gold.' " *Arctic Circular,* Vol. 17, No. 1, 1965-66 (pub. 1967), pp. 1-12.

BLAKE, E. Vale *ed* (1874). *Arctic experiences: containing Capt. George E. Tyson's wonderful drift on the ice-floe, a history of the Polaris expedition, the cruise of the Tigress and rescue of the Polaris survivors. To which is added a general Arctic chronology.* New York, Harper and Brothers, 1874.

BOAS, Franz (1884). "The Eskimo of Baffin Island." *Anthropological Society of Washing-*

ton. Transactions, Vol. 3, 1884, pp. 95-102.

BOAS, Franz (1884a). "A journey in Cumberland Sound and on the west shore of Davis Strait in 1883 and 1884." *American Geographical Society. Journal,* Vol. 16, 1884, pp. 242-72.

BOAS, Franz (1885). "Baffin Land. Geographischen Ergebnisse einer in den Jahren 1883 und 1884 ausgeführten Forschungsreise." *Petermanns Mitteilungen aus Justus Perthes' Geographischer Anstalt,* Ergänzungsband 17, 1884-85, No. 80, 1885.

BOAS, Franz (1885a). "Cumberland Sound and its Eskimos." *Popular Science Monthly,* Vol. 26, 1885, pp. 768-79.

BOAS, Franz (1885b). "Reise im Baffinlande 1883 und 1884." *Gesellschaft für Erdkunde zu Berlin. Verhandlungen,* Bd. 12, 1885, pp. 288-97.

BOAS, Franz (1885c). "Die Eskimos des Baffinlandes." *Deutsche Geographentag 5te, Hamburg, 1885. Verhandlungen,* pp. 101-12.

BOAS, Franz (1885d). "Die Sagen der Baffin-Land-Eskimos." *Zeitschrift für Ethnologie,* Vol. 17, 1885, pp. 161-66.

BOAS, Franz (1885e). "Die Wohnsitze und Wanderungen der Baffin-Land Eskimos." *Deutsche Geographische Blätter* (Bremen), Bd. 8, Heft 1, 1885, pp. 31-38.

BOAS, Franz (1887). "The geography and geology of Baffin Island." *Royal Society of Canada. Transactions,* Vol. 4, Sec. 4, 1887, pp. 75-78.

BOAS, Franz (1887a). "A year among the Eskimo." *American Geographical Society. Journal,* Vol. 19, 1887, pp. 383-402.

BOAS, Franz (1888). "The Central Eskimo." *Bureau of American Ethnology. Annual Report,* 6th, 1884-85 (pub. 1888), pp. 399-669.

BOAS, Franz (1888a). "Sagen der Eskimos von Baffin-Land." *Zeitschrift für Ethnologie,* Vol. 20, 1888, pp. 398-405.

BOAS, Franz (1894). "Eskimo-Dialekt des Cumberland-Sundes." *Anthropologische Gesellschaft in Wien. Mittheilungen,* Bd. 24, 1894 (Neue Folge Bd. 14), pp. 97-114.

BOAS, Franz (1901-07). "The Eskimo of Baffin Land and Hudson Bay." *American Museum of Natural History. Bulletin,* Vol. 15, Part 1, 1901, Part 2, 1907.

BODILLY, R.B. *ed* (1928). *The voyage of Captain Thomas James for the discovery of the North-west Passage, 1631.* London and Toronto, J. M. Dent & Sons, 1928.

BOISSEVAIN, Charles (1885). *Story of the life and aspirations of L. R. Koolemans Beynen (a young Dutch naval officer).* London, Sampson, Low, Marston, Searle, and Rivington, 1885.

BOON, T.C.B. (1962). *The Anglican Church from the Bay to the Rockies; a history of the ecclesiastical province of Rupert's Land and its dioceses from 1820 to 1950.* Toronto, Ryerson Press, 1962.

BOON, T.C.B. (1965). "William West Kirkby, first Anglican missionary to the Loucheux." *Beaver,* Outfit 295, Spring 1965, pp. 36-43.

BORDEN, K. Ethel (1961). "Northward 1903-04." *Canadian Geographical Journal,* Vol. 62, No. 1, January 1961, pp. 32-39.

BOSTOCK, Hugh Samuel *and* ROWLEY, Diana M.R. (1959). "Charles Camsell, C.M.G. 1876-1958." *Geographical Journal,* Vol. 125, Part 2, 1959, pp. 287-88.

BRAENDEGAARD, Jens Rasmus Jensen (1936). "Revisal of spiders from Ellesmereland collected by the Second Arctic expedition of the 'Fram'." *Norsk Entomologisk Tidsskrift,* Bd. 4, 1936, pp. 128-30.

BRAINARD, David Legge (1940). *Six came back; the Arctic adventure of David L. Brainard, edited by Bessie Rowland James.* Indianapolis and New York, Bobbs-Merrill Co., 1940.

BRAITHWAITE, John (1849). *Supplement to Captain Sir John Ross's narrative of a second voyage in the Victory, in search of a North-west*

Passage, containing the suppressed facts necessary to a proper understanding of the causes of the failure of the steam machinery of the Victory, and a just appreciation of Captain Sir John Ross's character as an officer and a man of science. London, printed by H. Silverlock, 1849.

BRETON, P. E. (1963). *Irish hermit of the Arctic: the life of Brother J. Patrick Kearney, O.M.I. Translated by J. S. Mullany.* Edmonton, Editions de l'Ermitage, 1963.

BRITISH MEDICAL JOURNAL (1910). "Inspector-General Coppinger, R.N." *British Medical Journal,* Vol. 1, 1910, p. 1090.

BRITTON, E. G. (1909). "Arctic mosses." *Bryologist,* Vol. 12, 1909, p. 106.

BROWN, Robert Neal Rudmose (1961). "Frederick William Beechey." *Polar Notes,* No. 3, 1961, pp. 41-43.

BRYANT, Henry G. (1892). "The Grand Falls of Labrador." *Century Magazine,* Vol. 44, No. 5, September 1892, pp. 643-55.

BRYANT, Henry G. (1894). "A journey to the Grand Falls of Labrador." *Geographical Society of Philadelphia. Bulletin,* Vol. 1, No. 2, March 1894, pp. 37-86.

BRYANT, Henry G. (1894). "The Peary auxiliary expedition, 1894." *Geographical Journal,* Vol. 4, 1894, pp. 531-34.

BRYANT, Henry G. (1895). "The Peary auxiliary expedition of 1894." *Geographical Society of Philadelphia. Bulletin,* Vol. 1, June 1895, pp. 141-215.

BRYCE, George (1883). "In memoriam: late A. K. Isbister, M.A., LL.B." *Historical and Scientific Society of Manitoba. Transactions,* No. 8, 1883, 4pp.

BUCKLAND, A. R. (n.d.). *John Horden, missionary bishop; a life on the shores of Hudson's Bay.* London, Sunday School Union, n.d.

BUGGE, Aage (1965). "Kallihirua, polar-eskimoen i Canterbury." *Grønland,* No. 5, 1965, pp. 161-75.

BUGGE, Aage (1966). "Polareskimoen i Canterbury. Supplerende oplysninger vedr. Kallihirua." *Grønland,* No. 1, 1966, pp. 17-22.

BURFORD, Robert (1850). *Description of summer and winter views of the polar regions, as seen during the expedition of Capt. James Clark Ross, KT. F.R.S. in 1848-9, now exhibiting at the Panorama, Leicester Square. Painted by the proprietor, Robert Burford, assisted by H. C. Selous, from drawings taken by Lieut Browne, of H.M.S. "Enterprise."* London, printed by W. J. Golbourn, 1850.

BURGESS, Helen (1968). "Tookoolito of Cumberland Sound." *North,* Vol. 15, No. 1, January-February 1968, pp. 40-43.

BURGESSE, J. A. (1947). "Jolliet on James Bay." *Beaver,* Outfit 278, December 1947, pp. 12-15.

BURPEE, Lawrence Johnstone *ed* (1907). "York Factory to the Blackfeet country. The journal of Anthony Hendry, 1754-55." *Royal Society of Canada. Proceedings and Transactions,* Third Series, Vol. 1, Section 2, 1907, pp. 307-64.

BURPEE, Lawrence Johnstone *ed* (1908). "An adventurer from Hudson Bay. Journal of Matthew Cocking from York Factory to the Blackfeet country 1772-73." *Royal Society of Canada. Proceedings and Transactions,* 3rd Series, Vol. 2, Section 2, 1908, pp. 89-121.

BURPEE, Lawrence Johnstone *ed* (1910). *Journal of the Yukon, 1847-48, by Alexander Hunter Murray. Edited with notes by L. J. Burpee. Published by authority of the Minister of Agriculture under the direction of the Archivist.* Ottawa, Government Printing Bureau, 1910. (Canada, Public Archives. Publications, No. 4).

BURPEE, Lawrence Johnstone (1935). *The search for the western sea; the story of the exploration of north-western America. Vols. 1 and 2.* Toronto, Macmillan Company of Canada, 1935.

BURPEE, Lawrence Johnstone (1945).

"Campbell of the Yukon." *Canadian Geographical Journal,* Vol. 30, No. 4, 1945, pp. 200-01.

BURPEE, Lawrence Johnstone (1946). "Samuel Hearne finds the Coppermine." *Canadian Geographical Journal,* Vol. 32, No. 3, March 1946, pp. 146-47.

BUSHNELL, Vivian C. (1956). "Marvin's ice shelf journey, 1906." *Arctic,* Vol. 9, No. 3, 1956, pp. 166-77.

BUSK, George (1880). "List of Polyzoa collected by Captain H. W. Feilden in the North Polar expedition; with descriptions of new species." *Linnean Society of London, Journal. Zoology,* Vol. 15, 1880, pp. 231-41.

BUTLER, Arthur Gardiner (1877). "Description of a new species of Argynnis from Arctic America." *Entomologists' Monthly Magazine,* Vol. 13, February 1877, p. 206.

CABOT, William Brooks (1912). *In northern Labrador.* Boston, Richard G. Badger, The Gorham Press, 1912.

CABOT, William Brooks (n.d.). *Labrador.* London, Heath Cranton, Ltd., n.d.

CALDER-MARSHALL, Arthur (1963). *The innocent eye; the life of Robert J. Flaherty.* London, W. H. Allen, 1963.

CALDERWOOD, W. L. (1935). "A boat journey in search of Franklin." *Scottish Geographical Magazine,* Vol. 51, No. 1, 1935, pp. 228-35.

CAMBRIDGE, Octavius Pickard (1877). "On some new and little-known spiders from the Arctic regions." *Annals and Magazine of Natural History,* Series 4, Vol. 20, October 1877, pp. 273-85.

CAMERON, Agnes Deans (1910). *The new north; being some account of a woman's journey through Canada to the Arctic.* New York and London, D. Appleton and Co., 1910.

CAMERON, Agnes Deans (1910a). "Canada's farthest north." *Geographical Journal,* Vol. 35, No. 6, June 1910, pp. 705-08.

CAMERON, Hector Charles (1952). *Sir Joseph Banks* K.B., F.R.S. *the autocrat of the philosophers.* London, Batchworth Press, 1952.

CAMPBELL, Gordon (1936). *Captain James Cook,* R.N., F.R.S. London, Hodder and Stoughton, 1936.

CAMPBELL, Marjorie Wilkins (1957). *The North West Company.* Toronto, Macmillan Company of Canada, 1957.

CAMPBELL, Robert (1967). "The discovery and exploration of the Pelly (Yukon) River." *North,* Vol. 14, No. 4, July-August 1967, pp. 9-15.

CAMSELL, Charles (1906). "Report on the Peel River and tributaries, Yukon and Mackenzie." *Geological Survey of Canada. Report,* No. 951, 1906. (In: *Geological Survey of Canada. Annual Report,* NS Vol. 16, 1904 (pub. 1906), Report CC).

CAMSELL, Charles (1916). "An exploration of the Tazin and Taltson Rivers North West Territories." *Geological Survey of Canada. Memoir* 84 (No. 69, Geological Series), 1916.

CAMSELL, Charles (1954). *Son of the north.* Toronto, Ryerson Press, 1954.

CANADA. Department of Marine and Fisheries (1886). "Report of the second Hudson's Bay Expedition under the command of Lieut. A. R. Gordon, R.N., 1885." *Canada. Parliament. Sessional Papers* (No. 11), Vol. 9, 1886.

CANADA. Department of Marine and Fisheries. (1887). *Report of the Hudson's Bay expedition, under the command of Lieut. A. R. Gordon,* R.N., *1886.* Ottawa, Department of Marine and Fisheries, 1887. (Canada. Parliament. Sessional Papers (No. 15), Vol. 14, 1887).

CANADA. Department of Marine and Fisheries (1898). *Report of the expedition to*

Hudson Bay and Cumberland Gulf in the steamship "Diana" under the command of William Wakeham . . . in the year 1897. Ottawa, Queen's Printer, 1898.

CANADA. Department of Marine and Fisheries (1911). *Report on the Dominion Government expedition to the northern waters and Arctic Archipelago of the D.G.S. "Arctic" in 1910.* Ottawa, Department of Marine and Fisheries, [1911?].

CANADA. Royal North-West Mounted Police (1904). "Report of Superintendent C. Constantine. Trip to Mackenzie River, with report of Sergeant Fitzgerald, Herschell Island Detachment." *Canada. Sessional Papers,* No. 28, 1904, pp. 43-54.

CANADA. Royal North-West Mounted Police (1905). "Report of Superintendent J. D. Moodie on service in Hudson Bay (per S.S. 'Neptune,' 1903-4)." *Canada. Sessional Papers,* No. 28, Part 4, 1905, pp. 1-12.

CANADA. Royal North-West Mounted Police (1907). "Patrol report, Constable L. E. Seller, Fullerton to Lyons Inlet." *Canada. Sessional Papers,* No. 28, 1907, pp. 116-27.

CANADA. Royal North-West Mounted Police (1908). "Report of Inspector A. M. Jarvis, C.M.G., on Wood Buffalo in the Mackenzie River district." *Canada. Sessional Papers,* No. 28, 1908, pp. 122-29.

CANADA. Royal North-West Mounted Police (1909). "Report of Sergeant D. McArthur patrol, Fullerton to Wager Bay." *Canada. Sessional Papers,* No. 28, 1909, pp. 265-67.

CANADA. Royal North-West Mounted Police (1910). "Patrol report Inspector E. A. Pelletier, Fort Saskatchewan, Alberta, to Chesterfield Inlet and Fullerton, Hudson Bay, and return to Regina via Churchill, Hudson Bay." *Canada. Sessional Papers,* No. 28, 1910, Appendix O.

CANADIAN ARCTIC EXPEDITION (1919-46). *Report of the Canadian Arctic Expedition 1913-18.* Vol. 3: *Insects.* Vol. 4: *Botany.* Vol. 5: *Botany.* Vol. 6: *Fishes and Tunicates.* Vol. 7: *Crustacea.* Vol. 8: *Mollusks, Echinoderms, Coelenterates, etc.* Vol. 9: *Annelids, parasitic worms, Protozoans, etc.* Vol. 10: *Plankton, hydrography, tides, etc.* Vol. 11: *Geology and geography.* Vol. 12: *The life of the Copper Eskimos.* Vol. 13: *Eskimo folk-lore.* Vol. 14: *Eskimo songs.* Vol. 15: *Eskimo language and technology.* Vol. 16: *Material culture of the Copper Eskimo.* Ottawa, King's Printer, 1919-46. (Vols. 1 and 2 were never published).

CANADIAN FIELD - NATURALIST (1920). "Obituary. John Macoun, 1831-1920. Assistant Director and Naturalist to the Geological Survey of Canada." *Canadian Field-Naturalist,* Vol. 34, No. 6, September 1920, pp. 110-14.

CANADIAN GEOGRAPHICAL JOURNAL (1934). "James Mackintosh Bell." *Canadian Geographical Journal,* Vol. 8, No. 4, April 1934, pp. 197-98.

CANADIAN RECORD OF SCIENCE (1893). "Obituary notice. Dr. John Rae, F.R.S., F.R.G.S." *Canadian Record of Science.* Vol. 5, No. 8, 1893, pp. 484-87.

CARRIÈRE, Gaston (1958). *Le roi des Betsiamites. Le père Charles Arnaud, O.M.I. (1826-1914).* Ottawa, Les Editions de l'Université d'Ottawa, 1958.

CARRIÈRE, Gaston (1963). *Explorateur pour le Christ. Louis Babel, O.M.I., 1826-1912.* Montréal, Rayonnement, 1963.

CARRIÈRE, Gaston (1964). "1864-1964. Glorieux centenaire d'un grand missionaire. Le père Henri Grollier, O.M.I." *La Bannière de Marie-Immaculée,* Année 72, 1964, pp. 40-44.

CARTER, H. J. (1877). "Arctic and Antarctic sponges." *Annals and Magazine of Natural History,* Series 4, Vol. 20, 1877, pp. 38-42.

CARY, Austin (1891). "Geological facts noted on Grand River, Labrador." *American Journal of Science,* Series 3, Vol. 42,

November 1891, pp. 419-21; correction, p. 516.

CARY, Austin (1892). "Exploration on Grand River, Labrador." *Journal of the American Geographical Society,* Vol. 24, No. 1 1892, pp. 1-17.

CASEBEER, Robert L. (1964). "Olaus Johan Murie, 1889-1963." *Journal of Wildlife Management,* Vol. 28, No. 1, 1964, pp. 191-92.

CASWELL, John Edwards (1977). "The RGS and the British Arctic Expedition, 1875-76." *Geographical Journal,* Vol. 143, Part 2, July 1977, pp. 200-10.

CHALMERS, J. W. (1960). *Fur trade governor; George Simpson 1820-1860.* Edmonton Institute of Applied Art Ltd., 1960.

CHALMERS, J. W. (1972). "Inland journey." *Beaver,* Outfit 303, No. 2, Autumn, 1972, pp. 52-58.

CHAPPELL, Edward (1817). *Narrative of a voyage to Hudson's Bay in His Majesty's Ship Rosamond, containing some account of the north-eastern coast of America and of the tribes inhabiting that remote region.* London, printed for J. Mawman, 1817.

CHARNOCK, John (1794-98). *Biographia navalis* . . . 6 vols. London, R. Faulder, 1794-98.

CHIMMO, William (1868). "A visit to the north-east coast of Labrador, during the autumn of 1867, by H.M.S., *Gannett,* Commander W. Chimmo, R.N." *Royal Geographical Society, Journal,* Vol. 38, 1868, pp. 258-81.

CHOQUE, Charles (1964). "Pierre Henri Grollier, O.M.I., 1826-1864." *Eskimo,* Vol. 67, Spring 1964, p. 3.

CREIGHTON, J.G.A. (1891). *Exploratory survey of southern Labrador.* [Ottawa], 1891. 43 pp. (Manuscript in the library of the Geological Survey of Canada, Ottawa.)

CHRISTY, Miller *ed* (1894). *The voyages of Captain Luke Foxe of Hull, and Captain Thomas James of Bristol, in search of a North-West Passage, in 1631-32 . . . 2 vols.* London, Hakluyt Society, 1894. (Works issued by the Hakluyt Society, First Series, Nos. 88-89.)

CHRISTY, Miller (1927). "Captain William Hawkeridge and his voyage in search of a North-west passage in 1625." *Mariner's Mirror,* Vol. 13, No. 1, January 1927, pp. 51-78.

CODY, H.A. (1908). *An apostle of the north. Memoirs of the Right Reverend William Carpenter Bompas,* D.D. London, Seeley and Co., 1908.

COLEMAN, Arthur P. (1921). "Northeastern part of Labrador, and New Quebec." *Geological Survey of Canada. Memoir 124* (No. 106, Geological Series), 1921.

COLLINSON, Richard (1855). "Account of the proceedings of H.M.S. Enterprise from Behring Strait to Cambridge Bay." *Royal Geographical Society. Journal,* Vol. 25, 1855, pp. 194-206.

COLLINSON, Richard (1867). *The three voyages of Martin Frobisher, in search of a passage to Cathaia and India by the North-West, A.D. 1576-78, reprinted from the first edition of Hakluyt's voyages, with selections from manuscript documents in the British Museum and State Paper Office.* London, Hakluyt Society, 1867. (First Series, No. 38. Reprinted by Burt Franklin, New York, n.d.).

COLLINSON, Richard (1889). *Journal of H.M.S. Enterprise, on the expedition in search of Sir John Franklin's ships by Behring Strait. 1850-55.* London, Sampson Low, Marston, Searle, & Rivington, 1889.

COMEAU, Napoleon Alexander (1915). "Report on the fisheries expedition to Hudson Bay in the auxiliary schooner *Burleigh.*" 1914. *In:* Melvill, J.C.D., *and others.* "Reports on fisheries investigations in Hudson and James bays and tributary waters in 1914." *Canada. Department of the Naval Service. Annual Report,* 1913-14, Appendix, pp. 69-85.

(Canada. Sessional Papers, No. 39a, 1915).

COMER, George (1906). "Whaling in Hudson Bay, with notes on Southampton Island." In: *Boas anniversary volume. Anthropological papers written in honor of Franz Boas . . . Presented to him on the twenty-fifth anniversary of his doctorate. . . .* New York, G.E. Stechert & Co., 1906, pp. 475-84.

COMER, George (1910). "A geographical description of Southampton Island and notes upon the Eskimo." *American Geographical Society of New York. Bulletin,* Vol. 42, 1910, pp. 84-90.

COMER, George (1913). "Additions to Captain Comer's map of Southampton Island." *American Geographical Society of New York. Bulletin,* Vol. 45, 1913, pp. 516-18.

CONINGHAM, William *ed* (n.d.). *The last journals of Captain Fitzjames,* R.N. *of the lost polar expedition.* Brighton, W. Pearce, n.d.

CONYBEARE, Crawford John Markland (1930). *A brief record of service by sea and land.* Printed for private circulation, 1930.

COOK, Frederick Albert (1911). *My attainment of the pole; being the record of the expedition that first reached the boreal center 1907-1909 with the final summary of the polar controversy.* New York, Polar Publishing Co., London, Arlen and Co., 1911.

COOK, Frederick Albert (1951). *Return from the pole. Edited, with an introduction, by Frederick J. Pohl.* New York, Pellegrini & Cudahy, 1951.

COOKE, Alan (1960). "A woman's way." *Beaver,* Outfit 291, Summer 1960, pp. 40-45.

COOKE, Alan (1964). "The exploration of New Quebec." *In:* Malaurie, Jean *and* Rousseau, Jacques *eds. Le Nouveau-Québec. Contribution à l'étude de l'occupation humaine.* Paris, La Haye, Mouton & Co., 1964. (Bibliothèque Arctique et Antarctique, Vol. 2).

COOKE, Alan (1969). *The Ungava venture of the Hudson's Bay Company, 1830-1843.* Ph.D. thesis, Cambridge, Eng., 1969.

COOKE, Alan *and* ROSS, W. Gillies (1969). "The drift of the whaler *Viewforth* in Davis Strait, 1835-36, from William Elder's journal." *Polar Record,* Vol. 14, No. 92, May 1969, pp. 581-91.

CORMACK, William Epps (1824). "Account of a journey across the island of Newfoundland." *Edinburgh Philosophical Journal,* Vol. 10, No. 19, 1824, pp. 156-62.

CORMACK, William Epps (1928). *Narrative of a journey across the island of Newfoundland in 1822, by W. E. Cormack. Centenary issue, edited by F.A. Bruton,* M.A. . . . London, New York, Toronto, Longmans, Green & Co., 1928.

CORNER, George W. (1972). *Doctor Kane of the Arctic seas.* Philadelphia, Temple University Press, 1972.

CUMMING, Peter A. *and* MICKENBERG, Neil H., *eds* (1972). *Native rights in Canada. 2nd ed.* Toronto, Indian-Eskimo Association of Canada in association with General Publishing Co. Ltd., 1972.

CURRAN, William Tees *and* ADAMS, Horace P. (n.d.). *Glimpses of northern Canada, the land of hidden treasure.* Montreal, Cambridge Corporation Ltd., n.d.

CURRAN, William Tees *and* CALKINS, Harold A. (1917). *In Canada's wonderful northland. A story of eight months of travel by canoe, motorboat, and dog-team on the northern rivers and along the New Quebec coast of Hudson Bay.* New York and London, G. P. Putnam's Sons, 1917.

CYRIAX, Richard J. (1939). *Sir John Franklin's last expedition; a chapter in the history of the Royal Navy.* London, Methuen, 1939.

CYRIAX, Richard J. (1942). "Sir James Clark Ross and the Franklin expedition." *Polar Record,* Vol. 3, No. 24, 1942, pp. 528-40.

CYRIAX, Richard J. (1944). "Captain Hall

and the so-called survivors of the Franklin expedition." *Polar Record,* Vol. 4, No. 28, July 1944, pp. 170-85.

CYRIAX, Richard J. (1951). "Recently discovered traces of the Franklin expedition." *Geographical Journal,* Vol. 117, Part 2, June 1951, pp. 211-14.

CYRIAX, Richard J. (1958). "The two Franklin expedition records found on King William Island." *Mariner's Mirror,* Vol. 44, No. 3, August 1958, pp. 179-89.

CYRIAX, Richard J. (1959). "Sir John Franklin. A note on the absence of records on the shores past which he sailed during his last voyage." *Scottish Geographical Magazine,* Vol. 75, No. 1, 1959, pp. 30-40.

CYRIAX, Richard J. (1964). "The voyage of H.M.S. *North Star,* 1849-50." *Mariner's Mirror,* Vol. 50, No. 4, 1964, pp. 307-18.

CYRIAX, Richard J. (1969). "The unsolved problem of the Franklin expedition records supposedly buried on King William Island." *Mariner's Mirror,* Vol. 55, No. 1, 1969, pp. 23-32.

CYRIAX, R.J. *and* JONES, A.G.E. (1954). "The papers in the possession of Harry Peglar, Captain of the Foretop, H.M.S. *Terror,* 1845." *Mariner's Mirror,* Vol. 40, No. 3, August 1954, pp. 186-95.

DALL, W.H. (1898). "Travels on the Yukon and in the Yukon Territory in 1866-1868." *In:* Trimmer, F. Mortimer *comp. The Yukon Territory. The narrative of W.H. Dall. . . .* London, Downey & Co., 1898, pp. 3-242.

DALY, Reginald A. (1902). "The geology of the northeast coast of Labrador." *Harvard University. Museum of Comparative Zoology, Bulletin,* Vol. 38, 1902, pp. 203-70. (Geological Series, Vol. 5, No. 5).

DANIELLS, Roy (1969). *Alexander Mackenzie and the north west.* London, Faber and Faber, 1969.

DAVIDSON, Gordon Charles (1918). *The North West Company.* Berkeley, University of California Press, 1918. (University of California. Publications in History, Vol. 7). (Reprinted by Russell and Russell, New York, 1967).

DAVIES, K.G. *ed* (1963). *Northern Quebec and Labrador journals and correspondence 1819-35. Edited by K.G. Davies, assisted by A.M. Johnson, with an introduction by Glyndwr Williams.* London, Hudson's Bay Record Society, 1963. (HBRS, Vol. 24).

DAVIES, K.G. *ed* (1965). *Letters from Hudson Bay 1703-40. Edited by K.G. Davies, assisted by A.M. Johnson, with an introduction by Richard Glover.* London, Hudson's Bay Record Society, 1965. (HBRS, Vol. 25).

DAVIS, C.H. *ed* (1876). *Narrative of the North Polar expedition. U.S. Ship Polaris, Captain Charles Francis Hall commanding.* Washington, D.C., Government Printing Office, 1876.

DAWSON, George Mercer (1888). "Report on an exploration in the Yukon district, N.W.T., and adjacent northern portion of British Columbia. 1887." *Canada. Geological Survey. Annual Report,* 1887 (pub. 1888), Part B.

[DAWSON, Henry P. *ed*] (1886). *Observations of the International Polar Expeditions, 1882-83. Fort Rae.* London, Trübner & Co., 1886.

DAWSON, L.S. (1885). *Memoirs of hydrography; including brief biographies of the principal officers who have served in H.M. Naval Surveying Service between the years 1750 and 1885.* 2 vols. Eastbourne, Henry W. Keay, 1885. (Reprinted in 1 vol. by Cornmarket Press, London, 1969).

DAWSON, Lionel (1946). *Lonsdale. The authorised life of Hugh Lowther, Fifth Earl of Lonsdale, K.G., G.C.V.O.* London, Odhams Press Ltd., 1946.

DEACON, M. *and* SAVOURS, A. (1976). "Sir George Strong Nares (1831-1915)." *Polar Record,* Vol. 18, No. 113, May 1976, pp. 127-41.

DEACON, Richard (1968). *John Dee; scientist, geographer, astrologer and secret agent to Elizabeth I.* London, Frederick Muller, 1968.

DEASE, Peter Warren *and* SIMPSON, Thomas (1840). "Arctic land expedition." *Nautical Magazine,* Vol. 9, 1840, pp. 439-45.

DELABARRE, Edmund Burke (1902). "Report of the Brown-Harvard expedition to Nachvak, Labrador in the year 1900." *Geographical Society of Philadelphia,* Vol. 3, No. 4, April 1902, pp. 65-212.

DELANGLEZ, Jean *ed* (1944). "The voyage of Louis Jolliet to Hudson Bay in 1679." *Mid-America,* Vol. 26, No. 3 [Vol. 15, n.s.], 1944, pp. 221-50.

DeRANCE, C.E. (1878). "The geology of the Arctic regions." *Manchester Geological Society. Transactions,* Vol. 14, 1877-78 (pub. 1878), pp. 441-47.

DICKIE, G. (1878). "On the Algae found during the Arctic expedition." *Linnean Society of London, Journal. Botany,* Vol. 17, 1878, pp. 6-12.

DICTIONARY OF CANADIAN BIOGRAPHY Vol. I. *1000 to 1700.* Toronto, University of Toronto Press, 1966. Vol. II. *1701 to 1740.* Toronto, University of Toronto Press, 1969. Vol. III. *1741 to 1770.* Toronto, University of Toronto Press, 1974. Vol. IV. *1771 to 1800.* In preparation. Vol. V. *1801 to 1820.* In preparation. Vol. IX. *1861 to 1870.* Toronto, University of Toronto Press, 1976. Vol. X. *1871 to 1880.* Toronto, University of Toronto Press, 1972.

DIETRICHSEN, O.C. (1931). "Otto Neumann Sverdrup." *Norsk Geografisk Tidsskrift,* Bd. 3, Hft. 5-8, 1931, pp. 285-89.

DOBBS, Arthur (1744). *An account of the countries adjoining to Hudson's Bay, in the north-west part of America. . . . With an abstract of Captain Middleton's journal, and observations upon his behaviour during his voyage and since his return. . . .* London, Printed for J. Robinson, 1744. (Reprinted by S.R. Publishers Ltd. and Johnson Reprint Corporation, 1967).

DODGE, Ernest S. (1961). *Northwest by sea.* New York, Oxford University Press, 1961.

DODGE, Ernest S. (1973). *The polar Rosses. John and James Clark Ross and their explorations.* London, Faber and Faber, 1973.

DOUGLAS, George Mellis (1914). *Lands forlorn; a story of an expedition to Hearne's Coppermine River.* New York and London, G.P. Putnam's Sons, 1914.

DOUGLAS, R. *and* WALLACE, J.N. (1926). *Twenty years of York Factory 1694-1714. Jeremie's account of Hudson strait and bay. Translated from the French edition of 1720 with notes and introduction.* Ottawa, Thorburn and Abbott, 1926.

DUNBABIN, Thomas (1963). "Captain Lyon in the Arctic." *Beaver,* Outfit 293, Spring 1963, pp. 45-51.

DUNCAN, Peter Martin *and* SLADEN, W. Percy (1877). "Report on the Echinodermata collected during the Arctic expedition, 1875-6." *Annals and Magazine of Natural History,* Series 4, Vol. 20, December 1877, pp. 447-70.

DUNCAN, Peter Martin *and* SLADEN, W. Percy (1881). *A memoir on the Echinodermata of the Arctic Sea to the west of Greenland.* London, J. Van Voorst, 1881.

DUNMORE, John (1965). *French explorers in the Pacific.* Vol. 1. *The eighteenth century.* Oxford, Clarendon Press, 1965.

DURAND, Elias *and others* (1864). "Enumeration of the Arctic plants collected by Dr. I.I. Hayes in his exploration of Smith's Sound, between parallels 78th and 82d, during the months of July, August and beginning of September, 1861. By E. Durand, Thomas P. James, and Samuel Ashmead." *Academy of Natural Sciences of Philadelphia. Proceedings,* 1863 (pub. 1864), pp. 93-96.

DYMOND, Joseph *and* WALES, William (1771). "Observations on the state of the air, winds, weather, &c made at Prince of Wales's Fort, on the north-west coast of Hudson's Bay, in the years 1768 and 1769." *Philosophical Transactions* (London, Royal Society), Vol. 60, 1770 (pub. 1771), pp. 137-77.

EAGAN, William Edward (1971). *Joseph Burr Tyrrell, 1858-1957.* London, Ont., University of Western Ontario Department of History, 1971. (PhD thesis. Canadian Theses on Microfilm No. 8549).

EASTON, Robert *and* BROWN, Mackenzie (1964). *Lord of beasts; the biography of Buffalo Jones.* London, André Deutsch, 1964.

EAVENSON, Howard N. (1949). *Map maker & Indian traders. An account of John Patten . . . Charles Swaine . . . Theodorus Swaine Drage.* Pittsburgh, University of Pittsburgh Press, 1949.

EAVENSON, Howard N. (1950). *Swaine and Drage. A sequel to Map maker & Indian traders.* Pittsburgh, University of Pittsburgh Press, 1950.

EIFRIG, Charles W. (1905). "Ornithological results of the Canadian *Neptune* expedition to Hudson Bay and northward, 1903-04." *Auk,* Vol. 22, 1905, pp. 233-41.

EKBLAW, Walter Elmer (1919). "The food-birds of the Smith Sound Eskimos." *Wilson Bulletin,* Vol. 31, March 1919, pp. 1-5.

EKBLAW, Walter Elmer (1927-28). "The material response of the Polar Eskimo to their far Arctic environment." *Association of American Geographers. Annals,* Vol. 17, 1927, pp. 147-98; Vol. 18, 1928, pp. 1-24.

ELLIS, Henry (1748). *A voyage to Hudson's Bay, by the 'Dobbs Galley' and 'California', in the years 1746 and 1747, for discovering a North West Passage; with an accurate survey of the coast and a short natural history of the country . . .* London, Whitridge, 1748.

ELLIS, Henry (1946a). "On Hudson's Bay in 1746." *Beaver,* Outfit 276, March 1946, pp. 40-43. [Extracts from Ellis 1748.]

ELLIS, Henry (1946b). "Eskimos of 1746." *Beaver,* Outfit 277, June 1946, pp. 30-33. [Extracts from Ellis 1748.]

EMERSON, Benjamin Kendall (1879). "On the geology of Frobisher Bay and Field Bay. Description of the collections made by C.F. Hall during his first expedition in the Arctic regions, 1860-'62." In: Nourse, J.E. *ed. Narrative of the second Arctic expedition made by Charles F. Hall . . . 1864-69.* Washington, D.C., Government Printing Office, 1879, pp. 553-83.

EMERSON, Benjamin Kendall (1905). "Notes on some rocks and minerals collected from north Greenland and Frobisher Bay." *American Geologist,* Vol. 35, February 1905, pp. 94-104.

EMMANUEL, Marthe (1959). *La France et l'exploration polaire; de Verrazano à La Pérouse, 1523-1788.* Paris, Nouvelles Editions Latines, 1959.

[ESPINOSA Y TELLO, José *ed*] (1802). *Relacion del viage hecho por las goletas Sutil y Mexicana en el ano de 1792, para reconocer el Estrecho de Fuca; con una introduccion en que se da noticia de las expediciones executadas anteriormente por los Espanoles en busca del paso del noroeste de la America.* Madrid, Imprenta Real, 1802.

ETHERIDGE, Robert (1878). "Palaeontology of the coasts of the Arctic lands visited by the late British expedition under Captain Sir George Nares." *Geological Society of London. Quarterly Journal,* Vol. 34, August 1878, pp. 568-639.

FAIRLEY, T.C. *and* ISRAEL, Charles E. (1957). *The true north; the story of Captain Joseph Bernier.* New York, St. Martin's Press, 1957.

FEILDEN, Henry Wemyss (1877). "Arctic molluscan fauna." *Zoologist,* Series 3, Vol. 1, October 1877, pp. 435-40.

FEILDEN, Henry Wemyss (1877a). "List of birds observed in Smith Sound and in the polar basin during the Arctic expedition of 1875-76." *Ibis,* Series 4, Vol. 1, October 1877, pp. 401-12.

FEILDEN, Henry Wemyss (1877b). "On the Mammalia of North Greenland and Grinnell Land." *Zoologist,* Series 3, Vol. 1, August-September 1877, pp. 313-21, 353-61.

FEILDEN, Henry Wemyss *and* JEFFREYS, J.G. (1877). "The post-Tertiary beds of Grinnell Land and North Greenland." *Annals and Magazine of Natural History,* Series 4, Vol. 20, December 1877, pp. 483-94.

FEILDEN, Henry Wemyss (1878). "Some remarks on interglacial epochs, in reference to fauna and flora existing at the present day in the northern hemisphere between the parallels of 81° and 83°N." *Royal Dublin Society. Scientific Proceedings,* NS, Vol. 2, October 1878, pp. 42-44.

FEILDEN, Henry Wemyss *and* DeRANCE, Charles Eugene (1878). "Geology of the coasts of the Arctic lands visited by the late British expedition under Captain Sir George Nares." *Geological Society of London. Quarterly Journal,* Vol. 34, August 1878, pp. 556-67.

FEILDEN, Henry Wemyss (1887). "On the zoology of Captain Markham's voyage to Hudson's Bay in the summer of 1886." *Norfolk and Norwich Naturalists' Society. Transactions,* Vol. 4, 1887, pp. 354-66.

FIDLER, Vera (1968). "A journey with the Chipewyans." *North,* Vol. 15, No. 6, November-December 1968, pp. 38-43.

FINNIE, R. S. (1963). "George Mellis Douglas." *Explorers Journal,* Vol. 41, No. 4, 1963, pp. 54-55.

[FISHER, Alexander] (1819). *Journal of a voyage of discovery, to the Arctic regions, performed between the 4th of April and the 18th of November, 1818, in His Majesty's Ship Alexander, Wm. Edw. Parry, Esq. Lieut. and Commander. By an officer of the Alexander.* London, Richard Phillips, [1819].

FISHER, Alexander (1821). *Journal of a voyage of discovery to the Arctic regions, in His Majesty's Ships Hecla and Griper, in the years 1819 & 1820.* London, Longman, Hurst, Rees, Orme, and Brown, 1821.

FITZGERALD, William G. (1895). "Rear-Admiral A.H. Markham, R.N., F.R.G.S." *Strand Magazine,* November 1895, pp. 547-59.

FLAHERTY, Robert Joseph (1918). "The Belcher Islands of Hudson Bay: their discovery and exploration." *Geographical Review.* Vol. 5, No. 6, June 1918, pp. 433-58.

FLAHERTY, Robert Joseph (1918a). "Two traverses across Ungava Peninsula, Labrador." *Geographical Review,* Vol. 6, No. 2, August 1918, pp. 116-32.

FLAHERTY, Robert Joseph (1923). "Wetalltook's Islands I-II." *World's Work,* Vol. 45, 1923, pp. 324-26, 422-33, 538-53.

FLAHERTY, Robert Joseph (1924). *My Eskimo friends, "Nanook of the north."* Garden City, N.Y., Doubleday, Page & Co., 1924.

FLEMING, Archibald Lang (1932). *Perils of the polar pack, or, the adventures of the Reverend E.W.T. Greenshield, KI.O.N., of Blacklead Island, Baffin Island.* Toronto, Missionary Society of the Church of England, and London, Society for Promoting Christian Knowledge, 1932.

FLEMING, Archibald Lang (1957). *Archibald the Arctic.* London, Hodder and Stoughton, 1957.

FLEMING, R. Harvey *ed* (1940). *Minutes of*

council, northern department of Rupert Land, 1821-31. Edited by R.H. Fleming . . . with an introduction by H.A. Innis. Toronto, Champlain Society for the Hudson's Bay Record Society, 1940. (HBRS, Vol. 3).

FLUHMANN, May (1976). *Second in command; a biography of Captain Francis Crozier, R.N., F.R.S., F.R.A.S.,* [Yellowknife], Government of the Northwest Territories, Department of Information, 1976.

FORBES, Alexander (1938). "Northernmost Labrador mapped from the air." *American Geographical Society. Special Publication,* No. 22, 1938.

FORSTER, Johann Reinhold (1786). *History of the voyages and discoveries made in the north. Translated from the German of John Reinhold Forster, I.U.D. and elucidated by several new and original maps.* London, Printed for G.G.J. and J. Robinson, 1786.

FOXE, Luke (1635). *North-west Fox, or, Fox from the North-west passage. . . .* London, Printed by B. Alsop and Tho. Fawcet, 1635. (Reprinted by S.R. Publishers Ltd. and Johnson Reprint Corporation, 1965).

FRANKLIN, John (1823). *Narrative of a journey to the shores of the Polar Sea, in the years 1819, 20, 21 and 22. . . . With an appendix on various subjects relating to science and natural history.* London, John Murray, 1823. (Reprinted by M.G. Hurtig Ltd., Edmonton, 1970).

FRANKLIN, John (1828). *Narrative of a second expedition to the shores of the polar sea in the years 1825, 1826, and 1827. . . . Including an account of the progress of a detachment to the eastward by John Richardson.* London, John Murray, 1828. (Reprinted by M.G. Hurtig Ltd., Edmonton, 1971).

FRASER, J. Keith (1957). "Tracing Ross across Boothia." *Canadian Geographer,* No. 10, 1957, pp. 40-60.

FREEMAN, Andrew A. (1961). *The case for Doctor Cook.* New York, Coward-McCann, 1961.

FRENCH, Francis Henry (1919). *Report of the Bathurst Inlet patrol, Royal Northwest Mounted Police, 1917-1918.* Ottawa, King's Printer, 1919.

GALBRAITH, John S. (1976). "The enigma of Sir George Simpson." *Beaver,* Outfit 306, No. 4, Spring 1976, pp. 4-9.

GARLINGTON, Ernest Albert (1883). *Report on Lady Franklin Bay expedition of 1883.* Washington City, Signal Office, 1883. (U.S. Army. Signal Corps. Signal Service Notes, No. 10).

GATHORNE-HARDY, G.M. (1921). *The Norse discoverers of America; the Wineland sagas. Translated and discussed by G.M. Gathorne-Hardy.* Oxford, Clarendon Press, 1921. (Reprinted, with a new preface by the author and a new introduction by Professor Gwyn Jones, 1970.)

GATHORNE-HARDY, G.M. (1922). "A recent journey to northern Labrador." *Geographical Journal,* Vol. 59, No. 3, March 1922, pp. 153-69.

GATHORNE-HARDY, Robert (1963). *Traveller's trio.* London [etc], Thomas Nelson and Sons, 1963.

GATHORNE-HARDY, Robert (1972). "Geoffrey Malcolm Gathorne-Hardy." *Polar Record,* Vol. 16, No. 102, September 1972, p. 447.

GELDART, Herbert D. (1887). "Notes on plants collected by Captain Markham, R.N., at Fort Churchill, Hudson's Bay, and West Digges Island, Hudson's Straits, in July and August, 1886." *Norfolk and Norwich Naturalists' Society. Transactions,* Vol. 4, 1887, pp. 354-66.

GEOGRAPHICAL JOURNAL (1893). "Dr. John Rae, F.R.S." *Geographical Journal,* Vol. 2, No. 3, September 1893, pp. 275-76.

GEOGRAPHICAL JOURNAL (1894). "Admiral Sir Edward Inglefield, K.C.B., D.C.L., F.R.S." *Geographical Journal,* Vol. 4,

No. 5, 1894, p. 464.

GEOGRAPHICAL JOURNAL (1897). "Admiral Sir George H. Richards, K.C.B., F.R.S." *Geographical Journal,* Vol. 9, No. 1, January 1897, pp. 97-98.

GEOGRAPHICAL JOURNAL (1900). "The Stein Arctic Expedition." *Geographical Journal,* Vol. 16, No. 6, December 1900, p. 689.

GEOGRAPHICAL JOURNAL (1905). "Sir Erasmus Ommanney and Sir James Donnet." *Geographical Journal,* Vol. 25, No. 2, February 1905, pp. 221-24.

GEOGRAPHICAL JOURNAL (1908). "Admiral Sir Leopold M'Clintock, K.C.B." *Geographical Journal,* Vol. 31, No. 1, January 1908, pp. 1-11.

GEOGRAPHICAL JOURNAL (1910). "Dr. W.B. Cheadle." *Geographical Journal,* Vol, 35, No. 6, June 1910, pp. 727-28.

GEOGRAPHICAL JOURNAL (1912). "George Borup." *Geographical Journal,* Vol. 39, No. 6, 1912, p. 618.

GEOGRAPHICAL JOURNAL (1912a). "Admiral Sir R. Vesey Hamilton." *Geographical Journal,* Vol. 40, No. 5, November 1912, pp. 570-72.

GEOGRAPHICAL JOURNAL (1915). "Vice-Admiral Sir George Strong Nares, K.C.B., F.R.S." *Geographical Journal,* Vol. 45, 1915, pp. 255-57.

GEOGRAPHICAL JOURNAL (1915a). "Warburton Pike." *Geographical Journal,* Vol. 46, No. 6, December 1915, pp. 478-79.

GEOGRAPHICAL JOURNAL (1916). "Death of Sir Clements Markham." *Geographical Journal,* Vol. 47, No. 3, March 1916, pp. 161-76.

GEOGRAPHICAL JOURNAL (1919). "Admiral Sir Albert Hastings Markham, K.C.B." *Geographical Journal,* Vol. 53, No. 1, January 1919, pp. 61-63.

GEOGRAPHICAL JOURNAL (1920). "The *Discovery* in Hudson Bay." *Geographical Journal,* Vol. 55, No. 4, April 1920, p. 326.

GEOGRAPHICAL JOURNAL (1921). "Obituary. Colonel H.W. Feilden, C.B." *Geographical Journal,* Vol. 58, No. 3, September 1921, pp. 247-48.

GEOGRAPHICAL JOURNAL (1922). "Sir Lewis Beaumont." *Geographical Journal,* Vol. 60, No. 2, 1922, pp. 161-62.

GEOGRAPHICAL JOURNAL (1931). "Captain Otto Sverdrup." *Geographical Journal,* Vol. 77, No. 1, January 1931, pp. 94-95.

GEOGRAPHICAL JOURNAL (1934). "Dr. J. Mackintosh Bell." *Geographical Journal,* Vol. 83, No. 6, June 1934, p. 535.

GEOGRAPHICAL JOURNAL (1935). "Major-General A.W. Greely, U.S.A." *Geographical Journal,* Vol. 86, No. 6, December 1935, pp. 563-64.

GEOGRAPHICAL MAGAZINE (1875). "Sherard Osborn." *Geographical Magazine,* Vol. 2, June 1875, pp. 161-69.

GEOGRAPHICAL REVIEW (1917). "Dr. Robert Bell." *Geographical Review,* Vol. 4, No. 2, August 1917, pp. 149-50.

GEOGRAPHICAL REVIEW (1933). "Henry Grier Bryant." *Geographical Review,* Vol. 23, No. 2, 1933, p. 338.

GEOGRAPHICAL REVIEW (1936). "General A.W. Greely." *Geographical Review,* Vol. 26, No. 1, January 1936, p. 161.

GEOGRAPHICAL REVIEW (1940). "Gunnar Isachsen." *Geographical Review,* Vol. 30, October 1940, pp. 690-91.

GEOGRAPHICAL REVIEW (1946). "David Legge Brainard." *Geographical Review,* Vol. 36, No. 4, 1936, p. 687.

GEOGRAPHICAL REVIEW (1949). "W. Elmer Ekblaw." *Geographical Review,* Vol. 39, No. 4, October 1949, p. 677.

GEOLOGICAL SOCIETY OF LONDON (1959). "Reginald Aldworth Daly." *Geological Society of London. Proceedings,* No. 1572, 1959, pp. 135-36.

GEOLOGICAL SURVEY OF CANADA (1899). *Summary report on the operations of the Geological Survey of the year 1898, by the director.* Ottawa, Geological Survey of Canada, 1899. (Geological Survey of Canada Annual Report, NS Vol. 11, 1898, [pub. 1901], Section A).

GEOLOGICAL SURVEY OF CANADA (1901). *Summary report on the operations of the Geological Survey for the year 1900, by the director.* Ottawa, Geological Survey of Canada, 1901. (Geological Survey of Canada. Annual Report, NS Vol. 13, 1900 [pub. 1903], Section A).

GEORGI, Johannes (1958). "Edward Sabine, ein grosser Geophysiker des 19. Jahrhundert." *Deutsche Hydrographische Zeitschrift,* Jahrg. 11, Heft 6, 1958, pp. 225-39.

GIBBONS, Russell W. (1956). *An historical evaluation of the Cook-Peary controversy; a critique of the acceptance by history and reference works of the claims of Robert Edwin Peary against those of Frederick Albert Cook as to the discovery of the North Pole, 1908-09.* [Ada, Ohio, PO Box 44], Russell W. Gibbons, 1956.

GIBBONS, Russell W. (1964). *Frederick Albert Cook, pioneer American polar explorer, commemorating the 100th anniversary of his birth, June 10, 1965.* Delaware, Dr. Frederick A. Cook Society, [1964?].

GIBBONS, Russell W. (1968). "Did Cook beat Peary to the pole?" *Science Digest,* Vol. 64, No. 3, September 1968, pp. 40-45.

GIBBONS, Russell W. (1970). "First atop the world: the great polar controversy." *Mankind; the Magazine of Popular History,* Vol. 2, No. 5, February 1970.

GIBSON, William (1937). "Sir John Franklin's last voyage. A brief history of the Franklin expedition and an outline of the researches which established the facts of its tragic outcome." *Beaver,* Outfit 268, No. 1, June 1937, pp. 44-75.

GILDER, William Henry (1881). *Schwatka's search; sledging in the Arctic in quest of the Franklin records.* London, Sampson Low, Marston, Searle, and Rivington; New York, C. Scribner's Sons, 1881.

GILPIN, James D. (1850). "Outline of the voyage of H.M.S. Enterprize and Investigator to Barrow Strait in search of Sir John Franklin." *Nautical Magazine,* Vol. 19, No. 1, 1850, pp. 8-9; No. 2, 1850, pp. 82-90; No. 3, 1850, pp. 160-70; No. 4, 1850, p. 230.

GLOVER, Richard (1947). "Sidelights on Sl Hearne." *Beaver,* Outfit 277, March 1947, pp. 10-14.

GLOVER, Richard (1951). "A note on John Richardson's 'Digression concerning Hearne's route.'" *Canadian Historical Review,* Vol. 32, No. 3, 1951, pp. 252-63.

GLOVER, Richard (1951a). "La Perouse on Hudson Bay." *Beaver,* Outfit 281, March 1951, pp. 42-46.

GLOVER, Richard *ed* (1958). *A journey from Prince of Wales's Fort in Hudson's Bay to the northern ocean, 1769, 1770, 1771, 1772. By Samuel Hearne, edited with an introduction by Richard Glover.* Toronto, Macmillan Company of Canada, 1958.

GLOVER, Richard *ed* (1962). *David Thompson's narrative 1784-1812. A new edition with added material, edited with an introduction and notes by Richard Glover.* Toronto, Champlain Society, 1962. (Publications of the Champlain Society, No. 40).

GODFREY, William C. (1857). *Godfrey's narrative of the last Grinnell Arctic exploring expedition in search of Sir John Franklin, 1853-4-5. With a biography of Dr. Elisha K. Kane.* Philadelphia, J.T. Lloyd & Co., 1857.

GOLDSON, William (1793). *Observations on the passage between the Atlantic and Pacific oceans, in two memoirs on the Straits of Anian, and the discoveries of de Fonte. Elucidated by a new and original map. To which is prefixed an historical abridgement of discoveries in the north of America.* Portsmouth, printed by W. Mowbray, 1793.

GOODSIR, Robert Anstruther (1850). *An Arctic voyage to Baffin's Bay and Lancaster Sound, in search of friends with Sir John Franklin.* London, John Van Voorst, 1850.

GORDON, Andrew Robertson (1885). "Report of the Hudson's Bay expedition under the command of Lieut. A.R. Gordon, R.N. 1884." *Canada. Department of Marine and Fisheries. Annual Report,* 17th, 1884 (pub. 1885), Appendix 30, pp. 189-228. (Canada. Parliament. Sessional Papers (No. 9), Vol. 6, 1885).

GORDON, Bryan C. (1973). "Archaeological discoveries on the Mackenzie Highway." *The Dispatch* (Ottawa, Department of Public Works), Vol. 18, No. 2, 1973, pp. 16-19, 24-25.

GOSCH, C.C.A. *ed* (1897). *Danish Arctic expeditions, 1605 to 1620. In two books. . . . Book II.- The expedition of Captain Jens Munk to Hudson's Bay in search of a North-west Passage in 1619-20.* London, Hakluyt Society, 1897. (Works issued by the Hakluyt Society, First Series, No. 97).

GOSLING, W.G. (1910). *Labrador; its discovery, exploration, and development.* London, Alston Rivers, Ltd., 1910.

GOUGH, Barry M. *ed* (1973). *To the Pacific and Arctic with Beechey. The journal of Lieutenant George Peard of H.M.S. "Blossom" 1825-1828.* Cambridge, Published for the Hakluyt Society at the University Press, 1973. (Works Issued by the Hakluyt Society, Second Series, No. 143).

GOULD, Albert T. (1946). "Captain Robert A. Bartlett." *Among the Deep Sea Fishers,* Vol. 44, No. 2, 1946, pp. 44-46.

GOULD, Rupert T. (1935). "Cheyne's proposed Arctic expedition, 1880." *Geographical Society of Philadelphia. Bulletin,* Vol. 33, No. 4, October 1935, pp. 99-105.

GOULD, S. (1917). *Inasmuch. Sketches of the beginnings of the Church of England in Canada in relation to the Indian and Eskimo races.* Toronto, Missionary Society of the Church of England in Canada, 1917. (Handbook No. 1).

GRANT, Madison (1902). "The caribou." *New York Zoological Society. Annual Report,* 7th, 1902, pp. 175-96.

GRAVES, Samuel H. (1908). *On the "White Pass" payroll.* Chicago, The Lakeside Press, 1908.

GREAT BRITAIN. Parliament (1848). *Return to an address of the Honourable The House of Commons, dated 21 March 1848; - for, copies of instructions to Captain Sir John Franklin, R.N., in reference to the Arctic expedition of 1845; to any officer or officers appointed by the Admiralty on any expedition in search of Captain Sir John Franklin, R.N.; and, copies or extracts of any proceedings and correspondence of the Admiralty in reference to Arctic expeditions, from 1845 to the present time, together with copies of charts illustrating the same.* Ordered, by The House of Commons, to be Printed, 13 April 1848. (Great Britain. Parliament. House of Commons, Sessional Papers, Accounts and Papers, 1847-48. Vol. 41, No. 264).

GREAT BRITAIN. Parliament (1848a). *Arctic expedition. Return to an address of the Honourable The House of Commons, dated 26 May 1848; - for, a "copy of the orders from the Lords Commissioners of the Admiralty, under which Captain Sir James Clark Ross, R.N., has proceeded on an expedition in search of Captain Sir John Franklin, R.N."* Ordered, by The House of Commons, to be printed, 7 June 1848. (Great Britain. Parliament. House of Commons. Sessional Papers. Accounts and Papers, 1847-48, Vol. 41, No. 386).

GREAT BRITAIN. Parliament (1849). *Arctic expedition (North Star). Expense of the estimates of the equipment of Her Majesty's Ship "North Star"; and copy of minute of the Board of Admiralty respecting the equipment of the*

"North Star", for the purpose of taking out a supply of provisions, &c. for the crews of Her Majesty's Ships "Investigator" and "Enterprise", employed on an expedition to the Arctic regions. Ordered, by the House of Commons, to be printed, 23 March 1849. (Great Britain. Parliament. House of Commons. Sessional Papers. Accounts and Papers, 1849, Vol. 31, No. 152.)

GREAT BRITAIN. Parliament (1849a). *Copy of a report from Sir John Richardson, dated Fort Confidence, Great Bear Lake, 16th September 1848, reporting his proceedings in search of Sir John Franklin's expedition.* Ordered, by the House of Commons, to be printed, 12 July 1849. (Great Britain. Parliament. House of Commons. Sessional Papers. Accounts and Papers, 1849, Vol. 32, No. 497).

GREAT BRITAIN. Parliament (1849b). *Arctic expeditions. Return to an order of The Honourable The House of Commons, dated 15 March 1849; - for, extracts "of any proceedings or correspondence of the Admiralty, in reference to the Arctic expeditions — (In continuation of Parliamentary Papers, Nos. 264 and 386, of session 1848, up to the present time.")* Ordered, by the House of Commons, to be Printed, 30 March 1849. (Great Britain. Parliament. House of Commons. Sessional Papers, Accounts and Papers, 1849, Vol. 32, No. 188).

GREAT BRITAIN. Parliament (1850a). *Arctic expedition. Return to an order of the Honourable The House of Commons, dated 5 February 1850; - for, copies "of any reports or statements from the officers employed in the Arctic expeditions" . . . "of any plan or plans of search . . ." "copy or extracts from any correspondence or proceedings of the Board of Admiralty . . ." "copies of the orders issued by the Board of Admiralty to the Captains Collinson, Kellett and Moore, and to Lieutenant Pullen; and, also, copy of the instructions given to Dr. Rae . . ." "of any reports made by any officer or officers employed in the late expeditions . . ." "and, of the latest chart of the Polar sea. . . ."* Ordered, by the House of Commons, to be Printed, 5 March 1850. (Great Britain. Parliament. House of Commons. Sessional Papers, Accounts and Papers, 1850, Vol. 35, No. 107).

GREAT BRITAIN. Parliament (1850b). *Arctic expedition. Return to an order of the Honourable The House of Commons, dated 12 April 1850; - for, an account "of the charge for the purchase, repair, and outfit of the vessels now fitting for the Arctic expedition."* Ordered, by The House of Commons, to be printed, 23 May 1850. (Great Britain. Parliament. House of Commons. Sessional Papers. Accounts and Papers, 1850, Vol. 35, No. 368).

GREAT BRITAIN. Parliament (1850c). *Arctic expeditions. Return to an address of the Honourable The House of Commons, dated 10 May 1850; - for, "Copies of instructions from the Admiralty to Captain Austin, R.N., C.B., and to any other officers in Her Majesty's Service engaged in Arctic expeditions, since the date of the last parliamentary return."* Ordered, by The House of Commons, to be printed, 31 May 1850. (Great Britain. Parliament. House of Commons. Sessional Papers, Reports and Papers, 1850, Vol. 35, No. 397).

GREAT BRITAIN. Parliament (1851). *Return to an address of the Honourable The House of Commons, dated 7 February 1851; - for, "copy or extracts from any correspondence or proceedings of the Board of Admiralty . . ." "copies of any instructions from the Admiralty to any officers in Her Majesty's service, engaged in Arctic expeditions . . ." "and, copy or extracts from any correspondence or communications from the Government of the United States . . . in relation to any search to be made on the part of the United States . . ."* Ordered, by the House of Commons, to be printed, 7 March 1851. (Great Britain. Parliament. House of Commons. Sessional Papers. Accounts and Papers, 1851, Vol. 33, No. 97).

GREAT BRITAIN. Parliament (1851a).

Arctic expeditions. Report of the committee appointed by the Lords Commissioners of the Admiralty to inquire into and report on the recent Arctic expeditions in search of Sir John Franklin, together with the minutes of evidence taken before the committee, and papers connected with the subject. London, Eyre and Spottiswoode, 1851. (Great Britain. Parliament. House of Commons. Sessional Papers. Accounts and Papers 1852, Vol. 50).

GREAT BRITAIN. Parliament (1852). *Arctic expedition. Further correspondence and proceedings connected with the Arctic expedition. Presented to both Houses of Parliament by command of Her Majesty.* London, Printed by George Edward Eyre and William Spottiswoode for Her Majesty's Stationery Office, 1852. (Great Britain. Parliament. House of Commons. Sessional Papers. Accounts and Papers, 1852, Vol. 50, No. 1449).

GREAT BRITAIN. Parliament (1852a). *Additional papers relative to the Arctic expedition under the orders of Captain Austin and Mr. William Penny.* London, Eyre and Spottiswoode, 1852. (Great Britain. Parliament. House of Commons. Sessional Papers. Accounts and Papers, 1852, Vol. 50, No. 1436).

GREAT BRITAIN. Parliament (1852b). *Return to an address of the Honourable The House of Commons, dated 19 November 1852; - for, "copies of any correspondence received at the Admiralty from Sir Edward Belcher's squadron . . ." "of any communications received at the Admiralty from Mr. Kennedy . . ." "of correspondence from Commander Inglefield . . ." "of any plans or suggestions of search . . ." "and, of any further correspondence on the subject of the Arctic regions. . . ."* Ordered, by The House of Commons, to be printed, 20 December 1852. (Great Britain. Parliament. House of Commons. Sessional Papers. Accounts and Papers, 1852-53, Vol. 60, No. 82).

GREAT BRITAIN. Parliament (1852c). *Arctic expedition. Return to an order of the Honourable The House of Commons, dated 6 April 1852; — for, copy "of the report of Dr. Rae, of the proceedings of the Arctic searching expedition under his command, since the 10th day of June last, with a sketch of his route."* Ordered, by the House of Commons, to be printed, 6 April 1852. (Great Britain. Parliament. House of Commons. Sessional Papers. Accounts and Papers, 1852, Vol. 50, No. 248).

GREAT BRITAIN. Parliament (1853). *Ship "Rattlesnake". Return to an order of The Honourable The House of Commons, dated 27 April 1853; - for, copies or extracts "of any minutes or other documents connected with the selection by the late Board of Admiralty of the ship 'Rattlesnake', in preference to a steamer, for the purpose of endeavouring to communicate with the 'Plover:' " "and, copies of instructions given by the Board of Admiralty to the captain of the 'Rattlesnake' ". Admiralty, 3 May 1853.* Ordered, by The House of Commons, to be printed, 4 May 1853. (Great Britain. Parliament. House of Commons. Sessional Papers. Accounts and Papers, 1852-53, Vol. 60, No. 444).

GREAT BRITAIN. Parliament (1854). *Papers relative to the recent Arctic expeditions in search of Sir John Franklin and the crews of H.M.S. "Erebus" and "Terror".* London, Eyre and Spottiswoode, 1854. (Great Britain. Parliament. House of Commons. Sessional Papers. Accounts and Papers, 1854, Vol. 42, No. 1725).

GREAT BRITAIN. Parliament (1854a). *Arctic expedition. Return to an order of the Honourable The House of Commons, dated 4 April 1854; - for, copies "of any instructions which either have been issued, or hereafter may be issued, during the present season, to the commanders of Her Majesty's Ships now engaged in the Arctic regions in the search for Sir John Franklin's expedition. Admiralty, 7 April 1854.* Ordered, by The House of Commons, to be printed, 7 April 1854. (Great Britain. Parliament. House of Commons. Sessional Papers. Accounts and Papers, 1854, Vol. 42, No 171).

GREAT BRITAIN. Parliament (1855). *Further papers relative to the recent Arctic expeditions in search of Sir John Franklin and the crews of H.M.S. "Erebus" and "Terror".* London, Eyre and Spottiswoode, 1855. (Great Britain. House of Commons. Sessional Papers. Accounts and Papers, 1854-55, Vol. 35, No. 1898).

GREAT BRITAIN. Parliament (1856). *Further papers relative to the recent Arctic expeditions in search of Sir J. Franklin, and the crews of Her Majesty's Ships "Erebus" and "Terror"; including the reports of Dr. Kane and Messrs Anderson and Stewart. . . . Presented to the House of Commons, 1856.* London, Printed by Harrison and Sons, 1856. (Great Britain. Parliament. House of Commons. Sessional Papers. Accounts and Papers, 1856, Vol. 41, No. 2124).

GREAT BRITAIN. Parliament (1877). *Arctic expedition, 1875-6. Journals and proceedings of the Arctic expedition, 1875-6, under the command of Captain Sir George S. Nares. Presented to both Houses of Parliament by Command of Her Majesty, 1877.* London, Her Majesty's Stationery Office, 1877. (C.-1636).

GREAT BRITAIN. Privy Council (1927). *In the matter of the boundary between the Dominion of Canada and the Colony of Newfoundland in the Labrador Peninsula, between the Dominion of Canada of the one part and the Colony of Newfoundland of the other part. . . .* 12 vols. London, Printed by W. Clowes and Sons Ltd.

GREELY, Adolphus Washington (1886). *Three years of Arctic service. An account of the Lady Franklin Bay expedition of 1881-84 and the attainment of the farthest north.* 2 vols. London, Richard Bentley and Son, 1886.

GREELY, Adolphus Washington (1888). *International polar expedition. Report on the proceedings of the United States expedition to Lady Franklin Bay, Grinnell Land.* 2 vols. Washington, D.C., Government Printing Office, 1888.

GREEN, FitzHugh (1928). "The Crocker Land Expedition; the story of the last extensive dog-sledge expedition in the Arctic." *Natural History,* Vol. 28, September-October 1928, pp. 463-75.

GRENFELL, Wilfred T. *and others* (1909). *Labrador, the country and the people.* New York, Macmillan, 1909.

GRIEG, James Alexanderson (1911). "*Ophiura griegi,* Strand, en varietet af *Ophiura sarsi,* Lütken." *Nytt Magasin for Naturvidenskapene,* Bd. 49, 1911, pp. 49-51.

GRIERSON, John (1960). *Sir Hubert Wilkins; enigma of exploration.* London, Robert Hale, 1960.

GRIFFITH, Richard (1953). *The world of Robert Flaherty.* New York, Duell, Sloan and Pearce; Boston, Little, Brown, 1953.

HAKLUYT, Richard (1965). *The principall navigations voiages & discoveries of the English nation. By Richard Hakluyt. Imprinted at London, 1589. A photo-lithographic facsimile with an introduction by David Beers Quinn and Raleigh Ashlin Skelton and with a new index by Alison Quinn.* 2 vols. Cambridge, Published for the Hakluyt Society and the Peabody Museum of Salem at the University Press, 1965. (Hakluyt Society Extra Series, No. 39).

HALKETT, Andrew (1905). "A naturalist in the frozen north." *Canadian Field-Naturalist,* Vol. 19, 1905, pp. 79-86, 104-09, 115-17.

HALL, Charles Francis (1864). *Life with the Esquimaux: the narrative of Captain Charles Francis Hall, of the whaling barque "George Henry", from the 29th May, 1860, to the 13th September, 1862 . . .* Vols. 1 and 2. London, Sampson, Low, Son, and Marston, 1864. (Also published in 1 vol., 1865; reprinted by M.G. Hurtig, Edmonton, 1970).

HALL, Charles Francis (1873). "Geographical discoveries in the Arctic regions." *American Geographical Society. Journal,* Vol. 3, 1873, pp. 216-21.

HALLIDAY, E.M. (1961). "Captain Cook's

American." *American Heritage*, Vol. 13, No. 1, 1961, pp. 60-72, 84-87.

HAMILTON, Walter R. (1964). *The Yukon story*. Vancouver, Mitchell Press, 1964.

HANBURY, David Theophilus (1900). "A journey from Chesterfield Inlet to Great Slave Lake, 1898-9." *Geographical Journal*, Vol. 16, July 1900, pp. 63-77.

HANBURY, David Theophilus (1903). "Through the Barren Ground of north-eastern Canada to the Arctic coast." *Geographical Journal*, Vol. 22, August 1903, pp. 178-91.

HANBURY, David Theophilus (1904). *Sport and travel in the northland of Canada*. London, Edward Arnold, 1904.

HANSEN, Thorkild (1970). *North west to Hudson Bay; the life and times of Jens Munk. Translated by James McFarlane and John Lynch*. London, Collins, 1970.

HANSON, Earl Parker (1941). *Stefansson: prophet of the north*. New York and London, Harper and Brothers, 1941.

HANSSEN, Helmer (1936). *Voyages of a modern Viking*. London, George Routledge & Sons Ltd., 1936.

HANTZSCH, Bernhard Adolf (1928-29). "Contribution to the knowledge of the avifauna of north-eastern Labrador." *Canadian Field-Naturalist*, Vol. 42, January 1928, pp. 2-9; February 1928, pp. 33-40; April 1928, pp. 87-94; May 1928, pp. 123-25; September 1928, pp. 146-48; October 1928, pp. 172-77; November 1928, pp. 201-07; December 1928, pp. 221-27; January 1929, pp. 11-18; February 1929, pp. 31-34; March 1929, pp. 52-59.

HANTZSCH, Bernhard Adolf (1930). "Eskimo stone graves in north-eastern Labrador and the collection of anthropological material from them." *Canadian Field-Naturalist*, Vol. 44, 1930, pp. 180-82.

HANTZSCH, Bernhard Adolf (1931-32). "Contributions to the knowledge of extreme north-eastern Labrador." *Canadian Field-Naturalist*, Vol. 45, No. 3, March 1931, pp. 49-55; No. 4, April 1931, pp. 85-90; No. 5, May 1931, pp. 115-18; No. 6, September 1931, pp. 143-46; No. 7, October 1931, pp. 169-74; No. 8, November 1931, pp. 194-98; No. 9, December 1931, pp. 222-24; Vol. 46, No. 1, January 1932, pp. 7-12; No. 2, February 1932, pp. 34-36; No. 3, March 1932, pp. 56-63; No. 4, April 1932, pp. 84-89; No. 5, May 1932, pp. 112-16; No. 6, September 1932, pp. 143-45; No. 7, October 1932, pp. 153-62.

HARPER, Francis (1915). "The Athabasca-Great Slave Lake Expedition, 1914." *Geological Survey of Canada. Summary Report*, 1914 (pub. 1915), pp. 159-63.

HARPER, Francis (1931). "Amphibians and reptiles of the Athabasca and Great Slave Lakes region." *Canadian Field-Naturalist*, Vol. 45, March 1931, pp. 68-70.

HARPER, Francis (1931a). "Some plants of the Athabasca and Great Slave Lake region." *Canadian Field-Naturalist*, Vol. 45, May 1931, pp. 97-107.

HARPER, Francis (1932). "Mammals of the Athabasca and Great Slave Lakes region." *Journal of Mammalogy*, Vol. 13, February 1932, pp. 19-36.

HARPER, J. Russell (1967). *William G.R. Hind (1833-1888): a Confederation painter in Canada*. Windsor, Ont., Willistead Art Gallery, 1967.

HARRADON, H.D. (1942). "William John Peters, 1863-1942." *Terrestrial Magnetism and Atmospheric Electricity*, Vol. 47, No. 3, 1942, pp. 187-93.

HARRISON, Alfred H. (1908). *In search of a polar continent*. London, Edward Arnold, 1908.

HARRISON, Alfred H. (1908a). "In search of an Arctic continent." *Geographical Journal*, Vol. 31, 1908, pp. 277-87.

HART, Henry Chichester (1880). "On the botany of the British polar expedition of 1875-6." *Journal of Botany*, Vol. 9, 1880, pp.

52-56, 70-79, 111-15, 141-45, 177-82, 204-08, 235-42, 303-06.

HATTERSLEY-SMITH, G. (1976). "The British Arctic Expedition, 1875-76." *Polar Record,* Vol. 18, No. 113, May 1976, pp. 117-26.

HAUGHTON, Samuel (1862). "On fossils brought home from the Arctic regions in 1859, by Capt. Sir F.L. M'Clintock." *Royal Dublin Society. Journal*, Vol. 3, 1860 (pub. 1862), pp. 53-58.

HAWKES, Ernest W. (1916). "The Labrador Eskimo." *Geological Survey of Canada. Memoir* 91 (No. 14, Anthropological Series), 1916.

HAYES, Isaac Israel (1860). *An Arctic boat journey in the autumn of 1854.* Boston, Brown and Taggard; London, Sampson Low, Son and Co., 1860.

HAYES, Isaac Israel (1862). "Lecture on Arctic explorations." *Smithsonian Institution. Annual Report*, 1861 (pub. 1862), pp. 149-60.

HAYES, Isaac Israel (1867a). *The open polar sea: a narrative of a voyage of discovery towards the north pole, in the schooner "United States".* London, Sampson Low, Son and Marston, 1867.

HAYES, Isaac Israel (1867b). *Physical observations in the Arctic seas, made on the west coast of north Greenland, the vicinity of Smith Strait and the west side of Kennedy Channel, during 1860 and 1861. Reduced and discussed at the expense of the Smithsonian Institution, by Charles A. Schott.* Washington, D.C., Smithsonian Institution, 1867. (Smithsonian Contributions to Knowledge, Vol. 15, Art. 5; Smithsonian Institution Publication 196).

HEARNE, Samuel (1795). *A journey from Prince of Wales's Fort in Hudson's Bay, to the northern ocean undertaken by order of the Hudson's Bay Company, for the discovery of copper mines, a North West Passage, etc. in the years 1769, 1770, 1771 & 1772.* London, A. Strahan and T. Cadell, 1795. (Reprinted by N. Israel, Amsterdam, and Da Capo Press, New York, 1968).

HEATON, J.R. (1975). Francis Skead, Hydrogapher. *Africana Notes and News* (Johannesburg), Vol. 21, No. 7, 1975, pp. 281-85.

HEER, Oswald (1878). "Notes on fossil plants discovered in Grinnell Land by Captain H.W. Feilden, naturalist of the English North Polar expedition." *Geological Society of London. Quarterly Journal*, Vol. 34, February 1878, pp. 66-72.

HENRY, Alexander (1969). *Travels and adventures in Canada and the Indian territories between the years 1760 and 1776. By Alexander Henry, fur trader. Edited with notes . . . by James Bain.* Edmonton, M.G. Hurtig Ltd., 1969. (First published 1809).

HENSON, Matthew A. (1969). *A black explorer at the North Pole. An autobiographical report by the Negro who conquered the top of the world with Admiral Robert E. Peary.* New York, Walker and Co., 1969. (Originally published in 1912 as *A Negro explorer at the North Pole*).

HILLER, James K. (1966). "The Moravian expedition to Labrador, 1752." *Newfoundland Quarterly*, Vol. 65, No. 2, 1966, pp. 19-22.

HILLER, James K. (1971). "The Moravians in Labrador, 1771-1805." *Polar Record*, Vol. 15, No. 99, September 1971, pp. 839-54.

HIND, Henry Youle (1863). *Explorations in the interior of the Labrador peninsula, the country of the Montagnais and Nasquapee Indians.* Vols. 1 and 2. London, Longman, Green, Longman, Roberts, & Green, 1863.

HIND, Henry Youle (1864). "An exploration of the Moisie River, to the edge of the table-land of the Labrador peninsula." *Royal Geographical Society. Journal*, Vol. 34, 1864, pp. 82-87.

HINDS, Margery (1971-72). "A governor general goes north." *Beaver*, Outfit 302,

No. 1, Summer 1971, pp. 14-19; Outfit 303, No. 3, Winter 1972, pp. 48-53.

HOBBS, William Herbert (1936). *Peary.* New York, Macmillan, 1936.

HOCQUART, Gilles (1733). "Ordinance of Intendant Hocquart, 23 May 1733, re limits of traité de Tadoussac."
In: GREAT BRITAIN. Privy Council. *In the matter of the boundary between the Dominion of Canada and the Colony of Newfoundland in the Labrador Peninsula between the Dominion of Canada of the one part, and the Colony of Newfoundland of the other part. Volume VII of joint appendix, containing part XIX. . . .* London, Printed by William Clowes and Sons, Ltd., 1927.

HODGSON, Maurice (1974). "Bellot and Kennedy: a contrast in personalities." *Beaver,* Outfit 305, No. 1, Summer 1974, pp. 55-58.

HOEL, Adolf (1940). "Gunnar Isachsen." *Norsk Geografisk Tidsskrift,* Bd. 8, Heftel-2, 1940, pp. 1-5.

HOFFMAN, Bernard G. (1961). *Cabot to Cartier. Sources for a historical enthnography of northeastern North America 1497-1550.* Toronto, University of Toronto Press, 1961. (Reprinted 1968).

HOHN, E.O. (1963). "Roderick MacFarlane of Anderson River and Fort." *Beaver,* Outfit 294, Winter 1963, pp. 22-29.

HOLAND, Hjalmar R. (1930). "An early Norwegian fur trader of the Canadian northwest." *Studies and Records* (Northfield, Minn., Norwegian-American Historical Association), Vol. 5, 1930.

HOLLAND, Clive Anthony (1970). "William Penny, 1809-92: Arctic whaling master." *Polar Record,* Vol. 15, No. 94, January 1970, pp. 25-43.

HOLMAN, John P. (1951). "Robert Joseph Flaherty." *Explorers Journal,* Vol. 29, No. 4, 1951, pp. 38-40.

HOLME, Randle F. (1888). "A journey in the interior of Labrador, July to October 1887." *Royal Geographical Society. Proceedings,* NS, Vol. 10, No. 4, April 1888, pp. 189-205.

HOOKER, *Sir* Joseph Dalton (1857). "On some collections of Arctic plants, chiefly made by Dr. Lyall, Dr. Anderson, Herr Miertsching, and Mr. Rae, during the expeditions in search of Sir John Franklin, under Sir John Richardson, Sir Edward Belcher, and Sir Robert M'Clure." *Journal of the Proceedings of the Linnean Society. Botany,* Vol. 1, 1857, pp. 114-19.

HOOKER, *Sir* William Jackson (1829-40). *Flora Boreali-americana; or, the botany of the northern parts of British America: compiled principally from the plants collected by Dr. Richardson & Mr. Drummond on the late northern expeditions under command of Captain Sir John Franklin,* R.N. . . . 2 vols. London, H.G. Bohn, [1829]-1840.

HOOKER, *Sir* William Jackson *and* ARMOTT, G.A.W. (1841). *The botany of Captain Beechey's voyage; comprising an account of the plants collected by Messrs Lay and Collie, during the voyage to the Pacific and Behring's Strait* . . . London, H.G. Bohn, 1841. (Reprinted 1965).

HOOPER, William Hulme (1853). *Ten months among the tents of the Tuski, with incidents of an Arctic boat expedition in search of Sir John Franklin, as far as the Mackenzie River and Cape Bathurst.* London, John Murray, 1853.

HOPWOOD, Victor G. (1968). "David Thompson, mapmaker and mythmaker." *Canadian Literature,* No. 38, Autumn 1968, pp. 5-17.

HOPWOOD, Victor G. *ed* (1971). *David Thompson. Travels in western North America, 1784-1812.* Toronto, Macmillan Company of Canada, 1971.

HOUSTON, C. Stuart *ed* (1974). *To the Arctic by canoe, 1819-1821. The journal and paintings of Robert Hood, midshipman with Franklin.* Montreal and London, Arctic Institute of North America and McGill-Queen's University Press, 1974.

HOUSTON, C. Stuart (1974a). "Birds by Hood." *Beaver,* Outfit 305, No. 1, Summer 1974, pp. 4-11.

HOUSTON, C. Stuart (1974b). "The journals and paintings of Robert Hood." *Arctic,* Vol. 27, No. 4, December 1974, pp. 251-55.

HOWAY, F.W. (1930). "A list of trading vessels in the maritime fur trade 1785-1794." *Royal Society of Canada. Transactions,* 3rd Series, Vol. 24, Section 2, 1930, pp. 111-34.

HUBBARD, Mina Benson (1906). "My explorations in unknown Labrador." *Harper's Monthly Magazine,* Vol. 112, No. 672, May 1906, pp. 813-23.

HUBBARD, Mina Benson (1906a). "Labrador, from Lake Melville to Ungava Bay." *American Geographical Society of New York. Bulletin,* Vol. 38, 1906, pp. 529-39.

HUBBARD, Mina Benson (1908). *A woman's way through unknown Labrador; an account of the exploration of the Nascaupee and George rivers.* London, John Murray, 1908.

HUDSON'S BAY COMPANY (1). *The book of ships' movements, 1719-1929.* Hudson's Bay Company Archives, C.4/1.
Note: although this manuscript and Hudson's Bay Company (2) were found sufficient for compiling entries on supply voyages in this list, they supply only very basic information on each voyage. However, log books of many of the voyages are also available in the archives of the Hudson's Bay Company.

HUDSON'S BAY COMPANY (2). *Ships on Hudson's Bay Company business sailing between England and York Fort 1670-1900.* Unpublished typescript. Hudson's Bay Company Archives.

HUISH, Robert (1835). *The last voyage of Capt. Sir John Ross, R.N. to the Arctic regions; for the discovery of a North West Passage; performed in the years 1829-30-31-32 and 33. . . .* London, John Saunders, 1835.

HUISH, Robert (1836). *A narrative of the voyages and travels of Captain Beechey . . . to the Pacific and Behring's Straits; performed in the years 1825, 26, 27 and 28. For the purpose of cooperating with the expeditions under Captains Parry and Franklin. And of Captain Back, R.N., to the Thlew-ee-choh River & the Arctic sea . . .* London, W. Wright, [1836?].

HULTON, P.H. (1961). "John White's drawings of Eskimos." *Beaver,* Outfit 292, Summer 1961, pp. 16-20.

HURD, Marjorie (1950). "David Legge Brainard, 1856-1946." *Appalachia,* Vol. 48, 1950, p. 95.

ILLUSTRATED LONDON NEWS (1848). "Expedition in search of Sir John Franklin. H.M.S. 'Plover.' " *Illustrated London News,* 22 January 1848, p. 39.

INGLEFIELD, Edward Augustus (1853). *A summer search for Sir John Franklin; with a peep into the polar basin. . . . With short notices, by Professor Dickie, on the botany, and by Dr. Sutherland, on the meteorology and geology.* London, Thomas Harrison, 1853.

INGLEFIELD, Edward Augustus (1853a). "Report on the return of the Isabel from the Arctic regions." *Royal Geographical Society. Journal,* Vol. 23, 1853, pp. 136-45.

INGSTAD, Helge (1969). *Westward to Vinland: the discovery of pre-Columbian Norse house-sites in North America. Translated from the Norwegian by Erik J. Friis.* London, Jonathan Cape, 1969.

INNES-LILLINGSTON, F.G. (1876). *The land of the white bear: being a short account of the "Pandora's" voyage during the summer of 1875.* Portsmouth, J. Griffin and Co., London, Simpkin, Marshall and Co., 1876.

INNIS, Harold Adams (1927). "The North

West Company." *Canadian Historical Review,* Vol. 8, 1927, pp. 308-21.

INNIS, Harold Adams (1928). "Peter Pond in 1780." *Canadian Historical Review,* Vol. 9, 1928, p. 333.

INNIS, Harold Adams (1930). *Peter Pond, fur trader and adventurer.* Toronto, Irwin & Gordon, 1930.

INNIS, Harold Adams (1956). *The fur trade in Canada. An introduction to Canadian economic history. Revised ed.* Toronto, University of Toronto Press, 1956.

IREDALE, Thomas (1938). "William Anderson - ornithologist." *Emu,* Vol. 38, Part 1, 1938, pp. 60-62.

IRMINGER, O. (1880). "Tolken Carl Petersen." *Geografisk Tidsskrift,* Bd. 4, 1880, p. 82.

ISACHSEN, Fridtjof (1961). "Christian Leden, 17. juli 1882 - 19 november 1957." *Norsk Geografisk Tidsskrift,* Bd. 17, Hft. 5-8, 1959-60 (pub. 1961), pp. 339-40.

ISACHSEN, Gunnerius Ingvald (1902). "En kort oversigt over den anden norske polarfaerds geografiske arbejde." *Geografisk Tidsskrift,* Bd. 16, 1901-02 (pub. 1902), pp. 193-94.

ISACHSEN, Gunnerius Ingvald (1931). "Otto Sverdrup, 30/10 1854 - 26/11 1930." *Norsk Geografisk Tidsskrift,* Bd. 3, Hft. 5-8, 1931, pp. 290-96.

ISACHSEN, Gunnerius Ingvald (1938). "Godfred Hansen." *Norsk Geografisk Tidsskrift,* Bd. 7, Hefte 1, 1938, pp. 1-4.

ISBISTER, Alexander Kennedy (1845). "Some account of Peel River, N. America." *Royal Geographical Society. Journal,* Vol. 15, 1845, pp. 332-45.

JACKSON, Melvin H. (1963). "The Labrador landfall of John Cabot: the 1497 voyage reconsidered." *Canadian Historical Review,* Vol. 44, No. 2, June 1963, pp. 122-41.

JACOBI, A. (1913). "Bernhard Hantzsch." *Journal für Ornithologie,* Jahrg. 61, Nr. 1, 1913, pp. 161-67.

JAMES, James Alton (1962). *The first scientific exploration of Russian America and the purchase of Alaska.* Evanston and Chicago, Northwestern University, 1942. (Northwestern University Studies in the Social Sciences, No. 4).

JEFFREYS, John Gwyn (1877). "The post-Tertiary fossils procured by the late Arctic expedition, with notes on some of the recent or living Mollusca from the same expedition." *Annals and Magazine of Natural History,* Series 4, Vol. 20, September 1877, pp. 229-42.

JEFFREYS, J.G. (1879). "On some of the Mollusca procured during the Arctic expedition of the 'Fox' in 1858 and 1859." *Royal Dublin Society. Scientific Proceedings,* NS, Vol. 2, 1879, pp. 125-28.

JENNESS, Diamond (1917). "The Copper Eskimos." *Geographical Review,* Vol. 4, 1917, pp. 81-91.

JENNESS, Diamond (1928). *The people of the twilight.* New York, Macmillan, 1928.

JENNESS, Diamond (1961). "R.M. Anderson (1877-1961)." *Arctic,* Vol. 14, No. 4, 1961, p. 268.

JOHNSON, Robert E. (1976). *Sir John Richardson; Arctic explorer, natural historian, naval surgeon.* London, Taylor & Francis, 1976.

JONES, A.G.E. (1950). "The voyage of HMS *Cove,* Captain James Clark Ross, 1835-36." *Polar Record,* Vol. 5, No. 40, July 1950, pp. 543-56.

JONES, A.G.E. (1955). "Frederick John Hornby." *Mariner's Mirror,* Vol. 41, No. 4, November 1955, pp. 303-07.

JONES, A.G.E. (1958). "Robert Shedden and the *Nancy Dawson.*" *Mariner's Mirror,* Vol. 44, No. 2, May 1958, pp. 137-39.

JONES, A.G.E. (1969). "Captain Robert

Martin: a Peterhead whaling master in the nineteenth century." *Scottish Geographical Magazine,* Vol. 85, No. 3, December 1969, pp. 196-202.

JONES, A.G.E. (1975). "Lieutenant Robert Hood, R.N. 1797-1821." *Musk-ox,* No. 16, 1975, pp. 67-68.

JONES, A.G.E. *and* CYRIAX, Richard J. (1953). "Lieutenant Edward Griffiths and the Franklin expedition." *Mariner's Mirror,* Vol. 39, No. 3, August 1953, pp. 178-86.

JONES, Gwyn (1964). *The Norse Atlantic saga: being the Norse voyages of discovery and settlement to Iceland, Greenland, America.* London, Oxford University Press, 1964.

JONES, Gwyn (1968). *A history of the Vikings.* London, New York, Toronto, Oxford University Press, 1968.

JUNGERSEN, Hector Frederik Estrup (1916). *Alcyonarian and madreporarian corals in the Museum of Bergen, collected by the Fram-Expedition 1898-1902 and by the "Michael Sars" 1900-1906.* Bergen, J. Grieg, 1916. (Bergen Museum. Aarbok, No. 6, 1915-16).

KANE, Elisha Kent (1854). *The U.S. Grinnell expedition in search of Sir John Franklin: a personal narrative.* London, Sampson Low, Son, and Co., 1854.

KANE, Elisha Kent (1856). *Arctic explorations: the Second Grinnell Expedition in search of Sir John Franklin, 1853, '54, '55. 2 vols.* Philadelphia, Childs and Peterson; London, Trübner and Co., 1856.

KANE, Elisha Kent (1858). *Magnetical observations in the Arctic seas by Elisha Kent Kane. Made during the second Grinnell Expedition in search of Sir John Franklin. . . .* Washington, D.C. Smithsonian Institution, 1858. (Smithsonian Contributions to Knowledge, Vol. 10, Art. 3. Smithsonian Institution Publication No. 97).

KANE, Elisha Kent (1859). *Meteorological observations in the Arctic seas, made during the Second Grinnell Expedition in search of Sir John Franklin. . . .* Washington, D.C., Smithsonian Institution, 1859. (Smithsonian Contributions to Knowledge, Vol. 11, Art. 5. Smithsonian Institution Publication No. 104).

KANE, Elisha Kent (1860). *Astronomical observations in the Arctic seas, made during the Second Grinnell Expedition in search of Sir John Franklin.* Washington, D.C., Smithsonian Institution, 1860. (Smithsonian Contributions to Knowledge, Vol. 12, Art 2. Smithsonian Institution Publication No. 129).

KANE, Elisha Kent (1860a). *Tidal observations in the Arctic seas, made during the Second Grinnell Expedition in search of Sir John Franklin. . . .* Washington, D.C., Smithsonian Institution, 1860. (Smithsonian Contributions to Knowledge, Vol. 13, Art. 2. Smithsonian Institution Publication No. 130).

KANN, Leopold (1903). *Eineinhalb Jahre in Nordpolarregionen.* Vienna, Verlag von Moritz Perles, 1903.

KEELE, Joseph (1906). "Report on the upper Stewart River region, Yukon." *Geological Survey of Canada. Report,* No. 943, 1906. (In: *Geological Survey of Canada. Annual Report,* NS, Vol. 16, 1904 [pub. 1906], Report C).

KEELE, Joseph (1910). "A reconnaissance across the Mackenzie Mountains and on the Pelly, Ross, and Gravel rivers, Yukon, and North West Territories." *Canada. Department of Mines. Geological Survey Branch.* Report No. 1097, 1910.

KEEVIL, J.J. (1933). "William Anderson, 1748-1778, Master Surgeon, Royal Navy." *Annals of Medical History* (New York), NS, Vol. 5, No. 6, 1933, pp. 511-24.

[KEEVIL, J.J.] (1936). "Sir Alexander Armstrong." *Royal Naval Medical Service. Journal,* Vol. 22, No. 4, October 1936, pp. 275-84.

KEEVIL, J.J. (1943). "Robert McCormick, R.N. The stormy petrel of naval medicine." *Royal Naval Medical Service. Journal,* Vol. 29, No. 1, January 1943, pp. 36-62.

KENDALL, Lace (1963). *Elisha Kent Kane: Arctic challenger.* Philadelphia, Macrae Smith, 1963.

KENNEDY, William (1853). *A short narrative of the second voyage of the Prince Albert in search of Sir John Franklin.* London, W.H. Dalton, 1853.

KENNEY, G.I. (1971). "Charting Labrador's rivers." *North,* Vol. 18, No. 3, May-June 1971, pp. 10-16.

KENNEY, James F. *ed* (1932). *The founding of Churchill. Being the journal of Captain James Knight, Governor-in-Chief in Hudson Bay, from the 14th of July to the 13th of September, 1717.* Toronto, etc., J.M. Dent and Sons Ltd., 1932.

KENYON, Walter A. (1961). "Old Fort Albany relics." *Beaver,* Outfit 292, Summer 1961, pp. 21-23.

KENYON, Walter A. (1965). "The 'Old House' at Albany." *Beaver,* Outfit 296, Autumn 1965, pp. 48-52.

KENYON, Walter A. (1973). " 'Old House' at Albany." *Beaver,* Outfit 304, No. 3, Winter 1973, pp. 48-53.

KENYON, Walter A. (1974). "Charlton Island." *Beaver,* Outfit 305, No. 1, Summer 1974, pp. 24-31.

KENYON, W.A. *and* TURNBULL, J.R. (1971). *The battle for James Bay 1686.* Toronto, Macmillan Company of Canada, 1971.

KERR, J. Lennox (1959). *Wilfred Grenfell, his life and work.* London [etc], George G. Harrap & Co., 1959.

KIEFFER, Jean Jacques (1926). "Chironomiden der 2. Fram-Expedition (1898-1902)." *Norsk Entomologisk Tidsskrift,* Bd. 2, 1925 (pub. 1926), pp. 78-89.

KING, Richard (1836). *Narrative of a journey to the shores of the Arctic Ocean in 1833, 1834 and 1835, under the command of Capt. Back, R.N.* 2 vols. London, Richard Bentley, 1836.

KIRK, Edwin (1925). "Notes on an early collection of Paleozoic fossils from Ellesmereland." *American Journal of Science,* Series 5, Vol. 10, November 1925, pp. 445-47.

KITSON, Arthur (1907). *Captain James Cook, R.N., F.R.S.; "the circumnavigator."* London, John Murray, 1907.

KLUTSCHAK, Heinrich W. (1881). *Als Eskimo unter den Eskimos. Eine Schilderung der Erlebnisse der Schwatka'schen Franklin-Aufsuchungs-Expedition in den Jahren 1878-80.* Wien, etc., A. Hartleben's Verlag, 1881.

KNOX, H.G. (1957). "Alexander Kennedy Isbister." *Historical and Scientific Society of Manitoba.* Series 3, No. 12, 1957, pp. 17-28.

KOHLMEISTER, Benjamin *and* KMOCH, George (1814). *Journal of a voyage from Okkak, on the coast of Labrador, to Ungava Bay, westward of Cape Chudleigh; undertaken to explore the coast, and visit the Esquimaux in that unknown region.* London, Brethren's Society for the Furtherance of the Gospel among the Heathen, 1814.

KOCK, W.J. de *ed* (1968). *Dictionary of South African biography.* Vol. 1. Cape Town, Nasionale Boekhandel Bpk, 1968.

KOOLEMANS BEYNEN, L.R. (1876). *De reis der Pandora naar de Noordpoolgewesten, in den zomer van 1875.* Amsterdam, C.F. Stemler, 1876.

KROEBER, A.L. *and others* (1943). "Franz Boas, 1858-1942." *American Anthropological Association. Memoirs,* No. 61, 1943. (*American Anthropologist,* NS Vol. 45, No. 3, Pt 2).

KUMLIEN, Ludwig (1879). "Contributions to the natural history of Arctic America, made in connection with the Howgate polar expedition, 1877-78." *Bulletin of the United States National Museum,* No. 15, 1879.

LAMAR, William H. *and* ELLIS, Frank W.

(1884). *Physical observations during the Lady Franklin Bay expedition of 1883.* Washington, D.C., Government Printing Office, 1884. (U.S. Army. Signal Corps. Signal Service Notes, No. 14).

LAMB, G.F. (1956). *Franklin – happy voyager. Being the life and death of Sir John Franklin.* London, Ernest Benn, 1956.

LAMB, W. Kaye *ed* (1960). *The letters and journals of Simon Fraser.* Toronto, Macmillan Company of Canada, 1960.

LAMB, W. Kaye *ed* (1970). *The journals and letters of Sir Alexander Mackenzie.* Cambridge, published for the Hakluyt Society at the University Press, 1970. (Hakluyt Society Extra Series No. 41).

LANAUZE, Charles D. (1918). "A police patrol in the North-west Territories of Canada." *Geographical Journal,* Vol. 51, No. 5, May 1918, pp. 316-23.

LANDSVERK, O.G. (1961). *The Kensington runestone; a reappraisal of the circumstances under which the stone was discovered.* Glendale California, Church Press, 1961.

LANMAN, Charles (1885). *Farthest north; or, the life and explorations of Lieutenant James Booth Lockwood, of the Greely Arctic expedition.* New York, D. Appleton and Co., 1885.

[LA PÉROUSE, Jean François Galaup de] (1797). *Voyage de La Pérouse autour du monde, publié conformément au décret du 22 avril 1791, et rédigé par M.L.A. Milet-Mureau.* 4 vols. Paris, L'Imprimerie de la République, 1797.

LARSEN, Sofus (1921-22). "Hvad ved man om Gaspar Corte-Reals Opdagelsesrejser?" *Geografisk Tidsskrift,* Vol. 26, 1921-22, pp. 171-81.

LARSEN, Sofus (1925). "Nordamerikas Opdagelse 20 Aar før Columbus." *Geografisk Tidsskrift,* Vol. 28, No. 1, March 1925, pp. 88-110.

LAURSEN, Dan (1956). "Grønlaendere i forskningens tjeneste: II. Hans Hendrik fra Fiskenaesset." *Grønland,* Nr. 4, 1956, pp. 144-49.

LAWS, James (1967). "Dr. James Laws' journal of the Kane Relief Expedition of 1855." *Polar Notes,* No. 7, November 1967, pp. 1-24.

LAWSON, George (1888). "Remarks on the flora of the northern shores of America, with tabulated observations made by Mr. F.F. Payne on the seasonal development of plants at Cape Prince of Wales, Hudson Strait, during 1886." *Royal Society of Canada. Transactions,* Vol. 5, Section 4, 1887 (pub. 1888), pp. 207-12.

LEA, Alice (1917-18). "Some unpublished letters of Sir John Franklin, Sir John Richardson and others, written during the expeditions to North-west Canada for the purpose of exploration, 1819-22; and 1825-27." *Women's Canadian Historical Society of Toronto. Transactions,* Vol. 17, 1917-18, pp. 12-36.

LEARMONTH, L.A. (1970). "Sea gulls and flag poles." *Moccasin Telegraph,* Vol. 30, No. 1, Winter 1970, pp. 54-55.

LEBOURDAIS, Donat Marc (1963). *Stefansson, ambassador of the north.* Montreal, Harvest House, 1963.

LEDEN, Christian (1927). *Über Kiwatins Eisfelder; drei Jahre unter kanadischen Eskimos.* Leipzig, F.A. Brockhaus, 1927.

LEE, Herbert Patrick (1928). *Policing the top of the world.* London, John Lane the Bodley Head, 1928.

LEECHMAN, Douglas (1947). "Moravians to Ungava." *Beaver,* Outfit 278, September 1947, pp. 28-31.

LEECHMAN, John Douglas (1948). "McConnell's long trek." *Beaver,* Outfit 279, December 1948, pp. 12-15.

LEECHMAN, Douglas (1950). *Eskimo summer.* London, Museum Press, 1950.

LEFROY, *Sir* John Henry *and* RICHARDSON, *Sir* John (1855). *Magnetical and meteorological observations at Lake Athabasca and Fort Simpson, by Captain J.H. Lefroy . . . and at Fort Confidence, in Great Bear Lake, by Sir John Richardson.* London, published for H.M. Stationery Office by Longman, Brown, Green & Longmans, 1855.

LEITH, Charles Kenneth (1910). "An Algonkian basin in Hudson Bay, a comparison with the Lake Superior Basin." *Economic Geology,* Vol. 5, April 1910, pp. 227-46.

LEITH, Charles Kenneth *and* LEITH, Arthur T. (1912). *A summer and winter on Hudson Bay.* Madison, Wis., [Cartwell Printing Co.], 1912.

LEPAGE, Ernest *and* DUTILLY, Arthème (1945). "Retracing the route of Michaux's Hudson's Bay journey of 1792." *Revue de l'Université d'Ottawa,* January-March 1945.

LE VASSEUR, N. (1915). "La Baie d'Hudson; deux iles redécouvertes." *Société de Géographie de Québec. Bulletin,* Vol. 9, No. 3, 1915, pp. 140-49.

LEWIN, W. Henry (1911). *Did Peary reach the pole?* London, Simpkin, Marshall, Hamilton and Kent, 1911.

LEWIS, Arthur (1904). *The life and work of the Rev. E.J. Peck among the Eskimos.* London, Hodder and Stoughton, 1904.

LIDEGAARD, Mads (1968). "Hans Hendrik fra Fiskenaesset." *Gronland,* No. 8, 1968, pp. 249-56.

LIVERPOOL MERCURY (1820). [News item on whaling voyages of *Cumbrian* and *Earl Fauconberg,* 1820]. *Liverpool Mercury,* 6 October 1820.

LOOMIS, Chauncey C. (1972). *Weird and tragic shores; the story of Charles Francis Hall, explorer.* London, Macmillan, 1972.

LOTZ, J.R. (1961). *Whitehorse, Yukon Territory; a general introduction.* Ottawa, Department of Northern Affairs and National Resources, Industrial Division, Area and Community Planning Section, 1961.

LOTZ, J.R. (1962). "Northern Ellesmere Island: a study in the history of geographical discovery." *Canadian Geographer,* Vol. 6, Nos. 3-4, Winter 1962, pp. 151-61.

LOTZ, J.R. (n.d.). *The Dawson area; a regional monograph.* Ottawa, Department of Northern Affairs and National Resources, n.d. (Yukon Research Project Series, No. 2).

LOUDON, W.J. (1930). *A Canadian geologist.* Toronto, Macmillan Company of Canada Ltd., 1930.

LOW, Albert Peter (1888). *Report on explorations in James Bay and country east of Hudson Bay, drained by the Big, Great Whale and Clearwater rivers.* Montreal, William Foster Brown & Co., 1888. (Geological Survey of Canada. Annual Report, Vol. 3, 1887, Part J).

LOW, Albert Peter (1896). *Report on explorations in the Labrador peninsula along the East Main, Koksoak, Hamilton, Manicuagan and portions of other rivers in 1892-93-94-95.* Ottawa, Geological Survey of Canada, 1896. (Geological Survey of Canada Annual Report, Vol. 8, 1895, section L).

LOW, Albert Peter (1899). *Report on an exploration of part of the south shore of Hudson Strait and of Ungava Bay.* Ottawa, Geological Survey of Canada, 1899. (Geological Survey of Canada Annual Report, NS Vol. II, 1898 [pub. 1901], Section L).

LOW, Albert Peter (1902). *Report on an exploration of the coast of Hudson Bay from Cape Wolstenholme to the south end of James Bay.* Ottawa, Geological Survey of Canada, 1902. (Geological Survey of Canada Annual Report, NS Vol. 13, 1900 [pub. 1903], Section D).

LOW, Albert Peter (1903). *Report on the geology and physical character of the Nastapoka Islands, Hudson Bay.* Ottawa, King's Printer, 1903. (Geological Survey of Canada Annual Report, NS Vol. 13, 1900 [pub. 1903], Part DD).

LOW, Albert Peter (1906). *Report on the Dominion Government expedition to Hudson Bay and the Arctic Islands on board the* D.G.S. *Neptune 1903-1904.* Ottawa, Government Printing Bureau, 1906.

LOWER, Arthur Reginald Marsden (1915). "A report on the fish and fisheries of the west coast of James Bay." In: Melvill, J.C.D. *and others.* "Reports on fisheries investigations in Hudson and James bays and tributary waters in 1914." *Canada. Department of the Naval Service. Annual Report,* 1913-14, Appendix, pp. 29-67. (Canada. Sessional Papers, No. 39a, 1915).

LOWER, Arthur Reginald Marsden (1944). "By river to Albany." *Beaver,* Outfit 275, June 1944, pp. 16-19.

LUBBOCK, Adelaide (1968). *Owen Stanley* R.N. *1811-1850; Captain of the 'Rattlesnake'.* Melbourne and London, Heinemann, 1968.

LUBBOCK, Basil (1937). *The Arctic whalers.* Glasgow, Brown, Son & Ferguson, Ltd., 1937. (Reprinted 1955).

LUCAS, Frederick W. (1898). *The annals of the voyages of the brothers Nicolo and Antonio Zeno. . . .* London, Henry Stevens and Styles, 1898.

LYON, George Francis (1824). *The private journal of Captain G.F. Lyon, of* H.M.S. *Hecla, during the recent voyage of discovery under Captain Parry.* London, John Murray, 1824.

LYON, George Francis (1825). *A brief narrative of an unsuccessful attempt to reach Repulse Bay, through Sir Thomas Rowe's "Welcome", in His Majesty's Ship Griper, in the year mdcccxxiv.* London, John Murray, 1825.

LYSAGHT, A.M. (1971). *Joseph Banks in Newfoundland and Labrador, 1766. His diary, manuscripts and collections.* London, Faber and Faber, 1971.

McCLINTOCK, Francis Leopold *and* HAUGHTON, Samuel (1858). "Reminiscences of Arctic ice-travel in search of Sir John Franklin and his companions." *Royal Dublin Society Journal,* Vol. 1, 1856-57 (pub. 1858), pp. 183-250.

McCLINTOCK, Francis Leopold (1859). *The voyage of the 'Fox' in the Arctic seas. A narrative of the discovery of the fate of Sir John Franklin and his companions.* London, John Murray, 1859.

McCLINTOCK, *Sir* Francis Leopold (1861). "Surveys of HMS *Bulldog.*" *Royal Geographical Society. Proceedings,* Vol. 5, No. 2, January 1861, pp. 62-70.

McCLINTOCK, *Sir* Francis Leopold (1862). *Meteorological observations in the Arctic seas. By Sir Francis Leopold McClintock,* R.N., *made on board the Arctic searching yacht "Fox" in Baffin Bay and Prince Regent's Inlet in 1857, 1858, and 1859. Reduced and discussed . . . by Charles A. Schott.* Washington, D.C. Smithsonian Institution, 1862. (Smithsonian Contributions to Knowledge, Vol. 13, Art. 2.)

McCLURE, Robert John le Mesurier (1853). *The North-west Passage. Capt. M'Clure's despatches from Her Majesty's discovery ship, "Investigator", off Point Warren and Cape Bathurst.* London, John Betts, 1853.

McCLURE, Robert John Le Mesurier (n.d.). *The Arctic dispatches, containing an account of the discovery of the North-west Passage. . . .* London, J.D. Potter, n.d. (Reprinted from the *Nautical Magazine).*

McCONNELL, Richard George (1891). "Report on an exploration in the Yukon and Mackenzie basins, N.W.T." *Geological and Natural History Survey of Canada.* Annual Report, NS Vol. 4, 1888-89 (pub. 1891), Part D.

McCONNELL, Richard George (1906). The Macmillan River, Yukon District. *Geological Survey of Canada. Annual Report,* Vol. 15, 1902-03 (pub. 1906), pp. 22A-38A.

McCORMICK, Robert (1854). *Narrative of a*

boat expedition up the Wellington Channel in the year 1852, under the command of R. McCormick, R.N., F.R.C.S., *in* H.M.B. *"Forlorn Hope", in search of Sir John Franklin*, London, Printed by G.E. Eyre and William Spottiswoode, 1854.

McCORMICK, Robert (1884). *Voyages of discovery in the Arctic and Antarctic seas, and round the world: being personal narratives of attempts to reach the North and South Poles; and of an open-boat expedition up the Wellington Channel in search of Sir John Franklin and Her Majesty's Ships "Erebus" and "Terror", in Her Majesty's Boat "Forlorn Hope", under command of the author.* . . . 2 vols. London, Sampson Low, Marston, Searle, and Rivington, 1884.

McCOURT, Edward (1969). *The Yukon and Northwest Territories.* Toronto, Macmillan of Canada, 1969.

McDIARMID, F.A. (1923). "Geographical determinations of the Canadian Arctic Expedition." *Geographical Journal,* Vol. 62, No. 4, October 1923, pp. 293-302.

M'DONALD, Alexander (1841). *A narrative of some passages in the history of Eenoolooapik, a young Esquimaux, who was brought to Britain in 1839, in the ship "Neptune" of Aberdeen: an account of the discovery of Hogarth's Sound.* Edinburgh, Fraser & Co., 1841.

McDONALD, T.H. (1966). *Exploring the Northwest Territory. Sir Alexander Mackenzie's journal of a voyage by bark canoe from Lake Athabasca to the Pacific Ocean* [sic] *in the summer of 1789.* Norman, University of Oklahoma Press, 1966.

McDOUGALL, George Frederick (1857). *The eventful voyage of* H.M. *discovery ship "Resolute" to the Arctic regions in search of Sir John Franklin and the missing crews of* H.M. *discovery ships "Erebus" and "Terror", 1852, 1853, 1854.* . . . London, Longman, Brown, Green, Longmans, & Roberts, 1857.

MacFARLANE, Roderick (1890). "On an expedition down the Begh-ula or Anderson River." *Canadian Record of Science,* Vol. 4, January 1890, pp. 28-53.

McFEE, William (1928). *Sir Martin Frobisher.* London, John Lane the Bodley Head, 1928.

MACFIE, John (1970). "Severn House 1770." *Beaver,* Outfit 300, Spring 1970, pp. 42-49.

MacGAHAN, J.A. (1876). *Under the northern lights.* London, Sampson Low, Marston, Searle, and Rivington, 1876.

MacGREGOR, James G. (1952). *The land of twelve-foot Davis (a history of the Peace River country).* Edmonton, Applied Arts Products Ltd., 1952.

MacGREGOR, James G. (1966). *Peter Fidler: Canada's forgotten surveyor, 1769-1822.* Toronto, McClelland and Stewart, 1966.

McILRAITH, John (1868). *Life of Sir John Richardson,* C.B., LL.D., F.R.S. *Lond., Hon.* F.R.S. *Edin.; Inspector of Naval Hospitals and Fleets, &c. &c. &c.* London, Longmans, Green, & Co., 1868.

MacINNES, C.M. (1967). *Captain Thomas James and the North West Passage.* Bristol, Bristol Branch of the Historical Association, 1967.

MACK, G. Edmund (1938). "The Nascopie." *Beaver,* Outfit 269, No. 2, September 1938, pp. 5-9.

McKAY, Douglas *and* LAMB, W. Kaye (1938). "More light on Thomas Simpson." *Beaver,* Outfit 269, No. 2, September 1938, pp. 26-31.

McKEEVER, Harry P. (1968). "Simon Fraser." *B.C. Outdoors,* Vol. 24, No. 1, January-February 1968, pp. 48-53.

McKEEVOR, Thomas (1819). *A voyage to Hudson's Bay during the summer of 1812.* London, Richard Phillips & Co., 1819.

MACKENZIE, Alexander (1801). *Voyages from Montreal, on the River St. Laurence, through the continent of North America, to the frozen and Pacific oceans; in the years 1789 and 1793. With a preliminary account of the rise, progress, and present state of the fur trade of that country.* London, printed for T. Cadell, Jun, etc., 1801.

McKENZIE, W.G. (1970). "Able seaman John Nelson." *North,* Vol. 17, No. 5, September-October 1970, pp. 14-18.

McKENZIE, W.G. (1975). "The identity of Gateshead Island." *Polar Record,* Vol. 17, No. 110, May 1975, pp. 545-49.

[MACKINLAY, James] (1957). "A manuscript diary written on Mr. Warburton Pike's visit to Back River in 1890." *Arctic Circular,* Vol. 10, No. 3, 1957, pp. 32-50; No. 4, 1957, pp. 51-69.

McKINLAY, William Laird (1976). *Karluk; the great untold story of Arctic exploration.* London, Weidenfeld and Nicolson, 1976.

MacLEOD, Margaret Arnett *and* GLOVER, Richard *eds* (1971). "Franklin's first expedition as seen by the fur traders." *Polar Record,* Vol. 15, No. 98, May 1971, pp. 669-82.

MacMILLAN, Donald Baxter (1918). *Four years in the white north.* New York and London, Harper & Brothers, 1918.

MacMILLAN, Donald Baxter (1918). "New evidence that Cook did not reach the pole." *Geographical Review,* Vol. 5, 1918, pp. 140-41.

MacMILLAN, Donald Baxter (1918a). "Food supply of the Smith Sound Eskimos." *Natural History,* Vol. 18, March 1918, pp. 161-92.

[MacMILLAN, Donald Baxter] (1918b). "Notes on MacMillan's Ellesmere Island trip." *Geographical Review,* Vol. 5, March 1918, pp. 183-94.

MacMILLAN, Donald Baxter (1928). "Geographical report of the Crocker Land Expedition, 1913-1917." *American Museum of Natural History. Bulletin,* Vol. 56, 1928, pp. 379-435.

MacMILLAN, Donald Baxter (1934). *How Peary reached the Pole; the personal story of his assistant.* Boston and New York, Houghton Mifflin Co., 1934.

MacMILLAN, Donald Baxter (1955). "Matthew Henson." *Explorers Journal,* Vol. 33, No. 1, 1955, pp. 28-31, 93.

MACOUN, John (1892). "Mackenzie River notes." *Ottawa Naturalist,* Vol. 6, No. 9, November 1892, pp. 117-18.

MACOUN, John (1922). *Autobiography of John Macoun,* M.A. *Canadian explorer and naturalist. Assistant Director and Naturalist to the Geological Survey of Canada. 1831-1920.* Ottawa, Ottawa Field Naturalists' Club, 1922.

MAGNUSSON, Magnus *and* PALSSON, Hermann (1966). *The Vinland sagas: the Norse discovery of America. Graenlendinga saga and Eirik's saga.* New York, University Press, 1966.

MAINWARING, George Ernest (1931). *My friend the Admiral. The life, letters, and journals of Rear-Admiral James Burney,* FRS, *the companion of Captain Cook and friend of Charles Lamb.* London, George Routledge, 1931.

MAJOR, Richard Henry *ed* (1873). *The voyages of the Venetian brothers Nicolo & Antonio Zeno, to the northern seas in the XIVth century, comprising the lastest known accounts of the lost colony of Greenland; and of the Northmen in America before Columbus.* London, Hakluyt Society, 1873. (Works issued by the Hakluyt Society, First Series, No. 50).

MALTE, M.O. (1920). "Obituary. James Melville Macoun, C.M.G." *Canadian Field-Naturalist,* Vol. 34, No. 2, February 1920, pp. 38-40.

MARKHAM, Albert Hastings (1874). *Whaling cruise to Baffin's Bay and the Gulf of Boothia. And an account of the rescue of the crew of the "Polaris".* London, Sampson Low, Marston, Low, and Searle, 1874.

MARKHAM, Albert Hastings (1876). "On sledge travelling." *Royal Geographical Society. Proceedings,* Vol. 21, December 1876, pp. 110-20.

MARKHAM, Albert Hastings (1878). *The great frozen sea. A personal narrative of the voyage of the "Alert" during the Arctic expedition of 1875-6.* London, Daldy, Isbister & Co., 1878.

MARKHAM, Albert Hastings *ed* (1880). *The voyages and works of John Davis the navigator.* London, Hakluyt Society, 1880. (Works issued by the Hakluyt Society, First Series, No. 59).

MARKHAM, Albert Hastings (1917). *The life of Sir Clements R. Markham,* K.C.B., F.R.S. London, John Murray, 1917.

MARKHAM, Clements Robert *ed* (1881). *The voyages of William Baffin, 1612-1622.* London, Hakluyt Society, 1881. (Works issued by the Hakluyt Society, First Series, No. 63).

MARKHAM, Clements Robert (1889). *A life of John Davis, the navigator, 1550-1605, discoverer of Davis Straits.* London, George Philip & Son, 1889. ("The World's Great Explorers and Explorations" series).

MARKHAM, Clements Robert (1916). "Sir Allen Young." *Geographical Journal,* Vol. 47, No. 1, 1916, pp. 34-41.

MARKHAM, Clements Robert *ed* (1877). *The voyages of Sir James Lancaster, Kt., to the East Indies, with abstracts of journals of voyages to the East Indies, during the seventeenth century, preserved in the India Office, and the voyage of Captain John Knight (1606), to seek the North-west Passage.* London, Hakluyt Society, 1877. (Works issued by the Hakluyt Society, First Series, No. 56).

MARKHAM, Clements Robert (1880). "Expedition of Lieutenant F. Schwatka to King William Island." *Royal Geographical Society. Proceedings,* NS, Vol. 2, November 1880, pp. 658-62.

MARKHAM, *Sir* Clements Robert (1909). *Life of Admiral Sir Leopold McClintock,* K.C.B., D.C.L., LL.D., F.R.S., V.P.R.G.S. London, John Murray, 1909.

MARKHAM, M.E. *and* MARKHAM, F.A. (1927). *The life of Sir Albert Hastings Markham.* Cambridge, University Press, 1927.

MARQUIS, T.G. (1914). "The period of exploration." In: *Canada and its provinces, Vol. 21, Pacific province 1.* Toronto, 1914.

MARTIJN, Charles A. *and* ROGERS, Edward S. (1969). "Mistassini-Albanel; contributions to the prehistory of Quebec." *Centre d'Etudes Nordiques. Travaux Divers,* No. 29, 1969.

MASIK, August *and* HUTCHISON, Isobel Wylie (1935). *Arctic nights' entertainments; being the narrative of an Alaskan-Estonian digger, as told to Isobel Wylie Hutchison during the Arctic night of 1933-34 near Martin Point, Alaska.* London and Glasgow, Blackie and Son, 1935.

MASON, John Alden (1946). "Notes on the Indians of the Great Slave Lake area." *Yale University Publications in Anthropology,* No. 34, 1946.

MASSON, L.R. (1960). *Les bourgeois de la Compagnie du Nord-ouest. Récits de voyages, lettres et rapports inédits relatifs au nord-ouest canadien.* 2 vols. New York, Antiquarian Press, 1960. (First published 1889-90).

MATHEWS, Richard K. (1966). "Albany – pawn in Anglo-French conflict." *Beaver,* Outfit 297, Autumn 1966, pp. 49-55.

MATTOX, W.G. (1964). "Fort Nascopie on Petitsikapau Lake." *McGill Sub-Arctic Research Papers,* No. 18, August 1964, pp. 1-24. (McGill University. Geography Department. Publication No. 11).

MAURELLE, *Don* Francisco Antonio (1781). "Journal of a voyage in 1775 to explore the coast of America, northward of California." *In:* Barrington, Daines. *Miscellanies.* London, printed by J. Nichols, 1781.

MAURO, R.G. (1975). "Dillon Wallace of Labrador." *Beaver,* Outfit 306, No. 1, Summer 1975, pp. 50-57.

MEEK, Fielding Bradford (1865). "Preliminary notice of a small collection of fossils found by Dr. Hays [sic], on the west shore of Kennedy Channel, at the highest northern localities ever explored." *American Journal of Science,* Series 2, Vol. 40, July 1865, pp. 31-34.

MELVILL, James Cosmo Dobrée (1915). "Report on the east-coastal fisheries of James Bay." *In*: Melvill, J.C.D. *and others.* "Reports on fisheries investigations in Hudson and James bays and tributary waters in 1914." *Canada. Department of the Naval Service. Annual Report,* 1913-14, Appendix, pp. 3-28. (Canada. Sessional Papers, No. 39a, 1915).

MERBS, Charles F. (1969). "The Northwest Hudson Bay Thule Project." *Arctic Circular,* Vol. 19, No. 3, November 1969, pp. 46-50.

MERBS, Charles F. (1971). "An island of the dead; Sir Thomas Rowe's Welcome." *Beaver,* Outfit 301, Spring 1971, pp. 16-24.

MICHAUD, Louis G. (1843-65). *Biographie universelle (Michaud) ancienne et moderne.* 2nd ed. 45 vols. Paris, Madame C. Desplaces, Leipzig, F.A. Brockhaus, 1843-65.

MIDDLETON, Christopher (1743). *A vindication of the conduct of Captain Christopher Middleton, in a late voyage on board His Majesty's Ship the Furnace, for discovering a North-west Passage to the western American ocean. In answer to certain objections and aspersions of Arthur Dobbs Esq....* London, Printed by the Author's appointment, 1743. (Reprinted by S.R. Publishers Ltd. and Johnson Reprint Corporation, 1967).

MIERS, Edward John (1877). "Report on the Crustacea collected by the naturalists of the Arctic expedition in 1875-76." *Annals and Magazine of Natural History,* Series 4, Vol. 20, July-August 1877, pp. 52-66, 96-110.

MIERTSCHING, Johann August (1855). *Reise-Tagebuch des Missionars Joh. Aug. Miertsching, welcher als Dolmetscher die Nordpol-Expedition zur Aufsuchung Sir John Franklins auf dem Schiff Investigator begleitete. In den Jahren 1850 bis 1854.* Gnadau, im Verlag der Unitäts-Buchhandlung bei H.L. Menz und in Commission bei E. Kummer in Leipzig, 1855.

MILLER, Floyd (1963). *Ahdoolo! The biography of Matthew A. Henson.* New York, Dutton, 1963.

MILLS, Edwin (1946). "James W. Tyrrell, explorer." *Beaver,* Outfit 277, September 1946, pp. 38-41.

MILLWARD, A.E. *ed* (1930). *Southern Baffin Island. An account of exploration, investigation and settlement during the past fifty years. With an appendix "The crossing of Baffin Island to Foxe Basin by Bernard A. Hantzsch in 1910."* Ottawa, Department of the Interior, North West Territories and Yukon Branch, 1930.

MILTON, William Fitzwilliam *Viscount and* CHEADLE, Walter Butler (1865). *The North-west Passage by land. Being the narrative of an expedition from the Atlantic to the Pacific. Undertaken with the view of exploring a route across the continent to British Columbia through British territory by one of the northern passes in the Rocky Mountains.* London, Cassell, Petter, and Galpin, [1865].

MITCHELL, Hugh C. (1959). "Peary at the North Pole." *United States Naval Institute Proceedings,* Vol. 85, No. 4, 1959, pp. 64-72, illus.

MOCCASIN TELEGRAPH (1963). "Smellie, Thomas Farrar, 1881-1963." *Moccasin Telegraph,* Vol. 22, No. 2, Spring 1963, p. 51.

MOCCASIN TELEGRAPH (1970). "Store divisions and districts." *Moccasin Telegraph,* Vol. 30, No. 1, Winter 1970, pp. 85-129.

MOGG, William (1964). "The Arctic wintering of H.M.S. *Hecla* and *Fury* in Prince Regent Inlet, 1824-25." *Polar Record,* Vol. 12, No. 76, 1964, pp. 11-28.

MONGE, Alf *and* LANDSVERK, O.G. (1967). *Norse medieval cryptography in runic carvings.* Glendale, California, Norseman Press, 1967.

MONTGOMERY, Richard Gill (1932). *"Pechuck;" Lorne Knight's adventures in the Arctic.* New York, Dodd, Mead & Co., 1932.

MOODIE, J.D. (1907). "The Hudson Bay Expedition." *Canada. Department of Marine and Fisheries. Annual Report,* 39th, 1906 (pub. 1907), Appendix 18.

MOODIE, J.D. (1907a). "Report of superintendent J.D. Moodie, of voyage from Halifax to Churchill, Hudson Bay." *Canada. Sessional Papers,* No. 28, 1907, Appendix L.

MOORE, John Percy (1910). "The polychaetous annelids dredged in 1908 by Mr. Owen Bryant off the coasts of Labrador, Newfoundland, and Nova Scotia." *U.S. National Museum. Proceedings,* Vol. 37, 1910, pp. 133-46.

MORGAN, D. Parker *and others* (1907). "Rev. W. W. Kirkby, D.D. By D.P. Morgan, Joshua Kimber and Charles F. Canedy." *Churchman* (New York), 19 October 1907.

MORRISON, David R. (1968). *The politics of the Yukon Territory, 1898-1909.* Toronto, University of Toronto Press, 1968. (Canadian Studies in History and Government, No. 12).

MORTON, Arthur S. (1939). *A history of the Canadian west to 1870-71; being a history of Rupert's Land (the Hudson's Bay Company's territory) and of the North-West Territory (including the Pacific slope).* London, Thomas Nelson, [1939].

MORTON, Arthur S. (1944). *Sir George Simpson, overseas governor of the Hudson's Bay Company; a pen picture of a man of action.* Toronto, J.M. Dent, 1944.

MOSS, Edward L. (1878). *Shores of the polar sea; a narrative of the Arctic expedition of 1875-6. Illustrated by sixteen chromo-lithographs and numerous engravings from drawings made on the spot by the author.* London, Marcus Ward & Co., 1878.

MOSS, Edward L. (n.d.). *Polar sketches. Sixteen chromo-lithographs, from water-colour drawings made in the polar regions during the expedition of 1875-6, with descriptive notes.* London and Belfast, Marcus Ward & Co., n.d.

MOWAT, Farley (1958). *Coppermine journey; an account of a great adventure – selected from the journals of Samuel Hearne.* Boston, Little, Brown & Co., 1958.

MOWAT, Farley (1966). *Westviking; the ancient Norse in Greenland and North America.* London, Secker & Warburg, 1966.

MOWAT, Farley (1973). *Tundra. Selections from the great accounts of Arctic land voyages.* Toronto, McClelland and Stewart, 1973.

MUNN, Henry Toke (1919). "Southampton Island." *Geographical Journal,* Vol. 54, No. 1, July 1919, pp. 52-55, map.

MUNN, Henry Toke (1920). "Ice conditions in Davis Strait, 1919." *Geographical Journal,* Vol. 55, No. 2, February 1920, p. 155.

MUNN, Henry Toke (1932). *Prairie trails and Arctic by-ways.* London, Hurst and Blackett, Ltd., 1932.

MURCHISON, *Sir* Roderick Impey (1859). "Commander George Frederick Mecham, R.N." *Royal Geographical Society. Proceedings,* Vol. 3, No. 5, 1858-59 (pub. 1859), pp. 255-56.

MURCHISON, *Sir* Roderick Impey (1861). "Sir Geo. Simpson." *Royal Geographical Society. Proceedings,* Vol. 5, No. 4, 1860-61 (pub. 1861), pp. 163-64.

MURCHISON, *Sir* Roderick Impey (1861a). "Sir John Richardson, C.B." *Royal Geographical Society. Proceedings,* Vol. 5, No. 4, 1860-61 (pub. 1861), pp. 199-201.

MURRAY, T.B. (1857). *Kalli, the Esquimaux*

Christian. A memoir. 2nd ed. London, Society for Promoting Christian Knowledge, 1857.

NARES, George Strong (1876). *The official report of the recent Arctic expedition.* London, John Murray, 1876.

NARES, George Strong (1876a). "On the north circumpolar sea." *Royal Geographical Society. Proceedings,* Vol. 21, December 1876, pp. 96-106.

NARES, George Strong (1877). "On the navigation of Smith Sound, as a route to the polar sea." *Royal Geographical Society. Proceedings,* Vol. 21, March 1877, pp. 274-87.

NARES, *Sir* George Strong (1878). *Narrative of a voyage to the polar sea during 1875-6 in H.M. Ships 'Alert' and 'Discovery' . . . With notes on the natural history edited by H. W. Feilden . . .* 2 vols. London, Sampson Low, Marston, Searle, & Rivington, 1878.

NATHORST, Alfred Gabriel (1902). "Sverdrups polarexpedition 1898-1902." *Ymer,* Årg. 22, 1902, pp. 529-34.

NATVIG, Leif Reinhardt (1930). "Culiciden der 2 "Fram"-Expedition (1898-1902)." *Norsk Entomologisk Tidsskrift,* Bd. 2, 1930, pp. 358-59.

NAUTICAL MAGAZINE (1840). "Davis Strait whale fishery." *Nautical Magazine,* 1840, pp. 98-103.

NEATBY, Leslie H. (1958). *In quest of the North West Passage.* London, Constable & Co. Ltd., 1958.

NEATBY, Leslie H. (1959). "Joe and Hannah." *Beaver,* Outfit 290, Autumn 1959, pp. 16-21.

NEATBY, Leslie H. (1960). "McClure and the passage." *Beaver*, Outfit 291, Winter 1960, pp. 33-41.

NEATBY, Leslie H. (1961). "The Greely ordeal." *Beaver,* Outfit 292, Autumn 1961, pp. 4-11.

NEATBY, Leslie H. (1963). "Mr. Back of the Expedition." *Beaver,* Outfit 294, Summer 1963, pp. 13-21.

NEATBY, Leslie H. (1966). "Wrecked on the coast of Labrador." *Beaver,* Outfit 297, Autumn 1966, pp. 21-25.

NEATBY, Leslie H. *ed* (1967). *Frozen ships. The Arctic diary of Johann Miertsching 1850-1854. Translated and with introduction and notes by L.H. Neatby.* Toronto, Macmillan Company of Canada, 1967.

NEATBY, Leslie H. (1970). *The search for Franklin.* London, Arthur Barker, 1970.

NEATBY, Leslie H. (1975). "Hantzsch of Baffin Island." *Beaver,* Outfit 306, No. 3, Winter 1975, pp. 4-13.

NEUFELD, Peter Lorenz (1973). "Darrell." *North,* Vol. 20, No. 3, May-June 1973, pp. 34-36.

NEUMAYER, G.B. *and* BÖRGEN, C.N.J. *eds* (1886). *Die internationale Polarforschungen 1882-1883. Die Beobachtungs-Ergebnisse der deutschen Stationen. Band 1. Kingua-Fjord und die meteorologischen Stationen II. Ordnung in Labrador: Hebron, Okak, Nain, Zoar, Hoffenthal, Rama, sowie die magnetischen Observatorien in Breslau und Göttingen.* Berlin, Verlag von A. Asher, 1886.

NEUMAYER, Georg Balthasar *ed* (1890-91). *Die internationale Polarforschung 1882-1883. Die Deutschen Expeditionen und ihre Ergebnisse. Band 1. Geschichtlicher Theil, und in einem Anhange mehrere einzelne Abhandlungen physikalischen und sonstigen Inhalts. Band 2. Beschreibende Naturwissenschaften in einzelnen Abhandlungen.* Berlin, Verlag von A. Asher, 1890-91.

NICHOLSON, Norman L. (1954). *The boundaries of Canada, its provinces and territories.* Ottawa, Queen's Printer, 1954. (Geographical Branch Memoir 2).

NOICE, Harold (1924). *With Stefansson in the Arctic.* London, etc., George G. Harrap & Co., 1924.

NORSK GEOGRAFISK TIDSSKRIFT (1946). "Storker Storkerson 26/6/1883 — 22/3/1940." *Norsk Geografisk Tidsskrift,* Bd. 11, Hft 2, 1946, pp. 87-91.

NORSKE VIDENSKAPS-AKADEMI (1907-19). *Report of the second Norwegian Arctic Expedition in the "Fram" 1898-1902.* 4 vols. Kristiania, Videnskabs-selskabet i Kristiania, 1907-19.

NOURSE, J. E. *ed* (1879). *Narrative of the second Arctic expedition made by Charles F. Hall: his voyage to Repulse Bay, sledge journeys to the straits of Fury and Hecla and to King William's Land, and residence among the Eskimos during the years 1864-'69.* Washington, D.C., Government Printing Office, 1879.

NUTE, Grace Lee (1943). *Caesars of the wilderness. Médard Chouart, Sieur des Groseilliers and Pierre Esprit Radisson, 1618-1710.* New York and London, D. Appleton-Century Company, 1943.

NUTE, Grace Lee (1945). "Radisson and des Groseilliers." *Beaver,* Outfit 276, June 1945, pp. 36-41.

NUTE, Grace Lee (1948). "Paris to Peel's River in 1892." *Beaver,* Outfit 278, March 1948, pp. 19-23.

NUTTING, C.C. (1903). "Frank Russell." *Iowa Alumnus,* December 1903, [4pp].

OGILVIE, William (1890). *Exploratory survey of part of the Lewes, Tat-on-duc, Porcupine, Bell, Trout, Peel, and Mackenzie rivers . . . 1887-88.* Ottawa, Queen's Printer, 1890.

OGILVIE, William (1913). *Early days on the Yukon and the story of its gold finds.* London, John Lane, the Bodley Head, New York, Toronto, 1913.

OLESON, Tryggvi J. (1963). *Early voyages and northern approaches; 1000-1632.* Toronto, McClelland and Stewart, 1963. (Canadian Centenary Series, No. 1).

OLSEN, Ingvard (1961). "Grønlandsmissionaeren Christian Drachardt." *Atuagagdliutit/Grønlandsposten,* Nr. 13, 1961, pp. 17.

O'MEARA, E. (1862). "On some diatomaceous forms from the Arctic regions, collected during the late voyage of the 'Fox.' " *Royal Dublin Society. Journal,* Vol. 3, 1860 (pub. 1862), pp. 59-60.

O'NEILL, John Johnston (1916-17). "Geological reports, Canadian Arctic Expedition, 1915-16." *Geological Survey of Canada. Summary Report,* 1915 (pub. 1916), pp. 236-41; 1916 (pub. 1917), pp. 331-34.

ORMSBY, Margaret A. (1958). *British Columbia: a history.* Vancouver, Macmillan, 1958.

OSBORN, Sherard (1852). *Stray leaves from an Arctic journal; or, eighteen months in the polar regions, in search of Sir John Franklin's expedition, in the years 1850-51.* London, Longman, Brown, Green, and Longmans, 1852.

OSBORN, Sherard *ed* (1856). *The discovery of the North-west Passage by H.M.S. "Investigator," Capt. R. M'Clure, 1850, 1851, 1852, 1853, 1854 . . . from the logs and journals of Capt. Robert le M. M'Clure.* London, Longman, Brown, Green, Longmans, and Roberts, 1856.

OSBORN, Sherard (1856a). "Notes on the late Arctic expeditions." *Royal Geographical Society. Proceedings,* Vol. 1, June 1856, pp. 104-10.

OSBORN, Sherard (1858). "Remarks upon the amount of light experienced in high northern latitudes during the absence of the sun." *Royal Geographical Society. Journal,* Vol. 28, 1858, pp. 371-75.

OSBURN, Raymond C. (1932). "Biological and oceanographic conditions in Hudson Bay. 7. Bryozoa from Hudson Bay and Strait." In: *Canada. Biological Board. Contributions to Canadian Biology,* NS Vol. 17, 1932, pp. 361-76.

OTTAWA NATURALIST (1902). "Dr. Alfred R.C. Selwyn, C.M.G., F.R.S., Director,

Geological Survey of Canada, 1869-1894." *Ottawa Naturalist,* Vol. 16, No. 9, December 1902, pp. 171-77.

PACKARD, Alpheus Spring (1891). *The Labrador coast. A journal of two summer cruises to that region. With notes on its early discovery, on the Eskimo, on its physical geography, geology, and natural history.* New York, N.D.C. Hodges, and London, Kegan Paul, Trench, Trübner & Co., 1891.

PAINTER, George D. (1965). "The Tartar relation and the Vinland map; an interpretation." *In:* Skelton, R.A. *and others. The Vinland map and the Tartar relation.* New Haven and London, Yale University Press, 1965, pp. 241-62.

PARKER, Eric (n.d.). *Hesketh Prichard, D.S.O., M.C.; hunter, explorer, naturalist, cricketer, author, soldier. A memoir.* New York, E.P. Dutton & Co., n.d.

PARKES, Kenneth C. (1970). "In memoriam: Walter Edmond Clyde Todd." *Auk,* Vol. 87, No. 4, 1970, pp. 365-49.

PARNELL, C. (1941). "Ballantyne the brave." *Beaver,* Outfit 272, December 1941, pp. 4-6.

PARNELL, C. *and* KIRK, J.P. (1942). "Campbell of the Yukon." *Beaver,* Outfit 273, June 1942, pp. 4-6; September 1942, pp. 16-18; December 1942, pp. 23-27.

PARRY, Ann (1963). *Parry of the Arctic. The life story of Admiral Sir Edward Parry 1790-1855.* London, Chatto & Windus, 1963.

PARRY, Edward (1859). *Memoirs of Rear-Admiral Sir W. Edward Parry, Kt., F.R.S. etc., late Lieutenant-Governor of Greenwich Hospital. By his son . . .* 6th ed. London, Longman, Brown, Green, Longmans, and Roberts, 1859.

PARRY, Edward (1870). *Memorials of Charles Parry, Commander Royal Navy, by his brother. . . .* London, Strahan, 1870.

PARRY, William Edward (1821). *Journal of a voyage for the discovery of a North-West Passage from the Atlantic to the Pacific; performed in the years 1819-20, in His Majesty's Ships Hecla and Griper . . . with an appendix, containing the scientific and other observations.* London, John Murray, 1821.

PARRY, William Edward (1824). *Journal of a second voyage for the discovery of a North-West Passage from the Atlantic to the Pacific; performed in the years 1821-22-23, in His Majesty's Ships Fury and Hecla, under the orders of Captain William Edward Parry. . . .* London, John Murray, 1824.

PARRY, William Edward (1826). *Journal of a third voyage for the discovery of a North-west Passage from the Atlantic to the Pacific; performed in the years 1824-25, in His Majesty's Ships Hecla and Fury, under the orders of Captain William Edward Parry.* London, John Murray, 1826.

PARTRIDGE, Bellamy (1953). *Amundsen.* London, Robert Hale, 1953.

PATON, John (1825). *A journal of a voyage of discovery to the polar regions, in the year 1824, in His Majesty's Ship Griper, C. F. Lyon, Captain.* Paisley, Printed for the Author by J. Fraser, 1825.

PATTERSON, R.M. (1961). "The Nahany lands; J.M. McLeod's exploration in 1823 and 1824 of the South Nahanni River country." *Beaver,* Outfit 292, Summer 1961, pp. 40-47.

PAYNE, Frank F. (1888). "The mammals and birds of Prince of Wales Sound, Hudson's Strait." *Royal Canadian Institute. Proceedings,* Series 3, Vol. 5, 1887 (pub. 1888), pp. 111-23.

PEACOCK, F.W. (1960). "The Moravian mission in Labrador." *Canadian Geographical Journal,* Vol. 60, No. 5, May 1960, pp. 182-89.

PEAKE, Frank A. (1965). "William West Kirkby, missionary from Alaska to Florida." *Historical Magazine of the Episcopal Church,* Vol. 34, No. 3, 1965, pp. 265-76.

PEARY, Robert Edwin (1903). "Four years' Arctic exploration, 1898-1902." *Geographical Journal,* Vol. 22, 1903, pp. 646-72.

PEARY, Robert Edwin (1903a). "Report by R. E. Peary, C.E., U.S.N., on work done in the Arctic in 1898-1902." *American Geographical Society. Bulletin,* Vol. 35, 1903, pp. 496-534.

PEARY, Robert Edwin (1904). "Field work of the Peary Arctic Club 1898-1902." *Bulletin of the Geographical Society of Philadelphia,* Vol. 4, No. 1, January 1904, pp. 1-48.

PEARY, Robert Edwin (1904a). "North polar exploration: field work of the Peary Arctic Club, 1898-1902." *Smithsonian Institution. Annual Report,* 1903 (pub. 1904), pp. 427-57.

PEARY, Robert Edwin (1905). "Peary Arctic Club expedition, summer of 1905." *American Geographical Society of New York. Bulletin,* Vol. 37, October 1905, pp. 594-600.

PEARY, Robert Edwin (1907). *Nearest the pole. A narrative of the polar expedition of the Peary Arctic Club in the s.s. Roosevelt, 1905-1906.* London, Hutchinson & Co., 1907.

PEARY, Robert Edwin (1910). *The North Pole, its discovery in 1909 under the auspices of the Peary Arctic Club.* New York, Frederick A. Stokes, 1910.

PEARY, Robert Edwin (1910a). "Peary Arctic Club expedition to the North Pole, 1908-9." *Geographical Journal,* Vol. 36, August 1910, pp. 129-48.

PECK, Edmund James (n.d.). *Arctic experiences. Journal of the Rev. E. J. Peck from Wednesday, September 18th, 1901, to November 5th, 1902.* No publisher, n.d.

PERRY, B. Dale *and* CLARK, W. Dean (1971). "Fort Providence, N.W.T." *Musk-ox,* No. 8, 1971, pp. 1-13.

PETITOT, Emile (1875). "Géographie de l'Athabaskaw-Mackenzie et des grands lacs du bassin arctique." *Société de Géographie (Paris). Bulletin,* Sér 6, t. 10, 1875, pp. 5-42, 126-83, 242-90.

PETITOT, Emile (1875a). "Documents sur la géologie du bassin du Mac'Kenzie." *Société Géologique de France. Bulletin,* Ser. 3, t. 3, 1874-75 (pub. 1875), pp. 87-93, 611-12.

PETITOT, Emile (1875b). "Les Déné-Dindjiés." *International Congress of Americanists.* (1st, Nancy, 1875), Vol. 2, pub. 1875, pp. 13-37.

PETITOT, Emile (1886). *Traditions indiennes du Canada nord-ouest.* Paris, Maisonneuve Frères et C. Leclerc, 1886. (Les littératures populaires de toutes les nations, t. 23).

PETITOT, Emile (1887). *Les grands Esquimaux.* Paris, Librairie Plon, 1887.

PFEFFER, Georg Johann (1886). "Mollusken Krebse und Echinodermen von Cumberland-Sund nach der Ausbeute der deutschen Nordexpedition 1882 und 1883." *Jahrbuch der Hamburgischen Wissenschaftlichen Anstalten,* Jahrg 3, 1886, pp. 23-50.

PHILPOTS, Edward P. (1867). *An account of the wintering voyage of the ship Queen of Peterhead.* Peterhead, Scott and Walker ("Sentinel" office), 1867.

PHILPOTS, Edward P. (1869). "An account of the land in the vicinity of Cape Horsburgh, lat. 74°44'24", N., long. 79°W., and of the island discovered there." *Royal Geographical Society. Proceedings,* Vol. 13, 1869, pp. 372-75.

PIKE, Warburton (1892). *The barren ground of northern Canada.* London, Macmillan and Co., 1892.

PIKE, Warburton (1896). *Through the subarctic forest. A record of a canoe journey from Fort Wrangel to the Pelly Lakes and down the Yukon River to the Behring Sea.* London and New York, Edward Arnold, 1896.

POLAR RECORD (1931). "Obituary" [of Pelham Aldrich]. *Polar Record,* Vol. 1, No.

2, 1931, p. 39.

POLAR RECORD (1940). [Obituary of Gunnar Isachsen]. *Polar Record,* Vol. 3, No. 19, January 1940, pp. 208-09.

POLAR RECORD (1945). "James Williams Tyrrell." *Polar Record,* Vol. 4, No. 30, July 1945, p. 290.

POLAR RECORD (1949). "Ole Andreasen." *Polar Record,* Vol. 5, Nos. 37-38, 1949, p. 364.

POLAR RECORD (1954). "The Right Reverend Archibald Lang Fleming, first bishop of the Arctic." *Polar Record,* Vol. 7, No. 47, 1954, p. 85.

POLAR RECORD (1955). "Albert Peter Low." *Polar Record,* Vol. 7, No. 49, 1955, pp. 344-45.

POLAR RECORD (1957). "Helmer Julius Hanssen." *Polar Record,* Vol. 8, No. 55, 1957, p. 377.

POLAR RECORD (1958). "Joseph Burr Tyrrell." *Polar Record,* Vol. 9, No. 60, 1958, p. 271.

POLAR RECORD (1962). "Burt M. McConnell." *Polar Record,* Vol. 11, No. 70, 1962, p. 113.

POLAR RECORD (1970). "Walter Edmund Clyde Todd." *Polar Record,* Vol. 15, No. 96, September 1970, p. 361.

POLAR RECORD (1970a). "Diamond Jenness, C.C." *Polar Record,* Vol. 15, No. 96, September 1970, pp. 361-62.

POLUNIN, Nicholas (1949). *Arctic unfolding. Experiences and observations during a Canadian airborne expedition in northern Ungava, the Northwest Territories, and the Arctic Archipelago.* London, etc., Hutchinson & Co., 1949.

PORTER, Russell Williams (1898). "Frobisher Bay revisited." *American Geographical Society. Journal,* Vol. 30, 1898, pp. 97-110.

POWELL, D. M. (1970). "Western Line I (B.C.) District." *Moccasin Telegraph,* Vol. 30, No. 1, Winter 1970, pp. 126-28.

POWELL, Theodore (1961). *The long rescue; the story of the tragic Greely expedition.* London, W. H. Allen, 1961.

POWYS, Llewelyn (1927). *Henry Hudson.* London, John Lane, The Bodley Head, 1927.

PREBLE, Edward Alexander (1902). *A biological investigation of the Hudson Bay region.* Washington, D.C., Government Printing Office, 1902. (U.S. Department of Agriculture. Division of Biological Survey. North American Fauna, No. 22).

PREBLE, Edward Alexander (1908). *A biological investigation of the Athabasca-Mackenzie region.* Washington, D.C., Government Printing Office, 1908. (U.S. Department of Agriculture. Bureau of Biological Survey. North American Fauna, No. 27).

PRICHARD, H. Hesketh (1910). "Across Labrador from Nain to the George or Barren Grounds River." *Geographical Journal,* Vol. 36, No. 12, December 1910, pp. 691-92.

PRICHARD, H. Hesketh (1911). *Through trackless Labrador; with a chapter on fishing by G.M. Gathorne-Hardy.* London, Heinemann, 1911.

PULLEN, H. F. *ed* (1954). *The Pullen records.* Unpublished, 1954. (Copy in Scott Polar Research Institute).

PULLEN, William John Samuel (1947). "Pullen in search of Franklin." *Beaver,* Outfit 277, March 1947, pp. 40-43; Outfit 278, June 1947, pp. 22-25.

PURCHAS, Samuel (1906). *Hakluytus posthumus or Purchas his pilgrimes. Contayning a history of the world in sea voyages and lande travells by Englishmen and others.* In twenty volumes. Volumes 13 and 14. Glasgow, James MacLehose and Sons, 1906.

PUTNAM, George Palmer (1947). *Mariner of the north. The life of Captain Bob Bartlett.* New York, Duell, Sloan and Pearce, 1947.

QUAYLE, Eric (1967). *Ballantyne the brave; a Victorian writer and his family.* London, Rupert Hart-Davis, 1967.

RAE, John (1850). *Narrative of an expedition to the shores of the Arctic Sea in 1846 and 1847.* London, T. & W. Boone, 1850.

RAE, John (1852a). "Journey from Great Bear Lake to Wollaston Land." *Royal Geographical Society. Journal,* Vol. 22, 1852, pp. 73-82.

RAE, John (1852b). "Recent explorations along the south and east coast of Victoria Land." *Royal Geographical Society. Journal,* Vol. 22, 1852, pp. 82-96.

RAE, John (1855). "Arctic exploration, with information respecting Sir John Franklin's missing party." *Royal Geographical Society. Journal,* Vol. 25, 1855, pp. 246-56.

RAWLINS, Dennis (1970). "A retrospective critique of Peary's North Pole claim." *Polar Notes,* No. 10, October 1970, pp. 24-54.

RAWLINS, Dennis (1970a). "Peary and the North Pole: the lingering doubt." *United States Naval Institute Proceedings,* Vol. 96, No. 6/808, June 1970, pp. 32-41.

RAWLINS, Dennis (1973). *Peary at the North Pole; fact or fiction?* Washington, D.C. and New York, Robert B. Luce, 1973.

RAWLINSON, *Sir* Henry C. (1872). Dr. Berthold Seemann. *Royal Geographical Society. Proceedings,* Vol. 16, No. 4, 1872, pp. 310-11.

RICH, Edwin Ernest *ed* (1938). *Journal of occurrences in the Athabasca Department by George Simpson, 1820 and 1821, and report. Edited by E. E. Rich, with an introduction by Chester Martin.* Toronto, Published by the Champlain Society for the Hudson's Bay Record Society, 1938. (HBRS Vol. 1. Reprinted by Kraus Reprint Ltd., Nendeln, Liechtenstein, 1968).

RICH, Edwin Ernest *ed* (1939). *Colin Robertson's correspondence book, September 1817 to September 1822. Edited with an introduction by E. E. Rich . . . assisted by R. Harvey Fleming.* Toronto, Published by the Champlain Society for the Hudson's Bay Record Society, 1939. (HBRS Vol. 2).

RICH, Edwin Ernest *ed* (1941). *The letters of John McLoughlin from Fort Vancouver to the governor and committee. First series, 1825-28. Edited by E. E. Rich . . . with an introduction by W. Kaye Lamb.* Toronto, Published by the Champlain Society for the Hudson's Bay Record Society, 1941. (HBRS Vol. 4).

RICH, Edwin Ernest *ed* (1942). *Minutes of the Hudson's Bay Company 1671-1674. Edited by E. E. Rich . . . with an introduction by Sir John Clapham. . . .* Toronto, Published by the Champlain Society for the Hudson's Bay Record Society, 1942. (HBRS Vol. 5).

RICH, Edwin Ernest *ed* (1943). *The letters of John McLoughlin from Fort Vancouver to the governor and committee. Second series, 1839-44. Edited by E. E. Rich . . . with an introduction by W. Kaye Lamb.* Toronto, Published by the Champlain Society for the Hudson's Bay Record Society, 1943. (HBRS Vol. 6).

RICH, Edwin Ernest *ed* (1945). *Minutes of the Hudson's Bay Company 1679-1684. First part, 1679-82. Edited by E. E. Rich . . . with an introduction by G. N. Clark.* London, Hudson's Bay Record Society, 1945. (HBRS Vol. 8).

RICH, Edwin Ernest *ed* (1945). *Minutes of the Hudson's Bay Company 1679-1684. First part, 1679-82. Edited by E. E. Rich . . . with an introduction by G. N. Clark.* Toronto, Published by the Champlain Society for the Hudson's Bay Record Society, 1945. (HBRS Vol. 8).

RICH, Edwin Ernest *ed* (1946). *Minutes of the Hudson's Bay Company 1679-1684. Second part, 1682-84. Edited by E. E. Rich . . . with an introduction by G. N. Clark.* London, Hudson's Bay Record Society, 1946. (HBRS Vol. 9).

RICH, Edwin Ernest *ed* (1948). *Copy-book of letters outward &c. Begins 29th May, 1680, ends 5 July, 1687. Edited by E. E. Rich, assisted by A. M. Johnson, with an introduction by E. G. R. Taylor.* Toronto, Published by the Champlain Society for the Hudson's Bay Record Society, 1948. (HBRS Vol. 11).

RICH, Edwin Ernest *ed* (1949). *James Isham's observations on Hudsons Bay, 1743, and notes and observations on a book entitled A voyage to Hudsons Bay in the Dobbs Galley, 1749.* Toronto, Published by the Champlain Society for the Hudson's Bay Record Society, 1949. (HBRS Vol. 12).

RICH, Edwin Ernest *ed* (1951). *Cumberland House journals and inland journal 1775-82. First series. 1775-79. Edited by E. E. Rich, assisted by A. M. Johnson, with an introduction by Richard Glover.* London, Hudson's Bay Record Society, 1951. (HBRS Vol. 14).

RICH, Edwin Ernest *ed* (1952). *Cumberland House journals and inland journals 1775-82. Second series, 1779-82. Edited by E. E. Rich, assisted by A. M. Johnson, with an introduction by Richard Glover.* London, Hudson's Bay Record Society, 1952. (HBRS Vol. 15).

RICH, Edwin Ernest *ed* (1953). *John Rae's correspondence with the Hudson's Bay Company on Arctic exploration 1844-1855. Edited by E. E. Rich . . . assisted by A. M. Johnson . . . with an introduction by J. M. Wordie . . . and R. J. Cyriax.* London, Hudson's Bay Record Society, 1953. (HBRS Vol. 16).

RICH, Edwin Ernest (1954). "The Hudson's Bay Company and the Treaty of Utrecht." *Cambridge Historical Review*, Vol. 9, No. 2, 1954, pp. 183-203.

RICH, Edwin Ernest *ed* (1955). *A journal of a voyage from Rocky Mountain Portage in Peace River to the sources of Finlays Branch and North West ward in summer 1824 [by Samuel Black]. Edited by E. E. Rich, assisted by A. M. Johnson, with an introduction by R. M. Patterson.* London, Hudson's Bay Record Society, 1955. (HBRS Vol. 18).

RICH, Edwin Ernest *ed* (1956). *London correspondence inward from Eden Colvile, 1849-1852. Edited by E. E. Rich, assisted by A. M. Johnson, with an introduction by W. L. Morton.* London, Hudson's Bay Record Society, 1956. (HBRS Vol. 19).

RICH, Edwin Ernest (1958). *The history of the Hudson's Bay Company 1670-1870.* Volume I: *1670-1763.* London, Hudson's Bay Record Society, 1958. (HBRS Vol. 21).

RICH, Edwin Ernest (1959). *The history of the Hudson's Bay Company 1670-1870.* Volume II: *1763-1870.* London, Hudson's Bay Record Society, 1959. (HBRS Vol. 22).

RICH, Edwin Ernest (1966). *Montreal and the fur trade.* Montreal, McGill University Press, 1966.

RICH, Edwin Ernest (1967). *The fur trade and the northwest to 1857.* Toronto, McClelland and Stewart, 1967. (Canadian Centenary Series, No. 11).

RICHARDSON, John (1829-37). *Fauna boreali-americana; or the zoology of the northern parts of British America: containing descriptions of the objects of natural history collected in the late northern land expeditions, under command of Captain Sir John Franklin, R.N. By John Richardson . . . assisted by William Swainson . . . and William Kirby.* London, John Murray, 1829-37. 4 vols.

RICHARDSON, John (1836). "Digression concerning Hearne's route." *In:* Back, George. *Narrative of the Arctic land expedition to the mouth of the Great Fish River . . . in the years 1833, 1834, and 1835.* London, John Murray, 1836, pp. 144-55.

RICHARDSON, *Sir* John (1851). *Arctic searching expedition: a journal of a boat-voyage through Rupert's Land and the Arctic sea, in search of the discovery ships under command of Sir John Franklin. With an appendix on the physical geography of North America.* 2 vols. London, Longman, Brown, Green, and Longmans, 1851.

RINK, Hinrich Johannes (1877). "Om Grønlaenderen Hans Hendriks Deltagelse i Nordpolsexpeditionerne 1853-1876, under Kane, Hayes, Hall og Nares." *Geografisk Tidsskrift,* Bd. 1, 1877, pp. 185-92.

RIVETT-CARNAC, Charles (1973). "The establishment of the R.C.M.P. presence in the Northwest Territories and the Arctic." *Canadian Geographical Journal,* Vol. 86, No. 5, May 1973, pp. 155-67.

ROBINSON, Bradley (1947). *Dark companion.* New York, Robert M. McBride and Co., 1947.

ROBINSON, H. (1834). "Private journal kept on board HMS Favorite on the Newfoundland station." *Journal of the Royal Geographical Society,* Vol. 4, No. 2, 1834, pp. 207-20.

ROBINSON, M. J. *and* ROBINSON, J. L. (1946). "Exploration and settlement of Mackenzie District, N.W.T." *Canadian Geographical Journal,* Vol. 32, No. 6, June 1946, pp. 246-55; Vol. 33, No. 1, July 1946, pp. 43-49.

ROSENMÜLLER, M. (1913). "Bernhard Hantzsch und seine letzte Forschungsreise in Baffinland. Vorläufiger Bericht nach den Tagebüchern und sonstigen Aufzeichnungen des Forschers." *Verein Für Erdkunde zu Dresden, Mitteilungen,* Bd. 2, 1913, pp. 669-716.

ROSENTHAL, Eric *ed* (1966). *South African dictionary of national biography.* London and New York, Frederick Warne & Co., 1966.

ROSS, John (1819). *A voyage of discovery, made under the orders of the Admiralty, in His Majesty's Ships Isabellu and Alexander, for the purpose of exploring Baffin's Bay, and inquiring into the probability of a North-west Passage.* London, John Murray, 1819.

ROSS, *Sir* John (1835). *Narrative of a second voyage in search of a North-west Passage, and of a residence in the Arctic regions during the years 1829, 1830, 1831, 1832, 1833 . . . Including the reports of Commander, now Captain, James Clark Ross . . . and the discovery of the northern Magnetic Pole.* London, A. W. Webster, 1835.

ROSS, *Sir* John (1835a). *Appendix to the narrative of a second voyage in search of a Northwest Passage, and of a residence in the Arctic regions during the years 1829, 1830, 1831, 1832, 1833.* London, A. W. Webster, 1835.

ROSS, *Sir* John (1835b). *Explanation and answer to Mr. John Braithwaite's supplement to Captain Sir John Ross's narrative of a second voyage in the Victory, in search of a North-west Passage.* [London, printed for Sir John Ross, 1835].

ROSS, W. Gillies (1960). "Parry's second voyage." *History Today,* Vol. 10, No. 2, February 1960, pp. 100-05.

ROSS, W. Gillies (1971). *Hudson Bay whaling 1860-1915.* Ph.D. thesis, University of Cambridge, 1971. (Published as *Whaling and Eskimos: Hudson Bay 1860-1915* by National Museums of Canada, Ottawa, 1975).

ROSS, W. Gillies *and* BARR, William (1972). "Voyages in northwestern Hudson Bay (1720-1772) and discovery of the Knight relics on Marble Island." *Musk-ox,* No. 11, 1972, pp. 28-33.

ROSS, W. Gillies (1973). "Whaling in Hudson Bay. Part I - 1765 to 1772. The Marble Island whale fishery." *Beaver,* Outfit 303, No. 4, Spring 1973, pp. 4-11.

ROSS, W. Gillies (1973a). "Whaling in Hudson Bay. Part II - 1866-67. Whaling voyage of the 'Ocean Nymph.'" *Beaver,* Outfit 304, No. 1, Summer 1973, pp. 40-47.

ROSS, W. Gillies (1976). "Canadian sovereignty in the Arctic: the *Neptune expedition of 1903-04. Arctic,* Vol. 29, No. 2, June 1976, pp. 87-104.

ROUCH, Jules (1938). "Emile de Bray." *La Géographie* (Paris), Vol. 69, Nos. 5-6, 1938, pp. 257-63.

ROUCH, Jules (1945). "Deux officiers de marine français, Joseph Bellot et Emile de Bray à bord de S.M. Britannique dans les mers polaires (1852-54)." *France-Grande Bretagne. Bulletin des Relations Franco-Britanniques,* No. 194, March 1945, pp. 1-12.

ROUSSEAU, Jacques (1948). "Bataille de sextants autour du lac Mistassini." *L'Action Universitaire* (Montréal, Université de Montréal), An. 14, No. 2, January 1948, pp. 99-116.

ROUSSEAU, Jacques (1948a). "Le voyage d'André Michaux au lac Mistassini en 1792." *Mémoires du Jardin Botanique de Montréal/Memoirs of the Montreal Botanic Garden,* No. 3, 1948.

ROUSSEAU, Jacques (1950). "Les voyages du père Albanel au lac Mistassini et à la Baie James en 1672 et 1674." *Revue d'Histoire de l'Amérique Française,* Vol. 3, No. 4, March 1950.

ROUSSEAU, Jacques (1950a). "La mission politique du père Albanel à la baie d'Hudson." *Bulletin des Recherches Historiques,* Vol. 56, No. 4-6, April-June 1950, pp. 71-77.

ROWLEY, Graham (1977). "Bernhard Hantzsch: the probable cause of his death in Baffin Island in 1911." *Polar Record,* Vol. 18, No. 117, September 1977 (in press).

ROY, Pierre-Georges (1942). *Inventaire de pièces sur la côte de la Province de Québec.* Vol. 2. Québec, Archives de la Province de Québec, 1942.

ROYAL GEOGRAPHICAL SOCIETY (1860). "Dr. John Simpson, M.D., R.N." *Royal Geographical Society. Proceedings,* Vol. 4, No. 4, 1859-60 (pub. 1860), p. 138.

ROYAL GEOGRAPHICAL SOCIETY (1880). "Carl Petersen." *Royal Geographical Society. Proceedings,* NS Vol. 2, No. 8, August 1880, pp. 514-15.

ROYAL GEOGRAPHICAL SOCIETY (1882). "Commander Wyatt Rawson, R.N., F.R.G.S." *Royal Geographical Society. Proceedings,* NS Vol. 4, November 1882, pp. 697-700.

ROYAL GEOGRAPHICAL SOCIETY (1883). "Obituary. Vice-Admiral Sir Richard Collinson, K.C.B." *Royal Geographical Society. Proceedings,* NS Vol. 5, 1883, pp. 606-09.

ROYAL GEOGRAPHICAL SOCIETY (1927). "Capt. William Hawkeridge and the North-West Passage." *Geographical Journal,* Vol. 69, No. 4, April 1927, pp. 380-81.

ROYAL SOCIETY (1897). "Sir George Henry Richards, K.C.B." *Royal Society of London. Proceedings,* Vol. 60, 1897, pp. xxxii-xxxv.

ROYAL SOCIETY (1897a). "Dr. John Rae, LL.D." *Royal Society of London. Proceedings,* Vol. 60, 1897, pp. v-vii.

ROYAL SOCIETY (1915). "Sir George Nares, K.C.B., F.R.S., 1831-1915." *Royal Society of London. Proceedings,* Series A, Vol. 91, 1915, pp. lii-liii.

RUNDALL, Thomas (1849). *Narratives of voyages towards the North-west, in search of a passage to Cathay and India. 1496 to 1631. With selections from the early records of the Honourable the East India Company and from mss. in the British Museum.* London, Hakluyt Society, 1849. (Works issued by the Hakluyt Society, First Series, No. 5).

RUSSELL, Frank (1898). *Explorations in the far north. Being the report of an expedition under the auspices of the University of Iowa during the years 1892, '93 and '94.* [Iowa City?], The University, 1898.

RUSSELL, L.S. (1962). "Rudolph Martin Anderson." *Canadian Geographical Journal,* Vol. 64, No. 6, 1962, pp. 198-99.

RYDER, Carl Hartvig (1889-90). "Grønlaenderen Hans Hendrik." *Geografisk Tidskrift,* Bd. 10, Hefte 5-6, 1889-90, pp. 140-43.

SABINE, *Sir* Edward (1819a). "An account of the Esquimaux, who inhabit the west coast of Greenland, above the latitude

76°." *Quarterly Journal of Science, Literature and Art,* Vol. 7, 1819, pp. 72-94.

SABINE, *Sir* Edward (1819b). "Observations on the dip and variation of the magnetic needle, and on the intensity of the magnetic force; made during the late voyage in search of a North West Passage." *Royal Society of London. Philosophical Transactions,* 1819, pp. 132-44.

SABINE, *Sir* Edward (1819c). *Remarks on the account of the late voyage of discovery to Baffin's Bay published by Captain J. Ross,* R.N. London, John Booth, 1819.

SABINE, *Sir* Edward *ed* (1821). *The North Georgia gazette and winter chronicle.* London, John Murray, 1821.

SABINE, *Sir* Edward *and others* (1824). *A supplement to the appendix of Captain Parry's voyage for the discovery of a North-West Passage, in the years 1819-20. Containing an account of the subjects of natural history. By E. Sabine, William Kirby, John Edward Gray, Robert Brown, and Charles Konig.* London, John Murray, 1824.

SABINE, *Sir* Edward (1863). "Results of the hourly observations of the magnetic declination made by Sir F. L. M'Clintock, and the officers of the yacht 'Fox', at Port Kennedy, in the Arctic sea, in the winter of 1858-1859." *Royal Society of London. Philosophical Transactions,* Vol. 153, 1863, pp. 649-63.

SAINVILLE, Edouard de (1898). "Voyage à l'embouchure de la rivière Mackenzie." *Bulletin de la Société de Géographie* (Paris), Sér. 7, t. 19, 1898, pp. 291-307.

SALTER, John William (1853). "On Arctic Silurian fossils." *Geological Society of London. Quarterly Journal,* Vol. 9, June 1853, pp. 312-17.

SAVOIE, Donat *ed* (1970). *Les Amérindiens du nord-ouest canadien au 19e siècle selon Emile Petitot.* Volume 1: *Les Esquimaux Tchiglit.* Volume 2: *Les Indiens Loucheux, précédé d'une présentation générale des Indiens Dènèdindjié.* Canada. Department of Indian Affairs and Northern Development. Northern Science Research Group. Mackenzie Delta Research Project, Vols 9-10, 1970.

SCHAANNING, Hans Thomas Lange (1933). "Birds from Arctic North-America. Ornithological results of the Fram-Expedition 1898-1902 and the Gjoa-Expedition 1903-1907." *Nytt Magasin for Naturvidenskabene,* Bd. 73, 1933, pp. 137-65.

SCHEI, Per Elisaeus (1903). "Summary of geological results." *Geographical Journal,* Vol. 22, No. 1, July 1903, pp. 55-65.

SCHLAIKJER, Erich M. *and* SCHWARZ, Herbert F. (1949). "Walter Elmer Ekblaw." *Explorers Journal,* Vol. 27, No. 4, 1949, p. 42.

SCHLEY, Winfield S. *and* SOLEY, J.R. (1885). *The rescue of Greely.* New York, C. Scribner's; London, Sampson Low, Marston, Searle, and Rivington, 1885.

SCHLEY, Winfield S. (1887). *Report of Winfield S. Schley, Commander,* U.S. *Navy, commanding Greely Relief Expedition of 1884.* Washington, D.C., Government Printing Office, 1887.

SCHORGER, A.W. (1952). "Walter Elmer Ekblaw." *Auk,* Vol. 69, No. 1, 1952, pp. 122-23.

SCHWARZ, Herbert F. *ed* (1950). "John W. Goodsell." *Explorers Journal,* Vol. 28, No. 1, 1950, pp. 44-45.

SCHWARZ, Herbert F. *ed* (1958). "In memory. Clifford Hamilton Easton." *Explorers Journal,* Vol. 36, No. 4, December 1958, p. 33.

SCHWATKA, Frederick (1883). "The igloo of the Innuit." *Science,* Vol. 2, 1883, pp. 182, 216, 259, 304, 347.

SCHWATKA, Frederick (1884). "The implements of the igloo." *Science,* Vol. 4, 1884, pp. 81-85.

SCHWATKA, Frederick (1884a). "The Netschillik Innuits." *Science,* Vol. 4, 1884, pp. 543-45.

SCHWATKA, Frederick (1885). *Report of a military reconnaissance in Alaska, made in 1883.* Washington, D.C., Government Printing Office, 1885.

SCHWATKA, Frederick (1885a). *Along Alaska's great river. A popular account of the travels of the Alaska exploring expedition of 1883, along the great Yukon River, from its source to its mouth, in the British North-West Territory, and in the Territory of Alaska.* New York, Cassell & Co. Ltd., 1885.

SCIENCE (1960). "Obituary of Burt M. McConnell." *Science,* 21 October 1960, p. 1142.

SCOTTISH GEOGRAPHICAL MAGAZINE (1916). "Sir Clements Markham, K.C.B., F.R.S." *Scottish Geographical Magazine,* Vol. 32, No. 3, 1916, pp. 140-42.

SCOTTISH GEOGRAPHICAL MAGAZINE (1917). "Dr. Robert Bell, F.R.S." *Scottish Geographical Magazine,* Vol. 33, 1917, pp. 366-67.

SEEMANN, Berthold (1853). *Narrative of the voyage of H.M.S. Herald during the years 1845-51, under the command of Captain Henry Kellett, R.N., C.B.; being a circumnavigation of the globe, and three cruises to the Arctic regions in search of Sir John Franklin.* 2 vols. London, Reeve and Co., 1853.

SELWYN, Alfred R. C. (1877). "Report on exploration in British Columbia, by Alfred R. C. Selwyn in 1875." *Geological Survey of Canada. Report of Progress for 1875-76,* pub. 1877, pp. 28-232.

SETON, Ernest Thompson (1912). *The Arctic prairies.* London, Constable, 1912.

SEXÉ, Marcel (1923). *Two centuries of fur trading 1723-1923; romance of the Revillon family.* Paris, printed by Draeger Frères [for Revillon Frères], 1923.

SHAW, Alfred Victor (1945). "The Silliman Fossil Mount." *Mineralogist,* Vol. 13, 1945, pp. 212-14.

SHAW, Edward C. (1970). "Captain William Kennedy – an extraordinary Canadian." *Historical and Scientific Society of Manitoba, Transactions,* Series 3, No. 27, 1970, pp. 7-18.

SHAW, Margaret Mason (1958). *Canadian portraits: Tyrrell, Camsell, Cross, LaBine, geologists and prospectors.* Toronto, Clarke, Irwin, 1958.

SHERMAN, Orray Taft (1883). *Meteorological and physical observations on the east coast of British America.* Washington, D.C., Government Printing Office, 1883. (U.S. Army. Signal Corps. Professional Papers, No. 11).

SHIPLEY, Nan (1967). "Anne and Alexander Murray." *Beaver,* Outfit 298, Winter 1967, pp. 33-37.

SIMMONS, Herman Georg (1903). "Preliminary report on the botanical work of the second Norwegian Polar Expedition 1898-1902." *Nytt Magasin for Naturvidenskapene,* Bd. 41, 1903, pp. 223-38.

SIMPSON, Alexander (1845). *The life and travels of Thomas Simpson, the Arctic discoverer.* London, Richard Bentley, 1845.

SIMPSON, Thomas (1839). "Arctic discovery." *Nautical Magazine,* Vol. 8, 1839, pp. 564-65.

SIMPSON, Thomas (1843). *Narrative of the discoveries on the north coast of America; effected by the officers of the Hudson's Bay Company during the years 1836-39.* London, Richard Bentley, 1843.

SKELTON, R. A. (1965). "The Vinland map." *In:* Skelton, R.A. *and others. The Vinland map and the Tartar relation.* New Haven and London, Yale University Press, 1965, pp. 107-240.

SKELTON, R.A. *and others* (1965). *The Vinland map and the Tartar relation. By R.A. Skelton, Thomas E. Marston, and George D. Painter.* New Haven and London, Yale University Press, 1965.

SMITH, Alice Henderson (1919). "A historical enquiry into the efficacy of lime-juice for the prevention and cure of scurvy." *Journal of the Royal Army Medical Corps,* Vol. 32, No. 2, 1919, pp. 93-116; No. 3, 1919, pp. 188-208.

SMITH, D. Murray (1877). *Arctic expeditions from British and foreign shores from the earliest times to the expedition of 1875-76.* Southampton, Charles H. Calvert, 1877.

SMITH, Edgar A. (1877). "On the Mollusca collected during the Arctic expedition of 1875-76." *Annals and Magazine of Natural History,* Series 4, Vol. 20, August 1877, pp. 131-46.

SMITH, Ralph (1971). "Discovery of one of James Knight's ships at Marble Island: a preliminary report." *Musk-Ox,* No. 9, 1971, pp. 12-14.

SMITH, Ralph (1972). "Relics of James Knight." *Beaver,* Outfit 302, No. 4, Spring 1972, pp. 36-41.

SMITH, Ralph *and* BARR, William (1971). "Marble Island; a search for the Knight expedition, August 6-15, 1970." *Musk-Ox,* No. 8, 1971, pp. 40-46.

SNOW, William Parker (1851). *Voyage of the Prince Albert in search of Sir John Franklin: a narrative of every-day life in the Arctic seas.* London, Longman, Brown, Green, and Longmans, 1851.

SOLIS-COHEN, Bertha (1943). "An American search for the Northwest Passage." *Beaver,* Outfit 274, September 1943, pp. 24-27.

SOLIS-COHEN, Bertha (1952). "Philadelphia's expeditions to Labrador." *Pennsylvania History,* Vol. 19, No. 2, 1952, pp. 148-62.

SØRENSEN, Janus (1941). "En Danskers indsats i Franklin-tragediens opklaring." *Grønlandske Selskabs Aarsskrift,* 1941, pp. 45-99.

SPECK, Gordon (1963). *Samuel Hearne and the Northwest Passage.* Caldwell, Idaho, Caxton Printers, 1963.

STACKPOLE, Edouard A. *ed* (1965). *The long Arctic search; the narrative of Lieutenant Frederick Schwatka, U.S.A., 1878-1880, seeking the records of the lost Franklin expedition.* Mystic, Marine Historical Association, Inc., 1965.

STACKPOLE, Renny A. (1969). *American whaling in Hudson Bay 1861-1919.* Mystic, Munston Institute of American Maritime History, The Marine Historical Association, Inc., 1969.

STAFFORD, Marie Peary (1959). *Discoverer of the North Pole. The story of Robert E. Peary.* New York, William Morrow, 1959.

STAGER, John K. (1962). "Fur trading posts in the Mackenzie region up to 1850." *Canadian Association of Geographers. British Columbia Division. Occasional Papers in Geography,* Vol. 3, May 1962, pp. 37-46.

STAGER, John K. (1967). "Fort Anderson: the first post for trade in the western Arctic." *Geographical Bulletin,* Vol. 9, No. 1, 1967, pp. 45-56.

STANLEY, George F. G. *ed* (1955). *John Henry Lefroy in search of the magnetic north. A soldier-surveyor's letters from the north-west 1843-1844.* Toronto, Macmillan Company of Canada, 1955.

STEELE, Harwood (1936). *Policing the Arctic. The story of the conquest of the Arctic by the Royal Canadian (formerly North-West) Mounted Police.* London, Jarrolds, 1936.

STEEN, Aksel S. *ed* (1887). *Die Internationale Polarforschung 1882-1883. Beobachtungsergebnisse der norwegischen Polarstation Bossekop in Alten.* Christiania, Druck bei Grøndahl & Son, 1887.

STEFANSSON, Vilhjalmur (1908). "Notes on the theory and treatment of diseases among the Mackenzie River Eskimo." *Journal of American Folklore,* Vol. 21, 1908, pp. 43-45.

STEFANSSON, Vilhjalmur (1908a). "On the Mackenzie River." *American Geographical Society of New York. Bulletin,* Vol. 40, 1908, pp. 157-69.

STEFANSSON, Vilhjalmur (1908b). "Suitability of Eskimo methods of winter travel in scientific exploration." *American Geographical Society of New York. Bulletin,* Vol. 40, 1908, pp. 210-13.

STEFANSSON, Vilhjalmur (1908c). "Wintering among the Eskimos." *Harper's Magazine,* Vol. 117, 1908, pp. 38-48.

STEFANSSON, Vilhjalmur (1910). "An ethnologist in the Arctic." *Harper's Magazine,* Vol. 120, February 1910, pp. 455-63.

STEFANSSON, Vilhjalmur (1912). "Some details of Mr. Stefansson's geographical work." *American Geographical Society of New York. Bulletin,* Vol. 44, September 1912, pp. 660-63.

STEFANSSON, Vilhjalmur (1912a). "Work among the Arctic Eskimos." *Geological Survey of Canada. Summary Report,* 1911 (pub. 1912), pp. 389-90.

STEFANSSON, Vilhjalmur (1912-13). "My quest in the Arctic." *Harper's Magazine,* Vol. 126, December 1912-May 1913, pp. 3-13, 176-87, 348-59, 512-22, 671-84, 888-900.

STEFANSSON, Vilhjalmur (1913). *My life with the Eskimo.* New York, Macmillan, 1913.

STEFANSSON, Vilhjalmur (1913a). "Victoria Island and the surrounding seas." *American Geographical Society of New York. Bulletin,* Vol. 45, February 1913, pp. 93-106.

STEFANSSON, Vilhjalmur (1918). "The activities of the Canadian Arctic Expedition from October, 1916, to April, 1918." *Geographical Review,* Vol. 6, October 1918, pp. 354-69.

STEFANSSON, Vilhjalmur (1919). *The Stefansson-Anderson Arctic expedition.* New York, American Museum of Natural History, 1919. (American Museum of Natural History. Anthropological Papers, Vol. 14, Parts 1-2).

STEFANSSON, Vilhjalmur (1919a). "Solving the problem of the Arctic: a record of five years' exploration." *Harper's Magazine,* Vol. 138, 1919, pp. 557-90, 713-35; Vol. 139, 1919, pp. 35-47, 193-203, 386-98, 709-20.

STEFANSSON, Vilhjalmur (1921). "The Canadian Arctic Expedition of 1913 to 1918." *Geographical Journal,* Vol. 58, October 1921, pp. 283-305.

STEFANSSON, Vilhjalmur (1921a). *The friendly Arctic: the story of five years in polar regions.* New York, Macmillan, 1921.

STEFANSSON, Vilhjalmur (1923). *Hunters of the great north.* London, etc., George G. Harrap, 1923.

STEFANSSON, Vilhjalmur (1938). *The three voyages of Martin Frobisher in search of a passage to Cathay and India by the North-west, A.D. 1576-8. From the original 1578 text by George Best. Together with numerous other versions, additions, etc. Now edited, with preface, introduction, notes, appendixes and bibliography, by Vilhjalmur Stefansson.* London, Argonaut Press, 1938.

STEFANSSON, Vilhjalmur (1939). *The problem of Meighen Island. Intended as the third chapter but suppressed in the publication of Unsolved mysteries of the Arctic.* New York, privately printed for Mr Joseph Robinson, 1939.

STEFANSSON, Vilhjalmur (1956). *The fat of the land.* New York, Macmillan, 1956.

STEFANSSON, Vilhjalmur (1960). *Northwest to fortune. The search of western man for a commercially practical route to the Far East.* London, George Allen & Unwin, 1960.

STEFANSSON, Vilhjalmur (1964). *Discovery; the autobiography of Vilhjalmur Stefansson.* New York, McGraw Hill, 1964.

STEIN, R. (1909). "Bernhard Hantzsch's notes on northeastern Labrador." *American Geographical Society of New York. Bulletin,* Vol. 41, 1909, pp. 566-75.

STEPHENSON, Henry F. (1876). "The

winter quarters of the *Discovery.*" *Royal Geographical Society. Proceedings,* Vol. 21, December 1876, pp. 106-10.

STEVENS, Henry (1886). *The dawn of British trade in the East Indies as recorded in the court minutes of the East India Company, 1599-1603, containing an account of the formation of the company, the first adventure and Weymouth's voyage in search of the North-West Passage. With an introduction by Sir George Birdwood, Kt., C.S.I., M.D.* London, 1886.

STEVENS, R.P. (1863). "Report on the geological and mineralogical specimens collected by Mr. C. F. Hall in Frobisher Bay." *American Journal of Science,* Series 2, Vol. 35, March 1863, pp. 293-95.

STEVENSON, Alex (1966). "Penny's polar probe, 1850-51." *North,* Vol. 13, No. 6, November-December 1966, pp. 30-37.

STEVENSON, Alex (1967). "Apprentice voyage 1867." *Beaver,* Outfit 297, Spring 1967, pp. 50-55.

STEVENSON, Alex (1968). "Whalers' wait." *North,* Vol. 15, No. 5, September-October 1968, pp. 24-30.

STEVENSON, Alex (1969). "Lawless land." *North,* Vol. 16, No. 1, January-February 1969, pp. 22-30.

STEVENSON, Alex (1973). "The Robert Janes murder trial at Pond Inlet." *Beaver,* Outfit 304, No. 2, Autumn 1973, pp. 16-23.

STEWART, Elihu (1913). *Down the Mackenzie and up the Yukon in 1906.* London, John Lane, the Bodley Head, 1913.

STIMPSON, William (1864). "Synopsis of the marine invertebrata collected by the late Arctic expedition, under Dr. I.I. Hayes." *Academy of Natural Sciences of Philadelphia. Proceedings,* 1863 (pub. 1864), pp. 138-42.

STOBBE, Leslie H. (1969). "Integrated transportation system serves the Yukon." *British Columbia Business Journal,* Vol. 1, No. 4, June 1969, pp. 3-5.

STOCKTON, Charles H. (1890). "The Arctic cruise of the U.S.S. Thetis in the summer and autumn of 1889." *National Geographic Magazine,* Vol. 2, No. 3, 1890, pp. 171-98.

STONE, Andrew Jackson (1900). "Some results of a natural history journey to northern British Columbia, Alaska and the Northwest Territory." *American Museum of Natural History. Bulletin,* Vol. 13, 1900, pp. 31-62.

STONYHURST MAGAZINE (1933). "In memoriam. Alfred Henry Harrison (O.S. 1877)." *Stonyhurst Magazine,* No. 308, December 1933, p. 409.

STORM, Gustav (1889). *Studies on the Vineland voyages.* Copenhagen, 1889.

STOWELL, James Vinton (1954). *Ross Gilmore Marvin; scientist, teacher, explorer. A "tragedy in the Arctic".* Elvira, N.Y., Chemung County Historical Society, 1954.

STUPART, Robert Frederick (1887). "The Eskimo of Stupart Bay." *Royal Canadian Institute. Proceedings,* Series 3, Vol. 4, 1885-86 (pub. 1887), pp. 95-114.

SULLIVAN, Walter (1963). "Vilhjalmur Stefansson, 1879-1962." *Geographical Review,* Vol. 53, No. 2, 1963, pp. 287-91.

SUTHERLAND, Peter Cormack (1852). *Journal of a voyage in Baffin's Bay and Barrow Straits, in the years 1850-1851, performed by H.M. Ships "Lady Franklin" and "Sophia", under the command of Mr. William Penny, in search of the missing crews of H.M. Ships Erebus and Terror . . .* 2 vols. London, Longman, Brown, Green, and Longmans, 1852.

SUTHERLAND, Peter Cormack (1853). "On the geological and glacial phenomena of the coasts of Davis Strait and Baffin Bay." *Geological Society of London. Quarterly Journal,* Vol. 9, June 1853, pp. 296-312.

SUTHERLAND, Peter Cormack (1856). "On the Esquimaux." *Ethnological Society of London. Journal,* Vol. 4, 1856, pp. 193-214.

SUTTON, George Miksch (1932). "The exploration of Southampton Island, Hudson Bay . . . Part 1. Section 1, prefatory; section 2, introductory; section 3, bibliography of part 1 and part 2, section 2." *Carnegie Museum. Memoirs,* Vol. 12, Part 1, sections 1, 2, 3, March 1932.

SVERDRUP, Harald Ulrik (1959). "Roald Amundsen." *Arctic,* Vol. 12, No. 4, December 1959, pp. 221-36.

SVERDRUP, Otto Neumann (1903). "The Norwegian polar expedition in the Fram, 1898-1902." *Geographical Journal,* Vol. 22, 1903, pp. 38-56.

SVERDRUP, Otto Neumann (1903a). "The second Norwegian polar expedition in the Fram." *Scottish Geographical Magazine,* Vol. 19, 1903, pp. 337-53.

SVERDRUP, Otto Neumann (1904). *New land; four years in the Arctic regions. 2 vols.* London and New York, Longmans, Green, 1904. (Translation of *Nyt land; fire aar i arktiske agne.* Kristiania, 1903).

[SWAINE, Charles] (1748). *An account of a voyage for the discovery of a North-west Passage by Hudson's Streights, to the Western and Southern Ocean of America, performed in the year 1746 and 1747, in the ship California, Capt. Francis Smith, Commander. By the Clerk of the California.* Vols. 1 and 2. London, Printed and sold by Mr. Jolliffe, in St. James's-street; Mr. Corbett, in Fleet-street; and Mr. Clarke, under the Royal Exchange, 1748. (Reprinted by S.R. Publishers Ltd. and Johnson Reprint Corporation, 1968).

TASKER, Florence A. (1908). "A woman through husky-land." *Field and Stream,* February and March 1908, pp. 823-28, 937-43.

TAVENER, L.E. (1953). "William Mogg." *Nautical Magazine,* Vol. 170, 1953, pp. 141-44.

TAYLOR, E.G.R. (1931). "Master Hore's voyage of 1536." *Geographical Journal,* Vol. 77, No. 5, May 1931, pp. 469-70.

TAYLOR, E.G.R. (1966). *The mathematical practitioners of Hanoverian England, 1714-1840.* Cambridge, Cambridge University Press for the Institute of Navigation, 1966.

TERO, Richard D. *ed* (1973). "Alaska: 1779. Father Riobo's narrative." *Alaska Journal. History and Arts of the North – Quarterly,* Vol. 3, No. 2, Spring 1973, pp. 81-88.

THALBITZER, William (1951). *Two runic stones, from Greenland and Minnesota.* Washington, D.C., Smithsonian Institution, 1951. (Smithsonian Miscellaneous Collections, Vol. 116, No. 3).

THOMAS, Lowell (1961). *Sir Hubert Wilkins; his world of adventure.* London, Arthur Barker, 1961.

THOMSON, Don W. (1966). "The Yukon and William Ogilvie." *Canadian Geographical Journal,* Vol. 73, No. 6, 1966, pp. 194-97.

TILLOTSON, John (1869). *Adventures in the ice: a comprehensive summary of Arctic exploration, discovery, and adventure. Including experiences of Captain Penny, the veteran whaler.* London, James Hogg and Son, [1869].

TODD, Alden L. (1961). *Abandoned. The story of the Greely Arctic expedition 1881-1884.* New York, etc., McGraw Hill, 1961.

TODD, Walter Edmond Clyde (1963). *Birds of the Labrador peninsula and adjacent areas. A distributional list.* Toronto, Published in association with Carnegie Museum by University of Toronto Press, 1963.

TOLMACHOFF, Innokenty Pavlovich (1926). "Note on the Devonian fauna of Ellesmere Land." *American Journal of Science.* Series 5, Vol. 11, January 1926, pp. 47-57.

TOMLINSON, H.M. (1951). "Why James Bay?" *Bay,* Outfit 5, Shipment 1, 1951, pp. 6-9.

TOMPKINS, S.R. (1945). "Drawing the Alaska boundary." *Canadian Historical Review,* Vol. 26, March 1945, pp. 1-24.

TOWNSEND, Charles W. (1907). *A Labrador spring.* Boston, Dana Estes, 1907.

TOWNSEND, Charles W. *ed* (1911) *Captain Cartwright and his Labrador journal.* Boston, Dana Estes & Co., 1911.

TOWNSEND, Charles W. *and* ALLEN, G.M. (1907). "Birds of Labrador." *Boston Society of Natural History. Proceedings,* Vol. 33, July 1907, pp. 277-428.

TRAILL, H.D. (1896). *The life of Sir John Franklin,* R.N. London, John Murray, 1896.

TREMBLAY, Alfred (1921). *Cruise of the Minnie Maud. Arctic seas and Hudson Bay 1910-11 and 1912-13. Compiled and translated by A.B. Reader.* Quebec, Arctic Exchange and Publishing Ltd., 1921.

TRIMMER, F. Mortimer *comp* (1898). *The Yukon Territory. The narrative of W.H. Dall, leader of the expedition to Alaska in 1866-1868; the narrative of an exploration made in 1887 in the Yukon District by George M. Dawson; extracts from the report of an exploration made in 1896-1897 by Wm. Ogilvie.* London, Downey & Co., 1898.

TURNER, John Peter (1950). *The North-West Mounted Police 1873-1893.* 2 vols. Ottawa, King's Printer, 1950.

TURNER, Lucien McShan (1886). "List of the birds of Labrador." *U.S. National Museum. Proceedings,* Vol. 8, 1885 (pub. 1886), pp. 233-54.

TURNER, Lucien McShan (1888). "Physical and zoological character of the Ungava district, Labrador." *Royal Society of Canada Transactions,* Vol. 5, Section 4, 1887 (pub. 1888), pp. 79-83.

TURNER, Lucien McShan (1894). "Ethnology of the Ungava district, Hudson Bay territory." *Bureau of American Ethnology. Annual Report,* 11th, 1889-90 (pub. 1894), pp. 159-350.

TUTTLE, Charles Richard (1885). *Our north land: being a full account of the Canadian north-west and Hudson's Bay Route, together with a narrative of the experiences of the Hudson's Bay expedition of 1884, including a description of the climate, resources, and the characteristics of the native inhabitants between the 50th parallel and the Arctic Circle.* Toronto, C. Blackett Robinson, 1885.

TYLER, Winsor Marrett (1943). "In memoriam: Glover Morrill Allen." *Auk,* Vol. 60, No. 2, 1943, pp. 163-68.

TYRRELL, James Williams (1898). *Across the sub-Arctics of Canada. A journey of 3,200 miles by canoe and snowshoe through the Barren Lands.* London, T. Fisher Unwin, [1898].

TYRRELL, James Williams *and* FAIRCHILD, C.C. (1902). "Report on an exploratory survey between Great Slave Lake and Hudson Bay, districts of Mackenzie and Keewatin." *Canada. Department of the Interior. Annual Report,* 1901 (pub. 1902), Part 3, pp. 98-155, 209-329.

TYRRELL, Joseph Burr (1894). "An expedition through the Barren Lands of northern Canada." *Geographical Journal,* Vol. 4, 1894, pp. 437-50.

TYRRELL, Joseph Burr (1894a). "Notes on the Pleistocene of the North-West Territories of Canada, north-west and west of Hudson Bay. *Geological Magazine,* NS, Decade 4, Vol. 1, September 1894, pp. 394-99

TYRRELL, Joseph Burr (1895). "A second expedition through the Barren Lands of northern Canada." *Geographical Journal,* Vol. 6, 1895, pp. 438-48.

TYRRELL, Joseph Burr (1898). "Report on the Doobaunt, Kazan and Ferguson rivers and the north-west coast of Hudson Bay." *Geological Survey of Canada. Annual Report,* NS, Vol. 9, 1896 (pub. 1898), Report F.

TYRRELL, Joseph Burr (1909). "A story of a Franklin search expedition." *Royal Canadian Institute. Transactions,* Vol. 8, April 1909, pp. 393-402.

TYRRELL, Joseph Burr (1913). "Peter Fidler, trader and surveyor." *Royal Society of Canada. Transactions,* 3rd series, Vol. 7, Section 2, 1913.

TYRRELL, Joseph Burr (1931). *Documents relating to the early history of Hudson Bay, edited with an introduction and notes by J.B. Tyrrell.* Toronto, Champlain Society, 1931. (Publications of the Champlain Society, Vol. 18).

TYRRELL, Joseph Burr *ed* (1934). *Journals of Samuel Hearne and Philip Turnor.* Toronto, Champlain Society, 1934. (Publications of the Champlain Society, No. 21).

TYSON, George Emory (1879). *The cruise of the Florence, or, extracts from the journal of the preliminary Arctic expedition of 1877-'78. Edited by H.W. Howgate.* Washington, D.C., James J. Chapman, 1879.

UNITED STATES. Congress. House (1911). *Statement of Captain Robert E. Peary, United States Navy, before the Committee on Naval Affairs, House of Representatives.* Washington, D.C., Government Printing Office, 1911.

UNITED STATES. Congress. Senate (1884). *Proceedings of the "Proteus" Court of Inquiry on the Greely relief expedition of 1883.* Washington, D.C., Government Printing Office, 1884.

UPTON, L.F.S. (1968). "The wreck of the *Eldorado.*" *Beaver,* Outfit 299, Autumn 1968, pp. 27-31.

USHER, Peter J. (1971). *Fur trade posts of the Northwest Territories 1870-1970.* Canada. Department of Indian Affairs and Northern Development. Northern Science Research Group, NSRG 71-4, 1971.

VANCOUVER, George (1798). *A voyage of discovery to the North Pacific Ocean and round the world; in which the coast of north-west America has been carefully examined and accurately surveyed, in the years 1790-95, in the Discovery, sloop of war, and armed tender Chatham.* 3 vols. London, printed for G.G. and J. Robinson and J. Edwards, 1798.

VIANA, Francisco Javier de (1849). *Diario del viage explorador de las corbetas españolas "Descubierta" y "Atrevida", en los años de 1789 a 1794, llevado por el teniente de navio D. Francisco Javier de Viana, y ofrecido para su publicacion, en su original inedito, por el Sr. D. Francisco Javier de Viana, y demas hijos del autor.* Cerrito de la Victoria, Imprenta del Ejercito, 1849.

VIGNERAS, L.-A. (1970). "The voyage of Esteban Gomes from Florida to the Baccalos." *Terrae Incognitae,* Vol. 2, 1970, pp. 25-28.

VILLAREJO, Oscar M. *ed* (1965). *Dr. Kane's voyage to the polar lands. [By Johan Carl Christian Petersen, edited with an introduction by O.M. Villarejo].* Philadelphia, University of Pennsylvania Press, 1965.

VOORHIS, Ernest (1930). *Historic forts and trading posts of the French regime and of the English fur trading companies.* Ottawa, Department of the Interior, 1930.

WAGNER, Henry R. (1931). "Apocryphal voyages to the Northwest Coast of America." *Proceedings of the American Antiquarian Society,* New Series, Vol. 41, Part 1, 1931, pp. 179-234.

WAHLGREN, Erik (1959). "The case of the Kensington Rune Stone." *American Heritage,* Vol. 10, No. 3, April 1959, pp. 34-35, 101-05.

WALES, William (1770). "Astronomical observations made by order of the Royal Society, at Prince of Wales's Fort, on the north-west coast of Hudson's Bay." *Philosophical Transactions* (London, Royal Society), Vol. 59, 1769 (pub. 1770), pp. 467-88.

WALES, William. (1771). "Journal of a voyage, made by order of the Royal Society, to Churchill River, on the north-west coast of Hudson's Bay; of thirteen months residence in that country; and of the voyage back to England; in the years 1768 and 1769." *Philosophical Transactions* (London, Royal Society), Vol. 60, 1770 (pub. 1771), pp. 100-36.

WALKER, D. (1860). "Ornithological notes of the voyage of the 'Fox' in the Arctic seas." *Ibis,* Vol. 2, 1860, pp. 165-68.

WALKER, D. (1862). "Notes on the zoology of the last Arctic expedition under Capt. Sir F.L. M'Clintock." *Royal Dublin Society. Journal,* Vol. 3, 1860 (pub. 1862), pp. 61-77.

WALLACE, Dillon (1905). *The lure of the Labrador wild.* London, Hodder & Stoughton, 1905.

WALLACE, Dillon (1907). *The long Labrador trail.* Toronto, Fleming H. Revell Co., 1907.

WALLACE, Dillon (1932). "Crossing the Labrador barrens." *In:* Blossom, Frederick A. *ed. Told at the Explorers' Club; true tales of modern exploration.* London, etc., George G. Harrap, 1932.

WALLACE, J.N. (1928). "The explorer of Finlay River in 1824." *Canadian Historical Review,* Vol. 9, 1928, pp. 25-31.

WALLACE, J.N. (1929). *The wintering partners on Peace River from the earliest records to the union in 1821. With a summary of the Dunvegan journal, 1806.* Ottawa, Thorburn and Abbott, 1929.

WALLACE, W.S. *ed* (1932). *John McLean's notes of a twenty-five year's service in the Hudson's Bay territory.* Toronto, Champlain Society, 1932. (Publications of the Champlain Society, Vol. 19).

WALLACE, W.S. (1938). "Sir Henry Lefroy's journey to the north-west in 1843-4." *Royal Society of Canada. Proceedings and Transactions,* 3rd Series, Vol. 32, Section 2, 1938, pp. 67-96.

WALLACE, W.S. (1947). "The Nor'westers invade the Bay." *Beaver,* Outfit 277, March 1947, pp. 33-34.

WALLACE, W.S. (1952). "Notes of an 18th century northerner." *Beaver,* Outfit 283, June 1952, pp. 39-42.

WALLICH, George Charles (1860). *Notes on the presence of animal life at vast depths in the sea; with observations on the nature of the sea bed, as bearing on submarine telegraphy.* London, Taylor and Francis, 1860.

WALLICH, George Charles (1862). *The North-Atlantic sea-bed: comprising a diary of the voyage on board H.M.S. Bulldog, in 1860; and observations on the presence of animal life, and the formation and nature of organic deposits, at great depths in the ocean.* London, J. Van Voorst, 1862.

WALLIS, Helen *and others.* (1974). "The strange case of the Vinland map: a symposium [by] Helen Wallis, F.R. Maddison, G.D. Painter, D.B. Quinn, R.M. Perkins, G.R. Crone, A.D. Baynes-Cope, Walter C. and Lucy B. McCrone." *Geographical Journal,* Vol. 140, Part 2, June 1974, pp. 183-214.

WARMOW, Mathias (1859). "Extracts from Br. M. Warmow's journal of his residence in Cumberland Inlet, during the winter of 1857-1858." *Periodical Accounts Relating to the Missions of the Church of the United Brethren, Established among the Heathen,* Vol. 23, 1850-60, No. 242, March 1859, pp. 87-92.

WATT, Maud (1938). "Nascopie honeymoon." *Beaver,* Outfit 268, No. 4, March 1938, pp. 18-26.

WATSON, Robert (1929). "Chief trader Alexander Hunter Murray and Fort Youcon." *Beaver,* Outfit 260, No. 1, June 1929, pp. 211-13.

WEEKES, Mary *and* KING, William Cornwallis (1939). "Akaitcho, a link with Franklin." *Beaver,* Outfit 270, June 1939, pp. 25-27.

WEEMS, John Edward (1960). *Race for the Pole.* New York, Henry Holt, 1960.

WEEMS, John Edward (1967). *Peary, the explorer and the man, based on his personal papers.* London, Eyre and Spottiswoode, 1967.

WENTZEL, Willard Ferdinand (1821). *Account of Mackenzies River – with a Chart.* Unpublished manuscript, dated Fort Enterprize, 26 February 1821. Scott Polar Research Institute MS 802/1-2 (Xerox copy).

WENTZEL, Willard Ferdinand (1822). "Notice regarding the map of Mackenzie's River by Mr. W.F. Wentzel, of the North West Fur Company." *Memoirs of the Wernerian Natural History Society* (Edinburgh), Vol. 4, 1821-22 (pub. 1822), pp. 562-63.

[WENTZEL, Willard Ferdinand] (1823). "Notice of the attempts to reach the sea by Mackenzie's River, since the expedition of Sir Alexander Mackenzie." *Edinburgh Philosophical Journal,* Vol. 8, October 1822-April 1823 (pub. 1823), pp. 77-81.

WHALLEY, George (1962). *The legend of John Hornby.* London, John Murray, 1962.

WHITEAVES, Joseph F. (1899). "Recent discovery of rocks of the age of the Trenton Formation at Akpatok Island, Ungava Bay, Ungava." *American Journal of Science,* Series 4, Vol. 7, June 1899, pp. 433-34.

WHITELEY, W.H. (1964). "The establishment of the Moravian mission in Labrador and British policy, 1763-83." *Canadian Historical Review,* Vol. 45, No. 1, March 1964, pp. 29-50.

WHITFIELD, R.P. (1900). "Observations on and descriptions of Arctic fossils." *American Museum of Natural History. Bulletin,* Vol. 13, 1900, pp. 19-22.

WHITNEY, Caspar (1896). *On snow-shoes to the barren grounds. Twenty-eight hundred miles after musk-oxen and wood-bison.* London, Osgood, McIlvaine & Co., 1896.

WHITNEY, Harry (1910). *Hunting with the Eskimos. The unique record of a sportsman's year among the northernmost tribe – the big game hunting, the native life, and the battle for existence through the long Arctic night.* London, T. Fisher Unwin, 1910.

WHYMPER, Frederick (1869). *Travel and adventure in the Territory of Alaska, formerly Russian America – now ceded to the United States – and in various other parts of the North Pacific.* New York, Harper & Brothers, 1869.

WILD, Roland (1955). *Arctic command; the story of Smellie of the Nascopie.* Toronto, Ryerson Press, 1955.

WILLARD, Berton C. (1968). "Russell W. Porter – explorer." *Polar Notes,* No. 8, June 1968, pp. 69-84.

WILLIAMS, Glyndwr (1962). *The British search for the Northwest Passage in the eighteenth century.* London, Printed for the Royal Commonwealth Society by Longmans, 1962. (Imperial Studies series, No. 24).

WILLIAMS, Glyndwr (1963). "Captain Coats and exploration along the East Main." *Beaver,* Outfit 294, Winter 1963, pp. 4-13.

WILLIAMS, Glyndwr *ed* (1969). *Andrew Graham's observations on Hudson's Bay 1767-91. Edited by Glyndwr Williams, with an introduction by Richard Glover.* London, Hudson's Bay Record Society, 1969. (HBRS Vol. 27).

WILLIAMS, J. (1939). "The last voyage of the Stork." *Beaver,* Outfit 270, No. 2, September 1939, pp. 44-47.

WILLIAMSON, H. Anthony (1964). "The Moravian mission and its impact on the Labrador Eskimo." *Arctic Anthropology,* Vol. 2, No. 2, 1964, pp. 32-36.

WILLIAMSON, James A. (1937). *The voyages of John and Sebastian Cabot.* London, Published for the Historical Association by G. Bell and Sons, 1937. (Historical Association Pamphlet, No. 106).

WILLIAMSON, James A. (1962). *The Cabot voyages and Bristol discovery under Henry VII. By James A. Williamson. With the cartography of the voyages by R.A. Skelton.* Cambridge, Published for the Hakluyt Society at the University Press, 1962. (Hakluyt Society, Second Series, No. 120).

WILSON, Clifford P. (1938). "Letters from George Back." *Canadian Geographical Journal*, Vol. 17, No. 3, September 1938, pp. 130-35.

WILSON, Clifford P. (1947). "Founding Fort Yukon. . . ." *Beaver*, Outfit 278, June 1947, pp. 38-43.

WILSON, Clifford P. (1970). *Campbell of the Yukon.* Toronto, Macmillan Company of Canada, 1970.

WILSON, John Tuzo (1949). "New light on Hearne." *Beaver*, Outfit 280, June 1949, pp. 14-18.

WILSON, Malcolm (1973). "Sir John Ross's last expedition, in search of Sir John Franklin." *Musk-ox*, No. 13, 1973, pp. 3-11.

WOOD, Walter A. (1959). "George Hubert Wilkins." *Geographical Review*, Vol. 49, No. 3, 1959, pp. 410-16.

WOODBRIDGE, Dwight E. (1921). "Iron-ore deposits on the Belcher Islands." *Engineering and Mining Journal*, Vol. 112, No. 7, August 1921, pp. 251-54.

WOODS, Lawrence C. (1969). "W.E. Clyde Todd (1875-1969)." *Arctic*, Vol. 22, No. 4, December 1969, pp. 451-52.

WOODWARD, Frances J. (1948). "Edward Augustus Inglefield." *Polar Record*, Vol. 5, No. 35, 1948, pp. 189-92.

WOODWARD, Frances J. (1950). "Joseph René Bellot, 1826-53." *Polar Record*, Vol. 5, No. 39, 1950, pp. 398-407.

WOODWARD, Henry (1878). "Notes on some Arctic Silurian or Devonian (?) fossils from Beechey Island, brought home by the S.Y. "Pandora" in 1875, and from Port Dundas, Lancaster Sound, by an earlier expedition in 1853." *Geological Magazine*, New Series, Decade 2, Vol. 5, September 1878, pp. 385-90.

WOOLLACOTT, Arthur P. (1927). *Mackenzie and his voyageurs. By canoe to the Arctic and the Pacific 1789-93.* London and Toronto, J.M. Dent, 1927.

WORDIE, *Sir* James Mann (1944). "Captain William Adams and Dr. R.W. Gray." *Polar Record*, Vol. 4, No. 27, 1944, pp. 119-22.

WRAY, O.R. (1934). "In the footsteps of Samuel Hearne." *Canadian Geographical Journal,* Vol. 9, No. 3, September 1934, pp. 138-46.

WRIGHT, Noel (1949). *New light on Franklin.* Ipswich and London, W.S. Cowell Ltd., 1949.

WRIGHT, Noel (1959). *Quest for Franklin.* London, etc., Heinemann, 1959.

WRIGHT, Theon (1970). *The big nail; the story of the Cook-Peary feud.* New York, John Day Company, 1970.

YOUNG, Allen William (1875). *The search for Sir John Franklin. From the journal of Allen Young.* London and Portsmouth, J. Griffin & Co., 1875.

YOUNG, Allen William (1876). *Cruise of the 'Pandora'. From the private journal kept by Allen Young.* London, Printed by William Clowes and Sons, 1876.

YOUNG, Allen William (1879). *The two voyages of the 'Pandora' in 1875 and 1876.* London, Edward Stanford, 1879.

ZASLOW, Morris (1971). *The opening of the Canadian north, 1870-1914.* Toronto and Montreal, McClelland and Stewart, 1971.

Index

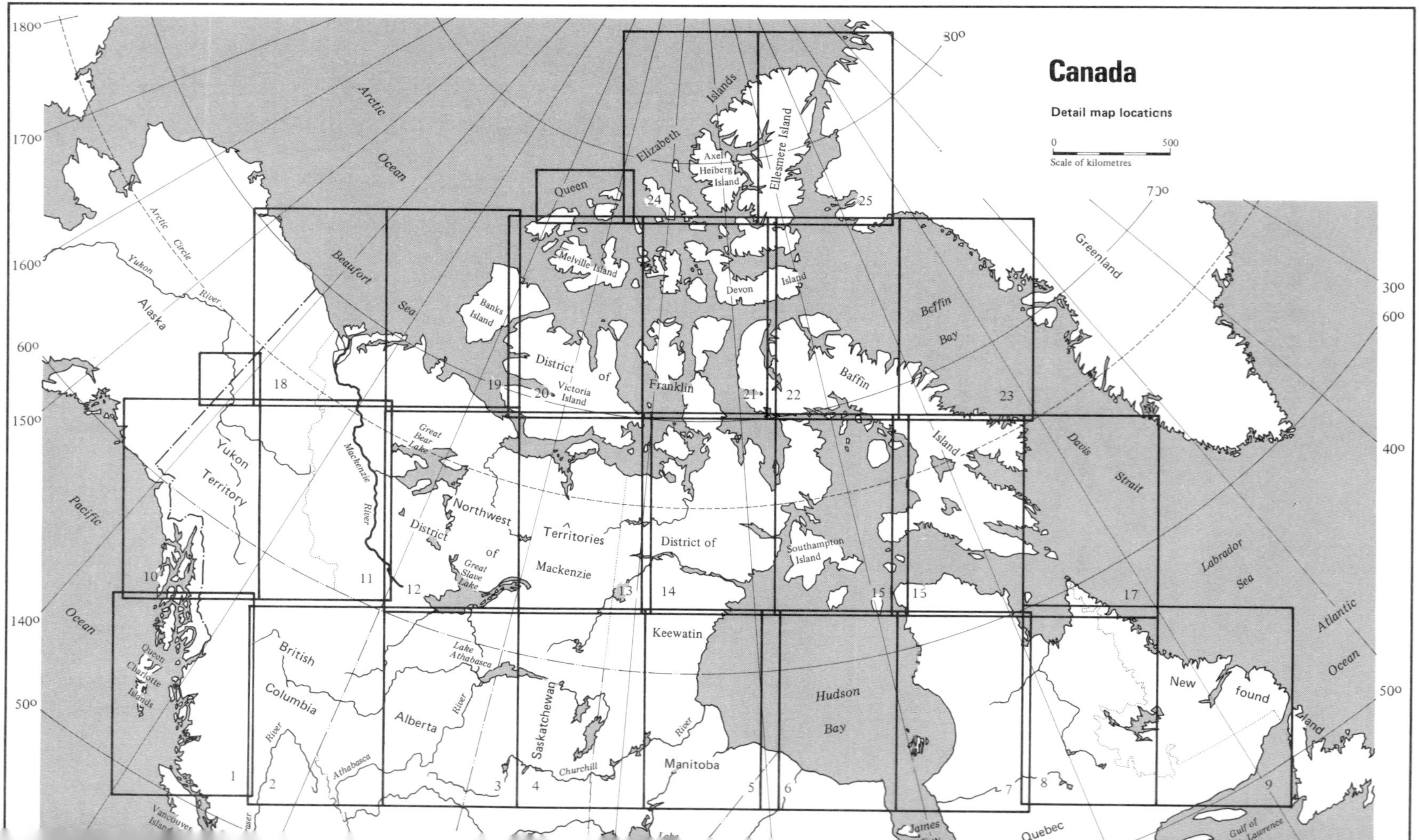
Canada
Detail map locations
0
500
Scale of kilometres
Arctic
Ocean
Beaufort
Sea
Pacific
Ocean
Alaska
Yukon
River
Arctic
Circle
Yukon
Territory
Mackenzie
River
Great
Bear
Lake
Great
Slave
Lake
Northwest
Territories
District
of
Mackenzie
District of
Franklin
District
of
Keewatin
Banks
Island
Victoria
Island
Melville Island
Queen
Elizabeth
Islands
Axel
Heiberg
Island
Ellesmere Island
Devon
Island
Baffin
Island
Southampton
Island
Baffin
Bay
Davis
Strait
Greenland
Labrador
Sea
Atlantic
Ocean
New
found
land
Gulf of
St. Lawrence
Quebec
James
Hudson
Bay
Manitoba
Saskatchewan
Churchill
Alberta
Lake
Athabasca
Athabasca
River
British
Columbia
Queen
Charlotte
Islands
Vancouver
Island
1
2
3
4
5
6
7
8
9
10
11
12
13
14
15
16
17
18
19
20
21
22
23
24
25
180°
170°
160°
60°
150°
140°
50°
30°
60°
40°
70°

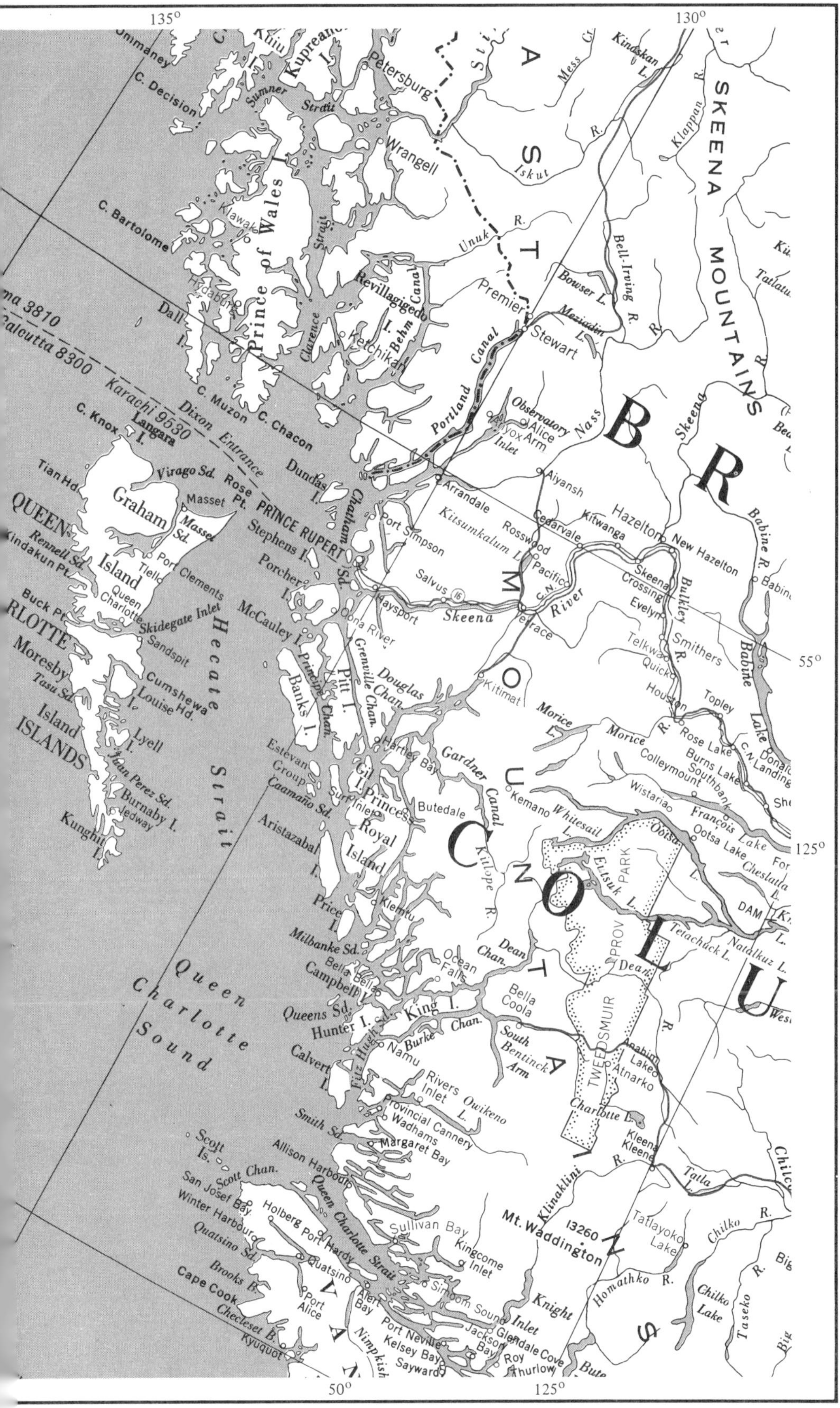
135°
130°
125°
55°
50°
Unimaney
C. Decision
Kuiu
Kupreanof I.
Petersburg
Sumner Strait
Wrangell
Prince of Wales I.
Klawak
C. Bartolome
Hydaburg
Dall I.
Clarence Strait
Revillagigedo I.
Behm Canal
Ketchikan
C. Muzon
C. Chacon
Dixon Entrance
Karachi 9530
Calcutta 8300
Langara I.
C. Knox
Tian Hd.
Virago Sd.
Rose Pt.
Masset
Graham Island
Masset Sd.
QUEEN
CHARLOTTE
ISLANDS
Rennell Sd.
Kindakun Pt.
Tlell
Port Clements
Queen Charlotte
Skidegate Inlet
Sandspit
Buck Pt.
Moresby Island
Tasu Sd.
Cumshewa Inlet
Louise I.
Lyell I.
Juan Perez Sd.
Burnaby I.
Jedway
Kunghit I.
Hecate Strait
PRINCE RUPERT
Dundas I.
Stephens I.
Porcher I.
Chatham Sd.
Port Simpson
McCauley I.
Oona River
Haysport
Salvus
Skeena River
Arrandale
Kitsumkalum L.
Rosswood
Aiyansh
Cedarvale
Kitwanga
Hazelton
New Hazelton
Pacific
Terrace
Skeena Crossing
Evelyn
Telkwa
Smithers
Quick
Houston
Bulkley R.
Topley
Rose Lake
Burns Lake
Colleymount
Southbank
Wistaria
François Lake
Ootsa Lake
Portland Canal
Observatory Inlet
Alice Arm
Anyox
Nass
Stewart
Premier
Meziadin L.
Bowser L.
Bell-Irving R.
Unuk R.
Iskut R.
Mess Cr.
Kinaskan L.
Klappan R.
Stikine
SKEENA MOUNTAINS
Skeena
Tatlatui
Babine R.
Babine Lake
Donald Landing
C.N.
Grenville Chan.
Principe Chan.
Pitt I.
Banks I.
Douglas Chan.
Kitimat
Morice L.
Morice R.
Hartley Bay
Gardner Canal
Kemano
Whitesail L.
Estevan Group
Gil I.
Caamaño Sd.
Surf Inlet
Princess Royal Island
Butedale
Aristazabal I.
Kitlope R.
Eutsuk L.
TWEEDSMUIR PROV PARK
Ootsa L.
Cheslatta L.
DAM
Natalkuz L.
Tetachuck L.
Klemtu
Price I.
Milbanke Sd.
Bella Bella
Campbell I.
Ocean Falls
Dean Chan.
Dean R.
Bella Coola
Queens Sd.
Hunter I.
King I.
Fitz Hugh Sd.
Burke Chan.
South Bentinck Arm
Namu
Calvert I.
Anahim Lake
Atnarko
Charlotte L.
Rivers Inlet
Owikeno L.
Provincial Cannery
Wadhams
Smith Sd.
Margaret Bay
Queen Charlotte Sound
Scott Is.
Scott Chan.
Allison Harbour
San Josef Bay
Winter Harbour
Holberg
Port Hardy
Queen Charlotte Strait
Sullivan Bay
Kingcome Inlet
Quatsino Sd.
Quatsino
Brooks B.
Cape Cook
Port Alice
Checleset B.
Kyuquot
Alert Bay
Simoom Sound
Port Neville
Kelsey Bay
Sayward
Jackson Bay
Glendale Cove
Roy
Thurlow
Nimpkish
Knight Inlet
Bute
Mt. Waddington
13260
Klinaklini R.
Kleena Kleene
Tatla L.
Tatlayoko Lake
Homathko R.
Chilko R.
Chilko Lake
Taseko R.
Chilcotin
B R I T I S H
C O L U M B I A
A L A S K A
M O U N T A I N S
V A N

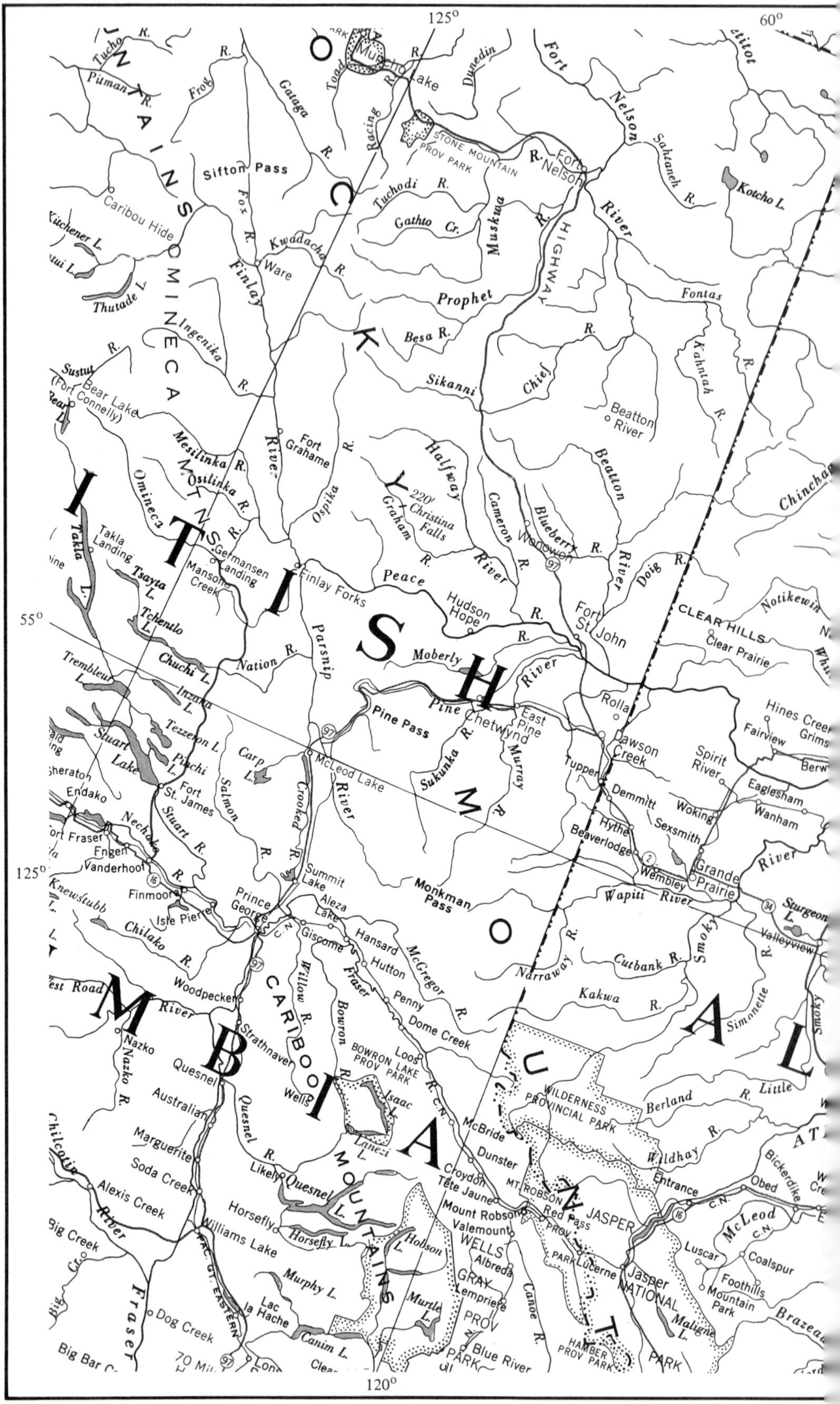

125°
60°
55°
125°
120°
Tucho R.
Pitman R.
Frog R.
Gataga R.
Toad R.
Muncho Lake
Racing R.
Dunedin
Fort Nelson River
Sahtaneh R.
Kotcho L.
Stone Mountain Prov Park
Fort Nelson
Sifton Pass
Caribou Hide
Kitchener L.
Thutade L.
Fox R.
Kwadacha R.
Ware
Finlay
Tuchodi R.
Gathto Cr.
Muskwa R.
Highway
Prophet
Fontas
Besa R.
Ingenika R.
Sustut R.
Bear Lake (Fort Connelly)
Bear L.
Sikanni
Chief R.
Kahntah R.
Beatton River
Beatton
Fort Grahame
Finlay River
Ospika R.
Halfway
Mesilinka R.
Osilinka R.
Omineca
220' Christina Falls
Graham R.
Cameron R.
Blueberry R.
Wonowon
Chinchaga
Takla Landing
Takla L.
Germansen Landing
Manson Creek
Tsayta L.
Finlay Forks
Peace
Hudson Hope
River
Doig R.
Fort St. John
Clear Hills
Clear Prairie
Notikewin
Tchentlo L.
Chuchi L.
Nation R.
Parsnip
Moberly
Trembleur L.
Inzana L.
Pine Pass
Pine
Chetwynd
East Pine
Rolla
Dawson Creek
Tupper
Hines Creek
Fairview
Spirit River
Tezzeron L.
Stuart Lake
Pinchi L.
Carp L.
McLeod Lake
Sukunka R.
Murray R.
Demmitt
Eaglesham
Woking
Wanham
Sheraton
Endako
Fort St. James
Salmon
Crooked R.
River
Hythe
Sexsmith
Beaverlodge
Grande Prairie
Wembley
Fort Fraser
Nechako
Stuart R.
Vanderhoof
Summit Lake
Wapiti River
Monkman Pass
Knewstubb
Finmoore
Prince George
Aleza Lake
Isle Pierre
Giscome
Hansard
Hutton
McGregor R.
Cutbank R.
Sturgeon L.
Valleyview
Chilako R.
Woodpecker
Cariboo
Willow R.
Fraser
Bowron
Penny
Narraway
Kakwa R.
Smoky R.
Simonette
West Road River
Nazko
Nazko R.
Strathnaver
Quesnel
Bowron Lake Prov Park
Dome Creek
Loos
Wilderness Provincial Park
Australian
Wells
Isaac L.
Berland R.
Little
Marguerite
Quesnel R.
Lanezi L.
McBride
Dunster
Wildhay R.
Likely
Quesnel L.
Croydon
Tête Jaune
Mt. Robson Prov Park
Entrance
Bickerdike
Obed
Chilcotin
Soda Creek
Alexis Creek
River
Williams Lake
Horsefly
Horsefly L.
Mount Robson
Red Pass
Jasper
C.N.
McLeod
Valemount
Wells Gray Prov Park
Luscar
Coalspur
Big Creek
Pac. Gt. Eastern
Hobson
Albreda
Lucerne
Jasper National Park
Foothills
Mountain Park
Murphy L.
Lempriere
Maligne L.
Brazeau
Lac la Hache
Murtle L.
Canoe R.
Fraser
Dog Creek
Canim L.
Blue River
Hamber Prov Park
Big Bar
70 Mile
British Columbia
Alberta
Rocky Mountains
Omineca Mountains
Cariboo Mountains

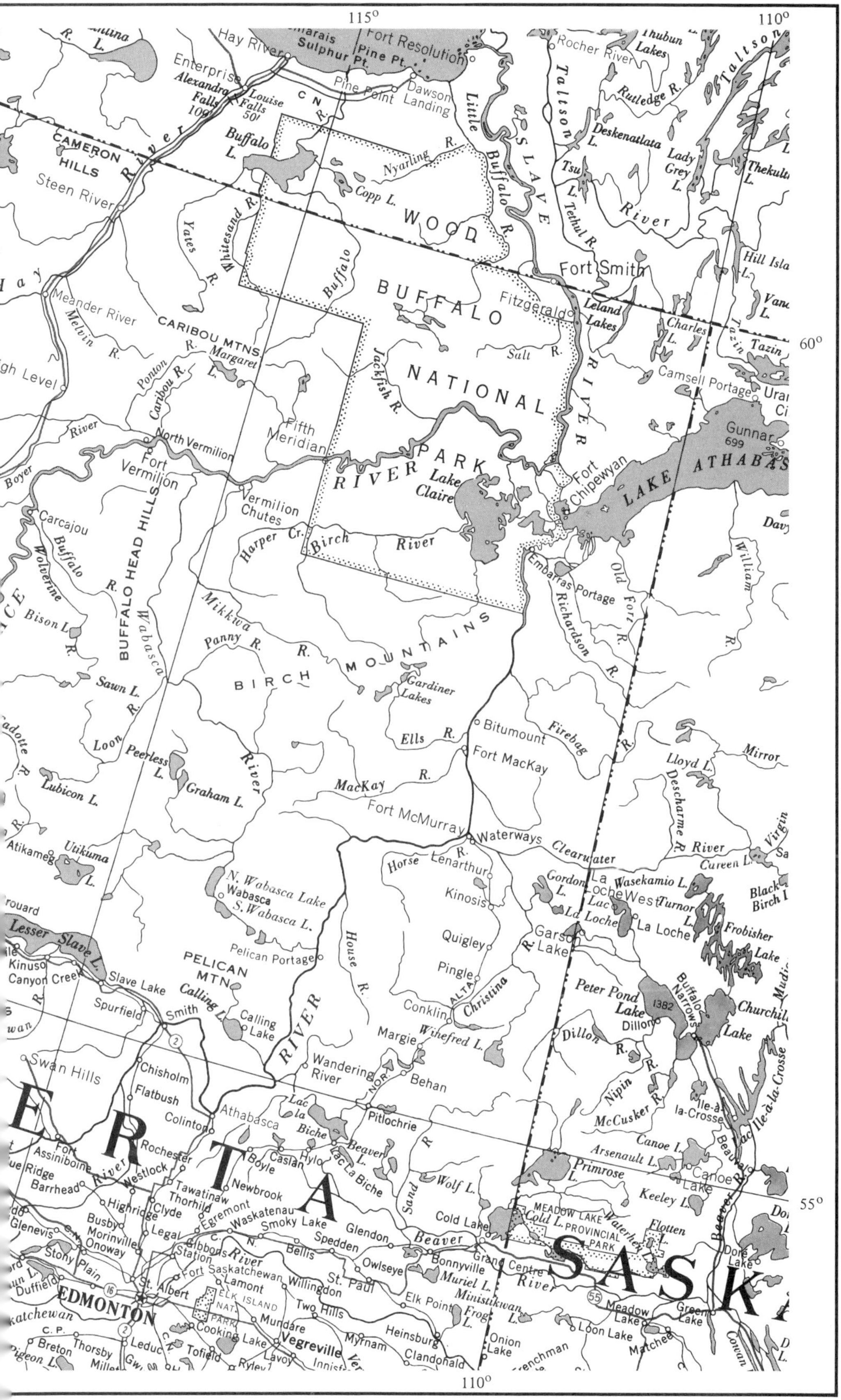
115°
110°
60°
55°
Hay River
Sulphur Pt.
Pine Pt.
Fort Resolution
Rocher River
Thubun Lakes
Taltson
Enterprise
Alexandra Falls 109'
Louise Falls 50'
Pine Point
Dawson Landing
C N R.
Little Buffalo R.
SLAVE
Taltson R.
Rutledge R.
Deskenatlata L.
Lady Grey L.
Thekulthili L.
CAMERON HILLS
Buffalo L.
Nyarling R.
Tsu L.
Tethul R.
River
Steen River
Copp L.
WOOD
Whitesand R.
Yates R.
Buffalo R.
Hill Island L.
Fort Smith
Hay
BUFFALO
Fitzgerald
Leland Lakes
Vander L.
Meander River
Melvin R.
CARIBOU MTNS
Ponton R.
Caribou R.
Margaret L.
Jackfish R.
Salt R.
Charles L.
Tazin L.
Tazin
High Level
NATIONAL
Camsell Portage
Uranium City
Fifth Meridian
RIVER
Gunnar 699
River
North Vermilion
Fort Vermilion
PARK
Lake Claire
Fort Chipewyan
LAKE ATHABASCA
Boyer
RIVER
Vermilion Chutes
Carcajou
Buffalo
Wolverine
Harper Cr.
Birch River
Davy
Embarras Portage
Old Fort R.
William R.
BUFFALO HEAD HILLS
Bison L.
Mikkwa R.
Richardson R.
Panny R.
MOUNTAINS
Wabasca
BIRCH
Gardiner Lakes
Sawn L.
Cadotte R.
Loon R.
Bitumount
Firebag R.
Peerless L.
Ells R.
Fort MacKay
Lloyd L.
Mirror
Lubicon L.
Graham L.
River
MacKay R.
Descharme R.
Fort McMurray
Waterways
Clearwater
River
Virgin
Atikameg
Utikuma L.
Horse R.
Lenarthur
Careen L.
N. Wabasca Lake
Wabasca
S. Wabasca L.
Kinosis
Gordon L.
La Loche West
Wasekamio L.
Turnor L.
Black Birch L.
Lac La Loche
La Loche
Frobisher Lake
Brouard
Lesser Slave L.
Quigley
Garson Lake
Christina R.
House R.
Pelican Portage
PELICAN MTN
Pingle
Peter Pond Lake
1382
Buffalo Narrows
Churchill Lake
Kinuso
Canyon Creek
Slave Lake
Calling L.
Conklin
ALTA
Dillon
Spurfield
Smith
Calling Lake
RIVER
Margie
Winefred L.
Dillon R.
Swan Hills
Chisholm
Wandering River
Behan
NOR
Nipin R.
McCusker R.
Île-à-la-Crosse
Lac Île-à-la-Crosse
Flatbush
Colinton
Athabasca
Lac la Biche
Pitlochrie
ERTA
Fort Assiniboine
Rochester
Beaver L.
Lac La Biche
Hylo
Caslan
Boyle
Canoe L.
Arsenault L.
Primrose L.
Beaver R.
Blue Ridge
Barrhead
River
Westlock
Tawatinaw
Thorhild
Newbrook
Sand R.
Wolf L.
Canoe Lake
Keeley L.
Highridge
Clyde
Egremont
A
Glenevis
Busby
Morinville
Legal
Waskatenau
Smoky Lake
Glendon
Cold Lake
MEADOW LAKE PROVINCIAL PARK
Cold L.
Waterhen
Flotten L.
Dore Lake
Onoway
Stony Plain
Gibbons
Station
River
Bellis
Spedden
Beaver
Bonnyville
Grand Centre
River
Fort Saskatchewan
Lamont
Owlseye
Muriel L.
Duffield
EDMONTON
St. Albert
ELK ISLAND NAT. PARK
Willingdon
St. Paul
Elk Point
Ministikwan L.
SASK
55
Meadow Lake
Green Lake
Saskatchewan
Two Hills
Frog L.
Loon Lake
C.P.
Mundare
Vegreville
Cooking Lake
Myrnam
Heinsburg
Onion Lake
Matchee
Cowan
Breton
Thorsby
Leduc
Tofield
Lavoy
Clandonald
Frenchman
Pigeon L.
Millet
Ryley
Innisfree
110°

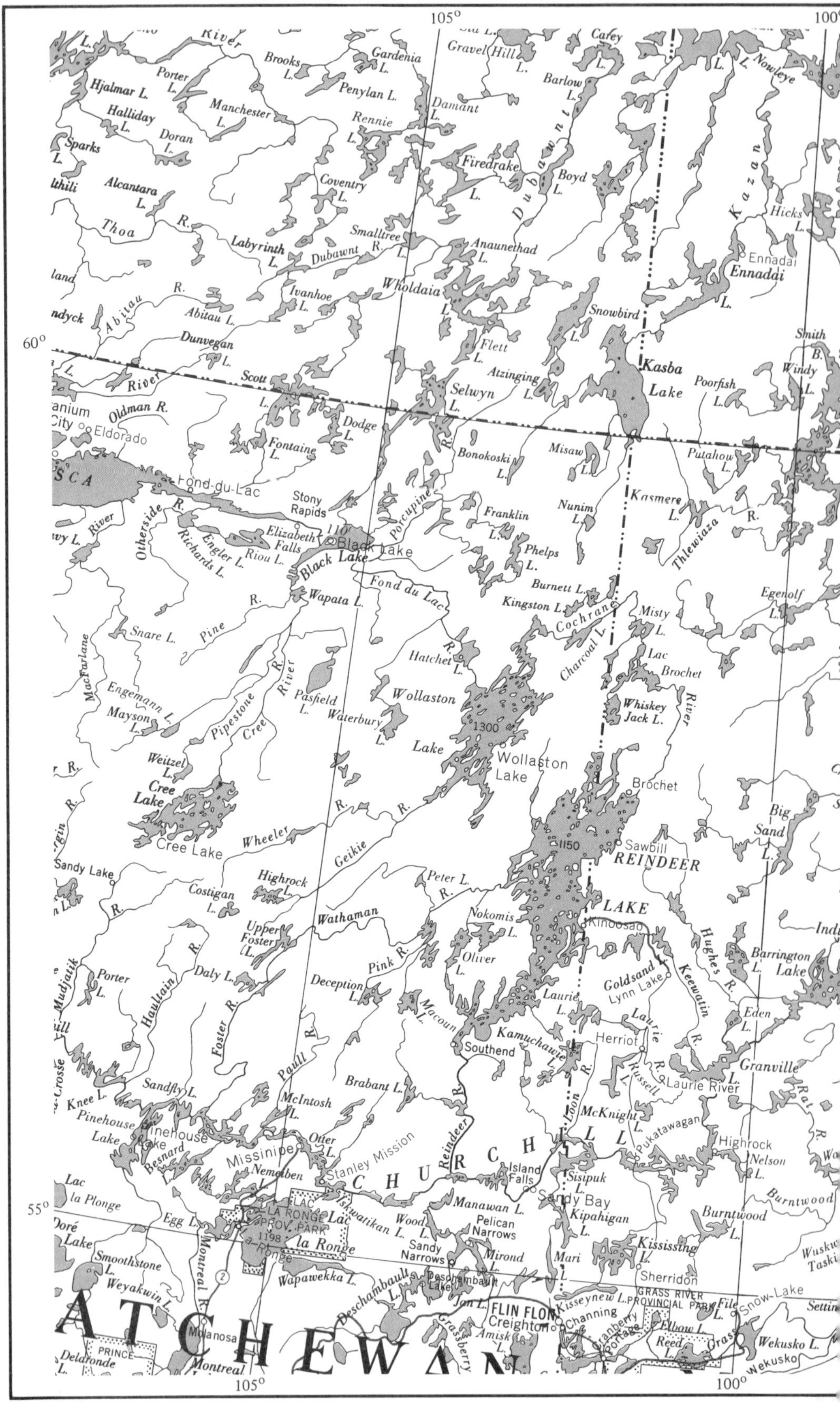
105°
100°
River
Brooks L.
Gardenia L.
Gravel Hill L.
Carey L.
Porter L.
Hjalmar L.
Halliday L.
Manchester L.
Penylan L.
Barlow L.
Nowleye
Doran L.
Rennie L.
Damant L.
Sparks L.
Dubawnt
Firedrake L.
Boyd L.
Kazan
Alcantara L.
Coventry L.
Hicks L.
Thoa R.
Labyrinth L.
Smalltree L.
Dubawnt R.
Anaunethad L.
Ennadai
Ennadai L.
Abitau R.
Ivanhoe L.
Wholdaia L.
Abitau L.
Snowbird L.
60°
Dunvegan L.
Flett L.
Smith B.
Kasba Lake
Windy L.
Scott L.
Atzinging L.
Poorfish L.
River
Selwyn L.
Oldman R.
Uranium City
Eldorado
Dodge L.
Fontaine L.
Bonokoski L.
Misaw L.
Putahow L.
Fond-du-Lac
Stony Rapids
Kasmere L.
Nunim L.
Franklin L.
Porcupine R.
Otherside R.
River
Elizabeth Falls
110
Black Lake
Thlewiaza R.
Engler L.
Riou L.
Richards L.
Black Lake
Phelps L.
Burnett L.
Fond du Lac
Kingston L.
Cochrane
Egenolf L.
Wapata L.
Misty L.
Pine R.
Snare L.
Charcoal L.
Hatchet R. L.
Lac Brochet
MacFarlane
Engemann L.
Wollaston
Whiskey Jack L.
River
Maysons L.
Pipestone
Cree River
R.
Pasfield L.
Waterbury L.
1300
Cree
Lake
Wollaston Lake
Weitzel L.
Cree Lake
Brochet
R.
R.
Big Sand L.
Cree Lake
Wheeler
Geikie
1150
Sawbill
REINDEER
Sandy Lake
Highrock L.
Peter L. R.
Costigan L.
LAKE
R.
Wathaman
Nokomis L.
Kinoosao
Upper Foster L.
Hughes R.
Barrington Lake
Daly L.
Oliver L.
Porter L.
Pink R.
Deception L.
Goldsand L.
Lynn Lake
Mudjatik
Hautain R.
Foster R.
Macoun L.
Laurie L.
Keewatin
Laurie R.
Eden L.
Kamuchawie L.
Herriot
Southend
Paull R.
Granville L.
Brabant L.
Reindeer R.
Russell L.
Laurie River
Loon R.
Sandfly L.
Knee L.
McIntosh L.
McKnight L.
Rat R.
Pinehouse Lake
Pinehouse Lake
Otter L.
Stanley Mission
CHURCHILL
Pukatawagan
Highrock
Besnard L.
Missinipe
Nemeiben
Island Falls
Sisipuk L.
Nelson L.
Lac la Plonge
Sandy Bay
Burntwood
55°
LAC LA RONGE PROV. PARK
Lac la Ronge
Manawan L.
Egg L.
Iskwatikan L.
Wood L.
Pelican Narrows
Kipahigan L.
Burntwood L.
Doré Lake
1198
La Ronge
Sandy Narrows
Mirond L.
Mari L.
Kississing L.
Smoothstone L.
Montreal R.
2
Wapawekka L.
Deschambault L.
Deschambault Lake
Sherridon
Weyakwin L.
GRASS RIVER PROVINCIAL PARK
File L.
Snow Lake
Jan L.
FLIN FLON
Kisseynew L.
Channing
Creighton
Cranberry Portage
Elbow L.
Deschambault
Grassberry
Amisk L.
Reed L.
Grass
Wekusko L.
Setting
ATCHEWAN
Molanosa
PRINCE
Delaronde L.
Montreal L.
Wekusko
105°
100°

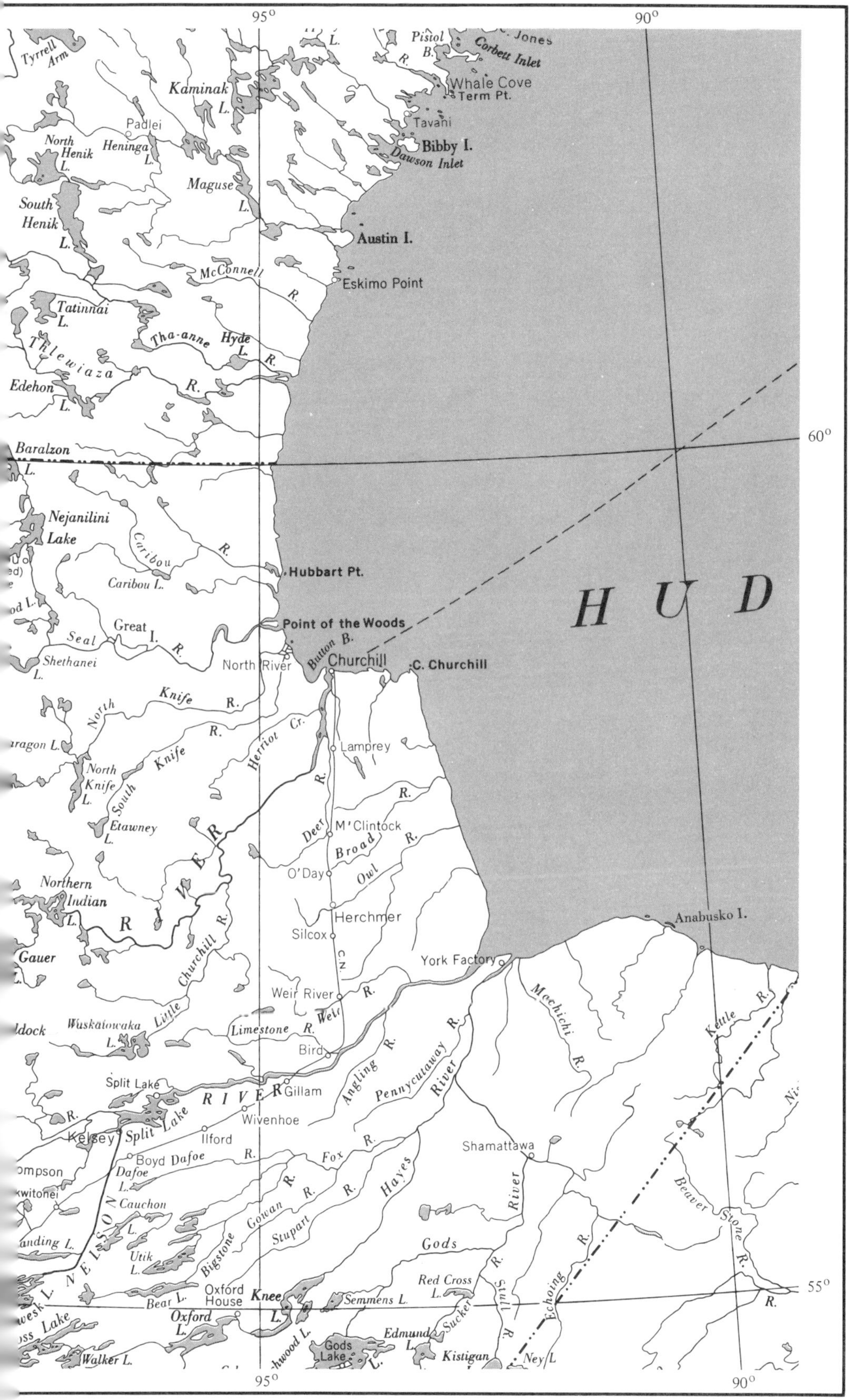

95°
90°
Tyrrell Arm
Kaminak L.
Pistol B.
Corbett Inlet
Whale Cove
Term Pt.
Tavani
Padlei
North Henik L.
Heninga L.
Bibby I.
Dawson Inlet
Maguse L.
South Henik L.
Austin I.
McConnell R.
Eskimo Point
Tatinnai L.
Tha-anne R.
Hyde L.
Thlewiaza R.
Edehon L.
60°
Baralzon L.
Nejanilini Lake
Caribou R.
Caribou L.
Hubbart Pt.
HUD
Point of the Woods
Great Seal R.
Button B.
Shethanei L.
North River
Churchill
C. Churchill
North Knife R.
South Knife R.
Herriot Cr.
Lamprey
North Knife L.
Etawney L.
Deer R.
M'Clintock
Broad R.
Owl R.
O'Day
Northern Indian L.
CHURCHILL RIVER
Herchmer
Silcox
C.N.
Anabusko I.
Gauer
Little Churchill R.
York Factory
Weir River
Weir R.
Limestone R.
Waskaiowaka L.
Bird
Machichi R.
Kettle R.
Angling R.
Pennycutaway River
Split Lake
NELSON RIVER
Gillam
Wivenhoe
Kelsey
Split Lake
Ilford
Boyd
Dafoe R.
Dafoe L.
Fox R.
Shamattawa
Cauchon L.
Gowan R.
Stupart R.
Hayes R.
River
Beaver Stone R.
Utik L.
Bigstone R.
Gods R.
Echoing R.
Red Cross L.
Sucker R.
Stull R.
55°
Bear L.
Oxford House
Knee L.
Semmens L.
Oxford L.
Edmund L.
Gods Lake
Kistigan
Ney L
Walker L.

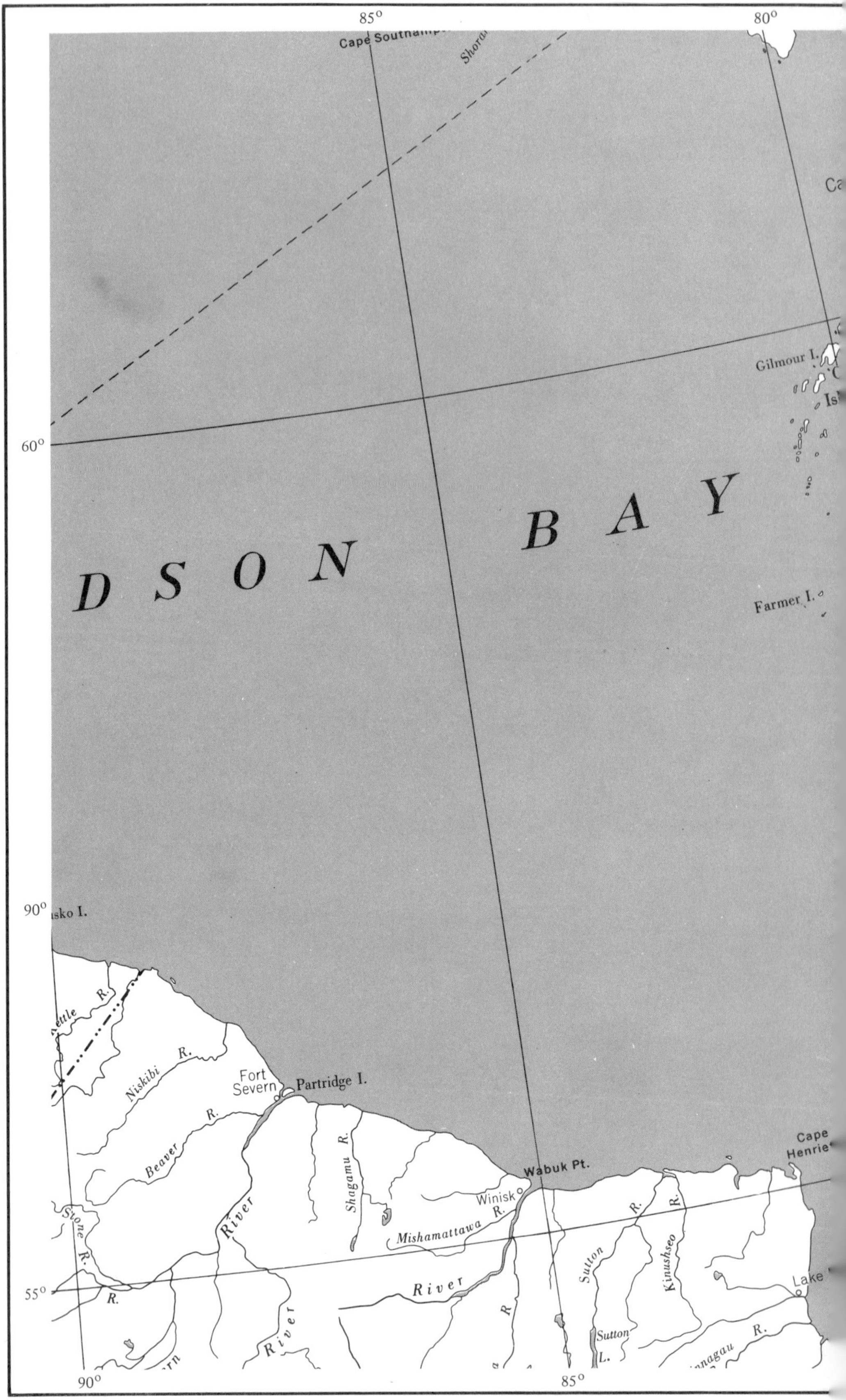
85°
80°
Cape Southampton
Gilmour I.
60°
D S O N B A Y
Farmer I.
90°
Fort Severn
Partridge I.
Niskibi R.
Beaver R.
Shagamu R.
Wabuk Pt.
Winisk
Cape Henrietta Maria
Mishamattawa R.
River
Stone R.
Sutton R.
Kinushseo R.
Sutton L.
Lake
55°
R.
85°

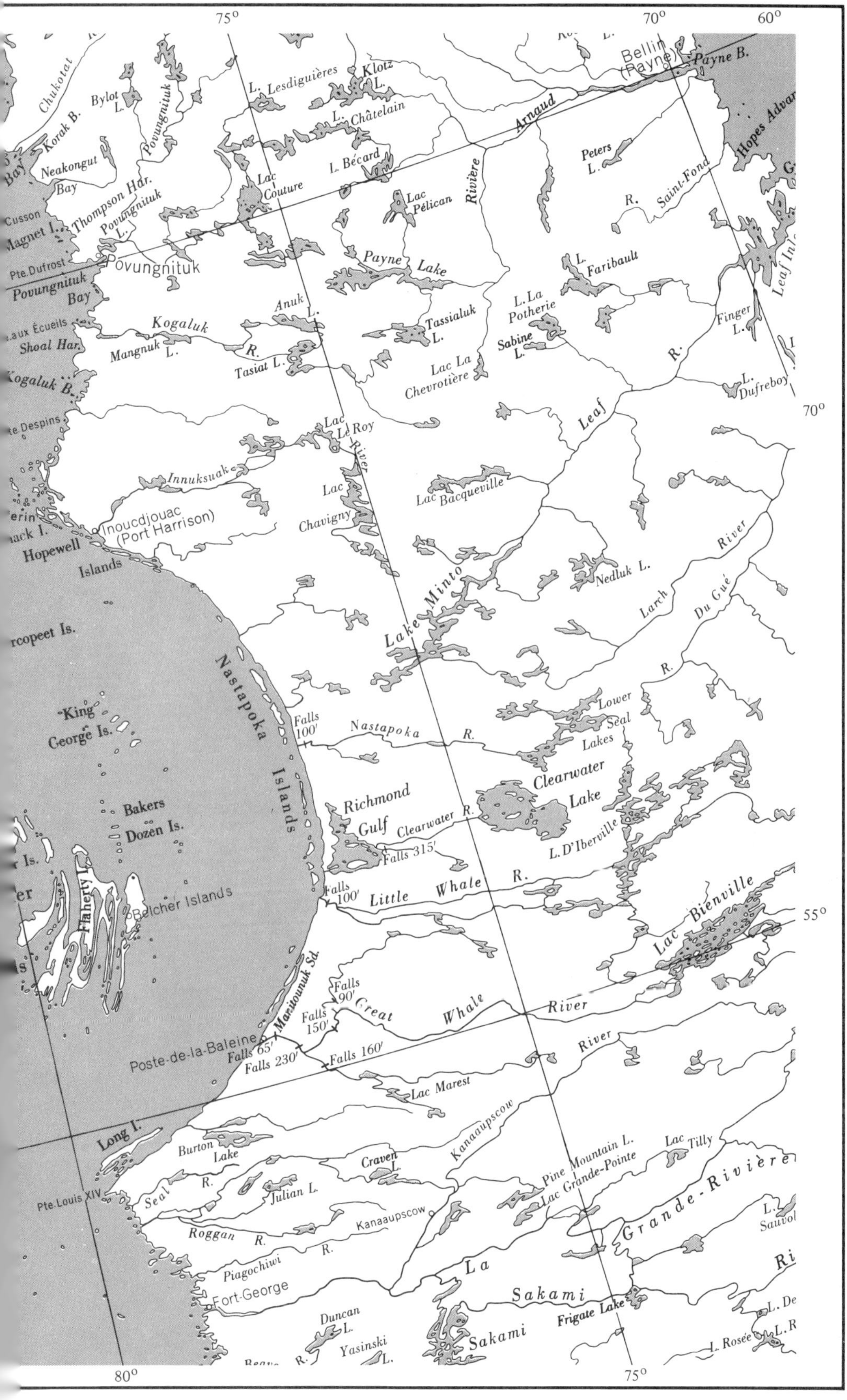
75°
70°
60°
Bellin (Payne)
Payne B.
Hopes Advan
Chukotat
Korak B.
Bylot L.
Povungnituk
L. Lesdiguières
Klotz L.
L. Châtelain
Arnaud
Neakongut Bay
L. Bécard
Lac Couture
Peters L.
Rivière
Saint-Fond
Thompson Har.
Povungnituk L.
Cusson
Magnet I.
Lac Pélican
R.
Pte. Dufrost
Povungnituk
Povungnituk Bay
Payne Lake
L. Faribault
Leaf Inl
Anuk L.
L. La Potherie
Kogaluk
Tassialuk L.
Finger L.
aux Écueils
Shoal Har.
Mangnuk L.
R.
Sabine L.
Tasiat L.
Lac La Chevrotière
R.
Kogaluk B.
L. Dufreboy
70°
Pte. Despins
Lac Le Roy
Leaf
River
Innuksuak
Lac Bacqueville
Lac Chavigny
Inoucdjouac (Port Harrison)
Hopewell Islands
River
Nedluk L.
Lake Minto
Larch
Du Gué
rcopeet Is.
R.
King George Is.
Nastapoka Islands
Falls 100′
Nastapoka R.
Lower Seal Lakes
Bakers Dozen Is.
Richmond Gulf
Clearwater R.
Clearwater Lake
Falls 315′
L. D'Iberville
Flaherty I.
Belcher Islands
Falls 100′
Little Whale R.
Lac Bienville
55°
Manitounuk Sd.
Falls 90′
Great Whale River
Falls 150′
Poste-de-la-Baleine
Falls 65′
Falls 230′
Falls 160′
River
Lac Marest
Long I.
Burton Lake
Kanaaupscow
Craven L.
Pine Mountain L.
Lac Tilly
Seal R.
Julian L.
Lac Grande-Pointe
Pte. Louis XIV
Grande-Rivière
L. Sauvol
Kanaaupscow
Roggan R.
Piagochiwi R.
La
Fort-George
Sakami
Frigate Lake
Duncan L.
Sakami
Yasinski L.
L. Rosée
80°
75°

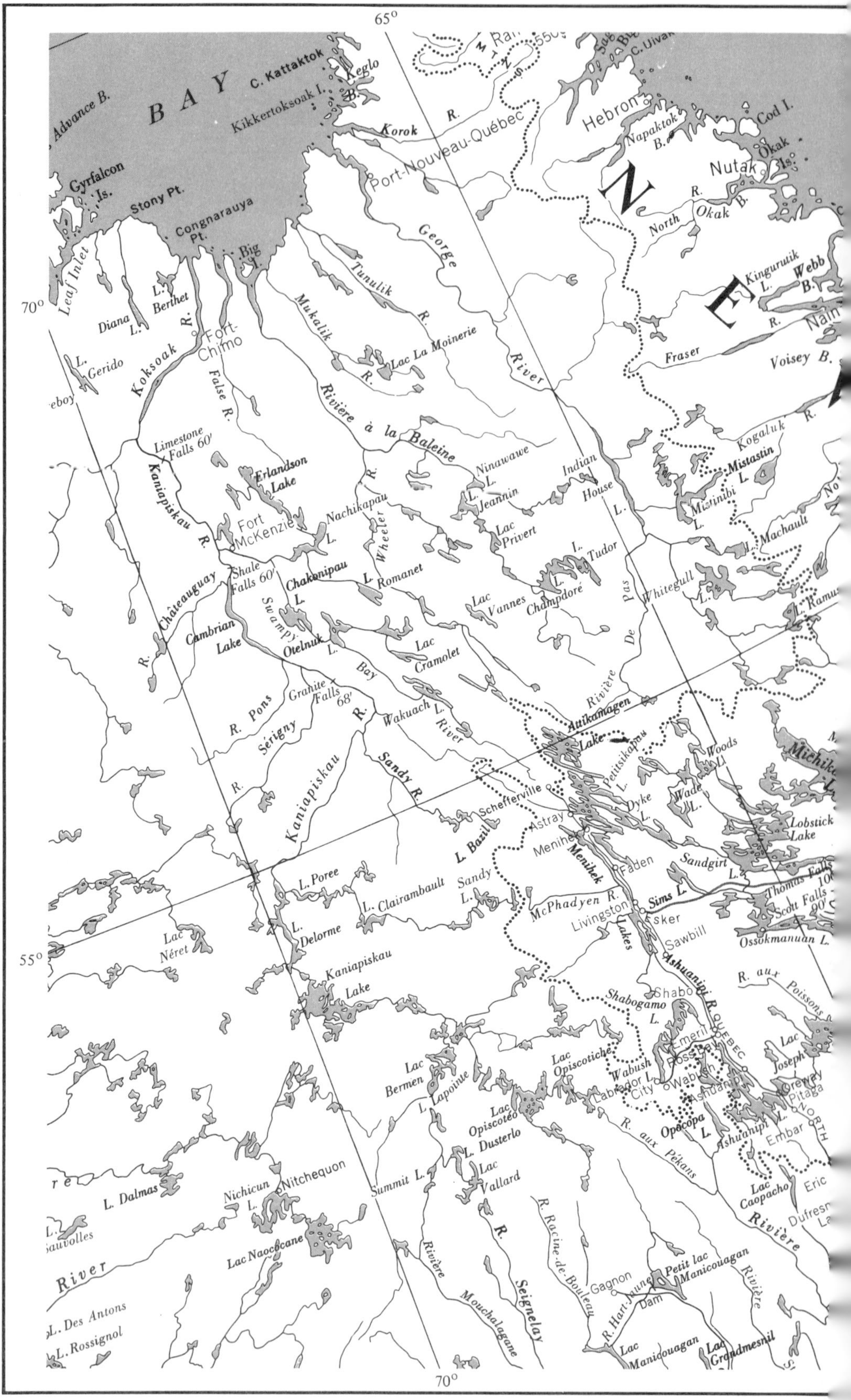
65°
70°
55°
70°
BAY
Advance B.
C. Kattaktok
Kikkertoksoak I.
Keglo B.
Korok R.
Gyrfalcon Is.
Stony Pt.
Congnarauya Pt.
Big I.
Port-Nouveau-Québec
George River
Hebron
C. Uivak
Napaktok B.
Cod I.
Okak Is.
Nutak
Okak B.
North R.
N
E
Kingurutik L.
Webb B.
Fraser R.
Nain
Voisey B.
Leaf Inlet
Diana L.
L. Berthet
Koksoak R.
Fort-Chimo
L. Gerido
Tunulik R.
Mukalik R.
Lac La Moinerie
False R.
Rivière à la Baleine
Limestone Falls 60'
Kaniapiskau R.
Erlandson Lake
Fort McKenzie
Nachikapau L.
Wheeler R.
Ninawawe L.
L. Jeannin
Lac Privert
Indian House L.
Kogaluk R.
Mistastin L.
Mistinibi L.
L. Machault
Shale Falls 60'
Chakonipau L.
L. Romanet
L. Tudor
L. Champdoré
Lac Vannes
De Pas
Whitegull L.
L. Ramusio
Châteauguay R.
Cambrian Lake
Swampy Bay
Otelnuk L.
Lac Cramolet
Granite Falls 68'
R. Pons
Sérigny R.
Wakuach L.
River
Rivière
Attikamagen Lake
Petitsikapau L.
Woods L.
Michikamau L.
Kaniapiskau R.
Sandy R.
Schefferville
Astray
Dyke L.
Wade L.
Lobstick Lake
L. Bazil
Menihek
Menihek Lakes
Faden
Sandgirt L.
Thomas Falls
Scott Falls 90'
L. Poree
L. Clairambault
Sandy L.
McPhadyen R.
Sims L.
Livingston
Esker
Sawbill
Ossokmanuan L.
Lac Néret
L. Delorme
Kaniapiskau Lake
Ashuanipi R.
R. aux Poissons
Shabogamo L.
Shabo
Emeril
QUEBEC
Lac Joseph
Ossokmanuan
Lac Bermen
L. Lapointe
Lac Opiscotiche
Wabush L.
Labrador City
Ross Bay
Wabush
Ashuanipi
Oreway
Pitaga
NORTH
Lac Opiscotéo
L. Dusterlo
R. aux Pékans
Opocopa L.
Ashuanipi L.
Embar
Lac Vallard
Summit L.
Nichicun L.
Nitchequon
L. Dalmas
Lac Caopacho
Eric
Dufresne Lake
Rivière
L. Sauvolles
Lac Naococane
R. Racine-de-Bouleau
Seignelay R.
Rivière Mouchalagane
Petit lac Manicouagan
Gagnon
R. Hart-Jaune
Dam
Rivière
River
L. Des Antons
L. Rossignol
Lac Manicouagan
Lac Grandmesnil

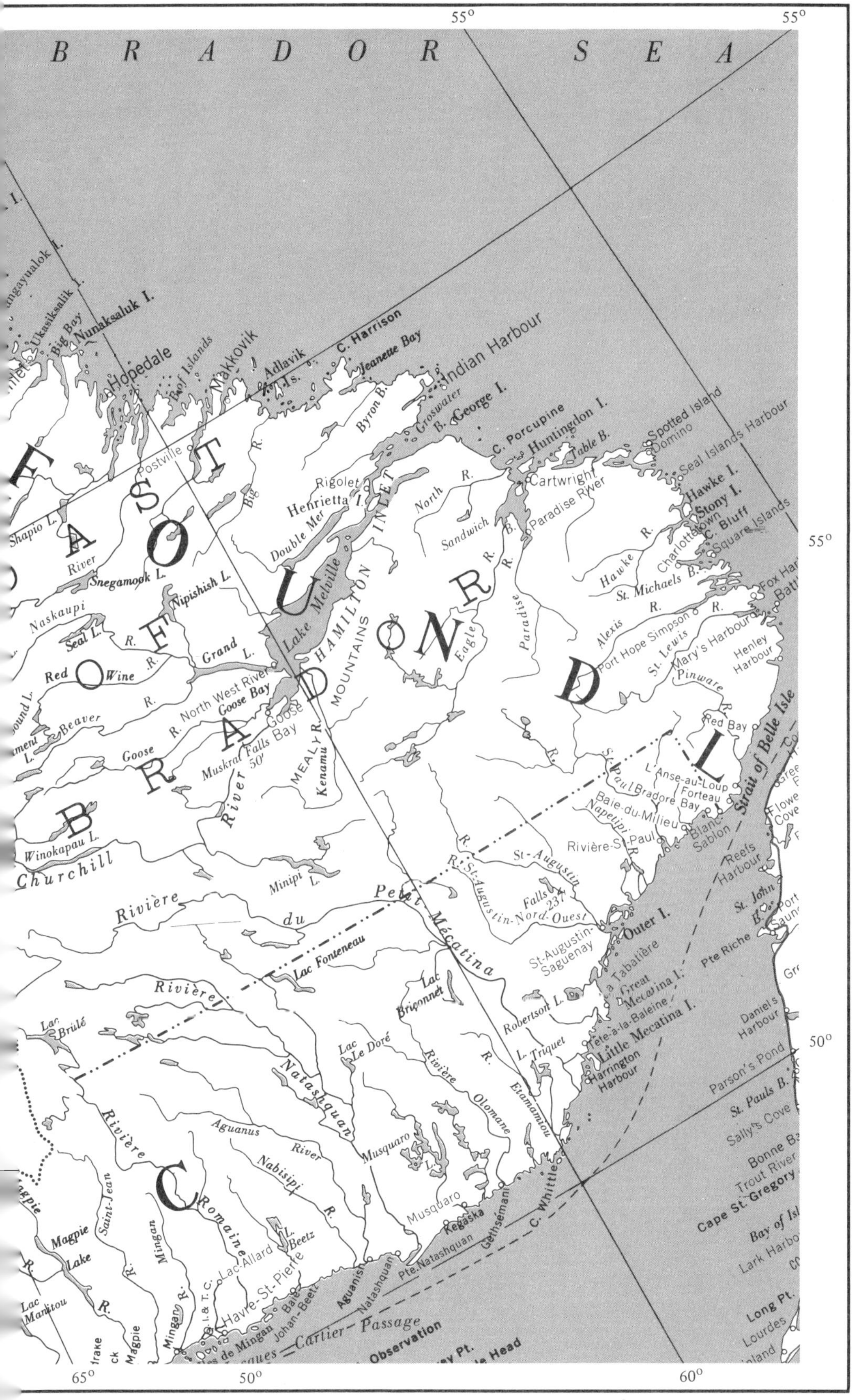

B R A D O R S E A
55°
55°
55°
50°
65°
50°
60°
L A B R A D O R
C O A S T
Hopedale
Nunaksaluk I.
Ukasiksalik I.
Big Bay
Makkovik
Adlavik Is.
C. Harrison
Jeanette Bay
Indian Harbour
Byron B.
Groswater B.
George I.
C. Porcupine
Huntingdon I.
Table B.
Spotted Island
Domino
Seal Islands Harbour
Cartwright
Paradise River
Hawke I.
Stony I.
C. Bluff
Charlottetown
Square Islands
St. Michaels B.
Fox Harbour
Battle
Port Hope Simpson
Mary's Harbour
St. Lewis R.
Henley Harbour
Pinware R.
Red Bay
L'Anse-au-Loup
Forteau
Bradore Bay
Baie-du-Milieu
Blanc Sablon
Rivière-St-Paul
Strait of Belle Isle
Reefs Harbour
St. John B.
Pte Riche
Daniel's Harbour
Parson's Pond
St. Pauls B.
Sally's Cove
Bonne Bay
Trout River
Cape St. Gregory
Bay of Islands
Lark Harbour
Long Pt.
Lourdes
Postville
Rigolet
Henrietta I.
Double Mer
Lake Melville
HAMILTON INLET
North R.
Sandwich B.
Eagle R.
Paradise R.
Hawke R.
Alexis R.
MEALY MOUNTAINS
Kenamu R.
Goose Bay
North West River
Goose Bay
Muskrat Falls 50'
Grand L.
Nipishish L.
Snegamook L.
Naskaupi R.
Seal L.
Red Wine R.
Beaver R.
Goose R.
Winokapau L.
Churchill River
Shapio L.
Minipi L.
Rivière du Petit Mécatina
St-Augustin R.
R. St-Augustin-Nord-Ouest
Falls 237
Outer I.
St-Augustin-Saguenay
La Tabatière
Great Mecatina I.
Tête-à-la-Baleine
Little Mecatina I.
Harrington Harbour
Robertson L.
L. Triquet
Lac Fonteneau
Lac Brûlé
Rivière Natashquan
Lac Brûlé
Lac Le Doré
Lac Briçonnet
Rivière Olomane
Etamamiou R.
Aguanus River
Nabisipi R.
Musquaro L.
Musquaro
Kegaska
Gethsémani
C. Whittle
Pte.Natashquan
Natashquan
Aguanish
L. Beetz
Lac-Allard
Havre-St-Pierre
Johan-Beetz
Baie Johan-Beetz
Îles de Mingan
Mingan
Mingan R.
Romaine
Rivière Saint-Jean
Magpie Lake
Magpie R.
Lac Manitou
Jacques Cartier Passage
Observation

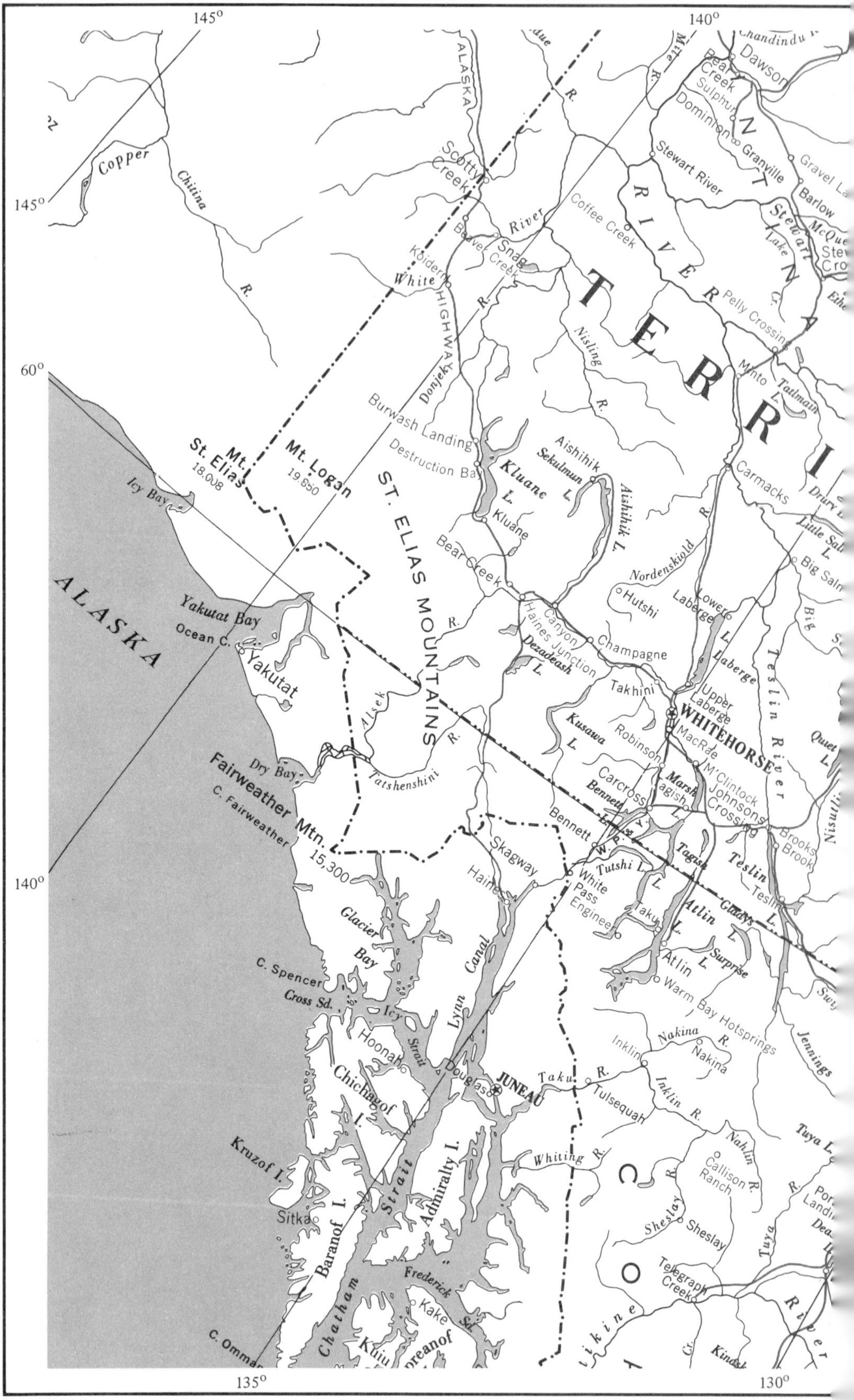
145°
140°
60°
135°
130°
Dawson
Bear Creek
Sulphur
Dominion
Granville
Stewart River
Gravel L.
Barlow
Stewart
Lake
Copper
Chitina
R.
ALASKA
Scotty Creek
River
Beaver Creek
Snag Creek
Koidern
White
Coffee Creek
RIVER
TERRI
HIGHWAY
R.
Donjek
Nisling
R.
Pelly Crossing
Minto
Tatlmain L.
Burwash Landing
Destruction Bay
Mt. St. Elias
18,008
Mt. Logan
19,850
Kluane L.
Kluane
Aishihik
Sekulmun L.
Aishihik L.
Carmacks
Little Salmon L.
Big Salmon
Icy Bay
ST. ELIAS MOUNTAINS
Bear Creek
Nordenskiold
Hutshi
Lower Laberge
Laberge
Canyon
Haines Junction
Champagne
Dezadeash L.
Yakutat Bay
Ocean C.
Yakutat
ALASKA
Alsek
Takhini
Upper Laberge
WHITEHORSE
Teslin River
Kusawa L.
Robinson
MacRae
M'Clintock
Johnsons Crossing
Dry Bay
Tatshenshini
R.
Fairweather Mtn.
C. Fairweather
15,300
Carcross
Bennett L.
Marsh L.
Tagish
Bennett
W. P. Y.
Tutshi L.
Tagish L.
Teslin L.
Teslin
Brooks Brook
Skagway
Haines
White Pass
Engineer
Taku
Atlin L.
Atlin
Gladys L.
Surprise L.
Glacier Bay
C. Spencer
Cross Sd.
Icy Strait
Lynn Canal
Warm Bay Hotsprings
Hoonah
Chichagof I.
Douglas
JUNEAU
Taku R.
Inklin
Nakina R.
Nakina
Inklin R.
Tulsequah
Jennings
Kruzof I.
Admiralty I.
Strait
Whiting R.
Nahlin R.
Callison Ranch
Tuya L.
C
Sitka
Baranof I.
Sheslay
Sheslay
R.
Tuya
O
Telegraph Creek
Chatham
Frederick Sd.
Kake
Kuiu
Kupreanof
C. Ommaney
Stikine
River
Cr.

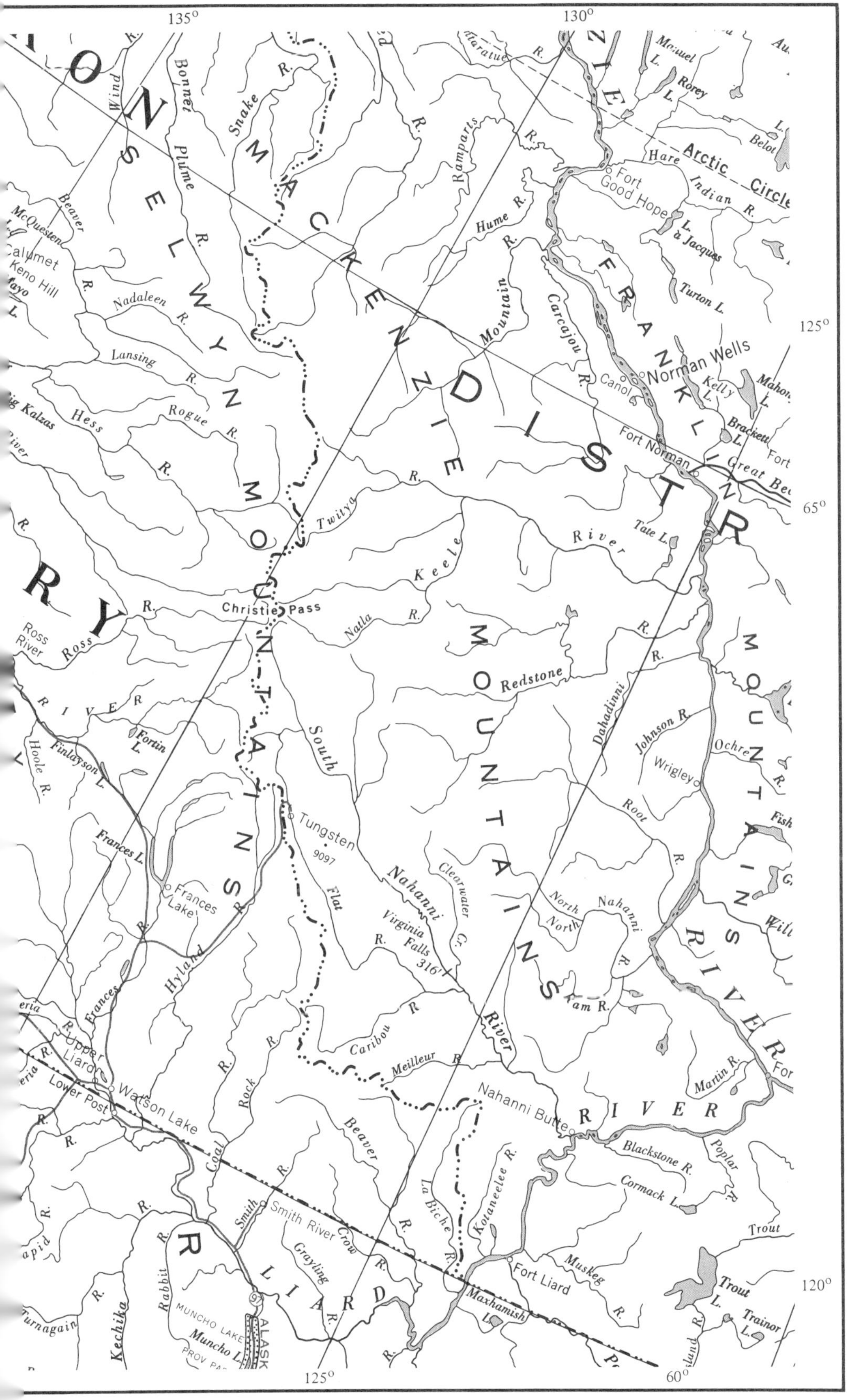
135°
130°
125°
65°
120°
60°
Arctic Circle
MACKENZIE DISTRICT
SELWYN MOUNTAINS
MACKENZIE MOUNTAINS
FRANKLIN MOUNTAINS
MACKENZIE RIVER
LIARD RIVER
Fort Good Hope
Norman Wells
Canol
Fort Norman
Great Bear
Hare Indian R.
L. à Jacques
Turton L.
Kelly L.
Brackett L.
Tate L.
Rorey L.
Belot L.
Ramparts R.
Hume R.
Mountain R.
Carcajou R.
Snake R.
Bonnet Plume R.
Wind R.
Beaver R.
McQuesten
Calumet
Keno Hill
Mayo L.
Nadaleen R.
Lansing R.
Hess R.
Rogue R.
Twitya
Keele R.
Natla R.
Christie Pass
Redstone R.
Dahadinni R.
Johnson R.
Wrigley
Ochre R.
Root R.
Ross River
Ross R.
Fortin L.
Finlayson L.
Hoole R.
Frances L.
Frances Lake
Frances R.
Hyland R.
South
Tungsten
9097
Flat R.
Nahanni
Clearwater Cr.
Virginia Falls 316'
North Nahanni R.
River
Caribou R.
Meilleur R.
Upper Liard
Lower Post
Watson Lake
Rock R.
Coal R.
Beaver R.
Nahanni Butte
La Biche R.
Kotaneelee R.
Blackstone R.
Poplar R.
Cormack L.
Martin R.
Trout
Trout L.
Trainor L.
Fort Liard
Muskeg R.
Maxhamish L.
Smith River
Smith
Crow R.
Grayling R.
Rabbit R.
Kechika
Turnagain
Muncho L.
MUNCHO LAKE PROV PARK
97
ALASKA

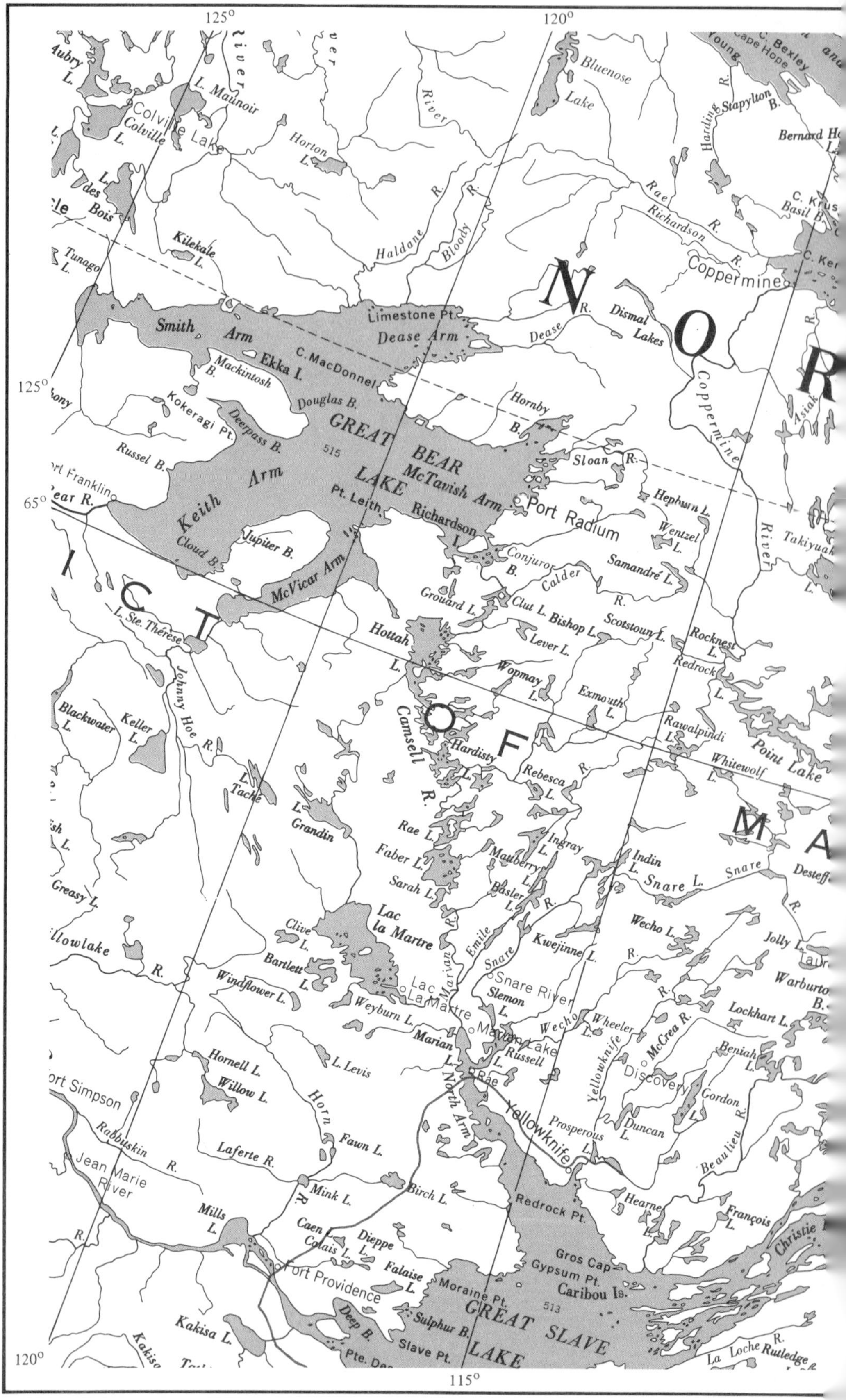

125°
120°
115°
65°
Aubry L.
L. Maunoir
Colville Lake
Colville L.
L. des Bois
Horton L.
Bluenose Lake
C. Bexley
Cape Hope
Young
Stapylton B.
Harding R.
Bernard H
Rae R.
Richardson R.
C. Krus
Basil B.
C. Ker
Coppermine
Haldane R.
Bloody R.
Kilekale L.
Tunago L.
N
O
R
Smith Arm
Limestone Pt.
Dease Arm
Dease R.
Dismal Lakes
Ekka I.
C. MacDonnel
Mackintosh B.
Douglas B.
Kokeragi Pt.
Deerpass B.
GREAT BEAR LAKE
515
Hornby B.
Coppermine
Asiak
Sloan R.
Russel B.
Keith Arm
McTavish Arm
Port Radium
Fort Franklin
Bear R.
Pt. Leith
Richardson I.
Hepburn L.
Wentzel L.
River
Takiyuak
Cloud B.
Jupiter B.
Conjuror B.
Calder R.
Samandré L.
I
C
T
McVicar Arm
Grouard L.
Clut L.
Bishop L.
Scotstown L.
L. Ste. Therese
Hottah L.
Lever L.
Rocknest L.
Redrock L.
Johnny Hoe R.
Wopmay L.
Exmouth L.
Blackwater L.
Keller L.
Camsell R.
O
Hardisty L.
F
Rawalpindi L.
Point Lake
Whitewolf L.
Rebesca L.
L. Taché
L. Grandin
Rae L.
M
A
Ingray L.
Greasy L.
Faber L.
Mattberry L.
Indin L.
Snare L.
Snare
Desteff
Sarah L.
Basler L.
R.
Lac la Martre
Clive L.
Wecho L.
Jolly L.
Yellowlake R.
Bartlett L.
Marian R.
Emile R.
Snare
Kwejinne L.
Warburton B.
Windflower L.
Lac La Martre
Snare River
Slemon L.
Weyburn L.
Wecho
Wheeler L.
McCrea R.
Lockhart L.
Marian L.
Marian Lake
Russell L.
Beniah L.
Hornell L.
L. Levis
Rae
Yellowknife
Discovery
Willow L.
North Arm
Gordon L.
Fort Simpson
Horn R.
Yellowknife
Prosperous L.
Duncan L.
Rabbitskin R.
Fawn L.
Beaulieu R.
Laferte R.
Jean Marie River
Mink L.
Birch L.
Redrock Pt.
Hearne L.
François L.
Mills L.
Caen L.
Dieppe L.
Christie
Colais L.
Gros Cap
Gypsum Pt.
Caribou Is.
Fort Providence
Falaise L.
Moraine Pt.
513
Kakisa L.
Deep B.
Sulphur B.
GREAT SLAVE LAKE
Slave Pt.
La Loche R.
Rutledge
120°
115°

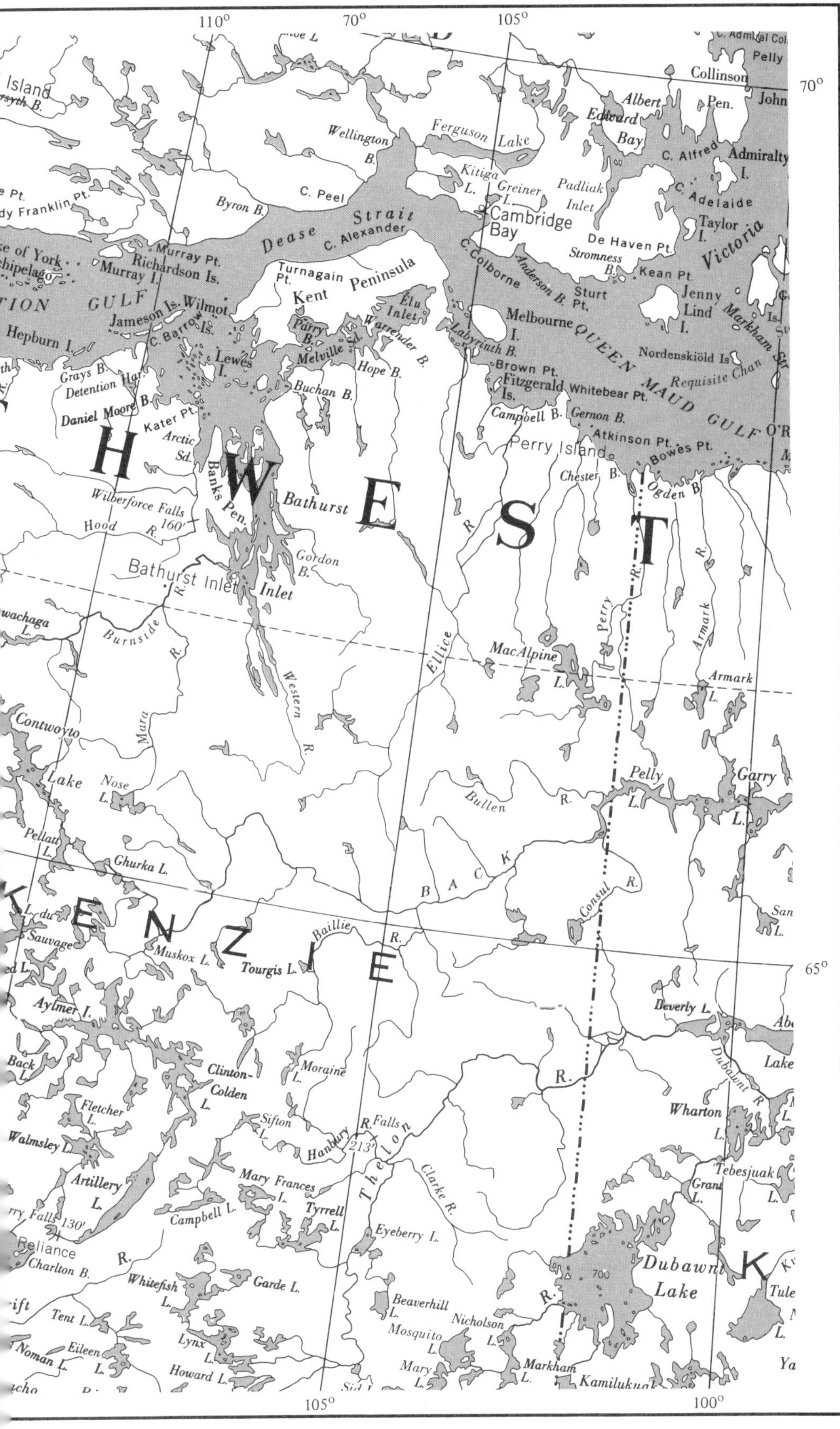
110°
70°
105°
70°
65°
105°
100°
Pelly
Collinson
Pen.
John
Albert
Edward
Bay
Wellington
B.
Ferguson
Lake
C. Alfred
Admiralty
I.
Kitiga
L.
Greiner
L.
Padliak
Inlet
C. Adelaide
C. Peel
Byron B.
Strait
Cambridge
Bay
Taylor
I.
Dease
C. Alexander
De Haven Pt.
Stromness
B.
Victoria
Kean Pt.
Murray Pt.
Richardson Is.
Murray I.
Turnagain
Pt.
Kent
Peninsula
C. Colborne
Anderson B.
Sturt
Pt.
Jenny
Lind
I.
GULF
Jameson Is.
Wilmot
Is.
C. Barrow
Parry
B.
Elu
Inlet
Warrender B.
Melbourne
I.
QUEEN
MAUD
GULF
Markham
Hepburn I.
Lewes
I.
Melville Sd.
Hope B.
Labyrinth B.
Nordenskiöld Is.
Requisite Chan
Grays B.
Detention Har.
Buchan B.
Brown Pt.
Fitzgerald
Is.
Whitebear Pt.
Daniel Moore B.
Kater Pt.
Campbell B.
Gernon B.
Arctic
Sd.
Atkinson Pt.
Perry Island
Bowes Pt.
H
W
E
S
T
Banks
Pen.
Bathurst
Chester
B.
Ogden B.
Wilberforce Falls
160′
Hood
R.
R.
R.
Gordon
B.
Bathurst Inlet
Inlet
R.
R.
Perry
Armark
Burnside
R.
Ellice
MacAlpine
L.
Armark
L.
Western
R.
Contwoyto
Lake
Mara
Nose
L.
Pelly
L.
Garry
L.
Bullen
R.
Pellatt
L.
Ghurka L.
B A C K
R.
Consul
L. du
Sauvage
E
N
Z
I
E
Baillie
R.
Muskox L.
Tourgis L.
Beverly L.
Aylmer L.
Lake
Dubawnt
R.
Back
L.
Moraine
L.
R.
Clinton-
Colden
L.
Wharton
L.
Fletcher
L.
Sifton
L.
Falls
213′
Walmsley L.
Hanbury
R.
Thelon
Clarke R.
Tebesjuak
Grant
L.
Artillery
L.
Mary Frances
L.
Campbell L.
Tyrrell
L.
Eyeberry L.
Falls
130′
Reliance
Charlton B.
R.
Whitefish
L.
Garde L.
700
Dubawnt
Lake
K
Tule
L.
Beaverhill
L.
Nicholson
L.
Tent L.
Mosquito
L.
R.
Lynx
L.
Eileen
L.
Noman L.
Howard L.
Mary
L.
Markham
L.
Kamilukuak

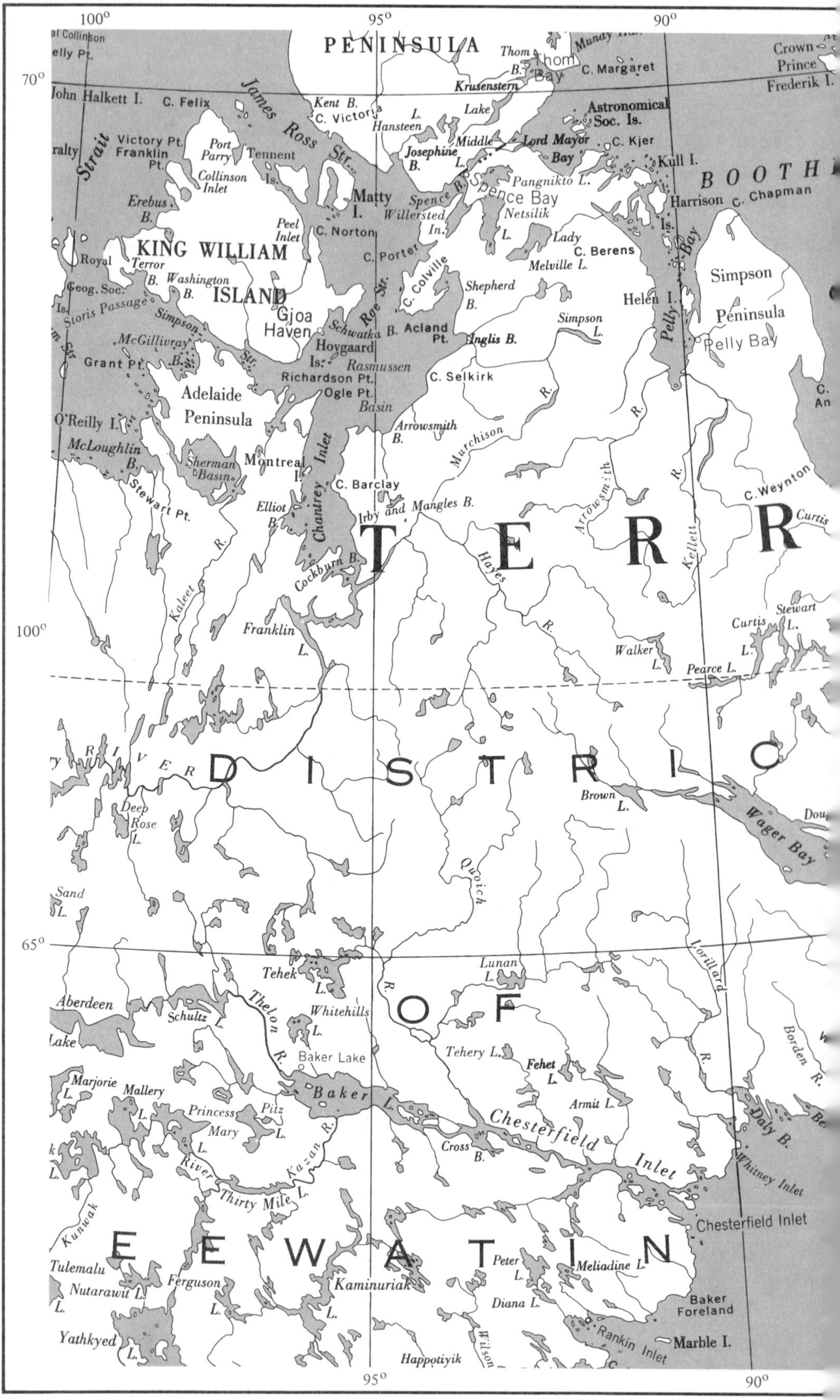

100°
95°
90°
70°
65°
PENINSULA
Thom B.
Thom Bay
C. Margaret
Crown Prince Frederik I.
John Halkett I.
C. Felix
James Ross Str.
Kent B.
C. Victoria
L. Hansteen
Krusenstern Lake
Astronomical Soc. Is.
Victory Pt.
Franklin Pt.
Strait
Port Parry
Tennent Is.
Collinson Inlet
Josephine B.
Middle L.
Lord Mayor Bay
C. Kjer
Kull I.
BOOTHI
Spence B.
Spence Bay
Pangnikto L.
Netsilik L.
Harrison Is.
C. Chapman
Erebus B.
Matty I.
Willersted In.
Peel Inlet
C. Norton
Lady Melville L.
C. Berens
KING WILLIAM ISLAND
Royal Geog. Soc. Is.
Terror B.
Washington B.
C. Porter
C. Colville
Shepherd B.
Simpson Peninsula
Helen I.
Pelly Bay
Storis Passage
Simpson Str.
Gjoa Haven
Schwatka B.
Rae Str.
Acland Pt.
Inglis B.
Simpson L.
Pelly Bay
McGillivray B.
Grant Pt.
Hoygaard Is.
Rasmussen Basin
Richardson Pt.
Ogle Pt.
C. Selkirk
Adelaide Peninsula
O'Reilly I.
Arrowsmith B.
Murchison R.
McLoughlin B.
Sherman Basin
Montreal I.
Chantrey Inlet
C. Barclay
Arrowsmith R.
Stewart Pt.
Elliot B.
Irby and Mangles B.
C. Weynton
Curtis
Kellett R.
Hayes R.
Cockburn B.
Kaleet R.
TERR
Franklin L.
Walker L.
Curtis L.
Stewart L.
Pearce L.
100°
RIVER
DISTRIC
Deep Rose L.
Brown L.
Wager Bay
Quoich
Sand L.
Lorillard
Tehek L.
Lunan L.
Aberdeen Lake
Schultz L.
Thelon R.
Whitehills L.
OF
Borden R.
Baker Lake
Tehery L.
Fehet L.
Marjorie L.
Mallery L.
Baker L.
Armit L.
Princess Mary L.
Pitz L.
Daly B.
Chesterfield Inlet
Cross B.
Kazan R.
Whitney Inlet
River
Thirty Mile L.
Chesterfield Inlet
Kunwak
EEWATIN
Tulemalu L.
Nutarawit L.
Ferguson L.
Kaminuriak L.
Peter L.
Meliadine L.
Diana L.
Baker Foreland
Yathkyed L.
Happotiyik
Wilson
Rankin Inlet
Marble I.
95°
90°

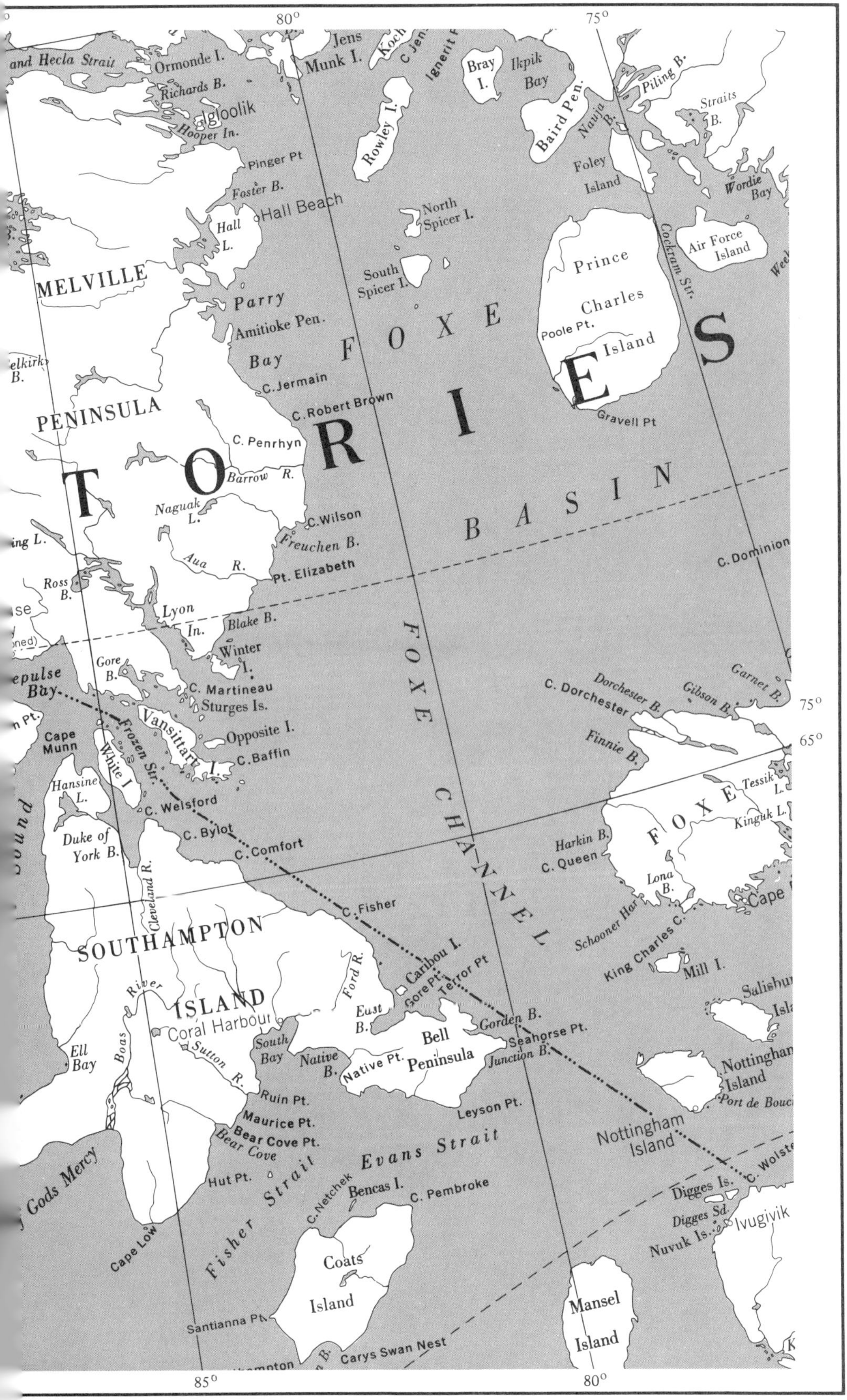

80°
75°
and Hecla Strait
Ormonde I.
Richards B.
Igloolik
Hooper In.
Jens Munk I.
Koch
Bray I.
Ikpik Bay
Rowley I.
Baird Pen.
Piling B.
Straits B.
Foley Island
Wordie Bay
Pinger Pt
Foster B.
Hall Beach
Hall L.
North Spicer I.
South Spicer I.
Air Force Island
Prince Charles Island
Cockram Str.
MELVILLE
PENINSULA
Parry Bay
Amitioke Pen.
FOXE BASIN
TORRIES
C. Jermain
C. Robert Brown
Poole Pt.
Gravell Pt
C. Penrhyn
Barrow R.
Naguak L.
C. Wilson
Freuchen B.
Aua R.
Pt. Elizabeth
C. Dominion
Ross B.
Lyon In.
Blake B.
Winter I.
Repulse Bay
Gore B.
C. Martineau
Sturges Is.
Vansittart I.
Frozen Str.
White I.
Opposite I.
C. Baffin
Cape Munn
Hansine L.
Duke of York B.
C. Welsford
C. Bylot
C. Comfort
FOXE CHANNEL
C. Dorchester
Dorchester B.
Gibson B.
Garnet B.
Finnie B.
75°
65°
Tessik L.
Kinguk L.
FOXE
Harkin B.
C. Queen
Lona B.
Cape
Cleveland R.
C. Fisher
SOUTHAMPTON ISLAND
Schooner Har.
King Charles C.
Mill I.
Caribou I.
Gore Pt.
Terror Pt
Ford R.
River
East B.
Coral Harbour
Bell Peninsula
Gorden B.
Seahorse Pt.
Junction B.
Ell Bay
Boas
South Bay
Sutton R.
Native B.
Native Pt.
Nottingham Island
Port de Bouc
Ruin Pt.
Maurice Pt.
Bear Cove Pt.
Bear Cove
Leyson Pt.
Evans Strait
Gods Mercy
Hut Pt.
Fisher Strait
C. Netchek
Bencas I.
C. Pembroke
Digges Is.
Digges Sd.
Nuvuk Is.
Ivugivik
Cape Low
Coats Island
Mansel Island
Santianna Pt.
Carys Swan Nest
85°
80°

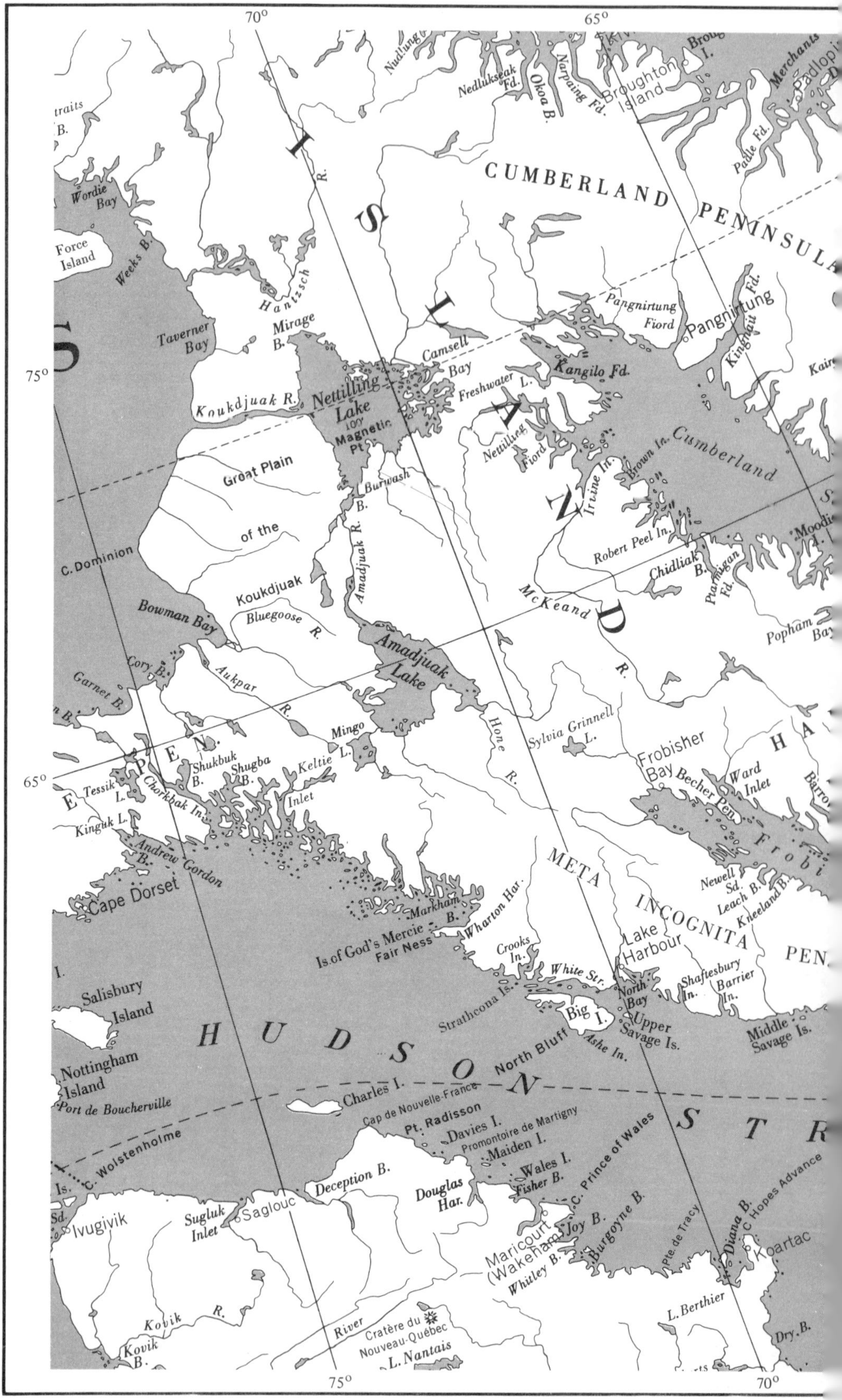

70°
65°
75°
65°
75°
70°
CUMBERLAND PENINSULA
Broughton Island
Merchants
Padle Fd.
Nedlukseak Fd.
Okoa B.
Napaing Fd.
Wordie Bay
Force Island
Weeks B.
Hantzsch
Mirage B.
Taverner Bay
Koukdjuak R.
Nettilling Lake
Magnetic Pt.
Camsell Bay
Freshwater L.
Pangnirtung Fiord
Pangnirtung
Kingnait Fd.
Kangilo Fd.
Nettilling Fiord
Irvine In.
Brown In.
Cumberland
Great Plain of the Koukdjuak
Burwash B.
Amadjuak R.
C. Dominion
Bowman Bay
Bluegoose R.
Robert Peel In.
Chidliak B.
Ptarmigan Fd.
McKeand R.
Popham Bay
Amadjuak Lake
Cory B.
Garnet B.
Aukpar R.
Mingo L.
Hone R.
Sylvia Grinnell L.
Frobisher Bay
Becher Pen.
Ward Inlet
Tessik L.
Chorkbak In.
Shukbuk B.
Shugba B.
Keltie Inlet
Kinguk L.
Andrew Gordon B.
Cape Dorset
Markham B.
Wharton Har.
Is. of God's Mercie
Fair Ness
Crooks In.
META INCOGNITA PEN.
Newell Sd.
Leach B.
Kneeland B.
Lake Harbour
White Str.
Strathcona Is.
Shaftesbury In.
Barrier In.
North Bay
Big I.
Upper Savage Is.
Ashe In.
Middle Savage Is.
Salisbury Island
Nottingham Island
Port de Boucherville
HUDSON STR
North Bluff
Charles I.
Cap de Nouvelle-France
Pt. Radisson
Davies I.
Promontoire de Martigny
Maiden I.
Wales I.
Fisher B.
C. Prince of Wales
C. Wolstenholme
Deception B.
Douglas Har.
Burgoyne B.
Joy B.
Pte. de Tracy
Diana B.
C. Hopes Advance
Koartac
Ivugivik
Sugluk Inlet
Saglouc
Maricourt (Wakeham)
Whitley B.
L. Berthier
Kovik R.
Kovik B.
River
Cratère du Nouveau-Québec
L. Nantais
Dry. B.

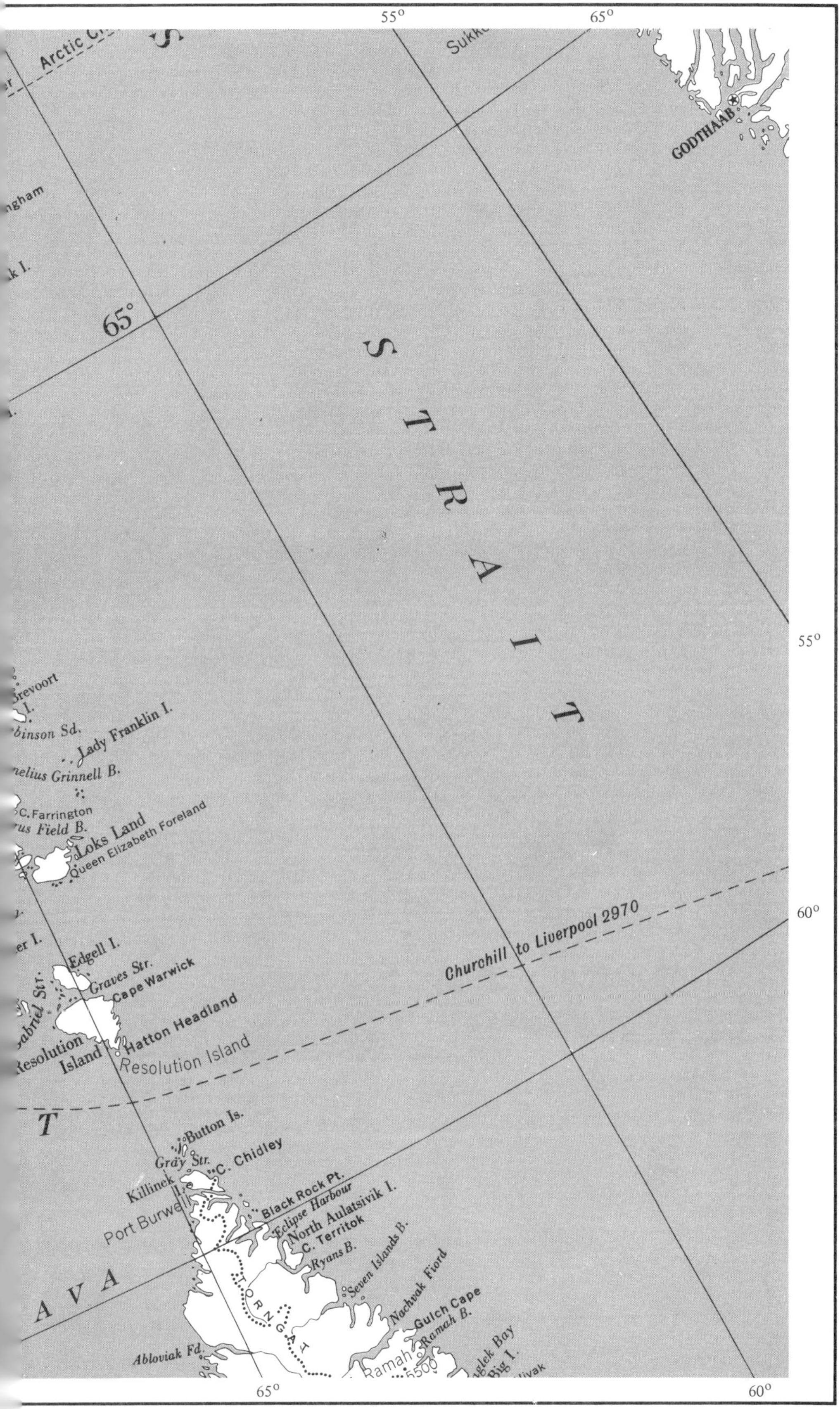

55°
65°
Arctic Ci
GODTHAAB
ngham
k I.
65°
S T R A I T
55°
Brevoort I.
obinson Sd.
Lady Franklin I.
nelius Grinnell B.
C. Farrington
rus Field B.
Loks Land
Queen Elizabeth Foreland
60°
Churchill to Liverpool 2970
er I.
Edgell I.
Graves Str.
Cape Warwick
abriel Str.
Hatton Headland
Resolution Island
Resolution Island
T
Button Is.
Gray Str.
C. Chidley
Killinek
Port Burwell
Black Rock Pt.
Eclipse Harbour
North Aulatsivik I.
C. Territok
Ryans B.
Seven Islands B.
Nachvak Fiord
Gulch Cape
Ramah B.
A V A
Abloviak Fd.
Ramah
5500
Big I.
65°
60°

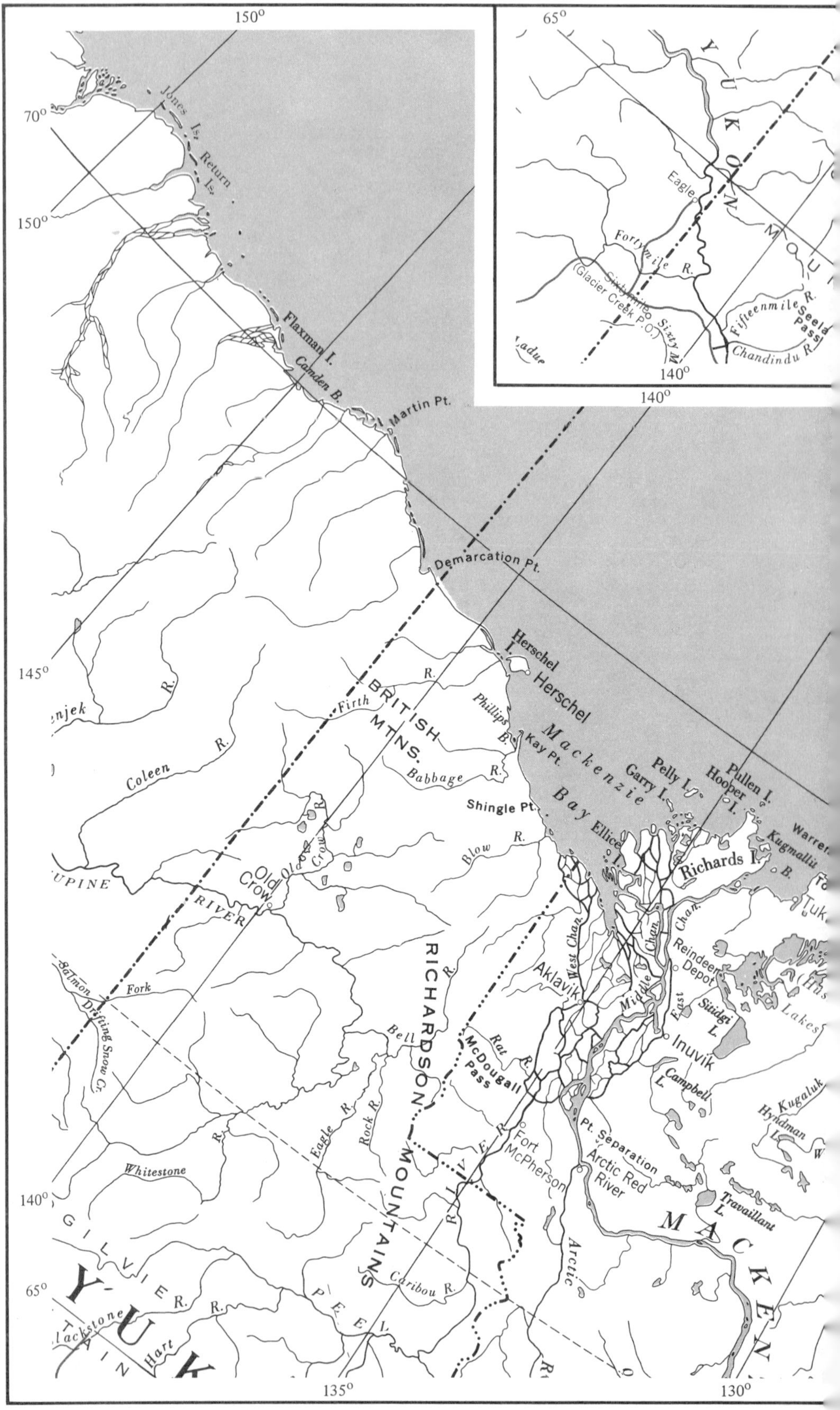

150°
70°
150°
145°
140°
65°
135°
130°
140°
65°
140°
140°
Jones Is.
Return Is.
Flaxman I.
Camden B.
Martin Pt.
Demarcation Pt.
Herschel I.
Herschel
Phillips B.
Kay Pt.
Mackenzie Bay
Shingle Pt.
Pelly I.
Garry I.
Hooper I.
Pullen I.
Ellice I.
Richards I.
Kugmallit B.
Warren
Tuk
BRITISH MTNS.
Firth R.
Babbage R.
Blow R.
Coleen R.
R.
UPINE RIVER
Old Crow
Old Crow R.
Salmon Fork
Drifting Snow Cr.
Whitestone
GILVIER R.
R.
Blackstone
Hart
YUK
RICHARDSON MOUNTAINS
Bell
Eagle R.
Rock R.
R.
R.
McDougall Pass
Rat R.
PEEL RIVER
Caribou R.
Fort McPherson
Aklavik
West Chan.
Middle Chan.
East Chan.
Reindeer Depot
Sitidgi L.
Inuvik
Campbell L.
Lakes
Pt. Separation
Arctic Red River
Arctic
Hyndman L.
Kugaluk
Travaillant L.
MACKEN
YUKON MOU
Eagle
Fortymile R.
Sixtymile (Glacier Creek P.O.)
Sixtymile
Fifteenmile R.
Seela Pass
Chandindu R.
Ladue

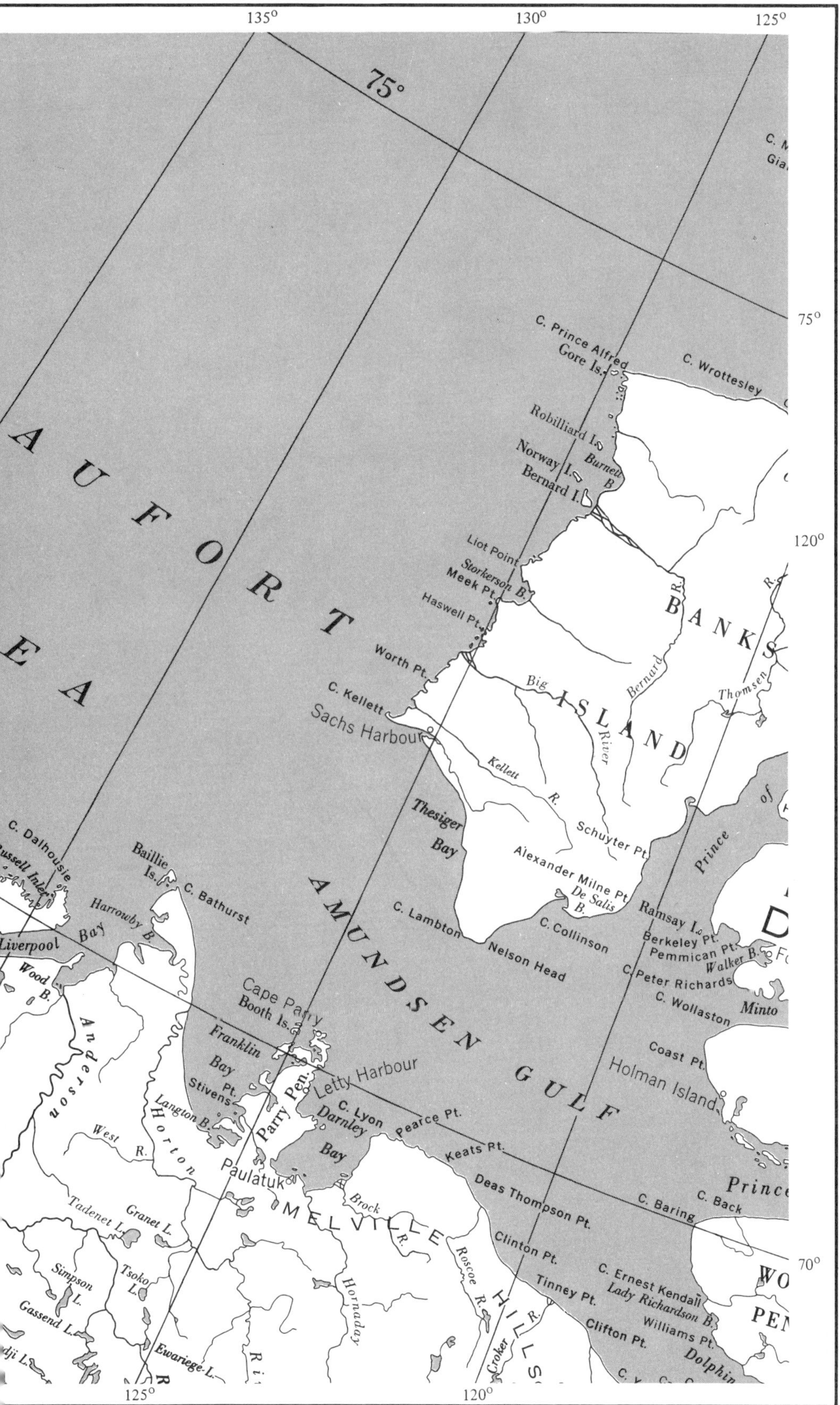
135°
130°
125°
75°
120°
70°
125°
120°
A U F O R T
E A
C. Prince Alfred
Gore Is.
C. Wrottesley
Robilliard Is.
Norway I.
Bernard I.
Burnett B.
Liot Point
Storkerson B.
Meek Pt.
Haswell Pt.
Worth Pt.
C. Kellett
Sachs Harbour
B A N K S
I S L A N D
Big
Bernard
River
Thomsen
Kellett
R.
Thesiger Bay
Schuyter Pt.
Alexander Milne Pt.
De Salis B.
C. Lambton
Nelson Head
C. Collinson
Ramsay I.
Berkeley Pt.
Pemmican Pt.
Walker B.
C. Peter Richards
C. Wollaston
Minto
Prince of
C. Dalhousie
Russell Inlet
Baillie Is.
C. Bathurst
Harrowby B.
Bay
Liverpool
Wood B.
A M U N D S E N   G U L F
Cape Parry
Booth Is.
Franklin Bay
Stivens Pt.
Langton B.
Parry Pen.
Letty Harbour
C. Lyon
Darnley Bay
Pearce Pt.
Keats Pt.
Coast Pt.
Holman Island
Anderson
Horton
West R.
Paulatuk
Deas Thompson Pt.
C. Baring
C. Back
Prince
M E L V I L L E
Brock R.
Roscoe R.
Hornaday
H I L L S
Clinton Pt.
Tinney Pt.
C. Ernest Kendall
Lady Richardson B.
Williams Pt.
Clifton Pt.
Dolphin
Tadenet L.
Granet L.
Simpson L.
Tsoko L.
Gassend L.
Ewariege L.
Croker R.

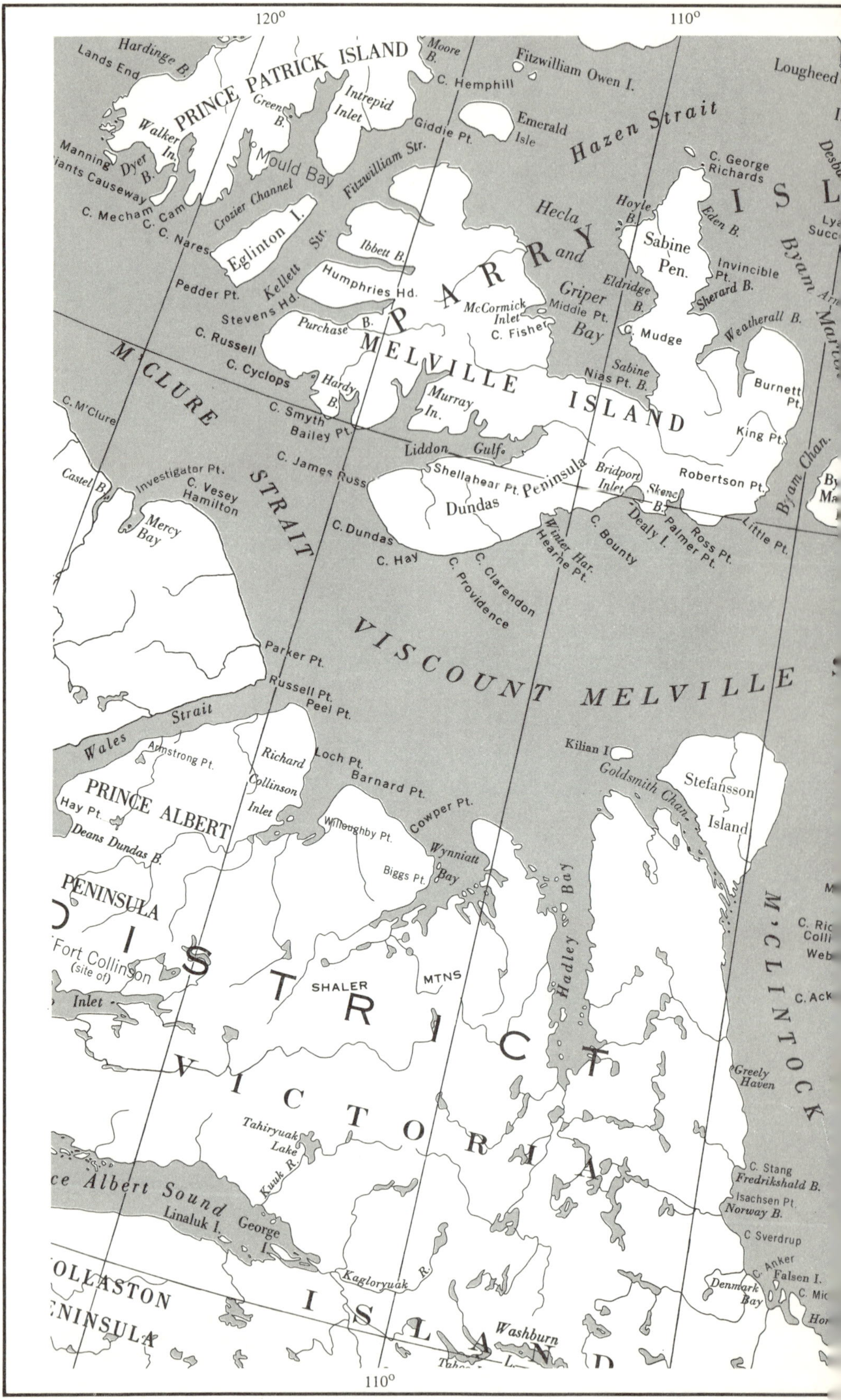

120°
110°
Hardinge B.
Lands End
PRINCE PATRICK ISLAND
Moore B.
Fitzwilliam Owen I.
Lougheed
C. Hemphill
Green B.
Intrepid Inlet
Walker In.
Manning
Dyer B.
Giants Causeway
Mould Bay
Giddie Pt.
Emerald Isle
Hazen Strait
Fitzwilliam Str.
C. George Richards
Crozier Channel
C. Mecham
C. Cam
C. Nares
Eglinton I.
Kellett Str.
Ibbett B.
Hecla and Griper Bay
Hoyle B.
Sabine Pen.
Eden B.
Byam Martin
PARRY
ISL
Humphries Hd.
Pedder Pt.
Stevens Hd.
Purchase B.
McCormick Inlet
C. Fisher
Middle Pt.
Eldridge B.
Invincible Pt.
Sherard B.
C. Mudge
Weatherall B.
C. Russell
MELVILLE ISLAND
C. Cyclops
Hardy B.
Murray In.
Sabine
Nias Pt. B.
Burnett Pt.
M'CLURE STRAIT
C. M'Clure
C. Smyth
Bailey Pt.
Liddon Gulf
King Pt.
C. James Ross
Shellabear Pt.
Dundas Peninsula
Bridport Inlet
Skene B.
Robertson Pt.
Byam Chan.
Castel B.
Investigator Pt.
C. Vesey Hamilton
Mercy Bay
C. Dundas
C. Hay
Winter Har.
Hearne Pt.
C. Bounty
Dealy I.
Palmer Pt.
Ross Pt.
Little Pt.
C. Providence
C. Clarendon
VISCOUNT MELVILLE
Parker Pt.
Russell Pt.
Peel Pt.
Wales Strait
Armstrong Pt.
Richard Collinson Inlet
Loch Pt.
Barnard Pt.
Kilian I.
Goldsmith Chan.
Stefansson Island
PRINCE ALBERT PENINSULA
Hay Pt.
Deans Dundas B.
Willoughby Pt.
Cowper Pt.
Wynniatt Bay
Biggs Pt.
Hadley Bay
M'CLINTOCK
Fort Collinson (site of)
Inlet
SHALER MTNS
DISTRICT
VICTORIA ISLAND
C. Ric
Colli
Web
C. Ack
Greely Haven
Tahiryuak Lake
Kuuk R.
Prince Albert Sound
C. Stang
Fredrikshald B.
Isachsen Pt.
Norway B.
Linaluk I.
George I.
C Sverdrup
WOLLASTON PENINSULA
Kagloryuak R.
C. Anker
Falsen I.
Denmark Bay
C. Mic
Washburn

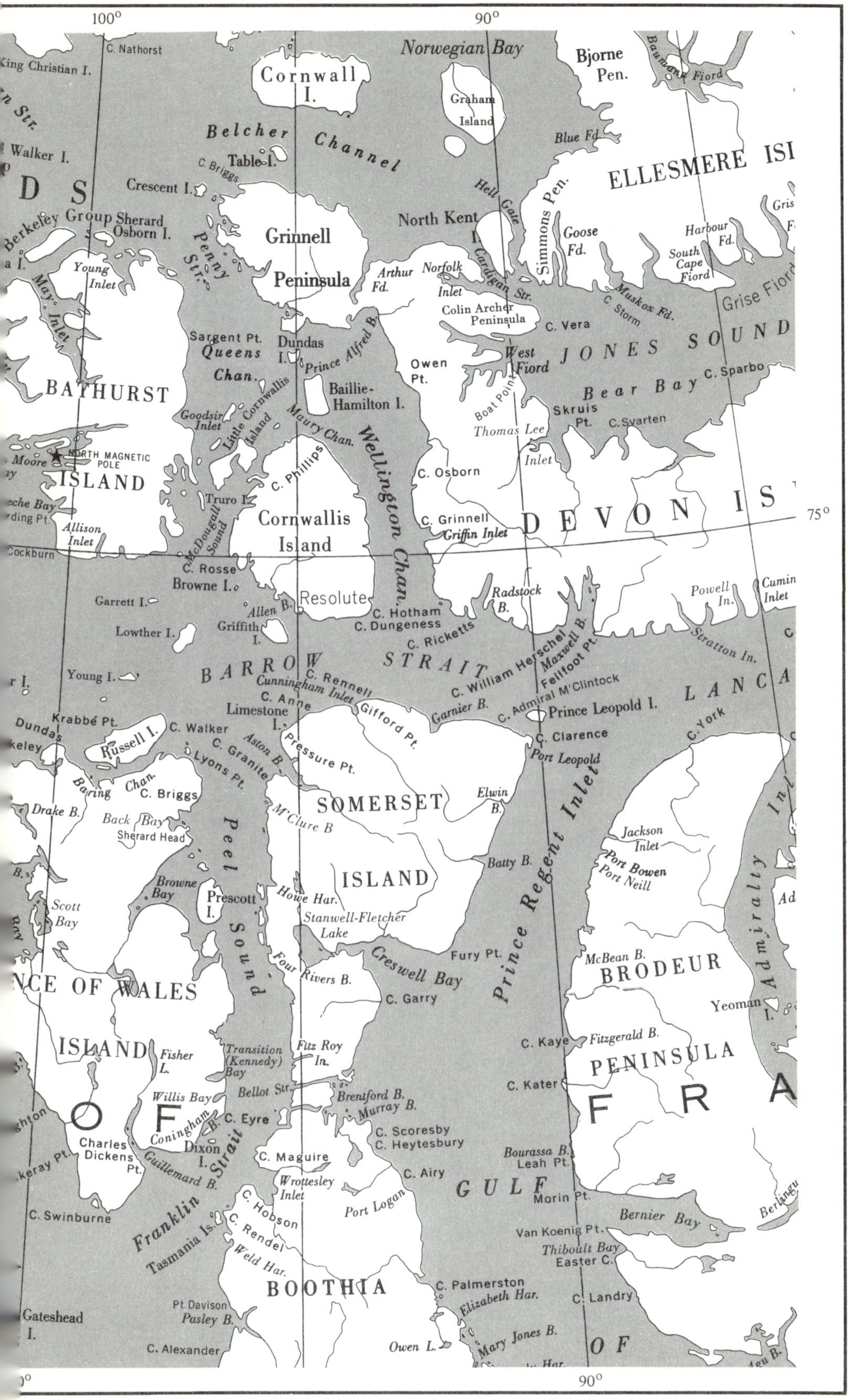
100°
90°
C. Nathorst
King Christian I.
Norwegian Bay
Cornwall I.
Graham Island
Bjorne Pen.
Baumann Fiord
Belcher Channel
Walker I.
Table I.
C. Briggs
Blue Fd.
ELLESMERE ISI
Crescent I.
Hell Gate
Simmons Pen.
Berkeley Group
Sherard Osborn I.
Penny Str.
Grinnell Peninsula
North Kent I.
Goose Fd.
Harbour Fd.
South Cape Fiord
Young Inlet
May Inlet
Arthur Fd.
Norfolk Inlet
Cardigan Str.
Muskox Fd.
C. Storm
Grise Fiord
Colin Archer Peninsula
C. Vera
Sargent Pt.
Queens Chan.
Dundas I.
Prince Alfred B.
Owen Pt.
West Fiord
JONES SOUND
BATHURST ISLAND
Little Cornwallis Island
Baillie-Hamilton I.
Bear Bay
C. Sparbo
Goodsir Inlet
Maury Chan.
Boat Point
Skruis Pt.
C. Svarten
Thomas Lee Inlet
Moore
NORTH MAGNETIC POLE
C. Phillips
Wellington Chan.
C. Osborn
Truro I.
Allison Inlet
McDougall Sound
Cornwallis Island
C. Grinnell
Griffin Inlet
DEVON IS
75°
Cockburn
C. Rosse
Browne I.
Garrett I.
Resolute
Radstock B.
Powell In.
Cumin Inlet
Allen B.
C. Hotham
C. Dungeness
Lowther I.
Griffith I.
C. Ricketts
Maxwell B.
Stratton In.
BARROW STRAIT
C. Rennell
C. William Herschel
Fellfoot Pt.
Young I.
Cunningham Inlet
C. Anne
C. Admiral M'Clintock
LANCA
Limestone I.
Gifford Pt.
Garnier B.
Prince Leopold I.
Krabbé Pt.
C. York
C. Walker
Dundas
Russell I.
C. Granite
Aston B.
Pressure Pt.
C. Clarence
Port Leopold
Lyons Pt.
Baring Chan.
C. Briggs
SOMERSET ISLAND
Elwin B.
Prince Regent Inlet
Drake B.
Back Bay
Sherard Head
M'Clure B
Jackson Inlet
Admiralty Inlet
Peel Sound
Port Bowen
Port Neill
Batty B.
Browne Bay
Scott Bay
Prescott I.
Howe Har.
Stanwell-Fletcher Lake
Four Rivers B.
Creswell Bay
Fury Pt.
McBean B.
BRODEUR PENINSULA
PRINCE OF WALES ISLAND
C. Garry
Yeoman I.
Transition (Kennedy) Bay
Fisher L.
Fitz Roy In.
C. Kaye
Fitzgerald B.
Bellot Str.
Willis Bay
C. Kater
Brentford B.
Murray B.
OF
C. Eyre
Coningham B.
FRA
C. Scoresby
C. Heytesbury
Charles Dickens Pt.
Dixon I.
Franklin Strait
C. Maguire
Bourassa B.
Leah Pt.
Guillemard B.
Wrottesley Inlet
C. Airy
GULF
Morin Pt.
C. Swinburne
Port Logan
C. Hobson
Bernier Bay
Tasmania Is.
C. Rendel
Van Koenig Pt.
Weld Har.
Thiboult Bay
Easter C.
BOOTHIA
C. Palmerston
Elizabeth Har.
C. Landry
Pt. Davison
Pasley B.
Gateshead I.
Mary Jones B.
C. Alexander
Owen L.
OF
90°

80°
70°
Makinson Inlet
Smith Bay
ISLAND
Starnes Fiord
Fd.
Lee Pt.
Clarence Head
C. Norton Shaw
Glacier Strait
Phillips Pt.
Coburg Island
King Edward Pt.
Lady Ann Strait
Ward Pt.
Belcher Pt.
Johnson Pt.
C. Parker
LAND
Philpots Island
75°
Cary Is.
Saunders I.
Thule
(Air Base)
Cape Atholl
Cape York
Bush
B A F
Croker Bay
C. Sherard
C. Warrender
C. Home
STER SOUND
C. Charles Yorke
C. Crauford
C. Joy
C. Hay
Maud Bight
C. Liverpool
C. Byam Martin
C. Walter Bathurst
C. Burney
C. Graham Moore
Navy Board Inlet
Canada Pt.
BYLOT ISLAND
Elwin Inlet
BORDEN
PENINSULA
Low Pt.
Strathcona Sd.
Arctic Bay
Adams Sd.
Pond Inlet
Pond Inlet
C. Weld
C. Macculloch
C. Coutts
Nova Zembla I.
Cape Jameson
(Ragged Pt)
Buchan Gulf
C. Cargenholm
C. Hunter
Eclipse Sound
Oliver Sd.
North Arm
Coutts Inlet
Tremblay Sd.
Milne Inlet
Tay Sd.
Paquet B.
Robertson R.
Moffet In.
Cambridge Fiord
Royal Society Fd.
Tromso Fd.
Dexterity Fd.
Clark Fd
NKLIN
BAFFI
Inuktorfik L.
Quartz L.
Nina Bang L.
Erichsen L.
Rowley R.
Conn Lake
Bieler Lake
Bell B.
Saputing L.
Gifford R.
Steensby In.
Neergaard L.
Grant-Suttie B.
Ege B.
Gifford Fiord
Murray Maxwell B.
Gillian
Flint L.
Autridge B.
70°
80°

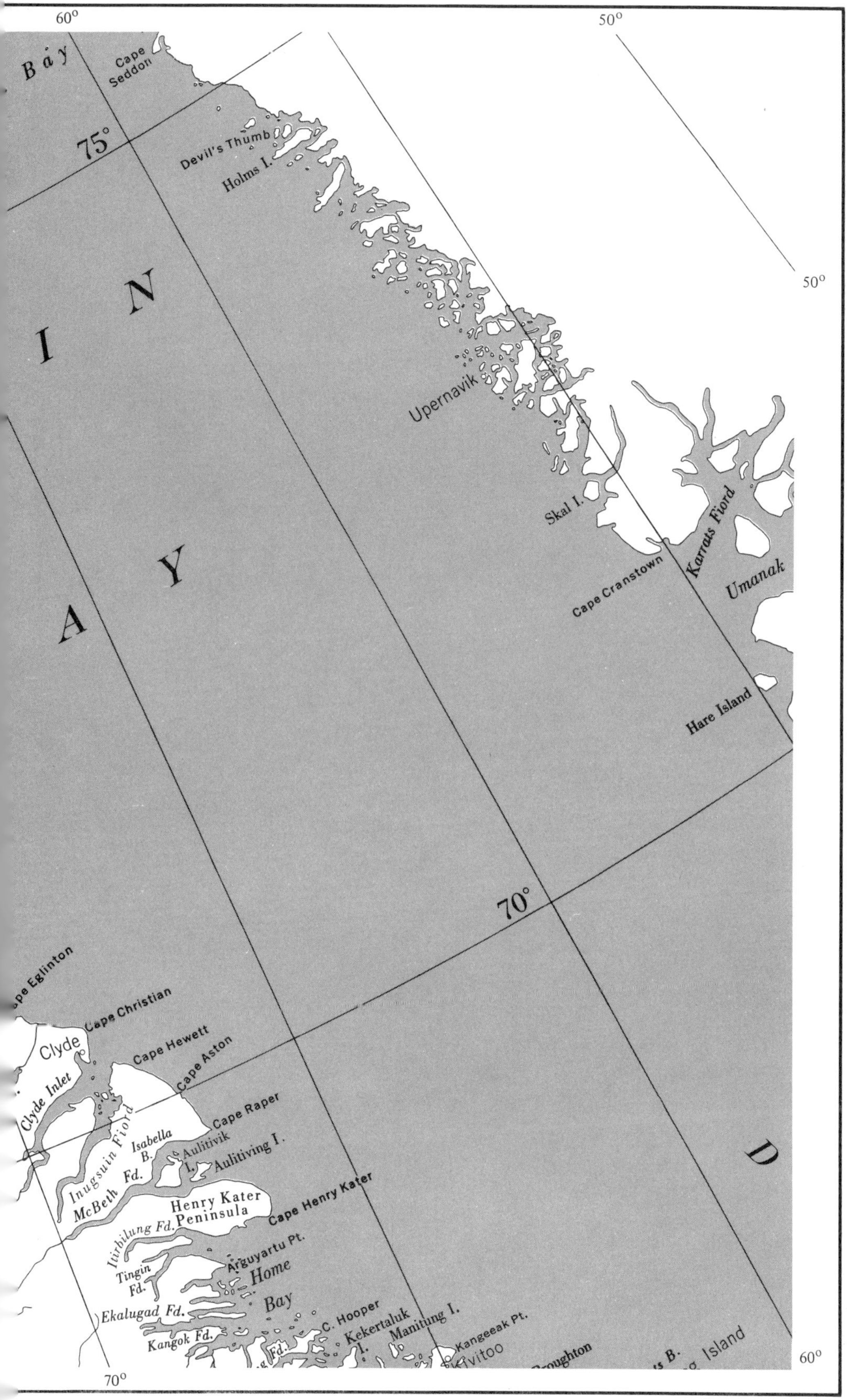
60°
50°
Bay
Cape Seddon
75°
Devil's Thumb
Holms I.
N
I
Upernavik
Skal I.
50°
Karrats Fiord
Umanak
Cape Cranstown
Y
A
Hare Island
70°
Cape Christian
Clyde
Clyde Inlet
Cape Hewett
Cape Aston
Cape Raper
Isabella B.
Aulitivik I.
Aulitiving I.
Inugsuin Fiord
McBeth Fd.
D
Henry Kater Peninsula
Cape Henry Kater
Itirbilung Fd.
Arguyartu Pt.
Tingin Fd.
Home Bay
Ekalugad Fd.
C. Hooper
Kekertaluk I.
Manitung I.
Kangeeak Pt.
Kangok Fd.
Island
60°
70°

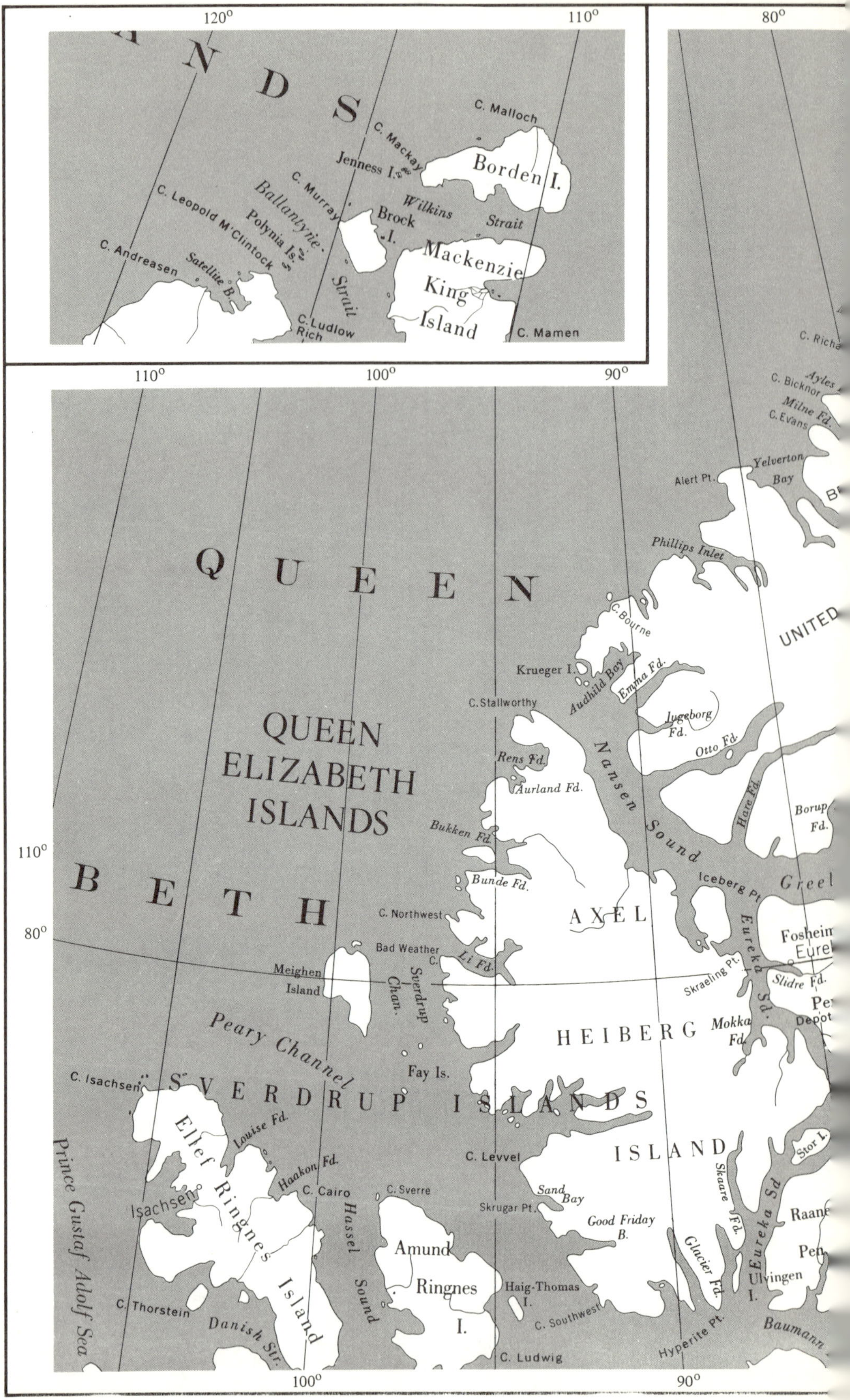

120°
110°
80°
N D S
C. Malloch
C. Mackay
Jenness I.
Borden I.
C. Murray
Ballantyne
C. Leopold M'Clintock
Polynia Is.
Wilkins
Strait
Brock I.
C. Andreasen
Satellite B.
Mackenzie King Island
Strait
C. Ludlow
Rich
C. Mamen
C. Richa
110°
100°
90°
Ayles
C. Bicknor
Milne Fd.
C. Evans
Yelverton Bay
Alert Pt.
Phillips Inlet
Q U E E N
C. Bourne
UNITED
Krueger I.
Audhild Bay
Emma Fd.
C. Stallworthy
Ingeborg Fd.
Otto Fd.
QUEEN ELIZABETH ISLANDS
Rens Fd.
Aurland Fd.
Nansen Sound
Hare Fd.
Borup Fd.
Bukken Fd.
110°
B E T H
Bunde Fd.
Iceberg Pt.
Greel
80°
C. Northwest
AXEL
Eureka Sd.
Fosheim
Eurel
Bad Weather C.
Li Fd.
Meighen Island
Sverdrup Chan.
Skraeling Pt.
Slidre Fd.
Pe
Depot
Peary Channel
HEIBERG
Mokka Fd.
Fay Is.
C. Isachsen
SVERDRUP ISLANDS
Louise Fd.
Ellef Ringnes Island
Stor I.
C. Levvel
ISLAND
Haakon Fd.
C. Cairo
C. Sverre
Sand Bay
Skaare Fd.
Eureka Sd
Raane Pen.
Prince Gustaf Adolf Sea
Isachsen
Hassel Sound
Skrugar Pt.
Good Friday B.
Amund Ringnes I.
Glacier Fd.
Ulvingen I.
C. Thorstein
Haig-Thomas I.
C. Southwest
Danish Str.
Hyperite Pt.
Baumann
C. Ludwig
100°
90°

70°
60°
50°
LINCOLN SEA
John Murray I.
Victoria Fd.
Sherard Osborn Fd.
Hunt
C. Nares
C. Columbia
Good Pt.
C. Colan
C. Hecla
James Ross B.
C. Joseph Henry
Black Cliffs B.
Markham Fd.
Clements Markham Inlet
Alert
C. Union
Disraeli Fd.
Robeson Channel
Newman Bay
RANGE
ELLESMERE ISLAND
STATES
RANGE
Lake Hazen
Lady Franklin Bay
C. Baird
Hall Basin
Archer Fd.
Judge Daly Promontory
Tanquary Fd.
Carl Ritter B.
Kennedy Channel
VICTORIA AND ALBERT MTNS
Rawlings Bay
C. Lawrence
C. Joseph Good
John Richardson Bay
C. Wilkes
C. M'Clintock
Scoresby B.
C. Fraser
Dobbin B.
C. Louis Napoleon
Kane Basin
C. Hawks
Copes B.
Woodward B.
Princess Marie B.
Victoria Head
Sawyer B.
Bache Pen.
Flagler B.
Knud Pen.
Buchanan Bay
Jokel Fd.
Thorvald Pen.
Pim I.
C. Sabine
Alexandra Fiord
Johan Pen.
C. Herschel
Smith Sound
Baird Inlet
C. Isabella
Cadogan Inlet
Paget Pt.
Thule (Qânâq)
Inglefield Bay
C. Dunsterville
C. Faraday
Northumberland I.
C. Mouat
C. Parry
80°
70°
50°
80°
60°